MATHEMATICAL
METHODS IN
CHEMICAL
ENGINEERING

MATHEMATICAL METHODS IN CHEMICAL ENGINEERING

ARVIND VARMA
Department of Chemical Engineering
University of Notre Dame

MASSIMO MORBIDELLI
Laboratorium für Technische Chemie
ETH Zürich

New York   Oxford
OXFORD UNIVERSITY PRESS
1997

OXFORD UNIVERSITY PRESS

Oxford   New York
Athens   Auckland   Bangkok   Bogotá   Bombay   Buenos Aires
Calcutta   Cape Town   Dar es Salaam   Delhi
Florence   Hong Kong   Istanbul   Karachi
Kuala Lumpur   Madras   Madrid   Melbourne
Mexico City   Nairobi   Paris   Singapore
Taipei   Tokyo   Toronto

 and associated companies in

Berlin   Ibadan

Copyright © 1997 by Oxford University Press, Inc.

Published by Oxford University Press, Inc.
198 Madison Avenue, New York, New York, 10016

Oxford is a registered trademark of Oxford University Press.

Library of Congress Cataloging-in-Publication Data

Varma, Arvind.
   Mathematical methods in chemical engineering / Arvind Varma,
Massimo Morbidelli.
      p.   cm.
   Includes bibliographical references and index.
   ISBN 0-19-509821-8
   1. Chemical engineering—Mathematics.   I. Morbidelli, Massimo.
II. Title.
TP149.V36   1996   95-26843 CIP
515'.14'02466—dc20

9 8 7 6 5 4 3 2

Printed in the United States of America
on acid-free paper

Dedicated to
our wives
Karen and Luisella
and children
Anita and Sophia
Melissa and Oreste
for their love and patience

CONTENTS

Chapter 2

FIRST-ORDER NONLINEAR ORDINARY DIFFERENTIAL EQUATIONS AND STABILITY THEORY

Chapter 3

THEORY OF LINEAR ORDINARY DIFFERENTIAL EQUATIONS

I. INITIAL VALUE PROBLEMS

II. BOUNDARY VALUE PROBLEMS

III. EIGENVALUE PROBLEMS

Chapter 4

SERIES SOLUTIONS AND SPECIAL FUNCTIONS

Chapter 5

FUNDAMENTALS OF PARTIAL DIFFERENTIAL EQUATIONS

Chapter 6

FIRST-ORDER PARTIAL DIFFERENTIAL EQUATIONS

Chapter 7

GENERALIZED FOURIER TRANSFORM METHODS FOR LINEAR PARTIAL DIFFERENTIAL EQUATIONS

Chapter 8

LAPLACE TRANSFORM

Chapter 9

PERTURBATION METHODS

PREFACE

Over the years, and especially since the 1960s, significant developments have occurred in chemical engineering through the applications of mathematics. These developments were supported by the available text and specialized books dealing with topics in applied mathematics, and some of these were written by, and for, chemical engineers.

Most chemical engineering departments offer a senior-graduate level course in mathematical methods. A variety of topics are covered in these courses. For some time, there has been need for a textbook which provides a mathematical background at a somewhat advanced level, along with applications to chemical engineering problems. The present book is an attempt in this direction.

We have included a relatively broad selection of mathematical topics in this book that we believe are essential for a modern chemical engineer to know, especially at the graduate level. With successful mastery of this material, students can more confidently undertake mathematical analysis in research and can better cope with other graduate level courses. In making these selections, however, we have had to delete a number of topics to limit the size of the book. Topics deleted include complex analysis, statistics, and numerical methods. Fortunately, for each of these topics good books are available in the market, and for numerical methods in particular there are several books available written specifically for chemical engineers. These sources can be used to supplement one's mathematical training.

This book can be used for either a one or two-semester course. In a two-semester course, the entire book can be covered comfortably. For a one-semester course, the instructor can choose the topics to be covered, since the individual chapters are essentially independent. Portions of the book have been used successfully as text by the authors at the University of Notre Dame, Politecnico di Milano and ETH-Zürich. In addition to chemical engineers, students from other engineering departments have frequently taken these courses. Professors William B. Russel of Princeton University and Gianni Astarita of the Universities of Naples and Delaware have also used portions of this book for their courses, and offered valued suggestions for improvements.

It is our pleasure to acknowledge here some debts of gratitude. To many readers, our enormous debt to Professor Neal R. Amundson of the University of Houston will become clear immediately. Several chapters bear an imprint of material offered in his course at the University of Minnesota, where one of us (AV) first learned about the beauty and power of mathematics applied to chemical engineering problems. Chapter 1 is essentially a summary of his book *Mathematical Methods in Chemical Engineering:*

Matrices and their Application, Prentice-Hall (1966). Indeed, we had hoped earlier that he would co-author the present book, but this was unfortunately not possible owing to his other commitments. We are also grateful to Professor Rutherford Aris of the University of Minnesota for his encouragement during the course of this writing.

The typing of the manuscript was largely the work of Mrs. Jeanne Davids, Mrs. Linda Phillips and Mrs. Sherry DePoy, although some early drafts were prepared by Mrs. Helen Deranek and Mrs. Miriam Oggioni. The illustrations were prepared by Mrs. Joanne Birdsell and Mrs. Ruth Quinn. To them, and to many students who took our courses and who helped with calculations behind some of the figures as well as galley proofs, we are grateful.

Our debts to our wives and children are too personal to describe here fully. They have suffered some neglect as we have struggled to complete the manuscript. It is to them that we dedicate this book, with love and affection.

Arvind Varma
Massimo Morbidelli

MATHEMATICAL METHODS IN CHEMICAL ENGINEERING

Chapter 1

MATRICES AND THEIR APPLICATION

1.1 DETERMINANTS AND ELEMENTARY PROPERTIES

A *determinant* of nth order is written in the form

$$D = \begin{vmatrix} a_{11} & a_{12} & \cdots & a_{1n} \\ a_{21} & a_{22} & \cdots & a_{2n} \\ \vdots & & & \\ a_{n1} & a_{n2} & \cdots & a_{nn} \end{vmatrix} = |a_{ij}| \tag{1.1.1}$$

and is a *square array* of n^2 elements a_{ij}, which may be real or complex numbers or functions. In the general element a_{ij}, i refers to the row number and j to the column number where the element is located.

The determinant is essentially a number or a function; it is defined as

$$D = \Sigma(-1)^h(a_{1l_1}a_{2l_2} \cdots a_{nl_n}) \tag{1.1.2}$$

where the summation is over *all* possible products of a_{ij}, each involving n elements, *one and only one* from each row and each column. Thus there exist n choices for the first element, $(n-1)$ choices for the second element, and so on.

For a particular determinant, there are $n!$ terms in the sum. In each term, the second subscripts j in a_{ij} will initially not be in the natural order $1, 2, 3, \ldots, n$, although all numbers from 1 to n will appear. The value of h is defined as the number of transpositions required to transform the sequence $l_1, l_2, \ldots, l_n$ into the order $1, 2, 3, \ldots, n$, where a transposition is an interchange of two numbers l_k and l_m.

The number h is not unique, but it can be shown to be always either even or odd. For example, consider the sequence

$$1 \quad 5 \quad 3 \quad 4 \quad 2$$

To put this sequence in its natural order, several alternative schemes are possible, such as

1	5	3	2	4		1	5	3	2	4		1	2	3	4	5
1	3	5	2	4		1	2	3	5	4			$h = 1$			
1	3	2	5	4		1	2	3	4	5						
1	2	3	5	4			$h = 3$									
1	2	3	4	5											
	$h = 5$														

and we see that in all of them, h is odd.

Consider the third-order determinant:

$$D = \begin{vmatrix} a_{11} & a_{12} & a_{13} \\ a_{21} & a_{22} & a_{23} \\ a_{31} & a_{32} & a_{33} \end{vmatrix}$$

There are $3! = 6$ terms in the sum

$$D = \Sigma(-1)^h \, a_{1l_1} a_{2l_2} a_{3l_3}$$

given by $a_{11}[a_{22}a_{33}, a_{23}a_{32}]$, $a_{12}[a_{21}a_{33}, a_{23}a_{31}]$, $a_{13}[a_{22}a_{31}, a_{21}a_{32}]$, where the algebraic signs have not yet been assigned. For each term, h can be calculated to yield

$$\begin{aligned} D = \; & +a_{11}a_{22}a_{33} - a_{11}a_{23}a_{32} \\ & -a_{12}a_{21}a_{33} + a_{12}a_{23}a_{31} \\ & -a_{13}a_{22}a_{31} + a_{13}a_{21}a_{32} \end{aligned} \qquad (1.1.3)$$

a result which can be verified easily by other methods (cf. section 1.3).

Multiplication of a determinant by a scalar quantity k is equivalent to multiplication of the elements of any row *or* any column by that scalar. Thus, for example,

$$k\begin{vmatrix} a_{11} & a_{12} & a_{13} \\ a_{21} & a_{22} & a_{23} \\ a_{31} & a_{32} & a_{33} \end{vmatrix} = \begin{vmatrix} a_{11} & ka_{12} & a_{13} \\ a_{21} & ka_{22} & a_{23} \\ a_{31} & ka_{32} & a_{33} \end{vmatrix} = \begin{vmatrix} ka_{11} & ka_{12} & ka_{13} \\ a_{21} & a_{22} & a_{23} \\ a_{31} & a_{32} & a_{33} \end{vmatrix}$$

This follows from expansion of the determinant in terms of its elements using eq. (1.1.2). Other elementary properties which can be obtained from the definition (1.1.2) and the scalar multiplication rule are as follows:

1. If all elements of any row (or any column) are zero, the determinant is zero.
2. The value of the determinant is unchanged if all its rows and columns are interchanged.
3. If two rows (or two columns) are interchanged, the sign of the determinant changes.
4. If all elements of two rows (or two columns) are equal, element by element, then the determinant is zero.

5. Two determinants may be added term by term if they are identical except for one row (or one column), as

$$
\begin{vmatrix}
a_{11} & a_{12} & a_{13} \\
a_{21} & a_{22} & a_{23} \\
a_{31} & a_{32} & a_{33}
\end{vmatrix}
+
\begin{vmatrix}
a_{11} & b_{12} & a_{13} \\
a_{21} & b_{22} & a_{23} \\
a_{31} & b_{32} & a_{33}
\end{vmatrix}
=
\begin{vmatrix}
a_{11} & a_{12} + b_{12} & a_{13} \\
a_{21} & a_{22} + b_{22} & a_{23} \\
a_{31} & a_{32} + b_{32} & a_{33}
\end{vmatrix}
$$

6. The most important property of a determinant involves multiplying one row (or column) by a constant and adding to another row (or column) element by element. Consider, for example, the determinant

$$
\begin{vmatrix}
a_{11} & a_{12} + ka_{11} & a_{13} \\
a_{21} & a_{22} + ka_{21} & a_{23} \\
a_{31} & a_{32} + ka_{31} & a_{33}
\end{vmatrix}
=
\begin{vmatrix}
a_{11} & a_{12} & a_{13} \\
a_{21} & a_{22} & a_{23} \\
a_{31} & a_{32} & a_{33}
\end{vmatrix}
+ k
\begin{vmatrix}
a_{11} & a_{11} & a_{13} \\
a_{21} & a_{21} & a_{23} \\
a_{31} & a_{31} & a_{33}
\end{vmatrix}
$$

due to property 5 and the definition of scalar multiplication. The second determinant on the rhs is zero, because columns 1 and 2 are identical (property 4). Thus

$$
\begin{vmatrix}
a_{11} & a_{12} & a_{13} \\
a_{21} & a_{22} & a_{23} \\
a_{31} & a_{32} & a_{33}
\end{vmatrix}
=
\begin{vmatrix}
a_{11} & a_{12} + ka_{11} & a_{13} \\
a_{21} & a_{22} + ka_{21} & a_{23} \\
a_{31} & a_{32} + ka_{31} & a_{33}
\end{vmatrix}
$$

7. A corollary of properties 4 and 6 is that if all elements of two rows (or columns) are in a constant ratio, element by element, then the determinant is zero.

8. If the elements of a determinant D are functions of a variable, say $a_{ij}(x)$, the derivative of D can be computed using eq. (1.1.2) as follows:

$$
\frac{dD}{dx} = \Sigma(-1)^h \left[\frac{da_{1l_1}}{dx} \, a_{2l_2} \cdots a_{nl_n} \right]
$$

$$
+ \Sigma(-1)^h \left[a_{1l_1} \, \frac{da_{2l_2}}{dx} \cdots a_{nl_n} \right] + \cdots
$$

$$
+ \Sigma(-1)^h \left[a_{1l_1} \, a_{2l_2} \cdots a_{n-1l_{n-1}} \, \frac{da_{nl_n}}{dx} \right]
$$

This is a sum of n determinants:

$$
\frac{dD}{dx} = \sum_{i=1}^{n} D'_i \tag{1.1.4}
$$

where

$$
D'_i =
\begin{vmatrix}
a_{11} & a_{12} & \cdots & a_{1n} \\
a_{21} & a_{22} & \cdots & a_{2n} \\
\vdots & & & \\
\dfrac{da_{i1}}{dx} & \dfrac{da_{i2}}{dx} & \cdots & \dfrac{da_{in}}{dx} \\
\vdots & & & \\
a_{n1} & a_{n2} & \cdots & a_{nn}
\end{vmatrix}
$$

1.2 MINORS AND COFACTORS

Given a determinant of order n, other determinants called *minors* may be formed by striking out an *equal* number of rows and columns. If, for example, in an nth-order determinant m rows and m columns are removed, what remains is a determinant of order $n - m$, and elements at the intersections form a determinant of order m. Thus canceling the second and fifth rows and also the second and fourth columns, the fifth-order determinant

$$\begin{vmatrix} a_{11} & a_{12} & a_{13} & a_{14} & a_{15} \\ a_{21} & a_{22} & a_{23} & a_{24} & a_{25} \\ a_{31} & a_{32} & a_{33} & a_{34} & a_{35} \\ a_{41} & a_{42} & a_{43} & a_{44} & a_{45} \\ a_{51} & a_{52} & a_{53} & a_{54} & a_{55} \end{vmatrix}$$

leads to the pair of determinants of orders 3 and 2, respectively:

$$\begin{vmatrix} a_{11} & a_{13} & a_{15} \\ a_{31} & a_{33} & a_{35} \\ a_{41} & a_{43} & a_{45} \end{vmatrix} \quad \text{and} \quad \begin{vmatrix} a_{22} & a_{24} \\ a_{52} & a_{54} \end{vmatrix}$$

These two determinants are called *complementary minors*. A specific complementary minor is that of a single element, called the *cofactor* of that element. The cofactor of element a_{ij} is defined as

$$A_{ij} = (-1)^{i+j}[\text{complement of } a_{ij}] \tag{1.2.1}$$

where the complement of a_{ij} is the determinant of order $(n - 1)$ obtained by striking out the ith row and the jth column of the original determinant.

Another special minor of a determinant of order n is the *principal minor* of order j, obtained by deleting $(n - j)$ rows and columns with the *same* index. Thus if the kth row is deleted, the kth column is also struck out. All the diagonal elements of a principal minor are also diagonal elements of the original determinant.

1.3 LAPLACE'S DEVELOPMENT OF A DETERMINANT

Although the value of a determinant may be obtained from the definition (1.1.2) as a sum of products, this is usually a difficult task because $n!$ terms are involved. Alternative methods are therefore sought and one of the most useful is that of Laplace (for a proof, see Bôcher, 1964, p. 24). In what follows, wherever the term *row* appears, the term *column* may be substituted as well.

Choose *any* row from the nth-order determinant, then Laplace's development implies that

$$D = \sum_{j=1}^{n} a_{ij} A_{ij}, \qquad i = 1, 2, \ldots, n \tag{1.3.1}$$

Let us now consider the quantity

$$E_{ik} = \sum_{j=1}^{n} a_{ij}A_{kj}, \qquad i, k = 1, 2, \ldots, n$$

Note that E_{ik} is the sum of products of elements of the ith row and the corresponding cofactors of elements of the kth row. Thus for $i = k$, we have $E_{ii} = D$. For $i \neq k$, E_{ik} is the expansion of a determinant in which the ith and kth rows are identical. Such a determinant with two identical rows is zero; thus $E_{ik} = 0$ for $i \neq k$.

Thus we have

$$\sum_{j=1}^{n} a_{ij}A_{kj} = \begin{cases} D, & i = k \\ 0, & i \neq k \end{cases} \qquad (1.3.2)$$

A_{kj} is called an *alien cofactor* of a_{ij} if $i \neq k$, and the expansion represented by E_{ik} for $i \neq k$ is called the expansion of D by alien cofactors.

1.4 SOLUTION OF A SYSTEM OF LINEAR ALGEBRAIC EQUATIONS

A large number of problems in applied mathematics eventually reduce to the solution of linear algebraic equations. For this reason, one frequently returns to this problem. We next develop *Cramer's rule*, a useful tool in many theoretical considerations, although it is never used for computation of large systems. Let us consider n equations in n unknowns, $x_1, x_2, \ldots, x_n$:

$$\begin{aligned} a_{11}x_1 + a_{12}x_2 + \cdots + a_{1n}x_n &= b_1 \\ a_{21}x_1 + a_{22}x_2 + \cdots + a_{2n}x_n &= b_2 \\ &\vdots \\ a_{n1}x_1 + a_{n2}x_2 + \cdots + a_{nn}x_n &= b_n \end{aligned} \qquad (1.4.1)$$

Multiply the first equation by A_{1j}, the second by $A_{2j}, \ldots,$ the nth by A_{nj} and add; this yields

$$x_1 \sum_{i=1}^{n} a_{i1}A_{ij} + x_2 \sum_{i=1}^{n} a_{i2}A_{ij} + \cdots + x_j \sum_{i=1}^{n} a_{ij}A_{ij} + \cdots + x_n \sum_{i=1}^{n} a_{in}A_{ij} = \sum_{i=1}^{n} b_iA_{ij}$$

However, from eq. (1.3.2), $\sum_{i=1}^{n} a_{ij}A_{ij} = D$, while the rest of the coefficients are zero, since they are an expansion of the determinant of the coefficient matrix by alien cofactors. Thus

$$x_j \sum_{i=1}^{n} a_{ij}A_{ij} = \sum_{i=1}^{n} b_iA_{ij}$$

which leads to Cramer's rule:

$$x_j = \frac{\sum_{i=1}^{n} b_iA_{ij}}{D} = \frac{D_j}{D}, \qquad D \neq 0, \quad j = 1, 2, \ldots, n \qquad (1.4.2)$$

where

$$D_j = \begin{vmatrix} a_{11} & a_{12} & \cdots & a_{1j-1} & b_1 & a_{1j+1} & \cdots & a_{1n} \\ a_{21} & a_{22} & \cdots & a_{2j-1} & b_2 & a_{2j+1} & \cdots & a_{2n} \\ \vdots & & & & & & & \\ a_{n1} & a_{n2} & \cdots & a_{nj-1} & b_n & a_{nj+1} & \cdots & a_{nn} \end{vmatrix} \qquad (1.4.3)$$

Let us now verify that x_j as given by eq. (1.4.2) is a solution of the original set. Substituting x_j into the ith of eqs. (1.4.1) gives

$$a_{i1}D_1 + a_{i2}D_2 + \cdots + a_{ij}D_j + \cdots + a_{in}D_n = b_iD \qquad (1.4.4)$$

We need to show that eq. (1.4.4) is an identity for *all* i. Expanding by elements of the jth column gives

$$D_j = b_1 A_{1j} + b_2 A_{2j} + \cdots + b_n A_{nj}$$

and substituting in eq. (1.4.4) yields

$$\begin{aligned} \text{lhs} &= a_{i1}[b_1 A_{11} + b_2 A_{21} + \cdots + b_n A_{n1}] \\ &\quad + a_{i2}[b_1 A_{12} + b_2 A_{22} + \cdots + b_n A_{n2}] + \cdots \\ &\quad + a_{in}[b_1 A_{1n} + b_2 A_{2n} + \cdots + b_n A_{nn}] \\ &= b_1 \left[\sum_{j=1}^{n} a_{ij} A_{1j} \right] \\ &\quad + b_2 \left[\sum_{j=1}^{n} a_{ij} A_{2j} \right] + \cdots \\ &\quad + b_i \left[\sum_{j=1}^{n} a_{ij} A_{ij} \right] + \cdots \\ &\quad + b_n \left[\sum_{j=1}^{n} a_{ij} A_{nj} \right] \end{aligned}$$

All terms in the square brackets, except for the coefficient of b_i, are zero due to eq. (1.3.2). Thus we have

$$\text{lhs} = b_iD = \text{rhs}$$

and hence eq. (1.4.4) is established as an identity.

Some Useful Results from Cramer's Rule:

1. If $D \neq 0$, then there exists a unique solution $\mathbf{x}$ given by eq. (1.4.2).
2. If $D \neq 0$ and *all* b_i are zero, then $\mathbf{x} = \mathbf{0}$.
3. If $D = 0$, then various possibilities arise; we may have no solutions, unique solutions, or infinite solutions.

These aspects are discussed further in section 1.10.

Remark.   As noted earlier, Cramer's rule is in fact never used for solving large systems of equations. The reason is as follows.

To solve a set of n linear algebraic equations in n unknowns, we need to evaluate a total of $(n + 1)$ determinants of order n. Now each determinant of order n has $n!$ terms in the sum, each term being a product of n elements. Since each product of n elements involves $(n - 1)$ multiplications, the evaluation of $(n + 1)$ determinants requires $(n + 1)(n - 1)(n!)$ multiplications. Because a division is computationally equivalent to a multiplication, by adding the n divisions ($x_j = D_j/D$), the total number of multiplications is $[(n + 1)(n - 1)(n!) + n]$. For large n, this leads to $\sim(n^2)(n!)$ multiplications. Thus, as an example, for $n = 100$, we have $(10^4)(100!) \simeq 10^{162}$ multiplications. Neglecting addition and bookkeeping time, and taking 10^{-8} sec for each multiplication, this means 10^{154} sec $\simeq 3 \times 10^{146}$ years!

For this reason, alternative methods have been developed to solve systems of linear algebraic equations. These techniques originate primarily from the ideas of Gauss elimination (cf. Kreyszig, 1993, chapter 19), and they prove useful in computer applications.

1.5   MATRICES, ELEMENTARY OPERATIONS AND PROPERTIES

1. An $(m \times n)$ *matrix* is an array of elements arranged in m rows and n columns as

$$\mathbf{A} = \begin{bmatrix} a_{11} & a_{12} & \cdots & a_{1n} \\ a_{21} & a_{22} & \cdots & a_{2n} \\ \vdots & & & \vdots \\ a_{m1} & a_{m2} & \cdots & a_{mn} \end{bmatrix}$$

2. A matrix which contains only a single column and m rows is called an *m-column vector*:

$$\mathbf{x} = \begin{bmatrix} x_1 \\ x_2 \\ \vdots \\ x_m \end{bmatrix}$$

   Similarly, a matrix that contains a single row and n columns is called an *n-row vector*:

$$y = [y_1 \quad y_2 \cdots y_n]$$

3. The transpose of a matrix $\mathbf{A}$ is indicated by $\mathbf{A}^T$, and it results from a *complete* interchange of rows and columns of $\mathbf{A}$. If $\mathbf{A}$ is an $(m \times n)$ matrix, $\mathbf{A}^T$ is an $(n \times m)$ matrix.

4. The transpose of a column vector is a row vector, and vice versa.

5. Two matrices are said to be equal if they contain the same number of rows *and* the same number of columns *and* if corresponding elements are equal.

6. Multiplication of a matrix $\mathbf{A}$ by a scalar k is defined as follows:

$$kA = k[a_{ij}] = [ka_{ij}] = \begin{bmatrix} ka_{11} & \cdots & ka_{1n} \\ \vdots & & \\ ka_{m1} & \cdots & ka_{mn} \end{bmatrix} \quad (1.5.1)$$

thus *every* element of $\mathbf{A}$ is multiplied by the scalar. Note the difference with scalar multiplication of a determinant, eq. (1.1.4).

7. Addition or subtraction of two matrices may be performed if they are of the same order, defining

$$\mathbf{C} = \mathbf{A} \pm \mathbf{B}$$

by

$$c_{ij} = a_{ij} \pm b_{ij} \quad (1.5.2)$$

Note that the addition or subtraction operation is *commutative*:

$$\mathbf{C} = \mathbf{A} + \mathbf{B} = \mathbf{B} + \mathbf{A}$$

8. Multiplication of matrices is possible only under special circumstances—that is, when they are *conformable*. Two matrices $\mathbf{A}$ and $\mathbf{B}$ are said to be conformable in the order $\mathbf{AB}$ if $\mathbf{A}$ has the same number of columns as $\mathbf{B}$ has rows. If $\mathbf{A}$ is an $(m \times n)$ matrix and $\mathbf{B}$ is an $(n \times p)$ matrix, then $\mathbf{C} = \mathbf{AB}$ is an $(m \times p)$ matrix whose elements are given as

$$c_{ij} = \sum_{k=1}^{n} a_{ik}b_{kj}, \quad i = 1, 2, \ldots, m \quad \text{and} \quad j = 1, 2, \ldots, p \quad (1.5.3)$$

It is necessary to distinguish between premultiplication of $\mathbf{B}$ by $\mathbf{A}$ to yield $\mathbf{AB}$, and postmultiplication of $\mathbf{B}$ by $\mathbf{A}$ to yield $\mathbf{BA}$. In general, $\mathbf{AB}$ may be defined but $\mathbf{BA}$ may not, due to nonconformability. Also, in general, even if $\mathbf{AB}$ and $\mathbf{BA}$ are defined, $\mathbf{AB} \neq \mathbf{BA}$.

Example 1. If $\mathbf{A}$ is of order $(m \times n)$ and $\mathbf{B}$ is of order $(n \times m)$, then $\mathbf{AB}$ is of order $(m \times m)$, while $\mathbf{BA}$ is of order $(n \times n)$. If $n \neq m$, the order of $\mathbf{AB} \neq$ order of $\mathbf{BA}$.

Example 2. Even if $m = n$, $\mathbf{AB} \neq \mathbf{BA}$ in general—for example

$$\begin{bmatrix} 2 & 1 \\ 0 & -1 \end{bmatrix}\begin{bmatrix} 1 & -1 \\ 2 & 1 \end{bmatrix} = \begin{bmatrix} 4 & -1 \\ -2 & -1 \end{bmatrix}$$

$$\begin{bmatrix} 1 & -1 \\ 2 & 1 \end{bmatrix}\begin{bmatrix} 2 & 1 \\ 0 & -1 \end{bmatrix} = \begin{bmatrix} 2 & 2 \\ 4 & 1 \end{bmatrix}$$

Thus the commutative law is *not* valid. The *distributive* and *associative* laws for multiplication are, however, valid:

$$\mathbf{A}[\mathbf{B} + \mathbf{C}] = \mathbf{AB} + \mathbf{AC}$$

$$\mathbf{A}(\mathbf{BC}) = (\mathbf{AB})\mathbf{C}$$

which may be shown by definitions of matrix multiplication and addition.

9. Multiplication of vectors follows in the same way.
   If

$$\mathbf{x} = \begin{bmatrix} x_1 \\ x_2 \\ \vdots \\ x_n \end{bmatrix}, \qquad \mathbf{y} = \begin{bmatrix} y_1 \\ y_2 \\ \vdots \\ y_n \end{bmatrix}$$

then

$$\mathbf{x}^T\mathbf{y} = \sum_{j=1}^{n} x_j y_j = \mathbf{y}^T\mathbf{x}$$

which is a scalar. Also,

$$\mathbf{x}^T\mathbf{x} = \sum_{j=1}^{n} x_j^2 = \|\mathbf{x}\|^2 \tag{1.5.4}$$

where $\|\mathbf{x}\|$ denotes the length of vector $\mathbf{x}$, and is called the *norm* of $\mathbf{x}$. On the other hand, note that

$$\mathbf{xy}^T = \begin{bmatrix} x_1 \\ x_2 \\ \vdots \\ x_n \end{bmatrix} [y_1 \quad y_2 \cdots y_n] = \begin{bmatrix} x_1y_1 & x_1y_2 & \cdots & x_1y_n \\ x_2y_1 & x_2y_2 & \cdots & x_2y_n \\ \vdots & \vdots & & \vdots \\ x_ny_1 & x_ny_2 & \cdots & x_ny_n \end{bmatrix}$$

is an $(n \times n)$ matrix.

10. Matrices provide a convenient shorthand notation for systems of linear algebraic equations. Thus the system

$$\begin{aligned} a_{11}x_1 + a_{12}x_2 + \cdots + a_{1n}x_n &= c_1 \\ a_{21}x_1 + a_{22}x_2 + \cdots + a_{2n}x_n &= c_2 \\ &\vdots \\ a_{m1}x_1 + a_{m2}x_2 + \cdots + a_{mn}x_n &= c_m \end{aligned} \tag{1.5.5}$$

is concisely written as

$$\mathbf{Ax} = \mathbf{c}, \tag{1.5.6}$$

where

$$\mathbf{A} = \begin{bmatrix} a_{11} & a_{12} & \cdots & a_{1n} \\ a_{21} & a_{22} & \cdots & a_{2n} \\ & \vdots & & \\ a_{m1} & a_{m2} & \cdots & a_{mn} \end{bmatrix}, \quad \mathbf{x} = \begin{bmatrix} x_1 \\ x_2 \\ \vdots \\ x_n \end{bmatrix}, \quad \text{and} \quad \mathbf{c} = \begin{bmatrix} c_1 \\ c_2 \\ \vdots \\ c_m \end{bmatrix} \tag{1.5.7}$$

1.6   SPECIAL TYPES OF MATRICES

1. A *square* matrix is one in which the number of rows equals the number of columns.

2. A *diagonal* matrix is a square matrix with all except the diagonal elements equal to zero, and some of these (but not all) may be zero as well.

3. the *identity* or *idem* matrix $\mathbf{I}_n$ is a diagonal matrix with all elements on the main diagonal equal to 1.

4. A *tridiagonal* (or *Jacobi*) matrix is a square matrix which may have nonzero terms only on the main diagonal and on the adjacent diagonals above and below it, as for example

$$
\begin{bmatrix}
a_1 & b_1 & 0 & \cdot & 0 \\
c_1 & a_2 & b_2 & & \vdots \\
0 & c_2 & a_3 & b_3 & \vdots \\
\vdots & \vdots & & c_{n-2} & a_{n-1} & b_{n-1} \\
0 & \cdot & \cdot & c_{n-1} & a_n
\end{bmatrix}
$$

5. The *null* or *zero* matrix, **0**, is one where all elements are zero.

6. A matrix where most of the elements are zero is called a *sparse* matrix.

7. A *symmetric* matrix is square, with $a_{ij} = a_{ji}$, for all values of i and j—that is, $\mathbf{A} = \mathbf{A}^T$.

8. A *skew-symmetric* matrix is also square and satisfies

$$\mathbf{A} = -\mathbf{A}^T$$

Thus $a_{ii} = 0$ for all values of i.

9. In a *triangular* matrix, $a_{ij} = 0$ for either $i > j$ or $i < j$, leading to *upper* and *lower* triangular matrices, respectively. For instance, the matrix

$$
\begin{bmatrix}
a_{11} & a_{12} & \cdots & a_{1n} \\
0 & a_{22} & \cdots & a_{2n} \\
0 & & \cdot & \cdot \\
\vdots & & \cdot & \vdots \\
0 & 0 & \cdots & a_{nn}
\end{bmatrix}
$$

is upper triangular.

1.7   THE ADJOINT AND INVERSE MATRICES

The *adjoint* matrix is defined only for a *square* matrix. Let A_{ij} be the cofactor of element a_{ij} in the determinant of a square matrix $\mathbf{A}$; then the adjoint matrix is defined as

$$\text{adj } \mathbf{A} = \begin{bmatrix} A_{11} & A_{21} & \cdots & A_{n1} \\ A_{12} & A_{22} & \cdots & A_{n2} \\ \vdots & \vdots & & \vdots \\ A_{1n} & A_{2n} & \cdots & A_{nn} \end{bmatrix} \tag{1.7.1}$$

where note that A_{ij} is in the transpose position of a_{ij}. From the definition of matrix multiplication, it follows that for the product $\mathbf{A}$ (adj $\mathbf{A}$), the (i, j) element is $\sum_{k=1}^{n} a_{ik} A_{jk}$. Thus from eq. (1.3.2), we have

$$\mathbf{A} \text{ (adj } \mathbf{A}) = D\mathbf{I} \tag{1.7.2}$$

and also

$$(\text{adj } \mathbf{A}) \mathbf{A} = D\mathbf{I} \tag{1.7.3}$$

where

$$D = |\mathbf{A}| = \det \mathbf{A}$$

Then for $D \neq 0$, we may define

$$\frac{1}{D} \text{ adj } \mathbf{A} = \mathbf{A}^{-1} \tag{1.7.4}$$

since in eqs. (1.7.2) and (1.7.3), $(1/D)$ adj $\mathbf{A}$ plays the role of an *inverse* matrix:

$$\mathbf{A}\mathbf{A}^{-1} = \mathbf{I} = \mathbf{A}^{-1}\mathbf{A} \tag{1.7.5}$$

and $\mathbf{I}$ plays the role of unity.

A matrix $\mathbf{A}$ is said to be *nonsingular* if $|\mathbf{A}| \neq 0$, and *singular* if $|\mathbf{A}| = 0$. Note that if $\mathbf{A}$ is singular, then $\mathbf{A}$ (adj $\mathbf{A}$) = $\mathbf{0}$. Thus a singular matrix has an adjoint, but *no* inverse.

We can now define various powers of square matrices. Thus, for example,

$$\mathbf{A}^2 = \mathbf{A}\mathbf{A}$$

and in general,

$$\mathbf{A}^n = \underbrace{\mathbf{A} \cdots \mathbf{A}}_{n \text{ times}} \tag{1.7.6}$$

It may be verified that if m and n are integers, then

$$\mathbf{A}^m \mathbf{A}^n = \mathbf{A}^{m+n} = \mathbf{A}^n \mathbf{A}^m \tag{1.7.7}$$

Thus the law of exponents holds for square matrices.

Also for a nonsingular $\mathbf{A}$:

$$\mathbf{A}^{-n} = (\mathbf{A}^{-1})^n$$

and

$$\mathbf{A}^m \mathbf{A}^{-n} = \underbrace{\mathbf{A} \cdots \mathbf{A}}_{m \text{ times}} \underbrace{\mathbf{A}^{-1} \cdots \mathbf{A}^{-1}}_{n \text{ times}} = \mathbf{A}^{m-n} \tag{1.7.8}$$

A direct calculation shows that

$$\mathbf{AI} = \mathbf{A} = \mathbf{IA}$$

Thus $\mathbf{I}$ is unity also in the sense that

$$\mathbf{A}^0 = \mathbf{I} \tag{1.7.9}$$

From the above, it is clear that we can define *matrix polynomials* as follows:

$$P_n(\mathbf{A}) = \sum_{j=0}^{n} a_j \mathbf{A}^j = a_0 \mathbf{I} + a_1 \mathbf{A} + a_2 \mathbf{A}^2 + \cdots + a_n \mathbf{A}^n \tag{1.7.10}$$

which is a square matrix of the same order as $\mathbf{A}$.

1.8 THE REVERSAL RULE FOR MATRIX TRANSPOSE AND INVERSE

Given two square matrices of the same order $\mathbf{A}$ and $\mathbf{B}$, it is sometimes required to calculate the inverse and transpose of their product. This is readily done, since by definition of the inverse

$$(\mathbf{AB})^{-1}(\mathbf{AB}) = \mathbf{I}$$

Postmultiplying by $\mathbf{B}^{-1}$ gives

$$(\mathbf{AB})^{-1}\mathbf{A} = \mathbf{B}^{-1}$$

and postmultiplying by $\mathbf{A}^{-1}$ yields

$$(\mathbf{AB})^{-1} = \mathbf{B}^{-1}\mathbf{A}^{-1} \tag{1.8.1}$$

Thus the inverse of a product of two matrices is the product of the inverses in the reverse order.

Similarly, it can be shown that

$$(\mathbf{AB})^T = \mathbf{B}^T\mathbf{A}^T \tag{1.8.2}$$

and therefore

$$(\mathbf{ABC})^T = \mathbf{C}^T\mathbf{B}^T\mathbf{A}^T \tag{1.8.3}$$

1.9 RANK OF A MATRIX

From a given matrix, many determinants may be formed by striking out whole rows and whole columns leaving a *square* array from which a determinant may be formed. For example, given the matrix

$$\begin{bmatrix} a_{11} & a_{12} & a_{13} & a_{14} & a_{15} \\ a_{21} & a_{22} & a_{23} & a_{24} & a_{25} \\ a_{31} & a_{32} & a_{33} & a_{34} & a_{35} \\ a_{41} & a_{42} & a_{43} & a_{44} & a_{45} \end{bmatrix}$$

striking out one column leaves a fourth-order determinant, and there are five such fourth-order determinants. Striking out two columns and one row leaves a third-order determinant, and there are 40 different such determinants. Many of these determinants may, of course, be zero.

Suppose that in a general $(m \times n)$ matrix all determinants formed by striking out whole rows and whole columns of order greater than r are zero, but there exists at least one determinant of order r which is nonzero; then the matrix is said to have *rank r*.

For example, in the matrix

$$\begin{bmatrix} -1 & 4 & 3 & 2 \\ 3 & -3 & -4 & 1 \\ 0 & 9 & 5 & 7 \end{bmatrix}$$

all third-order determinants are zero, but there are many second-order determinants which are nonzero. Therefore this matrix has rank two.

1.10   SOME IMPLICATIONS OF RANK

Consider the system of m linear algebraic equations in n unknowns:

$$a_{11}x_1 + a_{12}x_2 + \cdots + a_{1n}x_n = b_1$$
$$\vdots \qquad\qquad\qquad\qquad\qquad\qquad (1.10.1)$$
$$a_{m1}x_1 + a_{m2}x_2 + \cdots + a_{mn}x_n = b_m$$

or

$$\mathbf{Ax} = \mathbf{b} \qquad\qquad (1.10.2)$$

If $m = n$, then $\mathbf{A}$ is square and if $|\mathbf{A}| \neq 0$, then $\mathbf{A}^{-1}$ exists. Premultiplying both sides of eq. (1.10.2) by $\mathbf{A}^{-1}$, we get

$$\mathbf{A}^{-1}\mathbf{Ax} = \mathbf{A}^{-1}\mathbf{b}$$

so that the solution is given by

$$\mathbf{x} = \mathbf{A}^{-1}\mathbf{b} \qquad\qquad (1.10.3)$$

However, if $m \neq n$, then we may have either $m < n$ or $m > n$. If $m < n$, then there are less equations than unknowns and we may have *many* solutions. On the other hand, if $m > n$, we have more equations than unknowns, and so the possibility exists that the set of equations may be *incompatible*. All of this is effectively summarized in the ranks of the coefficient and the augmented matrices.

For the system (1.10.1), the *coefficient matrix* is defined as

$$\mathbf{A} = \begin{bmatrix} a_{11} & a_{12} & \cdots & a_{1n} \\ \vdots & & & \\ a_{m1} & a_{m2} & \cdots & a_{mn} \end{bmatrix}$$

while the *augmented matrix* is given by

$$
\text{aug } \mathbf{A} = \begin{bmatrix} a_{11} & a_{12} & \cdots & a_{1n} & b_1 \\ \vdots & & & & \\ a_{m1} & a_{m2} & \cdots & a_{mn} & b_m \end{bmatrix}
$$

Some general results, which are among the most important in linear algebra, can now be stated (for a proof, see Amundson, 1966, section 2.8).

General Results

1. The necessary and sufficient condition that a set of linear algebraic equations has solutions is that the rank of the coefficient matrix be the same as that of the augmented matrix.

2. If the rank of the matrix of coefficients is r and is the same as that of the augmented matrix, and if n is the number of unknowns, the values of $(n - r)$ unknowns may be arbitrarily assigned and the remaining r unknowns are then uniquely determined.

3. The necessary and sufficient condition that a set of homogeneous linear equations in n unknowns has other than the trivial solution is that the rank of the coefficient matrix be less than n.

1.11   EIGENVALUES AND EIGENVECTORS

Consider the system of equations

$$
\mathbf{A}\mathbf{x} = \lambda \mathbf{x} \tag{1.11.1}
$$

where $\mathbf{A}$ is a *square* matrix and λ is as yet an unspecified parameter. Then eq. (1.11.1) is equivalent to

$$
(\mathbf{A} - \lambda \mathbf{I})\mathbf{x} = \mathbf{0} \tag{1.11.2}
$$

where

$$
\mathbf{A} - \lambda \mathbf{I} = \begin{bmatrix} a_{11} - \lambda & a_{12} & a_{13} & \cdots & a_{1n} \\ a_{21} & a_{22} - \lambda & a_{23} & \cdots & a_{2n} \\ \vdots & \vdots & \vdots & & \vdots \\ a_{n1} & a_{n2} & a_{n3} & \cdots & a_{nn} - \lambda \end{bmatrix}
$$

Then if eq. (1.11.1) is to have a nontrivial solution, we *must* have

$$
|\mathbf{A} - \lambda \mathbf{I}| = 0 \tag{1.11.3}
$$

which implies that

$$
\begin{vmatrix} a_{11} - \lambda & a_{12} & \cdots & a_{1n} \\ a_{21} & a_{22} - \lambda & \cdots & a_{2n} \\ \vdots & \vdots & & \vdots \\ a_{n1} & a_{n2} & \cdots & a_{nn} - \lambda \end{vmatrix} = 0
$$

When expanded, this yields an nth-order polynomial equation in $\lambda : P_n(\lambda) = 0$. Since a polynomial of degree n has n zeros, $P_n(\lambda) = 0$ has n roots λ_i, $i = 1, 2, \ldots, n$. Even if all a_{ij} are real, some of the λ_i may be complex; and, if so, they occur in pairs.

The values of λ_i satisfying $P_n(\lambda) = 0$ are called *eigenvalues* of the matrix $\mathbf{A}$. Also, $(\mathbf{A} - \lambda\mathbf{I})$ is the *characteristic matrix*, $|\mathbf{A} - \lambda\mathbf{I}|$ is the *characteristic determinant*, and $|\mathbf{A} - \lambda\mathbf{I}| = 0$ is the *characteristic equation* or the *eigenvalue equation*.

We will primarily be concerned with cases where λ_i, $i = 1, 2, \ldots, n$, are distinct. Then $\mathbf{A}$ is said to have *simple* eigenvalues, and it follows that

$$P_n(\lambda_i) = 0, \quad P'_n(\lambda_i) \neq 0, \quad i = 1, 2, \ldots, n \tag{1.11.4}$$

Once the eigenvalues λ_i are determined, the set of equations

$$(\mathbf{A} - \lambda_i\mathbf{I})\,\mathbf{x} = 0 \tag{1.11.5}$$

will have nontrivial solutions, since the rank of $(\mathbf{A} - \lambda\mathbf{I})$ is less than n. The solution $\mathbf{x}_j$ which corresponds to a given eigenvalue λ_j is called an *eigenvector* and is said to belong to λ_j. $\mathbf{x}_j$ is determined only up to an arbitrary multiplicative constant and is *any* nonzero column of $\mathrm{adj}(\mathbf{A} - \lambda_j\mathbf{I})$.

In order to prove this last result, first note that the eigenvalues λ_j are determined from

$$|\mathbf{A} - \lambda\mathbf{I}| = \begin{vmatrix} a_{11} - \lambda & a_{12} & \cdots & a_{1n} \\ a_{21} & a_{22} - \lambda & \cdots & a_{2n} \\ \vdots & & & \\ a_{n1} & a_{n2} & \cdots & a_{nn} - \lambda \end{vmatrix} = 0$$

that is, $P_n(\lambda) = 0$. Then from the formula (1.1.4) for differentiating a determinant,

$$P'_n(\lambda) = \begin{vmatrix} -1 & 0 & 0 & \cdots & 0 \\ a_{21} & a_{22} - \lambda & a_{23} & \cdots & a_{2n} \\ \vdots & & & & \\ a_{n1} & a_{n2} & a_{n3} & \cdots & a_{nn} - \lambda \end{vmatrix}$$

$$+ \begin{vmatrix} a_{11} - \lambda & a_{12} & a_{13} & \cdots & a_{1n} \\ 0 & -1 & 0 & \cdots & 0 \\ \vdots & & & & \\ a_{n1} & a_{n2} & a_{n3} & \cdots & a_{nn} - \lambda \end{vmatrix}$$

$$+ \cdots \tag{1.11.6}$$

$$+ \begin{vmatrix} a_{11} - \lambda & a_{12} & a_{13} & \cdots & a_{1n} \\ \vdots & & & & \\ a_{n-1,1} & a_{n-1,2} & \cdots & a_{n-1,n-1} - \lambda & a_{n-1,n} \\ 0 & 0 & \cdots & 0 & -1 \end{vmatrix}$$

$$= (-1) \sum_{k-1}^{n} M_k$$

where M_k is a principal minor of $(\mathbf{A} - \lambda\mathbf{I})$ of order $(n - 1)$. Since for simple eigenvalues, from eq. (1.11.4), $P'_n(\lambda_j) \neq 0$, at least one of the principal minors M_j is not zero. Hence the rank of

$$\mathbf{B} = (\mathbf{A} - \lambda_j\mathbf{I}) \tag{1.11.7}$$

is $(n - 1)$, and then from Sylvester's law of nullity (cf. Amundson, 1966, page 26), the rank of $\mathrm{adj}(\mathbf{A} - \lambda_j\mathbf{I})$ equals one.

From eq. (1.11.2), $\mathbf{x}_j$ is a nontrivial solution of

$$(\mathbf{A} - \lambda_j\mathbf{I})\mathbf{x}_j = \mathbf{Bx}_j = \mathbf{0} \tag{1.11.8}$$

which may also be rewritten as

$$\sum_{k=1}^{n} b_{ik}x_k = 0, \qquad i = 1, 2, \ldots, n \tag{1.11.9}$$

where b_{ik} are the elements of $\mathbf{B}$. Since $|\mathbf{B}| = 0$, Laplace's development (1.3.2) leads to

$$\sum_{k=1}^{n} b_{ik}B_{mk} = 0 \qquad \text{for each } i \text{ and } m \tag{1.11.10}$$

where B_{mk} is the cofactor of the element b_{mk}.

Comparing eqs. (1.11.9) and (1.11.10), we have

$$x_k = B_{mk}, \qquad k = 1, 2, \ldots, n \tag{1.11.11}$$

for each m. Recalling definition (1.7.1) for the adjoint of a matrix, $\mathbf{x}_j$ thus is *any* nonzero column of $\mathrm{adj}(\mathbf{A} - \lambda_j\mathbf{I})$. In fact, since the rank of $\mathrm{adj}(\mathbf{A} - \lambda_j\mathbf{I})$ is one, its columns are constant multiples of each other.

Example. Given a matrix

$$\mathbf{A} = \begin{bmatrix} 1 & -1 \\ 2 & 4 \end{bmatrix}$$

find its eigenvalues and the corresponding eigenvectors.

The eigenvalue equation is $P_2(\lambda) = 0$—that is,

$$\begin{vmatrix} 1 - \lambda & -1 \\ 2 & 4 - \lambda \end{vmatrix} = 0$$

or

$$\lambda^2 - 5\lambda + 6 = 0$$

which has roots

$$\lambda_1 = 3 \quad \text{and} \quad \lambda_2 = 2$$

In order to determine the eigenvector $\mathbf{x}_1$ corresponding to λ_1, note that

$$\mathbf{A} - \lambda_1\mathbf{I} = \begin{bmatrix} -2 & -1 \\ 2 & 1 \end{bmatrix}$$

$$\text{adj}(\mathbf{A} - \lambda_1\mathbf{I}) = \begin{bmatrix} 1 & 1 \\ -2 & -2 \end{bmatrix}$$

and hence

$$\mathbf{x}_1 = \begin{bmatrix} 1 \\ -2 \end{bmatrix}$$

This can be verified by observing that

$$\mathbf{A}\mathbf{x}_1 = \begin{bmatrix} 1 & -1 \\ 2 & 4 \end{bmatrix} \begin{bmatrix} 1 \\ -2 \end{bmatrix} = \begin{bmatrix} 3 \\ -6 \end{bmatrix}$$

while

$$\lambda_1\mathbf{x}_1 = 3 \begin{bmatrix} 1 \\ -2 \end{bmatrix} = \begin{bmatrix} 3 \\ -6 \end{bmatrix}$$

Thus, indeed $\mathbf{A}\mathbf{x}_1 = \lambda_1\mathbf{x}_1$, as it should be.

For the eigenvector $\mathbf{x}_2$ corresponding to λ_2, we obtain

$$\mathbf{A} - \lambda_2\mathbf{I} = \begin{bmatrix} -1 & -1 \\ 2 & 2 \end{bmatrix}$$

$$\text{adj}(\mathbf{A} - \lambda_2\mathbf{I}) = \begin{bmatrix} 2 & 1 \\ -2 & -1 \end{bmatrix}$$

so that

$$\mathbf{x}_2 = \begin{bmatrix} 1 \\ -1 \end{bmatrix}$$

It is again easy to verity that $\mathbf{A}\mathbf{x}_2 = \lambda_2\mathbf{x}_2$

1.12 LINEAR INDEPENDENCE OF EIGENVECTORS

A set of vectors $\{\mathbf{z}_j\}$ is said to be *linearly independent* if the equations $c_1\mathbf{z}_1 + c_2\mathbf{z}_2 + \cdots + c_n\mathbf{z}_n = \Sigma_{j=1}^n c_j\mathbf{z}_j = \mathbf{0}$, where $\{c_j\}$ are constants, are satisfied *only* by the set $\mathbf{c} = \mathbf{0}$.

If $\mathbf{A}$ has simple eigenvalues λ_j (i.e., no repeated eigenvalues), then the set of eigenvectors $\{\mathbf{x}_j\}$ forms a linearly independent set. This property, along with the one discussed in the next section, has important consequences.

To see this, let $\mathbf{x}_i$ and $\mathbf{x}_j$ be eigenvectors which belong to λ_i and λ_j, respectively.

Then $\mathbf{x}_i$ and $\mathbf{x}_j$ are not constant multiples of each other, because if they were, then

$$\mathbf{x}_j = c\mathbf{x}_i \tag{1.12.1}$$

where c is a nonzero constant. Premultiplication by $\mathbf{A}$ yields

$$\mathbf{A}\mathbf{x}_j = c\mathbf{A}\mathbf{x}_i$$

which from eq. (1.11.5) is equivalent to

$$\lambda_j\mathbf{x}_j = c\lambda_i\mathbf{x}_i \tag{1.12.2.}$$

Also, multiplying both sides of eq. (1.12.1) by λ_j, we have

$$\lambda_j\mathbf{x}_j = c\lambda_j\mathbf{x}_i \tag{1.12.3}$$

Subtracting eq. (1.12.3) from eq. (1.12.2) yields

$$c(\lambda_i - \lambda_j)\mathbf{x}_i = \mathbf{0}$$

Since c is nonzero and $\lambda_i \neq \lambda_j$, this means that $\mathbf{x}_i = \mathbf{0}$, which is contrary to the hypothesis that $\mathbf{x}_i$ is an eigenvector. Thus it is clear that for different eigenvalues, the corresponding eigenvectors are *not* in simple proportion. Hence any two eigenvectors are linearly independent.

Suppose now, that of the n eigenvectors, *only r* are linearly independent, and for convenience let these be the first r. Then, by assumption, any of the other $(n - r)$ eigenvectors can be expressed as linear combinations of the first r, as follows:

$$\mathbf{x}_j = c_1\mathbf{x}_1 + c_2\mathbf{x}_2 + \cdots + c_r\mathbf{x}_r, \quad j = r + 1, \ldots, n \tag{1.12.4}$$

where the constants c_1 to c_r are not all zero. Premultiplying both sides by $\mathbf{A}$ gives

$$\mathbf{A}\mathbf{x}_j = c_1\mathbf{A}\mathbf{x}_1 + c_2\mathbf{A}\mathbf{x}_2 + \cdots + c_r\mathbf{A}\mathbf{x}_r$$

Since

$$\mathbf{A}\mathbf{x}_i = \lambda_i\mathbf{x}_i, \quad i = 1, \ldots, n$$

we have

$$\lambda_j\mathbf{x}_j = c_1\lambda_1\mathbf{x}_1 + c_2\lambda_2\mathbf{x}_2 + \cdots + c_r\lambda_r\mathbf{x}_r \tag{1.12.5}$$

Also, multiplying both sides of eq. (1.12.4) by λ_j yields

$$\lambda_j\mathbf{x}_j = c_1\lambda_j\mathbf{x}_1 + c_2\lambda_j\mathbf{x}_2 + \cdots + c_r\lambda_j\mathbf{x}_r \tag{1.12.6}$$

Subtracting eq. (1.12.6) from eq. (1.12.5) leads to

$$\mathbf{0} = c_1(\lambda_1 - \lambda_j)\mathbf{x}_1 + c_2(\lambda_2 - \lambda_j)\mathbf{x}_2 + \cdots + c_r(\lambda_r - \lambda_j)\mathbf{x}_r \tag{1.12.7}$$

Since all the eigenvalues are distinct, and not all c_j are zero, eq. (1.12.7) implies that there exists a set of constants, not all zero, such that the eigenvectors $\mathbf{x}_1, \mathbf{x}_2 \ldots, \mathbf{x}_r$ are linearly *dependent*. However, this is contrary to the earlier hypothesis that only $\mathbf{x}_1$, $\mathbf{x}_2, \ldots, \mathbf{x}_r$ are linearly independent. Thus the entire set of eigenvectors $\{\mathbf{x}_j\}, j = 1$ to n must be linearly independent.

1.13   THE BIORTHOGONALITY PROPERTY OF DISTINCT EIGENVECTORS

Corresponding to

$$\mathbf{A}\mathbf{x} = \lambda\mathbf{x}$$

another eigenvalue problem may also be formed:

$$\mathbf{y}^T\mathbf{A} = \eta\mathbf{y}^T \tag{1.13.1}$$

Taking transpose of both sides of eq. (1.13.1) gives

$$\mathbf{A}^T\mathbf{y} = \eta\mathbf{y} \tag{1.13.2}$$

The condition that eq. (1.13.2) has nontrivial solutions is

$$|\mathbf{A}^T - \eta\mathbf{I}| = 0$$

that is,

$$\begin{vmatrix} a_{11} - \eta & a_{21} & \cdots & a_{n1} \\ a_{12} & a_{22} - \eta & \cdots & a_{n2} \\ \vdots & \vdots & & \vdots \\ a_{1n} & a_{2n} & \cdots & a_{nn} - \eta \end{vmatrix} = 0$$

Since the value of a determinant remains the same when its rows and columns are interchanged, we have $\eta = \lambda$. Thus the two eigenvalue problems

$$\mathbf{A}\mathbf{x} = \lambda\mathbf{x} \quad \text{and} \quad \mathbf{y}^T\mathbf{A} = \eta\mathbf{y}^T$$

have the *same* eigenvalues λ_i, $i = 1$ to n. The eigenvectors, however, are *not* the same; $\mathbf{x}_j$ is one of the nonzero columns of adj($\mathbf{A} - \lambda_j\mathbf{I}$), while $\mathbf{y}_j$ is one of the nonzero columns of adj($\mathbf{A}^T - \lambda_j\mathbf{I}$). These are *not* the same unless $\mathbf{A} = \mathbf{A}^T$—that is, $\mathbf{A}$ is symmetric.

However, *columns* of adj($\mathbf{A}^T - \lambda_j\mathbf{I}$) are *rows* of adj($\mathbf{A} - \lambda_j\mathbf{I}$). Thus $\mathbf{y}_j^T$, called an *eigenrow*, is simply a nonzero row of adj($\mathbf{A} - \lambda_j\mathbf{I}$). This is very fortunate, because both eigenvectors $\mathbf{x}_j$ and $\mathbf{y}_j$ are obtained from the *same* calculation.

Similar to the set $\{\mathbf{x}_j\}$, the set of vectors $\{\mathbf{y}_j^T\}$ can also be shown to be linearly independent. We now show that these two sets possess an important property, that of *biorthogonality*. This means that each member of one set is orthogonal to each member of the other set, except for the one with which it has a common eigenvalue. The proof is quite straightforward, and it proceeds as follows.

By definition,

$$\mathbf{A}\mathbf{x}_j = \lambda_j\mathbf{x}_j \tag{1.13.3}$$

and

$$\mathbf{y}_i^T\mathbf{A} = \lambda_i\mathbf{y}_i^T, \qquad i \neq j \tag{1.13.4}$$

so that $\lambda_i \neq \lambda_j$. Premultiplying eq. (1.13.3) by $\mathbf{y}_i^T$ and postmultiplying eq. (1.13.4) by $\mathbf{x}_j$ gives

$$\mathbf{y}_i^T\mathbf{A}\mathbf{x}_j = \lambda_j\mathbf{y}_i^T\mathbf{x}_j$$

and

$$\mathbf{y}_i^T \mathbf{A} \mathbf{x}_j = \lambda_i \mathbf{y}_i^T \mathbf{x}_j$$

Since the lhs of both equations are now the same, subtracting yields

$$(\lambda_j - \lambda_i)\mathbf{y}_i^T \mathbf{x}_j = 0$$

Recalling that $\lambda_i \neq \lambda_j$, it follows that

$$\mathbf{y}_i^T \mathbf{x}_j = 0, \qquad i \neq j \tag{1.13.5a}$$

and since this is a *scalar*, we also have

$$\mathbf{x}_i^T \mathbf{y}_i = 0, \qquad i \neq j \tag{1.13.5b}$$

Let us now show that

$$\mathbf{y}_i^T \mathbf{x}_i \neq 0$$

For this, we will prove that given a set of n linearly independent vectors with n components, there exists no vector orthogonal to *all* members of the set *except* the zero vector. Let us assume that this is not true, and let $\mathbf{z}$ be such a vector, so that

$$\mathbf{z}^T \mathbf{x}_j = 0, \qquad \text{for all } j \tag{1.13.6}$$

Since

$$\mathbf{x}_j = \begin{bmatrix} x_{1j} \\ x_{2j} \\ \vdots \\ x_{nj} \end{bmatrix} \quad \text{and} \quad \mathbf{z} = \begin{bmatrix} z_1 \\ z_2 \\ \vdots \\ z_n \end{bmatrix}$$

eq. (1.13.6) is equivalent to the set of equations

$$z_1 x_{1j} + z_2 x_{2j} + \cdots + z_n x_{nj} = 0, \qquad j = 1, 2, \ldots, n$$

and, on writing these more fully,

$$\begin{aligned}
z_1 x_{11} + z_2 x_{21} + \cdots + z_n x_{n1} &= 0 \\
z_1 x_{12} + z_2 x_{22} + \cdots + z_n x_{n2} &= 0 \\
&\vdots \\
z_1 x_{1n} + z_2 x_{2n} + \cdots + z_n x_{nn} &= 0
\end{aligned} \tag{1.13.7}$$

Recalling that $\{\mathbf{x}_j\}$ are linearly independent, there exists *no* set of constants $\{c_j\}$, *not all zero*, such that

$$c_1 \mathbf{x}_1 + c_2 \mathbf{x}_2 + \cdots + c_n \mathbf{x}_n = \mathbf{0}$$

that is,

$$
c_1 \begin{bmatrix} x_{11} \\ x_{21} \\ \vdots \\ x_{n1} \end{bmatrix} + c_2 \begin{bmatrix} x_{12} \\ x_{22} \\ \vdots \\ x_{n2} \end{bmatrix} + \cdots + c_n \begin{bmatrix} x_{1n} \\ x_{2n} \\ \vdots \\ x_{nn} \end{bmatrix} = \begin{bmatrix} 0 \\ 0 \\ \vdots \\ 0 \end{bmatrix}
\tag{1.13.8}
$$

Since eq. (1.13.8) has *only* the trivial solution $\mathbf{c} = 0$, from condition 3 in "General Results" (section 1.10), we have

$$
\begin{vmatrix} x_{11} & x_{12} & \cdots & x_{1n} \\ x_{21} & x_{22} & \cdots & x_{2n} \\ \vdots & & & \\ x_{n1} & x_{n2} & \cdots & x_{nn} \end{vmatrix} \neq 0
$$

which is the same as the determinant of the coefficient matrix of the system (1.13.7). Thus $\mathbf{z} = \mathbf{0}$, which contradicts our earlier assumption.

However, since $\mathbf{y}_i^T \neq \mathbf{0}^T$, we have $\mathbf{y}_i^T \mathbf{x}_i \neq 0$

1.14   THE CASE OF A REAL SYMMETRIC MATRIX

If $\mathbf{A}$ is real and symmetric, then $\mathbf{A} = \mathbf{A}^T$ by definition, and

$$
\text{adj}(\mathbf{A} - \lambda_j \mathbf{I}) = \text{adj}(\mathbf{A}^T - \lambda_j \mathbf{I})
$$

so that $\{\mathbf{x}_j\}$ and $\{\mathbf{y}_j\}$ are identical, and from eqs. (1.13.5) we obtain

$$
\mathbf{x}_j^T \mathbf{x}_i = 0, \qquad i \neq j
\tag{1.14.1}
$$

Thus we have an *orthogonal* set of vectors $\{\mathbf{x}_j\}$—that is, each member of the set is orthogonal to every other member of the set. Also, the eigenvalues are *real*, because, if not, let

$$
\lambda_+ = \alpha + i\beta, \qquad \lambda_- = \alpha - i\beta
$$

be a pair of complex conjugate eigenvalues and let $\mathbf{x}_+$ and $\mathbf{x}_-$ be the corresponding eigenvectors. Then from eq. (1.14.1)

$$
\mathbf{x}_+^T \mathbf{x}_- = 0
$$

However, $\mathbf{x}_+$ and $\mathbf{x}_-$ are complex conjugates, because they are nonzero columns of $\text{adj}(\mathbf{A} - \lambda_\pm \mathbf{I})$. By denoting

$$
\mathbf{x}_\pm = [(a_1 \pm i b_1)(a_2 \pm i b_2) \cdots (a_n \pm i b_n)]^T
$$

we have

$$
\mathbf{x}_+^T \mathbf{x}_- = \sum_{i=1}^{n} (a_i^2 + b_i^2)
$$

which can only be zero for $\mathbf{a} = \mathbf{b} = \mathbf{0}$. However, $\mathbf{x}_\pm$ will then be zero vectors, which they cannot be because they are eigenvectors. Thus the eigenvalues of a real symmetric matrix are always real.

1.15 EXPANSION OF AN ARBITRARY VECTOR

As noted above, each square matrix $\mathbf{A}$ generates two sets of linearly independent vectors through the eigenvalue problems associated with it:

$$\mathbf{A}\mathbf{x} = \lambda\mathbf{x} \quad \text{and} \quad \mathbf{A}^T\mathbf{y} = \eta\mathbf{y}$$

and $\lambda_i = \eta_i$, $i = 1$ to n. The two sets of eigenvectors corresponding to $\{\lambda_j\}$, that is $\{\mathbf{x}_j\}$ and $\{\mathbf{y}_j\}$, are nonzero columns and nonzero rows, respectively, of $\mathrm{adj}(\mathbf{A} - \lambda_j\mathbf{I})$. Furthermore, $\{\mathbf{x}_j\}$ and $\{\mathbf{y}_j\}$ satisfy the biorthogonality relation

$$
\begin{aligned}
\mathbf{y}_i^T\mathbf{x}_j &= 0, \qquad i \neq j \\
&\neq 0, \qquad i = j
\end{aligned}
\tag{1.15.1}
$$

If we now consider an arbitrary vector $\mathbf{z}$ with the same number of components as vectors $\{\mathbf{x}_j\}$, we can expand it in terms of a linear combination of the set of eigenvectors $\{\mathbf{x}_j\}$:

$$\mathbf{z} = \alpha_1\mathbf{x}_1 + \alpha_2\mathbf{x}_2 + \cdots + \alpha_n\mathbf{x}_n = \sum_{j=1}^{n} \alpha_j\mathbf{x}_j \tag{1.15.2}$$

where the α_j are constants, yet to be determined. For this, premultiplying the lhs by $\mathbf{y}_i^T$ gives

$$
\begin{aligned}
\mathbf{y}_i^T\mathbf{z} &= \sum_{j=1}^{n} \alpha_j\mathbf{y}_i^T\mathbf{x}_j \\
&= \alpha_i\mathbf{y}_i^T\mathbf{x}_i
\end{aligned}
$$

from eq. (1.15.1). Thus we have

$$\alpha_i = \frac{\mathbf{y}_i^T\mathbf{z}}{\mathbf{y}_i^T\mathbf{x}_i} \tag{1.15.3}$$

and hence

$$\mathbf{z} = \sum_{j=1}^{n} \frac{\mathbf{y}_j^T\mathbf{z}}{\mathbf{y}_j^T\mathbf{x}_j} \mathbf{x}_j \tag{1.15.4}$$

If $\mathbf{A}$ is symmetric, then $\mathbf{y}_j = \mathbf{x}_j$, and

$$\mathbf{z} = \sum_{j=1}^{n} \frac{\mathbf{x}_j^T\mathbf{z}}{\mathbf{x}_j^T\mathbf{x}_j} x_j \tag{1.15.5}$$

The expansion of an arbitrary vector in terms of the eigenvectors of a matrix has important theoretical and practical implications. On the theoretical side, it implies that every finite n-dimensional vector space is *spanned* by the eigenvectors of *any* matrix of order n with simple eigenvalues. This is the basis of *spectral theory*, which connects solutions of equations in finite-dimensional vector spaces with those in infinite-dimensional vector spaces (or function spaces) as encountered in linear boundary value problems.

Two practical implications of these results are treated in the following sections.

1.16   APPLICATION TO SOLVING A SYSTEM OF LINEAR ALGEBRAIC EQUATIONS

Suppose we want to solve the set of n equations

$$\mathbf{Ax} = \mathbf{b} \tag{1.16.1}$$

in n unknowns $\mathbf{x}$, where $\mathbf{A}$ is a given nonsingular coefficient matrix with simple eigenvalues, and $\mathbf{b}$ is the known vector of constants. Since $|\mathbf{A}| \neq 0$, the set has a unique solution. Recalling that an arbitrary vector can be expanded in terms of the eigenvectors of any matrix, let

$$\mathbf{x} = \sum_{j=1}^{n} \alpha_j \mathbf{x}_j \tag{1.16.2}$$

and

$$\mathbf{b} = \sum_{j=1}^{n} \beta_j \mathbf{x}_j \tag{1.16.3}$$

where the eigenvalue problem for $\mathbf{A}$ is

$$\mathbf{Ax} = \lambda \mathbf{x} \tag{1.16.4}$$

and β_j are known from eq. (1.15.4). To find the solution of eqs. (1.16.1), we need to determine the constants α_j, $j = 1$ to n.

Substituting eqs. (1.16.2) and (1.16.3) in eq. (1.16.1) gives

$$\mathbf{A} \sum_{j=1}^{n} \alpha_j \mathbf{x}_j = \sum_{j=1}^{n} \beta_j \mathbf{x}_j \tag{1.16.5}$$

Since from eq. (1.16.4) the lhs equals $\sum_{j=1}^{n} \alpha_j \lambda_j \mathbf{x}_j$, eq. (1.16.5) can be rearranged as

$$\sum_{j=1}^{n} (\alpha_j \lambda_j - \beta_j) \mathbf{x}_j = \mathbf{0}$$

Now, since the set of eigenvectors $\{\mathbf{x}_j\}$ is linearly independent, there exists no set of constants $\{c_j\}$, not all zero, such that $\sum_{j=1}^{n} c_j \mathbf{x}_j = \mathbf{0}$. Thus

$$\alpha_j \lambda_j = \beta_j, \qquad j = 1, 2, \ldots, n \tag{1.16.6}$$

whence the unknown constants α_j are determined as

$$\alpha_j = \beta_j / \lambda_j, \qquad j = 1, 2, \ldots, n \tag{1.16.7}$$

Note that division by λ_j, as indicated in eq. (1.16.7), is permissible *only* if $\lambda_j \neq 0$, $j = 1$ to n. If $|\mathbf{A}| \neq 0$, then the eigenvalues of $\mathbf{A}$ are also all nonzero. This follows because if $|\mathbf{A}| \neq 0$ and $\lambda = 0$ is an eigenvalue, then eq. (1.16.4) implies

$$\mathbf{Ax} = \mathbf{0} \qquad (1.16.8)$$

where $\mathbf{x}$ is the eigenvector corresponding to the zero eigenvalue. Since nontrivial solutions of (1.16.8) can occur only if $|\mathbf{A}| = 0$, we must have $\mathbf{x} = \mathbf{0}$. However, $\mathbf{x}$ is an eigenvector; hence it cannot be the zero vector. Thus if $|\mathbf{A}| \neq 0$, $\lambda = 0$ is not an eigenvalue of $\mathbf{A}$.

The solution of eq. (1.16.1) is then given by

$$\mathbf{x} = \sum_{j=1}^{n} \frac{\beta_j}{\lambda_j} \mathbf{x}_j \qquad (1.16.9)$$

Similarly, the solution of the set of equations

$$\mathbf{Ax} - \gamma\mathbf{x} = \mathbf{b} \qquad (1.16.10)$$

is

$$\mathbf{x} = \sum_{j=1}^{n} \frac{\beta_j}{\lambda_j - \gamma} \mathbf{x}_j \qquad (1.16.11)$$

which exists only if γ is not an eigenvalue of $\mathbf{A}$.

1.17 APPLICATION TO SOLVING A SYSTEM OF FIRST-ORDER LINEAR ORDINARY DIFFERENTIAL EQUATIONS

A system of first-order linear ordinary differential equations (ODEs) with constant coefficients can also be solved using the idea of expansion of an arbitrary vector in terms of eigenvectors. The solution thus obtained forms the basis of *stability theory*, a topic which is considered in chapter 2.

Consider the system

$$\frac{dx_1}{dt} = a_{11}x_1 + a_{12}x_2 + \cdots + a_{1n}x_n + b_1$$

$$\frac{dx_2}{dt} = a_{21}x_1 + a_{22}x_2 + \cdots + a_{2n}x_n + b_2$$

$$\vdots \qquad\qquad\qquad (1.17.1)$$

$$\frac{dx_n}{dt} = a_{n1}x_1 + a_{n2}x_2 + \cdots + a_{nn}x_n + b_n$$

where the a_{ij} and b_i are constants. This can be written succinctly as

$$\frac{d\mathbf{x}}{dt} = \mathbf{Ax} + \mathbf{b} \qquad (1.17.2)$$

and let the initial conditions (ICs) be

$$\mathbf{x} = \mathbf{x}_0 \qquad \text{at } t = 0 \tag{1.17.3}$$

The object is to find how $\mathbf{x}$ varies with the independent variable t—that is, $\mathbf{x}(t)$. For this, we first consider the associated *homogeneous* equation:

$$\frac{d\mathbf{x}}{dt} = \mathbf{Ax} \tag{1.17.4}$$

Assume that the solution of eq. (1.17.4) has the form

$$\mathbf{x} = \mathbf{z}e^{\lambda t} \tag{1.17.5}$$

where $\mathbf{z}$ is an unknown vector of constants, and λ is also an as yet unknown constant. Substituting eq. (1.17.5) in eq. (1.17.4) gives

$$\mathbf{Az}e^{\lambda t} = \lambda \mathbf{z}e^{\lambda t}$$

and since $e^{\lambda t} \neq 0$, we have

$$\mathbf{Az} = \lambda \mathbf{z} \tag{1.17.6}$$

Since the assumed solution (1.17.5) requires $\mathbf{z} \neq \mathbf{0}$, eq. (1.17.6) implies that λ is an eigenvalue of $\mathbf{A}$ and $\mathbf{z}$ is the eigenvector which belongs to λ. Noting that $\mathbf{A}$ is a square matrix of order n, there are n values of λ satisfying

$$|\mathbf{A} - \lambda_j \mathbf{I}| = 0, \qquad j = 1, 2, \ldots, n$$

and corresponding to each of these is an eigenvector $\mathbf{z}_j$.

Thus there are n solutions of the form (1.17.5), and the general solution of the homogeneous equation (1.17.4) is then a linear combination of these:

$$\mathbf{x} = \sum_{j=1}^{n} c_j \mathbf{z}_j e^{\lambda_j t} \tag{1.17.7}$$

where the c_j are arbitrary constants.

A *particular solution* of eq. (1.17.2) can be readily obtained by assuming that it is a constant vector $\mathbf{k}$. Substituting $\mathbf{x} = \mathbf{k}$ in eq. (1.17.2) yields

$$\mathbf{0} = \mathbf{Ak} + \mathbf{b}$$

so that

$$\mathbf{k} = -\mathbf{A}^{-1}\mathbf{b} \tag{1.17.8}$$

Since eq. (1.17.7) is the general solution of the homogeneous equation (1.17.4), and (1.17.8) is a particular solution of eq. (1.17.2), the general solution of the non-homogeneous equation (1.17.2) is then

$$\mathbf{x} = \sum_{j=1}^{n} c_j \mathbf{z}_j e^{\lambda_j t} - \mathbf{A}^{-1}\mathbf{b} \tag{1.17.9}$$

where the c_j are constants, yet to be determined. It may be verified readily by direct substitution that eq. (1.17.9) indeed solves eq. (1.17.2). However, the ICs (1.17.3) have not been utilized yet, and when these are invoked the constants c_j are obtained as shown below. From eqs. (1.17.3) and (1.17.9), we have

$$\mathbf{x}_0 = \sum_{j=1}^{n} c_j \mathbf{z}_j - \mathbf{A}^{-1}\mathbf{b} \qquad (1.17.10)$$

which is a set of n equations in n unknowns c_j.

Since $\{\mathbf{z}_j\}$ form a linearly independent set, the rank of the system (1.17.10) is n, and the c_j can therefore be *uniquely* determined. For this, let $\{\mathbf{w}_j\}$ be the biorthogonal set of eigenvectors of $\mathbf{A}$ corresponding to $\{\mathbf{z}_j\}$. These are simply the nonzero rows of $\mathrm{adj}(\mathbf{A} - \lambda_j \mathbf{I})$ and satisfy

$$\begin{aligned} \mathbf{w}_k^T \mathbf{z}_j &= 0, & k \neq j \\ &\neq 0, & k = j \end{aligned} \qquad (1.17.11)$$

Premultiplying both sides of eq. (1.17.10) by $\mathbf{w}_k^T$ gives

$$\sum_{j=1}^{n} c_j \mathbf{w}_k^T \mathbf{z}_j = \mathbf{w}_k^T [\mathbf{x}_0 + \mathbf{A}^{-1}\mathbf{b}]$$

Owing to eq. (1.17.11), the only term in the sum on the lhs which is nonzero is for $j = k$; hence

$$c_j = \frac{\mathbf{w}_j^T [\mathbf{x}_0 + \mathbf{A}^{-1}\mathbf{b}]}{\mathbf{w}_j^T \mathbf{z}_j}, \qquad j = 1, 2, \ldots, n \qquad (1.17.12)$$

The unknown constants in the general solution (1.17.9) are thus determined, and the solution of eqs. (1.17.2) *with* ICs (1.17.3) is then given by

$$\mathbf{x} = \sum_{j=1}^{n} \frac{\mathbf{w}_j^T [\mathbf{x}_0 + \mathbf{A}^{-1}\mathbf{b}]}{\mathbf{w}_j^T \mathbf{z}_j} \mathbf{z}_j e^{\lambda_j t} - \mathbf{A}^{-1}\mathbf{b} \qquad (1.17.13)$$

If $\mathbf{A}$ is symmetric, the λ_j are all real from section 1.14, and $\mathbf{w}_j = \mathbf{z}_j$, then

$$\mathbf{x} = \sum_{j=1}^{n} \frac{\mathbf{z}_j^T [\mathbf{x}_0 + \mathbf{A}^{-1}\mathbf{b}]}{\mathbf{z}_j^T \mathbf{z}_j} \mathbf{z}_j e^{\lambda_j t} - \mathbf{A}^{-1}\mathbf{b} \qquad (1.17.14)$$

Whenever a "solution" is obtained, it is always a good practice to verify that it is indeed a solution. This can be checked by simply substituting it in the original differential equation and verifying that the resulting equation is identically satisfied, and also that the initial condition holds. It is left as an exercise to ascertain that both of the above statements are true for the solution represented by eq. (1.17.13).

1.18 SOME ASPECTS OF THE EIGENVALUE PROBLEM

As noted in section 1.11, in the system of n equations in n unknowns

$$\mathbf{A}\mathbf{x} = \lambda \mathbf{x}$$

nontrivial solutions **x** exist only for certain specific values of λ, called *eigenvalues*. These are the solutions of

$$|\mathbf{A} - \lambda\mathbf{I}| = 0$$

which upon expansion leads to a polynomial equation of degree n in λ:

$$P_n(\lambda) = 0$$

The polynomial $P_n(\lambda)$ must be written down explicitly in order to obtain numerical values of λ. For this, it is frequently helpful to consider the following expansion of $|\mathbf{A} - \lambda\mathbf{I}| = 0$, obtained from Laplace's development of the determinant:

$$P_n(\lambda) = (-1)^n\lambda^n + s_1\lambda^{n-1} + s_2\lambda^{n-2} + \cdots + s_{n-1}\lambda + |\mathbf{A}| = 0 \quad (1.18.1)$$

where

$$s_j = (-1)^{n-j}[\text{sum of all principal minors of order } j \text{ of } \mathbf{A}], \quad (1.18.2)$$

$$j = 1, 2, \ldots, (n-1)$$

and the principal minors have been defined in section 1.2.

Example. Let

$$\mathbf{A} = \begin{bmatrix} 1 & -4 & -1 \\ 2 & 0 & 5 \\ -1 & 1 & -2 \end{bmatrix}$$

and then from eq. (1.18.2), we have

$$s_1 = (-1)^{3-1}[1 + 0 - 2] = -1$$

and

$$s_2 = (-1)^{3-2}\left[\begin{vmatrix} 0 & 5 \\ 1 & -2 \end{vmatrix} + \begin{vmatrix} 1 & -1 \\ -1 & -2 \end{vmatrix} + \begin{vmatrix} 1 & -4 \\ 2 & 0 \end{vmatrix} \right] = 0$$

while

$$|\mathbf{A}| = -3$$

Thus the eigenvalue equation (1.18.1) is

$$P_3(\lambda) = -\lambda^3 - \lambda^2 - 3 = 0 \quad (1.18.3)$$

Alternatively, proceeding directly using expansion of $|\mathbf{A} - \lambda\mathbf{I}|$ by Laplace's development gives

$$\begin{vmatrix} 1-\lambda & -4 & -1 \\ 2 & -\lambda & 5 \\ -1 & 1 & -2-\lambda \end{vmatrix} = 0$$

or

$$(1 - \lambda)[\lambda^2 + 2\lambda - 5] + 4[-2\lambda + 1] - 1[-\lambda + 2] = 0$$

which leads to eq. (1.18.3).

Some useful relations between the various eigenvalues, λ_j can be obtained by rewriting eq. (1.18.1) in the equivalent form:

$$P_n(\lambda) = (-1)^n[(\lambda - \lambda_1)(\lambda - \lambda_2) \cdots (\lambda - \lambda_n)] \tag{1.18.4}$$

where the multiplier $(-1)^n$ arises because it is the coefficient of the leading term in eq. (1.18.1). Also, eq. (1.18.4) may be rearranged as

$$P_n(\lambda) = (-1)^n \left[\lambda^n - \lambda^{n-1} \sum_{j=1}^{n} \lambda_j + \right.$$

$$\left. \lambda^{n-2} \sum_{\substack{j,k=1 \\ j \neq k}}^{n} \lambda_j \lambda_k + \cdots + (-1)^n \prod_{j=1}^{n} \lambda_j \right] \tag{1.18.5}$$

Comparing the coefficients of the λ^{n-1} term in eqs. (1.18.1) and (1.18.5), we have

$$(-1)^{n+1} \sum_{j=1}^{n} \lambda_j = s_1 = (-1)^{n-1}[\text{sum of principal minors of order 1 of } \mathbf{A}]$$

This implies that

$$\sum_{j=1}^{n} \lambda_j = \text{tr } \mathbf{A} \tag{1.18.6}$$

where tr $\mathbf{A}$ is the *trace of* $\mathbf{A}$—that is, the sum of the diagonal terms of $\mathbf{A}$.

Similarly, comparing the constant terms in eqs. (1.18.1) and (1.18.5), we obtain

$$\prod_{j=1}^{n} \lambda_j = |\mathbf{A}| \tag{1.18.7}$$

Note that either eq. (1.18.1) or eq. (1.18.7) implies that if $|\mathbf{A}| = 0$, then $\lambda = 0$ is an eigenvalue of $\mathbf{A}$.

Example. Let

$$\mathbf{A} = \begin{bmatrix} a_{11} & a_{12} \\ a_{21} & a_{22} \end{bmatrix}$$

Then $|\mathbf{A} - \lambda\mathbf{I}| = 0$ is equivalent to

$$(a_{11} - \lambda)(a_{22} - \lambda) - a_{21}a_{12} = 0$$

which rearranges as

$$\lambda^2 - (a_{11} + a_{22})\lambda + (a_{11}a_{22} - a_{21}a_{12}) = 0$$

or

$$\lambda^2 - (\text{tr } \mathbf{A})\lambda + |\mathbf{A}| = 0$$

The two eigenvalues are given by

$$\lambda_\pm = \tfrac{1}{2}[(\text{tr } \mathbf{A}) \pm \sqrt{(\text{tr } \mathbf{A})^2 - 4|\mathbf{A}|}]$$

from which it is easily seen that

$$\lambda_+ + \lambda_- = \text{tr } \mathbf{A}$$

and

$$\lambda_+ \lambda_- = |\mathbf{A}|$$

in agreement with eqs. (1.18.6) and (1.18.7).

1.19   SOME OBSERVATIONS CONCERNING EIGENVALUES

Let $\mathbf{A}$ be a square matrix; then the eigenvalue problem is

$$\mathbf{Ax} = \lambda \mathbf{x} \qquad (1.19.1)$$

Premultiplying both sides by $\mathbf{A}$ gives

$$\mathbf{AAx} = \mathbf{A}^2 \mathbf{x} = \lambda \mathbf{Ax}$$
$$= \lambda^2 \mathbf{x}, \qquad \text{using eq. (1.19.1)}$$

Similarly,

$$\mathbf{A}^m \mathbf{x} = \lambda^m \mathbf{x}, \qquad m = 1, 2, \dots \qquad (1.19.2)$$

thus if λ is an eigenvalue of $\mathbf{A}$, λ^m is an eigenvalue of $\mathbf{A}^m$ where m is a positive integer. Also, premultiplying both sides of eq. (1.19.1) by $\mathbf{A}^{-1}$ yields

$$\mathbf{A}^{-1}\mathbf{Ax} = \lambda \mathbf{A}^{-1}\mathbf{x}$$

or

$$\mathbf{A}^{-1}\mathbf{x} = \frac{1}{\lambda}\mathbf{x}$$

and similarly,

$$\mathbf{A}^{-m}\mathbf{x} = \frac{1}{\lambda^m}\mathbf{x}, \qquad m = 1, 2, \dots \qquad (1.19.3)$$

which implies that if λ is an eigenvalue of $\mathbf{A}$, then $1/\lambda^m$ is an eigenvalue of $\mathbf{A}^{-m}$ for positive integers m. Thus

$$\mathbf{A}^j \mathbf{x} = \lambda^j \mathbf{x} \qquad (1.19.4)$$

where j is a positive *or* negative integer. Multiplying both sides of eq. (1.19.4) by a scalar α_j,

$$\alpha_j \mathbf{A}^j \mathbf{x} = \alpha_j \lambda^j \mathbf{x}$$

leads to

$$\sum_{j=-M}^{N} \alpha_j \mathbf{A}^j \mathbf{x} = \sum_{j=-M}^{N} \alpha_j \lambda^j \mathbf{x}$$

where M and N are positive integers. Hence we may state that

$$Q(\mathbf{A})\mathbf{x} = Q(\lambda)\mathbf{x} \qquad (1.19.5)$$

where $Q(\mathbf{A})$ is a matrix polynomial. Thus if λ is an eigenvalue of $\mathbf{A}$, $Q(\lambda)$ is an eigenvalue of $Q(\mathbf{A})$.

1.20 HAMILTON–CAYLEY THEOREM AND SOME USEFUL DEDUCTIONS

The Hamilton–Cayley (H–C) theorem is the building block for advanced topics in matrix analysis. We simply state the theorem without proof, which can be found elsewhere (cf. Amundson, 1966, section 5.11), and note some important results that follow from it.

The theorem states that: ''Every square matrix satisfies its own characteristic (or eigenvalue) equation.''

Now, given a square matrix $\mathbf{A}$ of order n, let the eigenvalue equation be

$$P_n(\lambda) = \lambda^n + a_1\lambda^{n-1} + a_2\lambda^{n-2} + \cdots + a_{n-1}\lambda + a_n = 0 \qquad (1.20.1)$$

Then according to the H–C theorem,

$$P_n(\mathbf{A}) = \mathbf{A}^n + a_1\mathbf{A}^{n-1} + a_2\mathbf{A}^{n-2} + \cdots + a_{n-1}\mathbf{A} + a_n\mathbf{I} = \mathbf{0} \qquad (1.20.2)$$

where $P_n(\mathbf{A})$ is a matrix polynomial of order n.

Example. Given the matrix

$$\mathbf{A} = \begin{bmatrix} 1 & -1 \\ 2 & 1 \end{bmatrix}$$

show that $\mathbf{A}$ satisfies its own characteristic equation.

The characteristic equation for $\mathbf{A}$ is

$$\begin{vmatrix} 1 - \lambda & -1 \\ 2 & 1 - \lambda \end{vmatrix} = 0$$

which, upon expansion, leads to

$$\lambda^2 - 2\lambda + 3 = 0$$

Also,

$$\mathbf{A}^2 = \begin{bmatrix} 1 & -1 \\ 2 & 1 \end{bmatrix}\begin{bmatrix} 1 & -1 \\ 2 & 1 \end{bmatrix} = \begin{bmatrix} -1 & -2 \\ 4 & -1 \end{bmatrix}$$

and then we have

$$\mathbf{A}^2 - 2\mathbf{A} + 3\mathbf{I} = \begin{bmatrix} -1 & -2 \\ 4 & -1 \end{bmatrix} - 2\begin{bmatrix} 1 & -1 \\ 2 & 1 \end{bmatrix} + 3\begin{bmatrix} 1 & 0 \\ 0 & 1 \end{bmatrix} = \mathbf{0}$$

Some useful deductions from the H–C theorem can now be made. Let

$$P_n(\lambda) = \lambda^n + a_1\lambda^{n-1} + a_2\lambda^{n-2} + \cdots + a_{n-1}\lambda + a_n = 0$$

be the characteristic equation of a square matrix $\mathbf{A}$. Then since from the H–C theorem,

$$P_n(\mathbf{A}) = \mathbf{A}^n + a_1\mathbf{A}^{n-1} + a_2\mathbf{A}^{n-2} + \cdots + a_{n-1}\mathbf{A} + a_n\mathbf{I} = \mathbf{0}$$

we have

$$\mathbf{A}^n = -[a_1\mathbf{A}^{n-1} + a_2\mathbf{A}^{n-2} + \cdots + a_{n-1}\mathbf{A} + a_n\mathbf{I}] \qquad (1.20.3)$$

Thus the nth power of a matrix can be expressed as a linear combination of the first $(n-1)$ powers of $\mathbf{A}$. Also, pre- or postmultiplying both sides of eq. (1.20.3) by $\mathbf{A}$ gives

$$\mathbf{A}^{n+1} = -[a_1\mathbf{A}^n + a_2\mathbf{A}^{n-1} + \cdots + a_{n-1}\mathbf{A}^2 + a_n\mathbf{A}]$$

which, by using eq. (1.20.3) for $\mathbf{A}^n$, leads to

$$\begin{aligned}
\mathbf{A}^{n+1} &= +a_1[a_1\mathbf{A}^{n-1} + a_2\mathbf{A}^{n-2} + \cdots + a_{n-1}\mathbf{A} + a_n\mathbf{I}] \\
&\quad - [a_2\mathbf{A}^{n-1} + \cdots + a_{n-1}\mathbf{A}^2 + a_n\mathbf{A}] \qquad (1.20.4) \\
&= b_1\mathbf{A}^{n-1} + b_2\mathbf{A}^{n-2} + \cdots + b_{n-1}\mathbf{A} + b_n\mathbf{I}
\end{aligned}$$

where $b_i = a_1 a_i - a_{i+1}$, $i = 1$ to $(n-1)$, and $b_n = a_1 a_n$.

Thus the $(n+1)$st power of $\mathbf{A}$ can also be expressed as a linear combination of the first $(n-1)$ powers of $\mathbf{A}$. It similarly follows that *any positive* integral power of $\mathbf{A}$ can be represented as a linear combination of $\mathbf{A}^i$, $i = 0, 1, 2, \ldots, (n-1)$.

Also, from eq. (1.20.3), if $a_n \neq 0$ (i.e., $|\mathbf{A}| \neq 0$), we have

$$\mathbf{I} = -\frac{1}{a_n}[\mathbf{A}^n + a_1\mathbf{A}^{n-1} + \cdots + a_{n-1}\mathbf{A}]$$

Now, pre- or postmultiplying by $\mathbf{A}^{-1}$ yields

$$\mathbf{A}^{-1} = -\frac{1}{a_n}[\mathbf{A}^{n-1} + a_1\mathbf{A}^{n-2} + \cdots + a_{n-2}\mathbf{A} + a_{n-1}\mathbf{I}] \qquad (1.20.5)$$

Similarly, we obtain

$$\begin{aligned}
\mathbf{A}^{-2} &= -\frac{1}{a_n}[\mathbf{A}^{n-2} + a_1\mathbf{A}^{n-3} + \cdots + a_{n-2}\mathbf{I} + a_{n-1}\mathbf{A}^{-1}] \\
&= c_1\mathbf{A}^{n-1} + c_2\mathbf{A}^{n-2} + \cdots + c_{n-1}\mathbf{A} + c_n\mathbf{I} \qquad (1.20.6)
\end{aligned}$$

using eq. (1.20.5) for $\mathbf{A}^{-1}$.

Thus it follows that *any negative* integral power of a nonsingular matrix $\mathbf{A}$ can also be expressed as a linear combination of $\mathbf{A}^i$, $i = 0, 1, 2, \ldots, (n-1)$.

1.21 SYLVESTER'S THEOREM

Sylvester's theorem builds on the Hamilton–Cayley theorem, and it is important both theoretically and in making computations. Consider a polynomial of degree $(n-1)$:

$$Q(x) = a_1 x^{n-1} + a_2 x^{n-2} + \cdots + a_{n-1}x + a_n$$

where the a_j are constants. We can rewrite the polynomial in Lagrangian form:

$$
\begin{aligned}
Q(x) = {} & c_1(x - \alpha_2)(x - \alpha_3) \cdots (x - \alpha_n) \\
& + c_2(x - \alpha_1)(x - \alpha_3) \cdots (x - \alpha_n) + \cdots \\
& + c_j(x - \alpha_1)(x - \alpha_2) \cdots (x - \alpha_{j-1})(x - \alpha_{j+1}) \cdots (x - \alpha_n) + \cdots \\
& + c_n(x - \alpha_1)(x - \alpha_2) \cdots (x - \alpha_{n-1})
\end{aligned}
$$

where $\alpha_j, j = 1, 2, \ldots, n$ are arbitrary scalars, while the constants c_j are related to the constants a_j. A convenient short form notation for $Q(x)$ is

$$
Q(x) = \sum_{j=1}^{n} c_j \prod_{k=1,j}^{n} (x - \alpha_k) \tag{1.21.1}
$$

where

$$
c_j = \frac{Q(\alpha_j)}{\prod_{k=1,j}^{n}(\alpha_j - \alpha_k)}
$$

and $\prod_{k=1,j}^{n} d_k$ denotes product of all terms d_k, for k varying from 1 to n, but $k \neq j$.

Now, consider the matrix polynomial of degree $(n - 1)$, where $\mathbf{A}$ is a square matrix of order n:

$$
\begin{aligned}
Q(\mathbf{A}) = {} & c_1(\mathbf{A} - \alpha_2\mathbf{I})(\mathbf{A} - \alpha_3\mathbf{I}) \cdots (\mathbf{A} - \alpha_n\mathbf{I}) \\
& + c_2(\mathbf{A} - \alpha_1\mathbf{I})(\mathbf{A} - \alpha_3\mathbf{I}) \cdots (\mathbf{A} - \alpha_n\mathbf{I}) + \cdots \\
& + c_j(\mathbf{A} - \alpha_1\mathbf{I})(\mathbf{A} - \alpha_2\mathbf{I}) \cdots (\mathbf{A} - \alpha_{j-1}\mathbf{I})(\mathbf{A} - \alpha_{j+1}\mathbf{I}) \cdots (\mathbf{A} - \alpha_n\mathbf{I}) + \cdots \\
& + c_n(\mathbf{A} - \alpha_1\mathbf{I})(\mathbf{A} - \alpha_2\mathbf{I}) \cdots (\mathbf{A} - \alpha_{n-1}\mathbf{I})
\end{aligned}
$$

$$
= \sum_{j=1}^{n} c_j \prod_{k=1,j}^{n} (\mathbf{A} - \alpha_k\mathbf{I}) \tag{1.21.2}
$$

The α_j were, so far, arbitrary; now let $\alpha_j = \lambda_j$, the eigenvalues of $\mathbf{A}$ and let $\mathbf{A}$ have simple eigenvalues. Then

$$
Q(\mathbf{A}) = \sum_{j=1}^{n} c_j \prod_{k=1,j}^{n} (\mathbf{A} - \lambda_k\mathbf{I}) \tag{1.21.3}
$$

Postmultiplying both sides by $\mathbf{x}_m$, the eigenvector corresponding to λ_m yields

$$
Q(\mathbf{A})\mathbf{x}_m = \sum_{j=1}^{n} c_j \left[\prod_{k=1,j}^{n} (\mathbf{A} - \lambda_k\mathbf{I}) \right] \mathbf{x}_m \tag{1.21.4}
$$

Since $\mathbf{A}\mathbf{x}_j = \lambda_j\mathbf{x}_j$, every term in the sum (1.21.4) contains the term $(\lambda_m - \lambda_m)$ *except* for the term which arises from $j = m$. Thus eq. (1.21.4) reduces to

$$
\begin{aligned}
Q(\mathbf{A})\mathbf{x}_m & = c_m \left[\prod_{k=1,m}^{n} (\mathbf{A} - \lambda_k\mathbf{I}) \right] \mathbf{x}_m \\
& = c_m(\lambda_m - \lambda_1)(\lambda_m - \lambda_2) \cdots (\lambda_m - \lambda_{m-1})(\lambda_m - \lambda_{m+1}) \cdots (\lambda_m - \lambda_n)\mathbf{x}_m \\
& = c_m \left[\prod_{k=1,m}^{n} (\lambda_m - \lambda_k) \right] \mathbf{x}_m
\end{aligned} \tag{1.21.5}
$$

However, from eq. (1.19.5) we have

$$Q(\mathbf{A})\mathbf{x}_m = Q(\lambda_m)\mathbf{x}_m$$

so that eq. (1.21.5) becomes

$$\left\{Q(\lambda_m) - c_m\left[\prod_{k=1,m}^{n}(\lambda_m - \lambda_k)\right]\right\}\mathbf{x}_m = \mathbf{0}$$

Since $\mathbf{x}_m \neq \mathbf{0}$, this implies that

$$c_m = \frac{Q(\lambda_m)}{\prod_{k=1,m}^{n}(\lambda_m - \lambda_k)}, \qquad m = 1, 2, \ldots, n \qquad (1.21.6)$$

So, if we represent a matrix polynomial $Q(\mathbf{A})$ in the Lagrangian form (1.21.3), the constants c_j are given by eq. (1.21.6); that is,

$$Q(\mathbf{A}) = \sum_{j=1}^{n} Q(\lambda_j)\frac{\prod_{k=1,j}^{n}(\mathbf{A} - \lambda_k\mathbf{I})}{\prod_{k=1,j}^{n}(\lambda_j - \lambda_k)} \qquad (1.21.7)$$

which is called *Sylvester's formula*. Recall that $\mathbf{A}$ is a square matrix of order n, and $Q(x)$ is a polynomial of degree $(n - 1)$ in x. However, since any polynomial $P(\mathbf{A})$ in $\mathbf{A}$ or $\mathbf{A}^{-1}$ can be expressed as a polynomial of degree $(n - 1)$ in $\mathbf{A}$, due to the H-C Theorem, Sylvester's formula in fact holds for an *arbitrary* polynomial in $\mathbf{A}$ or $\mathbf{A}^{-1}$.

Example. Given $\mathbf{A} = \begin{bmatrix} 1 & 3 \\ 5 & 1 \end{bmatrix}$, compute $\mathbf{A}^{100}$.
  $\mathbf{A}$ has eigenvalues $\lambda_1 = 1 + \sqrt{15}$, $\lambda_2 = 1 - \sqrt{15}$. From eq. (1.21.7), we have

$$\mathbf{A}^{100} = \lambda_1^{100}\frac{(\mathbf{A} - \lambda_2\mathbf{I})}{(\lambda_1 - \lambda_2)} + \lambda_2^{100}\frac{(\mathbf{A} - \lambda_1\mathbf{I})}{(\lambda_2 - \lambda_1)}$$

Thus, very simply, we have found $\mathbf{A}^{100}$!

The Sylvester's formula that is actually used in computations is slightly different from eq. (1.21.7), and it is based on the equality (see Amundson, 1966, section 5.14 for details)

$$\prod_{k=1,j}^{n}(\mathbf{A} - \lambda_k\mathbf{I}) = \text{adj}(\lambda_j\mathbf{I} - \mathbf{A})$$

Thus eq. (1.21.7) reduces to

$$Q(\mathbf{A}) = \sum_{j=1}^{n} Q(\lambda_j)\frac{\text{adj}(\lambda_j\mathbf{I} - \mathbf{A})}{\prod_{k=1,j}^{n}(\lambda_j - \lambda_k)} \qquad (1.21.8)$$

so that even the matrix multiplication indicated in eq. (1.21.7) does not have to be performed.

1.22   TRANSIENT ANALYSIS OF A STAGED ABSORPTION COLUMN

We consider here a specific problem which can be handled conveniently by matrix methods. It is an idealization of an important chemical engineering process, and it concerns the dynamics of a staged absorption column shown schematically in Figure 1.1. The purpose of such an operation is to selectively transfer one or more components from a gas stream fed to the column from the bottom, by absorbing them in a liquid stream fed at the top. The gas and liquid streams then pass countercurrently through the column, and they are intimately contacted by each of the N stages built therein for such contact.

   For simplicity, we shall consider that only one component is transferred from the gas to the liquid phase. We also invoke various assumptions to make the problem tractable, and there indeed are situations of practical importance where these are closely approximated. These various assumptions are as follows:

1. Each stage is an equilibrium contact stage; that is, the gas and the liquid streams *leaving* a stage are in equilibrium with each other.
2. The gas stream consists largely of inert gas, in which a solute, which has to be transferred to the liquid stream, is present. Similarly, the liquid stream is largely an inert absorbing liquid. The inert gas is assumed to be insoluble in the liquid, and the inert liquid is assumed to be nonvolatile.
3. Let y represent mass of solute per mass of inert gas in the gas stream, and let x be mass of solute per mass of inert liquid in the liquid stream. The streams leaving a stage are assumed to satisfy the relation

$$y = Kx \tag{1.22.1}$$

   where K is the Henry's law constant. This relationship holds, for example, for dilute solutions of SO_2 or of NH_3 in water.

   Now, let G and L represent the inert gas and liquid mass flow rates, respectively, and let h denote the mass holdup of inert liquid at each stage. For given values of x_0, y_{N+1}, N, G, L, h *and* an initial composition of liquid in the column, we wish to determine how the composition varies throughout the column as a function of time.

   The various streams entering and leaving the generic jth stage are shown schematically in Figure 1.2. The solute mass involves the following terms:

$$\text{Input} = Lx_{j-1} + Gy_{j+1}$$

$$\text{Output} = Lx_j + Gy_j$$

$$\text{Accumulation} = h \frac{dx_j}{d\theta}$$

where θ is time elapsed from column startup. Representing the principle of solute mass conservation as

$$\text{Input} = \text{Output} + \text{Accumulation}$$

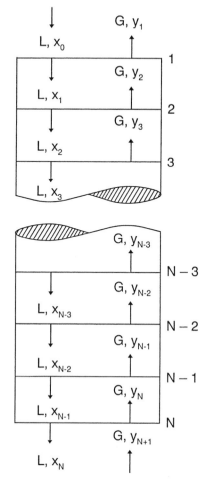

Figure 1.1 Schematic diagram of a staged absorption column.

we have

$$Lx_{j-1} + Gy_{j+1} = Lx_j + Gy_j + h\frac{dx_j}{d\theta}, \qquad j = 1, 2, \ldots, N \qquad (1.22.2)$$

Substituting eq. (1.22.1), this becomes

$$h\frac{dx_j}{d\theta} = Lx_{j-1} - (L + GK)x_j + GKx_{j+1}, \qquad j = 1, 2, \ldots, N$$

which reduces to the following set of ODEs with constant coefficients

$$\frac{d\mathbf{x}}{dt} = \mathbf{Ax} + \mathbf{b}, \qquad t > 0 \qquad (1.22.3)$$

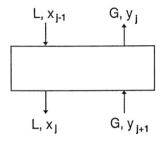

Figure 1.2 Input and output streams for stage j.

with the IC

$$\mathbf{x} = \mathbf{x}_0 \qquad \text{at } t = 0 \tag{1.22.4}$$

where

$$t = \frac{\theta}{h},$$

$$\mathbf{x} = \begin{bmatrix} x_1 \\ x_2 \\ \cdot \\ \cdot \\ \cdot \\ x_N \end{bmatrix}, \qquad \mathbf{b} = \begin{bmatrix} Lx_0 \\ 0 \\ \cdot \\ \cdot \\ \cdot \\ 0 \\ Gy_{N+1} \end{bmatrix}$$

and

$$\mathbf{A} = \begin{bmatrix} -(L + GK) & GK & 0 & 0 & \cdots & & & 0 \\ L & -(L + GK) & GK & 0 & \cdots & & & 0 \\ 0 & L & -(L + GK) & GK & 0 & \cdots & & 0 \\ \cdot & & & & & & & \cdot \\ \cdot & & & & & & & \cdot \\ 0 & & & & 0 & L & -(L + GK) & GK \\ 0 & & \cdots & & & 0 & L & -(L + GK) \end{bmatrix}$$

From eq. (1.17.13), the solution is given by

$$\mathbf{x}(t) = \sum_{j=1}^{N} \frac{\mathbf{w}_j^T[\mathbf{x}_0 + \mathbf{A}^{-1}\mathbf{b}]}{\mathbf{w}_j^T \mathbf{z}_j} \, \mathbf{z}_j \exp(\lambda_j t) - \mathbf{A}^{-1}\mathbf{b} \tag{1.22.5}$$

where λ_j, $j = 1$ to N, are the eigenvalues of the matrix $\mathbf{A}$, and $\mathbf{z}_j$ and $\mathbf{w}_j^T$ are the corresponding eigenvectors and eigenrows, respectively. Also, $\{\mathbf{z}_j\}$ and $\{\mathbf{w}_j\}$ form biorthogonal sets. In order to proceed further, we need to first find the eigenvalues λ_j.

Introducing the notation

$$\alpha = GK \quad \text{and} \quad \beta = L$$

the coefficient matrix can be written as

$$
\mathbf{A} =
\begin{bmatrix}
-(\alpha + \beta) & \alpha & 0 & 0 & \cdots & & & 0 \\
\beta & -(\alpha + \beta) & \alpha & 0 & \cdots & & & 0 \\
0 & \beta & -(\alpha + \beta) & \alpha & 0 & \cdots & & 0 \\
\vdots & & & & & & & \vdots \\
0 & & & & & \beta & -(\alpha + \beta) & \alpha \\
0 & & & \cdots & & 0 & \beta & -(\alpha + \beta)
\end{bmatrix}
\tag{1.22.6}
$$

and is a special form of the tridiagonal (or Jacobi) matrix. The eigenvalues satisfy the equation

$$
P_N(\lambda) = |\mathbf{A} - \lambda \mathbf{I}| = 0
$$

That is,

$$
P_N(\lambda) =
\begin{vmatrix}
-(\alpha + \beta + \lambda) & \alpha & 0 & 0 & \cdots & & 0 \\
\beta & -(\alpha + \beta + \lambda) & \alpha & 0 & \cdots & & 0 \\
0 & \beta & -(\alpha + \beta + \lambda) & \alpha & \cdots & & 0 \\
\vdots & & & & & & \vdots \\
0 & & & & \beta & -(\alpha + \beta + \lambda) & \alpha \\
0 & & & \cdots & 0 & \beta & -(\alpha + \beta + \lambda)
\end{vmatrix} = 0
\tag{1.22.7}
$$

For $N = 2$ we have

$$
P_2(\lambda) =
\begin{vmatrix}
-(\alpha + \beta + \lambda) & \alpha \\
\beta & -(\alpha + \beta + \lambda)
\end{vmatrix}
$$
$$
= (\alpha + \beta + \lambda)^2 - \alpha\beta
$$

For $N = 3$ we obtain

$$
P_3(\lambda) =
\begin{vmatrix}
-(\alpha + \beta + \lambda) & \alpha & 0 \\
\beta & -(\alpha + \beta + \lambda) & \alpha \\
0 & \beta & -(\alpha + \beta + \lambda)
\end{vmatrix}
$$
$$
= -(\alpha + \beta + \lambda)P_2(\lambda) + \alpha\beta(\alpha + \beta + \lambda)
$$

where the determinant is expanded by elements of the first column. Thus the recursion relationship

$$
P_n(\lambda) = -(\alpha + \beta + \lambda)P_{n-1}(\lambda) - \alpha\beta P_{n-2}(\lambda)
\tag{1.22.8}
$$

follows for $n \geq 2$, where

$$
P_0(\lambda) = 1 \quad \text{and} \quad P_1(\lambda) = -(\alpha + \beta + \lambda)
\tag{1.22.9}
$$

Note that eq. (1.22.8) is a *difference equation*, which may be solved by standard techniques (cf. Kelley and Peterson, 1991, section 3.3). For this, eq. (1.22.8) may be rearranged as

$$P_n(\lambda) + (\alpha + \beta + \lambda)P_{n-1}(\lambda) + \alpha\beta P_{n-2}(\lambda) = 0 \qquad (1.22.10)$$

Assuming that the solution is given by

$$P_n(\lambda) = \gamma^n \qquad (1.22.11)$$

substituting in eq. (1.22.10) leads to

$$\gamma^n + (\alpha + \beta + \lambda)\gamma^{n-1} + \alpha\beta\gamma^{n-2} = 0$$

Thus γ satisfies the quadratic equation

$$\gamma^2 + (\alpha + \beta + \lambda)\gamma + \alpha\beta = 0$$

which has roots

$$\gamma_{\pm} = \frac{-(\alpha + \beta + \lambda) \pm \delta}{2} \qquad (1.22.12)$$

where

$$\delta = \sqrt{(\alpha + \beta + \lambda)^2 - 4\alpha\beta} \qquad (1.22.12a)$$

Thus from eq. (1.22.11), the general solution is given by

$$P_n(\lambda) = c_1\gamma_+^n + c_2\gamma_-^n \qquad (1.22.13)$$

where c_1 and c_2 are constants, obtained by utilizing eq. (1.22.9) for two specific values of n: 0 and 1. This gives

$$c_1 + c_2 = 1 \quad \text{and} \quad c_1\gamma_+ + c_2\gamma_- = -(\alpha + \beta + \lambda)$$

so that

$$c_1 = \frac{\gamma_+}{\delta}, \qquad c_2 = \frac{-\gamma_-}{\delta}$$

Substituting in eq. (1.22.13) leads to

$$P_N(\lambda) = \frac{\gamma_+^{N+1} - \gamma_-^{N+1}}{\delta} \qquad (1.22.14)$$

which is a much simpler representation of $P_N(\lambda)$ than would have been obtained by a direct expansion of the determinant (1.22.7). Now, the eigenvalues λ satisfy $P_N(\lambda) = 0$; thus they are the solutions of the equation

$$\gamma_+^{N+1} = \gamma_-^{N+1}$$

That is

$$\left(\frac{\gamma_+}{\gamma_-}\right)^{N+1} = 1 \qquad (1.22.15)$$

The $(N + 1)$ roots of $z^{N+1} = 1$ are given by

$$z_k = \exp\left(\frac{i2\pi k}{N + 1}\right), \qquad k = 0, 1, 2, \ldots, N$$

where $i = \sqrt{-1}$. Substituting for $\gamma_\pm$, this means that the λ_k satisfy

$$\frac{\gamma_+}{\gamma_-} = \frac{-(\alpha + \beta + \lambda_k) + \delta_k}{-(\alpha + \beta + \lambda_k) - \delta_k} = \exp\left(\frac{i2\pi k}{N + 1}\right), \qquad k = 0, 1, \ldots, N \quad (1.22.16)$$

which we need to solve to yield λ_k explicitly. This requires some manipulation, described next in some detail.

Adding 1 to both sides of eq. (1.22.16) gives

$$\frac{-2(\alpha + \beta + \lambda_k)}{\Delta} = \exp\left(\frac{i2\pi k}{N + 1}\right) + 1$$

while subtracting 1 from both sides gives

$$\frac{2\delta_k}{\Delta} = \exp\left(\frac{i2\pi k}{N + 1}\right) - 1$$

where

$$\Delta = -(\alpha + \beta + \lambda_k) - \delta k$$

Dividing the above equations, we have

$$\frac{-\delta_k}{(\alpha + \beta + \lambda_k)} = \frac{\exp[i2\pi k/(N + 1)] - 1}{\exp[i2\pi k/N + 1)] + 1}$$

$$= i \tan\left(\frac{\pi k}{N + 1}\right)$$

$(1.22.17)$

and squaring both sides gives

$$\frac{(\alpha + \beta + \lambda_k)^2 - 4\alpha\beta}{(\alpha + \beta + \lambda_k)^2} = -\tan^2\left(\frac{\pi k}{N + 1}\right)$$

$$= 1 - \sec^2\left(\frac{\pi k}{N + 1}\right)$$

Thus we obtain

$$\frac{(\alpha + \beta + \lambda_k)^2}{4\alpha\beta} = \cos^2\left(\frac{\pi k}{N + 1}\right)$$

so that

$$\lambda_k = -(\alpha + \beta) \pm 2\sqrt{\alpha\beta} \cos\left(\frac{\pi k}{N + 1}\right), \qquad k = 0, 1, \ldots, N \quad (1.22.18)$$

Equation (1.22.18) represents a total of $2N + 2$ values of λ. However, $P_N(\lambda) = 0$ is a polynomial of degree N, which should have only N roots. Thus there are extraneous roots present, which have to be identified and discarded.

The index $k = 0$ should be excluded from eq. (1.22.18), because then $z_0 = 1$, which implies that $\gamma_+ = \gamma_- = \gamma$. Then the roots γ of the difference equation are not distinct, and the general solution cannot be written in the form (1.22.13). Also, since in this case

$$\gamma_+ = \gamma_- = \gamma$$

from eq. (1.22.12), we have

$$\delta = 0$$

implying that

$$(\alpha + \beta + \lambda)^2 = 4\alpha\beta \qquad (1.22.19)$$

and

$$\gamma = -\frac{\alpha + \beta + \lambda}{2} \qquad (1.22.20)$$

Instead of $P_n(\lambda) = \gamma^n$, the general solution of the difference equation (1.22.8) becomes

$$P_n(\lambda) = (c_1 + c_2 n)\gamma^n$$

Utilizing eqs. (1.22.9) and (1.22.20) to evaluate c_1 and c_2 yields

$$c_1 = 1, \qquad c_2 = -1 - \frac{\alpha + \beta + \lambda}{\gamma} = 1$$

which gives

$$P_n(\lambda) = (1 + n)\gamma^n$$

Since the eigenvalues λ satisfy $P_N(\lambda) = 0$, we have

$$(1 + N)\gamma^N = 0$$

whose solution is $\gamma = 0$, repeated N times. From eq. (1.22.20), this implies that

$$\lambda = -(\alpha + \beta)$$

However, this is incompatible with eq. (1.22.19), since α and β are both positive. Thus $k = 0$ is an extraneous root of eq. (1.22.18).

At this stage, the eigenvalues are therefore given by eq. (1.22.18) with $k = 1$, $2, \ldots, N$ only, and N extraneous roots have yet to be identified. For this, we observe that eq. (1.22.17),

$$\frac{\delta_k}{(\alpha + \beta + \lambda_k)} = -i \tan\left(\frac{\pi k}{N + 1}\right)$$

does *not* hold if the $+$ sign is taken in eq. (1.22.18). This eliminates the remaining extraneous roots, so that the eigenvalues are given by

$$\lambda_k = -(\alpha + \beta) - 2\sqrt{\alpha\beta}\,\cos\!\left(\frac{\pi k}{N+1}\right), \qquad k = 1, 2, \ldots, N \quad (1.22.21)$$

Note that the λ_k are *always* negative, because they are bounded above by the case where the cosine term equals -1; that is,

$$\lambda_k \leq -(\alpha + \beta) + 2\sqrt{\alpha\beta} = -(\sqrt{\alpha} - \sqrt{\beta})^2 < 0$$

From the general solution (1.22.5), this means that for all possible initial compositions $\mathbf{x}_0$ we obtain

$$\lim_{t\to\infty} \mathbf{x}(t) = \mathbf{x}_s = -\mathbf{A}^{-1}\mathbf{b}$$

and since the eigenvalues are all real, the approach to the steady-state $\mathbf{x}_s$ is always *nonoscillatory*.

To complete the solution $\mathbf{x}(t)$, we now need to determine $\mathbf{z}_j$, $\mathbf{w}_j$ and $\mathbf{A}^{-1}$. Recall that the eigenvectors $\mathbf{z}_j$ and the eigenrows $\mathbf{w}_j^T$ of $\mathbf{A}$ are simply the nonzero columns and rows of $\mathrm{adj}(\mathbf{A} - \lambda_j\mathbf{I})$, respectively. If we denote

$$\mathbf{B} = \mathbf{A} - \lambda_k\mathbf{I} = \begin{bmatrix} -(\alpha+\beta+\lambda_k) & \alpha & 0 & 0 & \cdots & & 0 \\ \beta & -(\alpha+\beta+\lambda_k) & \alpha & 0 & \cdots & & 0 \\ 0 & \beta & -(\alpha+\beta+\lambda_k) & \alpha & \cdots & & 0 \\ \vdots & & & & & & \vdots \\ 0 & & & & \beta & -(\alpha+\beta+\lambda_k) & \alpha \\ 0 & & \cdots & & 0 & \beta & -(\alpha+\beta+\lambda_k) \end{bmatrix}$$

then it is readily seen that the cofactors of the various elements of the first row and the first column of $\mathbf{B}$ are

$$B_{11} = P_{N-1}(\lambda_k), \quad B_{12} = -\beta P_{N-2}(\lambda_k), \qquad B_{13} = (-\beta)^2 P_{N-3}(\lambda_k), \cdots,$$
$$B_{1N} = (-\beta)^{N-1} P_0(\lambda_k)$$

$$B_{21} = -\alpha P_{N-2}(\lambda_k), \qquad B_{31} = (-\alpha)^2 P_{N-3}(\lambda_k), \cdots,$$
$$B_{N1} = (-\alpha)^{N-1} P_0(\lambda_k)$$

Thus we have

$$\mathbf{z}_k = \begin{bmatrix} P_{N-1}(\lambda_k) \\ -\beta P_{N-2}(\lambda_k) \\ (-\beta)^2 P_{N-3}(\lambda_k) \\ \vdots \\ (-\beta)^{N-1} P_0(\lambda_k) \end{bmatrix}, \qquad \mathbf{w}_k = \begin{bmatrix} P_{N-1}(\lambda_k) \\ -\alpha P_{N-2}(\lambda_k) \\ (-\alpha)^2 P_{N-3}(\lambda_k) \\ \vdots \\ (-\alpha)^{N-1} P_0(\lambda_k) \end{bmatrix} \qquad (1.22.22)$$

From eq. (1.22.13), we see that

$$P_n(\lambda) = \frac{\gamma_+^{n+1} - \gamma_-^{n+1}}{\delta}$$

where

$$\delta = \sqrt{(\alpha + \beta + \lambda)^2 - 4\alpha\beta}$$

Substituting the λ_k given by eqs. (1.22.21) yields

$$\delta_k = i2\sqrt{\alpha\beta}\,\sin\left(\frac{\pi k}{N + 1}\right)$$

and from eq. (1.22.12) we obtain

$$\gamma_\pm = \frac{-(\alpha + \beta + \lambda) \pm \delta}{2}$$

Using the above equations for λ_k and δ_k gives

$$\gamma_\pm = \sqrt{\alpha\beta}\,\exp\left(\frac{\pm i\pi k}{N + 1}\right)$$

which leads to

$$P_n(\lambda_k) = (\alpha\beta)^{n/2}\,\frac{\sin[\pi k(n + 1)/(N + 1)]}{\sin[\pi k/(N + 1)]} \qquad (1.22.23)$$

Thus the various elements of $\mathbf{z}_k$ and $\mathbf{w}_k$ may be explicitly written as

$$\mathbf{z}_k = \begin{bmatrix} (\alpha\beta)^{(N-1)/2}\{\sin[\pi kN/(N + 1)]\}/\phi_k \\ -\beta(\alpha\beta)^{(N-2)/2}\{\sin[\pi k(N - 1)/(N + 1)]\}/\phi_k \\ \vdots \\ (-\beta)^{N-1} \end{bmatrix} \qquad (1.22.24a)$$

$$\mathbf{w}_k = \begin{bmatrix} (\alpha\beta)^{(N-1)/2}\{\sin[\pi kN/(N + 1)]\}\phi_k \\ -\alpha(\alpha\beta)^{(N-2)/2}\{\sin[\pi k(N - 1)/(N + 1)]\}/\phi_k \\ \vdots \\ (-\alpha)^{N-1} \end{bmatrix} \qquad (1.22.24b)$$

where

$$\phi_k = \sin[\pi k/(N + 1)]$$

and the scalar product $\mathbf{w}_j^T\mathbf{z}_j$ required in the solution (1.22.5) is

$$\mathbf{w}_j^T\mathbf{z}_j = (\alpha\beta)^{N-1}\sum_{n=1}^{N}\left\{\frac{\sin[\pi jn/(N + 1)]}{\sin[\pi j/(N + 1)]}\right\}^2 \qquad (1.22.25)$$

The only important quantity now left to evaluate is $\mathbf{A}^{-1}$. By definition, this is given by

$$\mathbf{A}^{-1} = \frac{1}{D_N} \text{adj } \mathbf{A} = \frac{1}{D_N} \begin{bmatrix} A_{11} & A_{21} & \cdots & A_{N1} \\ A_{12} & A_{22} & \cdots & A_{N2} \\ \vdots & & & \\ A_{1N} & A_{2N} & \cdots & A_{NN} \end{bmatrix}$$

where $D_N = |\mathbf{A}|$, and A_{ij} is the cofactor of element a_{ij}. Note that the solution (1.22.5) requires $\mathbf{A}^{-1}\mathbf{b}$, and $\mathbf{b}$ has all elements zero, except the first ($= Lx_0$) and the Nth ($= Gy_{N+1}$). Thus, *only* the first and last columns of $\mathbf{A}^{-1}$ need to be evaluated.

First, let us calculate $D_N = |\mathbf{A}|$. Note that for $N = 2$ we obtain

$$D_2 = \begin{vmatrix} -(\alpha + \beta) & \alpha \\ \beta & -(\alpha + \beta) \end{vmatrix} = (\alpha + \beta)^2 - \alpha\beta$$

and for $N = 3$ we have

$$D_3 = \begin{vmatrix} -(\alpha + \beta) & \alpha & 0 \\ \beta & -(\alpha + \beta) & \alpha \\ 0 & \beta & -(\alpha + \beta) \end{vmatrix} = -(\alpha + \beta)D_2 + \alpha\beta(\alpha + \beta)$$

which generate the recursion relationship

$$D_n = -(\alpha + \beta)D_{n-1} - \alpha\beta D_{n-2} \qquad \text{for } n \geq 2 \qquad (1.22.26)$$

with

$$D_0 = 1 \quad \text{and} \quad D_1 = -(\alpha + \beta) \qquad (1.22.27)$$

The difference equation (1.22.26) may be solved in ways similar to those utilized for $P_n(\lambda)$; that is, letting $D_n = \psi^n$, which yields two values of ψ, we obtain

$$\psi_+ = -\beta \quad \text{and} \quad \psi_- = -\alpha$$

so that the general solution of eq. (1.22.26) is

$$D_n = c_1\psi_+^n + c_2\psi_-^n$$

The constants c_1 and c_2 are evaluated using eq. (1.22.27) as

$$c_1 = -\frac{\beta}{\alpha - \beta} \quad \text{and} \quad c_2 = \frac{\alpha}{\alpha - \beta}$$

so that

$$D_n = \frac{(-1)^{n+1}}{\alpha - \beta} [\beta^{n+1} - \alpha^{n+1}]$$

and

$$D_N = \frac{(-1)^{N+1}}{\alpha - \beta}[\beta^{N+1} - \alpha^{N+1}] \tag{1.22.28}$$

We now proceed to determine the first and last columns of $\mathbf{A}^{-1}$. From a direct calculation, it is seen that

$$A_{11} = D_{N-1}, \quad A_{12} = -\beta D_{N-2}, \quad A_{13} = (-\beta)^2 D_{N-3}, \dots, A_{1N} = (-\beta)^{N-1} D_0$$

$$A_{N1} = (-1)^{N+1}\alpha^{N-1}, \quad A_{N2} = (-1)^{N+3}(\alpha + \beta)\alpha^{N-2}, \dots,$$

$$A_{N,N-1} = (-1)^{2N-1}\alpha D_{N-2}, \quad A_{NN} = D_{N-1}$$

Thus the first and last columns of $\mathbf{A}^{-1}$ are

$$\begin{bmatrix} -[\beta^N - \alpha^N]/\phi \\ -\beta[\beta^{N-1} - \alpha^{N-1}]/\phi \\ -\beta^2[\beta^{N-2} - \alpha^{N-2}]/\phi \\ \vdots \\ -\beta^{N-2}[\beta^2 - \alpha^2]/\phi \\ -\beta^{N-1}[\beta - \alpha]/\phi \end{bmatrix} \quad \text{and} \quad \begin{bmatrix} -\alpha^{N-1}[\beta - \alpha]/\phi \\ -\alpha^{N-2}[\beta^2 - \alpha^2]/\phi \\ \vdots \\ -\alpha[\beta^{N-1} - \alpha^{N-1}]/\phi \\ -[\beta^N - \alpha^N]/\phi \end{bmatrix}$$

respectively, where

$$\phi = \beta^{N+1} - \alpha^{N+1}$$

We can then conclude that

$$\mathbf{A}^{-1}\mathbf{b} = -\begin{bmatrix} [Lx_0(\beta^N - \alpha^N) + Gy_{N+1}\alpha^{N-1}(\beta - \alpha)]/\phi \\ [Lx_0\beta(\beta^{N-1} - \alpha^{N-1}) + Gy_{N+1}\alpha^{N-2}(\beta^2 - \alpha^2)]/\phi \\ \vdots \\ [Lx_0\beta^{N-1}(\beta - \alpha) + Gy_{N+1}(\beta^N - \alpha^N)]/\phi \end{bmatrix} \tag{1.22.29}$$

and *all* elements in the solution (1.22.5) are now explicitly known.

1.23   REDUCTION OF A MATRIX TO DIAGONAL FORM

Given a square matrix $\mathbf{A}$, it is, in general, possible to find *several* pairs of matrices $\mathbf{P}$ and $\mathbf{Q}$ such that

$$\mathbf{PAQ} = \mathbf{D} \tag{1.23.1}$$

where $\mathbf{D}$ contains only diagonal elements, and $\mathbf{D}$ is called a *diagonal form* of $\mathbf{A}$.

Two matrices $\mathbf{S}$ and $\mathbf{T}$ related by

$$\mathbf{P}^{-1}\mathbf{SP} = \mathbf{T} \tag{1.23.2}$$

are said to be *similar*, and the transformation (1.23.2) is called a *similarity transformation*.

A similarity transformation has the important property that the eigenvalues of the transformed matrix are the *same* as those of the original matrix; thus in eq. (1.23.2), $\mathbf{S}$

and $\mathbf{T}$ have the same eigenvalues. To see this, let λ be the eigenvalues of $\mathbf{S}$; thus λ satisfy

$$|\mathbf{S} - \lambda\mathbf{I}| = 0 \tag{1.23.3}$$

Also, let η be the eigenvalues of $\mathbf{T}$, so that η satisfy

$$|\mathbf{T} - \eta\mathbf{I}| = 0 \tag{1.23.4}$$

Now, since $\mathbf{T} = \mathbf{P}^{-1}\mathbf{SP}$, we have

$$|\mathbf{P}^{-1}\mathbf{SP} - \eta\mathbf{I}| = 0$$

which may be rearranged as

$$|\mathbf{P}^{-1}(\mathbf{S} - \eta\mathbf{I})\mathbf{P}| = 0$$

However, since $|\mathbf{ABC}| = |\mathbf{A}||\mathbf{B}||\mathbf{C}|$, this becomes

$$|\mathbf{P}^{-1}||\mathbf{S} - \eta\mathbf{I}||\mathbf{P}| = 0$$

Also, $|\mathbf{P}^{-1}||\mathbf{P}| = 1$; hence

$$|\mathbf{S} - \eta\mathbf{I}| = 0 \tag{1.23.5}$$

which implies that $\eta = \lambda$.

We now consider the possibility of reducing a given square matrix $\mathbf{A}$ to a diagonal form by using a similarity transformation. This reduction is possible and proves useful in many applications of matrices.

Let the eigenvalue problem corresponding to $\mathbf{A}$ be

$$\mathbf{Ax} = \lambda\mathbf{x} \tag{1.23.6}$$

so that $\lambda_j, j = 1$ to n, are the eigenvalues of $\mathbf{A}$, and $\mathbf{x}_j$ are the corresponding eigenvectors. Also, let us define

$$\mathbf{X} = \begin{bmatrix} x_{11} & x_{12} & \cdots & x_{1n} \\ x_{21} & x_{22} & \cdots & x_{2n} \\ \vdots & & & \\ x_{n1} & x_{n2} & \cdots & x_{nn} \end{bmatrix} = [\mathbf{x}_1 \quad \mathbf{x}_2 \cdots \mathbf{x}_n] \tag{1.23.7}$$

as the matrix whose columns are the eigenvectors $\mathbf{x}_j$. Since these are linearly independent, $\mathbf{X}$ is nonsingular. Then by definition of matrix multiplication we obtain

$$\mathbf{AX} = \mathbf{A}[\mathbf{x}_1 \ \mathbf{x}_2 \ \cdots \ \mathbf{x}_n] = [\lambda_1\mathbf{x}_1 \ \lambda_2\mathbf{x}_2 \ \cdots \ \lambda_n\mathbf{x}_n]$$

$$= \begin{bmatrix} x_{11} & x_{12} & \cdots & x_{1n} \\ x_{21} & x_{22} & \cdots & x_{2n} \\ \vdots & \vdots & & \vdots \\ x_{n1} & x_{n2} & \cdots & x_{nn} \end{bmatrix} \begin{bmatrix} \lambda_1 & 0 & \cdots & 0 \\ 0 & \lambda_2 & \cdots & 0 \\ \vdots & & & \vdots \\ 0 & & \cdots & \lambda_n \end{bmatrix}$$

$$= \mathbf{X}\boldsymbol{\Lambda}$$

where Λ is a diagonal matrix, with the eigenvalues of $\mathbf{A}$ as the nonzero elements. This leads to

$$\mathbf{X}^{-1}\mathbf{A}\mathbf{X} = \Lambda \qquad (1.23.8)$$

Thus a similarity transformation using a matrix composed of the eigenvectors of $\mathbf{A}$ as columns produces a diagonal form in which the diagonal elements are the eigenvalues of $\mathbf{A}$.

Similarly, if $\mathbf{y}_j^T$ are the eigenrows of $\mathbf{A}$—that is, $\mathbf{y}$ satisfy

$$\mathbf{A}^T\mathbf{y} = \lambda\mathbf{y}$$

and $\mathbf{Y}^T$ is a matrix whose rows are the eigenrows $\mathbf{y}_j^T$—then

$$\mathbf{Y}^{-1}\mathbf{A}^T\mathbf{Y} = \Lambda \qquad (1.23.9)$$

as well.

A special case arises when $\mathbf{A}$ is symmetric, because then $\mathbf{A}^T = \mathbf{A}$ and

$$\mathbf{x}_j^T\mathbf{x}_i \begin{cases} = 0, & i \neq j \\ \neq 0, & i = j \end{cases} \qquad (1.23.10)$$

Then $\mathbf{X} = [\mathbf{x}_1 \quad \mathbf{x}_2 \cdots \mathbf{x}_n]$ and

$$\mathbf{X}^T\mathbf{X} = \begin{bmatrix} \mathbf{x}_1^T \\ \mathbf{x}_2^T \\ \vdots \\ \mathbf{x}_n^T \end{bmatrix} [\mathbf{x}_1 \quad \mathbf{x}_2 \cdots \mathbf{x}_n]$$

$$= \begin{bmatrix} \mathbf{x}_1^T\mathbf{x}_1 & \mathbf{x}_1^T\mathbf{x}_2 & \cdots & \mathbf{x}_1^T\mathbf{x}_n \\ \mathbf{x}_2^T\mathbf{x}_1 & \mathbf{x}_2^T\mathbf{x}_2 & \cdots & \mathbf{x}_2^T\mathbf{x}_n \\ \vdots & & & \vdots \\ \mathbf{x}_n^T\mathbf{x}_1 & \mathbf{x}_n^T\mathbf{x}_2 & \cdots & \mathbf{x}_n^T\mathbf{x}_n \end{bmatrix}$$

which, using eq. (1.23.10), reduces to

$$\mathbf{X}^T\mathbf{X} = \begin{bmatrix} \mathbf{x}_1^T\mathbf{x}_1 & 0 & \cdots & 0 \\ 0 & \mathbf{x}_2^T\mathbf{x}_2 & \cdots & 0 \\ \vdots & & & \vdots \\ 0 & & \cdots & \mathbf{x}_n^T\mathbf{x}_n \end{bmatrix}$$

Furthermore, if the eigenvectors $\mathbf{x}_j$ are *normalized* such that

$$\mathbf{x}_j^T\mathbf{x}_j = 1, \qquad j = 1, 2, \ldots, n$$

then

$$\mathbf{X}^T\mathbf{X} = \mathbf{I} \qquad (1.23.11)$$

Thus for a symmetric matrix $\mathbf{A}$ with normalized eigenvectors, we have

$$\mathbf{X}^{-1} = \mathbf{X}^T \qquad (1.23.12)$$

and the similarity transformation (1.23.8) becomes

$$\mathbf{X}^T \mathbf{A} \mathbf{X} = \mathbf{\Lambda} \tag{1.23.13}$$

Note that a matrix which satisfies eq. (1.23.12) is called an *orthogonal* matrix.

1.24 QUADRATIC FORMS

Given a second-order expression such as

$$
\begin{aligned}
G = \; & a_{11}x_1^2 + a_{12}x_1x_2 + \cdots + a_{1n}x_1x_n \\
& + a_{21}x_2x_1 + a_{22}x_2^2 + \cdots + a_{2n}x_2x_n \\
& + \cdots \\
& + a_{n1}x_nx_1 + a_{n2}x_nx_2 + \cdots + a_{nn}x_n^2 \\
& + b_1x_1 + b_2x_2 + \cdots b_nx_n \\
& + c
\end{aligned}
\tag{1.24.1}
$$

and then making the linear changes of variables,

$$x_j = y_j + c_j, \quad j = 1, 2, \ldots, n \tag{1.24.2}$$

it is possible to eliminate the linear terms completely by choosing c_j appropriately.
     For example, let

$$G = a_{11}x_1^2 + (a_{12} + a_{21})x_1x_2 + a_{22}x_2^2 + b_1x_1 + b_2x_2 + c$$

Then with the transformation (1.24.2) we have

$$
\begin{aligned}
G = \; & a_{11}y_1^2 + (a_{12} + a_{21})y_1y_2 + a_{22}y_2^2 \\
& + [2a_{11}c_1 + (a_{12} + a_{21})c_2 + b_1]y_1 + [c_1(a_{12} + a_{21}) + 2a_{22}c_2 + b_2]y_2 \\
& + [a_{11}c_1^2 + (a_{12} + a_{21})c_1c_2 + a_{22}c_2^2 + b_1c_1 + b_2c_2 + c]
\end{aligned}
\tag{1.24.3}
$$

The linear terms vanish if we choose c_1 and c_2 such that

$$2a_{11}c_1 + (a_{12} + a_{21})c_2 = -b_1$$

$$(a_{12} + a_{21})c_1 + 2a_{22}c_2 = -b_2$$

The condition for a solution to exist for all values of b_i is

$$
\Delta = \begin{vmatrix} 2a_{11} & a_{12} + a_{21} \\ a_{12} + a_{21} & 2a_{22} \end{vmatrix} \neq 0
$$

or

$$4a_{11}a_{22} \neq (a_{12} + a_{21})^2 \tag{1.24.4}$$

Thus, under the mild restriction (1.24.4) it is possible to always remove the linear terms. Then eq. (1.24.3) reduces to

$$Q = G - \alpha = a_{11}y_1^2 + (a_{12} + a_{21})y_1y_2 + a_{22}y_2^2 \tag{1.24.5}$$

where the constant term is denoted by α.

Thus, in general, any second-order expression such as eq. (1.24.1) can be reduced to

$$
\begin{aligned}
Q &= a_{11}x_1^2 + a_{12}x_1x_2 + \cdots + a_{1n}x_1x_n \\
&\quad + a_{21}x_2x_1 + a_{22}x_2^2 + \cdots + a_{2n}x_2x_n \\
&\quad + \cdots \\
&\quad + a_{n1}x_nx_1 + a_{n2}x_nx_2 + \cdots + a_{nn}x_n^2 \\
&= \mathbf{x}^T\mathbf{A}\mathbf{x}
\end{aligned}
\tag{1.24.6}
$$

where the coefficient matrix $\mathbf{A}$, without *any* loss of generality, may be taken as symmetric. The quantity Q contains only second-order terms and is called a *quadratic form*.

Note that given a quadratic form Q, the symmetric matrix $\mathbf{A}$ may be readily computed from the relation

$$
a_{ij} = \frac{1}{2}\frac{\partial^2 Q}{\partial x_i \partial x_j}
\tag{1.24.7}
$$

1.25 REDUCTION OF A QUADRATIC FORM TO THE CANONICAL FORM

As shown above, a quadratic form can be written in terms of a *symmetric* matrix $\mathbf{A}$ as

$$
Q = \mathbf{x}^T\mathbf{A}\mathbf{x}
\tag{1.25.1}
$$

Now, given a quadratic form, it is always possible to reduce it to a form which contains *only* a sum of squares, and this is called the *canonical* form of the quadratic form. To see how this is done, consider the change of variables.

$$
\mathbf{x} = \mathbf{C}\mathbf{y}
\tag{1.25.2}
$$

where $\mathbf{C}$ is nonsingular. Then from eq. (1.25.1), Q becomes

$$
\begin{aligned}
Q &= (\mathbf{C}\mathbf{y})^T\mathbf{A}(\mathbf{C}\mathbf{y}) \\
&= \mathbf{y}^T(\mathbf{C}^T\mathbf{A}\mathbf{C})\mathbf{y}
\end{aligned}
$$

y is not an eigenvectors

In section 1.23, it was shown that for a symmetric matrix $\mathbf{A}$ we obtain

$$
\mathbf{X}^T\mathbf{A}\mathbf{X} = \mathbf{\Lambda}
$$

Thus if we choose $\mathbf{C} = \mathbf{X}$, then

$$
\begin{aligned}
Q &= \mathbf{y}^T\mathbf{\Lambda}\mathbf{y} \\
&= \lambda_1 y_1^2 + \lambda_2 y_2^2 + \cdots + \lambda_n y_n^2
\end{aligned}
\tag{1.25.3}
$$

which is a canonical form, and the transformation which converts a quadratic form to the canonical form is

$$
\mathbf{x} = \mathbf{X}\mathbf{y}
\tag{1.25.4}
$$

1.26 DEFINITE QUADRATIC FORMS

Given a quadratic form

$$Q = \mathbf{x}^T \mathbf{A} \mathbf{x}$$

where $\mathbf{A}$ is symmetric, Q will be positive or negative, depending on $\mathbf{x}$.

If, however, $Q > 0$ for all $\mathbf{x} \neq \mathbf{0}$, and $Q = 0$ only for $\mathbf{x} = \mathbf{0}$, then Q is said to be *positive definite*. Similarly, if $Q < 0$ for all $\mathbf{x} \neq \mathbf{0}$ and $Q = 0$ only for $\mathbf{x} = \mathbf{0}$, then Q is *negative definite*.

Now, since Q may be reduced to the canonical form

$$Q = \lambda_1 y_1^2 + \lambda_2 y_2^2 + \cdots + \lambda_n y_n^2$$

it is obvious that a *necessary and sufficient* condition that Q be positive (negative) definite is that *all* the eigenvalues of $\mathbf{A}$ be positive (negative).

In many cases, it is relatively difficult to compute all the eigenvalues of $\mathbf{A}$. If the aim is only to show the definiteness of a quadratic form, then a result due to Sylvester is often helpful. Given the symmetric matrix $\mathbf{A}$ corresponding to the quadratic form Q, consider its special principal minors:

$$A_n = |\mathbf{A}| = \begin{vmatrix} a_{11} & a_{12} & \cdots & a_{1n} \\ a_{21} & a_{22} & \cdots & a_{2n} \\ \vdots & & & \vdots \\ a_{n1} & a_{n2} & \cdots & a_{nn} \end{vmatrix}, \quad A_{n-1} = \begin{vmatrix} a_{11} & a_{12} & \cdots & a_{1n-1} \\ a_{21} & a_{22} & \cdots & a_{2n-1} \\ \vdots & & & \\ a_{n-1,1} & a_{n-1,2} & \cdots & a_{n-1,n-1} \end{vmatrix}$$

$$\ldots, A_2 = \begin{vmatrix} a_{11} & a_{12} \\ a_{21} & a_{22} \end{vmatrix}, \quad A_1 = a_{11}$$

Then, the *necessary and sufficient* condition that Q be positive definite is that *all* the minors $A_1, A_2, \ldots, A_n$ be greater than zero (for a proof, see Gantmacher, 1959, Volume 1, chapter 10).

Also, note that Q is negative definite, if $-Q$ is positive definite. Thus the quadratic form

$$Q = a_{11} x_1^2 + 2a_{12} x_1 x_2 + a_{22} x_2^2$$

is negative definite, if and only if

$$a_{11} < 0 \quad \text{and} \quad a_{11} a_{22} > (a_{12})^2$$

1.27 MAXIMA AND MINIMA OF FUNCTIONS OF SEVERAL VARIABLES

From elementary calculus, we know that given a function f of a single variable x, the *necessary* condition that it has an extremum (a relative maximum *or* a relative minimum) at $x = a$ is that $f'(a) = 0$. The conditions for the extremum to be a maximum or a minimum are derived by considering the expansion of $f(x)$ in a Taylor series:

$$f(x) = f(x_0) + (x - x_0)f'(x_0) + \frac{(x - x_0)^2}{2!} f''(x_0)$$

$$+ \frac{(x - x_0)^3}{3!} f'''(x_0) + \cdots \quad (1.27.1)$$

If $x_0 = a$, and $x = a + h$, then

$$f(a + h) - f(a) = \frac{h^2}{2!} f''(a) + \frac{h^3}{3!} f'''(a) + \cdots \quad (1.27.2)$$

For a definite extremum, $[f(a + h) - f(a)]$ should be of constant sign for sufficiently small h. Thus we have

$$f''(a) < 0 \quad \text{if } x = a \text{ is a maximum}$$

and

$$f''(a) > 0 \quad \text{if } x = a \text{ is a minimum} \quad (1.27.3)$$

If now f is a function of several variables $x_1, x_2, \ldots, x_n$ the *necessary* condition that $\mathbf{x} = \mathbf{a}$ be an extremum is that

$$\frac{\partial f}{\partial x_j} = 0, \quad j = 1, 2, \ldots, N \quad (1.27.4)$$

at $\mathbf{x} = \mathbf{a}$. The Taylor series expansion for f around $\mathbf{a}$ with $\mathbf{x} = \mathbf{a} + \mathbf{h}$, is

$$f(\mathbf{a} + \mathbf{h}) = f(\mathbf{a}) + \left[\frac{\partial f}{\partial x_1}\bigg|_{\mathbf{a}} h_1 + \frac{\partial f}{\partial x_2}\bigg|_{\mathbf{a}} h_2 + \cdots + \frac{\partial f}{\partial x_n}\bigg|_{\mathbf{a}} h_n \right]$$

$$+ \frac{1}{2!} \left[\frac{\partial^2 f}{\partial x_1^2}\bigg|_{\mathbf{a}} h_1^2 + \frac{\partial^2 f}{\partial x_1 \partial x_2}\bigg|_{\mathbf{a}} h_1 h_2 + \cdots + \frac{\partial^2 f}{\partial x_1 \partial x_n}\bigg|_{\mathbf{a}} h_1 h_n \right.$$

$$+ \cdots$$

$$\left. + \frac{\partial^2 f}{\partial x_n \partial x_1}\bigg|_{\mathbf{a}} h_n h_1 + \frac{\partial^2 f}{\partial x_n \partial x_2}\bigg|_{\mathbf{a}} h_n h_2 + \cdots + \frac{\partial^2 f}{\partial x_n^2}\bigg|_{\mathbf{a}} h_n^2 \right] + \cdots$$

which is concisely written in symbolic operator form as

$$f(\mathbf{a} + \mathbf{h}) = f(\mathbf{a}) + \sum_{j=1}^{\infty} \frac{1}{j!} \left[\left(h_1 \frac{\partial}{\partial x_1} + \cdots + h_n \frac{\partial}{\partial x_n} \right)^j f(\mathbf{x}) \right]_{\mathbf{x}=\mathbf{a}} \quad (1.27.5)$$

Thus for all small $\mathbf{h}$, the quantity

$$Q = f(\mathbf{a} + \mathbf{h}) - f(\mathbf{a}) = \frac{1}{2!} \left[\left(h_1 \frac{\partial}{\partial x_1} + \cdots + h_n \frac{\partial}{\partial x_n} \right)^2 f(\mathbf{x}) \right]_{\mathbf{x}=\mathbf{a}}$$

should be of one sign: positive if $\mathbf{a}$ is a minimum, and negative if $\mathbf{a}$ is a maximum.
If we denote

$$\frac{\partial^2 f}{\partial x_i \partial x_j}\bigg|_{\mathbf{a}} = f_{ij}$$

then

$$
\begin{aligned}
Q = {} & f_{11}h_1^2 + f_{12}h_1h_2 + \cdots + f_{1n}h_1h_n \\
& + f_{21}h_2h_1 + f_{22}h_2^2 + \cdots + f_{2n}h_2h_n \\
& + \cdots \\
& + f_{n1}h_nh_1 + f_{n2}h_nh_2 + \cdots + f_{nn}h_n^2 \\
= {} & \mathbf{h}^T\mathbf{F}\mathbf{h}
\end{aligned}
\tag{1.27.6}
$$

which is a quadratic form, and $\mathbf{F}$ is the symmetric matrix $\{f_{ij}\}$, sometimes called the *Hessian* matrix of the function f. Thus whether $\mathbf{x} = \mathbf{a}$ is a maximum or a minimum depends on whether the quadratic form Q is negative definite or positive definite. Also, if Q is negative (positive) definite, then the matrix $\mathbf{F}$ is said to be negative (positive) definite.

1.28   FUNCTIONS DEFINED ON MATRICES

We have earlier defined finite polynomials on matrices, and shown by applying the H–C theorem that all such polynomials in $\mathbf{A}$ and $\mathbf{A}^{-1}$ can be reduced to a polynomial in $\mathbf{A}$ of degree $\leq (n - 1)$. We now consider *infinite* power series of matrices of the form

$$
a_0\mathbf{I} + a_1\mathbf{A} + a_2\mathbf{A}^2 + \cdots + a_n\mathbf{A}^n + \cdots
\tag{1.28.1}
$$

where $a_j, j = 1$ to n, are scalar coefficients. If we let

$$
\begin{aligned}
\mathbf{S}_m &= a_0\mathbf{I} + a_1\mathbf{A} + \cdots + a_m\mathbf{A}^m \\
&= \sum_{j=0}^{n} a_j\mathbf{A}^j
\end{aligned}
\tag{1.28.2}
$$

then $\mathbf{S}_m$ represents a partial sum, and we wish to consider its limit as $m \to \infty$. It should be apparent that if *each* element of $\mathbf{S}_m$ converges to a limit as $m \to \infty$, then

$$
\mathbf{S} = \lim_{m\to\infty} \mathbf{S}_m
\tag{1.28.3}
$$

exists.

If we consider the similarity transformation

$$
\mathbf{C} = \mathbf{T}^{-1}\mathbf{A}\mathbf{T}
\tag{1.28.4}
$$

then $\mathbf{A}$ and $\mathbf{C}$ are similar. From eq. (1.28.4) we have

$$
\mathbf{C}^2 = \mathbf{C}\mathbf{C} = (\mathbf{T}^{-1}\mathbf{A}\mathbf{T})(\mathbf{T}^{-1}\mathbf{A}\mathbf{T}) = \mathbf{T}^{-1}\mathbf{A}^2\mathbf{T}
$$

and in general we have

$$
\mathbf{C}^j = \mathbf{T}^{-1}\mathbf{A}^j\mathbf{T}
$$

where j is an integer. If a_j is a scalar, then

$$
a_j\mathbf{C}^j = a_j\mathbf{T}^{-1}\mathbf{A}^j\mathbf{T} = \mathbf{T}^{-1}(a_j\mathbf{A}^j)\mathbf{T}
$$

and

$$P_m(\mathbf{C}) = \sum_{j=0}^{m} a_j \mathbf{C}^j = \mathbf{T}^{-1}[P_m(\mathbf{A})]\mathbf{T} \qquad (1.28.5)$$

Thus if $\mathbf{A}$ and $\mathbf{C}$ are similar and $P_m(x)$ is any polynomial, then $P_m(\mathbf{A})$ and $P_m(\mathbf{C})$ are also similar.

Let

$$P_m(\mathbf{A}) = \mathbf{S}_m(\mathbf{A})$$

Then from eq. (1.28.5) we obtain

$$\mathbf{S}_m(\mathbf{C}) = \mathbf{T}^{-1}\mathbf{S}_m(\mathbf{A})\mathbf{T}$$

and the two infinite series

$$a_0\mathbf{I} + a_1\mathbf{A} + a_2\mathbf{A}^2 + \cdots$$

and

$$a_0\mathbf{I} + a_1\mathbf{C} + a_2\mathbf{C}^2 + \cdots$$

converge or diverge together. Let us now select

$$\mathbf{T} = \mathbf{X}$$

the matrix composed of the eigenvectors of $\mathbf{A}$; then from eq. (1.23.8) we obtain

$$\mathbf{X}^{-1}\mathbf{AX} = \boldsymbol{\Lambda}$$

and thus $\mathbf{A}$ and $\boldsymbol{\Lambda}$ are similar, with

$$\mathbf{S}_m(\boldsymbol{\Lambda}) = a_0\mathbf{I} + a_1\boldsymbol{\Lambda} + a_2\boldsymbol{\Lambda}^2 + \cdots + a_m\boldsymbol{\Lambda}^m \qquad (1.28.6)$$

Since $\boldsymbol{\Lambda}$ is a diagonal matrix,

$$\boldsymbol{\Lambda}^j = \begin{bmatrix} \lambda_1^j & 0 & \cdots & 0 \\ 0 & \lambda_2^j & \cdots & 0 \\ \vdots & & \ddots & \vdots \\ 0 & & \cdots & \lambda_n^j \end{bmatrix}$$

so that

$$\mathbf{S}_m(\boldsymbol{\Lambda}) = \begin{bmatrix} \sum_{j=0}^{m} a_j\lambda_1^j & 0 & \cdots & 0 \\ 0 & \sum_{j=0}^{m} a_j\lambda_2^j & \cdots & 0 \\ \vdots & \vdots & & \vdots \\ 0 & \cdots & & \sum_{j=0}^{m} a_j\lambda_n^j \end{bmatrix}$$

and the $\lim_{m \to \infty} \mathbf{S}_m(\mathbf{\Lambda})$ exists if each of the limits

$$\lim_{m \to \infty} \left[\sum_{j=0}^{m} a_j \lambda_k^j \right], \qquad k = 1, 2, \ldots, n$$

exists. Thus we need to consider convergence of the *scalar* power series

$$S(x) = \lim_{m \to \infty} S_m(x) = \sum_{j=0}^{\infty} a_j x^j$$

If *each* of the eigenvalues of $\mathbf{A}$ lies within the radius of convergence of this power series, then the matrix power series (1.28.1) will also converge.

Since

$$\exp(x) = 1 + x + \frac{x^2}{2!} + \cdots$$

converges for all x, the matrix series

$$\mathbf{I} + \mathbf{A} + \frac{\mathbf{A}^2}{2!} + \cdots$$

also converges for all $\mathbf{A}$, and is called $\exp(\mathbf{A})$. Similarly,

$$\mathbf{I} + \mathbf{A}t + \frac{\mathbf{A}^2 t^2}{2!} + \cdots = \mathbf{I} + (\mathbf{A}t) + \frac{(\mathbf{A}t)^2}{2!} + \cdots$$
$$= \exp(\mathbf{A}t) \tag{1.28.7}$$

and we can define other functions of matrices such as $\sin(\mathbf{A})$, $\cos(\mathbf{A})$, and so on, as well.

By differentiating both sides of eq. (1.28.7), we have

$$\frac{d}{dt}[\exp(\mathbf{A}t)] = \mathbf{A} + \frac{2\mathbf{A}^2 t}{2!} + \frac{3\mathbf{A}^3 t^2}{3!} + \cdots$$
$$= \mathbf{A}\left[\mathbf{I} + \mathbf{A}t + \frac{\mathbf{A}^2 t^2}{2!} + \cdots \right] \tag{1.28.8}$$
$$= \mathbf{A} \exp(\mathbf{A}t)$$
$$= \exp(\mathbf{A}t)\mathbf{A}$$

so that the derivative of the exponential matrix is analogous to that of a scalar exponential. Note also that from Sylvester's formula (1.21.8) we obtain

$$\exp(\mathbf{A}t) = \sum_{j=1}^{n} \exp(\lambda_j t) \frac{\mathrm{adj}(\lambda_j \mathbf{I} - \mathbf{A})}{\Pi_{i=1,j}^{n}(\lambda_j - \lambda_i)} \tag{1.28.9}$$

so that $\exp(\mathbf{A}t)$ can be readily evaluated.

1.29 AN ALTERNATIVE METHOD FOR SOLVING A SYSTEM OF FIRST-ORDER LINEAR ODES

For the system of first-order linear ODEs with constant coefficients

$$\frac{d\mathbf{x}}{dt} = \mathbf{A}\mathbf{x} + \mathbf{b} \qquad (1.29.1)$$

with ICs

$$\mathbf{x}(0) = \mathbf{x}_0 \qquad (1.29.2)$$

in section 1.17. we have obtained the solution $\mathbf{x}$ in terms of the eigenvalues and eigenvectors of $\mathbf{A}$. Another form of the solution can be obtained by appealing to the exponential function of the matrix.

The function

$$\mathbf{x} = \exp(\mathbf{A}t)\mathbf{c} \qquad (1.29.3)$$

where $\mathbf{c}$ is a constant vector, has the property that

$$\frac{d\mathbf{x}}{dt} = \mathbf{A}\,\exp(\mathbf{A}t)\mathbf{c} = \mathbf{A}\mathbf{x}$$

Hence eq. (1.29.3) is a solution of the homogeneous equation. Also

$$-\mathbf{A}^{-1}\mathbf{b}$$

is a particular solution of (1.29.1), so that the general solution is

$$\mathbf{x} = \exp(\mathbf{A}t)\mathbf{c} - \mathbf{A}^{-1}\mathbf{b} \qquad (1.29.4)$$

where $\mathbf{c}$ is yet not determined. This is obtained by satisfying the IC (1.29.2), which states that

$$\mathbf{x}_0 = c - \mathbf{A}^{-1}\mathbf{b}$$

thus leading to

$$\mathbf{x} = \exp(\mathbf{A}t)(\mathbf{x}_0 + \mathbf{A}^{-1}\mathbf{b}) - \mathbf{A}^{-1}\mathbf{b} \qquad (1.29.5)$$

The solution may also be obtained in a somewhat different manner, which resembles that of solving a single linear ODE. Equation (1.29.1) may be rewritten as

$$\frac{d\mathbf{x}}{dt} - \mathbf{A}\mathbf{x} = \mathbf{b}$$

or

$$\frac{d}{dt}\,[\exp(-\mathbf{A}t)\mathbf{x}] = \exp(-\mathbf{A}t)\mathbf{b}$$

Integrating both sides between 0 and t gives

$$\exp(-\mathbf{A}t)\mathbf{x} - \mathbf{x}_0 = -\mathbf{A}^{-1}[\exp(-\mathbf{A}t) - \mathbf{I}]\mathbf{b}$$

so that

$$\mathbf{x} = \exp(\mathbf{A}t)\mathbf{x}_0 - \exp(\mathbf{A}t)\ \mathbf{A}^{-1}[\exp(-\mathbf{A}t) - \mathbf{I}]\mathbf{b}$$
$$= \exp(\mathbf{A}t)(\mathbf{x}_0 + \mathbf{A}^{-1}\mathbf{b}) - \mathbf{A}^{-1}\mathbf{b} \tag{1.29.6}$$

since $\mathbf{A}^{-1}$ commutes with $\exp(\mathbf{A}t)$. Note that if in eq. (1.29.1), $\mathbf{b}$ is a function of t, then we have

$$\mathbf{x} = \exp(\mathbf{A}t)\mathbf{x}_0 + \int_0^t \exp[-\mathbf{A}(\tau - t)]\mathbf{b}(\tau)\ d\tau \tag{1.29.7}$$

by the above procedure.

1.30 SOME APPLICATIONS INVOLVING CHEMICAL REACTIONS

The methods of matrix analysis are well suited for solving problems which involve isothermal first-order chemical reactions. Two reactor types, the batch and the continuous-flow stirred tank reactor (CSTR), are considered here. In the first type, the reactants are placed in a stirred vessel and the reaction is allowed to proceed, with no further addition of the reactants or withdrawal of the products. In a CSTR, on the other hand, the reactants are fed continuously to a well-agitated tank, and the products are withdrawn continuously from it. In both reactor types, we assume ideal mixing so that species concentrations within the reactor are uniform in space, although they change as a function of time depending on the initial conditions (ICs).

For generality, let us consider the sequence of reversible reactions:

$$\begin{array}{ccccc} 1 & 2 & 3 & m \\ A_1 \rightleftharpoons A_2 \rightleftharpoons A_3 \rightleftharpoons \cdots \rightleftharpoons A_{m+1} \end{array} \tag{1.30.1}$$

where they all follow first-order kinetics, with rate given by

$$r_j = k_j a_j - k_j' a_{j+1}, \qquad j = 1, 2, \ldots, m \tag{1.30.2}$$

Here, k_j and k_j' are the rate constants of the forward and the backward jth reaction, respectively, and a_j represents the concentration of species A_j. If this sequence of reactions occurs in an isothermal *batch* reactor of constant volume V (see Figure 1.3), then the transient species mass balance

| Input | = | Output | + | Accumulation | (1.30.3) |

can be derived as follows.

For species A_1

Input	=	0		
Output	=	$r_1 V$	=	$(k_1 a_1 - k_1' a_2)V$
Accumulation	=	$\dfrac{d}{dt}(V a_1)$		

so that from eq. (1.30.3),

$$0 \quad = \quad (k_1 a_1 - k'_1 a_2) \quad + \quad \frac{da_1}{dt}$$

That is,

$$\frac{da_1}{dt} \quad = \quad -k_1 a_1 + k'_1 a_2 \tag{1.30.4}$$

For species A_j ($j = 2, 3, \ldots, m$)

Input	$=$	$r_{j-1} V$	$=$	$(k_{j-1} a_{j-1} - k'_{j-1} a_j) V$
Output	$=$	$r_j V$	$=$	$(k_j a_j - k'_j a_{j+1}) V$
Accumulation	$=$	$\dfrac{d}{dt} (V a_j)$		

which leads to

$$\frac{da_j}{dt} \quad = \quad k_{j-1} a_{j-1} - (k'_{j-1} + k_j) a_j + k'_j a_{j+1} \tag{1.30.5}$$

For species A_{m+1}

Input	$=$	$r_m V$	$=$	$(k_m a_m - k'_m a_{m+1}) V$
Output	$=$	0		
Accumulation	$=$	$\dfrac{d}{dt} (V a_{m+1})$		

and so we have

$$\frac{da_{m+1}}{dt} \quad = \quad k_m a_m - k'_m a_{m+1} \tag{1.30.6}$$

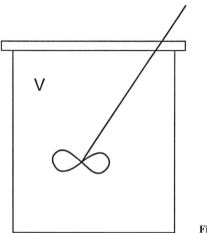

Figure 1.3 A batch reactor.

The balances (1.30.4)–(1.30.6) can be written concisely as

$$\frac{d\mathbf{a}}{dt} = \mathbf{K}\mathbf{a} \tag{1.30.7}$$

where

$$\mathbf{a}^T = [A_1 A_2 \cdots A_{m+1}]$$

and

$$\mathbf{K} = \begin{bmatrix} -k_1 & k_1' & 0 & 0 & 0 & \cdots & 0 \\ k_1 & -(k_2 + k_1') & k_2' & 0 & 0 & \cdots & 0 \\ 0 & k_2 & -(k_3 + k_2') & k_3' & 0 & \cdots & 0 \\ \vdots & & & & & & \vdots \\ 0 & & \cdots & & k_{m-1} & -(k_m + k_{m-1}') & k_m' \\ 0 & & \cdots & & & 0 & k_m & -k_m' \end{bmatrix} \tag{1.30.8}$$

The ICs are given by

$$\mathbf{a}(0) = \mathbf{a}_0 \tag{1.30.9}$$

which denotes the species concentrations in the reactor at start-up.

From eq. (1.17.13), we know that the solution of eqs. (1.30.7) and (1.30.9) is a linear combination of exponentials of $\lambda_j t$. We can then immediately conclude that

1. if all the eigenvalues of $\mathbf{K}$—that is, λ_j—have negative real parts, then $\mathbf{a} \to \mathbf{0}$ as $t \to \infty$, while

2. if *any* eigenvalue has a positive real part, then one or more of the species concentrations a_i become arbitrarily large as $t \to \infty$.

Both of these possibilities are clearly physically incorrect. The resolution of this problem occurs when it is realized that if all the rows of $\mathbf{K}$ are added up, say to the last row, then a row of zeros is produced. Thus $\mathbf{K}$ is singular, and by eq. (1.18.1), $\lambda = 0$ is an eigenvalue of $\mathbf{K}$. Let us denote the eigenvalues by

$$0, \lambda_1, \lambda_2, \ldots, \lambda_m$$

and the corresponding eigenvectors and eigenrows, respectively, by

$$\mathbf{z}_0, \{\mathbf{z}_j\}$$

and

$$\mathbf{w}_0^T, \{\mathbf{w}_j^T\}$$

Thus the solution (1.17.13) becomes

$$\mathbf{a} = \frac{\mathbf{w}_0^T a_0}{\mathbf{w}_0^T \mathbf{z}_0} \mathbf{z}_0 + \sum_{j=1}^m \frac{\mathbf{w}_j^T a_0}{\mathbf{w}_j^T \mathbf{z}_j} \mathbf{z}_j \exp(\lambda_j t) \tag{1.30.10}$$

so that *if* all the λ_j have negative real parts, then

$$\mathbf{a}_s = \lim_{t \to \infty} \mathbf{a} = \frac{\mathbf{w}_0^T \mathbf{a}_0}{\mathbf{w}_0^T \mathbf{z}_0} \mathbf{z}_0 \qquad (1.30.11)$$

We now show that this is indeed the case and that, in particular, all the λ_j are real as well.

The matrix **K** is real and tridiagonal and has the form

$$\mathbf{B} = \begin{bmatrix} a_1 & b_1 & 0 & \cdots & & 0 \\ c_1 & a_2 & b_2 & & & \vdots \\ 0 & c_2 & a_3 & b_3 & & \vdots \\ \vdots & & & & & \\ & & \cdots & c_{n-2} & a_{n-1} & b_{n-1} \\ 0 & & \cdots & & c_{n-1} & a_n \end{bmatrix} \qquad (1.30.12)$$

which occurs in a variety of staged processes—for example, the absorption column considered in section 1.22. In general, the determinant $|\mathbf{B} - \lambda\mathbf{I}|$ may be written as $P_n(\lambda)$. If we denote the determinant formed by omitting the last row and the last column as $P_{n-1}(\lambda)$, that formed by omitting the last two rows and columns as $P_{n-2}(\lambda)$, and so on, then it is easy to verify that

$$\begin{aligned} P_n(\lambda) &= (a_n - \lambda)P_{n-1}(\lambda) - b_{n-1}c_{n-1}P_{n-2}(\lambda) \\ P_{n-1}(\lambda) &= (a_{n-1} - \lambda)P_{n-2}(\lambda) - b_{n-2}c_{n-2}P_{n-3}(\lambda) \\ &\vdots \\ P_3(\lambda) &= (a_3 - \lambda)P_2(\lambda) - b_2c_2P_1(\lambda) \\ P_2(\lambda) &= (a_2 - \lambda)P_1(\lambda) - b_1c_1P_0(\lambda) \\ P_1(\lambda) &= (a_1 - \lambda)P_0(\lambda) \end{aligned} \qquad (1.30.13)$$

where we define $P_0(\lambda) = 1$. These recurrence relationships for the various $P_j(\lambda)$ are useful for evaluating the *real* eigenvalues, since if we assume a λ', we can compute the sequence

$$P_1(\lambda'), P_2(\lambda'), \ldots, P_{n-1}(\lambda'), P_n(\lambda')$$

and λ' is an eigenvalue of **B** if $P_n(\lambda') = 0$. Thus the eigenvalues can be found without writing the characteristic equation itself.

The sequence of polynomials $P_j(\lambda)$ forms a so-called Sturm sequence (cf. Amundson, 1966, section 5.18), which implies that the number of real zeros of $P_n(\lambda)$ in the range $a < \lambda < b$ is equal to the difference between the number of sign changes in the sequence for $\lambda = a$ and $\lambda = b$. Thus, let $V(a)$ be the number of sign changes in the sequence for $\lambda = a$, and let $V(b)$ be the same for $\lambda = b$; then

$$\begin{aligned} N &= \text{number of real roots of } P_n(\lambda) = 0 \text{ for } \lambda\varepsilon(a,b) \\ &= V(b) - V(a) \end{aligned} \qquad (1.30.14)$$

Recalling that the elements of matrix **B** are real, for λ large and positive, we have

$$P_0(\lambda) = 1 > 0$$
$$P_1(\lambda) = (a_1 - \lambda) < 0$$
$$P_2(\lambda) = (a_2 - \lambda)P_1(\lambda) - b_1 c_1 > 0$$
$$\vdots$$
$$P_{n-1}(\lambda) = (a_{n-1} - \lambda)P_{n-2}(\lambda) - b_{n-2}c_{n-2}P_{n-3}(\lambda) = [< 0, \quad n \text{ even}]$$
$$= [> 0, \quad n \text{ odd}]$$
$$P_n(\lambda) = (a_n - \lambda)P_{n-1}(\lambda) - b_{n-1}c_{n-1}P_{n-2}(\lambda) \quad = [> 0, \quad n \text{ even}]$$
$$= [< 0, \quad n \text{ odd}]$$

Thus there are n sign changes in the sequence. Similarly, for λ large and negative we have

$$P_0(\lambda) \quad = 1 > 0$$
$$P_1(\lambda) \quad > 0$$
$$P_2(\lambda) \quad > 0$$
$$\vdots$$
$$P_{n-1}(\lambda) > 0$$
$$P_n(\lambda) \quad > 0$$

and there are no sign changes in the sequence. Hence from eq. (1.30.14), we see that the characteristic equation $P_n(\lambda) = 0$ has n real roots. Since the matrix **B** of eq. (1.30.12) is of order n, we can then conclude that *all* the eigenvalues of **B** are *real*.

Consider now the same Sturmian sequence, with $\lambda = 0$, for the matrix **K** of eq. (1.30.8). Note that **K** is of order $(m + 1)$, with

$$a_1 = -k_1; \qquad a_j = -(k_j + k'_{j-1}) \quad \text{for } j = 2, 3, \ldots, m; \qquad a_{m+1} = -k'_m$$

and

$$b_j = k'_j; \qquad c_j = k_j \quad \text{for } j = 1, 2, \ldots, m \tag{1.30.15}$$

This along with eq. (1.30.13) leads to

$$P_0(0) = 1 > 0$$
$$P_1(0) = a_1 = -k_1 < 0$$
$$P_2(0) = a_2 P_1(0) - b_1 c_1$$
$$= (k_2 + k'_1)k_1 - k_1 k'_1$$
$$= k_1 k_2 = -k_2 P_1(0) > 0$$
$$P_3(0) = -k_3 P_2(0) < 0$$
$$\vdots$$

Thus $P_0, P_1, \ldots, P_m$ alternate in sign, and with $P_{m+1}(0) = 0$ there are a total of m sign changes. Since as noted above, there are no sign changes for λ large and negative, the matrix K has m negative real eigenvalues. The feature of the steady state represented by eq. (1.30.11) is therefore correct.

These m negative real eigenvalues must be found numerically. Once they have been determined, the computation of the eigenvectors is quite simple, for they are the non-trivial solutions of

$$(\mathbf{A} - \lambda \mathbf{I})\mathbf{x} = \mathbf{0}$$

which for the general tridiagonal matrix (1.30.12) may be written as

$$
\begin{aligned}
(a_1 - \lambda)x_1 + b_1 x_2 &= 0 \\
c_1 x_1 + (a_2 - \lambda)x_2 + b_2 x_3 &= 0 \\
c_2 x_2 + (a_3 - \lambda)x_3 + b_3 x_4 &= 0 \\
\vdots \\
c_{n-2} x_{n-2} + (a_{n-1} - \lambda)x_{n-1} + b_{n-1} x_n &= 0 \\
c_{n-1} x_{n-1} + (a_n - \lambda)x_n &= 0
\end{aligned}
\tag{1.30.16}
$$

These can be solved successively to give

$$
x_2 = -\frac{1}{b_1}(a_1 - \lambda)x_1
$$

$$
x_3 = -\frac{1}{b_2}[(a_2 - \lambda)x_2 + c_1 x_1]
\tag{1.30.17}
$$

$$
\vdots
$$

$$
x_n = -\frac{1}{b_{n-1}}[(a_{n-1} - \lambda)x_{n-1} + c_{n-2} x_{n-2}]
$$

and the last equation of the set (1.30.16) may be omitted or, alternatively, used as a consistency check. Thus for any eigenvalue λ, we can assume a nonzero value for x_1, and the remaining x_j are then computed from eqs. (1.30.17).

Similarly, the eigenrows $\mathbf{w}_j^T$ are obtained by following the same procedure as above and noting that they are the nontrivial solutions of

$$(\mathbf{A}^T - \lambda \mathbf{I})\mathbf{w} = \mathbf{0}$$

As discussed in section 1.29, an alternative procedure for solving eqs. (1.30.7) and (1.30.9) is to use the exponential function of a matrix. In this case, eq. (1.29.5) yields

$$\mathbf{a}(t) = \exp(\mathbf{K}t)\mathbf{a}_0 \tag{1.30.18}$$

where $\exp(\mathbf{K}t)$ is evaluated as indicated in eq. (1.28.9).

Let us now consider the same sequence of reactions (1.30.1) occurring in a series of n CSTRs, as shown in Figure 1.4, where the effluent of one becomes the influent of the next. The transient species mass balances can be derived in the same manner as for the batch reactor, by simply adding the contributions of the convective flow. This gives

$$V\frac{d\mathbf{a}_j}{dt} = q\mathbf{a}_{j-1} - q\mathbf{a}_j + V\mathbf{K}\mathbf{a}_j, \qquad j = 1, 2, \ldots, n \tag{1.30.19}$$

where q is the volumetric flow rate, V is the volume of each reactor in the series, $\mathbf{a}_j$ is the $(m + 1)$-dimensional vector representing the species concentrations leaving the jth reactor, and the matrix of rate constants $\mathbf{K}$ is again given by eq. (1.30.8). The ICs reflect

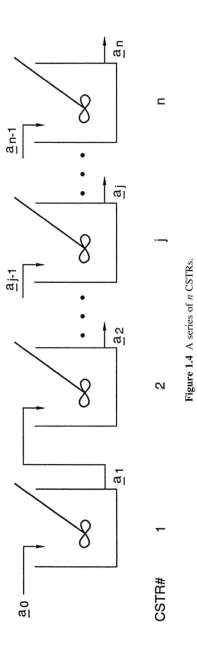

Figure 1.4 A series of n CSTRs.

the concentrations of the chemical species in each reactor at start-up conditions:

$$\mathbf{a}_j(0) = \mathbf{a}_{j0}, \qquad j = 1, 2, \ldots, n \tag{1.30.20}$$

If we let $\theta_h = V/q$, the average *residence (or holding) time* in each reactor, eq. (1.30.19) takes the form

$$\theta_h \frac{d\mathbf{a}_j}{dt} = \mathbf{a}_{j-1} - \mathbf{a}_j + \theta_h \mathbf{K} \mathbf{a}_j$$

or

$$\frac{d\mathbf{a}_j}{d\tau} - \mathbf{M}\mathbf{a}_j = \mathbf{a}_{j-1}, \qquad j = 1, 2, \ldots, n \tag{1.30.21}$$

where

$$\tau = \frac{t}{\theta_h}, \qquad \mathbf{M} = (\theta_h \mathbf{K} - \mathbf{I})$$

and $\mathbf{M}$ is not singular unless $1/\theta_h$ is an eigenvalue of $\mathbf{K}$. Since as shown above, the eigenvalues of $\mathbf{K}$ are all real and negative, except for one which is zero, this can never happen. Note that eq. (1.30.21) is a system of *differential-difference equations*, which can be solved sequentially.

From eq. (1.29.7), the solution for $j = 1$ in terms of the matrix exponential is

$$\mathbf{a}_1(\tau) = \exp(\mathbf{M}\tau)\mathbf{a}_{10} + \exp(\mathbf{M}\tau) \int_0^\tau \exp(-\mathbf{M}s)\mathbf{a}_0(s) \, ds$$

If $\mathbf{a}_0$, the vector of feed concentrations to the first reactor, is constant, then

$$\int_0^\tau \exp(-\mathbf{M}s)\mathbf{a}_0 \, ds = -\mathbf{M}^{-1} \exp(-\mathbf{M}s)\big|_0^\tau \mathbf{a}_0$$
$$= \mathbf{M}^{-1}[\mathbf{I} - \exp(-\mathbf{M}\tau)]\mathbf{a}_0$$

so that

$$\mathbf{a}_1(\tau) = \exp(\mathbf{M}\tau)\mathbf{a}_{10} + \mathbf{M}^{-1} \exp(\mathbf{M}\tau)\mathbf{a}_0 - \mathbf{M}^{-1}\mathbf{a}_0 \tag{1.30.22}$$

For $j = 2$, we have

$$\frac{d\mathbf{a}_2}{d\tau} - \mathbf{M}\mathbf{a}_2 = \mathbf{a}_1, \qquad \mathbf{a}_2(0) = \mathbf{a}_{20}$$

which leads to

$$\mathbf{a}_2(\tau) = \exp(\mathbf{M}\tau)\mathbf{a}_{20} + \exp(\mathbf{M}\tau) \int_0^\tau \exp(-\mathbf{M}s)\mathbf{a}_1(s) \, ds$$

$$= \exp(\mathbf{M}\tau)\mathbf{a}_{20} + \exp(\mathbf{M}\tau) \int_0^\tau [\mathbf{a}_{10} + \mathbf{M}^{-1}\mathbf{a}_0\tau - \mathbf{M}^{-1} \exp(-\mathbf{M}s)\mathbf{a}_0] \, ds$$

$$= \exp(\mathbf{M}\tau)\mathbf{a}_{20} + \exp(\mathbf{M}\tau)[\mathbf{a}_{10}\,\tau + \mathbf{M}^{-1}\mathbf{a}_0\tau + \mathbf{M}^{-2}\{\exp(-\mathbf{M}\tau) - \mathbf{I}) \, \mathbf{a}_0]$$
$$= \exp(\mathbf{M}\tau)(\mathbf{a}_{20} + \tau\mathbf{a}_{10}) + \exp(\mathbf{M}\tau)\mathbf{M}^{-2}[\mathbf{M}\tau - \mathbf{I}]\mathbf{a}_0 + \mathbf{M}^{-2}\mathbf{a}_0 \tag{1.30.23}$$

One can proceed similarly, and show by induction that

$$\mathbf{a}_k(\tau) = \exp(\mathbf{M}\tau) \sum_{j=1}^{k} \frac{\tau^{j-1}}{(j-1)!} \, \mathbf{a}_{k-j+1,0} - \qquad (1.30.24)$$

$$\exp(\mathbf{M}\tau)\,(-\mathbf{M})^{-k} \sum_{j=1}^{k} \frac{(-\mathbf{M}\tau)^{j-1}}{(j-1)!} \, \mathbf{a}_0 + (-\mathbf{M})^{-k}\mathbf{a}_0, \qquad k = 1, 2, \ldots, n$$

Note that since

$$\mathbf{M} = \theta_h \mathbf{K} - \mathbf{I}$$

from eq. (1.19.5), it follows that the eigenvalues of $\mathbf{M}$ are

$$\eta_j = \theta_h \lambda_{j-1}$$

where λ_j are the eigenvalues of $\mathbf{K}$. Thus η_j are also all real and negative.

References

N. R. Amundson, *Mathematical Methods in Chemical Engineering: Matrices and Their Application*, Prentice-Hall, Englewood Cliffs, NJ, 1966.

M. Bôcher, *Introduction to Higher Algebra*, Dover, New York, 1964.

F. R. Gantmacher, *The Theory of Matrices*, volumes 1 and 2, Chelsea, New York, 1959.

W. G. Kelley and A. C. Peterson, *Difference Equations: An Introduction with Applications*, Academic Press, San Diego, 1991.

E. Kreyszig, *Advanced Engineering Mathematics*, 7th ed., Wiley, New York, 1993.

Additional Reading

R. Bellman, *Introduction to Matrix Analysis*, 2nd ed., McGraw-Hill, New York, 1970.

J. N. Franklin, *Matrix Theory*, Prentice-Hall, Englewood Cliffs, NJ, 1968.

R. A. Frazer, W. J. Duncan, and A. R. Collar, *Elementary Matrices and Some Applications to Dynamics and Differential Equations*, Cambridge University Press, Cambridge, 1965.

F. G. Hildebrand, *Methods of Applied Mathematics*, 2nd ed., Prentice-Hall, Englewood Cliffs, NJ, 1965.

L. A. Pipes, *Matrix Methods for Engineering*, Prentice-Hall, Englewood Cliffs, NJ, 1963.

Problems

1.1 Evaluate the determinant

$$\begin{vmatrix} 2 & 5 & 1 & 4 \\ 1 & 0 & -1 & 2 \\ 2 & 1 & -4 & 1 \\ 0 & 2 & 4 & 8 \end{vmatrix}$$

(a) By Laplace's development, using elements of row 1, row 3, or column 2.

(b) By using the elementary operations to produce a number of zeros in rows and columns and then expanding.

1.2 Use Cramer's rule to solve the following systems of linear equations:

(a) $x - y + 3z = 2$
$2x - 3z = 1$
$x + y + z = 0$

(b) $2x - y + 3z - w = 1$
$x + 2y + z + 2w = 0$
$2x + 4y - z = -1$
$y - 2z + w = 2$

1.3 Show that the area of a triangle in the (x, y) plane, with vertices at the three points (x_i, y_i), $i = 1, 2,$ or 3, is given by

$$A = \frac{1}{2} \begin{vmatrix} x_1 & y_1 & 1 \\ x_2 & y_2 & 1 \\ x_3 & y_3 & 1 \end{vmatrix}$$

1.4 Show that in a three-dimensional space, the plane passing through the three points (x_i, y_i, z_i), $i = 1, 2,$ or 3, is given by the equation

$$\begin{vmatrix} x & y & z & 1 \\ x_1 & y_1 & z_1 & 1 \\ x_2 & y_2 & z_2 & 1 \\ x_3 & y_3 & z_3 & 1 \end{vmatrix} = 0$$

1.5 Show that

$$\begin{vmatrix} 1+c & 1 & 1 & 1 \\ 1 & 1+c & 1 & 1 \\ 1 & 1 & 1+c & 1 \\ 1 & 1 & 1 & 1+c \end{vmatrix} = c^4\left(1 + \frac{4}{c}\right)$$

1.6 Prove that

$$D = \begin{vmatrix} -1 & 0 & 0 & a \\ 0 & -1 & 0 & b \\ 0 & 0 & -1 & c \\ x & y & z & -1 \end{vmatrix} = 1 - ax - by - cz$$

Hence $D = 0$ is the equation of a plane.

1.7 Consider the two matrices

$$A = \begin{bmatrix} 1 & -1 & 3 \\ 2 & -3 & 1 \\ 4 & 1 & -2 \end{bmatrix}, \quad B = \begin{bmatrix} 5 & -1 & 2 \\ 1 & 4 & -3 \\ 3 & 2 & -1 \end{bmatrix}$$

(a) Compute the rank of $\mathbf{A}$ and $\mathbf{B}$.
(b) Compute $\mathbf{A}^{-1}$, $\mathbf{B}^{-1}$, adj $\mathbf{A}$ and adj $\mathbf{B}$.
(c) Compute $\mathbf{A}^{-2}$ and $\mathbf{B}^2$.
(d) Verify by direct computation that
$(\mathbf{AB})^{-1} = \mathbf{B}^{-1}\mathbf{A}^{-1}$
$(\mathbf{AB})^T = \mathbf{B}^T\mathbf{A}^T$

1.8 Given the set of equations

$$x_1 - 2x_2 + x_3 - x_4 = 1$$
$$3x_1 + x_2 - 3x_3 + 2x_4 = 2$$
$$x_1 + 5x_2 - 5x_3 + 4x_4 = 0$$
$$4x_1 - x_2 + 3x_3 - 9x_4 = 5$$
$$2x_1 + 17x_2 - 16x_3 + 13x_4 = -1$$

(a) Solve the system of equations.

(b) Solve the system constituted by only the first four equations above.

(c) Solve the above system, with the only change that in the rhs of the third equation, 0 is replaced by 1.

1.9 Solve the system of equations

$$2x_1 - 5x_2 + 6x_3 = -4$$
$$4x_1 + x_2 - 2x_3 = 0$$
$$3x_1 - 2x_2 + 2x_3 = -2$$
$$x_1 + 3x_2 - 4x_3 = 1$$

1.10 Solve the system of equations

$$x_2 - x_3 + 2x_4 = 1$$
$$3x_1 + x_2 + 3x_3 + x_4 = 2$$
$$-2x_1 + x_2 + 2x_3 + 3x_4 = 1$$
$$-7x_1 + 2x_3 + 3x_4 = -1$$

1.11 Solve the system of equations

$$x_1 + x_2 + 5x_3 - 3x_4 = 0$$
$$2x_1 - x_2 + 3x_3 + x_4 = 0$$
$$4x_1 + x_2 + 2x_3 + x_4 = 0$$

1.12 In the following systems of equations, for any real values of the parameters p and q, determine the existence of the solution and whether it can include arbitrary values for one or more of the unknowns.

(a) $px_1 - x_2 + 3x_3 = 1$
  $x_1 + x_2 - px_3 = q$
  $4x_1 + 3x_2 + x_3 = 0$

(b) $x_1 - x_2 + 2x_3 = 1$
  $-2x_1 + qx_2 - 4x_3 = 0$
  $3x_2 + x_3 = p$

1.13 Determine the real values of the parameters p and q for which the following homogeneous systems admit nontrivial solutions.

(a) $px_1 + 2x_2 - x_3 = 0$
  $2x_1 - x_2 + 3x_3 = 0$
  $qx_1 - 4x_2 + 2x_3 = 0$

(b) $px_1 + x_2 - 3x_3 = 0$
  $x_1 - 2x_2 - x_3 = 0$
  $x_1 - qx_2 + x_3 = 0$

1.14 Consider the radical polymerization of three monomer species: A, B, and C. Assuming that the reactivity of a growing radical chain is determined by the last monomer unit added, the propagation reactions for radical chains (A·) with terminal monomer unit of type A are given by

$$A\cdot + A \rightarrow A\cdot \qquad r = k_{pAA}A\cdot A$$
$$A\cdot + B \rightarrow B\cdot \qquad r = k_{pAB}A\cdot B$$
$$A\cdot + C \rightarrow C\cdot \qquad r = k_{pAC}A\cdot C$$

Similarly, we have three additional propagation reactions for each of the other two types of growing radical chains—that is, those terminating with monomer units of type B (i.e., B·) and C (ie., C·). Find the probability that a growing radical chain at steady state is of type A·, B·, or C·.

Note that the probabilities are defined as concentration ratios; for example, for radical of type A· we have $p_A = A\cdot/(A\cdot + B\cdot + C\cdot)$. In order to evaluate them, start by balancing separately radicals of type A·, B·, and C·.

1.15 In general, m equilibrium reactions occurring among n chemical species can be described in compact form by the system of linear equations

$$\sum_{i=1}^{n} v_{ij}A_i = 0, \qquad j = 1, 2, \ldots, m$$

where A_i represents the ith chemical species and v_{ij} is its stoichiometric coefficient in the jth reaction (positive for a product and negative for a reactant).

The above set of equations provides the stoichiometric constraint to the evolution of the reacting system. However, since the stoichiometry of chemical reactions may be written in various forms, we should consider the rank of the stoichiometric matrix in order to get the number of *independent reactions*. The number of *thermodynamic components*, which is defined as the minimum number of components needed to fully characterize the system equilibrium composition, is then obtained by subtracting from the total number of chemical species the number of stoichiometric constraints—that is, the independent reactions.

Determine the number of independent reactions and thermodynamic components for the kinetic scheme

$$C_2H_4 + 0.5O_2 \rightarrow C_2H_4O$$
$$C_2H_4 + 3O_2 \rightarrow 2CO_2 + 2H_2O$$
$$C_2H_4O + 2.5O_2 \rightarrow 2CO_2 + 2H_2O$$

which occurs in the epoxidation of ethylene to ethylene oxide.

1.16 In the combustion of methane in pure oxygen, the following chemical species are involved: CH_4, CH_3, O_2, O, OH, H_2O, CO, CO_2. Determine sets of independent reactions and of thermodynamic components.

Hint: A convenient way to attack this problem is to consider all the reactions of formation of each individual chemical species above from its atomic components. Next, using an appropriate number of such reactions, eliminate the atomic species, thus leading to a set of independent reactions.

1.17 Consider a fluid flowing in a straight tube of length L and diameter D. Based on physical arguments, it can be expected that the pressure drop ΔP is given by

$$\frac{\Delta P}{L} = f(D, \mu, \rho, v)$$

where μ and ρ are the fluid viscosity and density, respectively, while v is its linear velocity. Use dimensional analysis to find the form of the function above, and indicate the role of the Reynolds number, $Dv\rho/\mu$.

Hint: Assume that the function has the form $f(D, \mu, \rho, v) = kD^\alpha\mu^\beta\rho^\gamma v^\delta$, where k is a dimensionless constant and the exponents α to δ are determined by requiring that the dimensions of the lhs and rhs are identical.

1.18 The heat flux exchanged by a fluid flowing in a tubular pipe, with the surroundings, can be characterized through an appropriate heat transfer coefficient. In general, this is a function of the fluid viscosity μ, density ρ, specific heat C_p, and thermal conductivity k, as well as of the tube diameter D and fluid velocity v. Using dimensional analysis, show that this dependence can be represented in the form

$$\text{Nu} = k \, \text{Re}^a \text{Pr}^b$$

where

$$\text{Nu} = \frac{hD}{k}, \qquad \text{Re} = \frac{Dv\rho}{\mu}, \qquad \text{Pr} = \frac{C_p \mu}{k}$$

are dimensionless parameters, while the dimensionless constant k and exponents a and b are determined experimentally.

1.19 Compute the eigenvalues of matrices $\mathbf{A}$ and $\mathbf{B}$ defined in problem 1.7.

1.20 Compute a set of eigenvectors and a set of eigenrows for the matrices $\mathbf{A}$ and $\mathbf{B}$ defined in problem 1.7. For each matrix, show by direct computation that the two sets are biorthogonal.

1.21 Show by direct computation that for the symmetric matrix

$$\mathbf{C} = \begin{bmatrix} 1 & -1 & 2 \\ -1 & 3 & 0 \\ 2 & 0 & 2 \end{bmatrix}$$

  (a) the eigenvalues are real and
  (b) the eigenvectors form an orthogonal set of vectors.

1.22 Using the concept of arbitrary vector expansion, solve the systems of equations given in problem 1.2.

1.23 Solve the following system by the method of eigenvalues and eigenvectors:

$$\frac{dx_1}{dt} = 2x_1 + x_2 - 4$$

$$\frac{dx_2}{dt} = -4x_1 - 3x_2 + 7$$

with the IC

$$x_1 = 1 \quad \text{and} \quad x_2 = 2 \quad \text{at } t = 0.$$

1.24 In a batch reactor, two consecutive first-order reactions are taking place:

$$A \xrightarrow{k_1} B \xrightarrow{k_2} C$$

The mass balance leads to a set of linear ODEs describing concentrations of species A and B (i.e., C_A and C_B), which can be solved using the idea of vector expansion as discussed in section 1.17. Determine the solution and show that for $k_2 \gg k_1$, the system approaches pseudo-steady-state conditions for component B; that is, $C_B \cong k_1 C_A / k_2$.

1.25 Consider a plug-flow tubular reactor where the epoxidation of ethylene to ethylene oxide takes place, following the kinetic scheme described in problem 1.15. In excess of oxygen, all kinetics may be assumed first-order with respect to the hydrocarbon reactant. The steady-state mass balance leads to a system of two linear ODEs. Solve these equations to determine the concentration profiles of C_2H_4 and C_2H_4O along the reactor length. Determine the reactor length which maximizes the production of the desired product (C_2H_4O).

1.26 Repeat problem 1.19 using eq. (1.18.1) as the starting point.

1.27 For the matrices $\mathbf{A}$ and $\mathbf{B}$ in problem 1.7, find the eigenvalues of $\mathbf{A}^{-4}$ and $\mathbf{B}^4$. Check the obtained results by first determining the two matrices through direct multiplication and then computing the corresponding eigenvalues.

1.28 Consider the matrix $\mathbf{A}$ in problem 1.7.

    (a) Show by direct computation that following the Hamilton–Cayley theorem, $\mathbf{A}$ satisfies its own characteristic equation.

    (b) Using Sylvester's formula, compute the matrix

$$\mathbf{A}^5 - 2\mathbf{A}^4 + \mathbf{A}^{-3} - 2\mathbf{A} + 4\mathbf{I}$$

    and find its eigenvalues.

1.29 Find an orthogonal matrix $\mathbf{P}$ such that $\mathbf{P}^{-1}\mathbf{A}\mathbf{P}$ is diagonal and has as diagonal elements the eigenvalues of $\mathbf{A}$, where

$$\mathbf{A} = \begin{bmatrix} 3 & -1 & 1 \\ -1 & 5 & -1 \\ 1 & -1 & 3 \end{bmatrix}$$

    Verify your conclusions.

1.30 For the matrix

$$\mathbf{A} = \begin{bmatrix} 1 & 0 & 1 \\ 1 & -2 & 1 \\ 0 & 4 & -3 \end{bmatrix}$$

    find two different matrices which, through similarity transformations, produce the same diagonal form whose elements are the eigenvalues of $\mathbf{A}$.

1.31 Given the second-order expression

$$G = ax_1^2 + x_2^2 + x_3^2 + 2x_1x_2 - bx_2x_3 + 3x_1 - 1$$

    determine conditions under which it can be reduced to a quadratic form. Also, find the corresponding canonical form.

1.32 Determine for what values of c, if any,

$$x^2 + y^2 + z^2 \geq 2c(xy + yz + zx)$$

1.33 What conditions will ensure that

    (a) the function

$$f(x, y) = x^2 + y^2 + ax + by + cxy$$

        has a unique extremum?

    (b) the unique extremum is a minimum?

1.34 Compute the extrema of the function

$$f(x, y) = x^3 + 4x^2 + 3y^2 + 5x - 6y$$

1.35 For the matrix

$$\mathbf{A} = \begin{bmatrix} 2 & -2 \\ -1 & 3 \end{bmatrix}$$

    (a) Compute the eigenvalues of $\exp(\mathbf{A})$.

    (b) Compute the matrix $\exp(\mathbf{A}t)$.

1.36 Use the method described in section 1.29 to solve the systems of ODEs arising in problems 1.23, 1.24, and 1.25.

1.37 Consider the sequence in n CSTRs connected in series shown in Figure 1.4, where the volume of the jth tank is V_j (ft^3). A first-order isothermal reaction $A \xrightarrow{k} B$ occurs in the CSTRs, where the rate constant is k (sec^{-1}). Let q represent the specified flow rate (ft^3/sec) through the system, and let c_j be the molar concentration (mol/ft^3) of reactant A leaving—and hence in—the jth tank. The feed concentration of A is c_{A0} and is specified. The initial distribution of reactants in the tanks is

$$c_j(0) = \alpha_j, \qquad j = 1 \rightarrow n$$

(a) Develop the transient equations governing the concentration of reactant A in each tank. Write them in the form

$$\frac{d\mathbf{c}}{dt} = \mathbf{A}\mathbf{c} + \mathbf{b}, \qquad c(0) = \boldsymbol{\alpha}$$

where $\mathbf{c}^T = [c_1 c_2 \cdots c_n]$ and identify the matrix $\mathbf{A}$ and vector $\mathbf{b}$; t is time elapsed (sec) after start-up.

(b) Following the method developed in section 1.22 for the absorber problem, solve these equations to give $\mathbf{c}(t)$ explicitly.

Chapter 2

FIRST-ORDER NONLINEAR ORDINARY DIFFERENTIAL EQUATIONS AND STABILITY THEORY

2.1 SOME ELEMENTARY IDEAS

For a first-order ordinary differential equation (ODE) of the type

$$\frac{dy}{dx} = f(x, y) \tag{2.1.1}$$

with a prescribed initial condition (IC)

$$y = y_0 \quad \text{at } x = x_0 \tag{2.1.2}$$

where $f(x, y)$ is a *nonlinear* function of x and y, it is in general not possible to solve for y analytically. This is in contrast to *linear* ODEs, such as

$$\frac{dz}{dx} + a(x)z = b(x) \tag{2.1.3}$$

with IC

$$z = z_0 \quad \text{at } x = x_0 \tag{2.1.4}$$

which can be solved readily by using the familiar *integrating factor* approach, to yield

$$z(x) = z_0 \exp\left[\int_x^{x_0} a(t)\, dt\right] + \exp\left[-\int^x a(t)\, dt\right] \int_{x_0}^x b(s) \exp\left[\int^s a(t)\, dt\right] ds \tag{2.1.5}$$

It is straightforward to verify that eq. (2.1.5) is indeed the solution of the ODE (2.1.3) which satisfies the IC (2.1.4).

A nonlinear ODE can be solved analytically only for certain special cases. One such case is that of a *separable* equation:

$$\frac{dy}{dx} = \frac{g(x)}{f(y)} \tag{2.1.6}$$

where the solution follows by separating the variables x and y and then integrating.

A second case involves the so-called *exact* equations, which can be written in the form

$$P(x,\ y) + Q(x,\ y) \frac{dy}{dx} = \frac{d}{dx} [g\{x,\ y(x)\}] = 0 \tag{2.1.7}$$

The solution is given by

$$g\{x,\ y(x)\} = c \tag{2.1.8}$$

where c is an arbitrary constant. In order for an ODE to be described by the form (2.1.7), we require that the functions P and Q satisfy the condition derived below. From the first of eqs. (2.1.7), we have

$$\frac{dy}{dx} = -\frac{P(x,\ y)}{Q(x,\ y)} \tag{2.1.9}$$

while the second leads to

$$\frac{\partial g}{\partial x} + \frac{\partial g}{\partial y} \frac{dy}{dx} = 0$$

which rearranges as

$$\frac{dy}{dx} = \frac{-\partial g/\partial x}{\partial g/\partial y} \tag{2.1.10}$$

Comparing eqs. (2.1.9) and (2.1.10), we obtain

$$\frac{\partial P}{\partial y} = \frac{\partial Q}{\partial x} \tag{2.1.11}$$

the condition for eq. (2.1.7) to be exact. Also, a comparison of eqs. (2.1.6), (2.1.9), and (2.1.10) leads to the conclusion that a separable equation is simply a special case of an exact equation.

The final category which leads to analytical solutions is relatively small and involves substitutions which convert the nonlinear ODE into a linear or an exact equation. An example of this type is the *Bernoulli* equation:

$$\frac{dy}{dx} + a(x)y + b(x)y^n = 0 \tag{2.1.12}$$

where n is any number. For $n = 0$ or 1, eq. (2.1.12) is linear and can be solved as discussed above. For other values of n, if we change the dependent variable as

$$z = y^{1-n} \tag{2.1.13}$$

then

$$y = z^{1/(1-n)}$$

and

$$\frac{dy}{dx} = \frac{z^{n/(1-n)}}{(1-n)} \frac{dz}{dx}$$

Substituting in eq. (2.1.12) and dividing by $z^{n/(1-n)}$ yields

$$\frac{1}{(1-n)} \frac{dz}{dx} + a(x)z + b(x) = 0 \tag{2.1.14}$$

which is linear in the variable z and hence can be solved as discussed earlier. From eq. (2.11.13), the solution for y is then given by

$$y = z^{1/(1-n)} \tag{2.1.15}$$

A second example is the *Riccati* equation, which involves a quadratic in y:

$$\frac{dy}{dx} = a_1(x) + a_2(x)y + a_3(x)y^2 \tag{2.1.16}$$

There are two elementary cases, which are easy to solve. They occur for $a_3(x) = 0$ or $a_1(x) = 0$, leading to a linear equation or a Bernoulli equation ($n = 2$), respectively. For general $a_i(x)$, an analytical solution of eq. (2.1.16) cannot be obtained. This is not difficult to understand because the change of variable

$$y = -\frac{u'}{a_3 u} \tag{2.1.17}$$

reduces eq. (2.1.16) to

$$u'' - \left[\frac{a_3'}{a_3} + a_2 \right] u' + a_1 a_3 u = 0 \tag{2.1.18}$$

a second-order linear ODE with nonconstant coefficients, which cannot be solved in closed form for arbitrary functions $a_i(x)$. For such equations, in general, only series solutions can be obtained (see chapter 4).

In spite of this, it is indeed possible to solve some Riccati equations. This can be done when a particular solution, say $y_1(x)$, can be found by *any* means (usually by observation). If we let the general solution be

$$y(x) = y_1(x) + u(x) \tag{2.1.19}$$

then by substituting in eq. (2.1.16) it may be seen that $u(x)$ satisfies the equation

$$\frac{du}{dx} - [a_2 + 2y_1a_3]u - a_3u^2 = 0 \qquad (2.1.20)$$

This is a Bernoulli equation in u with $n = 2$ and can be solved using the substitution given in eq. (2.1.13).

In summary, although we have discussed some special cases of nonlinear ODEs where an analytical closed-form solution can be obtained, in general this is not possible. Thus numerical procedures are commonly adopted in order to obtain answers to specific problems; we do not discuss numerical techniques here, but, for example, see James et al. (1985) or Rice (1993). Before applying numerical procedures, we need to examine whether a solution in the mathematical sense exists and, if it does, whether it is unique. These questions are answered by the existence and uniqueness theorem, which is treated next.

2.2 EXISTENCE AND UNIQUENESS THEOREM FOR A SINGLE FIRST-ORDER NONLINEAR ODE

In considering the solutions of the nonlinear ODE

$$\frac{dy}{dx} = f(x, y) \qquad (2.2.1)$$

with IC

$$y = y_0 \qquad \text{at } x = x_0 \qquad (2.2.2)$$

the following result can be proven (cf. Ince, 1956, chapter 3; Coddington, 1989, chapter 5).

● **THEOREM**

Let $f(x, y)$ be a one-valued, continuous function of x and y in a domain $\mathscr{D}$ defined as

$$|x - x_0| \leq a, \qquad |y - y_0| \leq b$$

and shown as the rectangle $ABCD$ in Figure 2.1.

Let $|f(x, y)| \leq M$ in $\mathscr{D}$, and let

$$h = \min\left[a, \frac{b}{M}\right]$$

If $h < a$, then $\mathscr{D}$ is redefined as

$$|x - x_0| \leq h, \qquad |y - y_0| \leq b$$

and shown as the rectangle $A'B'C'D'$ in Figure 2.1. Furthermore, let $f(x, y)$ satisfy the *Lipschitz condition*, which implies that if (x, y) and (x, Y) are two points in $\mathscr{D}$, then

$$|f(x, y) - f(x, Y)| < K|y - Y| \qquad (2.2.3)$$

where K is a positive constant.

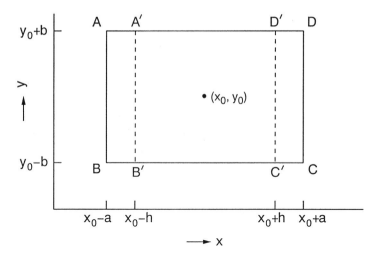

Figure 2.1 Definition of domain $\mathcal{D}$ for a case where $h = b/M < a$.

With these conditions satisfied, the existence theorem ensures that there *exists* a *unique continuous* function of x, say $y(x)$, defined for all x in the range $|x - x_0| \leq h$, which satisfies the ODE

$$\frac{dy}{dx} = f(x, y)$$

and the IC

$$y = y_0 \qquad \text{at } x = x_0$$

The theorem can be proved in more ways than one. The most common proof is by a *method of successive approximations*, also called *Picard iteration*. The proof derives from the fact that integrating both sides of eq. (2.2.1) with respect to x, as well as using eq. (2.2.2), leads to

$$\int_{y_0}^{y} dy = \int_{x_0}^{x} f(t, y(t)) \, dt$$

which gives

$$y(x) = y_0 + \int_{x_0}^{x} f(t, y(t)) \, dt \tag{2.2.4}$$

Equation (2.2.4) is an *integral equation* in y, since the unknown function y appears not only by itself but also within the integral. This equation can, however, be solved by a method of successive approximations, as shown next.

For x in the range $|x - x_0| \leq h$, consider the sequence of functions $y_1(x), y_2(x), \ldots$ defined as follows:

$$y_1(x) = y_0 + \int_{x_0}^{x} f(t, y_0) \, dt$$

$$y_2(x) = y_0 + \int_{x_0}^{x} f(t, y_1(t)) \, dt \qquad (2.2.5)$$

$$\vdots$$

$$y_n(x) = y_0 + \int_{x_0}^{x} f(t, y_{n-1}(t)) \, dt$$

We will show that

(a) as $n \to \infty$, the sequence $\{y_n(x)\}$ tends to a limit function, which is continuous in x,

(b) the limit function satisfies the ODE (2.2.1),

(c) the limit function equals y_0 when $x = x_0$, and

(d) it is the *only* (i.e., unique) continuous function having the above properties (b) and (c).

To prove these statements one by one, first note that

$$|y_1(x) - y_0| = \left| \int_{x_0}^{x} f(t, y_0) \, dt \right|$$

$$\leq \left| \int_{x_0}^{x} \left| f(t, y_0) \right| dt \right|$$

$$\leq M|x - x_0| \qquad (2.2.6)$$

$$\leq Mh$$

$$\leq b$$

If we now assume that

$$|y_{n-1}(x) - y_0| \leq b$$

so that the point (x, y_{n-1}) lies within $\mathcal{D}$, then

$$|f(t, y_{n-1}(t))| \leq M$$

and from eq. (2.2.5) we have

$$|y_n(x) - y_0| = \left| \int_{x_0}^{x} f(t, y_{n-1}(t)) \, dt \right|$$

$$\leq b$$

as before. This means that

$$|y_n(x) - y_0| \leq b \quad \text{for } |x - x_0| \leq h, \qquad n = 1, 2, \ldots \qquad (2.2.7)$$

Hence for $|x - x_0| \leq h$, the sequence of functions $\{y_n(x)\}$ lies in the domain $\mathcal{D}$.

Now, note from eq. (2.2.5) that

$$|y_1(x) - y_0| = \left| \int_{x_0}^{x} f(t, y_0) \, dt \right|$$

$$\leq \left| \int_{x_0}^{x} \left| f(t, y_0) \right| dt \right| \qquad (2.2.8)$$

$$\leq M|x - x_0|$$

Similarly, from eq. (2.2.5) we have

$$|y_2(x) - y_1(x)| = \left| \int_{x_0}^{x} [f(t, y_1(t)) - f(t, y_0)] \, dt \right|$$

$$\leq \left| \int_{x_0}^{x} \left| f(t, y_1(t)) - f(t, y_0) \right| dt \right|$$

$$< \left| \int_{x_0}^{x} K \left| y_1(t) - y_0 \right| dt \right|$$

using the Lipschitz condition (2.2.3). However, since from eq. (2.2.8),

$$|y_1(x) - y_0| \leq M|x - x_0|$$

we obtain

$$|y_2(x) - y_1(x)| < \frac{KM}{2!} |x - x_0|^2 \qquad (2.2.9)$$

Similarly, it may be shown that

$$|y_3(x) - y_2(x)| < \frac{MK^2}{3!} |x - x_0|^3 \qquad (2.2.10)$$

Assume now that the inequality

$$|y_j(x) - y_{j-1}(x)| < \frac{MK^{j-1}}{j!} |x - x_0|^j \qquad (2.2.11)$$

has been proven for $j = 1, 2, \ldots, n - 1$. Then from eq. (2.2.5) we have

$$|y_n(x) - y_{n-1}(x)| = \left| \int_{x_0}^{x} [f(t, y_{n-1}(t)) - f(t, y_{n-2}(t))] \, dt \right|$$

$$\leq \left| \int_{x_0}^{x} \left| f(t, y_{n-1}(t)) - f(t, y_{n-2}(t)) \right| dt \right|$$

$$< \left| \int_{x_0}^{x} K \left| y_{n-1}(t) - y_{n-2}(t) \right| dt \right|$$

$$< K \frac{MK^{n-2}}{(n - 1)!} \left| \int_{x_0}^{x} \left| t - x_0 \right|^{n-1} dt \right|$$

using (2.2.11) for $j = n - 1$. This leads to

$$|y_n(x) - y_{n-1}(x)| < \frac{MK^{n-1}}{n!} |x - x_0|^n$$

and hence inequality (2.2.11) holds for *all* $j \geq 1$.

Next, consider the identity

$$y_n(x) = y_0 + \sum_{j=1}^{n} [y_j(x) - y_{j-1}(x)] \tag{2.2.12}$$

For $|x - x_0| \leq h$, due to (2.2.11), the infinite series on the rhs obtained with $n = \infty$ converges *absolutely* and *uniformly*. Thus the limit function

$$y(x) = \lim_{n \to \infty} y_n(x) \tag{2.2.13}$$

exists. It is a continuous function of x in the internal $|x - x_0| \leq h$, since each term in the series (2.2.12) is a continuous function of x. This proves the above statement (a).

To prove statement (b), we need to show that $y(x)$ obtained from eq. (2.2.13) indeed satisfies the ODE (2.2.1). This would be easy if in the definition of $y_n(x)$ from eq. (2.2.5),

$$y_n(x) = y_0 + \int_{x_0}^{x} f(t, y_{n-1}(t))\, dt$$

we could let $n \to \infty$ on both sides and exchange the limit and integration operations. Then the lhs would give

$$\lim_{n \to \infty} y_n(x) = y(x)$$

from eq. (2.2.13), while the rhs would yield

$$y_0 + \lim_{n \to \infty} \int_{x_0}^{x} f(t, y_{n-1}(t))\, dt = y_0 + \int_{x_0}^{x} \lim_{n \to \infty} f(t, y_{n-1}(t))\, dt$$

$$= y_0 + \int_{x_0}^{x} f(t, y(t))\, dt$$

implying that

$$y(x) = y_0 + \int_{x_0}^{x} f(t, y(t))\, dt \tag{2.2.14}$$

which is just what is needed for the solution, as represented by eq. (2.2.4). This exchange of the limit and integration operations is, in fact, permitted from functional analysis, since $f(x, y)$ is bounded and the integral is over a finite range. Thus statement (b) follows.

It is clear by inspection of eq. (2.2.14) that at $x = x_0$, the limit function $y(x)$ indeed equals y_0, which is the content of statement (c).

Summarizing briefly, what we have shown so far is that under the restrictions of the theorem, if a sequence of functions $\{y_n(x)\}$ is defined as in eq. (2.2.5), then a continuous limit function of x exists, which satisfies both the ODE (2.2.1) and the IC (2.2.2). It should be evident that this method of successive approximations is a *constructive* one, since it provides a means of obtaining the solution $y(x)$, although for most functions $f(x, y)$ it is cumbersome to actually do so.

In this method, since the next approximation $y_j(x)$ is obtained directly from the previously computed approximation $y_{j-1}(x)$, it should be no surprise that the limit function $y(x)$ thus obtained is *unique*. However, this can also be proven by supposing that another *distinct* solution $Y(x)$ satisfying eq. (2.2.1) exists in $\mathscr{D}$, which also satisfies the IC (2.2.2). Thus we have

$$y(x) = y_0 + \int_{x_0}^{x} f(t, y(t)) \, dt \qquad (2.2.15a)$$

and also

$$Y(x) = y_0 + \int_{x_0}^{x} f(t, Y(t)) \, dt \qquad (2.2.15b)$$

Subtracting gives

$$|y(x) - Y(x)| = \left| \int_{x_0}^{x} [f(t, y(t)) - f(t, Y(t))] \, dt \right| \qquad (2.2.16)$$

We now need to consider separately the regions $x > x_0$ and $x < x_0$. In the interval $x > x_0$, eq. (2.2.16) yields

$$|y(x) - Y(x)| \leq \left| \int_{x_0}^{x} \left| f(t, y(t)) - f(t, Y(t)) \right| dt \right|$$

$$< K \int_{x_0}^{x} |y(t) - Y(t)| \, dt \qquad (2.2.17)$$

using the Lipschitz condition for f. If we denote

$$\delta(x) = \int_{x_0}^{x} |y(t) - Y(t)| \, dt \qquad (2.2.18)$$

then

$$\frac{d\delta}{dx} = |y(x) - Y(x)|$$

and eq. (2.2.17) leads to

$$\frac{d\delta}{dx} < K\delta$$

$$= K\delta - p(x) \qquad (2.2.19)$$

where $p(x) > 0$ for $(x - x_0) \leq h$. Also, by definition

$$\delta(x_0) = 0 \qquad (2.2.20)$$

The ODE (2.2.19) is linear, and its solution from eq. (2.1.5) is given by

$$\delta(x) = -\int_{x_0}^{x} p(t)e^{K(x-t)}\, dt$$

$$< 0 \qquad \text{for } x > x_0 \tag{2.2.21}$$

Following identical steps, for the region $x < x_0$ we have

$$\delta(x) > 0 \qquad \text{for } x < x_0 \tag{2.2.22}$$

However, by definition (2.2.18), if $y(x)$ and $Y(x)$ are *distinct* functions, then $\delta(x) > 0$ (<0) for $x > x_0$ ($<x_0$). This is a contradiction to eqs. (2.2.21) and (2.2.22). Thus

$$y(x) = Y(x) \tag{2.2.23}$$

which proves uniqueness of the solution—that is, statement (d).

2.3 SOME REMARKS CONCERNING THE EXISTENCE AND UNIQUENESS THEOREM

The two main assumptions in the theorem are that the function $f(x, y)$ is continuous and that it satisfies the Lipschitz condition in the domain $\mathcal{D}$. These two assumptions are independent of each other. Thus a function may be continuous but may not satisfy the Lipschitz condition [e.g., $f(x, y) = |y|^{1/2}$ for any $\mathcal{D}$ that contains $y = 0$]. Alternatively, it may be discontinuous and yet be Lipschitzian (e.g., a step function). We now examine the necessity of these two assumptions.

It should be apparent from the proof of the theorem that the continuity of $f(x, y)$ is not required. All the arguments in the proof remain valid if $f(x, y)$ is merely bounded, such that the integral

$$\int_{x_0}^{x} |f(t, y(t))|\, dt$$

exists. Thus $f(x, y)$ may well possess a finite number of discontinuities, and yet the solution $y(x)$ remains continuous.

Consider, for example, the initial value problem:

$$\frac{dy}{dx} = f(x, y) \tag{2.3.1}$$

with the IC

$$y = 1 \qquad \text{at } x = 0 \tag{2.3.2}$$

where

$$f(x, y) = \begin{cases} y(1 - x), & x > 0 \\ y(x - 1), & x < 0 \end{cases} \tag{2.3.3}$$

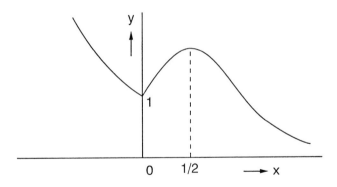

Figure 2.2 The solution $y(x)$ given by eq. (2.3.4).

is discontinuous at $x = 0$. However, the solution is *continuous*:

$$y(x) = \begin{cases} \exp[x - (x^2/2)], & x \geq 0 \\ \exp[(x^2/2) - x], & x \leq 0 \end{cases} \tag{2.3.4}$$

and is shown in Figure 2.2. Note that although $y(x)$ is continuous, it does not have a continuous derivative at $x = 0$ as implied by the discontinuity in $f(x, y)$.

The Lipschitz condition, on the other hand, is not necessary for the existence of the solution but is required to ensure its uniqueness. Thus, for example, in the problem

$$\frac{dy}{dx} = 3y^{2/3}, \qquad y(0) = 0$$

we have

$$f(x, y) = 3y^{2/3}$$

which is continuous for all (x, y). However, the functions

$$y_1(x) = x^3 \quad \text{and} \quad y_2(x) = 0$$

both satisfy the problem, and hence they are valid solutions. The reason for this behavior is that the function $3y^{2/3}$ is not Lipschitzian in any region that includes $y = 0$.

It is worth noting that if $f(x, y)$ has bounded first partial derivative with respect to y, then it also satisfies the Lipschitz condition. In order to prove this, note that the mean value theorem gives

$$f(x, y) - f(x, Y) = \frac{\partial f}{\partial y}(x, y^*)[y - Y], \qquad y^* \in (y, Y) \tag{2.3.5}$$

so that

$$|f(x, y) - f(x, Y)| \leq \left| \frac{\partial f}{\partial y}(x, y^*) \right| |y - Y|$$

$$< K|y - Y| \tag{2.3.6}$$

where K is the upper bound for $|\partial f/\partial y|$ in $\mathcal{D}$.

2.4 CONTINUOUS DEPENDENCE ON A PARAMETER OR THE INITIAL CONDITION

It should be evident that if in place of the ODE (2.2.1) with the IC (2.2.2), we consider the problem

$$\frac{dy}{dx} = f(x, y; \lambda)$$

$$y = y_0 \qquad \text{at } x = x_0$$

where λ is a parameter, the existence and uniqueness of the solution is again guaranteed as long as f is Lipschitz continuous in the variable y. We now show that besides these properties ensured for fixed values of λ and y_0, the solution possesses the additional feature of being continuously dependent on the parameter λ and the initial value y_0.

For this, consider the two problems

$$\frac{dy}{dx} = f(x, y; \lambda), \qquad y(x_0) = \delta_1 \tag{2.4.1}$$

$$\frac{dz}{dx} = f(x, z; \eta), \qquad z(x_0) = \delta_2 \tag{2.4.2}$$

where f satisfies a Lipschitz condition in y:

$$|f(x, y; \lambda) - f(x, Y; \lambda)| < K|y - Y| \tag{2.4.3}$$

Then eq. (2.2.4) applied to eqs. (2.4.1) and (2.4.2) leads to

$$y(x) = \delta_1 + \int_{x_0}^{x} f(t, y(t); \lambda) \, dt$$

and

$$z(x) = \delta_2 + \int_{x_0}^{x} f(t, z(t); \eta) \, dt$$

Subtracting gives

$$y(x) - z(x) = (\delta_1 - \delta_2) + \int_{x_0}^{x} [f(t, y(t); \lambda) - f(t, z(t); \eta)] \, dt$$

$$= (\delta_1 - \delta_2) + \int_{x_0}^{x} [\{f(t, y(t); \lambda) - f(t, z(t); \lambda)\}$$

$$+ \{f(t, z(t); \lambda) - f(t, z(t); \eta)\}] \, dt \tag{2.4.4}$$

We again need to consider separately the regions $x > x_0$ and $x < x_0$. For $x > x_0$, we obtain

$$|y(x) - z(x)| \leq |\delta_1 - \delta_2| + \int_{x_0}^{x} [|f(t, y(t); \lambda) - f(t, z(t); \lambda)|$$

$$+ |f(t, z(t); \lambda) - f(t, z(t); \eta)|] \, dt$$

Let us assume that the function f depends continuously on the parameter. Thus

$$|f(x, z(x); \lambda) - f(x, z(x); \eta)| < \varepsilon \qquad \text{in } \mathcal{D} \tag{2.4.5}$$

where ε is small if $|\lambda - \eta|$ is small. With this and eq. (2.4.3), we have

$$|y(x) - z(x)| < |\delta_1 - \delta_2| + \int_{x_0}^{x} [K|y(t) - z(t)| + \varepsilon] \, dt$$

$$< |\delta_1 - \delta_2| + \varepsilon(x - x_0) + K \int_{x_0}^{x} |y(t) - z(t)| \, dt \tag{2.4.6}$$

Let us denote

$$E(x) = \int_{x_0}^{x} |y(t) - z(t)| \, dt \tag{2.4.7}$$

Then

$$\frac{dE}{dx} = |y(x) - z(x)|$$

and (2.4.6) becomes the differential inequality

$$\frac{dE}{dx} - KE < |\delta_1 - \delta_2| + \varepsilon(x - x_0)$$

$$= |\delta_1 - \delta_2| + \varepsilon(x - x_0) - p(x) \tag{2.4.8}$$

with the IC

$$E = 0 \qquad \text{at } x = x_0 \tag{2.4.9}$$

by definition of E, and $p(x) > 0$.

Following the same approach employed in section 2.2, while solving for $\delta(x)$ defined by eq. (2.2.18), we obtain

$$E(x) = -\varepsilon \frac{(x - x_0)}{K} + \left[\frac{|\delta_1 - \delta_2|}{K} + \frac{\varepsilon}{K^2}\right] [e^{K(x-x_0)} - 1] - \int_{x_0}^{x} p(t) e^{K(x-t)} \, dt$$

Differentiating both sides with respect to x gives

$$\frac{dE}{dx} = |y(x) - z(x)|$$

$$= \frac{\varepsilon}{K} [e^{K(x-x_0)} - 1] + |\delta_1 - \delta_2| e^{K(x-x_0)} - p(x) - K \int_{x_0}^{x} p(t) e^{K(x-t)} \, dt$$

Since $p(x) > 0$, this leads to

$$|y(x) - z(x)| < \frac{\varepsilon}{K} [e^{K(x-x_0)} - 1] + |\delta_1 - \delta_2| e^{K(x-x_0)}, \qquad x > x_0 \tag{2.4.10}$$

If we repeat the steps from eq. (2.4.4) onward for $x < x_0$, the argument of the exponential on the rhs of inequality (2.4.10) reverses sign. Thus for *all* values of x, we have

$$|y(x) - z(x)| < \frac{\varepsilon}{K} [e^{K|x-x_0|} - 1] + |\delta_1 - \delta_2|e^{K|x-x_0|} \qquad (2.4.11)$$

This means that if the difference between λ and η is finite (so that the corresponding ε is also finite) and if the difference between the ICs δ_1 and δ_2 is finite, then the absolute difference between the solutions $y(x)$ and $z(x)$ remains bounded, for *all* $x \in \mathcal{D}$. This implies continuous dependence of the solution on a parameter and the IC. However, note that owing to the exponential on the rhs of inequality (2.4.11), the bound grows with increase of the independent variable, x. In unusual cases, for sufficiently large x, this can result in a substantial difference between $y(x)$ and $z(x)$, even though ε and $|\delta_1 - \delta_2|$ are both small. This, in fact, is a characteristic feature of systems which display *chaotic* behavior (see section 2.23).

Note that uniqueness of the solution for fixed values of the parameter and the IC follows immediately from (2.4.11), because ε and $|\delta_1 - \delta_2|$ are then identically zero.

2.5 SOME EXAMPLES

In this section we consider two examples. In the first we simply develop the successive approximations without concern for their domain of validity, while in the second we also examine this aspect.

Example 1. For the linear ODE

$$\frac{dy}{dx} = y \qquad (2.5.1)$$

with IC

$$y(0) = 1 \qquad (2.5.2)$$

construct the solution by successive approximations.

The analytic solution is readily obtained as

$$y = \exp(x) \qquad (2.5.3)$$

and let us now develop the successive approximations.

The function $f(x, y) = y$ is continuous and satisfies the Lipschitz condition (with $K = 1 + \varepsilon$, where $\varepsilon > 0$) for $\mathcal{D} = \{x \in (-\infty, \infty), y \in (-\infty, \infty)\}$. The successive approximations are given by

$$y_j(x) = 1 + \int_0^x y_{j-1}(t) \, dt, \qquad j = 1, 2, \dots \qquad (2.5.4)$$

Thus we have

$$y_1(x) = 1 + x$$

$$y_2(x) = 1 + \int_0^x (1 + t)\, dt = 1 + x + \frac{x^2}{2}$$

$$y_3(x) = 1 + \int_0^x \left(1 + t + \frac{t^2}{2}\right) dt = 1 + x + \frac{x^2}{2!} + \frac{x^3}{3!}$$

$$\vdots$$

and

$$y_n(x) = \sum_{j=0}^{n} \frac{x^j}{j!}$$

Taking the limit as $n \to \infty$ yields

$$y(x) = \lim_{n \to \infty} y_n(x) = \exp(x) \tag{2.5.5}$$

as expected.

Example 2. For the ODE

$$\frac{dy}{dx} = y^2 \tag{2.5.6}$$

with IC

$$y(0) = 1 \tag{2.5.7}$$

construct the successive approximations, and find the largest range of x for the solution.

The analytic solution can again be obtained easily by separation of variables:

$$y(x) = \frac{1}{1 - x} \tag{2.5.8}$$

which is clearly continuous for $-\infty < x < 1$, a range that contains the initial value of x (i.e., $x = 0$). A plot of eq. (2.5.8) is shown in Figure 2.3.

The nonlinear function $f(x, y) = y^2$ is continuous for all x and y. It also satisfies the Lipschitz condition for all y, since

$$|y_1^2 - y_2^2| = |(y_1 - y_2)(y_1 + y_2)| \le |y_1 + y_2| \cdot |y_1 - y_2|$$

so that $\max[2|y| + \varepsilon]$ in the chosen domain $\mathcal{D}$, where $\varepsilon > 0$, serves as the Lipschitz constant, K.

Before proceeding further, we must define a suitable domain $\mathcal{D}$. If we are ambitious, we may define $\mathcal{D}$ as

$$|x| < a = \infty, \qquad |y - 1| < b = \infty$$

However, in this case

$$M = \max_{\mathcal{D}} |f(x, y)| = \max_{|y-1| < \infty} |y^2| = \infty$$

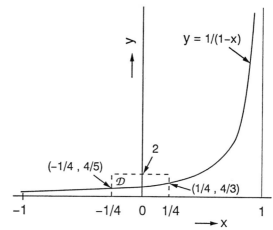

Figure 2.3 Plot of the exact solution and domain $\mathcal{D}$ for Example 2.

so that

$$h = \min\left[a, \frac{b}{M}\right] = 0$$

and we cannot even begin from the initial value of x—that is, $x_0 = 0$! So in this case, we have to be less ambitious.

Let us then define $\mathcal{D}$ as

$$|x| < a, \qquad |y - 1| < b \qquad\qquad (2.5.9)$$

in an effort to find the *maximum* range of validity in terms of x. Then

$$M = \max_{|y-1|<b} |y|^2 = (1 + b)^2$$

which leads to

$$h = \min\left[a, \frac{b}{(1 + b)^2}\right] \qquad\qquad (2.5.10)$$

It is easy to verify that the function

$$g(b) = \frac{b}{(1 + b)^2} \qquad\qquad (2.5.11)$$

attains a maximum value $\frac{1}{4}$ at $b = 1$; hence h is maximized with the choice $a \geq \frac{1}{4}$ and $b = 1$. Thus the *largest* domain that can be constructed around the point $(0, 1)$ is

$$\mathcal{D}: \qquad |x| \leq \tfrac{1}{4}, \quad |y - 1| \leq 1 \qquad\qquad (2.5.12)$$

and is shown in Figure 2.3.

Now, using the method of successive approximations, we have

$$y_j(x) = 1 + \int_0^x y_{j-1}^2(t)\, dt, \qquad y_0 = 1, \quad j = 1, 2, \ldots \qquad (2.5.13)$$

which lead to

$$y_1(x) = 1 + x$$

$$y_2(x) = \frac{2 + (1 + x)^3}{3}$$

$$y_3(x) = 1 + \frac{1}{9}\left[-1 + 4x + (1 + x)^4 + \frac{\{(1 + x)^7 - 1\}}{7}\right]$$

and further approximations become unwieldy. However, by comparing the exact value of y at the ends of the interval with y_3, we find that

$$\text{at } x = \tfrac{1}{4}: \qquad y_{\text{exact}} = \tfrac{4}{3} \quad \text{and} \quad y_3 = 1.331$$

while

$$\text{at } x = -\tfrac{1}{4}: \qquad y_{\text{exact}} = 0.8 \quad \text{and} \quad y_3 = 0.799$$

Thus y_3 is a rather good approximation to $y_{\text{exact}} = 1/(1 - x)$.

2.6 CONTINUATION OF SOLUTIONS

The exact continuous solution of Example 2 is shown in Figure 2.3 for the interval $-\infty < x < 1$, where the domain $\mathcal{D}$ (i.e., the region where a unique continuous solution can be proven to exist by the existence theorem) is also indicated. Since $\mathcal{D}$ is substantially smaller than the full interval, it appears that the conditions of the existence theorem restrict the actual domain of validity of the solution. This is true in many cases, but it should be apparent that we can continue the solution beyond both sides of $\mathcal{D}$ by taking new ICs

$$\text{at } x = \tfrac{1}{4}: \qquad y = \tfrac{4}{3}$$

and

$$\text{at } x = -\tfrac{1}{4}: \qquad y = \tfrac{4}{5}$$

for the ODE

$$\frac{dy}{dx} = y^2$$

For example, if we do this for $x_0 = \tfrac{1}{4}$, the largest domain $\mathcal{D}$ around $(\tfrac{1}{4}, \tfrac{4}{3})$ is found to be

$$\left|x - \tfrac{1}{4}\right| < \tfrac{3}{16}, \qquad \left|x - \tfrac{4}{3}\right| < \tfrac{4}{3}$$

so that existence is now proven for $-\tfrac{1}{4} < x < \tfrac{7}{16}$. Similarly, the solution can also be continued to the left of $x = -\tfrac{1}{4}$.

A natural question now arises: How far can the solution be continued? The answer is contained in the following result (cf. Brauer and Nohel, 1989, section 3.4; Coddington, 1989, chapter 5, section 7).

● THEOREM

Let $f(x, y)$ be continuous and bounded, and let it satisfy a Lipschitz condition in region $\mathcal{D}$. Then for every $(x_0, y_0) \in \mathcal{D}$, the ODE

$$\frac{dy}{dx} = f(x, y)$$

possesses a unique solution $y(x)$ satisfying $y(x_0) = y_0$, which can be extended in both directions of x_0 until it remains finite or until its graph meets the boundary of $\mathcal{D}$.

Hence it should thus be evident that in the case of Example 2, the solution $y = 1/(1 - x)$ can be extended over the interval $-\infty < x < 1$.

2.7 EXISTENCE AND UNIQUENESS THEOREM FOR A SYSTEM OF NONLINEAR ODEs

In previous sections we considered a single nonlinear ODE. However, the same conclusions also hold for a *system* of first-order ODEs

$$\frac{dy_1}{dx} = f_1(x, y_1, y_2, \ldots, y_n)$$

$$\frac{dy_2}{dx} = f_2(x, y_1, y_2, \ldots, y_n)$$

$$\vdots \qquad\qquad\qquad\qquad\qquad (2.7.1)$$

$$\frac{dy_n}{dx} = f_n(x, y_1, y_2, \ldots, y_n)$$

with ICs

$$y_i(x_0) = y_{i,0}, \qquad i = 1, 2, \ldots, n \qquad\qquad (2.7.2)$$

which may be written compactly as

$$\frac{d\mathbf{y}}{dx} = \mathbf{f}(x, \mathbf{y}) \qquad\qquad (2.7.3)$$

$$\mathbf{y}(x_0) = \mathbf{y}_0 \qquad\qquad (2.7.4)$$

The result, obtained by following the same steps as before, can be summarized as follows (cf. Ince, 1956, section 3.3; Brauer and Nohel, 1989, section 3.4).

● THEOREM

Let the functions f_i, $i = 1, 2, \ldots, n$, be single-valued and continuous in their $(n + 1)$ arguments in the domain $\mathcal{D}$ defined by

$$|x - x_0| \leq a, \quad |y_1 - y_{1,0}| \leq b_1, \ldots, |y_n - y_{n,0}| \leq b_n \qquad (2.7.5)$$

and let

$$|f_i(x, \mathbf{y})| \le M, \qquad i = 1, 2, \ldots, n \qquad (2.7.6)$$

in this domain. Also, define

$$h = \min\left[a, \frac{b_1}{M}, \cdots, \frac{b_n}{M} \right] \qquad (2.7.7)$$

and if $h < a$, restrict x to $|x - x_0| \le h$. Furthermore, let the functions f_i satisfy the Lipschitz condition in $\mathscr{D}$:

$$|f_i(x, y_1, \ldots, y_n) - f_i(x, Y_1, \ldots, Y_n)| < \sum_{j=1}^{n} K_j |y_j - Y_j|, \qquad i = 1, 2, \ldots, n \quad (2.7.8)$$

Then there exist unique, continuous functions of x, $y_i(x)$ satisfying eqs. (2.7.1) and (2.7.2). These functions are obtained from the sequences

$$y_{i,j}(x) = y_{i,0} + \int_{x_0}^{x} f_i(t, y_{1,j-1}(t), \ldots, y_{n,j-1}(t))\, dt, \qquad i = 1, 2, \ldots, n; \quad j = 1, 2, \ldots$$
$$(2.7.9)$$

in the limit as $j \to \infty$, and they are continuous with respect to the initial conditions $\mathbf{y}_0$ or parameters in $\mathbf{f}$. These solutions can be extended in both directions of x_0 until $\| \mathbf{y}(x) \|$ remains finite or until its graph meets the boundary of $\mathscr{D}$, where $\| \mathbf{y} \|$ denotes the norm of $\mathbf{y}$, defined by eq. (1.5.4).

This result is not only valuable by itself, but it can also be employed for nonlinear ODEs of higher order, such as

$$\frac{d^n y}{dx^n} = f\left(x, y, \frac{dy}{dx}, \ldots, \frac{d^{n-1}y}{dx^{n-1}} \right) \qquad (2.7.10)$$

with ICs

$$\text{at } x = x_0: \qquad y = y_0, \quad \frac{dy}{dx} = a_1, \ldots, \frac{d^{n-1}y}{dx^{n-1}} = a_{n-1} \qquad (2.7.11)$$

This is so because with the change of variables

$$\begin{aligned}
y &= y_1 \\
\frac{dy}{dx} &= \frac{dy_1}{dx} = y_2 \\
\frac{d^2 y}{dx^2} &= \frac{dy_2}{dx} = y_3 \\
&\vdots \\
\frac{d^{n-1} y}{dx^{n-1}} &= \frac{dy_{n-1}}{dx} = y_n
\end{aligned} \qquad (2.7.12)$$

eq. (2.7.10) reduces to the set of n first-order ODEs

$$\frac{dy_1}{dx} = y_2$$

$$\frac{dy_2}{dx} = y_3$$

$$\vdots \tag{2.7.13}$$

$$\frac{dy_{n-1}}{dx} = y_n$$

$$\frac{dy_n}{dx} = f(x, y_1, y_2, \ldots, y_n)$$

with ICs

$$\text{at } x = x_0: \quad y_1 = y_0, \quad y_2 = a_1, \ldots, y_n = a_{n-1} \tag{2.7.14}$$

Thus if f is continuous and bounded and satisfies a Lipschitz condition in $\mathcal{D}$, then a unique, continuous, and $(n-1)$ times continuously differentiable solution of eq. (2.7.10) exists, which also satisfies the IC (2.7.11).

2.8   QUALITATIVE BEHAVIOR OF NONLINEAR ODEs

A variety of important problems in science and engineering are *nonlinear* in nature. As discussed in section 2.1, except for some special cases they cannot be solved analytically. However, in many cases, a good deal of qualitative information about the solution can be extracted without actually solving the problem. Such information is of great value, because it not only provides insight into the problem, but also serves as a check on the numerical solution which is generally obtained using a computer.

As a motivation for this study, let us first consider two simple problems, one linear and the other nonlinear. It will become clear soon that nonlinear problems exhibit certain surprising features which are not found in linear ones.

The linear ODE

$$\frac{du}{dt} = -au + b = -a\left(u - \frac{b}{a}\right) \tag{2.8.1}$$

with IC

$$u(0) = u_0 \tag{2.8.2}$$

where a and b are constants, has the *unique* solution

$$u(t) = \frac{b}{a} + \left(u_0 - \frac{b}{a}\right)e^{-at} \tag{2.8.3}$$

If the problem arises physically, then $a > 0$, and let us restrict our discussion to this case. Then eq. (2.8.3) implies that as $t \to \infty$,

$$u(t) \to u_s = \frac{b}{a} \qquad (2.8.4)$$

where u_s refers to the *steady-state* (also called *equilibrium* or *critical*) value of u. Now, u_s could also be obtained directly from eq. (2.8.1), because in the steady state, by definition, u does not vary with t. Thus setting du/dt in eq. (2.8.1) equal to zero gives $u_s = b/a$.

The two points to note about the above linear problem are as follows:

1. The steady state is unique.
2. The steady state is approached asymptotically as $t \to \infty$.

A schematic plot of the solution given by eq. (2.8.3) is shown in Figure 2.4, where it may be noted that there is no relative extremum in the u profile for finite t. This follows by differentiating the solution (2.8.3), to give

$$\frac{du}{dt} = -a\left(u_0 - \frac{b}{a}\right)e^{-at} \begin{cases} < 0, & u_0 > b/a \\ = 0, & u_0 = b/a \\ > 0, & u_0 < b/a \end{cases} \qquad (2.8.5)$$

However, this conclusion can also be reached *directly* from eq. (2.8.1), without actually solving it, as follows. If $u_0 < b/a$, then the rhs of eq. (2.8.1) is > 0, implying that $du/dt|_{t=0} > 0$. Thus u increases with t, from its initial value u_0. However, the value of u can never exceed b/a since the rhs becomes progressively smaller (but positive) as u approaches b/a. Hence for $u_0 < b/a$, $u_0 < u(t) < b/a$ and $du/dt > 0$ for all finite $t > 0$. Similarly, for $u_0 > b/a$, we have $b/a < u(t) < u_0$ and $du/dt < 0$ for all finite $t > 0$. Note finally that if $u_0 = b/a$, then the rhs of eq. (2.8.1) is zero initially, and $u(t)$ remains at b/a for all $t > 0$.

A simple example which leads to eq. (2.8.1) is the cooling (or heating) of a fluid

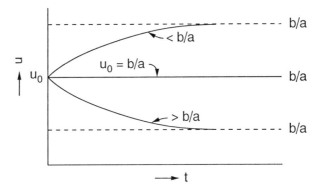

Figure 2.4 The solution of the ODE (2.8.1).

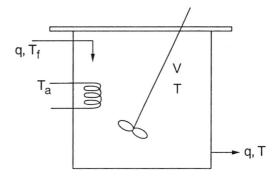

Figure 2.5 A continuous-flow stirred tank.

in a continuous-flow stirred tank (see Figure 2.5). In order to derive the energy balance equation, let q represent the volumetric fluid flow rate and let V denote the tank volume. Then an energy balance implies

$$\text{Input} = \text{Output} + \text{Accumulation} \tag{2.8.6}$$

where

$$\text{Input} = qC_p\rho T_f$$

$$\text{Output} = qC_p\rho T + UA(T - T_a)$$

and

$$\text{Accumulation} = VC_p\rho \frac{dT}{dt}$$

Here T is the fluid temperature; T_f and T_a are the fluid feed and coil temperatures, respectively; C_p and ρ are the heat capacity and density of the fluid, respectively; U is the heat transfer coefficient between the tank fluid and the coil; and A is the heat transfer area of the coil. Then eq. (2.8.6) leads to

$$VC_p\rho \frac{dT}{dt} = qC_p\rho(T_f - T) - UA(T - T_a)$$

or

$$\frac{dT}{dt} = -\left(\frac{q}{V} + \frac{UA}{VC_p\rho}\right)T + \left(\frac{q}{V}T_f + \frac{UAT_a}{VC_p\rho}\right) \tag{2.8.7}$$

By comparing eqs. (2.8.1) and (2.8.7), note that $a = [(q/V) + (UA/VC_p\rho)] > 0$, as observed earlier. The problem becomes complete when the IC is specified, which in this example reflects the fluid temperature in the tank at start-up ($t = 0$) conditions:

$$T(0) = T_0$$

Let us now consider a *nonlinear* ODE:

$$\frac{du}{dt} = f(u) = u(1 - u) \tag{2.8.8}$$

with IC

$$u(0) = u_0 \tag{2.8.9}$$

which arises in population dynamics. It is one of the simplest equations which describes the growth of a single species, having access to limited resources, and can be derived as follows (cf. Murray, 1990, section 1.1; Boyce and DiPrima, 1992, section 2.6).

Let u be the size of a population at time t. In the simplest model of population growth, it is assumed that the rate of growth of the population is proportional to its size; that is,

$$\frac{du}{dt} = \alpha u \tag{2.8.10}$$

where $\alpha > 0$ is a constant. This is readily solved to give

$$u(t) = u(0) \exp(\alpha t) \tag{2.8.11}$$

which implies exponential growth, sometimes referred to as *Malthus's law*.

The more realistic models respect the fact that growth is eventually limited by availability of resources. Thus we have

$$\frac{du}{dt} = ug(u) \tag{2.8.12}$$

where a suitable form for $g(u)$ is now determined. When the population size u is small, there are sufficient resources so the growth rate is essentially independent of resources. Thus

$$g(u) \simeq \alpha \tag{2.8.13}$$

a positive constant when u is small. However, as the population size increases, limited resources inhibit the growth rate. Thus $g(u)$ should decrease with u; that is,

$$\frac{dg}{du} < 0 \tag{2.8.14}$$

The simplest function $g(u)$ satisfying both conditions (2.8.13) and (2.8.14) is

$$g(u) = \alpha - \beta u \tag{2.8.15}$$

where α and β are positive constants. Thus eq. (2.8.12) takes the form

$$\frac{du}{dt} = u(\alpha - \beta u) \tag{2.8.16}$$

which is called the *logistic* or the *Verhulst equation*. For $\alpha = \beta = 1$, we obtain eq. (2.8.8).

The steady states of eq. (2.8.8) are obtained readily by setting $f(u) = 0$ as

$$u_{s1} = 0 \text{ and } u_{s2} = 1 \tag{2.8.17}$$

and we find that in contrast to the linear problem (2.8.1), the steady state in this case is *not* unique.

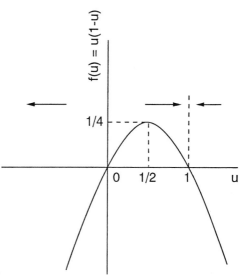

Figure 2.6 Plot of the function $f(u) = -u(u-1)$.

In order to develop and understand the solution, let us consider the shape of the function $f(u)$, shown in Figure 2.6. We observe that

$$f(u) \begin{cases} > 0, & u \in (0, 1) \\ = 0, & u = 0 \text{ or } 1 \\ < 0, & u < 0 \text{ or } > 1 \end{cases}$$

where we note that although from a physical viewpoint $u > 0$, we are also considering $u < 0$ for mathematical completeness.

From eq. (2.8.8), $du/dt = f(u)$. Thus if the IC $u_0 > 1$, du/dt will initially be negative, and u will decrease as t increases. This will continue until u equals 1, where u remains since du/dt then equals zero. Similarly, if $u_0 \in (0, 1)$, du/dt will be positive, and u will increase with t until it again reaches 1. However, if $u_0 < 0$, du/dt is negative and u decreases further as t increases. Finally, if $u_0 = 0$ *or* 1, then $du/dt = 0$ and u will not change with t. The directions of movement for u, for various intervals of u_0, are indicated by the arrows in Figure 2.6.

Thus we find that the steady-state $u_{s2} = 1$ is obtained from *all* ICs $u_0 > 0$, but the steady state $u_{s1} = 0$ is not obtained from any IC except $u_0 = 0$. It should therefore be evident that if the system governed by eq. (2.8.8) was operating at steady state (u_{s1} or u_{s2}) and was slightly perturbed away from it, it would return to u_{s2} but not to u_{s1}. In this sense, $u_{s2} = 1$ is *stable* while $u_{s1} = 0$ is *unstable*. We will return to these considerations in more detail later in this chapter.

Although we have reached these conclusions in a qualitative manner, they are confirmed by the analytic solution of eqs. (2.8.8) and (2.8.9), which can be found easily by the separation of variables:

$$\ln\left(\frac{u}{u_0}\right) - \ln\left(\frac{1-u}{1-u_0}\right) = t \qquad (2.8.18)$$

For the existence of the logarithms, this requires

$$u(t) \begin{cases} > 1 & \text{if } u_0 > 1 \\ \in (0, 1) & \text{if } u_0 \in (0, 1) \\ < 0 & \text{if } u_0 < 0 \end{cases} \qquad (2.8.19)$$

for all finite $t \geq 0$. The analytic solution is given by

$$u(t) = \begin{cases} \dfrac{1}{1 + \{(1 - u_0)/u_0\}e^{-t}}, & u_0 \in (0, 1) \text{ or } u_0 > 1 \\ \dfrac{-1}{\{(1 + |u_0|)/|u_0|\}e^{-t} - 1}, & u_0 < 0 \end{cases} \qquad (2.8.20)$$

and $u(t) = u_{si}$, if $u_0 = u_{si}$; $i = 1$ or 2. Thus $u(t) \to 1$ as $t \to \infty$, if $u_0 \in (0, 1)$ or $u_0 >$ 1. For the case $u_0 < 0$, it would again appear that $u(t) \to 1$ as $t \to \infty$, but this is *not* true. From the theorem on continuation of solutions (section 2.6), starting from $(u_0, 0)$, a continuous solution $(u(t), t)$ can be found only up to the point where $u(t)$ becomes unbounded. From eq. (2.8.20), it is clear that for $u_0 < 0$, $u(t)$ becomes unbounded at a finite $t = t^*$, since

$$\frac{1 + |u_0|}{|u_0|} > 1$$

and

$$e^{-t} \leq 1 \qquad \text{for } t \geq 0$$

Thus for a fixed $|u_0|$, there exists a finite $t = t^*$ such that

$$\left\{ \frac{1 + |u_0|}{|u_0|} \right\} e^{-t^*} = 1 \qquad (2.8.21)$$

hence

$$u(t^*) = -\infty$$

Note from eq. (2.8.21) that for $u_0 < 0$, t^* decreases as $|u_0|$ increases.

A schematic diagram of $u(t)$ for various u_0 values is shown in Figure 2.7, where

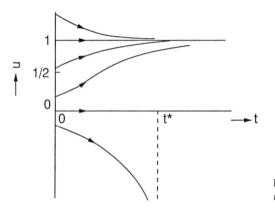

Figure 2.7 Solution of eq. (2.8.8) for various u_0 values.

it may be noted that an inflection point in the u profile exists only if $u_0 \in (0, \frac{1}{2})$. This occurs because at the inflection point

$$\frac{d^2u}{dt^2} = 0$$

while from eq. (2.8.8) we obtain

$$\frac{d^2u}{dt^2} = \frac{d}{dt}\,[f(u)]$$

$$= \frac{df}{du}\frac{du}{dt}$$

$$= f(u)\,\frac{df}{du}$$

and $df/du = (-2u + 1)$ vanishes only for $u = \frac{1}{2}$ (see Figure 2.6).

2.9 LINEARIZATION

In the nonlinear problem (2.8.8) considered in the previous section, even without solving the problem, we were able to conclude that all ICs $u_0 > 0$ led to the steady state $u_{s2} = 1$, while $u(t)$ tends to increase further if $u_0 < 0$. The *only* IC that led to $u_{s1} = 0$ was $u_0 = 0$. We thus concluded that the steady state $u_{s2} = 1$ was stable, while $u_{s1} = 0$ was unstable (note that the terms ''stable'' and ''unstable'' are used loosely here, and are made more precise in the next section). The problem (2.8.8) was relatively simple, and was even amenable to an analytic solution. This is typically not the case, particularly when a system of ODEs is involved. The question that arises then is whether something can be said about the character of the steady state(s) in more complex cases.

If we are interested in the behavior of the system in the immediate vicinity of the steady state, then one possibility is to consider the equations *linearized* about the steady state in question. For example, let us consider the problem

$$\frac{du}{dt} = f(u), \qquad u(0) = u_0 \tag{2.9.1}$$

The steady states, u_s, are obtained by setting $f(u) = 0$, thus they satisfy the nonlinear equation

$$f(u_s) = 0 \tag{2.9.2}$$

Let us define the *perturbation variable*:

$$x(t) = u(t) - u_s \tag{2.9.3}$$

which is a measure of the deviation of the solution $u(t)$ from the specific steady state u_s. Then

$$u(t) = u_s + x(t)$$

and eq. (2.9.1) takes the form

$$\frac{d}{dt}(u_s + x) = \frac{dx}{dt} = f(u_s + x) \tag{2.9.4}$$

with the IC

$$x(0) = u_0 - u_s = x_0 \tag{2.9.5}$$

If we expand $f(u_s + x)$ in a Taylor series about u_s, we obtain

$$f(u_s + x) = f(u_s) + \frac{df}{du}\bigg|_{u_s} x + \sum_{n=2}^{\infty} \frac{d^n f}{du^n}\bigg|_{u_s} \frac{x^n}{n!} \tag{2.9.6}$$

where $f(u_s) = 0$, by definition. If we are investigating the behavior of the system *near* the steady state, we may want to neglect terms involving powers of x greater than one in eq. (2.9.6), in the hope that if x is ''small,'' then x^n with $n > 1$ will be even smaller. If we do this, eq. (2.9.4) reduces to

$$\frac{dx}{dt} = f'(u_s)x \quad , \quad x(0) = x_0 \tag{2.9.7}$$

which has the solution

$$x(t) = x_0 \exp[f'(u_s)t] \tag{2.9.8}$$

Recall from eq. (2.9.3) that

$$x(t) = u(t) - u_s$$

is defined as the perturbation from the specific steady state u_s. Equation (2.9.8) clearly indicates that the ''small'' perturbation $x(t)$

1. decays in size as t increases, if $f'(u_s) < 0$,
2. grows in size as t increases, if $f'(u_s) > 0$, and
3. remains at its initial value x_0, if $f'(u_s) = 0$

It then appears that the steady state, in some sense, is

1. stable, if $f'(u_s) < 0$,
2. unstable, if $f'(u_s) > 0$, and
3. marginally stable, if $f'(u_s) = 0$.

For the nonlinear ODE (2.8.8) considered in the previous section,

$$f(u) = u(1 - u) \tag{2.9.9}$$

and the steady states are given by

$$u_s = 0 \text{ and } 1$$

Differentiating eq. (2.9.9) yields

$$f'(u) = 1 - 2u$$

Thus for

$$u_s = u_{s1} = 0,$$

we obtain

$$f'(u_s) = 1 > 0$$

while for

$$u_s = u_{s2} = 1$$

we have

$$f'(u_s) = -1 < 0$$

Hence $u_s = 0$ is expected to be unstable, while $u_s = 1$ should be stable.

A comparison with results from the previous section indicates that for this example, correct conclusions are reached for both steady states by the linearization procedure. However, as of now, one cannot be certain that linearization will always yield the correct behavior even in the vicinity of the steady states. In fact, it does, and this is the content of the *fundamental stability theorem*, to be considered in section 2.13. First, it is important to define various terms associated with the notion of stability more precisely.

2.10 DEFINITIONS OF STABILITY

The system of nonlinear ODEs

$$\frac{dy_i}{dt} = f_i(y_1, y_2, \ldots, y_n), \qquad i = 1, 2, \ldots, n \tag{2.10.1}$$

written compactly as

$$\frac{d\mathbf{y}}{dt} = \mathbf{f(y)} \tag{2.10.2}$$

where

$$\mathbf{y}^T = [y_1, y_2, \ldots, y_n]$$

and

$$\mathbf{f}^T = [f_1, f_2, \ldots, f_n]$$

and where the functions f_i depend solely on $\mathbf{y}$, and not explicitly on the independent variable t, is called an *autonomous* system. In the remainder of this chapter, we deal only with autonomous systems.

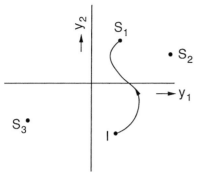

Figure 2.8 A phase plane, where S_i denote the steady states and I indicates a specific IC.

If the functions f_i are Lipschitz continuous in some region $\mathcal{D}$ of the real n-dimensional Euclidean space, then a unique solution of eq. (2.10.2), $\mathbf{y}(t)$, exists in $\mathcal{D}$ for the initial value problem (cf. section 2.2).

A constant vector $\mathbf{y}_s$ which satisfies

$$\mathbf{f}(\mathbf{y}_s) = \mathbf{0} \qquad (2.10.3)$$

is called a *steady state* (also *critical* or *equilibrium* point) of the system (2.10.2). It should be evident that multiple steady states may exist if more than one value of $\mathbf{y}_s$ satisfies eq. (2.10.3). Also, with the IC $\mathbf{y}(0) = \mathbf{y}_s$, the solution is $\mathbf{y}(t) \equiv \mathbf{y}_s$ for all $t \geq 0$.

Given an IC, it is often convenient to express the solution $\mathbf{y}(t)$ as a curve in the n-dimensional *phase space* of $\mathbf{y}$, and this curve is called a *trajectory*. Each steady state is a specific point in the phase space, and every point in the phase space is a potential IC $\mathbf{y}(0)$. Each IC thus generates a trajectory in the phase space, with the direction of increasing t indicated by an arrow. These aspects are shown in Figure 2.8 for a two-dimensional autonomous system possessing three steady states, where a particular trajectory is also shown.

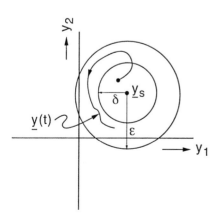

Figure 2.9 Definition of stability.

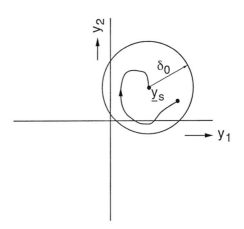

Figure 2.10 Definition of asymptotic stability.

With these preliminaries, we have the following definitions (cf. Brauer and Nohel, 1989, section 4.2).

Definition 1. A steady state $\mathbf{y}_s$ of eq. (2.10.2) is said to be *stable* if for each number $\varepsilon > 0$ we can find a number $\delta(\varepsilon) > 0$, such that if $\mathbf{y}(t)$ is any solution of eq. (2.10.2) with $\| \mathbf{y}_0 - \mathbf{y}_s \| < \delta(\varepsilon)$, then $\| \mathbf{y}(t) - \mathbf{y}_s \| < \varepsilon$ for all $t > 0$ (see Figure 2.9).

Recall that $\| \mathbf{x} \|$ denotes the norm of vector $\mathbf{x}$, defined by eq. (1.5.4).

Definition 2. A steady state is defined to be *unstable* if it is not stable.

Definition 3. A steady state $\mathbf{y}_s$ of eq. (2.10.2) is said to be *asymptotically stable* if it is stable and if there exists a number $\delta_0 > 0$ such that if $\mathbf{y}(t)$ is any solution of eq. (2.10.2) with $\| \mathbf{y}_0 - \mathbf{y}_s \| < \delta_0$, then $\lim_{t \to \infty} \mathbf{y}(t) = \mathbf{y}_s$ (see Figure 2.10).

Definition 4. A steady state $\mathbf{y}_s$ is said to be *globally asymptotically stable* if it is globally stable and if for all ICs $\mathbf{y}_0$, the solution $\mathbf{y}(t)$ of eq. (2.10.2) satisfies $\lim_{t \to \infty} \mathbf{y}(t) = \mathbf{y}_s$ (see Figure 2.11).

Note that global asymptotic stability implies that the steady state is *unique*.

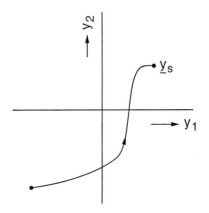

Figure 2.11 Definition of global asymptotic stability.

2.11 STABILITY OF HOMOGENEOUS LINEAR SYSTEMS WITH CONSTANT COEFFICIENTS

Before investigating the stability of nonlinear systems, it is important to first understand the dynamics of linear homogeneous systems with constant coefficients. Thus, let us consider

$$\frac{d\mathbf{y}}{dt} = \mathbf{A}\mathbf{y} \tag{2.11.1}$$

with ICs

$$\mathbf{y}(0) = \mathbf{y}_0 \tag{2.11.2}$$

where $\mathbf{A}$ is a real constant square matrix of order n, and $\mathbf{y}_s$ is an n-column vector.

If the eigenvalues of $\mathbf{A}$ are *distinct*, from eq. (1.17.13) we note that the solution $\mathbf{y}$ is represented by

$$\mathbf{y}(t) = \sum_{j=1}^{n} \frac{\mathbf{w}_j^T \mathbf{y}_0}{\mathbf{w}_j^T \mathbf{z}_j} \mathbf{z}_j e^{\lambda_j t} \tag{2.11.3}$$

where λ_j are the eigenvalues of $\mathbf{A}$, and $\mathbf{z}_j$ and $\mathbf{w}_j^T$ are the corresponding eigenvectors and eigenrows, respectively. It is then apparent from eq. (2.11.3) that if *all* eigenvalues λ_j are negative (if real) or have negative real parts (if complex), then

$$\lim_{t \to \infty} \mathbf{y}(t) = \mathbf{0} \tag{2.11.4}$$

Further, if $\mathbf{A}$ has at least one real eigenvalue that is positive or if it has at least one pair of complex eigenvalues with positive real parts, then due to the fact that for such eigenvalues the term $e^{\lambda_j t}$ increases without limit, the solution $\mathbf{y}$ will become unbounded as $t \to \infty$. In this case, we have

$$\lim_{t \to \infty} \mathbf{y}(t) = \pm\infty \tag{2.11.5}$$

Finally, if all eigenvalues of $\mathbf{A}$, except for one, are negative (if real) or have negative real parts (if complex), but there is either a real eigenvalue $\lambda = 0$, or there is a pair of purely imaginary eigenvalues $\lambda_\pm = \pm i\beta$, then

$$\lim_{t \to \infty} \mathbf{y}(t) = \frac{\mathbf{w}^T \mathbf{y}_0}{\mathbf{w}^T \mathbf{z}} \mathbf{z}, \qquad \lambda = 0 \tag{2.11.6a}$$

or

$$\lim_{t \to \infty} \mathbf{y}(t) = \frac{\mathbf{w}_+^T \mathbf{y}_0}{\mathbf{w}_+^T \mathbf{z}_+} \mathbf{z}_+ e^{i\beta t} + \frac{\mathbf{w}_-^T \mathbf{y}_0}{\mathbf{w}_-^T \mathbf{z}_-} \mathbf{z}_- e^{-i\beta t}, \qquad \lambda_\pm = \pm i\beta \tag{2.11.6b}$$

where $\mathbf{z}$ and $\mathbf{w}^T$ are the eigenvector and eigenrow, respectively corresponding to $\lambda = 0$, and $\mathbf{z}_\pm$ and $\mathbf{w}_\pm^T$ are the eigenvectors and eigenrows, respectively corresponding to $\lambda_\pm$. Since $\mathbf{z}_+$ and $\mathbf{z}_-$ *and* $\mathbf{w}_+^T$ and $\mathbf{w}_-^T$ are complex conjugates, it is easy to show that the solution (2.11.6b) is real and represents a self-sustained oscillation.

For the system (2.11.1), $\mathbf{y} = \mathbf{0}$ is a steady state. The above discussion indicates that when they are *distinct*, the eigenvalues of $\mathbf{A}$ determine completely whether the steady state $\mathbf{0}$ is reached as $t \to \infty$.

In fact, the same is also true when $\mathbf{A}$ has *repeated* eigenvalues. In order to prove this, let us assume that the characteristic equation

$$|\mathbf{A} - \lambda \mathbf{I}| = 0$$

has a root $\lambda = \rho$ repeated r times, so that ρ is an eigenvalue of *multiplicity r*. We then have two possibilities:

1. There are r linearly independent eigenvectors corresponding to $\lambda = \rho$, or
2. there are fewer than r such eigenvectors.

This issue is clarified by the following examples.

Example 1. For the matrix

$$\mathbf{A} = \begin{bmatrix} 1 & -1 \\ 1 & 3 \end{bmatrix}$$

determine the eigenvalues and the eigenvectors.

The eigenvalue problem is given by

$$\mathbf{A}\mathbf{x} = \lambda \mathbf{x} \tag{2.11.7}$$

so that the eigenvalues λ satisfy

$$|\mathbf{A} - \lambda \mathbf{I}| = 0$$

Thus we have

$$\begin{vmatrix} 1 - \lambda & -1 \\ 1 & 3 - \lambda \end{vmatrix} = 0$$

which yields

$$\lambda^2 - 4\lambda + 4 = 0$$

so that the eigenvalues are

$$\lambda_1 = \lambda_2 = 2$$

That is, the eigenvalue 2 has a multiplicity two. In order to determine the eigenvectors, from eq. (2.11.7) we have

$$(1 - \lambda)x_1 - x_2 = 0$$

and

$$x_1 + (3 - \lambda)x_2 = 0$$

For $\lambda = 2$, these equations reduce to

$$-x_1 - x_2 = 0$$

$$x_1 + x_2 = 0$$

which both yield the single condition

$$x_1 + x_2 = 0 \tag{2.11.8}$$

Thus either x_1 or x_2 may be chosen arbitrarily, and the other variable is obtained from eq. (2.11.8). Hence there is *only one* linearly independent eigenvector associated with the repeated eigenvalue:

$$\mathbf{x} = \begin{bmatrix} 1 \\ -1 \end{bmatrix} \tag{2.11.9}$$

Example 2. Determine the eigenvalues and the corresponding eigenvectors for the matrix

$$\mathbf{A} = \begin{bmatrix} 0 & 1 & 1 \\ 1 & 0 & 1 \\ 1 & 1 & 0 \end{bmatrix}$$

The eigenvalues satisfy the equation

$$\begin{bmatrix} -\lambda & 1 & 1 \\ 1 & -\lambda & 1 \\ 1 & 1 & -\lambda \end{bmatrix} = 0$$

Upon expansion, this leads to

$$-\lambda^3 + 3\lambda + 2 = 0$$

which has roots

$$\lambda_1 = 2, \ \lambda_2 = -1, \ \lambda_3 = -1 \tag{2.11.10}$$

Thus 2 is a simple eigenvalue, while -1 is an eigenvalue with multiplicity two.

The eigenvector corresponding to the simple eigenvalue $\lambda_1 = 2$ can be obtained by the procedure discussed in section 1.11—that is, by obtaining $\mathrm{adj}(\mathbf{A} - \lambda_1 \mathbf{I})$; this yields

$$\mathbf{x}_1 = \begin{bmatrix} 1 \\ 1 \\ 1 \end{bmatrix} \tag{2.11.11}$$

For the repeated eigenvalue $\lambda = -1$, the set of *three* equations

$$\mathbf{Ax} = \lambda \mathbf{x}$$

reduces to the *single* equation:

$$x_1 + x_2 + x_3 = 0 \tag{2.11.12}$$

Thus any two among x_1, x_2, and x_3 may be chosen arbitrarily, and the third is then obtained from eq. (2.11.12). For example, if we choose $x_1 = 0$ and $x_2 = 1$, then $x_3 = $

-1; alternatively, with the choice $x_1 = 1$ and $x_2 = 0$, we again have $x_3 = -1$. However, the two vectors

$$\mathbf{x}_2 = \begin{bmatrix} 0 \\ 1 \\ -1 \end{bmatrix} \quad \text{and} \quad \mathbf{x}_3 = \begin{bmatrix} 1 \\ 0 \\ -1 \end{bmatrix} \tag{2.11.13}$$

are linearly independent.

The difference between the two examples is that the matrix of Example 2 is symmetric, while that of Example 1 is not. In general, if $\mathbf{A}$ is a symmetric matrix of order n, then even though it possesses repeated eigenvalues, there exists a full set of n linearly independent eigenvectors. However, if $\mathbf{A}$ is not symmetric and has repeated eigenvalues, then a full set of n linearly independent eigenvectors may not exist (for further details, see Bellman, 1970, chapter 4, section 9 and chapter 11, section 10).

It is thus seen that the case where the matrix $\mathbf{A}$ has repeated eigenvalues is more difficult to analyze, because the linear independence of all the eigenvectors is not ensured. However, despite this, information about stability of the linear system can still be extracted, even though the proofs become cumbersome. For this analysis, we need the following upper triangularization theorem (cf. Bellman, 1969, chapter 1, section 10).

● **THEOREM 1**

Given a matrix $\mathbf{A}$, there exists a nonsingular matrix $\mathbf{T}$ such that

$$\mathbf{T}^{-1}\mathbf{A}\mathbf{T} = \begin{bmatrix} \lambda_1 & b_{12} & b_{13} & \cdots & b_{1n} \\ 0 & \lambda_2 & b_{23} & \cdots & b_{2n} \\ 0 & 0 & \lambda_3 & \cdots & b_{3n} \\ \vdots & \vdots & \vdots & \ddots & \vdots \\ 0 & 0 & \cdot & \cdots & \lambda_n \end{bmatrix} \tag{2.11.14}$$

where $\lambda_j, j = 1, 2, \ldots, n$, are the eigenvalues of $\mathbf{A}$, some perhaps repeated, and the matrix on the rhs is in upper-triangular form.

Proof

The proof is by induction. Let us first consider an arbitrary square matrix, $\mathbf{A}_2$ of order 2. Let λ_1 be an eigenvalue of $\mathbf{A}_2$, and $\mathbf{x}_1$ its associated eigenvector, determined by choosing any nonzero column of $\text{adj}(\mathbf{A}_2 - \lambda_1\mathbf{I})$. Now, form a square matrix $\mathbf{T}_2$, with first column equal to $\mathbf{x}_1$, while the second column is arbitrary, except that it maintains $\mathbf{T}_2$ as nonsingular. Thus we have

$$\mathbf{A}_2 = \begin{bmatrix} a_{11} & a_{12} \\ a_{21} & a_{22} \end{bmatrix} \tag{2.11.15}$$

and

$$\mathbf{T}_2 = \begin{bmatrix} x_{11} & c_1 \\ x_{21} & c_2 \end{bmatrix} \tag{2.11.16}$$

where

$$\mathbf{x}_1 = \begin{bmatrix} x_{11} \\ x_{21} \end{bmatrix}$$

and c_1 and c_2 are chosen such that $\mathbf{T}_2$ is nonsingular. The inverse of $\mathbf{T}_2$ is given by

$$\mathbf{T}_2^{-1} = \frac{1}{(c_2 x_{11} - c_1 x_{21})} \begin{bmatrix} c_2 & -c_1 \\ -x_{21} & x_{11} \end{bmatrix}$$

and with this we can evaluate the following similarity transform.

$$
\begin{aligned}
\mathbf{T}_2^{-1}\mathbf{A}_2\mathbf{T}_2 &= \frac{1}{(x_{11}c_2 - x_{21}c_1)} \begin{bmatrix} c_2 & -c_1 \\ -x_{21} & x_{11} \end{bmatrix} \begin{bmatrix} a_{11} & a_{12} \\ a_{21} & a_{22} \end{bmatrix} \begin{bmatrix} x_{11} & c_1 \\ x_{21} & c_2 \end{bmatrix} \\[6pt]
&= \frac{1}{(x_{11}c_2 - x_{21}c_1)} \begin{bmatrix} c_2 & -c_1 \\ -x_{21} & x_{11} \end{bmatrix} \begin{bmatrix} \lambda_1 x_{11} & a_{11}c_1 + a_{12}c_2 \\ \lambda_1 x_{21} & a_{21}c_1 + a_{22}c_2 \end{bmatrix} \\[6pt]
&= \frac{1}{(x_{11}c_2 - x_{21}c_1)} \begin{bmatrix} (c_2\lambda_1 x_{11} - c_1\lambda_1 x_{21}) & \begin{matrix} c_2(a_{11}c_1 + a_{12}c_2) \\ -c_1(a_{21}c_1 + a_{22}c_2) \end{matrix} \\ 0 & \begin{matrix} -x_{21}(a_{11}c_1 + a_{12}c_2) \\ +x_{11}(a_{21}c_1 + a_{22}c_2) \end{matrix} \end{bmatrix} \qquad (2.11.17) \\[6pt]
&= \begin{bmatrix} \lambda_1 & b_{12} \\ 0 & b_{22} \end{bmatrix} \\[6pt]
&= \begin{bmatrix} \lambda_1 & b_{12} \\ 0 & \lambda_2 \end{bmatrix}
\end{aligned}
$$

since b_{22} must equal λ_2, because a similarity transform preserves the eigenvalues. This proves the theorem for $n = 2$.

Let us assume that the theorem is valid for all square matrices of order n. This means that for any $\mathbf{A}_n$, we can find a corresponding $\mathbf{T}_n$ such that

$$\mathbf{T}_n^{-1}\mathbf{A}_n\mathbf{T}_n = \begin{bmatrix} \lambda_1 & b_{12} & \cdots & b_{1n} \\ 0 & \lambda_2 & \cdots & b_{2n} \\ \vdots & \vdots & & \vdots \\ 0 & 0 & \cdots & \lambda_n \end{bmatrix} \qquad (2.11.18)$$

If the theorem is valid, then we should be able to find a matrix $\mathbf{T}_{n+1}$ which will work for an arbitrary square matrix, $\mathbf{A}_{n+1}$ of order $(n + 1)$.

Let λ_1 be an eigenvalue of $\mathbf{A}_{n+1}$, and $\mathbf{x}_1$ be the corresponding eigenvector. Now choose n vectors $\mathbf{c}_j, j = 1, 2, \ldots, n$ such that

$$\mathbf{S}_1 = [\mathbf{x}_1 \quad \mathbf{c}_1 \quad \mathbf{c}_2 \cdots \mathbf{c}_n] \qquad (2.11.19)$$

is nonsingular. This leads to

$$\mathbf{A}_{n+1}\mathbf{S}_1 = \begin{bmatrix} \lambda_1 x_{11} & d_{12} & \cdots & d_{1,n+1} \\ \lambda_1 x_{21} & d_{22} & \cdots & d_{2,n+1} \\ \vdots & & & \\ \lambda_1 x_{n+1,1} & d_{n+1,2} & \cdots & d_{n+1,n+1} \end{bmatrix}$$

However, since $\mathbf{S}_1 \mathbf{E}_{n+1}$, where

$$
\mathbf{E}_{n+1} = \begin{bmatrix} \lambda_1 & e_{12} & \cdots & e_{1,n+1} \\ 0 & e_{22} & \cdots & e_{2,n+1} \\ \vdots & & & \\ 0 & e_{n+1,2} & \cdots & e_{n+1,n+1} \end{bmatrix}
\tag{2.11.20}
$$

has the same form as $\mathbf{A}_{n+1} \mathbf{S}_1$, we can choose $\mathbf{S}_1$ such that

$$
\mathbf{A}_{n+1} \mathbf{S}_1 = \mathbf{S}_1 \mathbf{E}_{n+1}
$$

which gives

$$
\mathbf{S}_1^{-1} \mathbf{A}_{n+1} \mathbf{S}_1 = \mathbf{E}_{n+1}
\tag{2.11.21}
$$

The matrix $\mathbf{E}_{n+1}$ may also be written in the partitioned form as

$$
\mathbf{E}_{n+1} = \begin{bmatrix} \lambda_1 & \mathbf{e}_n^T \\ \mathbf{0}_n & \mathbf{B}_n \end{bmatrix}
\tag{2.11.22}
$$

where

$$
\mathbf{e}_n^T = [e_{12} \quad e_{13} \cdots e_{1,n+1}]
$$

$$
\mathbf{B}_n = \begin{bmatrix} e_{22} & \cdots & e_{2,n+1} \\ \vdots & & \vdots \\ e_{n+1,2} & \cdots & e_{n+1,n+1} \end{bmatrix}
$$

and $\mathbf{0}_n$ is the n-dimensional null vector. Note that $\mathbf{B}_n$ is a square matrix of order n. Furthermore, since eq. (2.11.21) represents a similarity transformation, the eigenvalues of $\mathbf{A}_{n+1}$ and $\mathbf{E}_{n+1}$ are the same. Thus the eigenvalues of $\mathbf{B}_n$ are $\lambda_2, \lambda_3, \ldots, \lambda_{n+1}$, the remaining n eigenvalues of $\mathbf{A}_{n+1}$.

Recall that by assumption, every matrix of order n can be reduced to the form of eq. (2.11.18). In particular, $\mathbf{B}_n$ can be reduced to this form, and let the nonsingular matrix which does so be $\mathbf{T}_n$. Thus we have

$$
\mathbf{T}_n^{-1} \mathbf{B}_n \mathbf{T}_n = \begin{bmatrix} \lambda_2 & f_{12} & \cdots & f_{1n} \\ 0 & \lambda_3 & \cdots & f_{2n} \\ \vdots & \vdots & & \vdots \\ 0 & 0 & \cdots & \lambda_{n+1} \end{bmatrix}
\tag{2.11.23}
$$

Let $\mathbf{S}_{n+1}$ be the square matrix of order $n + 1$, generated as follows:

$$
\mathbf{S}_{n+1} = \begin{bmatrix} 1 & 0 & \cdots & 0 \\ 0 & & & \\ \vdots & & \mathbf{T}_n & \\ 0 & & & \end{bmatrix} = \begin{bmatrix} 1 & \mathbf{0}_n^T \\ \mathbf{0}_n & \mathbf{T}_n \end{bmatrix}
\tag{2.11.24}
$$

where $\mathbf{S}_{n+1}$ is also clearly nonsingular. Its inverse is given by

$$
\mathbf{S}_{n+1}^{-1} = \begin{bmatrix} 1 & \mathbf{0}_n^T \\ \mathbf{0}_n & \mathbf{T}_n^{-1} \end{bmatrix}
\tag{2.11.25}
$$

since

$$\mathbf{S}_{n+1}\mathbf{S}_{n+1}^{-1} = \begin{bmatrix} 1 & \mathbf{0}_n^T \\ \mathbf{0}_n & \mathbf{T}_n \end{bmatrix}\begin{bmatrix} 1 & \mathbf{0}_n^T \\ \mathbf{0}_n & \mathbf{T}_n^{-1} \end{bmatrix} = \begin{bmatrix} 1 & \mathbf{0}_n^T \\ \mathbf{0}_n & \mathbf{I}_n \end{bmatrix} = \mathbf{I}_{n+1}$$

where $\mathbf{I}_n$ is the identity matrix of order n.

From eqs. (2.11.21), (2.11.24), and (2.11.25) we have

$$\begin{aligned}
\mathbf{S}_{n+1}^{-1}(\mathbf{S}_1^{-1}\mathbf{A}_{n+1}\mathbf{S}_1)\mathbf{S}_{n+1} &= \begin{bmatrix} 1 & \mathbf{0}_n^T \\ \mathbf{0}_n & \mathbf{T}_n^{-1} \end{bmatrix}\begin{bmatrix} \lambda_1 & \mathbf{e}_n^T \\ \mathbf{0}_n & \mathbf{B}_n \end{bmatrix}\begin{bmatrix} 1 & \mathbf{0}_n^T \\ \mathbf{0}_n & \mathbf{T}_n \end{bmatrix} \\
&= \begin{bmatrix} \lambda_1 & \mathbf{e}_n^T \\ \mathbf{0}_n & \mathbf{T}_n^{-1}\mathbf{B}_n \end{bmatrix}\begin{bmatrix} 1 & \mathbf{0}_n^T \\ \mathbf{0}_n & \mathbf{T}_n \end{bmatrix} \\
&= \begin{bmatrix} \lambda_1 & \mathbf{e}_n^T\mathbf{T}_n \\ \mathbf{0}_n & \mathbf{T}_n^{-1}\mathbf{B}_n\mathbf{T}_n \end{bmatrix} \\
&= \begin{bmatrix} \lambda_1 & b_{12} & \cdots & b_{1,n+1} \\ 0 & \lambda_2 & \cdots & b_{2,n+1} \\ \vdots & \vdots & & \vdots \\ 0 & 0 & \cdots & \lambda_{n+1} \end{bmatrix}
\end{aligned} \tag{2.11.26}$$

which is the form required by the theorem. Also, note that

$$\mathbf{S}_{n+1}^{-1}(\mathbf{S}_1^{-1}\mathbf{A}_{n+1}\mathbf{S}_1)\mathbf{S}_{n+1} = (\mathbf{S}_1\mathbf{S}_{n+1})^{-1}\mathbf{A}_{n+1}(\mathbf{S}_1\mathbf{S}_{n+1})$$

Hence $\mathbf{T}_{n+1} = \mathbf{S}_1\mathbf{S}_{n+1}$ is the required matrix of order $(n+1)$ which converts $\mathbf{A}_{n+1}$ to an upper-triangular form. This completes the proof.

This result can be utilized to find solutions of the system

$$\frac{d\mathbf{y}}{dt} = \mathbf{A}\mathbf{y} \tag{2.11.1}$$

with ICs

$$\mathbf{y}(0) = \mathbf{y}_0 \tag{2.11.2}$$

for an arbitrary matrix $\mathbf{A}$ whose eigenvalues may well be repeated. Let $\mathbf{T}$ be the non-singular matrix which reduces $\mathbf{A}$ to an upper triangular form; that is,

$$\mathbf{T}^{-1}\mathbf{A}\mathbf{T} = \begin{bmatrix} \lambda_1 & b_{12} & \cdots & b_{1n} \\ 0 & \lambda_2 & \cdots & b_{2n} \\ \vdots & \vdots & & \vdots \\ 0 & 0 & \cdots & \lambda_n \end{bmatrix} \tag{2.11.27}$$

With the change of variables

$$\mathbf{y} = \mathbf{T}\mathbf{z} \tag{2.11.28}$$

eq. (2.11.1) reduces to

$$\mathbf{T}\frac{d\mathbf{z}}{dt} = \mathbf{A}\mathbf{T}\mathbf{z}$$

which yields

$$\frac{d\mathbf{z}}{dt} = (\mathbf{T}^{-1}\mathbf{A}\mathbf{T})\mathbf{z}$$

$$= \begin{bmatrix} \lambda_1 & b_{12} & \cdots & b_{1n} \\ 0 & \lambda_2 & \cdots & b_{2n} \\ \vdots & \vdots & & \vdots \\ 0 & 0 & \cdots & \lambda_n \end{bmatrix} \mathbf{z} \qquad (2.11.29)$$

Since from eq. (2.11.28),

$$\mathbf{z} = \mathbf{T}^{-1}\mathbf{y}$$

the IC for $\mathbf{z}$ is

$$\mathbf{z}(0) = \mathbf{T}^{-1}\mathbf{y}_0 \qquad (2.11.30)$$

It is straightforward to solve for all the z_j, starting with z_n, since eq. (2.11.29) is equivalent to

$$\frac{dz_1}{dt} = \lambda_1 z_1 + b_{12}z_2 + \cdots + b_{1n}z_n$$

$$\frac{dz_2}{dt} = \lambda_2 z_2 + b_{23}z_3 + \cdots + b_{2n}z_n$$

$$\vdots$$

$$\frac{dz_{n-1}}{dt} = \lambda_{n-1}z_{n-1} + b_{n-1,n}z_n$$

$$\frac{dz_n}{dt} = \lambda_n z_n$$

The last equation directly gives

$$z_n = z_n(0)e^{\lambda_n t}$$

so that the ODE for z_{n-1} takes the form

$$\frac{dz_{n-1}}{dt} = \lambda_{n-1}z_{n-1} + b_{n-1,n}z_n(0)e^{\lambda_n t} \qquad (2.11.31)$$

If λ_{n-1} and λ_n are distinct, then the solution is given by

$$z_{n-1} = \frac{\alpha}{\lambda_n - \lambda_{n-1}} e^{\lambda_n t} + \left[z_{n-1}(0) - \frac{\alpha}{\lambda_n - \lambda_{n-1}} \right] e^{\lambda_{n-1} t} \qquad (2.11.32a)$$

where $\alpha = b_{n-1,n}z_n(0)$. Alternatively, if $\lambda_{n-1} = \lambda_n = \lambda$, then we have

$$z_{n-1} = [z_{n-1}(0) + \alpha t]e^{\lambda t} \qquad (2.11.32b)$$

In general, an eigenvalue λ of multiplicity k will give rise to solutions with components of the form $p_{k-1}(t)e^{\lambda t}$, where $p_{k-1}(t)$ is a polynomial of degree $(k-1)$ at most. However, if λ has a negative real part, then

$$\lim_{t\to\infty} p_{k-1}(t)e^{\lambda t} = 0 \qquad (2.11.33)$$

for all finite k. Thus we have the following important stability result.

● THEOREM 2

The necessary and sufficient condition that all solutions of

$$\frac{d\mathbf{y}}{dt} = \mathbf{A}\mathbf{y}$$

with IC

$$\mathbf{y}(0) = \mathbf{y}_0$$

tend to zero as $t \to \infty$, is that the real parts of all the eigenvalues of $\mathbf{A}$ should be negative. This is true whether $\mathbf{A}$ has simple or repeated eigenvalues.

2.12   A COROLLARY OF THE UPPER-TRIANGULARIZATION THEOREM

A result which is useful in analyzing the stability character of nonlinear systems can be derived as a corollary of the upper-triangularization theorem proved in the previous section.

 COROLLARY

In Theorem 1 of section 2.11, the matrix $\mathbf{T}$ may be chosen such that $\Sigma_{i,j}|b_{ij}|$ is made less than any prescribed positive constant.

Proof
Given an arbitrary matrix $\mathbf{A}$, suppose that we have already found a matrix $\mathbf{S}$ such that $\mathbf{S}^{-1}\mathbf{A}\mathbf{S}$ is upper-triangular, with the eigenvalues of $\mathbf{A}$ on the main diagonal; that is,

$$\mathbf{S}^{-1}\mathbf{A}\mathbf{S} = \begin{bmatrix} \lambda_1 & d_{12} & \cdots & d_{1n} \\ 0 & \lambda_2 & \cdots & d_{2n} \\ \vdots & & & \vdots \\ 0 & 0 & \cdots & \lambda_n \end{bmatrix} \qquad (2.12.1)$$

Consider now the diagonal matrix

$$\mathbf{E} = \begin{bmatrix} \varepsilon & 0 & 0 & \cdots & 0 \\ 0 & \varepsilon^2 & 0 & \cdots & 0 \\ \vdots & \vdots & & & \vdots \\ 0 & 0 & 0 & \cdots & \varepsilon^n \end{bmatrix} \qquad (2.12.2)$$

where ε is a positive constant. Then we have

$$
\mathbf{E}^{-1} = \begin{bmatrix} \varepsilon^{-1} & 0 & 0 & \cdots & 0 \\ 0 & \varepsilon^{-2} & 0 & \cdots & 0 \\ \vdots & \vdots & & & \vdots \\ 0 & 0 & 0 & \cdots & \varepsilon^{-n} \end{bmatrix} \tag{2.12.3}
$$

and

$$
\mathbf{E}^{-1}(\mathbf{S}^{-1}\mathbf{A}\mathbf{S})\mathbf{E} = \begin{bmatrix} \lambda_1 & d_{12}\varepsilon & \cdots & & d_{1n}\varepsilon^{n-1} \\ 0 & \lambda_2 & \cdots & & d_{2n}\varepsilon^{n-2} \\ \vdots & \vdots & & & \vdots \\ 0 & 0 & \cdots & 0 & \lambda_{n-1} & d_{n-1,n}\varepsilon \\ 0 & 0 & & \cdots & & \lambda_n \end{bmatrix} \tag{2.12.4}
$$

If ε is chosen to be arbitrarily small, but *not* zero, then each element above the main diagonal can be made arbitrarily small. Note that ε cannot be chosen as zero, because then $\mathbf{E}^{-1}$ would not exist. If we define

$$
\mathbf{T} = \mathbf{S}\mathbf{E} \tag{2.12.5}
$$

then

$$
\mathbf{T}^{-1}\mathbf{A}\mathbf{T} = (\mathbf{E}^{-1}\mathbf{S}^{-1})\mathbf{A}(\mathbf{S}\mathbf{E}) = \begin{bmatrix} \lambda_1 & b_{12} & \cdots & b_{1n} \\ 0 & \lambda_2 & \cdots & b_{2n} \\ \vdots & \vdots & & \vdots \\ 0 & 0 & \cdots & \lambda_n \end{bmatrix} \tag{2.12.6}
$$

is upper-triangular, with $\Sigma_{i,j}|b_{ij}|$ arbitrarily small.

Remark. It should be evident that if the eigenvalues of $\mathbf{A}$ are distinct, then all b_{ij} can be made equal to zero. This is done following eq. (1.23.8), by taking $\mathbf{T} = \mathbf{X}$, the matrix with the eigenvectors of $\mathbf{A}$ as columns.

2.13 NONLINEAR SYSTEMS: THE FUNDAMENTAL STABILITY THEOREM

In this section we discuss a result of central importance for analyzing the dynamics of autonomous nonlinear systems (cf. Bellman, 1969, chapter 4; Brauer and Nohel, 1989, section 4.4). This result derives from the original independent work of the mathematicians Liapunov and Poincaré near the turn of the twentieth century.

● THEOREM

Given the autonomous nonlinear system

$$
\frac{d\mathbf{y}}{dt} = \mathbf{f}(\mathbf{y}) \tag{2.13.1}
$$

a steady state $\mathbf{y}_s$, obtained from $\mathbf{f}(\mathbf{y}_s) = \mathbf{0}$, is

(a) asymptotically stable if the corresponding linearized system is asymptotically stable
or

(b) unstable if the corresponding linearized system is unstable. No conclusion can be
derived about the nature of the steady state if the corresponding linearized system is
marginally stable.

Proof
First let us construct the linearized system corresponding to the nonlinear system (2.13.1),
where linearization is performed around *a* steady state $\mathbf{y}_s$ whose stability character we wish
to investigate.

If we define

$$x_i(t) = y_i(t) - y_{is}, \qquad i = 1, 2, \ldots, n \tag{2.13.2}$$

as the perturbation variables, which are a measure of the deviation from the steady state
in question, then by Taylor series expansion of $\mathbf{f(y)}$ around $\mathbf{y}_s$ we obtain

$$f_i(\mathbf{y}) = f_i(\mathbf{y}_s) + \left[\frac{\partial f_i}{\partial y_1}\bigg|_{\mathbf{y}_s} (y_1 - y_{1s}) + \frac{\partial f_i}{\partial y_2}\bigg|_{\mathbf{y}_s} (y_2 - y_{2s}) + \cdots + \frac{\partial f_i}{\partial y_n}\bigg|_{\mathbf{y}_s} (y_n - y_{ns})\right]$$

$$+ \frac{1}{2!}\left[\frac{\partial^2 f_i}{\partial y_1^2}\bigg|_{\mathbf{y}_s} (y_1 - y_{1s})^2 + \frac{\partial^2 f_i}{\partial y_1 \partial y_2}\bigg|_{\mathbf{y}_s} (y_1 - y_{1s})(y_2 - y_{2s}) + \cdots + \frac{\partial^2 f_i}{\partial y_n^2}\bigg|_{\mathbf{y}_s}\right.$$

$$\left. (y_n - y_{ns})^2 \right]$$

+ higher-order terms

(2.13.3)

If we truncate the series after the linear terms—that is, those involving $(y_i - y_{is})$—the
remainder of the series is simply

$$r_i = \frac{1}{2!} \mathbf{x}^T \mathbf{Q}_i(\mathbf{y}^*(t))\mathbf{x}, \qquad i = 1, 2, \ldots, n \tag{2.13.4}$$

where

$$\mathbf{Q}_i(\mathbf{y}^*(t)) = \begin{bmatrix} \dfrac{\partial^2 f_i}{\partial y_1^2} & \dfrac{\partial^2 f_i}{\partial y_1 \partial y_2} & \cdots & \dfrac{\partial^2 f_i}{\partial y_1 \partial y_n} \\ \vdots & \vdots & & \vdots \\ \dfrac{\partial^2 f_i}{\partial y_n \partial y_1} & \dfrac{\partial^2 f_i}{\partial y_n \partial y_2} & \cdots & \dfrac{\partial^2 f_i}{\partial y_n^2} \end{bmatrix}_{\mathbf{y}^*(t)} \tag{2.13.5}$$

is the symmetric Hessian matrix of second-order partial derivatives of f_i evaluated at
$\mathbf{y}^*(t) \in [\mathbf{y}(t), \mathbf{y}_s]$
Since we have

$$\frac{d\mathbf{y}}{dt} = \frac{d}{dt}(\mathbf{x} + \mathbf{y}_s) = \frac{d\mathbf{x}}{dt} \tag{2.13.6}$$

the nonlinear system (2.13.1) is *exactly* equivalent to

$$\frac{d\mathbf{x}}{dt} = \mathbf{Ax} + \mathbf{r(x)} \tag{2.13.7}$$

where

$$
\mathbf{A} = \begin{bmatrix} \dfrac{\partial f_1}{\partial y_1} & \dfrac{\partial f_1}{\partial y_2} & \cdots & \dfrac{\partial f_1}{\partial y_n} \\[2mm] \dfrac{\partial f_2}{\partial y_1} & \dfrac{\partial f_2}{\partial y_2} & \cdots & \dfrac{\partial f_2}{\partial y_n} \\[2mm] \vdots & & & \\[2mm] \dfrac{\partial f_n}{\partial y_1} & \dfrac{\partial f_n}{\partial y_2} & \cdots & \dfrac{\partial f_n}{\partial y_n} \end{bmatrix}_{\mathbf{y}_s} \quad \text{and} \quad \mathbf{r} = \begin{bmatrix} r_1 \\ r_2 \\ \vdots \\ r_n \end{bmatrix} \tag{2.13.8}
$$

Note that $\mathbf{A}$ is a constant matrix of the first-order partial derivatives of f_i, evaluated at the steady state $\mathbf{y}_s$, and is called the *Jacobian* or *stability matrix*. The system of ODEs (2.13.7) is often referred to as an *almost linear system*, since

$$
\lim_{\|\mathbf{x}\| \to 0} \frac{\|\mathbf{r}(\mathbf{x})\|}{\|\mathbf{x}\|} = 0 \tag{2.13.9}
$$

The linearized system corresponding to the nonlinear system, from eq. (2.13.7), is simply

$$
\frac{d\mathbf{w}}{dt} = \mathbf{A}\mathbf{w} \tag{2.13.10}
$$

The reason why the fundamental stability theorem is such a strong result is that the stability character of eq. (2.13.9), if it can be definitely established (i.e., either asymptotic stability *or* instability of $\mathbf{w} = \mathbf{0}$), is enough to tell us about the stability character of the steady state in question, $\mathbf{y}_s$ of the original nonlinear system (2.13.1). Precisely, if $\mathbf{w} = \mathbf{0}$ is an asymptotically stable (unstable) steady state of eq. (2.13.10), then $\mathbf{y}_s$ is also asymptotically stable (unstable). If, however, $\mathbf{w} = \mathbf{0}$ is a marginal stable steady state of eq. (2.13.9), then no conclusion can be drawn about the stability character of $\mathbf{y}_s$, that is, it may be asymptotically stable, unstable, or marginally stable, and further analysis is required (see section 2.20).

Let $\mathbf{T}$ be the matrix which reduces $\mathbf{A}$ to an upper-triangular form, with terms above the main diagonal arbitrarily small. Then with

$$
\mathbf{x} = \mathbf{T}\mathbf{w} \tag{2.13.11}
$$

eq. (2.13.7) reduces to

$$
\frac{d\mathbf{w}}{dt} = (\mathbf{T}^{-1}\mathbf{A}\mathbf{T})\mathbf{w} + \mathbf{T}^{-1}\mathbf{r}(\mathbf{T}\mathbf{w})
$$

which on writing more fully yields

$$
\frac{dw_1}{dt} = \lambda_1 w_1 + b_{12}w_2 + \cdots + b_{1n}w_n + \sum_{l=1}^{n}\sum_{m=1}^{n} p_{1lm}w_l w_m
$$

$$
\frac{dw_2}{dt} = \lambda_2 w_2 + b_{23}w_3 + \cdots + b_{2n}w_n + \sum_{l=1}^{n}\sum_{m=1}^{n} p_{2lm}w_l w_m \tag{2.13.12}
$$

$$
\vdots
$$

$$
\frac{dw_n}{dt} = \qquad\qquad\qquad\quad \lambda_n w_n + \sum_{l=1}^{n}\sum_{m=1}^{n} p_{nlm}w_l w_m
$$

where the b_{jk} are arbitrarily small, while some p_{jlm} may be arbitrarily large.

Now let us introduce the positive definite function:

$$V(t) = \sum_{j=1}^{n} |w_j|^2 \tag{2.13.13}$$

where

$$|w_j|^2 = w_j w_j^*$$

with w_j^* being the complex conjugate of w_j. Since some of the eigenvalues may be complex, let us denote them

$$\lambda_j = \alpha_j + i\beta_j, \qquad j = 1, 2, \ldots, n$$

so that

$$\lambda_j^* = \alpha_j - i\beta_j$$

where α_j and β_j are real.

Recall from Theorem 2 of section 2.11 that the linearized system (2.13.10) is asymptotically stable if and only if all the α_j are negative. Let us assume first that this is the case; then

$$\alpha_j \leq \tilde{\alpha} < 0, \qquad j = 1, 2, \ldots, n \tag{2.13.14}$$

so that $\tilde{\alpha}$ is the real part of the eigenvalue with the largest real part. Differentiating eq. (2.13.13) gives

$$\begin{aligned}
\frac{dV}{dt} &= \frac{d}{dt}\left[\sum_{j=1}^{n} w_j w_j^*\right] \\
&= \sum_{j=1}^{n}\left[w_j^* \frac{dw_j}{dt} + w_j \frac{dw_j^*}{dt}\right] \\
&= \sum_{j=1}^{n}\left[\lambda_j w_j w_j^* + w_j^* \sum_{k=j+1}^{n} b_{jk} w_k + w_j^* \sum_{l=1}^{n}\sum_{m=1}^{n} p_{jlm} w_l w_m \right. \\
&\qquad \left. + \lambda_j^* w_j w_j^* + w_j \sum_{k=j+1}^{n} b_{jk}^* w_k^* + w_j \sum_{l=1}^{n}\sum_{m=1}^{n} p_{jlm}^* w_l^* w_m^*\right]
\end{aligned} \tag{2.13.15}$$

utilizing the fact that if z_1 and z_2 are complex numbers, then

$$(z_1 z_2)^* = z_1^* z_2^*$$

The rhs of eq. (2.13.15) is clearly real since the terms appear in complex conjugate pairs. We will replace these pairs by terms that are larger, thus changing the equality to an upper-bound inequality. The *first* one reduces to

$$\begin{aligned}
(\lambda_j + \lambda_j^*) w_j w_j^* &= 2\alpha_j |w_j|^2 \\
&\leq 2\tilde{\alpha}\,|w_j|^2
\end{aligned}$$

so that

$$\begin{aligned}
\sum_{j=1}^{n} (\lambda_j + \lambda_j^*) w_j w_j^* &= \sum_{j=1}^{n} 2\alpha_j |w_j|^2 \\
&\leq 2\tilde{\alpha} \sum_{j=1}^{n} |w_j|^2 \\
&\leq 2\tilde{\alpha}V
\end{aligned} \tag{2.13.16}$$

Since for a complex number z we have

$$[z + z^*] \leq 2|z|$$

the *second* pair on the rhs of eq. (2.13.15) yields

$$\sum_{j=1}^{n}\left[w_j^* \sum_{k=j+1}^{n} b_{jk}w_k + w_j \sum_{k=j+1}^{n} b_{jk}^* w_k^*\right] = \sum_{j=1}^{n}\sum_{k=j+1}^{n}[b_{jk}w_j^*w_k + b_{jk}^*w_jw_k^*]$$

$$\leq 2 \sum_{j=1}^{n}\sum_{k=j+1}^{n}|b_{jk}w_j^*w_k|$$

$$\leq 2 \sum_{j=1}^{n}\sum_{k=1}^{n}|b_{jk}w_j^*w_k|$$

and since for complex numbers we have $|z_1 z_2| = |z_1||z_2|$, we obtain

$$\leq 2 \sum_{j=1}^{n}\sum_{j=k}^{n}|b_{jk}||w_j||w_k| \qquad (2.13.17)$$

Let us denote

$$|b_{jk}| \leq \tilde{b} > 0, \qquad j, k = 1, 2, \ldots, n \qquad (2.13.18)$$

so that

$$2 \sum_{j=1}^{n}\sum_{k=1}^{n}|b_{jk}||w_j||w_k| \leq 2\,\tilde{b} \sum_{j=1}^{n}|w_j| \sum_{k=1}^{n}|w_k| \qquad (2.13.19)$$

Furthermore, from the Cauchy–Schwarz inequality for real numbers:

$$\sum_{i=1}^{n} c_i d_i \leq \left[\sum_{i=1}^{n} c_i^2\right]^{1/2}\left[\sum_{i=1}^{n} d_i^2\right]^{1/2} \qquad (2.13.20)$$

by taking $c_i = 1$ and $d_i = |w_i|$, we have

$$\sum_{i=1}^{n}|w_i| \leq n^{1/2}\left[\sum_{i=1}^{n}|w_i|^2\right]^{1/2}$$

$$\leq n^{1/2}V^{1/2} \qquad (2.13.21)$$

Now using eqs. (2.13.17), (2.13.19), and (2.13.21), the second pair on the rhs of eq. (2.13.15) is

$$\leq 2\tilde{b}nV \qquad (2.13.22)$$

Finally, for the *third* pair:

$$\sum_{j=1}^{n}\left[w_j^* \sum_{l=1}^{n}\sum_{m=1}^{n} p_{jlm}w_lw_m + w_j \sum_{l=1}^{n}\sum_{m=1}^{n} p_{jlm}^* w_l^*w_m^*\right]$$

$$= \sum_{j=1}^{n}\sum_{l=1}^{n}\sum_{m=1}^{n}[p_{jlm}w_j^*w_lw_m + p_{jlm}^*w_jw_lw_m]$$

Repeating the same steps as above for the second pair gives

$$\leq 2Mn^{3/2}V^{3/2} \qquad (2.13.23)$$

where

$$|p_{jlm}| \leq M \geq 0, \qquad j, l, m = 1, 2, \dots, n \tag{2.13.24}$$

As noted earlier, by the choice of ε in defining $\mathbf{T}$, b_{ij} and hence $\bar{b}$ can be made arbitrarily small. However, once an ε is chosen, M is fixed and may be a large number.

Substituting eqs. (2.13.16), (2.13.22), and (2.13.23) in eq. (2.13.15) leads to

$$\frac{dV}{dt} \leq 2[\tilde{\alpha} + \bar{b}n + Mn^{3/2}V^{1/2}]V = [-\gamma + \delta V^{1/2}]V \tag{2.13.25}$$

where

$$\gamma = -2(\tilde{\alpha} + \bar{b}n) \quad \text{and} \quad \delta = 2Mn^{3/2} \geq 0 \tag{2.13.26}$$

are constants. Note that with $\tilde{\alpha} < 0$ because of the assumed asymptotic stability of the linearized system, and $\bar{b}$ arbitrarily small, we have

$$\gamma > 0 \tag{2.13.27}$$

We now need to solve the differential inequality (2.13.25) to determine $V(t)$. If $V(0)$ is sufficiently small, then

$$[-\gamma + \delta V(0)^{1/2}] < 0$$

so that

$$\left.\frac{dV}{dt}\right|_0 < 0 \tag{2.13.28}$$

and initially V tends to decrease, keeping $[-\gamma + \delta V^{1/2}] < 0$. Note that by definition (2.13.13), $V(t) \geq 0$, and it equals zero *if and only if* $\mathbf{w} = \mathbf{0}$.

In the region near $t = 0$, the differential inequality (2.13.25) rearranges as

$$dt \leq \frac{dV}{[-\gamma + \delta V^{1/2}]V}$$

which upon integration from 0 to t gives

$$t \leq \int_{V(0)}^{V} \frac{dV}{[-\gamma + \delta V^{1/2}]V} \tag{2.13.29}$$

The integral on the right can be readily evaluated by the change of variables:

$$y^2 = V$$

so that $2ydy = dV$, and

$$\int_{V(0)}^{V} \frac{dV}{[-\gamma + \delta V^{1/2}]V} = \int_{y(0)}^{y} \frac{2dy}{[-\gamma + \delta y]y}$$

$$= \frac{2}{\gamma} \ln\left[\frac{-\gamma + \delta y}{y}\right]\Bigg|_{y(0)}^{y}$$

Substituting in eq. (2.13.29) gives

$$\frac{\gamma t}{2} \leq \ln\left[\frac{-\gamma + \delta V^{1/2}}{V^{1/2}} \frac{V^{1/2}(0)}{-\gamma + \delta V^{1/2}(0)}\right]$$

which can be rearranged to yield

$$V^{1/2} \leq \frac{\gamma V^{1/2}(0)\exp(-\gamma t/2)}{\gamma - \delta V^{1/2}(0)[1 - \exp(-\gamma t/2)]} \tag{2.13.30}$$

Thus we have

$$\lim_{t \to \infty} V^{1/2}(t) \leq 0$$

However, by definition, $V(t) \geq 0$, and it equals zero *only if* $\mathbf{w} = \mathbf{0}$, which implies that

$$\lim_{t \to \infty} \mathbf{w}(t) = \mathbf{0} \tag{2.13.31}$$

Hence we have shown that the nonlinear system is asymptotically stable if the corresponding linearized system is asymptotically stable. This completes the proof for the first part of the theorem.

The second part of the theorem states that if the linearized system is unstable, the nonlinear system will also be unstable. To show this, assume that the linearized system is unstable. Thus, some eigenvalues of $\mathbf{A}$ will have positive real parts. Let $\lambda_1, \lambda_2, \ldots, \lambda_r$ have positive real parts, and let the remaining $\lambda_{r+1}, \ldots, \lambda_n$ have zero or negative real parts. Define the function

$$W(t) = \sum_{j=1}^{r} |w_j|^2 - \sum_{j=r+1}^{n} |w_j|^2 \tag{2.13.32}$$

Since

$$\frac{d}{dt}|w_j|^2 = w_j^* \frac{dw_j}{dt} + w_j \frac{dw_j^*}{dt}$$

from (2.13.12) as before, we have

$$\frac{d}{dt}\left[\sum_{j=1}^{r} |w_j|^2\right] = \sum_{j=1}^{r} 2\mathrm{Re}(\lambda_j)|w_j|^2$$

$$+ \sum_{j=1}^{r}\sum_{k=j+1}^{n} [b_{jk}w_j^*w_k + b_{jk}^*w_jw_k^*]$$

$$+ \sum_{j=1}^{r}\sum_{l=1}^{n}\sum_{m=1}^{n} [p_{jlm}w_j^*w_lw_m + p_{jlm}^*w_jw_l^*w_m^*] \tag{2.13.33a}$$

$$\frac{d}{dt}\left[\sum_{j=r+1}^{n} |w_j|^2\right] = \sum_{j=r+1}^{n} 2\mathrm{Re}(\lambda_j)|w_j|^2$$

$$+ \sum_{j=r+1}^{n}\sum_{k=j+1}^{n} [b_{jk}w_j^*w_k + b_{jk}^*w_jw_k^*]$$

$$+ \sum_{j=r+1}^{n}\sum_{l=1}^{n}\sum_{m=1}^{n} [p_{jlm}w_j^*w_lw_m + p_{jlm}^*w_jw_l^*w_m^*] \tag{2.13.33b}$$

The terms in the square brackets in the equations above are real since they are a pair of complex conjugates. Similar to the first part of the theorem, we replace these terms by bounds which in this case are lower bounds. For this, note that

$$
\begin{aligned}
+ \sum_{j=1}^{r} \sum_{k=j+1}^{n} [b_{jk}w_j^* w_k + b_{jk}^* w_j w_k^*] &\geq -2 \sum_{j=1}^{r} \sum_{k=j+1}^{n} |b_{jk}w_j^* w_k| \\
&\geq -2 \sum_{j=1}^{r} \sum_{k=1}^{n} |b_{jk}w_j^* w_k| \\
&\geq -2 \sum_{j=1}^{r} \sum_{k=1}^{n} |b_{jk}||w_j||w_k| \\
&\geq -2\tilde{b} \sum_{j=1}^{r} |w_j| \sum_{k=1}^{n} |w_k| \\
&\geq -2\tilde{b}n^{1/2}V^{1/2} \sum_{j=1}^{r} |w_j| \qquad (2.13.34)
\end{aligned}
$$

where in the last step the Cauchy–Schwarz inequality (2.13.21) has been used. Similarly, for the corresponding term in eq. (2.13.33b), we have

$$
\sum_{j=r+1}^{n} \sum_{k=j+1}^{n} [b_{jk}w_j^* w_k + b_{jk}^* w_j w_k^*] \leq 2\tilde{b}n^{1/2}V^{1/2} \sum_{j=r+1}^{n} |w_j| \qquad (2.13.35)
$$

In an analogous manner, the last terms of eqs. (2.13.33) yield

$$
\sum_{j=1}^{r} \sum_{l=1}^{n} \sum_{m=1}^{n} [p_{jlm}w_j^* w_l w_m + p_{jlm}^* w_j w_l^* w_m^*] \geq -2MnV \sum_{j=1}^{r} |w_j| \qquad (2.13.36)
$$

and

$$
\sum_{j=r+1}^{n} \sum_{l=1}^{n} \sum_{m=1}^{n} [p_{jlm}w_j^* w_l w_m + p_{jlm}^* w_j w_l^* w_m^*] \leq 2MnV \sum_{j=r+1}^{n} |w_j| \qquad (2.13.37)
$$

Using the above inequalities, eqs. (2.13.32) and (2.13.33) lead to

$$
\begin{aligned}
\frac{dW}{dt} &\geq \sum_{j=1}^{r} 2\mathrm{Re}(\lambda_j)|w_j|^2 - \sum_{j=r+1}^{r} 2\mathrm{Re}(\lambda_j)|w_j|^2 \\
&\quad -2\tilde{b}n^{1/2}V^{1/2} \sum_{j=1}^{n} |w_j| - 2MnV \sum_{j=1}^{n} |w_j| \\
&\geq \sum_{j=1}^{r} 2\mathrm{Re}(\lambda_j)|w_j|^2 - 2n^{1/2}V^{1/2}(\tilde{b} + Mn^{1/2}V^{1/2}) \sum_{j=1}^{n} |w_j|
\end{aligned}
$$

since $\mathrm{Re}(\lambda_j) \leq 0$ for $j = r + 1, \ldots, n$. Again, using the Cauchy–Schwarz inequality (2.13.21) gives

$$
\frac{dW}{dt} \geq \sum_{j=1}^{r} 2\mathrm{Re}(\lambda_j)|w_j|^2 - 2nV(\tilde{b} + Mn^{1/2}V^{1/2}) \qquad (2.13.38)
$$

The proof now proceeds by contradiction. Thus let us assume that although the linearized system is unstable, the nonlinear system is stable. Stability of the nonlinear system implies that by selecting $V(0)$ sufficiently small, $V(t)$ can be maintained as small as desired. Thus the second term in the inequality (2.13.38) can be neglected. If we define

$$\tilde{\lambda} = \min_{j=1 \to r} \text{Re}(\lambda_j) > 0 \qquad (2.13.39)$$

then (2.13.38) reduces to

$$\frac{dW}{dt} \geq 2\tilde{\lambda} \sum_{j=1}^{r} |w_j|^2$$

$$\geq 2\tilde{\lambda} \left[\sum_{j=1}^{r} |w_j|^2 - \sum_{j=r+1}^{n} |w_j|^2 \right]$$

$$\geq 2\tilde{\lambda} W$$

Upon integration, this leads to

$$W(t) \geq W(0) \exp[2\tilde{\lambda}t] \qquad (2.13.40)$$

Since $\tilde{\lambda} > 0$, it is evident that for sufficiently large t, $W(t)$ cannot be made arbitrarily small. Furthermore, by definition,

$$V(t) = \sum_{j=1}^{n} |w_j|^2 \geq \sum_{j=1}^{r} |w_j|^2 - \sum_{j=r+1}^{n} |w_j|^2 = W(t)$$

so that $V(t)$ can also not be made arbitrarily small. Thus we have a contradiction to the assumption that the nonlinear system is stable. Hence we can conclude that if the linearized system is unstable, the nonlinear system is also unstable. This completes the proof of the theorem.

2.14 STABILITY OF TWO-DIMENSIONAL AUTONOMOUS SYSTEMS

In the next several sections, we study two-dimensional autonomous systems in some detail. A number of examples from science and engineering fit in this category. For these systems, detailed analysis can be carried out, which leads to a rather complete understanding of the dynamic behavior.

Thus, consider the pair of ODEs

$$\frac{dy_1}{dt} = f_1(y_1, y_2) \qquad (2.14.1a)$$

$$\frac{dy_2}{dt} = f_2(y_1, y_2) \qquad (2.14.1b)$$

where f_i, in general, are nonlinear functions of $\mathbf{y}$ and let the ICs be

$$\mathbf{y}(0) = \mathbf{y}_0 \qquad (2.14.2)$$

We also assume appropriate conditions on f_i (see section 2.7), so that the solution of eqs. (2.14.1) and (2.14.2) is unique.

The *steady states* $\mathbf{y}_s$ of the system (2.14.1) are obtained from the solution of

$$\mathbf{f}(\mathbf{y}) = \mathbf{0} \tag{2.14.3}$$

where we note that there may be *multiple* $\mathbf{y}_s$ values satisfying eq. (2.14.3).

Let $\mathbf{y}_s$ be a specific steady state of eq. (2.14.1), and we now investigate its asymptotic stability. For this, from the fundamental stability theorem, we need to consider the corresponding system linearized about the steady state being examined. If

$$x_1 = y_1 - y_{1s} \quad \text{and} \quad x_2 = y_2 - y_{2s} \tag{2.14.4}$$

represent perturbations from the steady state $\mathbf{y}_s$, then the linearized system is given by

$$\frac{dx_1}{dt} = a_{11}x_1 + a_{12}x_2$$

$$\frac{dx_2}{dt} = a_{21}x_1 + a_{22}x_2$$

This is written compactly as

$$\frac{d\mathbf{x}}{dt} = \mathbf{A}\mathbf{x} \tag{2.14.5}$$

where

$$a_{ij} = \frac{\partial f_i}{\partial y_j}\bigg|_{\mathbf{y}_s}$$

are constants, and $\mathbf{A}$ is the Jacobian matrix. The eigenvalues of $\mathbf{A}$ satisfy the equation

$$|\mathbf{A} - \lambda\mathbf{I}| = 0$$

That is,

$$\begin{vmatrix} a_{11} - \lambda & a_{12} \\ a_{21} & a_{22} - \lambda \end{vmatrix} = 0$$

Upon expansion this gives

$$\lambda^2 - (a_{11} + a_{22})\lambda + (a_{11}a_{22} - a_{21}a_{12}) = 0$$

and recognizing the coefficients yields

$$\lambda^2 - (\text{tr } \mathbf{A})\lambda + |\mathbf{A}| = 0 \tag{2.14.6}$$

The two eigenvalues are thus given by

$$\lambda_{1,2} = \tfrac{1}{2}[\text{tr } \mathbf{A} \pm \sqrt{(\text{tr } \mathbf{A})^2 - 4|\mathbf{A}|}] \tag{2.14.7}$$

and may be real, purely imaginary, or a pair of complex conjugate numbers.

We now establish conditions under which it can be stated definitively that $\mathbf{y}_s$, the steady state being examined, is stable to small perturbations. From section 2.11, this requires that both eigenvalues be negative if they are real, or have a negative real part if they are complex.

Let us first consider the case where both λ_1 and λ_2 are *real*. This means that the coefficients of $\mathbf{A}$ satisfy the condition

$$(\mathrm{tr}\ \mathbf{A})^2 > 4|\mathbf{A}|$$

It is evident that two real numbers λ_1 and λ_2 are both negative if and only if their sum is negative and their product is positive; that is,

$$\lambda_1 + \lambda_2 < 0 \tag{2.14.8a}$$

and

$$\lambda_1\lambda_2 > 0 \tag{2.14.8b}$$

Using eq. (2.14.7) for the definitions, eq. (2.14.8a) implies that

$$\mathrm{tr}\ \mathbf{A} < 0 \tag{2.14.9a}$$

while eq. (2.14.8b) gives

$$|\mathbf{A}| > 0 \tag{2.14.9b}$$

On the other hand, if λ_1 and λ_2 are a pair of *complex* conjugates, then the coefficients of $\mathbf{A}$ satisfy

$$(tr\ A)^2 < 4|A|$$

which already implies that

$$|\mathbf{A}| > 0 \tag{2.14.10a}$$

The real part of λ_1 and λ_2 is negative if

$$\mathrm{tr}\ \mathbf{A} < 0 \tag{2.14.10b}$$

Note that conditions (2.14.9) and (2.14.10) are the *same*. Thus, regardless of whether the eigenvalues are real or complex, the steady state being examined is stable to small perturbations if and only if the two conditions

$$\mathrm{tr}\ \mathbf{A} < 0 \quad \text{and} \quad |\mathbf{A}| > 0 \tag{2.14.11}$$

are satisfied simultaneously.

2.15 CLASSIFICATION OF THE STEADY STATE FOR TWO-DIMENSIONAL AUTONOMOUS SYSTEMS

In this section we consider in detail the nature of the zero steady-state solution of the two-dimensional linear system:

$$\frac{d\mathbf{x}}{dt} = \mathbf{A}\mathbf{x} \tag{2.15.1}$$

where

$$\mathbf{A} = \begin{bmatrix} a_{11} & a_{12} \\ a_{21} & a_{22} \end{bmatrix} \tag{2.15.2}$$

is a matrix of constants. It is evident from the previous section that eq. (2.15.1) represents the linearized system corresponding to the nonlinear system (2.14.1). We are interested in establishing the precise behavior of $\mathbf{x}(t)$.

The nature of $\mathbf{x}(t)$ depends critically on the eigenvalues of $\mathbf{A}$, given by eq. (2.14.7). In general, the following cases are possible:

1. Real and unequal eigenvalues
   a. both negative
   b. both positive
   c. opposite in sign
2. Real and equal eigenvalues
3. Complex eigenvalues
   a. real part negative
   b. real part positive
4. Pure imaginary eigenvalues.

We will examine each of these cases in the x_1–x_2 *phase plane*. Although we will analyze only the *linear* system (2.15.1), let us digress briefly to first identify some general characteristics which apply also to nonlinear systems. Thus for the two-dimensional *nonlinear* system

$$\frac{d\mathbf{x}}{dt} = \mathbf{f}(\mathbf{x}) \tag{2.15.3}$$

each point in the phase plane is a potential IC for the system, as noted in section 2.10. The solution $\mathbf{x}(t)$ of eq. (2.15.3), with the IC $\mathbf{x}(0)$, generates a *trajectory* with the direction of increasing t shown by the arrow (see Figure 2.12).

If the functions $f_1(x_1, x_2)$ and $f_2(x_1, x_2)$ satisfy the conditions of the existence and uniqueness theorem (section 2.7), then each IC—that is, each point in the phase plane—gives rise to a unique trajectory. This leads to the following important result.

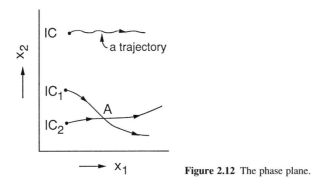

Figure 2.12 The phase plane.

● **THEOREM**

The trajectories do not cross in the phase plane.

The proof is simple. Let us assume that two trajectories do cross at a point such as A, as shown in Figure 2.12. Then at A, which represents a specific IC, there are two possible trajectories. This is a contradiction to the existence and uniqueness theorem and is thus impossible.

We now investigate the detailed behavior of the two-dimensional linear system (2.15.1)

$$\frac{d\mathbf{x}}{dt} = \mathbf{Ax}$$

in the phase plane. The steady state of the linear system is clearly $\mathbf{x} = \mathbf{0}$—that is, the origin of the phase plane. We will consider individually the various cases for the eigenvalues discussed above.

Case 1a. Real, Unequal, and Negative Eigenvalues, $\lambda_2 < \lambda_1 < 0$
From eq. (1.17.9), the solution of eq. (2.15.1) is given by

$$\mathbf{x}(t) = c_1 \mathbf{z}_1 e^{\lambda_1 t} + c_2 \mathbf{z}_2 e^{\lambda_2 t}$$

where $\mathbf{z}_i$ is the eigenvector corresponding to eigenvalue λ_i, and the constants c_i are determined by the ICs. Let us denote the two eigenvectors as

$$\mathbf{z}_1 = \begin{bmatrix} z_{11} \\ z_{21} \end{bmatrix} \quad \text{and} \quad \mathbf{z}_2 = \begin{bmatrix} z_{12} \\ z_{22} \end{bmatrix}$$

so that in expanded form we have

$$x_1(t) = c_1 z_{11} e^{\lambda_1 t} + c_2 z_{12} e^{\lambda_2 t} \tag{2.15.4a}$$

$$x_2(t) = c_1 z_{21} e^{\lambda_1 t} + c_2 z_{22} e^{\lambda_2 t} \tag{2.15.4b}$$

Consider now the specific IC which gives $c_2 = 0$. For this IC,

$$x_1(t) = c_1 z_{11} e^{\lambda_1 t} \quad \text{and} \quad x_2(t) = c_1 z_{21} e^{\lambda_1 t}$$

which imply that

$$\frac{x_2(t)}{x_1(t)} = \frac{z_{21}}{z_{11}} \tag{2.15.5}$$

a constant. Similarly, for the IC which yields $c_1 = 0$, we have

$$x_1(t) = c_2 z_{12} e^{\lambda_2 t} \quad \text{and} \quad x_2(t) = c_2 z_{22} e^{\lambda_2 t}$$

so that

$$\frac{x_2(t)}{x_1(t)} = \frac{z_{22}}{z_{12}} \tag{2.15.6}$$

another constant. Thus we conclude that with these two special ICs, the trajectories in the phase plane are two straight lines which are called *characteristic trajectories*. Since both eigenvalues λ_1 and λ_2 are negative, the characteristic trajectories have directions pointing toward the origin—that is, the steady state—as shown in Figure 2.13.

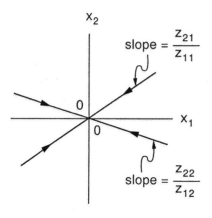

Figure 2.13 The characteristic trajectories.

Furthermore, since $\lambda_2 < \lambda_1 < 0$, for sufficiently large values of t and for *all* ICs, eq. (2.15.4) yields

$$x_1(t) \sim c_1 z_{11} e^{\lambda_1 t} \quad \text{and} \quad x_2(t) \sim c_1 z_{21} e^{\lambda_1 t} \tag{2.15.7}$$

except for the specific IC which makes $c_1 = 0$. Thus for all ICs and sufficiently large t, we have

$$\frac{x_2}{x_1} \sim \frac{z_{21}}{z_{11}} \tag{2.15.8}$$

Hence *all* trajectories (except one) eventually have the slope of the characteristic trajectory corresponding to λ_1, the eigenvalue closer to zero, and ultimately approach it as shown in Figure 2.14. An origin of this type is called a *node* or sometimes an *improper node*. Since the eigenvalues are negative, it is asymptotically stable.

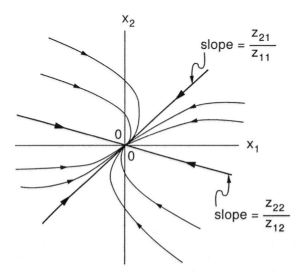

Figure 2.14 A stable node.

From eq. (2.14.7) and the considerations of section 2.14, in this case of real, unequal, and negative eigenvalues we have

$$\text{tr }\mathbf{A} < 0, \qquad |\mathbf{A}| > 0 \tag{2.15.9}$$

and

$$\Delta = (\text{tr }\mathbf{A})^2 - 4|\mathbf{A}| > 0 \tag{2.15.10}$$

Case 1b. Real, Unequal, and Positive Eigenvalues, $\lambda_2 > \lambda_1 > 0$

Real eigenvalues require that $\Delta > 0$. For both λ_1 and λ_2 to be positive as well, we must have in addition

$$\text{tr }\mathbf{A} > 0 \quad \text{and} \quad |\mathbf{A}| > 0 \tag{2.15.11}$$

Following the procedure as above for case 1a leads to the conclusion that the phase plane has the same form as shown in Figure 2.14, with the only difference that the arrows point outward, away from the origin. An origin of this type is again a node, but this time it is unstable. In this case, note that all ICs *close* to the origin emanate with slope equal to that of the characteristic trajectory corresponding to λ_1, the eigenvalue closer to zero.

Case 1c. Real Eigenvalues of Opposite Sign, $\lambda_1 < 0 < \lambda_2$

As in case 1a, eq. (2.15.4) states

$$x_1(t) = c_1 z_{11} e^{\lambda_1 t} + c_2 z_{12} e^{\lambda_2 t}$$

$$x_2(t) = c_1 z_{21} e^{\lambda_1 t} + c_2 z_{22} e^{\lambda_2 t}$$

Thus the two straight-line characteristic trajectories again have slopes given by the same expressions—that is, eqs. (2.15.5) and (2.15.6)—but the directions of the arrows are now as shown in Figure 2.15. In order to understand this, consider, for example, the IC for which $c_1 = 0$. In this case,

$$x_1 = c_2 z_{12} e^{\lambda_2 t} \quad \text{and} \quad x_2 = c_2 z_{22} e^{\lambda_2 t}$$

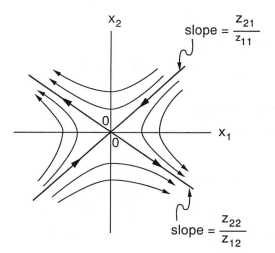

Figure 2.15 A saddle point.

Hence

$$\frac{x_2}{x_1} = \frac{z_{22}}{z_{12}}$$

which generates the characteristic trajectory corresponding to eigenvalue λ_2. However, since $\lambda_2 > 0$, *both* x_1 and x_2 increase in magnitude as t increases, so the arrows point away from the origin.

Also, since $\lambda_1 < 0$, for sufficiently large t, *all* ICs eventually lead to the solution

$$x_1 \sim c_2 z_{12} e^{\lambda_2 t} \quad \text{and} \quad x_2 \sim c_2 z_{22} e^{\lambda_2 t}$$

so that

$$\frac{x_2}{x_1} = \frac{z_{22}}{z_{12}}$$

the characteristic trajectory corresponding to eigenvalue λ_2.

An origin of this type is called a *saddle point*, and as apparent from the phase plane, it is inherently unstable. Note that eigenvalues which are real but opposite in sign require that

$$\Delta > 0 \quad \text{and} \quad |\mathbf{A}| < 0 \qquad (2.15.12)$$

Case 2. Real and Equal Eigenvalues, $\lambda_1 = \lambda_2 = \lambda$
From eq. (2.14.7), the two eigenvalues are real and equal if

$$\Delta = (\text{tr } \mathbf{A})^2 - 4|\mathbf{A}| = 0 \qquad (2.15.13)$$

and then

$$\lambda_1 = \lambda_2 = (\text{tr } \mathbf{A})/2 \qquad (2.15.14)$$

In this case, as discussed in section 2.11, the two eigenvectors may, in general, not be linearly independent. For this reason, the expression

$$\mathbf{x} = \sum_{j=1}^{2} c_j \mathbf{z}_j e^{\lambda_j t}$$

is *not* a valid solution of

$$\frac{d\mathbf{x}}{dt} = \mathbf{A}\mathbf{x}$$

and a different strategy is required.

For this, writing the equations in full, we have

$$\frac{dx_1}{dt} = a_{11}x_1 + a_{12}x_2 \qquad (2.15.15a)$$

$$\frac{dx_2}{dt} = a_{21}x_1 + a_{22}x_2 \qquad (2.15.15b)$$

with ICs

$$x_1(0) = x_{10}, \qquad x_2(0) = x_{20} \qquad (2.15.16)$$

These two first-order ODEs are equivalent to a single second-order ODE, which can be readily derived as follows. Differentiating eq. (2.15.15a) gives

$$\frac{d^2x_1}{dt^2} = a_{11}\frac{dx_1}{dt} + a_{12}\frac{dx_2}{dt}$$

$$= a_{11}\frac{dx_1}{dt} + a_{12}[a_{21}x_1 + a_{22}x_2]$$

$$= a_{11}\frac{dx_1}{dt} + a_{12}a_{21}x_1 + a_{22}\left[\frac{dx_1}{dt} - a_{11}x_1\right]$$

which on rearrangement yields

$$\frac{d^2x_1}{dt^2} - (\text{tr }\mathbf{A})\frac{dx_1}{dt} + |\mathbf{A}|x_1 = 0 \tag{2.15.17}$$

If we let

$$x_1(t) = e^{\lambda t}$$

then λ satisfies

$$\lambda^2 - (\text{tr }\mathbf{A})\lambda + |\mathbf{A}| = 0$$

which gives the familiar expression (2.14.7). However, since $\lambda_1 = \lambda_2 = \lambda$, the general solution of eq. (2.15.17) is given by

$$x_1(t) = (\alpha + \beta t)e^{\lambda t} \tag{2.15.18a}$$

where α and β are constants. From eq. (2.15.15a),

$$x_2(t) = \frac{1}{a_{12}}\left[\frac{dx_1}{dt} - a_{11}x_1\right]$$

$$= \left[\frac{\beta + \alpha(\lambda - a_{11})}{a_{12}} + \frac{\beta(\lambda - a_{11})}{a_{12}}t\right]e^{\lambda t} \tag{2.15.18b}$$

and the constants α and β can now be evaluated by using the ICs (2.15.16). This gives

$$\alpha = x_{10}$$

and

$$\beta = a_{12}x_{20} - (\lambda - a_{11})x_{10}$$

Substituting these in eq. (2.15.18) yields

$$x_1(t) = [x_{10} + \{a_{12}x_{20} - x_{10}(\lambda - a_{11})\}t]e^{\lambda t} \tag{2.15.19a}$$

and

$$x_2(t) = \left[x_{20} + (\lambda - a_{11})\left\{x_{20} - \frac{x_{10}}{a_{12}}(\lambda - a_{11})\right\}t\right]e^{\lambda t} \tag{2.15.19b}$$

and hence their ratio is

$$\frac{x_2(t)}{x_1(t)} = \frac{x_{20} + [(\lambda - a_{11})/a_{12}][a_{12}x_{20} - x_{10}(\lambda - a_{11})]t}{x_{10} + [a_{12}x_{20} - x_{10}(\lambda - a_{11})]t} \tag{2.15.20}$$

SUBCASE 2.1. $a_{11} \neq a_{22}$. In this subcase, from eq. (2.15.14) we note that

$$\lambda - a_{11} = \frac{a_{22} - a_{11}}{2} \neq 0$$

Thus as $t \to \infty$, from eq. (2.15.20) we have

$$\frac{x_2(t)}{x_1(t)} \to \frac{\lambda - a_{11}}{a_{12}} = \frac{a_{22} - a_{11}}{2a_{12}} \tag{2.15.21}$$

which represents the constant eventual slope of all trajectories.

It may be seen from eq. (2.15.19) that the straight-line characteristic trajectory is obtained from the specific IC given by

$$\frac{x_{20}}{x_{10}} = \frac{\lambda - a_{11}}{a_{12}} = \frac{a_{22} - a_{11}}{2a_{12}} \tag{2.15.22}$$

since this yields

$$\frac{x_2(t)}{x_1(t)} = \frac{x_{20}}{x_{10}} \tag{2.15.23}$$

for *all* values of t. Note that the rhs expressions of eqs. (2.15.21)–(2.15.23) are identical.

In this subcase, the phase plane has the form shown in Figure 2.16. The origin is again called an *improper node*, but has only one independent eigenvector and hence only one characteristic trajectory. Finally, it is asymptotically stable if tr $\mathbf{A} < 0$ and unstable if tr $\mathbf{A} > 0$.

SUBCASE 2.2. $a_{11} = a_{22}$ AND $a_{12} = a_{21} = 0$. If we let $a_{11} = a_{22} = a$, then from eq. (2.15.15) the set of ODEs for this subcase reduces to

$$\frac{dx_1}{dt} = ax_1$$

$$\frac{dx_2}{dt} = ax_2$$

along with ICs

$$x_1(0) = x_{10}, \qquad x_2(0) = x_{20}$$

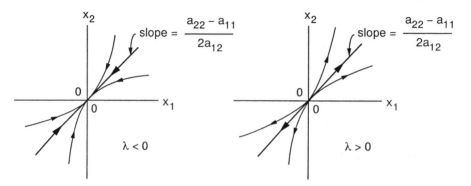

Figure 2.16 An improper node, one independent eigenvector.

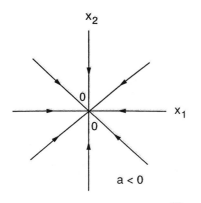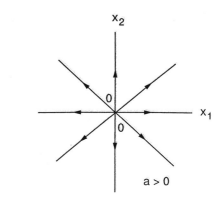

Figure 2.17 A proper node.

These equations have the solution

$$x_1(t) = x_{10}e^{at} \quad \text{and} \quad x_2(t) = x_{20}e^{at}$$

which yields

$$\frac{x_2(t)}{x_1(t)} = \frac{x_{20}}{x_{10}} \tag{2.15.24}$$

a constant. It is clear then that the phase plane consists *only* of straight lines, as shown in Figure 2.17. An origin of this type is called a *proper node*. This case is characterized by two linearly independent eigenvectors, in spite of the fact that the eigenvalues are repeated. The reason for this behavior as discussed in section 2.11 is that the matrix **A** is symmetric.

Remark. Note that the subcase $a_{11} = a_{22}$, with $a_{12} = a_{21} \neq 0$, is not feasible if the two eigenvalues are real and equal.

Case 3. Complex Eigenvalues with Nonzero Real Part

When the two eigenvalues of **A** are complex, they are a complex conjugate pair and may be written as

$$\lambda_\pm = \tfrac{1}{2}[\text{tr } \mathbf{A} \pm \sqrt{(\text{tr } \mathbf{A})^2 - 4|\mathbf{A}|}] \tag{2.15.25}$$

$$= \alpha \pm i\beta$$

For the eigenvalues to be complex, we must have

$$\Delta = (\text{tr } \mathbf{A})^2 - 4|\mathbf{A}| < 0$$

so that

$$|\mathbf{A}| > \frac{(\text{tr } \mathbf{A})^2}{4} > 0 \tag{2.15.26}$$

and stability of the origin, following eq. (2.15.9), requires that

$$\text{tr } \mathbf{A} = 2\alpha < 0$$

The general solution of eq. (2.15.1),

$$\frac{d\mathbf{x}}{dt} = \mathbf{A}\mathbf{x}$$

with the IC

$$\mathbf{x}(0) = \mathbf{x}_0$$

is then given by

$$\mathbf{x}(t) = c_+\mathbf{z}_+e^{\lambda_+ t} + c_-\mathbf{z}_-e^{\lambda_- t} \tag{2.15.27}$$

where

$$c_\pm = \frac{\mathbf{w}_\pm^T \mathbf{x}_0}{\mathbf{w}_\pm^T \mathbf{z}_\pm}$$

and $\mathbf{z}_\pm$, $\mathbf{w}_\pm^T$ are the eigenvectors and eigenrows corresponding to the eigenvalues $\lambda_\pm$, respectively.

Note that since $\lambda_\pm$ are complex conjugates, so are $\mathbf{z}_\pm$ and $\mathbf{w}_\pm^T$. In order to see this for $\mathbf{z}_\pm$, for example, recall that $\mathbf{z}_+$ is a nonzero column of $\mathrm{adj}(\mathbf{A} - \lambda_+\mathbf{I})$, while $\mathbf{z}_-$ is a nonzero column of $\mathrm{adj}(\mathbf{A} - \lambda_-\mathbf{I})$. Since the adjoints are complex conjugates, $\mathbf{z}_\pm$ must also be a complex conjugate pair. If we denote

$$\mathbf{w}_\pm^T = [a \pm ib \quad c \pm id] \quad \text{and} \quad \mathbf{z}_\pm = \begin{bmatrix} e \pm if \\ g \pm ih \end{bmatrix}$$

then it is easy to verity that $c_\pm$ are also complex conjugates and, for the sake of brevity, may be written as

$$c_\pm = \gamma \pm i\delta$$

Substituting the expressions for $c_\pm$, $z_\pm$ and $\lambda_\pm$ in eq. (2.15.27) yields

$$\mathbf{x} = \begin{bmatrix} x_1 \\ x_2 \end{bmatrix} = (\gamma + i\delta)\begin{bmatrix} e + if \\ g + ih \end{bmatrix}e^{(\alpha+i\beta)t} + (\gamma - i\delta)\begin{bmatrix} e - if \\ g - ih \end{bmatrix}e^{(\alpha-i\beta)t}$$

which, recalling that

$$\cos\theta = \frac{e^{i\theta} + e^{-i\theta}}{2} \quad \text{and} \quad \sin\theta = \frac{e^{i\theta} - e^{-i\theta}}{2i}$$

simplifies to

$$x_1(t) = 2e^{\alpha t}[(\gamma e - \delta f)\cos\beta t - (\delta e + \gamma f)\sin\beta t] \tag{2.15.28a}$$

and

$$x_2(t) = 2e^{\alpha t}[\gamma g - \delta h)\cos\beta t - (\delta g + \gamma h)\sin\beta t] \tag{2.15.28b}$$

Thus $x_1(t)$ and $x_2(t)$ are *periodic functions*, with period $= 2\pi/\beta$. They oscillate with t, and the oscillations

grow with t if $\alpha > 0$

and

decay with t if $\alpha < 0$

The phase plane behavior is then as shown in Figure 2.18, and the origin is called a *focus* or *spiral*. It is asymptotically stable or unstable, depending on whether $\alpha < 0$ or > 0, respectively.

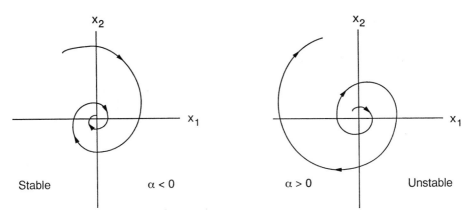

Figure 2.18 A focus or spiral.

DIRECTION OF ROTATION. Whether stable or unstable, the direction of rotation for the focus can be either clockwise or counterclockwise. The conditions for determining the direction can be obtained readily, by converting to polar coordinates. Thus consider eq. (2.15.1), written fully as

$$\frac{dx_1}{dt} = a_{11}x_1 + a_{12}x_2 \tag{2.15.29a}$$

$$\frac{dx_2}{dt} = a_{21}x_1 + a_{22}x_{12} \tag{2.15.29b}$$

Introducing polar coordinates (see Figure 2.19)

$$r = (x_1^2 + x_2^2)^{1/2} \quad \text{and} \quad \phi = \tan^{-1}(x_2/x_1)$$

yields

$$\begin{aligned}
\frac{d\phi}{dt} &= \frac{d}{dt}\left[\tan^{-1}\left(\frac{x_2}{x_1}\right)\right] \\
&= \left[x_1\frac{dx_2}{dt} - x_2\frac{dx_1}{dt}\right]/(x_1^2 + x_2^2) \\
&= Q(x_1, x_2)/(x_1^2 + x_2^2)
\end{aligned} \tag{2.15.30}$$

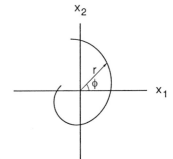

Figure 2.19 The polar coordinates.

where

$$Q(x_1, x_2) = a_{21}x_1^2 + (a_{22} - a_{11})x_1x_2 - a_{12}x_2^2 \tag{2.15.31}$$

is a *quadratic form*. This may be rewritten as

$$Q(x_1, x_2) = [x_1 \quad x_2] \begin{bmatrix} a_{21} & (a_{22} - a_{11})/2 \\ (a_{22} - a_{11})/2 & -a_{12} \end{bmatrix} \begin{bmatrix} x_1 \\ x_2 \end{bmatrix}$$

Hence Q is positive definite if and only if (cf. section 1.26)

$$a_{21} > 0 \tag{2.15.32a}$$

and

$$(a_{22} - a_{11})^2 + 4a_{12}a_{21} < 0 \tag{2.15.32b}$$

However, the latter condition *always* holds when **A** has purely imaginary or complex eigenvalues, because in such cases

$$\Delta = (\text{tr } \mathbf{A})^2 - 4|\mathbf{A}| < 0$$

which substituting for tr **A** and $|\mathbf{A}|$ yields eq. (2.15.32b).

Thus $Q(x_1, x_2)$ is positive (negative) definite if and only if $a_{21} > 0$ (<0). From eq. (2.15.30), we then conclude that

$$\text{if } a_{21} > 0, \quad \text{then } \frac{d\phi}{dt} > 0, \text{ implying } \textit{counterclockwise} \text{ rotation} \tag{2.15.33a}$$

while

$$\text{if } a_{21} < 0, \quad \text{then } \frac{d\phi}{dt} < 0, \text{ leading to a } \textit{clockwise} \text{ rotation.} \tag{2.15.33b}$$

Note that if $a_{21} = 0$, then

$$\Delta = (a_{22} - a_{11})^2 > 0$$

and the eigenvalues are real:

$$\lambda_1 = a_{11}, \qquad \lambda_2 = a_{22}$$

so that there is *no* rotation at all!

Case 4. Purely Imaginary Eigenvalues

This is merely a subcase of case 3. Since

$$\lambda_{\pm} = \frac{1}{2} [\text{tr } \mathbf{A} \pm \sqrt{(\text{tr } \mathbf{A})^2 - 4|\mathbf{A}|}]$$

purely imaginary eigenvalues result when

$$\text{tr } \mathbf{A} = 0 \quad \text{and} \quad |\mathbf{A}| > 0 \tag{2.15.34}$$

With $\lambda_{\pm} = \pm i\beta$, from eqs. (2.15.28) we have

$$x_1(t) = 2[(\gamma e - \delta f) \cos \beta t - (\delta e + \gamma f) \sin \beta t]$$

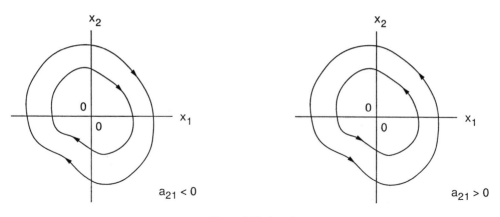

Figure 2.20 A center.

and

$$x_2(t) = 2[(\gamma g - \delta h) \cos \beta t - (\delta g + \gamma h) \sin \beta t]$$

Thus $x_1(t)$ and $x_2(t)$ are purely oscillatory functions, of period $2\pi/\beta$. The phase plane is of the form shown in Figure 2.20, and the origin is called a *center*, which is marginally stable.

Table 2.1 summarizes cases discussed in this section.

Note that the two eigenvalues of matrix $\mathbf{A}$ can be computed directly from tr $\mathbf{A}$ and $|\mathbf{A}|$ as follows:

$$\lambda_{\pm} = \tfrac{1}{2}[\text{tr } \mathbf{A} \pm \sqrt{\Delta}]$$

where

$$\Delta = (\text{tr } \mathbf{A})^2 - 4|\mathbf{A}|$$

Hence it is easy to translate the stability information obtained in this section, and summarized in Table 2.1, to the (tr $\mathbf{A}$, $|\mathbf{A}|$)-plane, as shown in Figure 2.21.

TABLE 2.1
Summary of Type and Stability Character of Origin

Eigenvalues	Type	Stability Character
$\lambda_2 < \lambda_1 < 0$	Improper node	Asymptotically stable
$\lambda_2 > \lambda_1 > 0$	Improper node	Unstable
$\lambda_1 = \lambda_2 = \lambda$	Proper or improper node	Asymptotically stable if $\lambda > 0$
		Unstable if $\lambda > 0$
$\lambda_1 < 0 < \lambda_2$	Saddle point	Unstable
$\lambda_{\pm} = \alpha \pm i\beta$	Focus or spiral	Asymptotically stable if $\alpha < 0$
		Unstable if $\alpha > 0$
$\lambda_{\pm} = \pm i\beta$	Center	Marginally stable

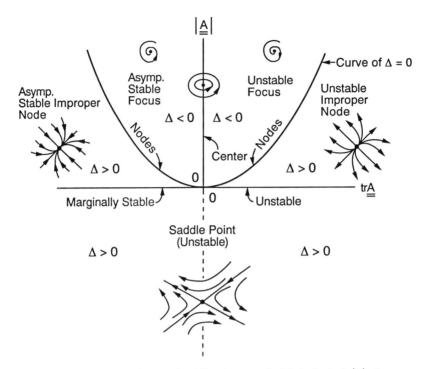

Figure 2.21 Summary of type and stability character of origin in the (tr $\mathbf{A}$, $|\mathbf{A}|$) plane.

2.16 EXAMPLE 1: POPULATION DYNAMICS OF TWO COMPETING SPECIES

In the next two sections, we consider the application of phase plane analysis to two classic examples. The first, examined here, is an extension of the problem treated in section 2.8, which involved the growth of a single species with limited resources. A conceptually simple generalization concerns the case of two species competing for a common limited food supply (cf. Murray, 1990, section 3.5; Boyce and DiPrima, 1992, section 9.4). Examples include two types of fish in a pond that do not prey on each other but compete for the available food, or two types of bacteria competing for the same nutrient.

Let x and y represent populations of the two species at time t. From the derivation of eq. (2.8.16), if y is absent, then the growth of species x is given by

$$\frac{dx}{dt} = x(\alpha_1 - \beta_1 x)$$

while if x is absent, then the evolution of y follows

$$\frac{dy}{dt} = y(\alpha_2 - \beta_2 y)$$

Now if species y is present and competes for the same resources as x, then the growth rate of x decreases further. Similarly, the presence of x inhibits the growth rate of y. The simplest account of such interference is provided by the equations

$$\frac{dx}{dt} = x(\alpha_1 - \beta_1 x - \gamma_1 y) \tag{2.16.1a}$$

and

$$\frac{dy}{dt} = y(\alpha_2 - \beta_2 y - \gamma_2 x) \tag{2.16.1b}$$

where α_i, β_i, and γ_i are positive constants. Equations (2.16.1) may be written as

$$\frac{dx}{dt} = x\Delta_1(x, y) \quad \text{and} \quad \frac{dy}{dt} = y\Delta_2(x, y) \tag{2.16.2}$$

where

$$\Delta_1(x, y) = \alpha_1 - \beta_1 x - \gamma_1 y \tag{2.16.3a}$$

and

$$\Delta_2(x, y) = \alpha_2 - \beta_2 y - \gamma_2 x \tag{2.16.3b}$$

Note that the biological interpretation of constants β and γ is that the β_i represent the inhibitory effect each species has on its *own* growth rate (i.e., *self-inhibition*), while the γ_i are a measure of the inhibiting effect one species has on the growth rate of the *other* species (i.e., *interaction*).

2.16.1   The Steady States

The steady states of the system (2.16.2) need to be determined first. These are *simultaneous* solutions of

$$x_s \Delta_1(x_s, y_s) = 0 \tag{2.16.4a}$$

and

$$y_s \Delta_2(x_s, y_s) = 0 \tag{2.16.4b}$$

Now, eq. (12.16.4a) is satisfied for

$$x_s = 0 \quad \text{or} \quad x_s = \frac{\alpha_1 - \gamma_1 y_s}{\beta_1}$$

while eq. (2.16.4b) requires that

$$y_s = 0 \quad \text{or} \quad y_s = \frac{\alpha_2 - \gamma_2 x_s}{\beta_2}$$

Hence there are four steady states:

S1. $x_s = 0$, $y_s = 0$

S2. $x_s = 0$, $y_s = \dfrac{\alpha_2}{\beta_2}$

S3. $x_s = \dfrac{\alpha_1}{\beta_1}$, $y_s = 0$

S4. $x_s = \dfrac{\alpha_1\beta_2 - \alpha_2\gamma_1}{\beta_1\beta_2 - \gamma_1\gamma_2}$, $y_s = \dfrac{\alpha_2\beta_1 - \alpha_1\gamma_2}{\beta_1\beta_2 - \gamma_1\gamma_2}$

where the coexistence steady state S4 arises from the intersection of the two straight lines $\Delta_1(x, y) = 0$ and $\Delta_2(x, y) = 0$ in the (x, y) plane (*if* an intersection is possible). Since we are dealing with species populations, only the first quadrant of the (x, y) plane is of interest.

2.16.2 Global Analysis

We now analyze the dynamic behavior of the system. For this, it is instructive to first plot separately the lines $\Delta_1 = 0$ and $\Delta_2 = 0$ in the (x, y) plane—that is, the x and y *nullclines*, respectively. Since

$$\frac{dx}{dt} = x\Delta_1(x, y) \quad \text{and} \quad \frac{dy}{dt} = y\Delta_2(x, y)$$

the signs of Δ_1 and Δ_2 in various sections of the plane, and the directions of some specific trajectories, can be identified as shown in Figure 2.22. Superimposing the two parts of Figure 2.22, we get *four* possibilities for the relative dispositions of the two lines as shown in Figure 2.23, where (a) occurs when the nullcline $\Delta_2 = 0$ is above Δ_1.

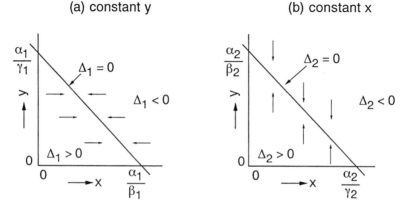

Figure 2.22 The x and y nullclines.

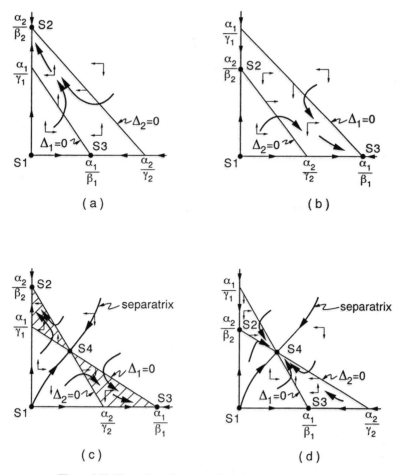

Figure 2.23 The various phase portraits for two competing species.

= 0, and (*b*) results when $\Delta_1 = 0$ is above $\Delta_2 = 0$. Cases (*c*) and (*d*) result *only* when the fourth steady state is in the first quadrant—that is,

$$x_{s4} > 0 \quad \text{and} \quad y_{s4} > 0$$

Then, (*c*) occurs if the slope of the nullcline $\Delta_2 = 0$ is less than that of $\Delta_1 = 0$, and (*d*) when the converse is true. In terms of the parameters α_i, β_i and γ_i, this means first that

$$x_{s4} = \frac{\alpha_1\beta_2 - \alpha_2\gamma_1}{\beta_1\beta_2 - \gamma_1\gamma_2} > 0 \quad \text{and} \quad y_{s4} = \frac{\alpha_2\beta_1 - \alpha_1\gamma_2}{\beta_1\beta_2 - \gamma_1\gamma_2} > 0 \qquad (2.16.5)$$

Case (*c*) is then the possible configuration when the slope of $\Delta_2 = 0$ is

$$-\frac{\gamma_2}{\beta_2} < -\frac{\beta_1}{\gamma_1}$$

and the slope of $\Delta_1 = 0$ is

$$\gamma_1\gamma_2 > \beta_1\beta_2 \tag{2.16.6}$$

which means that the interaction term $\gamma_1\gamma_2$ is larger than the self-inhibition term $\beta_1\beta_2$. Similarly, (*d*) results when

$$\gamma_1\gamma_2 < \beta_1\beta_2 \tag{2.16.7}$$

which implies that the self-inhibition term $\beta_1\beta_2$ dominates over the interaction term $\gamma_1\gamma_2$.

Combining the directions of the specific trajectories shown in Figure 2.22, the net direction of movement for all initial conditions (ICs) can then be identified. This has been done in Figure 2.23, from which the following *global* conclusions can be drawn for each of the possibilities (*a*) to (*d*).

Case a.
1. The steady state S1: $x_s = 0$, $y_s = 0$ is unstable and is maintained only if the IC is (0, 0).
2. The steady state S3: $x_s = \alpha_1/\beta_1$, $y_s = 0$ is attained only with those ICs for which $y_0 = 0$; that is, only the ICs on the x axis lead to this steady state.
3. All other ICs lead to the steady state S2: $x_s = 0$, $y_s = \alpha_2/\beta_2$.

Thus for case *a*, except when the species y is absent initially, it is the one that survives eventually and species x dies out.

Case b.
1. The steady state S1: $x_s = 0$, $y_s = 0$ is unstable, and is maintained only if the IC is (0, 0).
2. The steady state S2: $x_s = 0$, $y_s = \alpha_2/\beta_2$ is attained only with those ICs which lie on the y axis.
3. All other ICs lead to the steady state S3: $x_s = \alpha_1/\beta_1$, $y_s = 0$.

Hence for case *b*, except when the species x is initially not present, it is the eventual survivor and species y is eliminated.

Case c.
1. The steady state S1: $x_s = 0$, $y_s = 0$ is unstable and is maintained only with the IC (0, 0).
2. The coexistence steady state S4 is unstable and is maintained only if the IC coincides with S4.
3. If x is absent initially, then we have the following for the steady state S2: $x_s = 0$, $y_s = \alpha_2/\beta_2$ is attained as $t \to \infty$. Similarly, for the steady state S3 the following holds: $x_s = \alpha_1/\beta_1$, $y_s = 0$ is attained as $t \to \infty$ when y is not present initially.
4. All ICs in the lower shaded triangle lead to S3, while those in the upper shaded triangle lead to S2.

When we examine the directions of some of the trajectories in this case, it becomes clear that the first quadrant is divided up into two portions by a *separatrix*. All ICs above the separatrix eventually lead to the steady state S_2, while those below lead to steady state S_3.

Thus each pure species steady state has a finite *domain of attraction* (also called the *region of asymptotic stability*, or *RAS*). The coexistence steady state is unstable; that is, the two species are unable to exist together.

Case d.
1. The steady state S1: $x_s = 0$, $y_s = 0$ is unstable and is maintained only with the IC (0, 0).
2. All ICs on the x axis lead to the steady state S3: $x_s = \alpha_1/\beta_1$, $y_s = 0$.
3. All ICs on the y axis lead to the steady state S2: $x_s = 0$, $y_s = \alpha_2/\beta_2$.
4. All other ICs lead to the coexistence steady state S4.

Hence for this case, except for the x and y axes, the entire first quadrant of the (x, y) plane is the RAS of the coexistence steady state.

The preceding conclusions are *global* in nature; that is, we did not limit ourselves to considerations "near" a steady state. We now perform linearization around each of the steady states, to examine their type. The stability character of the various steady states (i.e., whether stable) is already known to us, and this will simply be confirmed by the linearized analysis.

2.16.3   Linearized Analysis

From eq. (2.14.5), the linearized system corresponding to the nonlinear system (2.16.1) is given by

$$\frac{d\boldsymbol{\xi}}{dt} = \mathbf{A}\boldsymbol{\xi} \tag{2.16.8}$$

where

$$\boldsymbol{\xi} = \begin{bmatrix} x - x_s \\ y - y_s \end{bmatrix}$$

is the vector of perturbations from the steady state (x_s, y_s), and

$$\mathbf{A} = \begin{bmatrix} \alpha_1 - 2\beta_1 x - \gamma_1 y & -\gamma_1 x \\ -\gamma_2 y & \alpha_2 - 2\beta_2 y - \gamma_2 x \end{bmatrix}_s \tag{2.16.9}$$

is the Jacobian or stability matrix, evaluated at the steady state. As noted earlier, there are four steady states S1–S4, and we now examine the eigenvalues of $\mathbf{A}$ for each of them.

Steady State S1

$$x_s = 0, \qquad y_s = 0$$

Substituting the coordinates for S1 in eq. (2.16.9), the matrix $\mathbf{A}$ reduces to

$$\mathbf{A} = \begin{bmatrix} \alpha_1 & 0 \\ 0 & \alpha_2 \end{bmatrix}$$

whose eigenvalues are

$$\lambda_1 = \alpha_1 > 0 \quad \text{and} \quad \lambda_2 = \alpha_2 > 0$$

Since the two eigenvalues are real and positive, following the classification described in section 2.15, the steady state S1 is an *unstable node*.

Steady State S2

$$x_s = 0, \qquad y_s = \frac{\alpha_2}{\beta_2}$$

In this case, the Jacobian matrix is

$$A = \begin{bmatrix} \alpha_1 - \dfrac{\gamma_1 \alpha_2}{\beta_2} & 0 \\[2mm] -\dfrac{\gamma_2 \alpha_2}{\beta_2} & -\alpha_2 \end{bmatrix}$$

which has eigenvalues given by the diagonal elements

$$\lambda_1 = \alpha_1 - \frac{\gamma_1 \alpha_2}{\beta_2} \quad \text{and} \quad \lambda_2 = -\alpha_2 < 0$$

Both eigenvalues are real and λ_2 is also negative. However, since λ_1 may be positive or negative, we have the following possibilities for the steady state S2:

If $\alpha_1 \beta_2 < \gamma_1 \alpha_2$, then it is a *stable node* (cases *a* and *c*).

If $\alpha_1 \beta_2 > \gamma_1 \alpha_2$, then it is an *unstable saddle point* (cases *b* and *d*).

Steady State S3

$$x_s = \frac{\alpha_1}{\beta_1}, \qquad y_s = 0$$

For this steady state, we have

$$A = \begin{bmatrix} -\alpha_1 & \dfrac{-\alpha_1 \gamma_1}{\beta_1} \\[2mm] 0 & \alpha_2 - \dfrac{\alpha_1 \gamma_2}{\beta_1} \end{bmatrix}$$

and the eigenvalues are again given by the diagonal elements

$$\lambda_1 = \alpha_2 - \frac{\alpha_1 \gamma_2}{\beta_1} \quad \text{and} \quad \lambda_2 = -\alpha_1 < 0$$

Similarly to S2, we have the following possibilities for the steady state S3:

If $\alpha_2 \beta_1 < \alpha_1 \gamma_2$, then it is a *stable node* (cases *b* and *c*).

If $\alpha_2 \beta_1 > \alpha_1 \gamma_2$, then it is an *unstable saddle point* (cases *a* and *d*).

Steady State S4

$$x_s = \frac{\alpha_1\beta_2 - \alpha_2\gamma_1}{\beta_1\beta_2 - \gamma_1\gamma_2}, \qquad y_s = \frac{\alpha_2\beta_1 - \alpha_1\gamma_2}{\beta_1\beta_2 - \gamma_1\gamma_2}$$

In the case of the coexistence steady state, the matrix $\mathbf{A}$ is given by

$$\mathbf{A} = \begin{bmatrix} -\beta_1 x_s & -\gamma_1 x_s \\ -\gamma_2 y_s & -\beta_2 y_s \end{bmatrix}$$

for which

$$\text{tr } \mathbf{A} = -(\beta_1 x_s + \beta_2 y_s) < 0$$

and

$$|\mathbf{A}| = (\beta_1\beta_2 - \gamma_1\gamma_2)x_s y_s$$

may be either positive or negative. It is then evident from Figure 2.21 that S4 is an *unstable saddle point* if

$$\beta_1\beta_2 < \gamma_1\gamma_2 \quad (\text{case } c)$$

Also, since

$$\Delta = (\text{tr } \mathbf{A})^2 - 4|\mathbf{A}| = (\beta_1 x_s + \beta_2 y_s)^2 - 4(\beta_1\beta_2 - \gamma_1\gamma_2)x_s y_s$$
$$= (\beta_1 x_s - \beta_2 y_s)^2 + 4\gamma_1\gamma_2 x_s y_s > 0$$

the steady state S4 is a *stable node* if

$$\beta_1\beta_2 > \gamma_1\gamma_2 \quad (\text{case } d)$$

The above conclusions regarding the linearized stability character of each of the steady states are consistent with the global analysis described earlier. The linearized analysis provides additional information about the type of the steady state (i.e., node, saddle point, etc.) and a precise description of the trajectories near the steady state. A comparison of the canonical forms of the trajectories for the various types of steady states in Figure 2.21, with the shapes of the trajectories in Figure 2.23, also indicates a consistency. Thus, for example, it is evident that steady state S4 is a saddle point in Figure 2.23c, while it is a stable node in the case of Figure 2.23d.

Finally, it is worth summarizing that the coexistence steady state S4 is stable *only* when the interaction term $(\gamma_1\gamma_2)$ between the two competing species is smaller than the self-inhibition term $(\beta_1\beta_2)$. In *all* other cases, the species cannot coexist in a stable manner, and either one or the other is eventually annihilated.

2.17 EXAMPLE 2: A NONADIABATIC CONTINUOUS-FLOW STIRRED TANK REACTOR

The problem of chemical reactor stability has received considerable attention in the chemical engineering literature since the early 1950s. The simplest flow reactor is a

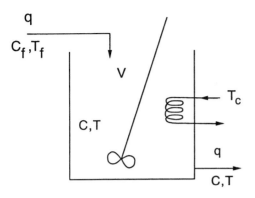

Figure 2.24 A nonadiabatic CSTR.

continuous-flow stirred tank reactor (CSTR), in which the reactor contents are kept well-mixed. The species concentrations and temperature therefore do not vary with position in the reactor, and these values are the same as in the effluent stream. As we shall see in this and later sections (2.21.2 and 2.23.3), this simple system can exhibit rather complex and interesting behavior.

Consider, as shown in Figure 2.24, a nonadiabatic CSTR in which the reaction A → products occurs. Let C represent the concentration (mol/cm^3) of reactant A, T the temperature (K), and $f(C, T)$ the rate of reaction (mol/cm^3-sec) by which A disappears. The transient mass and energy balances then lead to the pair of ODEs

$$V \frac{dC}{dt} = q(C_f - C) - Vf(C, T) \qquad (2.17.1)$$

$$VC_p\rho \frac{dT}{dt} = qC_p\rho(T_f - T) - Ua(T - T_c) + (-\Delta H)Vf(C, T) \qquad (2.17.2)$$

with ICs

$$C = C_0 \quad \text{and} \quad T = T_0 \qquad \text{at } t = 0 \qquad (2.17.3)$$

The notation used is as follows:

a	heat transfer area (cm^2)
C_p	heat capacity of reaction mixture (cal/g-K)
q	volumetric flow rate (cm^3/sec)
t	time elapsed after start-up (sec)
T_c	coolant temperature (K)
U	heat transfer coefficient (cal/sec-cm^2-K)
V	reactor volume (cm^3)
$(-\Delta H)$	heat of reaction (cal/mol)
ρ	density of reaction mixture (g/cm^3)

and the subscript f denotes feed value.

These balance equations can be cast into dimensionless form by introducing the following quantities:

$$u = \frac{C}{C_f}, \quad v = \frac{T}{T_f}, \quad \theta = \frac{V}{q}, \quad \tau = \frac{t}{\theta}, \quad r(u, v) = \frac{f(C, T)}{f(C_f, T_f)} \quad (2.17.4)$$

which transform eq. (2.17.1) to

$$\frac{du}{d\tau} = 1 - u - \frac{\theta f(C_f, T_f)}{C_f} r(u, v)$$

Thus a dimensionless parameter reflecting reactor size arises, and if we denote it by

$$\phi = \frac{\theta f(C_f, T_f)}{C_f} \quad (2.17.5)$$

then the mass balance becomes

$$\frac{du}{d\tau} = 1 - u - \phi r(u, v)$$

Similarly, the energy balance (2.17.2) yields

$$\frac{dv}{d\tau} = 1 - v - \frac{Ua}{qC_p\rho}(v - v_c) + \frac{(-\Delta H)\theta f(C_f, T_f)}{C_p\rho T_f} r(u, v)$$

which gives rise to two additional dimensionless parameters

$$\beta = \frac{(-\Delta H)C_f}{C_p\rho T_f} \quad \text{and} \quad \delta = \frac{Ua}{qC_p\rho} \quad (2.17.6)$$

representing the heat of reaction intensity and the extent of cooling, respectively. The dimensionless energy balance then takes the form

$$\frac{dv}{d\tau} = 1 - v - \delta(v - v_c) + \beta\phi r(u, v)$$

In summary, the dimensionless problem is represented by the autonomous system

$$\frac{du}{d\tau} = 1 - u - \phi r(u, v) \qquad\qquad = f_1(u, v) \quad (2.17.7)$$

$$\frac{dv}{d\tau} = (1 + \delta v_c) - (1 + \delta)v + \beta\phi r(u, v) = f_2(u, v) \quad (2.17.8)$$

with ICs

$$u = u_0 \quad \text{and} \quad v = v_0 \quad \text{at } \tau = 0 \quad (2.17.9)$$

Since $f_i(u, v)$, $i = 1$ or 2, are infinitely continuously differentiable functions of u and v, we are ensured (see section 2.7) that the above initial value problem has a unique solution.

2.17.1 The Steady States

The steady states are obtained by setting the time derivatives equal to zero. Thus they are simultaneous solutions of the two algebraic equations

$$1 - u_s - \phi r(u_s, v_s) = 0$$

and

$$(1 + \delta v_c) - (1 + \delta)v_s + \beta\phi r(u_s, v_s) = 0$$

where the subscript s refers to steady state. These two equations, however, can be combined into one by noting that multiplying the first by β and adding to the second yields

$$\beta(1 - u_s) + (1 + \delta v_c) - (1 + \delta)v_s = 0$$

so that

$$u_s = \frac{1 + \beta + \delta v_c - (1 + \delta)v_s}{\beta} \qquad (2.17.10)$$

Hence u_s may be replaced in terms of v_s, and we now only need to solve

$$(1 + \delta)v_s - (1 + \delta v_c) = \beta\phi r\left(\frac{1 + \beta + \delta v_c - (1 + \delta)v_s}{\beta}, v_s\right)$$

$$= \beta\phi g(v_s) \qquad (2.17.11)$$

to determine the steady states. Once the steady-state temperature v_s has been computed, the corresponding steady-state concentration u_s can be obtained from eq. (2.17.10).

    In physical terms, the left- and right-hand sides of eq. (2.17.11) represent the rates of heat removal (by flow and heat transfer to the cooling medium) and heat generation (by reaction), respectively, at steady state. Thus the equation states that the rates of heat generation and removal must be equal if a steady state is to exist. With this in mind, we may rewrite eq. (2.17.11) as

$$R(v_s) = \frac{1}{\phi}\left[v_s - \frac{1 + \delta v_c}{1 + \delta}\right] = \frac{\beta}{1 + \delta} g(v_s) = G(v_s) \qquad (2.17.12)$$

where R and G are the heat removal and heat generation functions, respectively.

    Taking advantage of the relationship (2.17.10) between u_s and v_s, and noting that the dimensionless reactant concentration satisfies the bounds

$$0 \le u_s \le 1$$

it can be seen that for the exothermic ($\beta > 0$) case of interest here, the dimensionless steady-state temperature must lie in the range

$$v_{lb} = \frac{1 + \delta v_c}{1 + \delta} \le v_s \le \frac{1 + \beta + \delta v_c}{1 + \delta} = v_{ub} \qquad (2.17.13)$$

where lb and ub represent the lower and upper bound, respectively.

2.17.2 Analysis of Steady-State Multiplicity

A complication that arises sometimes, which also leads to interesting phenomena, is that eq. (2.17.2) may have *multiple* solutions. To understand this feature, let us focus on the specific case of a first-order reaction, for which

$$f(C, T) = k_0 C \exp\left(\frac{-E}{R_g T}\right)$$

where k_0 is the frequency factor (sec^{-1}), E is the activation energy (cal/g mol), and R_g the universal gas constant (cal/g mol-K). From eq. (2.17.4) we have

$$r(u, v) = u \exp\left\{\gamma\left(1 - \frac{1}{v}\right)\right\}$$

where

$$\gamma = \frac{E}{R_g T_f}$$

is the dimensionless activation energy. Equation (2.17.12) then takes the form

$$R(v_s) = \frac{1}{\phi}\left[v_s - v_{lb}\right] = \frac{\beta}{1 + \delta}\left[\frac{1 + \beta + \delta v_c - (1 + \delta)v_s}{\beta}\right]\exp\left\{\gamma\left(1 - \frac{1}{v_s}\right)\right\}$$

$$= (v_{ub} - v_s)\exp\left\{\gamma\left(1 - \frac{1}{v_s}\right)\right\} = G(v_s)$$

This may be rearranged further as

$$\frac{1}{\phi} = \frac{G(v_s)}{v_s - v_{lb}} = F(v_s) \tag{2.17.14}$$

which implies that the values of v_s are obtained as intersections of the *constant* line $1/\phi$ with the function $F(v_s)$. Note that only those intersections which are in the range

$$v_{lb} \le v_s \le v_{ub}$$

are of interest. Here, the function F has the following properties:

1. $F(v) > 0$ for $v_{lb} \le v < v_{ub}$.
2. $F(v_{lb}) = +\infty$.
3. $F(v_{ub}) = 0$ and $F'(v_{ub}) < 0$.

Thus if $F(v)$ is a function which decreases monotonically with increasing v, then it is evident geometrically that only one intersection with a constant line is possible. This is the situation shown in Figure 2.25.

Hence *unique* steady states result for *all* values of ϕ if and only if

$$\underset{v_{lb} \le v \le v_{ub}}{\text{Sup}} \frac{dF}{dv} \le 0 \tag{2.17.15}$$

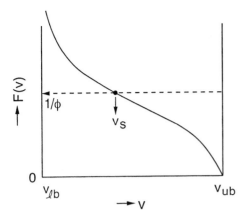

Figure 2.25 Form of the function $F(v)$ for unique steady states for all ϕ.

It should be evident from Figure 2.25 that in such cases the higher the value of ϕ (directly proportional to θ, the reactor residence time), the higher the steady-state temperature v_s; it also should be evident from eq. (2.17.10) that the higher the value of ϕ, the higher the corresponding reactant conversion $(1 - u_s)$. Since

$$F(v) = \frac{(v_{ub} - v) \exp\left\{\gamma\left(1 - \dfrac{1}{v}\right)\right\}}{v - v_{lb}}$$

differentiating once gives

$$F'(v) = \left[(v - v_{lb})\left\{-1 + (v_{ub} - v)\frac{\gamma}{v^2}\right\} - (v_{ub} - v)\right] \frac{\exp\left\{\gamma\left(1 - \dfrac{1}{v}\right)\right\}}{(v - v_{lb})^2}$$

so that $F'(v) \leq 0$ if and only if

$$(v - v_{lb})[-v^2 + \gamma(v_{ub} - v)] - (v_{ub} - v)v^2 \leq 0$$

which can be rearranged as

$$(\gamma + v_{ub} - v_{lb})v^2 - \gamma(v_{lb} + v_{ub})v + \gamma v_{lb}v_{ub} \geq 0 \qquad (2.17.16)$$

The lhs is a quadratic in v, with roots given by

$$v_\pm = \frac{\gamma(v_{lb} + v_{ub}) \pm \sqrt{\gamma^2(v_{lb} + v_{ub})^2 - 4\gamma v_{lb}v_{ub}(\gamma + v_{ub} - v_{lb})}}{2(\gamma + v_{ub} - v_{lb})}$$

The discriminant can be rewritten as $\beta\gamma\overline{\Delta}/(1 + \delta)^2$, where

$$\overline{\Delta} = \beta\gamma - \frac{4(1 + \delta v_c)(1 + \beta + \delta v_c)}{1 + \delta} \qquad (2.17.17)$$

Hence it is clear that if the physicochemical parameters are such that

$$\overline{\Delta} \leq 0 \qquad (2.17.18)$$

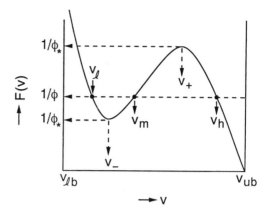

Figure 2.26 Form of the function $F(v)$ when multiple steady states exist.

then eq. (2.17.16) will *always* be satisfied. Furthermore, since $F'(v) \leq 0$ in such cases, unique steady states will result for *all* values of ϕ.

Alternatively, if the parameters are such that $\overline{\Delta} > 0$, then since eq. (2.17.16) may be rewritten for $F'(v) \leq 0$ as

$$(\gamma + v_{ub} - v_{lb})[(v - v_+)(v - v_-)] \geq 0$$

$F(v)$ has a local minimum at v_- and a local maximum at v_+ as shown in Figure 2.26, where $v_\pm$ are simplified as

$$v_\pm = \frac{\gamma(1 + \beta + 2\delta v_c) \pm \sqrt{\beta\gamma\overline{\Delta}}}{2[\beta + \gamma(1 + \delta)]} \qquad (2.17.19)$$

It then follows from eq. (2.17.14) that multiple steady states exist if and only if (see Figure 2.26)

$$F(v_-) \leq \frac{1}{\phi} \leq F(v_+)$$

which upon rearrangement gives

$$\phi_* = \frac{1}{F(v_+)} \leq \phi \leq \frac{1}{F(v_-)} = \phi^* \qquad (2.27.20)$$

while unique steady states are ensured if

$$\text{either } \phi < \phi_* \quad \text{or} \quad \phi > \phi^* \qquad (2.17.21)$$

Hence three steady states exist for $\phi_* < \phi < \phi^*$, two for either $\phi = \phi_*$ or $\phi = \phi^*$, and only one otherwise. For $\phi_* < \phi < \phi^*$, the three steady states may be labeled v_l, v_m, and v_h corresponding to low, middle, and high conversions, respectively. Note that along the low-conversion ($v < v_-$) and high-conversion ($v > v_+$) *branches*, an increase in the residence time τ (and thus ϕ) causes an increase in the reactor conversion; however, the converse holds for the middle branch. The critical values of ϕ—that is, ϕ_* and ϕ^*—are termed *bifurcation* points because two distinct branches of solutions originate at these points.

Remark. The strategy utilized above to analyze steady-state multiplicity for first-order reactions also works for other reaction kinetics (e.g., *n*th order), although the analysis becomes somewhat more complicated. Isothermal reactions for which the reaction rate versus concentration functionality is not monotone increasing [e.g., $r(c) = kC/(1 + KC)^2$ for sufficiently large K] can also yield multiple steady states. For a discussion of these issues, see Morbidelli et al. (1987).

Note that in terms of the heat generation and removal functions, G and R defined in eq. (2.17.12), the case of multiple steady states has the form shown in Figure 2.27. It is apparent from the figure that for a unique steady state or for the low- and high-conversion steady states when more than one exist, we have

$$R'(v_s) > G'(v_s) \qquad (2.17.22)$$

That is, at steady state, the slope of the heat removal function is greater than that of the heat generation function. However, the converse is true for the middle-conversion steady state.

Finally, it can be shown that unique steady states exist for all ϕ if the minimum value of

$$H(v) = G(v) - (v - v_{1b})\, G'(v) \qquad (2.17.23)$$

is non-negative. This condition is an equivalent form of that given by eq. (2.17.15).

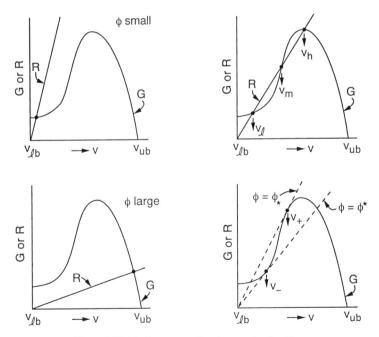

Figure 2.27 The heat removal and generation functions.

Geometrically, this means that the tangent drawn from any point on the G curve meets the ordinate axis at v_{lb} with a nonzero intercept.

2.17.3   Linearized Stability Analysis

In order to examine asymptotic stability of the steady states, we study the linearized system. Thus let

$$x_1(\tau) = u(\tau) - u_s \quad \text{and} \quad x_2(\tau) = v(\tau) - v_s \qquad (2.17.24)$$

represent perturbations from the steady state. Then, a linearization of eqs. (2.17.7) and (2.17.8) around the steady state yields the system

$$\frac{dx_1}{d\tau} = -x_1 - \phi(r_u x_1 + r_v x_2)$$

$$\frac{dx_2}{d\tau} = -(1 + \delta)x_2 + \beta\phi\{r_u x_1 + r_v x_2\}$$

where

$$r_u = \left.\frac{\partial r}{\partial u}\right|_s \quad \text{and} \quad r_v = \left.\frac{\partial r}{\partial v}\right|_s$$

are partial derivatives of the rate function, r, evaluated at the steady state. The Jacobian matrix is therefore

$$\mathbf{A} = \begin{bmatrix} -1 - \phi r_u & -\phi r_v \\ \beta\phi r_u & \beta\phi r_v - (1 + \delta) \end{bmatrix} \qquad (2.17.25)$$

for which

$$\begin{aligned} \operatorname{tr}\mathbf{A} &= -1 - \phi r_u + \beta\phi r_v - (1 + \delta) \\ &= -(2 + \delta) + \phi(\beta r_v - r_u) \end{aligned} \qquad (2.17.26a)$$

and

$$\begin{aligned} |\mathbf{A}| &= \{-1 - \phi r_u\}\{\beta\phi r_v - (1 + \delta)\} + \beta\phi^2 r_u r_v \\ &= (1 + \delta) - \phi\{\beta r_v - (1 + \delta)r_u\} \end{aligned} \qquad (2.17.26b)$$

From eq. (2.14.11), asymptotic stability of the steady state is ensured if

$$\operatorname{tr}\mathbf{A} < 0 \quad \text{and} \quad |\mathbf{A}| > 0$$

The first condition, $\operatorname{tr}\mathbf{A} < 0$, requires that

$$\phi(\beta r_v - r_u) < 2 + \delta \qquad (2.17.27a)$$

while the second, $|\mathbf{A}| > 0$, is satisfied if

$$\phi\{\beta r_v - (1 + \delta)r_u\} < 1 + \delta \qquad (2.17.27b)$$

Thus for a steady state to be asymptotically stable, these two conditions must be satisfied. However, if the reactor is *adiabatic*, then $\delta = 0$ by definition, and these conditions reduce to

$$\phi(\beta r_v - r_u) < 2 \tag{2.17.28a}$$

and

$$\phi(\beta r_v - r_u) < 1 \tag{2.17.28b}$$

respectively. Since tr $\mathbf{A} < 0$ is then implied by $|\mathbf{A}| > 0$, we reach the important conclusion that for an adiabatic reactor, the *only* condition required for asymptotic stability of a steady state is $|\mathbf{A}| > 0$.

The condition $|\mathbf{A}| > 0$ also has a physical interpretation which is now developed. From the definition (2.17.12):

$$\left.\frac{dR}{dv}\right|_s = \text{slope of the heat removal function at the steady state}$$

$$= \frac{1}{\phi}$$

Similarly,

$$\left.\frac{dG}{dv}\right|_s = \text{slope of the heat generation function at the steady state}$$

$$= \frac{\beta}{1 + \delta} \left.\frac{dg}{dv}\right|_s$$

$$= \frac{\beta}{1 + \delta} \left[\left.\frac{\partial r}{\partial u}\right|_s \left\{ -\frac{(1 + \delta)}{\beta} \right\} + \left.\frac{\partial r}{\partial v}\right|_s \right]$$

$$= \frac{\beta}{1 + \delta} r_v - r_u$$

Thus, we have

Slope of the heat removal function > Slope of the heat generation function

if and only if

$$\frac{1}{\phi} > \frac{\beta}{1 + \delta} r_v - r_u$$

or

$$\phi\{\beta r_v - (1 + \delta)r_u\} < 1 + \delta$$

which is the same as the condition (2.17.27b) which ensures $|\mathbf{A}| > 0$. For this reason, the condition $|\mathbf{A}| > 0$ is sometimes referred to as the *slope condition*. Note that in Figure 2.25 the entire curve satisfies $|\mathbf{A}| > 0$. Similarly, in Figures 2.26 and 2.27 the low- and high-conversion branches satisfy $|\mathbf{A}| > 0$, whereas on the middle-conversion branch we have $|\mathbf{A}| < 0$. Based on the above, we can summarize that in the adiabatic ($\delta = 0$) case the slope condition is necessary and sufficient for asymptotic stability of a steady state. In the nonadiabatic ($\delta > 0$) case, however, it is only one of the two conditions

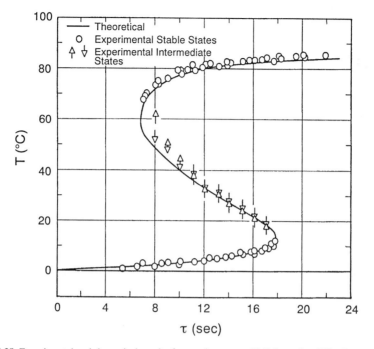

Figure 2.28 Experimental and theoretical results for steady-state multiplicity and stability for an adiabatic CSTR (Vejtasa and Schmitz, 1970).

required. When three steady states exist, the middle-conversion steady state is definitely unstable (saddle point) since it violates the slope condition. Furthermore:

1. If the reactor is adiabatic, then the unique steady states and the high- and low-conversion steady states are asymptotically stable (node or focus).

2. If the reactor is nonadiabatic, then the unique, and the high- and low-conversion steady states have a possibility of being unstable. They are asymptotically stable if tr **A** < 0; otherwise they are unstable.

Before concluding this example, it is worth noting that a variety of studies which link results of theoretical predictions with experimental observations have been reported in the literature. A pioneering work in this field is that of Vejtasa and Schmitz (1970), who carried out a second-order reaction between sodium thiosulfate and hydrogen peroxide in an adiabatic CSTR. As seen in Figure 2.28, experimental results for both steady-state multiplicity and stability closely match theoretical predictions.

2.18 LIAPUNOV'S DIRECT METHOD

In section 2.13, we discussed in detail that, owing to the fundamental stability theorem, stability character of a steady state for an almost linear system can be obtained by examining the corresponding linearized system. However, this analysis has limitations. For example, no conclusion can be derived if the linearized system is marginally stable. This will be the case if for the linearized system the largest eigenvalue (if real) is zero *or* if the real part of the eigenvalue having the largest real part (if complex) is zero. The latter case occurs frequently in systems which exhibit oscillatory behavior. Also, even for those cases where the steady state is asymptotically stable, no estimate of the size of the region of asymptotic stability (RAS) can be obtained from the fundamental stability theorem. This occurs because the theorem provides only *local* information in the vicinity of the steady state itself.

It is in this context that *Liapunov's direct method* (sometimes called *second method*) becomes important (cf. LaSalle and Lefschetz, 1961; Boyce and DiPrima, 1992, section 9.6). It is a direct method because it does not require any knowledge about the solution of the set of ODEs. Instead, it proceeds by constructing a suitable scalar function which can be regarded as related to the ''energy'' of the system. It then seeks to determine whether the energy decreases, indicating stability, or increases, implying instability.

In order to develop this notion further, consider the autonomous system

$$\frac{d\mathbf{x}}{dt} = \mathbf{f}(\mathbf{x}) \tag{2.18.1}$$

where $\mathbf{x}$ is an n-dimensional vector, and let

$$\mathbf{f}(\mathbf{0}) = \mathbf{0} \tag{2.18.2}$$

so that $\mathbf{x} = \mathbf{0}$ is an isolated critical point of the system. If we now consider a scalar function, $V(\mathbf{x})$, defined on some domain $\mathcal{D}$ containing the origin, then V is called *positive definite* if $V(\mathbf{0}) = 0$ and $V(\mathbf{x}) > 0$ for all other $\mathbf{x} \in \mathcal{D}$. Similarly, if $V(\mathbf{0}) = 0$ and $V(\mathbf{0}) < 0$ for all other $\mathbf{x} \in \mathcal{D}$, then V is said to be *negative definite*. Alternatively, if the inequalities $>$ and $<$ are replaced by $\geq$ and $\leq$, then $V(\mathbf{x})$ is defined to be *positive semidefinite* and *negative semidefinite*, respectively.

If $V(\mathbf{x})$ is a continuous function with continuous first partial derivatives, then we have

$$\frac{dV}{dt} = \dot{V} = \frac{\partial V}{\partial x_1}\frac{dx_1}{dt} + \frac{\partial V}{\partial x_2}\frac{dx_2}{dt} + \cdots + \frac{\partial V}{\partial x_n}\frac{dx_n}{dt}$$

and using eq. (2.18.1)

$$\frac{dV}{dt} = \frac{\partial V}{\partial x_1} f_1 + \frac{\partial V}{\partial x_2} f2 + \cdots + \frac{\partial V}{\partial x_n} f_n$$
$$= (\nabla V)^T \mathbf{f} \tag{2.18.3}$$

where ∇ is the gradient operator.

We can now state two theorems of Liapunov, where the first deals with stability and the second deals with instability of an isolated critical point.

● **THEOREM 1**

Consider the autonomous system (2.18.1), and assume that it has an isolated critical point at the origin. In a domain $\mathcal{D}$ containing the origin, if there exists a continuous function $V(\mathbf{x})$ with continuous first partial derivatives, which is positive definite and for which $\dot{V}$ given by eq. (2.18.3) is negative definite, then the origin is an asymptotically stable critical point. If all else is the same as above, but $\dot{V}$ is only negative semidefinite, then the origin is a stable critical point.

● **THEOREM 2**

Assume that the autonomous system (2.18.1) has an isolated critical point at the origin. Also, in a domain $\mathcal{D}$ containing the origin, let $V(\mathbf{x})$ be a continuous function with continuous first partial derivatives. Furthermore, let $V(\mathbf{0}) = 0$ and assume that in every neighborhood of the origin, there exists at least one point where V is positive. Then if $\dot{V}$ is positive definite in $\mathcal{D}$, the origin is an unstable critical point.

The proofs are straightforward and can be found in various places (cf. LaSalle and Lefschetz, 1961; Boyce and DiPrima, 1992, section 9.6). The results can be understood intuitively quite easily. For example, in the case of Theorem 1, the equation $V(\mathbf{x}) = c$, a constant, defines the surface of a sphere in the n-dimensional $\mathbf{x}$ space. The numerical value of c can be regarded as a measure of the distance from the surface to the origin. If $\dot{V} < 0$, then V (i.e., the distance between a point on the surface and the origin) will decrease continuously, and the system will eventually reach the origin (i.e., asymptotic stability).

The function V in the theorems above is called a *Liapunov function*. Note that if a suitable Liapunov function can be found, the first theorem permits one to construct a region of asymptotic stability around the critical point without having to solve the ODEs. This is a strong characteristic feature of Liapunov's direct method, and one that gives the method its directness. However, it should be noted that the RAS predicted by a Liapunov function typically tends to be conservative and does not include the entire region of stability.

The two theorems above give sufficient conditions for stability or instability. However, they are not necessary. In addition, the inability to find a suitable Liapunov function does not mean that one does not exist. The theorems themselves do not provide a procedure to construct Liapunov functions, although much work has been done in this direction [see Gurel and Lapidus (1969) for a survey]. The most common Liapunov functions are

$$V = \mathbf{x}^T \mathbf{x} \tag{2.18.4a}$$

and

$$V = \mathbf{x}^T \mathbf{P} \mathbf{x} \tag{2.18.4b}$$

where $\mathbf{P}$ is a positive definite matrix.

We now illustrate the use of Liapunov's direct method by two examples.

Example 1. Consider the system

$$\frac{dx_1}{dt} = -x_1^3 - 2x_1 x_2^2 \tag{2.18.5a}$$

$$\frac{dx_2}{dt} = -x_2 + x_1^2 x_2 \tag{2.18.5b}$$

Clearly, the origin is a unique steady state of the system, and the Jacobian matrix evaluated at the origin is

$$\mathbf{J} = \begin{bmatrix} 0 & 0 \\ 0 & -1 \end{bmatrix}$$

which has eigenvalues 0 and -1. Since the linearized system is marginally stable, we are unable to conclude about the stability character of the nonlinear system.

Consider now the Liapunov function

$$V = ax_1^2 + x_2^2 \tag{2.18.6}$$

where a is yet to be determined. V is positive definite when $a > 0$. Differentiating yields

$$\dot{V} = 2ax_1\dot{x}_1 + 2x_2\dot{x}_2$$

or

$$-\dot{V} = 2ax_1^4 + 2x_1^2 x_2^2(2a - 1) - 2x_2^2 \tag{2.18.7}$$

so that $\dot{V}$ is negative definite for $a \geq \frac{1}{2}$. Thus from Theorem 1 the origin is asymptotically stable. Since the origin is a unique critical point, it is globally asymptotically stable and its RAS is the entire (x_1, x_2) plane.

Example 2.

$$\frac{dx_1}{dt} = x_1^3 - x_2^3 \tag{2.18.8a}$$

$$\frac{dx_2}{dt} = 2x_1 x_2^2 + 4x_1^2 x_2 + 2x_2^3 \tag{2.18.8b}$$

Again, the origin is a unique steady state of the system, with Jacobian matrix

$$\mathbf{J} = \begin{bmatrix} 0 & 0 \\ 0 & 0 \end{bmatrix}$$

which has a repeated eigenvalue $\lambda = 0$. Hence we are again unable to assess the stability character from the linearized system.

Consider the Liapunov function

$$V = ax_1^2 + x_2^2 \tag{2.18.9}$$

where $a > 0$ is as yet an unknown constant. Then

$$\dot{V} = 2ax_1^4 + 2x_1 x_2^3(2 - a) + 8x_1^2 x_2^2 + 4x_2^4$$

With the choice

$$a = 2$$

we have

$$V > 0 \quad \text{and} \quad \dot{V} > 0 \tag{2.18.10}$$

Hence, following Theorem 2, the origin is unstable.

Example 3. Consider the autonomous system

$$\frac{dx_1}{dt} = x_2 \qquad\qquad = f_1(x_1, x_2) \tag{2.18.11a}$$

$$\frac{dx_2}{dt} = -2bx_1 - 3x_1^2 - ax_2 = f_2(x_1, x_2) \tag{2.18.11b}$$

which has two critical points, the origin 0 and $A: (-2b/3, 0)$. The Jacobian matrix is

$$\mathbf{J} = \begin{bmatrix} 0 & 1 \\ -2b - 6x_1 & -a \end{bmatrix}$$

At the origin, it satisfies

$$\text{tr } \mathbf{J} = -a < 0, \qquad |\mathbf{J}| = 2b > 0$$

which implies asymptotic stability. The eigenvalues may be real or complex, so that the origin is a stable node or focus, respectively.

For the point A we have

$$\mathbf{J} = \begin{bmatrix} 0 & 1 \\ 2b & -a \end{bmatrix}$$

leading to

$$\text{tr } \mathbf{J} = -a < 0, \qquad |\mathbf{J}| = -2b < 0$$

indicating an unstable saddle point. For the nonlinear system (2.18.11) the trajectories can be computed *numerically* and are shown in Figure 2.29 for a specific combination of a and b values. The hatched area is the RAS for the origin.

Let us now proceed to construct a RAS by using Liapunov's direct method. Consider the Liapunov function

$$V(x_1, x_2) = bx_1^2 + x_1^3 + \frac{x_2^2}{2} \tag{2.18.12}$$

then

$$\frac{dV}{dt} = -ax_2^2$$

$$< 0 \tag{2.18.13}$$

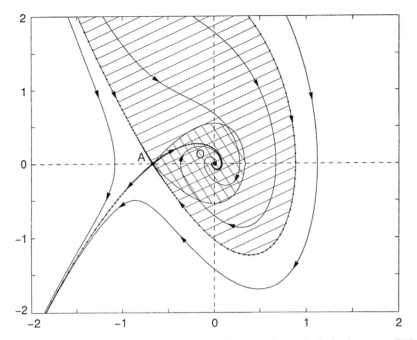

Figure 2.29 Some trajectories and the RAS (hatched area) computed numerically for the system (2.18.11), as well as the RAS (crosshatched area) predicted by the Liapunov function; $a = b = 1$.

as long as $x_2 \neq 0$. Note that

$$\frac{dx_2}{dx_1} = -a - \frac{x_1(2b + 3x_1)}{x_2}$$

$$\rightarrow \pm \infty$$

as $x_2 \rightarrow 0$. Hence, except at the origin, the trajectories are deflected away from the line $x_2 = 0$, so $x_2 \neq 0$ is not a limitation for $\dot{V} < 0$.

It is clear from eq. (2.18.12) that for sufficiently large and negative x_1, $V < 0$. Also, since the point A is unstable, the RAS for the origin certainly cannot include this point. Since

$$V = K$$

(where K is a positive constant) defines a closed curve surrounding the origin, the largest region enclosed by the curve that can be hoped for is constructed by using the coordinates of the point A in eq. (2.18.12). This yields

$$K = \left(\frac{-2b}{3}\right)^2 \left[b - \frac{2b}{3}\right] = \frac{4b^3}{27}$$

The RAS is thus bounded by the curve

$$V = bx_1^2 + x_1^3 + \frac{x_2^2}{2} = \frac{4b^3}{27} \tag{2.18.14}$$

and is shown as the crosshatched region in Figure 2.29. The plot of eq. (2.18.14) is facilitated by rearranging it as

$$x_2^2 = 2\left\{ \frac{4b^3}{27} - x_1^2(b + x_1) \right\} \tag{2.18.15}$$

Differentiating once gives

$$\frac{dx_2^2}{dx_1} = -2x_1(2b + 3x_1)$$

which implies that the curve has extreme points at $x_1 = 0$ and $-2b/3$. The second derivative

$$\frac{d^2 x_2^2}{dx_1^2} = -4b - 12x_1$$

is negative at $x_1 = 0$ and positive at $x_1 = -2b/3$, indicating a maximum and minimum for x_2^2, respectively. Substituting these x_1 values in eq. (2.18.15) yields

$$x_2 \Big|_{x_1=0} = \pm \left(\frac{8b^3}{27} \right)^{1/2}$$

and

$$x_2 \Big|_{x_1=-2b/3} = 0$$

Note that the RAS of the origin, as determined by the Liapunov function (2.18.12), is smaller than the actual RAS determined numerically. In fact, this occurs frequently and is a limitation of the method. However, it should also be noted that the smaller RAS is obtained *analytically* with relatively little effort.

2.19 INTRODUCTORY EXAMPLES OF PERIODIC SOLUTIONS

Many physical and chemical systems display an *oscillatory* or *periodic* behavior. In such cases the dynamics does not lead to any particular steady state, but rather the system oscillates repeatedly with a fixed period. Thus for the autonomous system

$$\frac{d\mathbf{x}}{dt} = \mathbf{f}(\mathbf{x}) \tag{2.19.1}$$

we have

$$\mathbf{x}(t) = \mathbf{x}(t + T) \tag{2.19.2}$$

where $T > 0$ is the period. Examples of this behavior include the classical predator–prey systems (cf. Murray, 1990), a nonisothermal CSTR with a positive-order reaction or an isothermal CSTR with an autocatalytic reaction (cf. Gray and Scott, 1990), and the van der Pol equation which describes current flow in an electric circuit with nonlinear damping (cf. Minorsky, 1962). In this section we consider two simple examples which display oscillatory behavior.

Example 1. Discuss the dynamic behavior of the system

$$\frac{dx}{dt} = x - y - x^3 - xy^2 = f_1(x, y) \tag{2.19.3a}$$

$$\frac{dy}{dt} = x + y - x^2y - y^3 = f_2(x, y) \tag{2.19.3b}$$

The steady states are obtained by solving the equations

$$x_s - y_s - x_s^3 - x_sy_s^2 = 0$$

$$x_s + y_s - x_s^2y_s - y_s^3 = 0$$

Multiplying the first equation by y_s, multiplying the second by x_s, and then subtracting the first from the second yields

$$x_s^2 + y_s^2 = 0$$

Thus (0, 0) is the unique steady state of the system (2.19.3).

In the linearized analysis, the Jacobian matrix is

$$\mathbf{J} = \begin{bmatrix} f_{1x} & f_{1y} \\ f_{2x} & f_{2y} \end{bmatrix}_s = \begin{bmatrix} 1 & -1 \\ 1 & 1 \end{bmatrix}$$

which has eigenvalues $1 \pm i$, implying that the origin (0, 0) is an unstable focus. Also, since $j_{21} = 1 > 0$, the direction of rotation for the focus is counterclockwise. Hence the phase plane for the linearized system has the form shown in Figure 2.30. The linearized analysis tells us only this much. From this analysis, it is not clear what happens as the trajectories spiral away from the origin. For example, do they continue to move away indefinitely, eventually going to infinity, or does something else happen?

In certain cases, such as this, additional information about the full nonlinear system can be obtained by converting to polar coordinates:

$$x = r \cos \phi, \qquad y = r \sin \phi \tag{2.19.4}$$

so that

$$\frac{dx}{dt} = \cos \phi \frac{dr}{dt} - r \sin \phi \frac{d\phi}{dt} \tag{2.19.5a}$$

and

$$\frac{dy}{dt} = \sin \phi \frac{dr}{dt} + r \cos \phi \frac{d\phi}{dt} \tag{2.19.5b}$$

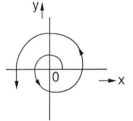

Figure 2.30 The nature of trajectories from the linearized analysis.

Introducing eq. (2.19.14) in the rhs of eq. (2.19.3) and using eq. (2.19.5), we obtain

$$\cos\phi\,\frac{dr}{dt} - r\sin\phi\,\frac{d\phi}{dt} = r\cos\phi - r\sin\phi - r^3\cos\phi$$

$$\sin\phi\,\frac{dr}{dt} + r\cos\phi\,\frac{d\phi}{dt} = r\cos\phi + r\sin\phi - r^3\sin\phi$$

which can be viewed as two equations in the two unknowns dr/dt and $d\phi/dt$. The solution is simply

$$\frac{dr}{dt} = r(1 - r^2) \qquad (2.19.6a)$$

and

$$\frac{d\phi}{dt} = 1 \qquad (2.19.6b)$$

Equation (2.19.6b) has the solution

$$\phi(t) = \phi(0) + t \qquad (2.19.7)$$

which implies that the phase angle ϕ increases with t; hence it has counterclockwise motion as also obtained earlier by the linearized analysis.

For r, from eq. (2.19.6a) the possible "steady states" are $r = 0$ and 1, where the latter describes a circle in the x–y plane. It is clear that for $0 < r < 1$, $dr/dt > 0$, while $dr/dt < 0$ for $r > 1$. Combining this information with eq. (2.19.7), we can conclude that all trajectories spiral in a counterclockwise manner toward the unit circle. Initial conditions within the circle spiral toward the circle from within, while those that originate from outside the circle approach it from the outside, as shown in Figure 2.31.

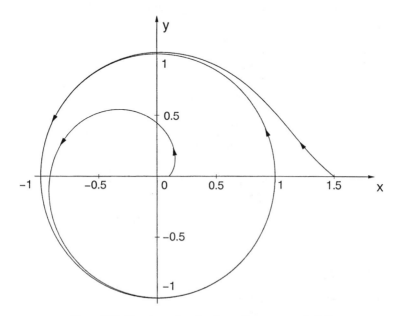

Figure 2.31 The phase plane for the nonlinear system (2.19.3).

The exact solution of eq. (2.19.6a) can be easily obtained by separation of variables:

$$r(t) = \frac{1}{\{1 + [1/r^2(0) - 1] \exp(-2t)\}^{1/2}} \tag{2.19.8}$$

where $r(0)$ denotes the initial value of r.

The periodic solution [in this example, $r = 1$] is also called a *limit cycle*. Note that since trajectories approach the limit cycle as $t \to \infty$, it is a *stable limit cycle*, and we have *orbital stability*. In some cases the trajectories approach the limit cycle from one side, but not the other. In such cases we have *one-sided stability*. In other cases a limit cycle may not be approached at all, and trajectories on both sides are repelled away. Then we say that the periodic solution is *unstable*.

Example 2. In this example we consider the Lotka–Volterra model of a predator–prey system. In such a system the predator species (e.g., foxes) feeds upon the prey species (e.g., rabbits), while the latter utilize a different food source. If we denote the populations of the prey and predator species by x_1 and x_2, respectively, then the Lotka–Volterra model is constructed by the following assumptions (cf. Murray, 1990, section 3.1; Boyce and DiPrima, 1992, section 9.5).

1. If the predator is absent, then the prey grows at a rate proportional to its population; that is,

$$\frac{dx_1}{dt} = ax_1, \quad a > 0 \qquad \text{when } x_2 = 0$$

2. If the prey is absent, the predator population decays in size; that is,

$$\frac{dx_2}{dt} = -bx_2, \quad b > 0 \qquad \text{when } x_1 = 0$$

3. When both the predator and prey are present, their interactions are proportional to the product of their populations. Each interaction promotes growth of the predator and inhibits growth of the prey.

With these assumptions we have the following model equations:

$$\frac{dx_1}{dt} = ax_1 - \alpha x_1 x_2 = x_1(a - \alpha x_2) = f_1(x_1, x_2) \tag{2.19.9a}$$

$$\frac{dx_2}{dt} = -bx_2 + \beta x_1 x_2 = x_2(-b + \beta x_1) = f_2(x_1, x_2) \tag{2.19.9b}$$

where α and β are also both positive and represent the effect of the interaction between the two species on their growth rates.

The critical points of the system are solutions of the simultaneous equations

$$f_1(x_1, x_2) = 0 \quad \text{and} \quad f_2(x_1, x_2) = 0$$

that is, the points $(0, 0)$ and $(b/\beta, a/\alpha)$. Let us first carry out linearized stability analysis, which yields the Jacobian matrix

$$\mathbf{A} = \begin{bmatrix} f_{1x_1} & f_{1x_2} \\ f_{2x_1} & f_{2x_2} \end{bmatrix}_s = \begin{bmatrix} a - \alpha x_2 & -\alpha x_1 \\ \beta x_2 & -b + \beta x_1 \end{bmatrix}_s$$

For the critical point $(0, 0)$ we have

$$\mathbf{A} = \begin{bmatrix} a & 0 \\ 0 & -b \end{bmatrix}$$

which has eigenvalues $\lambda_1 = a > 0$ and $\lambda_2 = -b < 0$. Hence the origin is an unstable saddle point.

For the critical point $(b/\beta, a/\alpha)$ the Jacobian matrix takes the form

$$\mathbf{A} = \begin{bmatrix} 0 & -\alpha b/\beta \\ \beta a/\alpha & 0 \end{bmatrix}$$

and its eigenvalues are purely imaginary, $\lambda = \pm i\sqrt{ab}$. Hence the point $(b/\beta, a/\alpha)$ is a center, with counterclockwise rotation since $a_{21} > 0$, and we cannot establish the stability character of the full system near this point.

Since we are dealing with populations, only the first quadrant of the (x_1, x_2) plane is of interest. Let us now identify the directions of the trajectories in various parts of this region. For this, it is convenient to rewrite eqs. (2.19.9) as

$$\frac{dx_1}{dt} = x_1 \Delta_1 \qquad\qquad (2.19.10a)$$

$$\frac{dx_2}{dt} = x_2 \Delta_2 \qquad\qquad (2.19.10b)$$

where

$$\Delta_1 = a - \alpha x_2 \quad \text{and} \quad \Delta_2 = -b + \beta x_1 \qquad\qquad (2.19.11)$$

The signs of Δ_1 and Δ_2, as well as the directions of the trajectories, are as shown in Figure 2.32. This analysis confirms the counterclockwise rotation of trajectories and indicates a periodic solution. However, the latter cannot be confirmed by this analysis since the trajectories could spiral inward to the critical point $(b/\beta, a/\alpha)$ or outward to infinity.

In this example we can obtain an exact representation of the trajectories in the phase plane. Dividing eq. (2.19.9b) by (2.19.9a) yields

$$\frac{dx_2}{dx_1} = \frac{x_2(-b + \beta x_1)}{x_1(a - \alpha x_2)}$$

where the variables can be separated to give

$$\frac{a - \alpha x_2}{x_2} dx_2 = \frac{-b + \beta x_1}{x_1} dx_1$$

Integration leads to

$$a \ln x_2 - \alpha x_2 = -b \ln x_1 + \beta x_1 + K \qquad\qquad (2.19.12)$$

where K is a constant. It can be shown (see problem 2.24) that for each K (fixed by the ICs), eq. (2.19.12) represents a closed curve in the (x_1, x_2) plane, surrounding the critical

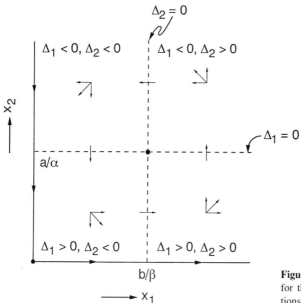

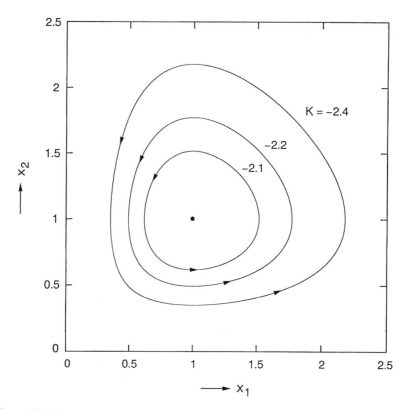

Figure 2.32 The direction of trajectories for the Lotka–Volterra system of equations.

Figure 2.33 Solution of the Lotka–Volterra equations (2.19.9) for various ICs; $a = b = \alpha = \beta = 1$.

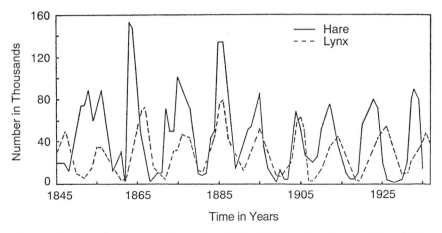

Figure 2.34 Fluctuations in the number of pelts sold by the Hudson Bay Company (Odum, 1953).

point $(b/\beta, a/\alpha)$, as shown in Figure 2.33. Hence eqs. (2.19.9) exhibit a periodic solution, which varies as the ICs are varied. This is a situation somewhat different from Example 1, where there was only one periodic solution that attracted trajectories from both within and without. In the present example the periodic solution is *marginally stable* since each IC generates its own limit cycle which neither grows nor decays with time.

An interesting example of a predator–prey interaction is provided by the data on the Canadian lynx–snowshoe hare populations reported in the trading records of the Hudson Bay Company from about 1845 until the 1930s (Odum, 1953). As seen in Figure 2.34, the species populations oscillate with a period of about 10 years. However, it should be noted that this is not a pure predator–prey system owing to additional complicating factors, such as epidemic diseases and fur trappers (for details see Gilpin, 1973).

2.20 THE HOPF BIFURCATION THEOREM

In the previous section we saw two examples of periodic solutions. In this section we present a general result which describes the possible situations that give birth to periodic solutions. Before going on to the general result, let us first consider a simple motivational example (Seydel, 1988, Section 2.6) which is similar to Example 1 of section 2.19.

For the system

$$\frac{dx}{dt} = -y + x(\mu - x^2 - y^2) \tag{2.20.1a}$$

$$\frac{dy}{dt} = x + y(\mu - x^2 - y^2) \tag{2.20.1b}$$

where μ is a real constant, it can be shown easily that $(0, 0)$ is the only critical point for all values of μ. The Jacobian matrix

$$\mathbf{J} = \begin{bmatrix} \mu & -1 \\ 1 & \mu \end{bmatrix} \tag{2.20.2}$$

has eigenvalues $\mu \pm i$. Hence it is clear from the classification of section 2.15 that the origin is a stable focus for $\mu < 0$ and an unstable focus for $\mu > 0$, with a counter-clockwise rotation.

As in the previous example, we can construct the periodic solution by converting to polar coordinates:

$$x = r \cos \phi \quad \text{and} \quad y = r \sin \phi$$

which leads to

$$\frac{d\phi}{dt} = 1 \tag{2.20.3a}$$

and

$$\frac{dr}{dt} = r(\mu - r^2) \tag{2.20.3b}$$

Hence it is clear that for $r = \mu^{1/2}$, $dr/dt = 0$. Thus for any $\mu > 0$, there is a periodic solution with radius in the phase plane equal to $\mu^{1/2}$. Further, since

$$\frac{dr}{dt} > 0 \qquad \text{for } r < \mu^{1/2} \tag{2.20.4a}$$

and

$$\frac{dr}{dt} < 0 \qquad \text{for } r > \mu^{1/2} \tag{2.20.4b}$$

the limit cycle is stable and all trajectories are attracted to it.

A summary of these results is shown in Figure 2.35. It may be seen that at $\mu = \mu_0 = 0$, there is an *exchange of stability*, from a stable critical point to a stable limit cycle. The stable limit cycle surrounds the unstable critical point and has an amplitude which grows as $(\mu - \mu_0)^{1/2}$. A bifurcation of this type, from an equilibrium to a periodic solution, is called a *Hopf bifurcation*. Note that at the bifurcation the Jacobian matrix has a pair of purely imaginary eigenvalues.

We now state the Hopf bifurcation theorem. Note that although the basic result was known by earlier mathematicians (notably Poincaré and Andronov), the bifurcation from an equilibrium state to a limit cycle is commonly referred to as a Hopf bifurcation; it is named after E. Hopf, who proved the result for the general n-dimensional case in 1942 (for proof, see Hassard et al., 1981, chapter 1; Marsden and McCracken, 1976, section 3).

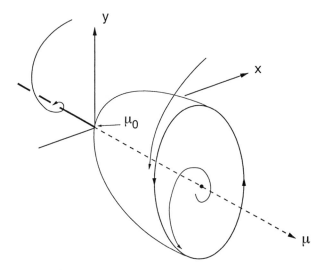

Figure 2.35 Hopf bifurcation from a stable equilibrium to a stable limit cycle, as the parameter μ crosses μ_0.

● **THEOREM**

Consider the n-dimensional autonomous system

$$\frac{d\mathbf{x}}{dt} = \mathbf{f}(\mathbf{x}, \mu) \tag{2.20.5}$$

which depends upon a parameter μ.

(a) Let $\mathbf{f}(\mathbf{x}_0, \mu_0) = 0$, so that $\mathbf{x}_0$ is the equilibrium solution corresponding to the parameter value μ_0.

(b) Furthermore, let the Jacobian matrix $\mathbf{f}_\mathbf{x}(\mathbf{x}_0, \mu_0)$ have a simple pair of purely imaginary eigenvalues $\lambda(\mu_0) = \pm i\omega_0$, with the remaining eigenvalues having strictly negative real parts.

(c) Finally, let the real part of λ indeed pass through zero as μ crosses μ_0—that is,

$$\frac{d}{d\mu} \text{Re}(\lambda) \bigg|_{\mu_0} \neq 0 \tag{2.20.6}$$

Then there is a birth of periodic solutions as μ crosses μ_0 (emergence in one direction of crossing and disappearance in the opposite direction). The period of the oscillatory solution at birth (i.e., a zero-amplitude oscillation) is $T_0 = 2\pi/\omega_0$.

The *direction of bifurcation* can be evaluated by calculating the algebraic sign of a specific function of μ, at $\mu = \mu_0$. This is generally a difficult task, and Hassard et al. (1981) have provided recipes and numerical techniques for this purpose. For two-dimensional systems, Friedrichs (1965) has described a detailed formulation for computing this function [also see Poore (1973) for a more explicit version of this formu-

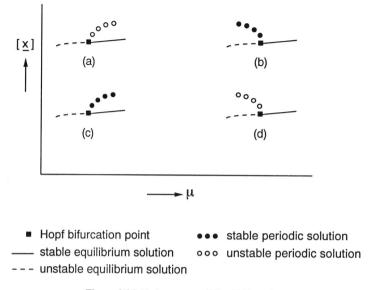

Figure 2.36 Various types of Hopf bifurcation.

lation]. The *stability* of the periodic solution (i.e., orbital stability) is related to the direction of bifurcation. The typical situation is that near the Hopf point, unstable periodic solutions surround stable equilibrium points (*subcritical bifurcation*) while stable periodic solutions surround unstable equilibria (*supercritical bifurcation*). Examples of these, with both stable and unstable periodic solutions emerging as the parameter μ increases or decreases beyond μ_0, are shown in Figure 2.36, where [**x**] denotes some measure of the solution.

Finally, it should be noted that Hopf bifurcation is a *local* theory. It provides results only *near* the bifurcation point, as the parameter μ just crosses μ_0.

2.21 APPLICATIONS OF THE HOPF BIFURCATION THEOREM

We now consider two examples involving applications of the Hopf bifurcation theorem. Both examples deal with chemical reactions. In the first example, dynamics of a first-order reaction in a nonadiabatic CSTR is investigated. The second example involves the dynamic behavior of the Belousov–Zhabotinskii reaction in an isothermal batch reactor. In both cases the analysis provides substantial insight into the physicochemical behavior of the systems.

2.21.1 The Nonadiabatic CSTR

This example was treated previously in section 2.17. For the analysis here, which closely follows Poore (1973), it is more convenient to adopt a somewhat different form of the dimensionless equations. Hence, starting with eqs. (2.17.1) and (2.17.2), for the case of a first-order irreversible reaction

$$f(C, T) = k_0 C \, \exp\left(-\frac{E}{R_g T}\right) \tag{2.21.1}$$

define the dimensionless quantities

$$x_1 = \frac{T - T_f}{T_f} \frac{E}{R_g T_f}, \qquad x_2 = \frac{C_f - C}{C_f}, \qquad \theta = \frac{V}{q}, \qquad \tau = \frac{t}{\theta}$$

$$x_{1c} = \frac{T_c - T_f}{T_f} \frac{E}{R_g T_f}, \qquad \gamma = \frac{E}{R_g T_f}, \qquad \mathrm{Da} = k_0 e^{-\gamma} \theta \tag{2.21.2}$$

$$B = \frac{(-\Delta H) C_f}{C_p \rho T_f} \frac{E}{R_g T_f}, \qquad \delta = \frac{Ua}{q C_p \rho}$$

Equations (2.17.1) and (2.17.2) then reduce to

$$\frac{dx_1}{d\tau} = -x_1 - \delta(x_1 - x_{1c}) + B \, \mathrm{Da}(1 - x_2) \exp\left(\frac{x_1}{1 + x_1/\gamma}\right) \tag{2.21.3a}$$

$$\frac{dx_2}{d\tau} = -x_2 + \mathrm{Da}(1 - x_2) \exp\left(\frac{x_1}{1 + x_1/\gamma}\right) \tag{2.21.3b}$$

Let us now focus exclusively on the case $\gamma \to \infty$, which simplifies the analysis considerably. Poore (1973) and Uppal et al. (1974) have shown that finiteness of γ does not change the qualitative picture that emerges. With this, the argument of the exponential becomes simplified, and we have

$$\frac{dx_1}{d\tau} = -x_1 - \delta(x_1 - x_{1c}) + B \, \mathrm{Da}(1 - x_2) \exp(x_1) = F_1(x_1, x_2) \tag{2.21.4a}$$

$$\frac{dx_2}{d\tau} = -x_2 + \mathrm{Da}(1 - x_2) \exp(x_1) = F_2(x_1, x_2) \tag{2.21.4b}$$

Analysis of Steady-State Behavior. We now need to analyze the steady-state behavior of the system (2.21.4). The procedure is identical to that carried out in section 2.17, so that we can expedite the investigation. If we denote the steady-state solution to be (a_1, a_2), then

$$-a_1 - \delta(a_1 - x_{1c}) + B \, \mathrm{Da}(1 - a_2) \exp(a_1) = 0 \tag{2.21.5a}$$

$$-a_2 \qquad\qquad + \mathrm{Da}(1 - a_2) \exp(a_1) \; = 0 \tag{2.21.5b}$$

Multiplying eq. (2.21.5b) by B and subtracting from the first gives

$$-a_1 - \delta(a_1 - x_{1c}) + B a_2 = 0$$

so that

$$a_1 = \frac{B a_2 + \delta x_{1c}}{1 + \delta} \tag{2.21.6}$$

and we may replace temperature (a_1) in terms of conversion (a_2). Thus eq. (2.21.5b) yields

$$\text{Da} = \frac{a_2}{(1 - a_2) \exp[(Ba_2 + \delta x_{1c})/(1 + \delta)]} = g(a_2) \qquad (2.21.7)$$

which gives the steady-state conversion. Note that it satisfies the bounds $0 \leq a_2 \leq 1$. Analyzing for the shape of g over the range of interest, we have

$$g(0) = 0, \qquad g(1) = +\infty$$

and

$$\frac{dg}{da_2} = [Ba_2^2 - Ba_2 + (1 + \delta)] \frac{\exp[-(Ba_2 + \delta x_{1c})/(1 + \delta)]}{(1 + \delta)(1 - a_2)^2}$$

$$= 0$$

when

$$Ba_2^2 - Ba_2 + (1 + \delta) = 0 \qquad (2.21.8)$$

This quadratic has roots given by

$$m_1 = \frac{1}{2} - \frac{1}{2}\sqrt{1 - \frac{4(1 + \delta)}{B}} \quad \text{and} \quad m_2 = \frac{1}{2} + \frac{1}{2}\sqrt{1 - \frac{4(1 + \delta)}{B}} \qquad (2.21.9)$$

If

$$B \leq 4(1 + \delta) = f_1(\delta) \qquad (2.21.10)$$

then g has the shape shown in Figure 2.37a, and unique steady states exist for all Da values. On the other hand, if

$$B > 4(1 + \delta) \qquad (2.21.11)$$

then with the notation

$$\text{Da}_i = \frac{m_i}{(1 - m_i) \exp[(Ba_i + \delta x_{1c})/(1 + \delta)]}, \qquad i = 1, 2 \qquad (2.21.12)$$

we have

　　unique steady states for $\text{Da} > \text{Da}_1$ or $\text{Da} < \text{Da}_2$
　　two steady states for $\text{Da} = \text{Da}_1$ or $\text{Da} = \text{Da}_2$
　　three steady states for $\text{Da}_2 < \text{Da} < \text{Da}_1$

Linearized Stability Analysis.　If we define the perturbations from steady states as

$$y_1 = x_1 - a_1 \quad \text{and} \quad y_2 = x_2 - a_2$$

then the corresponding linearized system is

$$\frac{d\mathbf{y}}{d\tau} = \mathbf{Ay} \qquad (2.21.13)$$

The Jacobian matrix is given by

$$\mathbf{A} = \begin{bmatrix} \dfrac{\partial F_1}{\partial x_1} & \dfrac{\partial F_1}{\partial x_2} \\ \dfrac{\partial F_2}{\partial x_1} & \dfrac{\partial F_2}{\partial x_2} \end{bmatrix}_s$$

$$= \begin{bmatrix} -1 - \delta + Ba_2 & -\dfrac{Ba_2}{1 - a_2} \\ a_2 & \dfrac{-1}{1 - a_2} \end{bmatrix}$$

where eq. (2.21.5b) has been used. From the definitions,

$$\text{tr } \mathbf{A} = -1 - \delta + Ba_2 - \frac{1}{1 - a_2}$$

$$= -\frac{1}{(1 - a_2)} [Ba_2^2 - (B + 1 + \delta)a_2 + (2 + \delta)] \qquad (2.21.14a)$$

$$|\mathbf{A}| = \frac{1}{(1 - a_2)} [Ba_2^2 - Ba_2 + (1 + \delta)]$$

$$= \frac{B}{(1 - a_2)} [(a_2 - m_1)(a_2 - m_2)] \qquad (2.21.14b)$$

and the sign of $|\mathbf{A}|$ on various branches in Figure 2.37 is readily established.

Stability requires that tr $\mathbf{A} < 0$ and $|\mathbf{A}| > 0$. From the classification of section 2.15, we can make the following observations:

1. When there is a unique steady state for all Da (Figure 2.37a), then it can be either a node, focus, or center, but *never* a saddle point.
2. When multiple steady states may exist (Figure 2.37b), then
   a. For $\text{Da}_2 < \text{Da} < \text{Da}_1$, the middle steady state is a saddle point, while the low- and high-conversion steady states are either node, focus, or center.
   b. For $\text{Da} = \text{Da}_1$ or $\text{Da} = \text{Da}_2$, $|\mathbf{A}| = 0$. The two eigenvalues of $\mathbf{A}$ are 0 and tr $\mathbf{A}$.
   c. For $\text{Da} < \text{Da}_2$ or $\text{Da} > \text{Da}_1$, the unique steady state is either a node, focus, or center.

Since we know the variation of $|\mathbf{A}|$ along $g(a_2)$, we now need to examine how tr $\mathbf{A}$ behaves along the same curve. For this, let us examine the roots of tr $\mathbf{A} = 0$.

Roots of tr $A = 0$. From eq. (2.21.14a),

$$\text{tr } \mathbf{A} = -\frac{B}{(1 - a_2)} [(a_2 - s_1)(a_2 - s_2)]$$

where

$$s_1, s_2 = \frac{(B + 1 + \delta) \pm \sqrt{(B + 1 + \delta)^2 - 4B(2 + \delta)}}{2B}, \qquad s_1 < s_2 \quad (2.21.15)$$

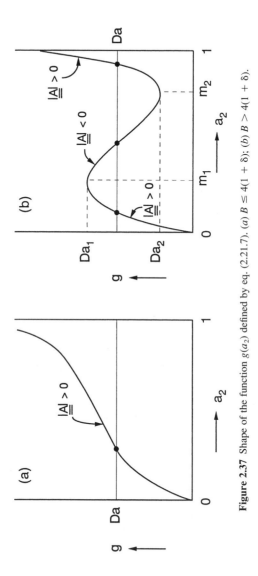

Figure 2.37 Shape of the function $g(a_2)$ defined by eq. (2.21.7). (a) $B \leq 4(1 + \delta)$; (b) $B > 4(1 + \delta)$.

170

are roots of tr $\mathbf{A} = 0$. If $(B + 1 + \delta)^2 - 4B(2 + \delta) < 0$, then s_1 and s_2 are a pair of complex conjugates and tr $\mathbf{A} < 0$. Alternatively, s_1 and s_2 are real if

$$(B + 1 + \delta)^2 - 4B(2 + \delta) > 0$$

which rearranges as

$$(B - B_+)(B - B_-) > 0$$

where

$$B_\pm = (3 + \delta) \pm 2\sqrt{(2 + \delta)} \qquad (2.21.16)$$

Note that $0 < B_- < B_+$, so that s_1 and s_2 are real if either $0 < B < B_-$ or $B > B_+$. However, for $0 < B < B_-$, we have $0 < 1 < s_1 < s_2$, so s_1 and s_2 are outside the range of interest. Thus $0 < s_1 < s_2 < 1$ *only* for

$$B > B_+ = (3 + \delta) + 2\sqrt{2 + \delta} = f_2(\delta) \qquad (2.21.17)$$

Finally, let us denote

$$\mathrm{Da}_3 = g(s_1) \quad \text{and} \quad \mathrm{Da}_4 = g(s_2) \qquad (2.21.18)$$

When s_1 and s_2 are real, then tr $\mathbf{A} > 0$ for a_2 in the range $s_1 < a_2 < s_2$, while tr $\mathbf{A} < 0$ for $0 < a_2 < s_1$ or $s_2 < a_2 < 1$.

Steady-State Response Diagrams. The object now is to consider *simultaneously* the various steady-state multiplicity and stability possibilities. In doing this, Poore (1973) exhaustively classified six possible regions of the (B, δ) plane (Figure 2.38), in each of which the reactor exhibits a different type of response, with certain specific types of temperature versus conversion phase plane behavior (Figure 2.39).

The classification of the regions is made by plotting three curves on the (B, δ) plane, as shown in Figure 2.38. The functions

$$B = 4(1 + \delta) = f_1(\delta) \qquad (2.21.10)$$

and

$$B = (3 + \delta) + 2\sqrt{2 + \delta} = f_2(\delta) \qquad (2.21.17)$$

were developed earlier. The third function

$$B = (1 + \delta)^3/\delta = f_3(\delta) \qquad (2.21.19)$$

arises when establishing the relative locations of m_i where $|\mathbf{A}| = 0$ and of s_i where tr $\mathbf{A} = 0$.

The six regions are defined as follows ($\delta \geq 0$ and $B \geq 0$):

I. $\quad 0 \leq B \leq \min[f_1(\delta), f_2(\delta)]$

II. $\quad$ IIa. $\quad f_1(\delta) \leq B \leq f_2(\delta), 0 \leq \delta \leq \dfrac{7}{9}$

$\qquad\quad$ IIb. $\quad f_2(\delta) \leq B \leq f_3(\delta), 0 \leq \delta \leq \dfrac{\sqrt{5} - 1}{2}$

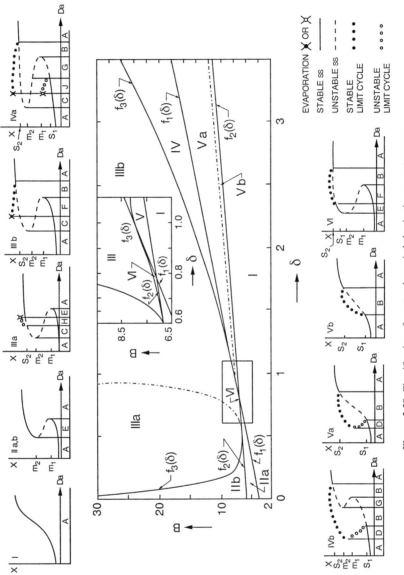

Figure 2.38 Classification of reactor dynamic behavior in parameter space.

172

III. $B \geq f_3(\delta)$

IV. $f_1(\delta) \leq B \leq f_3(\delta), \delta \geq 1$

V. $f_2(\delta) \leq B \leq f_1(\delta), \delta \geq \dfrac{7}{9}$

VI. $f_2(\delta) \leq B \leq f_3(\delta), \dfrac{\sqrt{5}-1}{2} \leq \delta \leq \dfrac{7}{9}$

        $f_1(\delta) \leq B \leq f_3(\delta), \dfrac{7}{9} \leq \delta \leq 1$

In each region the nature and relative locations of m_i and s_i can be identified precisely as shown below. It is worth noting that these conclusions are reached after very careful (essentially heroic!) analysis and require tedious manipulations.

I. m_1, m_2 complex and s_1, s_2 either complex or > 1.

II.   IIa. $0 < m_1 < m_2 < 1$; s_1, s_2 complex
      IIb. $0 < m_1 < s_1 < s_2 < m_2 < 1$

III. $0 < m_1 < s_1 < m_2 < s_2 < 1$

IV. $0 < s_1 < m_1 < m_2 < s_2 < 1$

V. m_1, m_2 complex and $0 < s_1 < s_2 < 1$

VI. $0 < m_1 < m_2 < s_1 < s_2 < 1$.

Based on the relative locations of m_i and s_i, the reactor steady-state response in each region can be identified; this is shown in the conversion (a_2) versus Damköhler number (Da) response diagrams surrounding the central (B, δ) plane in Figure 2.38. The stability character of each point on these diagrams is also established and is indicated by either ——— or --- portions, implying stable or unstable critical points, respectively.

Hopf Bifurcation and Periodic Solutions.   We are now in a position to apply the Hopf bifurcation theorem of section 2.20. At all the s_i points identified in the response diagrams

      Region III at s_2
      Region IV at s_1 and s_2
      Region V at s_1 and s_2, and
      Region VI at s_1 and s_2

(note that only those s_i are considered which are *not* in the range $m_1 < s_i < m_2$), we have tr $\mathbf{A} = 0$ and $|\mathbf{A}| > 0$. Hence at these points, the Jacobian matrix has a pair of purely imaginary eigenvalues (center), and these are the only eigenvalues of $\mathbf{A}$ since the system is two-dimensional.

The only condition in the theorem remaining to be verified is that given by eq.

(2.20.6). For a two-dimensional system, in the case of complex eigenvalues we have

$$\text{Re}(\lambda) = \tfrac{1}{2}\text{tr } \mathbf{A}$$

from eq. (2.14.7). Hence this condition becomes

$$\left. \frac{d(\text{tr } \mathbf{A})}{d \text{ Da}} \right|_{\text{Da}_0} = \text{tr } \mathbf{C}_0 \neq 0 \qquad (2.21.20)$$

where

$$\mathbf{C}_0 = \left. \frac{d\mathbf{A}}{d \text{ Da}} \right|_{\text{Da}_0}$$

and Da_0 is the critical value of Da corresponding to either s_1 or s_2.

Now, from eq. (2.21.13) we obtain

$$\mathbf{A} = \begin{bmatrix} -1 - \delta + Ba_2 & -\dfrac{Ba_2}{1 - a_2} \\[2mm] a_2 & -\dfrac{1}{1 - a_2} \end{bmatrix} \qquad (2.21.21)$$

so that

$$\mathbf{C}_0 = \left. \frac{d\mathbf{A}}{d \text{ Da}} \right|_{\text{Da}_0} = \begin{bmatrix} B & -\dfrac{B}{(1 - a_2^0)^2} \\[2mm] 1 & \dfrac{1}{(1 - a_2^0)^2} \end{bmatrix} \left(\frac{da_2}{d \text{ Da}} \right)_{\text{Da}_0} \qquad (2.21.22)$$

where a_2^0 is either s_1 or s_2. From eq. (2.21.5b) we have

$$a_2 = \text{Da}(1 - a_2)e^{a_1}$$

Differentiating this with respect to Da and using eq. (2.21.6) to relate a_1 and a_2 lead to

$$\frac{da_2}{d \text{ Da}} = \frac{(1 + \delta)a_2}{\text{Da}\left[\dfrac{(1 + \delta) - Ba_2(1 - a_2)}{1 - a_2} \right]}$$

Evaluation at $\text{Da} = \text{Da}_0$ and using eq. (2.21.14b) yields

$$\left(\frac{da_2}{d \text{ Da}} \right)_{\text{Da}_0} = \frac{(1 + \delta)a_2^0}{\text{Da}_0|\mathbf{A}|}$$

Hence from eq. (2.21.22) we have

$$\text{tr } \mathbf{C}_0 = \left[B - \frac{1}{(1 - a_2^0)^2} \right] \frac{(1 + \delta)a_2^0}{\text{Da}^0|\mathbf{A}|} \qquad (2.21.23)$$

The sign of tr $\mathbf{C}_0$ is determined by

$$B(1 - a_2^0)^2 - 1, \qquad a_2^0 = s_1 \text{ or } s_2$$

since $|\mathbf{A}| > 0$ at s_1 and s_2. Using eq. (2.21.15) for s_1 and s_2, it can be shown that

$$\text{tr } \mathbf{C}_0 \begin{cases} > 0 & \text{for } a_2^0 = s_1 \\ < 0 & \text{for } a_2^0 = s_2 \end{cases} \qquad (2.21.24)$$

Hence tr $\mathbf{C}_0 \neq 0$; following the theorem, bifurcation to periodic solutions does indeed occur at s_1 and s_2.

As noted at the end of the theorem is section 2.20, the *direction of bifurcation* at Da_0 (i.e., whether the periodic solution originates with increasing or decreasing values of Da) is established by the algebraic sign of a function. This function is obtained after a substantial analysis (for details see Poore, 1973) to yield

$$\delta^1 = \frac{Ba^2b^2}{8\omega_0^4\text{tr } \mathbf{C}_0} [\omega_0^2(b-1) + (2b - Ba) - (2b - Ba)^2] \qquad (2.21.25)$$

where

$$a = s_i, \, i = 1 \text{ or } 2; \qquad b = Bs_i - 1 - \delta; \qquad \omega_0^2 = b(Ba^2 - b)$$

Then, bifurcation of periodic solutions at s_i occurs

for $Da > Da_0$ if $\delta^1 > 0$ and
for $Da < Da_0$ if $\delta^1 < 0$

This establishes the directions of bifurcation in the response diagrams of Figure 2.38. The stability of the bifurcating periodic solutions follows the principle outlined at the end of the theorem in section 2.20. The filled and unfilled circles in the response diagrams indicate stable and unstable periodic solutions, respectively, and their distance to the steady-state curve represents the maximum amplitude of the oscillation.

As noted earlier, Hopf bifurcation theory provides only local information about the emerging periodic solutions—that is, in the immediate vicinity of the Hopf points at conversions s_1 and s_2. The response diagrams in Figure 2.38 have been taken from Uppal et al. (1974), who extended the Poore (1973) results to the case of finite activation energy (γ) and completely defined the CSTR dynamics by following the branches of periodic solutions up to their completion using numerical integration. The variety of dynamics is summarized in the nine phase portraits of Figure 2.39, where the letters A to J identify the portrait corresponding to the indicated Da regions of the response diagrams.

Several interesting dynamic features of the CSTR become evident from this analysis. For example, for a set of parameters lying in region IIIb of Figure 2.38, if we decrease the Da from large values, then the system goes from a stable high-conversion steady state to a stable limit cycle (transition A → B) whose amplitude increases up to a specific Da value. There, the limit cycle evaporates and the system settles down to a stable low-conversion steady state (transition F → C). Another interesting feature is the occurrence of nested limit cycles (portraits D or J), where an unstable limit cycle surrounds a stable steady state and is itself enveloped by a large-amplitude stable limit cycle. In this case, all ICs within the unstable limit cycle lead to the steady state, while those outside it lead to the outer stable limit cycle.

CASE			A	B	C	D	E	F	G	H	J
NODE or FOCUS	STABLE		1	0	1	1	2	1	0	2	1
	UNSTABLE		0	1	1	0	0	1	2	0	1
SADDLE POINT			0	0	1	0	1	1	1	1	1
LIMIT CYCLES	STABLE		0	1	0	1	0	1	1	0	1
	UNSTABLE		0	0	0	1	0	0	0	1	1
TOTAL INVARIANTS			1	2	3	3	3	4	4	4	5

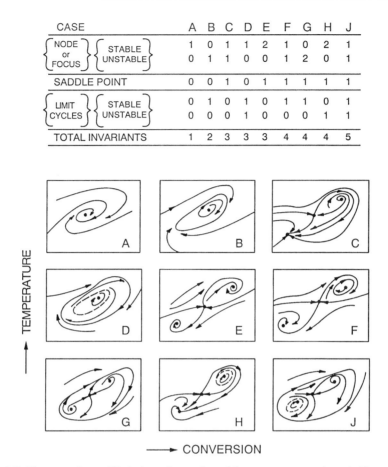

TEMPERATURE

CONVERSION

Figure 2.39 Phase portraits possible in the various regions of the parameter space shown in Figure 2.38.

2.21.2   The Belousov–Zhabotinskii Reaction

The occurrence of isothermal oscillations in chemically reacting systems is a fascinating subject which has attracted the attention of many scientists in recent years. Although there were isolated previous reports of oscillatory reactions, the first systematic work in this area appears to be that of the Russian scientists Belousov and Zhabotinskii, starting in the 1950s. This reaction, popularly known as the *BZ reaction*, has many variations, the most common involving oxidation of malonic acid by bromate in acid medium catalyzed by the redox couple Ce^{3+}/Ce^{4+}. In this case the color of the solution changes periodically from yellow to colorless and back to yellow.

The chemistry of this reaction is complex and many detailed kinetic schemes have been proposed in the literature. However, the sequence of steps

$$A + Y + 2H \rightarrow X + P \qquad (2.21.26)$$

$$X + Y + H \rightarrow 2P \qquad (2.21.27)$$

$$A + X + H \rightarrow 2X + 2Z \tag{2.21.28}$$

$$2X \qquad\qquad \rightarrow A + P + H \tag{2.21.29}$$

$$B + Z \qquad \rightarrow fY \tag{2.21.30}$$

known as the *Oregonator*, after the University of Oregon researchers Field and Noyes, describes the basic features of the system (cf. Tyson, 1985). Here, $A = BrO_3^-$, $H = H^+$, and $B =$ malonic acid are the reactants, P is a product, $X = HBrO_2$, $Y = Br^-$, and $Z = Ce^{4+}$ are the intermediate species which exhibit the oscillating behavior, and f is a stochiometric coefficient. The rate of each reaction step is determined by the law of mass action.

It is worth noting that reaction (2.21.28) is autocatalytic, since a single molecule of X produces two molecules of the same species. This provides the internal feedback for the system which is the source of its rich dynamic behavior.

Based on the above kinetic scheme, the equations describing the Oregonator model are

$$\frac{dX}{dt} = k_1 H^2 AY - k_2 HXY + k_3 HAX - 2k_4 X^2 \tag{2.21.31}$$

$$\frac{dY}{dt} = -k_1 H^2 AY - k_2 HXY + fk_5 BZ \tag{2.21.32}$$

$$\frac{dZ}{dt} = 2k_3 HAX - k_5 BZ \tag{2.21.33}$$

where the reactant concentrations A, B, and H are assumed constant in time. Since an inlet stream is not included, strictly speaking these equations are a model neither of a CSTR (open system) nor of a batch reactor (closed system). However, this model can be used to study the oscillations of the BZ reaction in a batch reactor because they occur on a time scale which is much faster than that of consumption of the reactants. In fact, by monitoring the behavior of a batch reactor up to reaction completion, we obtain a qualitative picture of the bifurcation behavior of the system with the reactant concentrations as the bifurcation parameters. Because of the argument given above, at each instant of time we observe the approximate stationary behavior of the system corresponding to the specific values of the reactant concentrations prevailing at that time.

In order to make the equations dimensionless, we introduce the following quantities:

$$x = \frac{X}{X_0}, \qquad y = \frac{Y}{Y_0}, \qquad z = \frac{Z}{Z_0}, \qquad \tau = \frac{t}{t_0}$$

$$a = \frac{A}{A_0}, \qquad b = \frac{B}{B_0}, \qquad \varepsilon = \frac{k_5 B_0}{k_3 HA_0} \tag{2.21.34}$$

$$\delta = \frac{2k_4 k_5 B_0}{k_2 k_3 H^2 A_0}, \qquad q = \frac{2k_1 k_4}{k_2 k_3}, \qquad h = 2f$$

where the subscript zero indicates reference conditions, defined as

$$X_0 = \frac{k_3 H A_0}{2k_4}, \qquad Y_0 = \frac{k_3 A_0}{k_2}$$

$$Z_0 = \frac{(k_3 H A_0)^2}{k_4 k_5 B_0}, \qquad t_0 = \frac{1}{k_5 B_0} \tag{2.21.35}$$

Equations (2.21.31) to (2.21.33) then lead to the classical form of the Oregonator model:

$$\varepsilon \frac{dx}{d\tau} = qay - xy + ax - x^2 \tag{2.21.36}$$

$$\delta \frac{dy}{d\tau} = -qay - xy + hbz \tag{2.21.37}$$

$$\frac{dz}{d\tau} = ax - bz \tag{2.21.38}$$

A further model reduction is achieved by considering typical values of the kinetic parameters, as reported in Table 2.2. From these, it is clear that the time constant of the variable y is much smaller than that of the other two—that is, $\delta \ll \varepsilon < 1$. Thus we can assume pseudo-steady state for y, and by setting the rhs of eq. (2.21.37) equal to zero we obtain

$$y_0 = \frac{hbz}{(x + qa)} \tag{2.21.39}$$

Substituting in eqs. (2.21.36) and (2.21.38) yields the reduced Oregonator model:

$$\varepsilon \frac{dx}{d\tau} = ax - x^2 - hbz \frac{(x - qa)}{(x + qa)} \tag{2.21.40}$$

$$\frac{dz}{d\tau} = ax - bz \tag{2.21.41}$$

The equilibrium point (x_0, z_0) is obtained by solving the cubic equation

$$ax_0 - x_0^2 - hax_0 \frac{(x_0 - qa)}{(x_0 + qa)} = 0 \tag{2.21.42}$$

TABLE 2.2
Kinetic Parameter Values for the Oregonator Model (Tyson, 1985)

$A_0 = 1$ M	$B_0 = 1$ M	$Z_0 = 3.2 \times 10^{-2}$ M	$H = 0.8$ M
$X_0 = 2 \times 10^{-3}$ M	$Y_0 = 10^{-5}$ M	$t_0 = 1$ sec	$q = 8 \times 10^{-4}$
	$\varepsilon = 0.12$	$\delta = 6 \times 10^{-4}$	

which yields the unique positive value

$$x_0 = a\phi \quad \text{and} \quad z = \frac{a^2\phi}{b} \tag{2.21.43}$$

where

$$\phi = \tfrac{1}{2}\{(1 - h - q) + [(1 - h - q)^2 + 4q(1 + h)]^{1/2}\}$$

is a positive constant which does not depend upon the concentrations a and b.

For the Jacobian matrix of the linearized system

$$\mathbf{J} = \begin{bmatrix} \dfrac{a}{\varepsilon}\left[1 - 2\phi - \dfrac{2qh\phi}{(\phi + q)^2}\right] & -\dfrac{hb(\phi - q)}{\varepsilon(\phi + q)} \\ a & -b \end{bmatrix} \tag{2.21.44}$$

we have

$$\operatorname{tr}\mathbf{J} = \frac{a}{\varepsilon}\left[1 - 2\phi - \frac{2qh\phi}{(\phi + q)^2}\right] - b \tag{2.21.45}$$

and

$$|\mathbf{J}| = \frac{ab}{\varepsilon}\left[2\phi - 1 + \frac{2qh\phi}{(\phi + q)^2} + \frac{h(\phi - q)}{(\phi + q)}\right] \tag{2.21.46}$$

Using eq. (2.21.42), the rhs of eq. (2.21.46) can be rearranged to show that

$$|\mathbf{J}| = \phi + \frac{2qh\phi}{(\phi + q)^2} > 0 \tag{2.21.47}$$

Thus the equilibrium point is either a focus or a node, but not a saddle point. It is asymptotically stable if $\operatorname{tr}\mathbf{J} < 0$ and is unstable for $\operatorname{tr}\mathbf{J} > 0$.

Let us now investigate the onset of oscillatory behavior by using the Hopf bifurcation theorem, taking the concentration a as the bifurcation parameter. For a two-dimensional system, the second condition of the theorem requires $\operatorname{tr}\mathbf{J} = 0$; that is, the equilibrium point is a center and $\mathbf{J}$ has a pair of purely imaginary eigenvalues. This can be cast in the form

$$a = b\varepsilon\left[1 - 2\phi - \frac{2qh\phi}{(\phi + q)^2}\right]^{-1} \tag{2.21.48}$$

For a given b, this equation provides the bifurcation value of the parameter a, denoted by a_0.

The only condition in the theorem which remains to be verified is that given by eq. (2.20.6). For a two-dimensional system, in the case of complex eigenvalues we have

$$\operatorname{Re}(\lambda) = \tfrac{1}{2}\operatorname{tr}\mathbf{J} \tag{2.21.49}$$

from eq. (2.14.7). Hence this condition becomes

$$\left.\frac{d(\text{tr } \mathbf{J})}{da}\right|_{a_0} \neq 0 \tag{2.21.50}$$

Differentiating eq. (2.21.45) gives

$$\left.\frac{d(\text{tr } \mathbf{J})}{da}\right|_{a_0} = \frac{1}{\varepsilon}\left[1 - 2\phi - \frac{2qh\phi}{(\phi + q)^2}\right] = \frac{b}{a_0} \neq 0 \tag{2.21.51}$$

Thus following the theorem, bifurcation to periodic solutions does indeed occur at a_0, which is shown in Figure 2.40a as a function of the stoichiometric coefficient h for the

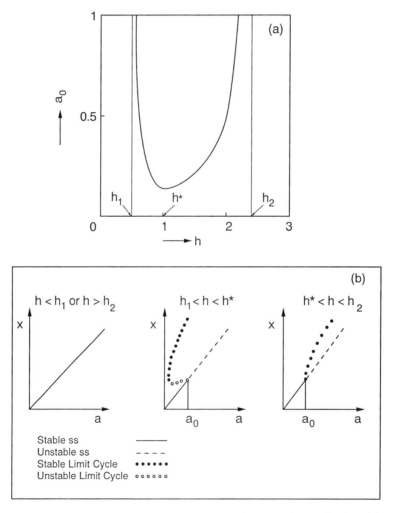

Figure 2.40 (a) The value of the concentration a at the Hopf bifurcation point as a function of the stoichiometric coefficient h. Parameter values are as in Table 2.2 and $b = 1$. (b) Steady-state response diagrams representing the concentration of the intermediate species, x, as a function of the reactant concentration, a.

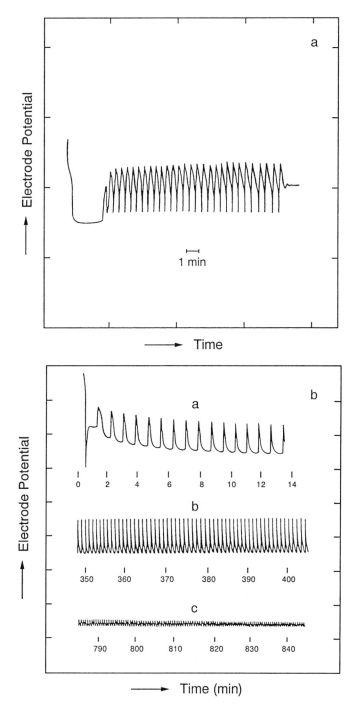

Figure 2.41 Oscillations of the potential of the bromide-selective electrode during the BZ reaction in a batch reactor (Ruoff and Noyes, 1989). (*a*) Subcritical Hopf bifurcation. (*b*) Supercritical Hopf bifurcation.

181

parameter values of Table 2.2. It is apparent from eq. (2.21.45) that above the curve we have tr $\mathbf{J} > 0$ (unique unstable equilibrium point), while below it, as well as for $h < h_1$ and $h > h_2$, we have tr $\mathbf{J} < 0$ (unique stable equilibrium point).

The determination of the direction of bifurcation at a_0 is based on the algebraic sign of a rather complicated function. Similar to the previous example involving a nonadiabatic CSTR, we report here only the conclusion of this analysis (for details see Mazzotti et al., 1995): For values of the stoichiometric coefficient in the interval $h_1 < h < h^*$ the Hopf bifurcation point is subcritical, while for $h^* < h < h_2$ it is supercritical.

Thus for any given value of h, we can fully characterize the behavior of the system using the steady-state response diagrams shown in Figure 2.40b. They illustrate the intermediate species concentration (x) as a function of reactant concentration (a) using the same notation adopted earlier in Figure 2.38.

As noted above, owing to the widely different time scales of oscillations and reactant depletion, the evolution of a batch reaction in time is a qualitative image of the bifurcation pattern when the concentration a is taken as the bifurcation parameter. This may be represented by a vertical line in Figure 2.40a, where the concentration a decreases while h remains constant. The experimental data shown in Figure 2.41 represent the potential measured by a bromide-selective electrode as a function of time. In agreement with the results in Figure 2.40a, these experiments confirm the existence of oscillatory behavior which may expire through either subcritical ($h_1 < h < h^*$) or supercritical ($h^* < h < h_2$) Hopf bifurcation.

2.22   LINEAR ANALYSIS FOR SYSTEMS OF DIMENSION LARGER THAN TWO

As seen in section 2.15, a comprehensive linear stability analysis can be carried out for two-dimensional systems. For systems of larger dimension, the analysis becomes difficult and it is not easy to visualize the geometric behavior. In fact, as we shall see in the next section, nonlinear systems of dimension larger than two can give rise to rather exotic dynamics.

As far as the linear stability analysis is concerned, recall from the fundamental stability theorem (section 2.13) that asymptotic stability of a steady state is ensured if all eigenvalues of the Jacobian matrix are negative (if real) or have negative real parts (if complex). It is then useful to have a convenient test or criterion which gives information about the nature of the eigenvalues. Such a test was developed by Routh and Hurwitz (see Gantmacher, 1989, chapter XV, section 6).

The Routh–Hurwitz Criterion for Eigenvalues.   Consider the real polynomial

$$P_n(\lambda) = a_0\lambda^n + a_1\lambda^{n-1} + \cdots + a_{n-1}\lambda + a_n, \qquad a_0 > 0 \qquad (2.22.1)$$

and the following n determinants

$$\Delta_1 = a_1$$

$$\Delta_2 = \begin{vmatrix} a_1 & a_3 \\ a_0 & a_2 \end{vmatrix}$$

$$\Delta_3 = \begin{vmatrix} a_1 & a_3 & a_5 \\ a_0 & a_2 & a_4 \\ 0 & a_1 & a_3 \end{vmatrix}$$

(2.22.2)

$$\Delta_4 = \begin{vmatrix} a_1 & a_3 & a_5 & a_7 \\ a_0 & a_2 & a_4 & a_6 \\ 0 & a_1 & a_3 & a_5 \\ 0 & a_0 & a_2 & a_4 \end{vmatrix}$$

$$\vdots$$

$$\Delta_n = \begin{vmatrix} a_1 & a_3 & a_5 & a_7 & \cdots & 0 \\ a_0 & a_2 & a_4 & a_6 & \cdots & 0 \\ 0 & a_1 & a_3 & a_5 & \cdots & 0 \\ 0 & a_0 & a_2 & a_4 & \cdots & 0 \\ \vdots & & & & & \vdots \\ 0 & \cdot & \cdot & \cdot & & a_n \end{vmatrix}$$

where note that the jth order determinant Δ_j contains elements $a_1, a_2, \ldots, a_j$ on the main diagonal.

The *necessary and sufficient* condition that all roots of P_n have negative real parts is that all the determinants $\Delta_1, \Delta_2, \ldots, \Delta_n$ be positive.

For example, consider the second-degree polynomial

$$P_2(\lambda) = a_0\lambda^2 + a_1\lambda + a_2, \qquad a_0 > 0 \tag{2.22.3}$$

Then

$$\Delta_1 = a_1$$

and

$$\Delta_2 = \begin{vmatrix} a_1 & 0 \\ a_0 & a_2 \end{vmatrix} = a_1 a_2$$

The Routh–Hurwitz conditions are

$$a_1 > 0 \quad \text{and} \quad a_2 > 0 \tag{2.22.4}$$

Hence for the roots of the quadratic (2.22.3) to have negative real parts, all the coefficients a_j must have the same sign.

Let us confirm this to be the case by considering the matrix

$$\mathbf{A} = \begin{bmatrix} a_{11} & a_{12} \\ a_{21} & a_{22} \end{bmatrix}$$

then the eigenvalues satisfy the equation

$$\lambda^2 - \text{tr } \mathbf{A}\lambda + |\mathbf{A}| = 0$$

We know from eq. (2.14.11) that both roots have negative real parts if and only if

$$\text{tr } \mathbf{A} < 0 \quad \text{and} \quad |\mathbf{A}| > 0$$

which matches with the condition (2.22.4).

Note that it is necessary and sufficient *only* for second-degree polynomials that all coefficients are of the same sign in order to ensure that all roots have negative real parts. For example, if we consider the cubic polynomial

$$P_3(\lambda) = a_0\lambda^3 + a_1\lambda^2 + a_2\lambda + a_3, \qquad a_0 > 0 \tag{2.22.5}$$

then the Routh–Hurwitz conditions are

$$\Delta_1 = a_1 > 0, \qquad \Delta_2 = \begin{vmatrix} a_1 & a_3 \\ a_0 & a_2 \end{vmatrix} > 0 \quad \text{and} \quad \Delta_3 = \begin{vmatrix} a_1 & a_3 & 0 \\ a_0 & a_2 & 0 \\ 0 & a_1 & a_3 \end{vmatrix} > 0$$

which require that

$$a_1 > 0$$

$$\Delta_2 = a_1 a_2 - a_0 a_3 > 0$$

and

$$\Delta_3 = a_3 \Delta_2 > 0$$

Hence we definitely need that all coefficients a_j, $j = 0 \rightarrow 3$, are positive, but also in addition that $\Delta_2 > 0$.

Finally, it may be noted that there is a redundancy in the Routh–Hurwitz conditions and that fewer conditions are in fact required to test for the character of the roots. These were developed by Linard and Chipart and are discussed in the book by Gantmacher (1989, chapter XV, section 13).

2.23 COMPLEX DYNAMICS

In our study so far, we have investigated only those cases where the transients lead either to a stable steady state or to a periodic solution. This is the extent of complexity possible in a system governed by two nonlinear ODEs. However, in some cases of practical import the dynamic behavior is complex and appears to be essentially random. This topic is of great current interest in applied mathematics, and we introduce it here by the use of two examples.

2.23.1 Nonlinear Difference Equations

A first-order nonlinear difference equation

$$x_{n+1} = f(x_n ; \mu) \tag{2.23.1}$$

where μ is a constant, represents a one-dimensional *map* (cf. May, 1976). It can be viewed as a finite difference analog of a nonlinear ODE. In biological context, it may represent how the population of a species in the nth generation, x_n, changes when going to the next generation, x_{n+1}. A *fixed point* of the equation, x^*, satisfies

$$x^* = f(x^*; \mu) \tag{2.23.2}$$

and arises from the intersection(s) of the function $f(x; \mu)$ with the 45° line, $x = x$. An example of a map, with successive values of x_n converging to the fixed point x^*, is shown in Figure 2.42.

If eq. (2.23.1) were linear, then we would have

$$x_{n+1} = \mu x_n \tag{2.23.3}$$

with the unique fixed point $x^* = 0$. It would have the solution

$$
\begin{aligned}
x_{n+1} &= \mu x_n \\
&= \mu(\mu x_{n-1}) = \mu^2 x_{n-1} \\
&\ \vdots \\
&= \mu^{n+1} x_0
\end{aligned}
\tag{2.23.4}
$$

which is *asymptotically stable* if $|\mu| < 1$; that is,

$$\lim_{n \to \infty} x_n = 0 = x^*$$

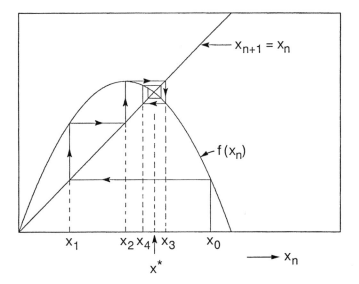

Figure 2.42 Iterates of $x_{n+1} = f(x_n)$, showing convergence to the fixed point x^*.

If $|\mu| > 1$, then the fixed-point is *unstable*. Finally, for $|\mu| = 1$, we have *marginal stability*, since

$$x_n = x_0 \qquad \text{if } \mu = 1$$

and

$$x_n = (-1)^n x_0 \qquad \text{if } \mu = -1$$

In the latter case, x_n oscillates between $\pm x_0$.

for the nonlinear difference eq. (2.23.1), the local stability of x^* is evaluated by linearizing $f(x)$ around x^* and then applying the above theory for the linear case. Thus x^* is locally stable if

$$\left. \left| \frac{df}{dx} \right| \right|_{x^*} < 1 \tag{2.23.5}$$

and it is unstable if

$$\left. \left| \frac{df}{dx} \right| \right|_{x^*} > 1 \tag{2.23.6}$$

A classic example of eq. (2.23.1) is the *logistic difference equation* (cf. Jackson, 1989, section 4.2; Doherty and Ottino, 1988):

$$x_{n+1} = \mu x_n (1 - x_n) = f(x_n; \mu), \qquad 0 \le x_n \le 1 \tag{2.23.7}$$

which is a finite difference analog of eq. (2.8.16). This transformation has the x and μ dependence as shown in Figure 2.43. Specifically,

$$f(0; \mu) = 0 = f(1; \mu) \tag{2.23.8}$$

and f exhibits a maximum, f_{max} at $x = 0.5$. For $0 < \mu \le 4$ we have $0 < f_{max} \le 1$, so that f maps the interval $[0, 1]$ into itself.

The fixed points of eq. (2.23.7), x^*, are determined by solving

$$x^* = \mu x^* (1 - x^*) \tag{2.23.9}$$

and yield

$$x^* = 0 \quad \text{and} \quad x^* = 1 - \frac{1}{\mu} \tag{2.23.10}$$

The nontrivial fixed point is in the $[0, 1]$ range of interest only for $\mu > 1$. Since

$$\frac{df}{dx} = \mu(1 - 2x)$$

at the fixed point $x^* = 0$, we have

$$\left. \frac{df}{dx} \right|_0 = \mu$$

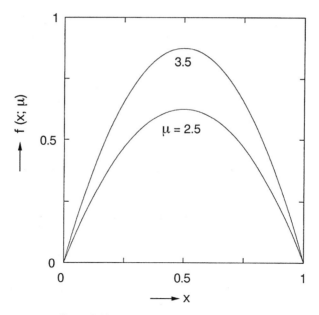

Figure 2.43 The function $f(x; \mu) = \mu x(1 - x)$.

and applying the stability criterion (2.23.5) shows that $x^* = 0$ is stable for $0 < \mu < 1$ and unstable for $\mu > 1$. For the nontrivial fixed point,

$$\left.\frac{df}{dx}\right|_{1-1/\mu} = 2 - \mu$$

and application of criterion (2.23.5) yields that $x^* = 1 - (1/\mu)$ is stable for

$$|2 - \mu| < 1 \quad \text{or} \quad 1 < \mu < 3 \tag{2.23.11}$$

and unstable for $\mu > 3$.

Before proceeding further, we need to introduce the concept of a *sequence of mappings*. For the map

$$x_{n+1} = f(x_n)$$

we have

$$x_1 = f(x_0)$$

$$x_2 = f(x_1) = f(f(x_0)) = f^2(x_0)$$

and similarly,

$$x_n = f(x_{n-1}) = f^2(x_{n-2}) = \cdots = f^n(x_0) \tag{2.23.12}$$

where f^j simply means j applications of the map f. Also, similar to fixed points of the map f, we also have fixed points of the sequence map f^j. It is evident that fixed points of f will also be fixed points of the sequence map. However, there can also be additional fixed points for the latter, satisfying

$$x = f^j(x) \quad \text{and} \quad x \neq f^k(x) \qquad \text{for } k < j \qquad (2.23.13)$$

and are called *period-j points*. Clearly, a fixed point of f is simply a period-one point.

Let us now return to the example map (2.23.7) and consider values of $\mu > \mu_1 = 3$. Here, as noted earlier, both the fixed points of f—that is, 0 and $1 - (1/\mu)$—are unstable. As seen in Figure 2.44a for $\mu = 3.2$, the sequence of iterates settles down to alternating values of x_1^* and x_2^*, which are fixed points of f^2 but not of f (see Figure 2.44b). Hence we say that at $\mu = 3$, the period-one cycle has *bifurcated* to a period-two cycle. Note that f^2 has a total of four fixed points: two unstable ones which are also fixed points of f, and two additional ones, x_1^* and x_2^*.

Following eqs. (2.23.5) and (2.23.6), the period-two fixed-points are stable if

$$\left| \frac{df^2}{dx} \right|_{x_k^*} = < 1, \qquad k = 1 \text{ or } 2 \qquad (2.23.14)$$

However, it turns out that the stability character of these points is *identical*, since

$$\left. \frac{df^2}{dx} \right|_{x_1^*} = \left. \frac{df^2}{dx} \right|_{x_2^*} \qquad (2.23.15)$$

In order to show this, consider any point x_0. Then

$$x_1 = f(x_0), \qquad x_2 = f(x_1) = f^2(x_0)$$

so that

$$\left. \frac{df^2}{dx} \right|_{x_0} = \left. \frac{df}{dx} \right|_{x_1} \cdot \frac{dx_1}{dx_0} = \left. \frac{df}{dx} \right|_{x_1} \cdot \left. \frac{df}{dx} \right|_{x_0}$$

For a period-two cycle (see Figure 2.44)

$$x_0 = x_2^* \quad \text{and} \quad x_1 = x_1^*$$

and thus we have

$$\left. \frac{df^2}{dx} \right|_{x_2^*} = \left. \frac{df}{dx} \right|_{x_1^*} \cdot \left. \frac{df}{dx} \right|_{x_2^*} \qquad (2.23.16)$$

Repeating the above procedure, but starting at x_1^* rather than at x_0, leads to the same eq. (2.23.16) for $df^2/dx|_{x_1^*}$, hence eq. (2.23.15) follows.

If we examine Figure 2.44, we note that while f has one extreme point (a maximum), f^2 has three extreme points (two maxima and one minimum). At the critical value of $\mu = \mu_1 = 3$, the curve f^2 is tangent to the 45° line at the previously stable fixed point of f(i.e., $x = 1 - 1/\mu$). At $\mu = \mu_1 = 3$, this fixed point becomes unstable and gives rise to two stable period-two fixed-points x_1^* and x_2^*.

As μ increases further, the function f^2 develops larger maxima and a smaller minimum, and its slope at the fixed points increases. Finally, a value of $\mu = \mu_2 = 1 + 6^{1/2} \simeq 3.45$ is reached where eq. (2.23.14) no longer holds, and the period-two fixed

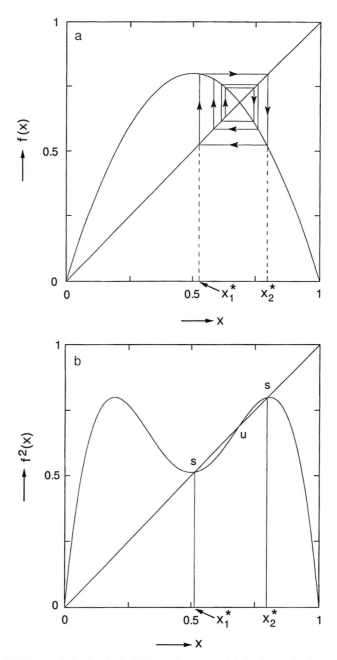

Figure 2.44 Period-two solution for the logistic equation at $\mu = 3.2$. (*a*) Iterates leading to a period-two cycle. (*b*) The function $f^2(x)$.

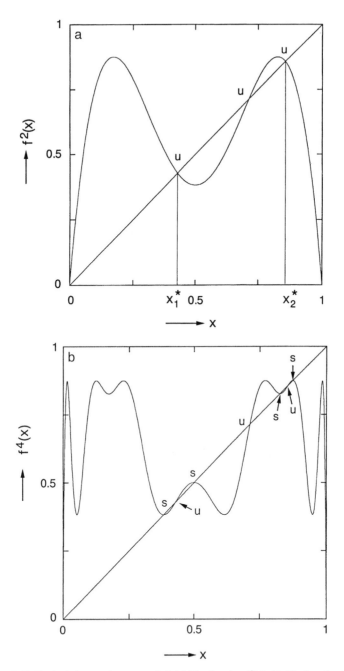

Figure 2.45 The logistic equation at $\mu = 3.5$. (*a*) The function $f^2(x)$. (*b*) The function $f^4(x)$.

points become unstable simultaneously. This is shown in Figure 2.45a where all three fixed points of f^2 are unstable. The two fixed points of f^2 which become unstable at μ_2 each give birth to two stable fixed points of f^4 (Figure 2.45b), which is now the fundamental function governing the process.

Similar to eq. (2.23.16), it is straightforward to show that

$$\left.\frac{df^n}{dx}\right|_{x_j^*} = \prod_{k=1}^{n} \left.\frac{df}{dx}\right|_{x_k^*}, \qquad j = 1, 2, \ldots, n \qquad (2.23.17)$$

where x_k^*, $k = 1 \to n$, are the period-n points of $f(x)$—that is, the n fixed points of $f^n(x)$. Thus the slope of f^n is the *same* at all its fixed points. Since stability of x_k^* follows if

$$\left|\frac{df^n}{dx}\right|_{x_k^*} < 1 \qquad (2.23.18)$$

the period-n points are either all stable or all unstable.

This period-doubling process continues as μ increases further, with each successive doubling giving rise to a stable period-2^n cycle in $f(x)$ of higher order. The bifurcation values of μ (i.e., μ_1, μ_2, μ_3, . . .) get closer, accumulating at $\mu_\infty = 3.5699 \ldots$, as shown in Table 2.3. The bifurcation diagram for this period-doubling cascade is shown in Figure 2.46.

It was observed by Feigenbaum that if we define

$$\delta_n = \frac{\mu_n - \mu_{n-1}}{\mu_{n+1} - \mu_n} \qquad (2.23.19)$$

then as n increases, δ_n rapidly approaches a constant value $\delta = 4.6692016. \ldots$ He also studied period-doubling for the map

$$x_{n+1} = \mu \sin(\pi x_n) \qquad (2.23.20)$$

and again found the same δ value, leading to the belief that it is a *universal* value for all functions with a single maximum. A very readable account of this discovery and its wide implications is given in Feigenbaum (1980).

As complex as what we have described so far may seem, it is still not the end of the story! As μ increases beyond μ_∞, additional complications occur (see detailed description in May, 1976). These include an infinite number of fixed points with different

TABLE 2.3
Bifurcation Values of μ_n for the Logistic Map

n	μ_n	n	μ_n
1	3	5	3.568 759
2	3.449 499	6	3.569 692
3	3.544 090	7	3.569 891
4	3.564 407	8	3.569 934
		∞	3.569 946

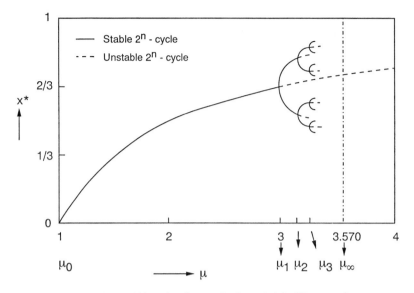

Figure 2.46 The bifurcation diagram for the period-doubling cascade.

periodicities, and they also include an infinite number of fixed points with different *aperiodic* (although bounded) solutions. At $\mu = 3.6786\ldots$ the first *odd-period* cycle appears, whose period is initially very long. As μ continues to increase, cycles with smaller odd periods occur, until at last at $\mu = 1 + 8^{1/2} = 3.8284\ldots$ a cycle of period 3 appears. Beyond this point, there are cycles with every integer period, as well as an uncountable number of asymptotically aperiodic trajectories—a situation aptly described as *chaotic*.

One feature of a chaotic solution is extreme *sensitivity* to initial conditions. This is illustrated in Figure 2.47, where two solutions of eq. (2.23.7) for $\mu = 3.9$, with $x_0 = 0.5$ and 0.502, are shown for various values of n. The two solutions are essentially the same until about $n = 15$, but subsequently, although they wander about in approximately the same set of values, their graphs are quite dissimilar. Finally, it may be noted that although dynamics such as that in Figure 2.47 appears to be the result of a *random* process, it is in fact the response of a very simple *deterministic* system. Studies such as this point out that even simple nonlinear systems can have very complicated dynamic behavior, which may well reflect irregularities experienced in nature.

2.23.2 The Lorenz Equations

In 1963, E. N. Lorenz published a landmark paper which has become a classic in the field of complex dynamics. In this paper he considered a set of three remarkably simple first-order ODEs and showed that when a certain parameter exceeds a critical value, the dynamic behavior is chaotic; that is, it exhibits a sensitive dependence on the initial conditions. The equations have their origin in hydrodynamic flow resulting when a liquid layer is confined between two flat plates, and the lower plate is maintained at a

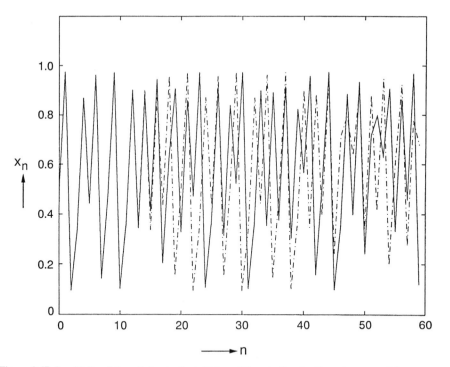

Figure 2.47 Sensitivity of the solution to the initial conditions in the chaotic regime, $\mu = 3.9$; ——$x_0 = 0.5$, —·—$x_0 = 0.502$.

temperature higher than the upper plate. The Lorenz equations are a simplified version of this physical problem (for details of the simplification see Lorenz, 1963) and are written as

$$\frac{dx}{dt} = -\sigma(x - y)$$

$$\frac{dy}{dt} = rx - y - xz \qquad (2.23.21)$$

$$\frac{dz}{dt} = xy - bz$$

where σ, r, and b are three real and positive parameters. In these equations, t represents dimensionless time, x is proportional to the intensity of the convective motion, y is proportional to the temperature difference between the ascending and descending currents, and z is proportional to the distortion of the vertical temperature profile from linearity. The parameter σ is the Prandtl number, b is a positive constant, and r is ratio of the Rayleigh number, Ra (for a given fluid and fixed geometry, it is a measure of the temperature difference between the plates), to its critical value, Ra_c, at which convection sets in. The definitions of all the variables and parameters may be found in the

Lorenz paper (1963). A great deal is known about the Lorenz equations and is available in a variety of sources, most notably in the book by Sparrow (1982). Doherty and Ottino (1988) provide a good summary of the results.

The equilibrium or critical points of the system are obtained by setting the rhs of eq. (2.23.21) equal to zero. It is easy to establish that they are three in number.

S1: $x_s = 0, \quad y_s = 0, \quad z_s = 0;$ valid for all r

S2: $x_s = \sqrt{b(r-1)}, \quad y_s = \sqrt{b(r-1)}, \quad z_s = r-1; \qquad r > 1$

S3: $x_s = -\sqrt{b(r-1)}, \quad y_s = -\sqrt{b(r-1)}, \quad z_s = r-1; \qquad r > 1$

$$(2.23.22)$$

The critical point S1 corresponds to no convection (i.e., pure thermal conduction through the liquid layer), while S2 and S3 correspond to convection and occur only for $r > 1$ as expected from the definition of r. The bifurcation diagram for eqs. (2.23.21) is shown in Figure 2.48.

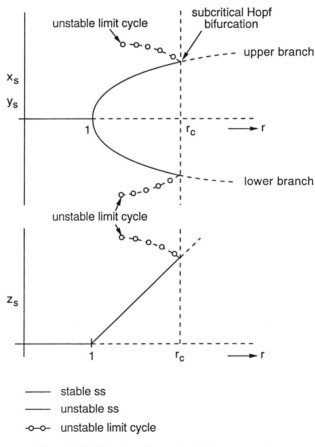

Figure 2.48 Bifurcation diagram for the Lorenz equations.

We now investigate linearized stability of the critical points. The Jacobian matrix is given by

$$J = \begin{bmatrix} -\sigma & \sigma & 0 \\ r - z & -1 & -x \\ y & x & -b \end{bmatrix}_s \qquad (2.23.23)$$

For the critical point S1, the eigenvalues of J are given by

$$\lambda_1 = -b < 0$$
$$\lambda_\pm = \tfrac{1}{2}[-(1 + \sigma) \pm \sqrt{(1 + \sigma)^2 + 4\sigma(r - 1)}] \qquad (2.23.24)$$

so that λ_1 and λ_- are *always* real and negative. The eigenvalue λ_+ is always real, but it is < 0 for $0 < r < 1$, $= 0$ for $r = 1$, and > 0 for $r > 1$. Hence the critical point S1 is asymptotically stable for $0 < r < 1$, loses stability at $r = 1$, and is unstable for $r > 1$. The onset of convection occurs at $r = 1$ (i.e., Ra $=$ Ra$_c$), where the system acquires the two additional critical points S2 and S3.

It turns out that the eigenvalue equation for *both* S2 and S3 ($r > 1$) is

$$\lambda^3 + (\sigma + b + 1)\lambda^2 + b(r + \sigma)\lambda + 2\sigma b(r - 1) = 0 \qquad (2.23.25)$$

To investigate stability character, we can apply the Routh–Hurwitz conditions. From eq. (2.22.5), all the eigenvalues have negative real parts if and only if

$$r[\sigma - (b + 1)] < \sigma(\sigma + b + 3) \qquad (2.23.26)$$

If

$$\sigma < (b + 1) \qquad (2.23.27)$$

then condition (2.23.26) is always satisfied, implying that S2 and S3 are asymptotically stable for all $r > 1$. However, if

$$\sigma > (b + 1) \qquad (2.23.28)$$

then from condition (2.23.26), S2 and S3 are asymptotically stable for

$$r < \frac{\sigma(\sigma + b + 3)}{\sigma - (b + 1)} = r_c \qquad (2.23.29)$$

and unstable for

$$r > r_c \qquad (2.23.30)$$

We will focus on the case $\sigma > (b + 1)$ from now on. It is clear from the above analysis that in this case *all* critical points are unstable for $r > r_c$.

It is straightforward to show that at $r = r_c$ the change of stability character for the critical points S2 and S3 occurs through the real part of a pair of complex conjugate eigenvalues changing sign (i.e., condition 2 of the Hopf bifurcation theorem, section 2.20). The argument is as follows. Since the coefficients of eq. (2.23.25) are *all* positive, Descartes rule of signs implies that there are *no* real and positive eigenvalues and that there is at least one real and negative eigenvalue. Thus when the stability character changes at $r = r_c$, it cannot occur by a real eigenvalue becoming positive.

In order to be certain that Hopf bifurcation does indeed occur at $r = r_c$, we need to verify condition c of the theorem in section 2.20. It is shown by Marsden and McCracken (1976, page 143) that

$$\left. \frac{d \, \text{Re } \lambda(r)}{dr} \right|_{r_c} = \frac{b[\sigma - (b + 1)]}{2[b(r_c + \sigma) + (\sigma + b + 1)^2]}$$

$$> 0$$

for $\sigma > b + 1$. Hence Hopf bifurcation is ensured at $r = r_c$. The direction of bifurcation is, as usual, more difficult to establish. However, it can be shown (Marsden and McCracken, 1976, pages 143–147) that it is subcritical; that is, unstable periodic solutions exist for $r < r_c$ (see Figure 2.48).

For $r > r_c$ we have no stable critical points and no stable periodic solutions (at least those of small amplitude originating from $r = r_c$). Lorenz (1963) showed two important results which ensure that something interesting happens in this region. The first is that all trajectories are bounded in the phase space, and the second is that for all $r > 0$, the flow of trajectories is *volume-contracting*; that is, all trajectories eventually approach a certain limiting set of points which has zero volume. For $r > r_c$, this set has a rather complicated structure and is called a *strange attractor*.

Numerical calculations indicate a *chaotic behavior* for the solution, as shown in Figure 2.49. Note that for these specific σ and b values, which have been used exten-

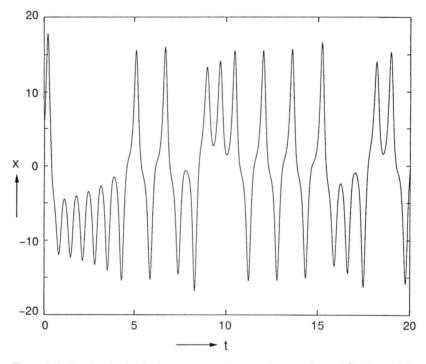

Figure 2.49 Chaotic behavior for the Lorenz equations; $r = 28$, $\sigma = 10$, $b = 8/3$, IC = (6,6,6).

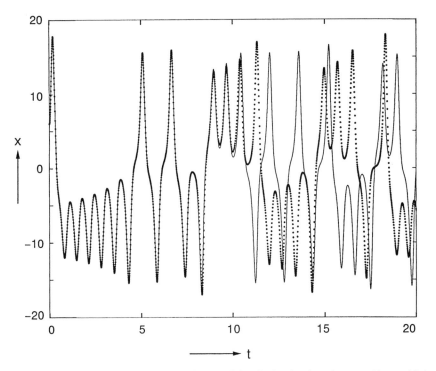

Figure 2.50 Sensitivity of the solution to the initial conditions in the chaotic regime; $r = 28$, $\sigma = 10$, $b = 8/3$; ——IC = (6,6,6), . . . IC = (6.01, 6,6).

sively in the literature since Lorenz (1963), $r_c = 470/19 \simeq 24.737$. As is characteristic of chaotic behavior, the solution displays an extreme *sensitivity* to the initial conditions. In Figure 2.50 the plot of x versus t for two nearby initial conditions indicates divergence for $t \gtrsim 10$ (recall similar behavior in Figure 2.47). The general structure in the phase space, however, remains essentially the same and is shown in Figure 2.51. Note that the trajectories actually do not cross (owing to the existence and uniqueness theorem) but appear to, because of the two-dimensional nature of the projections. Lorenz studied eq. (2.23.21) as a model for weather predictions. Based on calculations such as those in Figure 2.50, he noted the difficulty of making detailed long-range weather forecasts.

Chaotic behavior in the Lorenz equations can also occur for $r < r_c$, since here we have (at least near r_c) unstable periodic solutions surrounding stable critical points and the trajectories are bounded. The range of complex dynamic behavior possible for the Lorenz equations is indeed large and is determined generally by computations. For a fixed $\sigma = 10$, Jackson (1990, page 189) summarizes the wide variety of possibilities in the (b, r) plane.

Finally, while we have used the Lorenz equations as a simple example (it is historically the first) which leads to complex dynamic behavior, there are many other examples available in the literature. A class of these is the so-called Rossler models.

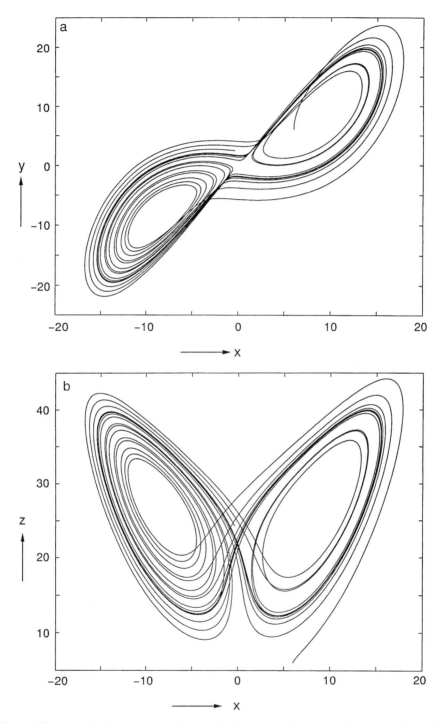

Figure 2.51 A numerically computed solution to the Lorenz equations; $r = 28$, $\sigma = 10$, $b = 8/3$. (*a*) Projection on the (x, y) plane. (*b*) Projection on the (x, z) plane.

These and other examples are reported by Jackson (1990). Note that a minimum of three nonlinear ODEs is required for chaotic behavior. The two examples treated in section 2.21 also exhibit chaos, when a third state variable is included. In a nonadiabatic CSTR, this occurs in the case of two reactions (Kahlert et al., 1981; Jorgensen et al., 1984), while for the Belousov–Zhabotinskii reaction, it requires more complex kinetic models (cf. Gyorgyi and Field, 1991).

References

R. Bellman, *Stability Theory of Differential Equations*, Dover, New York, 1969.

R. Bellman, *Introduction to Matrix Analysis*, 2nd ed., McGraw-Hill, New York, 1970.

W. E. Boyce, and R. C. DiPrima, *Elementary Differential Equations and Boundary Value Problems*, 5th ed., Wiley, New York, 1992.

F. Brauer and J. A. Nohel, *The Qualitative Theory of Ordinary Differential Equations*, Dover, New York, 1989.

E. A. Coddington, *An Introduction to Ordinary Differential Equations*, Dover, New York, 1989.

M. F. Doherty and J. M. Ottino, "Chaos in Deterministic Systems: Strange Attractors, Turbulence, and Applications in Chemical Engineering," *Chemical Engineering Science*, **43**, 139–183 (1988).

M. J. Feigenbaum, "Universal Behavior in Nonlinear Systems," *Los Alamos Science*, **1**, 4–27 (1980); reprinted in *Physica* **7D**, 16–39 (1983).

K. O. Friedrichs, *Advanced Ordinary Differential Equations*, Gordon and Breach, New York, 1965.

F. R. Gantmacher, *The Theory of Matrices*, volume 2, Chelsea, New York, 1989.

M. E. Gilpin, "Do Hares Eat Lynx?" *American Naturalist*, **107**, 727—730 (1973).

P. Gray and S. K. Scott, *Chemical Oscillations and Instabilities: Non-linear Chemical Kinetics*, Clarendon Press, Oxford, 1990.

O. Gurel and L. Lapidus, "A Guide to the Generation of Liapunov Functions," *Industrial and Engineering Chemistry*, **61**(3), 30 (1969).

L. Gyorgyi and R. J. Field, "Simple Models of Deterministic Chaos in the Belousov–Zhabotinsky Reaction," *Journal of Physical Chemistry*, **95**, 6594–6602 (1991).

B. D. Hassard, N. D. Kazarinoff, and Y-H. Wan, *Theory and Applications of Hopf Bifurcation*, Cambridge University Press, Cambridge, 1981.

E. L. Ince, *Ordinary Differential Equations*, Dover, New York, 1956.

E. A. Jackson, *Perspectives of Nonlinear Dynamics*, Cambridge University Press, Cambridge: Vol. 1, 1989; Vol. 2, 1990.

M. L. James, G. M. Smith, and J. C. Wolford, *Applied Numerical Methods for Digital Computation*, 3rd ed., Harper & Row, New York, 1985.

D. V. Jorgensen, W. W. Farr, and R. Aris, "More on the Dynamics of a Stirred Tank with Consecutive Reactions," *Chemical Engineering Science*, **39**, 1741–1752 (1984).

C. Kahlert, O. E. Rössler, and A. Varma, "Chaos in a Continuous Stirred Tank Reactor with Two Consecutive First-Order Reactions, One Exo-, One Endothermic," in *Modeling of Chemical Reaction Systems*, K. H. Ebert, P. Deuflhard, and W. Jäger, eds., Springer-Verlag, Berlin, 1981, pp. 355–365.

J. LaSalle and S. Lefschetz, *Stability by Liapunov's Direct Method with Applications*, Academic Press, New York, 1961.

E. N. Lorenz, "Deterministic Nonperiodic Flow," *Journal of Atmospheric Sciences*, **20**, 130–141 (1963).

J. E. Marsden and M. McCracken, *The Hopf Bifurcation and Its Applications*, Springer-Verlag, New York, 1976.

R. M. May, "Simple Mathematical Models with Very Complicated Dynamics," *Nature*, **261**, 459–467 (1976).

M. Mazzotti, M. Morbidelli, and G. Serravalle, "Bifurcation Analysis of the Oregonator Model in the 3-D Space: Bromate/Malonic Acid/Stoichiometric Coefficient," *Journal of Physical Chemistry*, **99**, 4501–4511 (1995).

N. Minorsky, *Nonlinear Oscillations*, Van Nostrand, Princeton, NJ, 1962.

M. Morbidelli, A. Varma, and R. Aris, "Reactor Steady-State Multiplicity and Stability," in *Chemical Reaction and Reactor Engineering*, J. J. Carberry and A. Varma, eds., Marcel Dekker, New York, 1987, chapter 15.

J. D. Murray, *Mathematical Biology*, Springer-Verlag, Berlin, 1990.

E. P. Odum, *Fundamentals of Ecology*, Saunders, Philadelphia, 1953.

A. B. Poore, "A Model Equation Arising from Chemical Reactor Theory," *Archives of Rational Mechanics and Analysis*, **52**, 358–388 (1973).

J. R. Rice, *Numerical Methods, Software, and Analysis*, 2nd ed., Academic Press, New York, 1993.

P. Ruoff and R. M. Noyes, "Exceptionally Large Oxygen Effect in the Belousov–Zhabotinskii Reaction," *Journal of Physical Chemistry*, **93**, 7394–7398 (1989).

R. Seydel, *From Equilibrium to Chaos: Practical Bifurcation and Stability Analysis*, Elsevier, New York, 1988.

C. Sparrow, *The Lorenz Equations: Bifurcations, Chaos, and Strange Attractors*, Springer-Verlag, New York, 1982.

J. J. Tyson, "A Quantitative Account of Oscillations, Bistability, and Traveling Waves in the Belousov–Zhabotinskii Reaction," in *Oscillations and Traveling Waves in Chemical Systems*, R. J. Field and M. Burger, eds., Wiley, New York, 1985, pp. 93–144.

A. Uppal, W. H. Ray, and A. B. Poore, "On the Dynamic Behavior of Continuous Stirred Tank Reactors," *Chemical Engineering Science*, **29**, 967–985 (1974).

S. A. Vejtasa and R. A. Schmitz, "An Experimental Study of Steady State Multiplicity and Stability in an Adiabatic Stirred Reactor," *AIChE Journal*, **16**, 410–419 (1970).

Problems

2.1 Find the general solution of the following ODEs.

(a) $y' + 3y = e^{-x}$

(b) $y' + 2xy = xe^{-x^2}$

(c) $y' = x^2 y$

(d) $y' = \dfrac{e^{x-y}}{1 + e^x}$

(e) $y' = \dfrac{3x^2 - 2xy}{x^2 - 2y}$

(f) $y' = -\dfrac{3y + e^x \sin y}{3x + e^x \cos y}$

(g) $y' = \dfrac{y}{x} + \dfrac{1}{y}$

(h) $xy' - y + xy^2 = 0$

(i) $y' = (x + y)^2$

(j) $y' = -\dfrac{1}{x^2} - \dfrac{2y}{x} + 2y^2$

2.2 The parallel reaction scheme

$$A \xrightarrow{k_1} B \qquad A \xrightarrow{k_2} C$$

where the first reaction is first order and the second is nth order, is carried out in an isothermal batch reactor. Determine an expression for the conversion of reactant A as a function of time.

2.3 Determine whether the following functions satisfy the Lipschitz condition over the domain indicated. If they do, then also determine the appropriate Lipschitz constant.

(a) $f(x, y) = xy^2$; $|x| \le 1$, $|y| \le 1$
(b) $f(x, y) = xy^2$; $|x| \le 1$, $|y| < \infty$
(c) $f(x, y) = x^2 e^{-xy^2}$; $0 \le x \le 1$, $|y| < \infty$
(d) $f(x, y) = y^{1/2}$; $|x| \le 1$, $0 \le y \le 1$
(e) $f(x, y) = y^{1/2}$; $|x| \le 1$, $1 \le y \le 2$

2.4 Consider the problem

$$\frac{dy}{dx} = 2x(y + 1); \qquad y(0) = 0$$

over the domain $|x| \le 1$, $|y| \le 2$.

(a) Determine the analytic solution $y(x)$.
(b) Determine the successive approximations, $y_j(x)$, $j = 1 \rightarrow 4$. From these, by induction, find $\lim_{j \to \infty} y_j(x)$ and show that it is the same as that obtained in part (a).
(c) Compare $y(1)$ with $y_4(1)$.

2.5 Consider the problem

$$\frac{dy}{dx} = 1 + y^2; \qquad y(0) = 0$$

(a) Determine the analytic solution, and then determine the domain over which it exists.
(b) Construct the initial largest range of x over which the solution is ensured to exist from the existence and uniqueness theorem.
(c) Find the first three successive approximations, $y_j(x)$.
(d) Compare y and y_3 at the extremities of the range determined in part (b).

2.6 For the system of equations

$$\frac{dy_1}{dx} = 1 + y_2$$

$$\frac{dy_2}{dx} = y_1^2$$

along with the ICs

$$y_1(0) = 0 \quad \text{and} \quad y_2(0) = 0$$

compute the first three successive approximations.

2.7 Confirm that eq. (2.8.20) represents the solution of eqs. (2.8.8) and (2.8.9).

2.8 For the problem

$$\frac{du}{dt} = f(u) = u(1 - u)(u - 2) \qquad \text{at } t = 0, \quad u = u_0 \in (-\infty, \infty)$$

repeat the analysis carried out in the text for eq. (2.8.8). Thus:

(a) Determine the steady states.

(b) Carry out a qualitative analysis by plotting $f(u)$ versus u, and investigate the behavior as $t \to \infty$ for various ICs.

(c) Confirm that linearization about the steady states yields correct information about their stability character for small perturbations.

(d) Confirm results of part (b) by the analytic solution.

2.9 Repeat the above problem for $f(u) = u^2(1 - u)$. Do you find something unusual about one of the steady states, in the sense that ICs less than or greater than this steady-state value lead to different equilibrium values? The case where ICs on one side of the steady state are attracted to it while those on the other side are repelled is called *one-sided stability*.

2.10 The dynamics of a certain process is described by the ODE

$$a\frac{dx}{dt} = 2x - 3x^2 + x^3$$

where $a > 0$ is a constant. Determine the steady states and whether they are asymptotically stable or unstable.

2.11 For each of the problems below,

1. Determine the eigenvalues and eigenvectors.
2. Classify the origin regarding its type and stability character.
3. Sketch a few trajectories in the phase plane.

(a) $\dfrac{dx}{dt} = \begin{bmatrix} 1 & -2 \\ 3 & -4 \end{bmatrix} \mathbf{x}$

(b) $\dfrac{dx}{dt} = \begin{bmatrix} -1 & 2 \\ -3 & 4 \end{bmatrix} \mathbf{x}$

(c) $\dfrac{dx}{dt} = \begin{bmatrix} 1 & 3 \\ 4 & 2 \end{bmatrix} \mathbf{x}$

(d) $\dfrac{dx}{dt} = \begin{bmatrix} -6 & 8 \\ -2 & 2 \end{bmatrix} \mathbf{x}$

(e) $\dfrac{dx}{dt} = \begin{bmatrix} 1 & 1 \\ -5 & -3 \end{bmatrix} \mathbf{x}$

(f) $\dfrac{dx}{dt} = \begin{bmatrix} 1 & -2 \\ 1 & 3 \end{bmatrix} \mathbf{x}$

(g) $\dfrac{dx}{dt} = \begin{bmatrix} 2 & -6 \\ 1 & -2 \end{bmatrix} \mathbf{x}$

2.12 For each of the following systems, determine the real steady states and discuss their type and stability character.

(a) $\dfrac{dx_1}{dt} = x_1 + x_2^2,$ $\qquad\qquad\dfrac{dx_2}{dt} = x_1 + x_2$

(b) $\dfrac{dx_1}{dt} = x_1 - x_1^2 - x_1x_2,$ $\qquad\dfrac{dx_2}{dt} = 3x_2 - x_1x_2 - 2x_2^2$

(c) $\dfrac{dx_1}{dt} = -2x_1 - x_2 + 2,$ $\qquad\dfrac{dx_2}{dt} = x_1x_2$

(d) $\dfrac{dx_1}{dt} = x_2,$ $\qquad\qquad\dfrac{dx_2}{dt} = -2bx_1 - 3x_1^2 - ax_2$

where a and b are positive constants.

2.13 For the system of equations

$$\frac{dx}{dt} = x$$

$$\frac{dy}{dt} = x^3 - 2y$$

(a) Determine the real steady states, their type, and their stability character.
(b) In the phase plane for the linearized system, identify the characteristic trajectories. Also sketch a few other trajectories.
(c) Confirm the results of part (a) by integrating the nonlinear system for dy/dx. Sketch a few trajectories for the nonlinear system as well.

2.14 A single-plate absorption column (see section 1.22) is described by the following equation:

$$h\frac{dx}{d\theta} = Lx_f + Gy_f - Lx - Gy$$

where x and y are mole fractions of the solute in the liquid and gas phases, respectively, θ is time, h is the liquid holdup, and L and G are the liquid and gas flow rates, respectively. The quantities h, L, G, x_f, and y_f are prescribed constants, with the last two representing compositions of the two feed streams. Furthermore, let the equilibrium relation be

$$y = \frac{\alpha x}{1 + (\alpha - 1)x}$$

where $\alpha > 0$ is the constant relative volatility.
Determine the number of steady states and their stability character.

2.15 A two-plate absorption column is described by

$$h\frac{dx_1}{d\theta} = Lx_f - Lx_1 - Gy_1 + Gy_2$$

$$h\frac{dx_2}{d\theta} = Lx_1 - Lx_2 - Gy_2 + Gy_f$$

where x_i and y_i refer to the solute mole fractions in the liquid and gas streams, respectively, leaving the ith plate, and the remaining notation is the same as in the problem above. The equilibrium relation is also the same:

$$y_i = \frac{\alpha x_i}{1 + (\alpha - 1)x_i}, \qquad i = 1 \text{ or } 2$$

where $\alpha > 0$. Without determining the steady states, examine the system for asymptotic stability.

2.16 Consider the set of eqs. (2.16.1) describing the population dynamics of two competing species, and assume that $\beta_1\beta_2 - \gamma_1\gamma_2 = 0$.

(a) Find all steady states of the system. Observe that the result depends on whether $\alpha_2\beta_1 - \alpha_1\gamma_2 = 0$.

(b) If $\alpha_2\beta_1 > \alpha_1\gamma_2$, determine the type and stability character of each steady state.

(c) Repeat part (b) for the case $\alpha_2\beta_1 < \alpha_1\gamma_2$.

(d) Analyze the nature of trajectories in the phase plane when $\alpha_2\beta_1 = \alpha_1\gamma_2$.

2.17 A reaction with the bimolecular Langmiur–Hinshelwood kinetic expression

$$f(C) = \frac{kC}{(1 + KC)^2}$$

is carried out in an *isothermal* CSTR, where C is reactant concentration; k and K are constants.

(a) Cast the problem in dimensionless form, by introducing the dimensionless quantities

$$u = \frac{C}{C_f}, \qquad r(u) = f(C)/f(C_f), \qquad \theta = \frac{V}{q}, \qquad \sigma = KC_f \quad \text{and} \quad Da = k\theta$$

where V is reactor volume, q is volumetric flow rate, and the subscript f refers to feed conditions.

(b) Determine the condition on the physicochemical parameters which ensures a unique steady state for *all* values of the Damköhler number, Da.

(c) If the condition in part (b) above is not satisfied, then determine the region where multiple steady states exist.

(d) Analyze the steady states for linearized stability.

2.18 In a nonadiabatic CSTR (section 2.17), consider a first-order reaction which is *endothermic* (i.e., $\beta < 0$).

(a) Develop the bounds on the steady state temperature.

(b) Show that the steady state is always unique.

(c) Show that the steady state is always asymptotically stable.

2.19 Recall that the notation in sections 2.17 and 2.21.1 is somewhat different. Starting with eqs. (2.21.3) for finite values of γ, derive the condition which ensures unique steady states for all values of Da. Show that this condition is the same as that given by eq. (2.17.18).

2.20 In a nonadiabatic CSTR (section 2.17), consider the case of an *n*th-order exothermic reaction. Repeat the analysis for unique and multiple steady states.

2.21 In the problems below,

1. Determine stability character of the origin by linearization.

2. Consider a Liapunov function of the type $V = ax_1^2 + bx_2^2$, where a and b are positive constants. How should these constants be chosen so that the origin has the stability character indicated?

(a) $\dfrac{dx_1}{dt} = -x_1^3 + x_1x_2^2, \quad \dfrac{dx_2}{dt} = -2x_1^2x_2 - x_2^3$ (asymptotically stable)

(b) $\dfrac{dx_1}{dt} = x_1^3 - x_2^3, \quad \dfrac{dx_2}{dt} = 2x_1x_2^2 + 4x_1^2x_2 + 2x_2^3$ (unstable)

2.22 The generalized motion of a damped pendulum is described by the *Lienard equation*

$$\frac{d^2u}{dt^2} + \frac{du}{dt} + g(u) = 0$$

where u is the angle between the pendulum rod and the downward vertical. In the classic case, $g(u) = \sin u$ whereas in general, $g(0) = 0$, $g(u) > 0$ for $0 < u < k$ and $g(u) < 0$ for $-k < u < 0$; that is, $ug(u) > 0$ for $-k < u < k$ and $u \neq 0$.

(a) Letting $x_1 = u$, $x_2 = du/dt$, show that $(0, 0)$ is an asymptotically stable critical point of the system.

(b) In this case a choice for the Liapunov function is

$$V(x_1, x_2) = \frac{x_2^2}{2} + \int_0^{x_1} g(s) \, ds, \qquad -k < x_1 < k$$

where the terms on the rhs represent kinetic and potential energy, respectively. Show that V is positive definite and that the origin is a stable critical point.

It is interesting to note that although we fully expect it, we are unable to ensure asymptotic stability of the origin with this particular form of V. Alternative forms of V are required for this purpose (see, for example, Brauer and Nohel, 1989, page 202).

2.23 Show that the system

$$\frac{dx}{dt} = -y + \frac{xf(r)}{r}$$

$$\frac{dy}{dt} = x + \frac{yf(r)}{r}$$

where $r = (x^2 + y^2)^{1/2}$, has periodic solutions which correspond to the roots of the function $f(r)$. Also, determine the direction of motion on the closed trajectories in the phase plane.

2.24 Show that in the (x_1, x_2) plane, eq. (2.19.12) represents a family of concentric closed curves surrounding the critical point $(b/\beta, a/\alpha)$.

2.25 In the Lotka–Volterra model, it is assumed that when the predator is absent, the prey population grows without bound. This assumption can be improved by allowing the prey to follow a logistic equation [see eq. (2.8.16)] in the absence of the predator. Thus in place of eqs. (2.19.9), consider the system

$$\frac{dx_1}{dt} = x_1(a - \gamma x_1 - \alpha x_2)$$

$$\frac{dx_2}{dt} = x_2(-b + \beta x_1)$$

where a, b, α, β, and γ are all positive constants. Determine all the critical points and discuss their stability character. Discuss the predator and prey population dynamics for various ICs.

2.26 Consider the two-dimensional system

$$\frac{dx}{dt} = \mu x + y - xy^2$$

$$\frac{dy}{dt} = -x + \mu y$$

Show that the system exhibits a Hopf bifurcation from the zero solution $x = y = 0$ at $\mu = 0$.

2.27 The *van der Pol equation*

$$\frac{d^2 x}{dt^2} - \mu(1 - x^2)\frac{dx}{dt} + x = 0$$

where μ is a constant, arises in electrical circuit theory. With the change of variables

$$x_1 = x, \qquad x_2 = \frac{dx}{dt}$$

it reduces to the pair of equations

$$\frac{dx_1}{dt} = x_2$$

$$\frac{dx_2}{dt} = -x_1 + \mu(1 - x_1^2)x_2$$

which has a unique critical point $(0, 0)$.

(a) Determine the stability character of the critical point as a function of the parameter μ.

(b) Are there values of μ where Hopf bifurcation occurs?

2.28 In the case of the nonadiabatic CSTR (section 2.21), develop the details leading to classification of the (B, δ) plane into regions I–VI, as shown in Figure 2.38.

2.29 The classic development of the nonadiabatic CSTR (section 2.21) assumes that the intrinsic thermal and material time constants are the same. In general, this may not be true. Then the only change that occurs in the system of eqs. (2.21.4) is that the lhs of eq. (2.21.4a) becomes Le $dx_1/d\tau$, where Le $\neq 1$ is the Lewis number [cf. W. H. Ray and S. P. Hastings, *Chemical Engineering Science*, **35**, 589 (1980)]. Obviously, steady-state multiplicity features are not influenced by Le; however, stability and oscillatory behavior are affected. Determine the critical value of the Lewis number (Le$_c$) such that for Le $>$ Le$_c$, Hopf bifurcation to periodic solutions is not possible.

2.30 The features of steady-state multiplicity, stability, and oscillatory behavior present for a first-order reaction occurring in a nonadiabatic CSTR can also be found in the case of an *isothermal* CSTR for reactions with complex kinetics. A well-known example involves the reaction scheme

$$A + 2B \xrightarrow{k_1} 3B \qquad \text{rate} = k_1 ab^2$$

$$B \xrightarrow{k_2} C \qquad \text{rate} = k_2 b$$

where the first step exhibits cubic autocatalysis [P. Gray and S. K. Scott, *Chemical Engineering Science*, **39**, 1087 (1984)]. Consider the case where the feed contains pure A

(a) Write down the transient equations governing the concentrations of species A and B. Reduce the equations to a dimensionless form.

(b) Analyze the system for steady-state multiplicity as a function of the reactor residence time.

(c) Analyze the steady states for linearized stability character.

(d) Determine the critical residence time at which Hopf bifurcation to periodic solutions occurs.

2.31 Determine the Hopf bifurcation set for the full Oregonator model (2.21.36) to (2.21.38), taking the concentration a as the bifurcation parameter.

2.32 A well-known model proposed to describe chemical oscillations is based on the so-called *Brusselator* kinetic scheme, named after the University of Brussels where it was first conceived [I. Prigogine and R. Lefever, *J Chem. Phys.*, **48**, 1695 (1968)]:

$$A \rightarrow X$$

$$B + X \rightarrow D + Y$$

$$Y + 2X \rightarrow 3X$$

$$X \rightarrow E$$

where the two reactants A and B are converted to products D and E through a sequence of irreversible reactions, which involve two intermediate species X and Y. Assume constant concentrations for species A and B, and reaction orders equal to the stoichiometric coefficients (that is, elementary steps).

(a) Derive the dimensionless transient equations governing the concentrations of the intermediate species X and Y.

(b) Determine the steady state solutions.

(c) Find the stability character of the steady states.

(d) Derive the Hopf bifurcation condition.

2.33 Consider the system
$$x' = x + y - x(x^2 + y^2)$$
$$y' = -x + y - y(x^2 + y^2)$$
$$z' = -z$$

(a) Show that the origin $(0, 0, 0)$ is the unique critical point of the system.

(b) By computing the eigenvalues of the linearized system explicitly, show that the origin is unstable.

(c) Confirm the results of part (b) by using the Routh–Hurwitz conditions.

2.34 Repeat the details of section 2.23.1 for the *cubic map* (Gray and Scott, 1990, section 13.1)
$$x_{n+1} = \mu x_n (1 - x_n)^2$$
where $\mu > 0$ is a tunable parameter.

2.35 The difference equation
$$x_{n+1} = x_n \exp[r(1 - x_n)]$$
is another realistic model for population dynamics, because it describes simple exponential growth at small population sizes and a tendency to decrease at large sizes. The steepness of the nonlinear behavior is tuned by the parameter $r > 0$ [R. M. May and G. F. Oster, *American Naturalist*, **110**, 573 (1976)].

(a) Determine the fixed point of the system.

(b) Determine the range $0 < r < r_1$ for which the fixed point is stable.

(c) Show that as r increases beyond r_1, bifurcation to a period-two cycle occurs.

(d) Successive period doublings occur as r increases further. Determine the values of r_n for stable period-2^n cycles, for $n = 1, 2,$ and 3.

Note. As in the case of the logistic difference equation (2.23.7), a limit point of the successive period doublings exists at $r_\infty = 2.6924 \cdots$. As r increases further, the first odd-period cycle appears at $r = 2.8332 \cdots$, followed by a cycle of period three at $r = 3.1024 \cdots$, beyond which there is chaos.

2.36 Develop the details for the various steps in the analysis of the Lorenz equations (section 2.23.2)

2.37 Perhaps the simplest differential equation system which leads to chaos is that conceived by Rossler (*Zeitschrift für Naturforschung*, **31a**, 259 (1976)):
$$\frac{dx}{dt} = -(y + z)$$

$$\frac{dy}{dt} = x + ay$$

$$\frac{dz}{dt} = b + z(x - c)$$

where a, b, and c are constants, and there is only a single quadratic nonlinearity! By numerical simulations, show that for the parameter set $a = b = 0.2$ and $c = 5.7$ the behavior is chaotic.

Chapter 3

I. INITIAL VALUE PROBLEMS

3.1 DEFINITIONS, LINEARITY, AND SUPERPOSITION

A general nth-order linear ordinary differential equation (ODE) can be written as

$$a_0(x)\frac{d^n y}{dx^n} + a_1(x)\frac{d^{n-1}y}{dx^{n-1}} + \cdots + a_{n-1}(x)\frac{dy}{dx} + a_n(x)y = -f(x) \quad (3.1.1)$$

where the coefficients $a_j(x)$ are continuous over the closed interval $x \in [a, b]$, and $a_0(x) \neq 0$ for $x \in [a, b]$. If $f(x) = 0$, eq. (3.1.1) is said to be *homogeneous*; otherwise it is *nonhomogeneous*.

If we denote

$$y = y_1$$
$$\frac{dy}{dx} = \frac{dy_1}{dx} = y_2$$
$$\frac{d^2 y}{dx^2} = \frac{dy_2}{dx} = y_3$$
$$\vdots$$
$$\frac{d^{n-1}y}{dx^{n-1}} = \frac{dy_{n-1}}{dx} = y_n$$

then from eq. (3.1.1) we have

$$\frac{d^n y}{dx^n} = \frac{dy_n}{dx} = -\frac{1}{a_0(x)} [f(x) + a_1 y_n + a_2 y_{n-1} + \cdots + a_n y_1]$$

Thus the nth-order ODE (3.1.1) reduces to a set of n coupled first-order ODEs of the form

$$\frac{d\mathbf{y}}{dx} = \mathbf{f}(x, \mathbf{y}) \tag{3.1.2}$$

We recall from chapter 2 that, when the ICs are provided, unique solutions of nonlinear equations of the form (3.1.2) are guaranteed to exist if the f_j have continuous partial derivatives with respect to all y_j. For the linear problem at hand, with the mild restrictions on $a_j(x)$ given above, the latter is ensured. Thus a unique solution for the ODE (3.1.1) exists for all $x \in [a, b]$, provided that the values of

$$y, \frac{dy}{dx}, \ldots, \frac{d^{n-1}y}{dx^{n-1}}$$

are given at some point $x_0 \in [a, b]$.

Although we have used the term *linear* above, in connection with eq. (3.1.1), let us define what is meant by linearity of a differential operator. Consider

$$L[y] = \left[a_0 \frac{d^n}{dx^n} + a_1 \frac{d^{n-1}}{dx^{n-1}} + \cdots + a_{n-1} \frac{d}{dx} + a_n \right](y) \tag{3.1.3}$$

Then $L[y]$ is linear if

$$L[c_1 y_1 + c_2 y_2] = c_1 L[y_1] + c_2 L[y_2] \tag{3.1.4}$$

where c_1 and c_2 are constants. It is easy to verify that the differential operator of eq. (3.1.1) is linear.

Solutions of homogeneous linear ODEs satisfy the important property of *superposition*, which can be stated as follows. If y_1 and y_2 are solutions of

$$L[y] = 0, \quad \text{i.e.,} \quad L[y_1] = 0 \quad \text{and} \quad L[y_2] = 0$$

then

$$y = c_1 y_1 + c_2 y_2 \tag{3.1.5}$$

where c_1 and c_2 are arbitrary constants, is also a solution of

$$L[y] = 0$$

This follows directly from the definition of linearity, which implies that

$$L[y] = L[c_1 y_1 + c_2 y_2] = c_1 L[y_1] + c_2 L[y_2] = 0$$

since y_1 and y_2 satisfy

$$L[y_1] = 0 \text{ and } L[y_2] = 0$$

Thus the sum of any finite number of solutions $L[y] = 0$ is also a solution.

Definition. The functions

$$f_1(x), f_2(x), \ldots, f_n(x)$$

are said to be *linearly dependent* in the interval $x \in [a, b]$ if there exist n constants c_1, $c_2, \ldots, c_n$ not all zero, such that

$$c_1 f_1(x) + c_2 f_2(x) + \cdots + c_n f_n(x) = 0 \tag{3.1.6}$$

for *all* $x \in [a, b]$. If a set of functions $\mathbf{f}(x)$ is not linearly dependent, then it is said to be *linearly independent*. It follows from above that in this case, the only set of constants c satisfying eq. (3.1.6) is $c = 0$.

Example 1. For $r_1 \neq r_2$, the functions $e^{r_1 x}$ and $e^{r_2 x}$ are linearly independent.
        The proof follows by contradiction. Thus, let us assume that there exist c_1 and c_2, both not equal to zero, such that

$$c_1 e^{r_1 x} + c_2 e^{r_2 x} = 0$$

or

$$c_1 + c_2 e^{(r_2 - r_1)x} = 0 \tag{3.1.7}$$

Differentiating once gives

$$c_2(r_2 - r_1)\, e^{(r_2 - r_1)x} = 0 \tag{3.1.8}$$

Since $r_1 \neq r_2$, eq. (3.1.8) implies that

$$c_2 = 0$$

and then from eq. (3.1.7) we have

$$c_1 = 0$$

However, this contradicts our earlier hypothesis. Hence $e^{r_1 x}$ and $e^{r_2 x}$ are linearly independent functions.

Example 2. The functions $\sin x$ and $\cos x$ are linearly independent. Proceeding as above, let us assume that they are not. Then we have

$$c_1 \sin x + c_2 \cos x = 0$$

with c_1 and c_2 both not equal to zero. Differentiating gives

$$c_1 \cos x - c_2 \sin x = 0$$

which together with the above equation leads to

$$\begin{bmatrix} \sin x & \cos x \\ \cos x & -\sin x \end{bmatrix} \begin{bmatrix} c_1 \\ c_2 \end{bmatrix} = \begin{bmatrix} 0 \\ 0 \end{bmatrix} \tag{3.1.9}$$

Since this is a set of linear homogeneous equations, nontrivial c_1 and c_2 exist if and only if

$$D = \begin{vmatrix} \sin x & \cos x \\ \cos x & -\sin x \end{vmatrix} = 0$$

However, $D = -1$. Hence $\sin x$ and $\cos x$ are linearly independent functions.

3.2   THE FUNDAMENTAL SET OF SOLUTIONS

An important concept in the solution of linear ODEs is that involving a *fundamental set of solutions*, which essentially means a set of linearly independent solutions of the ODE (cf. Ince, 1956, chapter 5).

● **THEOREM 1**

Given an nth-order linear ODE: $L[y] = 0$, there exist n linearly independent solutions in the interval $x \in [a, b]$.

Proof
Consider the following n solutions of $L[y] = 0$:

$$f_1(x), f_2(x), \ldots, f_n(x) \tag{3.2.1}$$

with the specific ICs

$$\begin{aligned}
f_1(x_0) &= 1, \quad f_1^{(1)}(x_0) = 0 = f_1^{(2)}(x_0) = \cdots = f_1^{(n-1)}(x_0) \\
f_2(x_0) &= 0, \quad f_2^{(1)}(x_0) = 1, \quad f_2^{(2)}(x_0) = 0 = \cdots = f_2^{(n-1)}(x_0) \\
&\vdots \\
f_n(x_0) &= 0 = f_n^{(1)}(x_0) = \cdots = f_n^{(n-2)}(x_0), \quad f_n^{(n-1)}(x_0) = 1
\end{aligned} \tag{3.2.2}$$

and $x_0 \in [a, b]$. Note that $f_j^{(k)}$ means the kth derivative of f_j.

Each of these solutions exists and is unique. What we would like to show is that this set of n solutions is linearly independent as well. As is usual in such cases, the proof is by contradiction.

Assume that the set is linearly dependent. Thus, by assumption, we can find constants $c_1, \ldots, c_n$ *not all zero* such that

$$c_1 f_1(x) + c_2 f_2(x) + \cdots + c_n f_n(x) = 0; \qquad x \in [a, b] \tag{3.2.3}$$

Differentiating successively $(n - 1)$ times gives

$$\begin{aligned}
c_1 f_1^{(1)} + c_2 f_2^{(1)} + \cdots + c_n f_n^{(1)} &= 0 \\
c_1 f_1^{(2)} + c_2 f_2^{(2)} + \cdots + c_n f_n^{(2)} &= 0 \\
&\vdots \\
c_1 f_1^{(n-1)} + c_2 f_2^{(n-1)} + \cdots + c_n f_n^{(n-1)} &= 0; \qquad x \in [a, b]
\end{aligned} \tag{3.2.4}$$

In particular, at $x = x_0$ we have

$$\begin{aligned}
c_1 f_1(x_0) + c_2 f_2(x_0) + \cdots + c_n f_n(x_0) &= 0 \\
c_1 f_1^{(1)}(x_0) + c_2 f_2^{(1)}(x_0) + \cdots + c_n f_n^{(1)}(x_0) &= 0 \\
&\vdots \\
c_1 f_1^{(n-1)}(x_0) + c_2 f_2^{(n-1)}(x_0) + \cdots + c_n f_n^{(n-1)}(x_0) &= 0
\end{aligned} \tag{3.2.5}$$

which, utilizing eq. (3.2.2), reduce to

$$\begin{aligned}
c_1 &= 0 \\
c_2 &= 0 \\
&\vdots \\
c_n &= 0
\end{aligned} \tag{3.2.6}$$

However, this is a contradiction to the earlier assumption. Thus the set of n solutions of $L[y] = 0$:

$$f_1(x), f_2(x), \ldots, f_n(x)$$

which satisfy the specific initial conditions (ICs) (3.2.2) is linearly independent. This specific set of solutions is called the *fundamental set of solutions* of $L[y] = 0$. The reason for this name is the following result.

● **THEOREM 2**

Let $f_1, f_2, \ldots, f_n$ be n solutions of $L[y] = 0$ satisfying the ICs (3.2.2). Then if $g(x)$ is *any* solution of $L[y] = 0$ in $x \in [a, b]$, with ICs

$$g(x_0) = \alpha_1, \ g^{(1)}(x_0) = \alpha_2, \ldots, g^{(n-1)}(x_0) = \alpha_n \qquad (3.2.7)$$

we have

$$g(x) = \alpha_1 f_1(x) + \alpha_2 f_2(x) + \cdots + \alpha_n f_n(x) \qquad (3.2.8)$$

Proof
Consider the function

$$\phi(x) = \alpha_1 f_1(x) + \alpha_2 f_2(x) + \cdots + \alpha_n f_n(x)$$

By superposition, $\phi(x)$ is a solution of $L[y] = 0$. Also, since f_j satisfy the specific ICs (3.2.2), it follows that

$$\phi(x_0) = \alpha_1$$
$$\phi^{(1)}(x_0) = \alpha_2$$
$$\vdots$$
$$\phi^{(n-1)}(x_0) = \alpha_n$$

Since the solution of $L[y] = 0$ with ICs (3.2.7) is unique, we must have

$$\phi(x) = g(x)$$

Thus, once the fundamental set of solutions of $L[y] = 0$ has been determined, the solution of $L[y] = 0$ with *arbitrary* ICs can be immediately written down.

Example

$$L[y] = \frac{d^2 y}{dx^2} + y = 0, \qquad x \in [0, 1]$$

Find the fundamental set of solutions $f_1(x)$ and $f_2(x)$. Let $x_0 = 0$. Then f_1 and f_2 satisfy the ICs

$$f_1(0) = 1, \qquad f_1^{(1)}(0) = 0$$
$$f_2(0) = 0, \qquad f_2^{(1)}(0) = 1$$

The general solution of $L[y] = 0$ is given by

$$y = c_1 \cos x + c_2 \sin x$$

where c_1 and c_2 are constants, different for f_1 and f_2, and are determined next. For f_1, we have

$$f_1(0) = 1 = c_1, \qquad f_1^{(1)}(0) = 0 = c_2$$

Thus

$$f_1 = \cos x$$

Similarly, for f_2

$$f_2(0) = 0 = c_1, \qquad f_2^{(1)}(0) = 1 = c_2$$

so that

$$f_2 = \sin x$$

Thus from eq. (3.2.8) the solution of $L[y] = 0$ with ICs

$$y(0) = \alpha_1, \ y^{(1)}(0) = \alpha_2$$

is given by

$$y = \alpha_1 \cos x + \alpha_2 \sin x$$

3.3 THE WRONSKIAN DETERMINANT AND LINEAR INDEPENDENCE OF SOLUTIONS

Given *any* n solutions of $L[y] = 0$: $\phi_1, \phi_2, \ldots, \phi_n$, the determinant

$$W(x) = \begin{vmatrix} \phi_1 & \phi_2 & \cdots & \phi_n \\ \phi_1^{(1)} & \phi_2^{(1)} & \cdots & \phi_n^{(1)} \\ \vdots & & & \\ \phi_1^{(n-1)} & \phi_2^{(n-1)} & \cdots & \phi_n^{(n-1)} \end{vmatrix} \tag{3.3.1}$$

is called the *Wronskian* determinant corresponding to these solutions. The value of the Wronskian indicates whether the n solutions are linearly independent.

● **THEOREM**

If $\phi_1(x), \phi_2(x), \ldots, \phi_n(x)$ are n solutions of $L[y] = 0$, then they are linearly dependent *if and only if* $W(x) = 0$ for all $x \in [a, b]$.

Proof
To show the *if* part, let $W(x) = 0$, so that

$$\begin{vmatrix} \phi_1 & \phi_2 & \cdots & \phi_n \\ \phi_1^{(1)} & \phi_2^{(1)} & \cdots & \phi_n^{(1)} \\ \vdots & & & \\ \phi_1^{(n-1)} & \phi_2^{(n-1)} & \cdots & \phi_n^{(n-1)} \end{vmatrix} = 0; \ x \ [a, b] \tag{3.3.2}$$

This means that each column is a linear combination of the other columns. Thus there exists a set of constants c_j, not all zero, such that

$$c_1\phi_1 + c_2\phi_2 + \cdots + c_n\phi_n = 0$$
$$\vdots \tag{3.3.3}$$
$$c_1\phi_1^{(n-1)} + c_2\phi_2^{(n-1)} + \cdots + c_n\phi_n^{(n-1)} = 0; \qquad x \in [a, b]$$

and the first of these equations directly implies that the ϕ_j are linearly dependent.

To show the *only if* part, assume that the ϕ_j are linearly dependent. Then there is a set of constants c_j, not all zero, such that

$$c_1\phi_1 + c_2\phi_2 + \cdots + c_n\phi_n = 0; \qquad x \in [a, b] \tag{3.3.4}$$

Differentiating $(n - 1)$ times, we have

$$c_1\phi_1^{(1)} + c_2\phi_2^{(1)} + \cdots + c_n\phi_n^{(1)} = 0$$
$$c_1\phi_1^{(2)} + c_2\phi_2^{(2)} + \cdots + c_n\phi_n^{(2)} = 0$$
$$\vdots \tag{3.3.5}$$
$$c_1\phi_1^{(n-1)} + c_2\phi_2^{(n-1)} + \cdots + c_n\phi_n^{(n-1)} = 0$$

These equations constitute a set of n homogeneous linear algebraic equations in n unknowns c_j. Since $\mathbf{c}$ is nontrivial, the determinant of the coefficient matrix must be zero. Thus

$$W(x) = 0; \qquad x \in [a, b] \tag{3.3.6}$$

The Wronskian determinant provides a practical test to determine the linear dependence or independence of a set of n solutions of the nth-order ODE, $L[y] = 0$.

For example, in the previous section we saw that the ODE

$$L[y] = \frac{d^2y}{dx^2} + y = 0; \qquad x \in (0, 1)$$

has the fundamental set

$$f_1 = \cos x, \qquad f_2 = \sin x$$

The Wronskian determinant for this set is given by

$$W(x) = \begin{vmatrix} f_1 & f_2 \\ f_1^{(1)} & f_2^{(1)} \end{vmatrix} = \begin{vmatrix} \cos x & \sin x \\ -\sin x & \cos x \end{vmatrix} = 1$$

Thus $\sin x$ and $\cos x$ are linearly independent functions.

A Practical Test. We saw above that given a homogeneous linear ODE, the value of $W(x)$ corresponding to a set of solutions over the interval $x \in [a, b]$ determines whether the set is linearly independent. A more convenient test for linear independence is now formulated.

Let us consider the Wronskian determinant given by

$$W(x) = \begin{vmatrix} \phi_1 & \phi_2 & \cdots & \phi_n \\ \phi_1^{(1)} & \phi_2^{(1)} & \cdots & \phi_n^{(1)} \\ \vdots & & & \\ \phi_1^{(n-1)} & \phi_2^{(n-1)} & \cdots & \phi_n^{(n-1)} \end{vmatrix}$$

Its derivative using eq. (1.1.4) is

$$\frac{dW}{dx} = \begin{vmatrix} \phi_1 & \phi_2 & \cdots & \phi_n \\ \vdots & & & \\ \phi_1^{(n-2)} & \phi_2^{(n-2)} & \cdots & \phi_n^{(n-2)} \\ \phi_1^{(n)} & \phi_2^{(n)} & \cdots & \phi_n^{(n)} \end{vmatrix} \quad \begin{array}{l} + (n - 1) \text{ other determinants,} \\ \text{which are all } zero \end{array}$$

Since all the ϕ_j are solutions of $L[y] = 0$, they satisfy

$$a_0 y^{(n)} + a_1 y^{(n-1)} + \cdots + a_{n-1} y^{(1)} + a_n y = 0$$

where $a_i = a_i(x)$ and $a_0(x) \neq 0$ for $x \in [a, b]$. This leads to

$$y^{(n)} = -\sum_{j=1}^{n} \frac{a_j(x)}{a_0(x)} y^{(n-j)}$$

and we have

$$\frac{dW}{dx} = \begin{vmatrix} \phi_1 & \phi_2 & \cdots & \phi_n \\ \phi_1^{(1)} & \phi_2^{(1)} & \cdots & \phi_n^{(1)} \\ \vdots & & & \\ \phi_1^{(n-2)} & \phi_2^{(n-2)} & \cdots & \phi_n^{(n-2)} \\ -\sum_{j=1}^{n} \frac{a_j}{a_0} \phi_1^{(n-j)} & -\sum_{j=1}^{n} \frac{a_j}{a_0} \phi_2^{(n-j)} & \cdots & -\sum_{j=1}^{n} \frac{a_j}{a_0} \phi_n^{(n-j)} \end{vmatrix}$$

Further, recall that the value of a determinant remains unchanged if a row or column is multiplied by a scalar and added to another row or column. Thus, multiplying the first row by a_n/a_0, the second row by $a_{n-1}/a_0, \ldots$, and the $(n - 1)$st row by a_2/a_0 and then adding all these to the nth row gives

$$\frac{dW}{dx} = \begin{vmatrix} \phi_1 & \phi_2 & \cdots & \phi_n \\ \phi_1^{(1)} & \phi_2^{(1)} & \cdots & \phi_n^{(1)} \\ \vdots & & & \\ \phi_1^{(n-2)} & \phi_2^{(n-2)} & \cdots & \phi_n^{(n-2)} \\ -\frac{a_1}{a_0} \phi_1^{(n-1)} & -\frac{a_1}{a_0} \phi_2^{(n-1)} & \cdots & -\frac{a_1}{a_0} \phi_n^{(n-1)} \end{vmatrix}$$

$$= -\frac{a_1(x)}{a_0(x)} W$$

Integrating, we obtain

$$W(x) = W(x_0) \exp\left[-\int_{x_0}^{x} \frac{a_1(t)}{a_0(t)} dt\right] \tag{3.3.7}$$

Thus if $W(x) = 0$ for *any* $x_0 \in [a, b]$, then $W(x) = 0$ for *all* $x \in [a, b]$.

If the ϕ_j are the fundamental set, then $W(x_0) = 1$ by definition, and eq. (3.3.7) reduces to

$$W(x) = \exp\left[-\int_{x_0}^{x} \frac{a_1(t)}{a_0(t)} dt\right] \neq 0 \tag{3.3.8}$$

3.4 NONHOMOGENEOUS EQUATIONS

We now consider the solution of nonhomogeneous linear ODEs of the form

$$L[y] = a_0(x)y^{(n)} + a_1(x)y^{(n-1)} + \cdots + a_{n-1}(x)y^{(1)} + a_n(x)y = -f(x) \quad (3.4.1)$$

● **THEOREM 1**

If $f_1(x), f_2(x), \ldots, f_n(x)$ is a fundamental set of solutions of the homogeneous equation $L[y] = 0$ and if $\psi(x)$ is a particular solution of $L[y] = -f$, then the solution of $L[y] = -f$ can be written as

$$y(x) = \psi(x) + \sum_{j=1}^{n} c_j f_j(x) \tag{3.4.2}$$

where c_j are arbitrary constants.

Proof
By the linearity of $L[\ ]$ and the principle of superposition, we see that

$$\begin{aligned}
L[y] &= L\left[\psi + \sum_{j=1}^{n} c_j f_j\right] \\
&= L[\psi] + \sum_{j=1}^{n} c_j L[f_j] \\
&= -f
\end{aligned}$$

● **THEOREM 2**

If ψ_1 and ψ_2 are two particular solutions of the nonhomogeneous equation $L[y] = -f$, that is, $L[\psi_1] = -f$ and $L[\psi_2] = -f$, then ψ_1 and ψ_2 differ by *at most* a linear combination of the fundamental set of $L[y] = 0$.

Proof
From Theorem 1, we have

$$\begin{aligned}
y &= \psi_1 + \sum_{j=1}^{n} c_j f_j \\
&= \psi_2 + \sum_{j=1}^{n} d_j f_j
\end{aligned}$$

Subtracting gives

$$0 = \psi_1 - \psi_2 + \sum_{j=1}^{n} (c_j - d_j)f_j$$

and hence

$$\psi_1 - \psi_2 = \sum_{j=1}^{n} (d_j - c_j)f_j$$

● **THEOREM 3**

If ψ_1 is a particular solution of $L[y] = -f_1$ and ψ_2 is a particular solution of $L[y] = -f_2$, then

$$\psi = \alpha\psi_1 + \beta\psi_2$$

is a particular solution of

$$L[y] = -(\alpha f_1 + \beta f_2)$$

Proof
Let $\psi = \alpha\psi_1 + \beta\psi_2$; then by linearity of $L[\ ]$ we obtain

$$L[y] = \alpha L[\psi_1] + \beta L[\psi_2] = -\alpha f_1 - \beta f_2 = -(\alpha f_1 + \beta f_2)$$

Thus, summarizing the results obtained so far, in order to solve nonhomogeneous problems, we need to find

(a) a fundamental set of $L[y] = 0$, and
(b) a particular solution of $L[y] = -f$.

3.5 THE METHOD OF VARIATION OF PARAMETERS

This is a general method of obtaining *particular* solutions of nonhomogeneous ODEs. There is some arbitrariness in the procedure, so at the end we have to confirm that we in fact have a solution. Consider the problem

$$L[y] = -f \qquad (3.5.1)$$

where

$$L[y] = a_0(x)y^{(n)} + a_1(x)y^{(n-1)} + \cdots + a_{n-1}(x)y^{(1)} + a_n(x)y$$

$a_j(x)$ are continuous for $x \in [a, b]$, and $a_0(x) \neq 0$ for all $x \in [a, b]$.
Let us assume that there exists a particular solution of the form

$$y_p(x) = \sum_{j=1}^{n} \mu_j(x)f_j(x) \qquad (3.5.2)$$

where $\mu_j(x)$ are as yet undetermined, and $\{f_j(x)\}$ is a fundamental set of $L[y] = 0$. Note that the $\mu_j(x)$ cannot all be constants, because then eq. (3.5.2) will represent a solution of $L[y] = 0$ rather than one of $L[y] = -f$.

Differentiating eq. (3.5.2) gives

$$y_p^{(1)} = \sum_{j=1}^n \mu_j^{(1)} f_j + \sum_{j=1}^n \mu_j f_j^{(1)}$$

If we set

$$\sum_{j=1}^n \mu_j^{(1)} f_j = 0 \tag{3.5.3}$$

then we have

$$y_p^{(1)} = \sum_{j=1}^n \mu_j f_j^{(1)} \tag{3.5.4}$$

Differentiating once again,

$$y_p^{(2)} = \sum_{j=1}^n \mu_j^{(1)} f_j^{(1)} + \sum_{j=1}^n \mu_j f_j^{(2)}$$

and setting

$$\sum_{j=1}^n \mu_j^{(1)} f_j^{(1)} = 0 \tag{3.5.5}$$

leads to

$$y_p^{(2)} = \sum_{j=1}^n \mu_j f_j^{(2)} \tag{3.5.6}$$

Proceeding similarly, we obtain

$$y_p^{(3)} = \sum_{j=1}^n \mu_j^{(1)} f_j^{(2)} + \sum_{j=1}^n \mu_j f_j^{(3)} \qquad \text{and set} \quad \sum_{j=1}^n \mu_j^{(1)} f_j^{(2)} = 0 \tag{3.5.7}$$

$$\vdots$$

$$y_p^{(n-1)} = \sum_{j=1}^n \mu_j^{(1)} f_j^{(n-2)} + \sum_{j=1}^n \mu_j f_j^{(n-1)} \quad \text{and set} \quad \sum_{j=1}^n \mu_j^{(1)} f_j^{(n-2)} = 0 \tag{3.5.8}$$

and finally

$$y_p^{(n)} = \sum_{j=1}^n \mu_j^{(1)} f_j^{(n-1)} + \sum_{j=1}^n \mu_j f_j^{(n)} \tag{3.5.9}$$

Note that in the equation above, we cannot set $\sum_{j=1}^n \mu_j^{(1)} f_j^{(n-1)} = 0$ as we did previously for the lower-order derivatives. The reason is that otherwise, this equation along with eqs. (3.5.3), (3.5.5), (3.5.7) and (3.5.8) would constitute a system of n linear homogeneous algebraic equations in the n unknowns $\mu_j^{(1)}$. Since the determinant of the coefficient matrix of this system is the Wronskian $W(x)$, it is nonzero, which would lead to the unique solution $\mu_j^{(1)} = 0$. This implies that the μ_j would be constants, which is not permissible as noted earlier.

Although we cannot set $\sum_{j=1}^{n} \mu_j^{(1)} f_j^{(n-1)} = 0$, an expression for this sum can be derived by substituting the various $y_p^{(j)}$ obtained above in $L[y_p] = -f$, as shown next.

$$
\begin{aligned}
L[y_p] &= a_0 y_p^{(n)} + a_1 y_p^{(n-1)} + \cdots + a_{n-1} y_p^{(1)} + a_n y_p \\
&= a_0 \sum_{j=1}^{n} \mu_j^{(1)} f_j^{(n-1)} + \left[a_0 \sum_{j=1}^{n} \mu_j f_j^{(n)} + a_1 \sum_{j=1}^{n} \mu_j f_j^{(n-1)} + \cdots \right. \\
&\quad \left. + a_{n-1} \sum_{j=1}^{n} \mu_j f_j^{(1)} + a_n \sum_{j=1}^{n} \mu_j f_j \right] \\
&= a_0 \sum_{j=1}^{n} \mu_j^{(1)} f_j^{(n-1)} + \left[\sum_{j=1}^{n} \mu_j \{ a_0 f_j^{(n)} + a_1 f_j^{(n-1)} + \cdots \right. \\
&\quad \left. + a_{n-1} f_j^{(1)} + a_n f_j \} \right] \\
&= a_0 \sum_{j=1}^{n} \mu_j^{(1)} f_j^{(n-1)} + \sum_{j=1}^{n} \mu_j L[f_j] \\
&= a_0 \sum_{j=1}^{n} \mu_j^{(1)} f_j^{(n-1)} \qquad\qquad (3.5.10)
\end{aligned}
$$

since $L[f_j] = 0$. Recalling that $L[y_p] = -f$, we have

$$
\sum_{j=1}^{n} \mu_j^{(1)} f_j^{(n-1)} = -\frac{f(x)}{a_0(x)} \qquad\qquad (3.5.11)
$$

Now, eqs. (3.5.3), (3.5.5), (3.5.7), (3.5.8), and (3.5.11) provide a system of n linear algebraic equations in n unknowns $\mu_j^{(1)}(x)$:

$$
\begin{aligned}
\mu_1^{(1)} f_1 + \mu_2^{(1)} f_2 + \cdots + \mu_n^{(1)} f_n &= 0 \\
\mu_1^{(1)} f_1^{(1)} + \mu_2^{(1)} f_2^{(1)} + \cdots + \mu_n^{(1)} f_n^{(1)} &= 0 \\
&\vdots \\
\mu_1^{(1)} f_1^{(n-2)} + \mu_2^{(1)} f_2^{(n-2)} + \cdots + \mu_n^{(1)} f_n^{(n-2)} &= 0 \\
\mu_1^{(1)} f_1^{(n-1)} + \mu_2^{(1)} f_2^{(n-1)} + \cdots + \mu_n^{(1)} f_n^{(n-1)} &= -\frac{f(x)}{a_0(x)}
\end{aligned}
$$

The solution, from Cramer's rule, is given by

$$
\mu_j^{(1)}(x) = \frac{1}{W(x)}
\begin{vmatrix}
f_1 & f_2 & \cdots & f_{j-1} & 0 & f_{j+1} & \cdots & f_n \\
f_1^{(1)} & f_2^{(1)} & \cdots & f_{j-1}^{(1)} & 0 & f_{j+1}^{(1)} & \cdots & f_n^{(1)} \\
\vdots & & & & & & & \\
f_1^{(n-1)} & f_2^{(n-1)} & \cdots & f_{j-1}^{(n-1)} & \dfrac{-f(x)}{a_0(x)} & f_{j+1}^{(n-1)} & \cdots & f_n^{(n-1)}
\end{vmatrix}
$$

$$
= \frac{D_j(x)}{W(x)} \qquad\qquad (3.5.12)
$$

which on integration gives

$$\mu_j(x) = \int_{x_0}^{x} \frac{D_j(t)}{W(t)} \, dt \qquad (3.5.13)$$

where $\mu_j(x_0)$ has been arbitrarily set equal to 0. Note that a nonzero value for $\mu_j(x_0)$ would lead, from eq. (3.5.2), to simply a linear combination of the fundamental set of $L[y] = 0$.

Introducing the determinant

$$W_j(t) = \begin{vmatrix} f_1 & f_2 & \cdots & f_{j-1} & 0 & f_{j+1} & \cdots & f_n \\ f_1^{(1)} & f_2^{(1)} & \cdots & f_{j-1}^{(1)} & 0 & f_{j+1}^{(1)} & \cdots & f_n^{(1)} \\ \vdots & & & & & & \\ f_1^{(n-2)} & f_2^{(n-2)} & \cdots & f_{j-1}^{(n-2)} & 0 & f_{j+1}^{(n-2)} & \cdots & f_n^{(n-2)} \\ f_1^{(n-1)} & f_2^{(n-1)} & \cdots & f_{j-1}^{(n-1)} & 1 & f_{j+1}^{(n-1)} & \cdots & f_n^{(n-1)} \end{vmatrix} \qquad (3.5.14)$$

and substituting it in eq. (3.5.13) yields

$$\mu_j(x) = \int_{x_0}^{x} \frac{-f(t)}{a_0(t)} \frac{W_j(t)}{W(t)} \, dt$$

From eq. (3.5.2) the particular solution is then given by

$$\begin{aligned} y_p(x) &= \sum_{j=1}^{n} \mu_j(x) f_j(x) \\ &= \sum_{j=1}^{n} f_j(x) \int_{x_0}^{x} \frac{-f(t)}{a_0(t)} \frac{W_j(t)}{W(t)} \, dt \\ &= \int_{x_0}^{x} \frac{-f(t)}{a_0(t)} \frac{\sum_{j=1}^{n} f_j(x) W_j(t)}{W(t)} \, dt \end{aligned} \qquad (3.5.15)$$

where the integration and summation operations have been exchanged.

Noting that

$$\sum_{j=1}^{n} f_j(x) W_j(t) = \tilde{W}(x, t) = \begin{vmatrix} f_1(t) & f_2(t) & \cdots & f_n(t) \\ f_1^{(1)}(t) & f_2^{(1)}(t) & \cdots & f_n^{(1)}(t) \\ \vdots & & & \\ f_1^{(n-2)}(t) & f_2^{(n-2)}(t) & \cdots & f_n^{(n-2)}(t) \\ f_1(x) & f_2(x) & \cdots & f_n(x) \end{vmatrix} \qquad (3.5.16)$$

eq. (3.5.15) reduces to

$$y_p(x) = \int_{x_0}^{x} \frac{-f(t)}{a_0(t)} \frac{\tilde{W}(x, t)}{W(t)} \, dt \qquad (3.5.17)$$

which is the desired particular solution of $L[y] = -f$.

The expression $\bar{W}(x, t)/a_0(t)W(t)$ is called the *one-sided Green's function* of the operator $L[\ ]$. Note that it is a function *only* of $L[\ ]$ and not of the nonhomogeneous term $-f(x)$. Once it is determined, the particular solution for an *arbitrary* $f(x)$ can be found directly from eq. (3.5.17).

Some steps of the procedure utilized above to derive the particular solution (3.5.17) involve arbitrariness. These include eqs. (3.5.3), (3.5.5), and so on. For this reason, we now need to verify that eq. (3.5.17) is indeed a particular solution of $L[y] = -f$. In doing so, we will also show that the particular solution (3.5.17) satisfies the homogeneous ICs:

$$y_p(x_0) = 0, \quad y_p^{(1)}(x_0) = 0, \ldots, \quad y_p^{(n-1)}(x_0) = 0 \tag{3.5.18}$$

It is apparent by observation that $y_p(x_0) = 0$, since the integration limits in eq. (3.5.17) are then identical. To evaluate the higher derivatives, we will need to differentiate the integral. For this, we use *Leibnitz formula*, which states that the derivative of the integral

$$I(x) = \int_{\alpha_1(x)}^{\alpha_2(x)} F(x, t) \, dt \tag{3.5.19a}$$

is given by

$$\frac{dI}{dx} = \int_{\alpha_1(x)}^{\alpha_2(x)} \frac{\partial F(x, t)}{\partial x} \, dt + \left[F(x, \alpha_2(x)) \frac{d\alpha_2}{dx} - F(x, \alpha_1(x)) \frac{d\alpha_1}{dx} \right] \tag{3.5.19b}$$

Thus differentiating eq. (3.5.17) gives

$$y_p^{(1)}(x) = \int_{x_0}^{x} \frac{-f(t)}{a_0(t)W(t)} \frac{\partial \bar{W}(x, t)}{\partial x} \, dt + \left[\frac{-f(x)}{a_0(x)} \right] \frac{\bar{W}(x, x)}{W(x)}$$

However, since by definition

$$\bar{W}(x, t) = \begin{vmatrix} f_1(t) & f_2(t) & \cdots & f_n(t) \\ f_1^{(1)}(t) & f_2^{(1)}(t) & \cdots & f_n^{(1)}(t) \\ \vdots & & & \\ f^{(n-2)}(t) & f_2^{(n-2)}(t) & \cdots & f_n^{(n-2)}(t) \\ f_1(x) & f_2(x) & \cdots & f_n(x) \end{vmatrix}$$

it follows that

$$\bar{W}(x, x) = 0$$

since the first and last rows are identical. This leads to

$$y_p^{(1)}(x) = \int_{x_0}^{x} \frac{-f(t)}{a_0(t)W(t)} \frac{\partial \bar{W}(x, t)}{\partial x} \, dt \tag{3.5.20}$$

and hence $y_p^{(1)}(x_0) = 0$.

Also note from the definition of derivative of a determinant (1.1.4) that

$$\frac{\partial^j \tilde{W}(x, t)}{\partial x^j} = \begin{vmatrix} f_1(t) & f_2(t) & \cdots & f_n(t) \\ f_1^{(1)}(t) & f_2^{(1)}(t) & \cdots & f_n^{(1)}(t) \\ \vdots & & & \\ f_1^{(n-2)}(t) & f_2^{(n-2)}(t) & \cdots & f_n^{(n-2)}(t) \\ f_1^{(j)}(x) & f_2^{(j)}(x) & \cdots & f_n^{(j)}(x) \end{vmatrix}$$

which implies that

$$\frac{\partial^j \tilde{W}(x, x)}{\partial x^j} = 0 \qquad \text{for } j = 0, 1, \ldots, (n - 2) \qquad (3.5.21)$$

since the $(j + 1)$st and last rows are identical.

Differentiating eq. (3.5.20), we obtain

$$y_p^{(2)}(x) = \int_{x_0}^{x} \frac{-f(t)}{a_0(t)W(t)} \frac{\partial^2 \tilde{W}(x, t)}{\partial x^2} \, dt + \left[\frac{-f(x)}{a_0(x)W(x)} \right] \frac{\partial \tilde{W}(x, x)}{\partial x}$$

which, using eq. (3.5.21), reduces to

$$y_p^{(2)}(x) = \int_{x_0}^{x} \frac{-f(t)}{a_0(t)W(t)} \frac{\partial^2 \tilde{W}(x, t)}{\partial x^2} \, dt \qquad (3.5.22)$$

and thus

$$y_p^{(2)}(x_0) = 0$$

Following this procedure successively yields

$$y_p^{(n-1)}(x) = \int_{x_0}^{x} \frac{-f(t)}{a_0(t)W(t)} \frac{\partial^{n-1} \tilde{W}(x, t)}{\partial x^{n-1}} \, dt + \left[\frac{-f(x)}{a_0(x)W(x)} \right] \frac{\partial^{n-2} \tilde{W}(x, x)}{\partial x^{n-2}}$$

Because the last term is zero, we have

$$y_p^{(n-1)}(x) = \int_{x_0}^{x} \frac{-f(t)}{a_0(t)W(t)} \frac{\partial^{n-1} \tilde{W}(x, t)}{\partial x^{n-1}} \, dt \qquad (3.5.23)$$

which implies that

$$y_p^{(n-1)}(x_0) = 0$$

With this last result, we have completed the proof of eq. (3.5.18).

We will now show that eq. (3.5.17) satisfies $L[y] = -f$. Note that differentiating eq. (3.5.23) gives

$$y_p^{(n)}(x) = \int_{x_0}^{x} \frac{-f(t)}{a_0(t)W(t)} \frac{\partial^n \tilde{W}(x, t)}{\partial x^n} \, dt + \left[\frac{-f(x)}{a_0(t)} \right] \qquad (3.5.24)$$

since $\partial^{n-1} \tilde{W}(x, x)/\partial x^{n-1} = W(x)$.

Utilizing the integral expressions for $y_p^{(j)}(x)$ just derived [i.e., eqs. (3.5.17), (3.5.20), and (3.5.22) to (3.5.24)], we obtain

$$L[y_p] = a_0(x)y_p^{(n)} + a_1(x)y_p^{(n-1)} + \cdots a_{n-1}(x)y_p^{(1)} + a_n(x)y_p$$

$$= -f(x) + \int_{x_0}^x \frac{-f(t)}{a_0(t)W(t)} \left[a_0(x)\frac{\partial^n \bar{W}(x,t)}{\partial x^n} + a_1(x)\frac{\partial^{n-1}\bar{W}(x,t)}{\partial x^{n-1}} + \cdots \right.$$

$$\left. + a_{n-1}(x)\frac{\partial \bar{W}(x,t)}{\partial x} + a_n(x)\bar{W}(x,t) \right] dt$$

However, the term in the square brackets can be simplified as

$$\begin{vmatrix} f_1(t) & f_2(t) & \cdots & f_n(t) \\ f_1^{(1)}(t) & f_2^{(1)}(t) & \cdots & f_n^{(1)}(t) \\ \vdots & & & \\ f_1^{(n-2)}(t) & f_2^{(n-2)}(t) & \cdots & f_n^{(n-2)}(t) \\ L[f_1] & L[f_2] & \cdots & L[f_n] \end{vmatrix}$$

which is zero because each term in the last row is zero, since $\{f_j\}$ is a fundamental set of solutions of the homogeneous equation $L[y] = 0$. We thus have the desired result that

$$L[y_p] = -f \tag{3.5.25}$$

so that the solution (3.5.17) is indeed a particular solution.

Finally, let us consider the case where $\{f_j\}$ is the fundamental set of solutions of $L[y] = 0$; that is, they satisfy the ICs

$$f_j^{(i-1)}(x_0) = \delta_{ij}$$

where

$$\delta_{ij} = \begin{cases} 1, & i = j \\ 0, & i \neq j \end{cases}$$

is the Kronecker delta. Then it follows from Theorem 2 of section 3.2 and Theorem 1 of section 3.4 that

$$y(x) = \sum_{j=1}^n \alpha_j f_j(x) + \int_{x_0}^x \frac{-f(t)}{a_0(t)} \frac{\bar{W}(x,t)}{W(t)} dt \tag{3.5.26}$$

is the unique solution of the ODE

$$L[y] = -f(x)$$

with ICs

$$y^{(j-1)}(x_0) = \alpha_j, \qquad j = 1, 2, \ldots, n$$

where α_j are prescribed constants.

Example. Using the one-sided Green's function, find a particular solution of

$$L[y] = y^{(3)} - y^{(1)} = x \qquad (3.5.27)$$

Recall from eq. (3.5.17) that a particular solution of

$$L[y] = -f$$

is given by

$$y_p = \int_{x_0}^{x} \frac{-f(t)}{a_0(t)} \frac{\bar{W}(x, t)}{W(t)} \, dt$$

where $W(x)$ is the Wronskian determinant corresponding to any fundamental set of solutions $\{f_j\}$ of the homogeneous equation $L[y] = 0$. Thus we need to first find a fundamental set of solutions of

$$y^{(3)} - y^{(1)} = 0$$

By the approach that is familiar, taking $y = e^{mx}$, the equation above leads to

$$m^3 - m = 0$$

This gives

$$m = 0, 1, -1$$

and hence

$$f_1 = 1, \qquad f_2 = e^x, \qquad f_3 = e^{-x} \qquad (3.5.28)$$

is a fundamental set of solutions of $L[y] = 0$. This can be verified by evaluating the Wronskian determinant

$$W(x) = \begin{vmatrix} f_1(x) & f_2(x) & f_3(x) \\ f_1^{(1)}(x) & f_2^{(1)}(x) & f_3^{(1)}(x) \\ f_1^{(2)}(x) & f_2^{(2)}(x) & f_3^{(2)}(x) \end{vmatrix}$$

$$= \begin{vmatrix} 1 & e^x & e^{-x} \\ 0 & e^x & -e^{-x} \\ 0 & e^x & e^{-x} \end{vmatrix}$$

$$= 2$$

which is nonzero. Further,

$$\bar{W}(x, t) = \begin{vmatrix} f_1(t) & f_2(t) & f_3(t) \\ f_1^{(1)}(t) & f_2^{(1)}(t) & f_3^{(1)}(t) \\ f_1(x) & f_2(x) & f_3(x) \end{vmatrix}$$

$$= \begin{vmatrix} 1 & e^t & e^{-t} \\ 0 & e^t & -e^{-t} \\ 1 & e^x & e^{-x} \end{vmatrix}$$

$$= -2 + e^{t-x} + e^{-t+x}$$

The particular solution is then

$$y_p(x) = \int_{x_0}^{x} \frac{t}{2} (-2 + e^{t-x} + e^{-t+x}) \, dt$$

$$= \frac{1}{2} [-2 + (x_0^2 - x^2) + (1 - x_0)e^{x_0-x} + (1 + x_0)e^{x-x_0}]$$

(3.5.29)

3.6 THE INVERSE OPERATOR

Consider the simple problem of a well-stirred surge tank (see Figure 3.1) which is utilized to damp out the fluctuations of, for example, temperature of a feed stream. If V is the tank volume, q is the volumetric flow rate, and C_p and ρ are the mass heat capacity and density, respectively, then the energy balance yields the ODE:

$$qC_p\rho T_i = qC_p\rho T + V \frac{d}{dt} (\rho C_p T)$$

Assuming that the physical properties are constant, this becomes

$$\theta \frac{dT}{dt} + T = T_i(t)$$

(3.6.1)

where $\theta = V/q$ is the tank holding time. Furthermore, if the initial temperature in the tank (i.e., an IC) is given, then we can readily solve for $T(t)$.

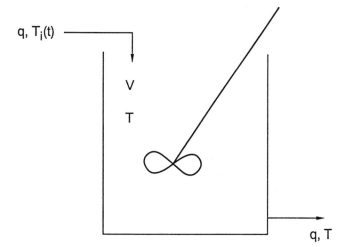

Figure 3.1 A well-stirred tank.

Note that the mathematical problem is a differential operator:

$$L[\ ] = \left(\theta \frac{d}{dt} + 1\right)[\ ] \tag{3.6.2}$$

operating on the *output* T to give the *input* T_i. This is just the opposite of what nature does: It operates on the input T_i to give the output T. The *natural* operator is thus the *inverse* of the mathematical operator, $L[\ ]$. With the IC $T(0) = 0$, this ODE can be solved by the familiar integrating factor approach [cf. eq. (2.1.5)] to yield the solution

$$T(t) = \int_0^t \frac{e^{-(t-\tau)/\theta}}{\theta} T_i(\tau)\ d\tau \tag{3.6.3}$$

The rhs may be viewed as an integral operator

$$L^{-1}[\ ] = \int_0^t \frac{e^{-(t-\tau)/\theta}}{\theta} [\ ]\ d\tau \tag{3.6.4}$$

acting on the input T_i; hence $L^{-1}[\ ]$ is the inverse or the natural operator for this system.
For the nth-order nonhomogeneous initial value problem

$$L[y] = -f \qquad x \in [a, b]$$
$$y^{(j)}(x_0) = 0 \qquad j = 0, 1, \ldots, (n-1) \tag{3.6.5}$$

the solution from eq. (3.5.26) will be just the particular solution:

$$y = \int_{x_0}^x \frac{-f(t)}{a_0(t)} \frac{\tilde{W}(x, t)}{W(t)}\ dt = L^{-1}[-f] \tag{3.6.6}$$

where the functions $\{f_j\}$ used to define W and $\tilde{W}$ are the fundamental set of $L[y] = 0$. The inverse operator

$$L^{-1}[\ ] = \int_{x_0}^x \frac{[\ ]}{a_0(t)} \frac{\tilde{W}(x, t)}{W(t)}\ dt \tag{3.6.7}$$

is the natural operator, and it is *linear* as well.
Furthermore, from the definitions of $L[\ ]$ and $L^{-1}[\ ]$ we have

$$L[y] = -f \quad \text{and} \quad L^{-1}[-f] = y$$

This leads to

$$L^{-1}L[y] = y$$

which implies that

$$L^{-1}L[\ ] = I[\ ] \tag{3.6.8}$$

defined as the *identity* operator. Similarly,

$$LL^{-1}[-f] = -f$$

which yields

$$LL^{-1}[\ ] = L^{-1}L[\ ] = I[\ ] \tag{3.6.9}$$

thus indicating that the operators $L[\ ]$ and $L^{-1}[\ ]$ are commutative.

3.7 THE ADJOINT EQUATION AND OPERATOR

Given a linear differential operator, an important concept is that of the corresponding *adjoint* operator (cf. Ince, 1956, chapter 5). This concept has significance when solving boundary value and eigenvalue problems. It will become clear later that these problems are readily solvable *only* for specific types of differential operators whose adjoints satisfy a certain property.

To see how the adjoint equation and the adjoint operator arise, let us consider the first-order linear ODE

$$L[y] = a_0(x)y^{(1)} + a_1(x)y = -f(x) \tag{3.7.1}$$

and ask the question, Is it possible that the lhs is a total derivative of something? If it were so, we could directly integrate the ODE to solve for y.

For example, if

$$a_0(x) = 0 \quad \text{and} \quad a_1(x) = 1$$

then we have

$$L[y] = xy^{(1)} + y = \frac{d}{dx}(xy) = -f$$

If the IC is $y(x_0) = y_0$, then integrating once and utilizing the IC gives

$$xy = -\int_{x_0}^{x} f(t)\, dt + x_0 y_0$$

and hence

$$y = -\frac{1}{x}\int_{x_0}^{x} f(t)\, dt + \frac{x_0 y_0}{x}$$

is the desired solution.

In general, however, the lhs will not be the total derivative of anything. Then we ask the question, Is it possible to find a function v such that

$$vL[y] = v\{a_0(x)y^{(1)} + a_1(x)y\} \tag{3.7.2}$$

is the derivative of something? In order to answer this, let us consider the identity

$$\{va_0 y\}^{(1)} = va_0 y^{(1)} + y\{va_0\}^{(1)}$$

which yields

$$va_0 y^{(1)} = \{va_0 y\}^{(1)} - y\{va_0\}^{(1)} \tag{3.7.3}$$

Utilizing this, eq. (3.7.2) becomes

$$\{va_0 y\}^{(1)} + y[-\{va_0\}^{(1)} + a_1 v] = -vf = vL[y] \tag{3.7.4}$$

If we *choose* v such that it satisfies

$$-\{va_0\}^{(1)} + a_1 v = 0 \tag{3.7.5}$$

then y is obtained by solving

$$\{va_0y\}^{(1)} = -vf \tag{3.7.6}$$

for which the lhs is indeed a total derivative. It can be seen that the function v is the familiar *integrating factor* for the first-order ODE (3.7.1). The first-order ODE (3.7.5) which defines it is called the *adjoint equation*, and the differential operator

$$L^*[\ ] = -\{a_0(x)[\ ]\}^{(1)} + a_1(x)[\ ] \tag{3.7.7}$$

is said to be *adjoint* to the operator $L[\ ]$. Note from eqs. (3.7.4) and (3.7.7) that

$$vL[y] - yL^*[v] = \frac{d}{dx}[va_0y] \tag{3.7.8}$$

Let us now consider the second-order linear ODE

$$L[y] = a_0y^{(2)} + a_1y^{(1)} + a_2y = -f \tag{3.7.9}$$

We again seek a function v such that $vL[y]$ given by

$$vL[y] = va_0y^{(2)} + va_1y^{(1)} + va_2y = -vf \tag{3.7.10}$$

becomes a total derivative. Now, as in eq. (3.7.3):

$$va_1y^{(1)} = \{va_1y\}^{(1)} - y\{va_1\}^{(1)}$$

Similarly,

$$va_0y^{(2)} = va_0\{y^{(1)}\}^{(1)}$$
$$= \{va_0y^{(1)}\}^{(1)} - y^{(1)}\{va_0\}^{(1)}$$
$$= \{va_0y^{(1)}\}^{(1)} - [\{(va_0)^{(1)}y\}^{(1)} - y\{va_0\}^{(2)}]$$

Upon substituting in eq. (3.7.10), these lead to

$$vL[y] = \frac{d}{dx}[va_0y^{(1)} + y\{va_1 - (va_0)^{(1)}\}]$$
$$+ y[\{va_0\}^{(2)} - \{va_1\}^{(1)} + va_2] = -vf \tag{3.7.11}$$

If we *choose* v such that

$$L^*[v] = \{va_0\}^{(2)} - \{va_1\}^{(1)} + va_2 = 0 \tag{3.7.12}$$

(the *adjoint* equation), then the lhs of eq. (3.7.11) is a total derivative.

The operator

$$L^*[\ ] = \{a_0[\ ]\}^{(2)} - \{a_1[\ ]\}^{(1)} + a_2[\ ] \tag{3.7.13}$$

is the adjoint operator of

$$L[\ ] = a_0[\ ]^{(2)} + a_1[\ ]^{(1)} + a_2[\ ]$$

and from eqs. (3.7.11) and (3.7.12) we have

$$vL[y] - yL^*[v] = \frac{d}{dx}[va_0y^{(1)} + y\{va_1 - (va_0)^{(1)}\}] \tag{3.7.14}$$

Definition. If $L[\ ] = L^*[\ ]$, then the operator $L[\ ]$ is said to be a *self-adjoint* operator.

For the second-order operator

$$L[u] = a_0 u^{(2)} + a_1 u^{(1)} + a_2 u \qquad (3.7.15)$$

from eq. (3.7.13), the adjoint operator is

$$L^*[v] = \{a_0 v\}^{(2)} - \{a_1 v\}^{(1)} + a_2 v$$
$$= a_0 v^{(2)} + \{2a_0^{(1)} - a_1\} v^{(1)} + \{a_0^{(2)} - a_1^{(1)} + a_2\} v$$

Then $L[\ ]$ will be self-adjoint if and only if

$$2a_0^{(1)} - a_1 = a_1 \quad \text{and} \quad a_0^{(2)} = a_1^{(1)}$$

However, both of these conditions are satisfied with

$$a_1 = a_0^{(1)} \qquad (3.7.16)$$

Thus we have the important result that the second-order operator (3.7.15) is self-adjoint if and only if eq. (3.7.16) applies.

Note that the second-order operator

$$L[y] = \frac{d}{dx}\left\{p(x)\,\frac{dy}{dx}\right\} + q(x)y \qquad (3.7.17)$$

which arises commonly in problems involving transport phenomena, is intrinsically self-adjoint. On the other hand, since the leading order terms in eqs. (3.7.1) and (3.7.7) have opposite signs, a first-order operator can *never* be self-adjoint.

We now discuss an important result which applies only to second-order operators.

● **THEOREM**

Every second-order linear differential operator can be made self-adjoint.

Proof
Let us consider

$$L[y] = a_0 y^{(2)} + a_1 y^{(1)} + a_2 y$$

along with the condition

$$a_1(x) \neq a_0^{(1)}(x)$$

so that as it stands, $L[\ ]$ is not self-adjoint. The claim is that it can *always* be made self-adjoint.

For this, consider the modified operator

$$L_1[y] = g(x)L[y] = ga_0 y^{(2)} + ga_1 y^{(1)} + ga_2 y \qquad (3.7.18)$$

where g is as yet an unknown function. If the new operator $L_1[\ ]$ is to be self-adjoint, following eq. (3.7.16), we must have

$$\{ga_0\}^{(1)} = ga_1$$

or

$$g^{(1)}a_0 + ga_0^{(1)} = ga_1$$

Rearranging the above equation as

$$\frac{g^{(1)}}{g} = -\frac{a_0^{(1)}}{a_0} + \frac{a_1}{a_0}$$

and integrating once,

$$\ln g = -\ln a_0 + \int^x \frac{a_1(t)}{a_0(t)}\, dt$$

we obtain

$$g(x) = \frac{1}{a_0(x)} \exp\left\{ \int^x \frac{a_1(t)}{a_0(t)}\, dt \right\} \tag{3.7.19}$$

up to an arbitrary multiplying constant. So although $L[\ ]$ by itself is not self-adjoint, we know the function $g(x)$ to multiply it with, so that the new operator

$$L_1[\ ] = gL[\ ]$$

is self-adjoint. This completes the proof.

We now derive the adjoint operator corresponding to the nth-order linear differential operator:

$$L[y] = a_0 y^{(n)} + a_1 y^{(n-1)} + \cdots + a_{n-1} y^{(1)} + a_n y \tag{3.7.20}$$

Multiplying both sides by v gives

$$vL[y] = v\{a_0 y^{(n)} + a_1 y^{(n-1)} + \cdots + a_{n-1} y^{(1)} + a_n y\} \tag{3.7.21}$$

Using the same manipulations as before with first- and second-order operators, we obtain

$$va_n y = y(a_n v)$$
$$va_{n-1} y^{(1)} = \{va_{n-1} y\}^{(1)} - \{va_{n-1}\}^{(1)} y$$
$$\vdots$$
$$va_1 y^{(n-1)} = \{va_1 y^{(n-2)}\}^{(1)} - \{va_1\}^{(1)} y^{(n-2)}$$
$$= \{va_1 y^{(n-2)}\}^{(1)} - \{(va_1)^{(1)} y^{(n-3)}\}^{(1)} + \{va_1\}^{(2)} y^{(n-3)}$$
$$\vdots$$
$$\vdots = \{va_1 y^{(n-2)} - (va_1)^{(1)} y^{(n-3)} + \cdots + (-1)^{n-2}(va_1)^{(n-2)} y\}^{(1)}$$
$$+ (-1)^{n-1}\{va_1\}^{(n-1)} y$$
$$va_0 y^{(n)} = \{va_0 y^{(n-1)}\}^{(1)} - \{va_0\}^{(1)} y^{(n-1)}$$
$$= \{va_0 y^{(n-1)}\}^{(1)} - \{(va_0)^{(1)} y^{(n-2)}\}^{(1)} + \{va_0\}^{(2)} y^{(n-2)}$$
$$\vdots$$
$$\vdots = \{va_0 y^{(n-1)} - (va_0)^{(1)} y^{(n-2)} + \cdots + (-1)^{n-1}(va_0)^{(n-1)} y\}^{(1)}$$
$$+ (-1)^n \{va_0\}^{(n)} y$$

Substituting all these in eq. (3.7.21) gives

$$vL[y] = y[va_n - \{va_{n-1}\}^{(1)} + \cdots + (-1)^{n-1}\{va_1\}^{(n-1)} + (-1)^n\{va_0\}^{(n)}]$$

$$+ \begin{bmatrix} va_{n-1}y + \{va_{n-2}\}y^{(1)} - \{va_{n-2}\}^{(1)}y \\ \vdots \\ + \{va_0\}y^{(n-1)} - \{va_0\}^{(1)}y^{(n-2)} + \cdots + (-1)^{n-1}\{va_0\}^{n-1}y \end{bmatrix}^{(1)}$$

$$= yL^*[v] + \frac{d}{dx}\pi(v, y) \qquad (3.7.22)$$

where

$$L^*[\ ] = [\ ]a_n - \{[\ ]a_{n-1}\}^{(1)} + \cdots + (-1)^{n-1}\{[\ ]a_1\}^{(n-1)} + (-1)^n\{[\ ]a_0\}^{(n)}$$

is the operator adjoint to $L[\ ]$, and $\pi(v, y)$ is called the *bilinear concomitant*. Equation (3.7.22) rearranged as

$$vL[y] - yL^*[v] = \frac{d}{dx}\pi(v, y) \qquad (3.7.23)$$

is called the *Lagrange identity*.

In solving boundary value problems, we will be interested in the integral of eq. (3.7.23) over a finite interval $[a, b]$. This gives

$$\int_a^b \{vL[y] - yL^*[v]\}\, dx = \pi(v, y)\Big|_a^b \qquad (3.7.24)$$

which is called *Green's formula*.

II. BOUNDARY VALUE PROBLEMS

3.8 THE FREDHOLM ALTERNATIVE

Many problems in science and engineering are such that conditions are specified not at one point (as in initial value problems) but rather over a surface—that is, at two or more points. Examples include heating, cooling, or drying of solids, simultaneous diffusion and reaction in a catalyst pellet, structural problems, and so on. Such problems are eventually formulated as

$$L[y] = -f, \qquad x \in (a, b) \qquad (3.8.1)$$

where $L[\ ]$ is an nth-order linear differential operator and $-f$ is the nonhomogeneous term. To determine the solution y completely, we need n subsidiary conditions. In general, these may involve the function y and its first $(n - 1)$ derivatives at both ends of the interval $[a, b]$. The general *linear* boundary conditions (BCs) thus are

$$\alpha_{10}y(a) + \alpha_{11}y^{(1)}(a) + \cdots + \alpha_{1,n-1}y^{(n-1)}(a)$$
$$+ \beta_{10}y(b) + \beta_{11}y^{(1)}(b) + \cdots + \beta_{1,n-1}y^{(n-1)}(b) = \gamma_1$$
$$\alpha_{20}y(a) + \alpha_{21}y^{(1)}(a) + \cdots + \alpha_{2,n-1}y^{(n-1)}(a)$$
$$+ \beta_{20}y(b) + \beta_{21}y^{(1)}(b) + \cdots + \beta_{2,n-1}y^{(n-1)}(b) = \gamma_2 \qquad (3.8.2)$$
$$\vdots$$
$$\alpha_{n0}y(a) + \alpha_{n1}y^{(1)}(a) + \cdots + \alpha_{n,n-1}y^{(n-1)}(a)$$
$$+ \beta_{n0}y(b + \beta_{n1}y^{(1)}(b) + \cdots + \beta_{n,n-1}y^{(n-1)}(b) = \gamma_n$$

where α_{ij}, β_{ij}, and γ_i are prescribed constants. These BCs may be compactly rewritten as

$$U_i[y] = \gamma_i, \qquad i = 1, 2, \ldots, n \qquad (3.8.3)$$

where $U_i[\ ]$ is the linear *boundary operator*:

$$U_i[\ ] = \alpha_{i0}[\ ](a) + \alpha_{i1}[\ ]^{(1)}(a) + \cdots + \alpha_{i,n-1}[\ ]^{(n-1)}(a) \qquad (3.8.4)$$
$$+ \beta_{i0}[\ ](b) + \beta_{i1}[\ ]^{(1)}(b) + \cdots + \beta_{i,n-1}[\ ]^{(n-1)}(b)$$

From eq. (3.4.2) the general solution of eq. (3.8.1) is given by

$$y = \sum_{j=1}^{n} c_j f_j(x) + y_p(x) \qquad (3.8.5)$$

where $\{f_j\}$ is a fundamental set of solutions of the homogeneous equation $L[y] = 0$, and y_p is a particular solution of $L[y] = -f$. The unknown constants $\{c_j\}$ are determined by satisfying the BCs (3.8.3):

$$U_i[y] = U_i\left[\sum_{j=1}^{n} c_j f_j(x) + y_p(x)\right]$$
$$= \sum_{j=1}^{n} c_j U_i[f_j(x)] + U_i[y_p(x)]$$
$$= \gamma_i$$

If we let

$$U_i[f_j(x)] = U_{ij} \qquad (3.8.6a)$$

and

$$\gamma_i - U_i[y_p(x)] = \mu_i \qquad (3.8.6b)$$

the unknown constants $\{c_j\}$ satisfy the system of linear algebraic equations:

$$U_{11}c_1 + U_{12}c_2 + \cdots + U_{1n}c_n = \mu_1$$
$$\vdots \qquad\qquad\qquad\qquad (3.8.7)$$
$$U_{n1}c_1 + U_{n2}c_2 + \cdots + U_{nn}c_n = \mu_n$$

A unique solution of eq. (3.8.7) exists if and only if

$$
|\mathbf{U}| =
\begin{vmatrix}
U_{11} & U_{12} & \cdots & U_{1n} \\
U_{21} & U_{22} & \cdots & U_{2n} \\
\vdots & & & \\
U_{n1} & U_{n2} & \cdots & U_{nn}
\end{vmatrix}
\neq 0
$$

If *all* γ_i equal zero *and* the nonhomogeneous term f equals zero—that is, we have a homogeneous ODE with homogeneous BCs—then $\mu_i = 0$. This along with $|\mathbf{U}| \neq 0$ would imply that $\mathbf{c} = \mathbf{0}$, and then from eq. (3.8.5) we have $y(x) = 0$.

Thus we can summarize the important result called *Fredholm alternative*: The nonhomogeneous boundary value problem given by eqs. (3.8.1) and (3.8.2) has a *unique* solution *if and only if* the corresponding homogeneous problem has only the *trivial* solution.

Example 1. Consider the nonhomogeneous boundary value problem

$$
L[y] = \frac{d^2y}{dx^2} = x, \qquad x \in (0, 1) \tag{3.8.8}
$$

$$
y(0) = 0 \;\; = y(1) \tag{3.8.9}
$$

The corresponding *homogeneous problem*

$$
L[y] = \frac{d^2y}{dx^2} = 0, \qquad x \in (0, 1) \tag{3.8.10}
$$

$$
y(0) = 0 \quad y(1) \tag{3.8.11}
$$

has the general solution

$$
y = c_1 x + c_2
$$

where c_1 and c_2 are arbitrary constants, which are determined by satisfying the BCs as follows:

$$
c_1 = 0 \quad \text{and} \quad c_2 = 0
$$

Thus the homogeneous problem has only the trivial solution $y = 0$. The Fredholm alternative ensures that the nonhomogeneous problem will have a unique solution. Let us confirm that this is the case.

A fundamental set for the homogeneous problem is

$$
f_1 = x, \qquad f_2 = 1
$$

A particular solution may be obtained either from the one-sided Green's function approach of section 3.5 or by the familiar *method of undetermined coefficients*. Proceeding by the latter, assume that the particular solution has the form

$$
y_p = d_1 x^3 + d_2 x^2
$$

Then we have

$$
\frac{dy_p}{dx} = 3d_1 x^2 + 2d_2 x
$$

and

$$\frac{d^2y_p}{dx^2} = 6d_1x + 2d_2$$

Substituting these expressions in the ODE (3.8.8) yields

$$6d_1x + 2d_2 = x$$

which, by comparing terms on both sides, gives

$$d_1 = \tfrac{1}{6} \quad \text{and} \quad d_2 = 0$$

The general solution of the nonhomogeneous ODE (3.8.8) is then

$$y = c_1x + c_2 + \frac{x^3}{6}$$

where the constants c_1 and c_2 are determined from the BCs (3.8.9) as

$$c_1 = -\tfrac{1}{6} \quad \text{and} \quad c_2 = 0$$

Hence

$$y = \frac{x}{6}(-1 + x^2)$$

is the unique solution of the nonhomogeneous problem.

Example 2. Let us now consider the same ODE as in the previous example:

$$L[y] = \frac{d^2y}{dx^2} = x, \qquad x \in (0, 1) \tag{3.8.12}$$

but with different BCs:

$$y^{(1)}(0) = 0 = y^{(1)}(1) \tag{3.8.13}$$

The corresponding homogeneous problem

$$\frac{d^2y}{dx^2} = 0, \qquad x \in (0, 1)$$

$$y^{(1)}(0) = 0 = y^{(1)}(1)$$

has the solution

$$y = a$$

where a is an *arbitrary* constant.

For the nonhomogeneous ODE (3.8.12), as in Example 1, the general solution is

$$y = c_1x + c_2 + \frac{x^3}{6} \tag{3.8.14}$$

where c_1 and c_2 are constants to be determined by satisfying the BCs (3.8.13). Differentiating eq. (3.8.14) yields

$$y^{(1)} = c_1 + \frac{x^2}{2}$$

Applying the BC at $x = 0$, we obtain

$$c_1 = 0$$

which implies that

$$y^{(1)} = \frac{x^2}{2}$$

At $x = 1$, this gives

$$y^{(1)}(1) = \tfrac{1}{2} \neq 0$$

which contradicts the given BC at $x = 1$.

Thus we see that, as dictated by the Fredholm alternative, since the corresponding homogeneous problem does not have *only* the trivial solution, there is *no* solution for the nonhomogeneous problem!

3.9   DISCUSSION OF BOUNDARY CONDITIONS

Here we take a diversion to discuss boundary conditions (BCs) and how they arise. In reality, all BCs come from nature, and not from mathematics. To derive them, the same principles that are utilized in arriving at the ODE itself are used. Their correctness, like that of the differential equation model, can be verified only by comparison with experiment.

As an example, consider the problem of steady-state heat conduction in an infinite slab shown in Figure 3.2. Since the other dimensions are infinite, the flow of heat is in the x-direction only. Let the temperature of the fluid in which the slab is immersed be

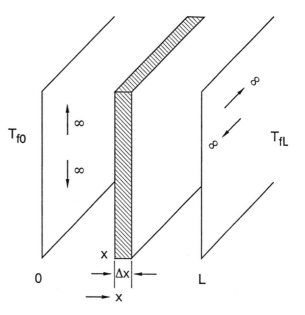

Figure 3.2 Heat conduction in an infinite slab.

T_{f0} and T_{fL}, on the lhs and rhs, respectively. Further, the temperature within the slab is denoted by $T(x)$, and the slab thermal conductivity is denoted by k. Now, consider a differential element of thickness Δx, located at x. Then by Fourier's law of heat conduction, we have

$$\text{Flux of heat } in \text{ at } x \qquad = -k \left.\frac{dT}{dx}\right|_x$$

$$\text{Flux of heat } out \text{ at } x + \Delta x = -k \left.\frac{dT}{dx}\right|_{x+\Delta x}$$

In the steady state, the principle of conservation of energy states that the two fluxes must be equal; thus

$$k \left.\frac{dT}{dx}\right|_{x+\Delta x} - k \left.\frac{dT}{dx}\right|_x = 0$$

Dividing through by Δx and taking the limit as $\Delta x \to 0$ yields the ODE:

$$\frac{d}{dx}\left(k \frac{dT}{dx}\right) = 0, \qquad x \in (0, L) \tag{3.9.1}$$

If we consider a somewhat more difficult problem which in addition involves internal heat *generation* (perhaps due to electrical heaters embedded, or due to chemical reaction) at rate Q, then for the differential element we have

$$\text{Flux of heat } in \text{ at } x \qquad = -k \left.\frac{dT}{dx}\right|_x + Q\Delta x$$

$$\text{Flux in heat } out \text{ at } x + \Delta x = -k \left.\frac{dT}{dx}\right|_{x\ +\ \Delta x}$$

By the same arguments as above, we obtain

$$\frac{d}{dx}\left(k \frac{dT}{dx}\right) = -Q, \qquad x \in (0, L) \tag{3.9.2}$$

where Q may well be a function of position.

If the thermal conductivity k is constant, eq. (3.9.2) becomes

$$\frac{d^2T}{dx^2} = -\frac{Q}{k}, \qquad x \in (0, L) \tag{3.9.3}$$

Note that arriving at the relevant mathematical problem (in this case an ODE) is the model-building step. In this step, a physical problem has been translated into a mathematical problem. In the present example, this required using the ideas of

1. conservation of energy, and
2. phenomenological law of heat conduction—that is, Fourier law.

From the way the ODE has been derived, it should be apparent that it is valid for $x \in (0, L)$. To solve the problem, we need to provide conditions which reflect how the slab interacts with the surroundings; these are called *boundary conditions* (BCs).

Consider the right end of the slab, at $x = L$:

$$\text{Flux of heat } \textit{out}, \text{ from the viewpoint of an observer } \textit{in} \text{ the slab } = -k \left. \frac{dT}{dx} \right|_{L_-}$$

$$\text{Flux of heat } \textit{out}, \text{ from the viewpoint of an observer } \textit{outside} \text{ the slab}$$
$$= h_L(T|_{L_+} - T_{fL})$$

following Newton's law of cooling, where h_L is the heat transfer coefficient. Since the two observers are witnessing the same fact, the two fluxes must be equal. Thus

$$-k \left. \frac{dT}{dx} \right|_{L_-} = h_L(T|_{L_+} - T_{fL})$$

and since the surface has no thickness, L_+ and L_- are the same location (i.e., L); thus

$$-k \frac{dT}{dx} = h_L(T - T_{fL}), \qquad x = L \qquad (3.9.4)$$

is the BC at $x = L$.

to derive the BC at $x = 0$, we repeat the same considerations for the left end of the slab:

$$\text{Flux of heat } \textit{in}, \text{ from the viewpoint of an observer } \textit{in} \text{ the slab } = -k \left. \frac{dT}{dx} \right|_{0_+}$$

$$\text{Flux of heat } \textit{in}, \text{ from the viewpoint of an observer } \textit{outside} \text{ the slab}$$
$$= h_0(T_{f0} - T|_{0_-})$$

and thus

$$-k \frac{dT}{dx} = h_0(T_{f0} - T), \qquad x = 0 \qquad (3.9.5)$$

is the BC at $x = 0$. Note the sign difference between the BCs at $x = 0$ and $x = L$.

Boundary conditions of the type (3.9.4) and (3.9.5) are called *third-kind, natural,* or *Robin* BCs. There are two special cases of these BCs which are worth considering.

Case 1. $h \to \infty$
This means that we have good external convection, so the heat arriving at the ends of the slab (from within) is removed immediately. Then $k/h \to 0$, and we have

$$T = T_{f0}, \qquad x = 0 \qquad (3.9.6)$$
$$T = T_{fL}, \qquad x = L \qquad (3.9.7)$$

which are called *first-kind* or *Dirichlet* BCs.

Case 2. $h \to 0$

This implies that the ends are insulated, and thus the heat arriving at the ends of the slab (from within) cannot escape. This leads to

$$\frac{dT}{dx} = 0, \qquad x = 0 \tag{3.9.8}$$

$$\frac{dT}{dx} = 0, \qquad x = L \tag{3.9.9}$$

termed *second-kind* or *Neumann* BCs.

Note that in a given problem we may have a combination of BCs. Thus, for example, the left end of the slab may be insulated while the right end may have good convection. The point is that what the applicable BC is, will depend entirely on the physical situation maintained outside the slab (i.e., h equal to zero, finite, or infinite).

3.10 SELF-ADJOINT PROBLEMS

There are two entities which constitute a boundary value problem: the differential operator and the BCs. By the method discussed in section 3.7, given a differential operator $L[\ ]$ the adjoint operator $L^*[\ ]$ can be found. Now Green's formula (3.7.24) states that

$$\int_a^b \{vL[y] - yL^*[v]\}\ dx = \pi(v, y)\big|_a^b \tag{3.10.1}$$

We now examine, for a given set of *homogeneous* BCs for the original operator $L[\ ]$, how the BCs for the adjoint operator $L^*[\ ]$ should be chosen so that

$$\pi(v, y)\big|_a^b = 0$$

that is,

$$\int_a^b \{vL[y] - yL^*[v]\}\ dx = 0 \tag{3.10.2}$$

The BCs for the adjoint operator thus chosen are called *adjoint* BCs.

We can address this issue for every $L[\ ]$, but let us in particular consider

$$L[y] = \frac{d}{dx}\left[p(x)\frac{dy}{dx}\right] + q(x)y, \qquad x \in (a, b) \tag{3.10.3}$$

a second-order operator on which we will focus from now on almost exclusively. We know from section 3.7 that $L[\ ]$ is self-adjoint; thus

$$L^*[v] = \frac{d}{dx}\left[p(x)\frac{dv}{dx}\right] + q(x)v$$

From eq. (3.10.1) we obtain

$$\int_a^b \{vL[y] - yL^*[v]\} \, dx = \pi(v, y)\big|_a^b$$

$$= p(b)[v(b)y'(b) - y(b)v'(b)]$$

$$- p(a)[v(a)y'(a) - y(a)v'(a)] \qquad (3.10.4)$$

which is obtained either directly from the definition of $\pi(v, y)$ in eq. (3.7.22) or, alternatively, from the particular $L[\ ]$ being considered as follows:

$$\int_a^b \{vL[y] - yL^*[v]\} \, dx = \int_a^b \left\{ v\left[\frac{d}{dx}\left(p(x)\frac{dy}{dx}\right) + q(x)y \right] \right.$$

$$\left. - y\left[\frac{d}{dx}\left(p(x)\frac{dv}{dx}\right) + q(x)v \right] \right\} dx$$

$$= \int_a^b \left[v\frac{d}{dx}\left(p(x)\frac{dy}{dx}\right) - y\frac{d}{dx}\left(p(x)\frac{dv}{dx}\right) \right] dx$$

$$= \left[vp(x)\frac{dy}{dx} - yp(x)\frac{dv}{dx} \right]_a^b$$

which gives eq. (3.10.4).

Let us now consider the homogeneous versions of the three types of BCs for y and in each case investigate what the BCs for v should be so that the rhs of eq. (3.10.4) is zero.

First-Kind BCs

$$y(a) = 0 \quad \text{and} \quad y(b) = 0 \qquad (3.10.5)$$

From eq. (3.10.4) we obtain

$$\int_a^b \{vL[y] - yL^*[v]\} \, dx = p(b)y'(b)v(b) - p(a)y'(a)v(a)$$

$$= 0$$

if we choose as the BCs for v:

$$v(a) = 0 \quad \text{and} \quad v(b) = 0 \qquad (3.10.6)$$

that is, the *same* BCs as for y.

Second-Kind BCs

$$y'(a) = 0 \quad \text{and} \quad y'(b) = 0 \qquad (3.10.7)$$

From eq. (3.10.4) we obtain

$$\int_a^b \{vL[y] - yL^*[v]\} \, dx = -p(b)y(b)v'(b) + p(a)y(a)v'(a)$$

$$= 0$$

if we choose as the BCs for v:

$$v'(a) = 0 \quad \text{and} \quad v'(b) = 0 \tag{3.10.8}$$

which are again the same BCs as for y.

Third-Kind BCs

$$-k(a)y'(a) + h(a)y(a) = 0$$
$$k(b)y'(b) + h(b)y(b) = 0 \tag{3.10.9}$$

From eq. (3.10.4) we obtain

$$\int_a^b \{vL[y] - yL^*[v]\}\, dx = p(b)[v(b)y'(b) - y(b)v'(b)]$$
$$- p(a)[v(a)y'(a) - y(a)v'(a)]$$

Replacing $y'(a)$ and $y'(b)$ in terms of $y(a)$ and $y(b)$, respectively, using the BCs (3.10.9) for y, we obtain

$$\int_0^b \{vL[y] - yL^*[v]\}\, dx = p(b)v(b)\left[-\frac{h(b)}{k(b)}y(b)\right] - p(b)y(b)v'(b)$$
$$- p(a)v(a)\left[\frac{h(a)}{k(a)}y(a)\right] + p(a)y(a)v'(a)$$
$$= -\frac{p(b)y(b)}{k(b)}[k(b)v'(b) + h(b)v(b)]$$
$$- \frac{p(a)y(a)}{k(a)}[-k(a)v'(a) + h(a)v(a)]$$
$$= 0$$

with the choice of BCs for v:

$$-k(a)v'(a) + h(a)v(a) = 0$$
$$k(b)v'(b) + h(b)v(b) = 0 \tag{3.10.10}$$

the same BCs as for y.

We thus find that for the general self-adjoint second-order operator (3.10.3) together with *homogeneous* versions of the three types of BCs, we obtain

$$\int_a^b \{vL[y] - yL^*[v]\}\, dx = 0$$

if we choose the *same* BCs for $L^*[\ ]$ as for $L[\ ]$. In this case we say that the full problem—that is, the differential operator along with the BCs—is self-adjoint. It can be readily verified that the adjoint BCs are the same as the original BCs even if the original BCs are mixed—that is, of one kind (first, second, or third) at one end and another kind at the other end.

3.11   SOLUTION OF NONHOMOGENEOUS EQUATIONS

In this section we develop the Green's function for boundary value problems, which can be used to solve problems with nonhomogeneous terms (cf. Roach, 1982 and Greenberg, 1971).

Specifically, we analyze cases where the nonhomogeneity arises *only* in the ODE and not in the BCs. The case of nonhomogeneous BCs is treated in section 3.12.

3.11.1   The Green's Function

Consider the second-order nonhomogeneous ODE

$$L[y] = \frac{d}{dx}\left(p(x)\frac{dy}{dx}\right) + q(x)y = -f(x), \qquad x \in (a, b) \tag{3.11.1}$$

for which the operator $L[\ ]$ is self-adjoint, along with self-adjoint *homogeneous* BCs:

$$-k_a \frac{dy}{dx} + h_a y = 0, \qquad x = a \tag{3.11.2a}$$

$$k_b \frac{dy}{dx} + h_b y = 0, \qquad x = b \tag{3.11.2b}$$

Let $\phi_1(x)$ and $\phi_2(x)$ be a fundamental set of $L[y] = 0$; then from eq. (3.5.17)

$$y_p(x) = \int_a^x \frac{-f(t)}{p(t)} \frac{\tilde{W}(x, t)}{W(t)}\, dt \tag{3.11.3}$$

is a particular solution, where

$$W(x) = \begin{vmatrix} \phi_1(x) & \phi_2(x) \\ \phi_1'(x) & \phi_2'(x) \end{vmatrix}$$

is the Wronskian determinant corresponding to the fundamental set of solutions, and

$$\tilde{W}(x, t) = \begin{vmatrix} \phi_1(t) & \phi_2(t) \\ \phi_1(x) & \phi_2(x) \end{vmatrix}$$

The general solution of the ODE

$$L[y] = -f(x), \qquad x \in (a, b)$$

can then be written as

$$y(x) = c_1\phi_1(x) + c_2\phi_2(x) + y_p(x) \tag{3.11.4}$$

where c_1 and c_2 are constants to be evaluated using the BCs. Differentiating yields

$$y' = c_1\phi_1' + c_2\phi_2' + y_p'$$

$$= c_1\phi_1' + c_2\phi_2' + \int_a^x \frac{-f(t)}{p(t)W(t)} \begin{vmatrix} \phi_1(t) & \phi_2(t) \\ \phi_1'(x) & \phi_2'(x) \end{vmatrix} dt$$

$$+ \left[\frac{-f(x)}{p(x)}\right] \frac{\tilde{W}(x, x)}{W(x)}$$

in which by observation the last term is zero.

We now apply the BC (3.11.2a) at $x = a$:

$$-k_a[c_1\phi_1'(a) + c_2\phi_2'(a)] + h_a[c_1\phi_1(a) + c_2\phi_2(a)] = 0$$

and the BC (3.11.2b) at $x = b$:

$$k_b[c_1\phi_1'(b) + c_2\phi_2'(b)] + k_b \int_a^b \frac{-f(t)}{p(t)W(t)} \begin{vmatrix} \phi_1(t) & \phi_2(t) \\ \phi_1'(b) & \phi_2'(b) \end{vmatrix} dt$$

$$+ h_b[c_1\phi_1(b) + c_2\phi_2(b)] + h_b \int_a^b \frac{-f(t)}{p(t)W(t)} \begin{vmatrix} \phi_1(t) & \phi_2(t) \\ \phi_1(b) & \phi_2(b) \end{vmatrix} dt = 0$$

Rearranging these two equations in the two unknowns c_1 and c_2, we have

$$[-k_a\phi_1'(a) + h_a\phi_1(a)]c_1 + [-k_a\phi_2'(a) + h_a\phi_2(a)]c_2 = 0 \qquad (3.11.5)$$

$$[k_b\phi_1'(b) + h_b\phi_1(b)]c_1 + [k_b\phi_2'(b) + h_b\phi_2(b)]c_2$$

$$= \int_a^b \frac{f(t)}{p(t)W(t)} \begin{vmatrix} \phi_1(t) & \phi_2(t) \\ k_b\phi_1'(b) + h_b\phi_1(b) & k_b\phi_2'(b) + h_b\phi_2(b) \end{vmatrix} dt$$
$$(3.11.6)$$

Note that so far no restrictions have been placed on $\phi_1(x)$ and $\phi_2(x)$, *except* that they are a fundamental set of solutions of $L[y] = 0$—that is, they satisfy

$$L[\phi_1] = 0, \qquad L[\phi_2] = 0$$

and their Wronskian determinant $W(x) \neq 0$. In particular, both ϕ_1 and ϕ_2 contain two arbitrary constants each, which can be evaluated by providing suitable conditions. For this we choose

$$\phi_1 \text{ such that it satisfies the BC at } x = a$$
$$(3.11.7)$$
$$\phi_2 \text{ such that it satisfies the BC at } x = b$$

Thus ϕ_1 and ϕ_2 still have one arbitrary constant left in each, which we will determine later. Using conditions (3.11.7), we have

$$-k_a\phi_1'(a) + h_a\phi_1(a) = 0$$

and

$$k_b\phi_2'(b) + h_b\phi_2(b) = 0$$

Thus from eq. (3.11.5)

$$[-k_a\phi_2'(a) + h_a\phi_2(a)]c_2 = 0$$

which implies that

$$c_2 = 0 \qquad (3.11.8)$$

since

$$-k_a \phi_2'(a) + h_a \phi_2(a) \neq 0$$

The latter condition holds, because otherwise ϕ_2 would satisfy homogeneous BCs at both ends $x = a$ and $x = b$, and since it is a solution of the homogeneous equation $L[\phi] = 0$, it would then be merely a trivial solution.

From eq. (3.11.6) we obtain

$$[k_b \phi_1'(b) + h_b \phi_1(b)]c_1 = \int_a^b \frac{f(t)}{p(t)W(t)} \begin{vmatrix} \phi_1(t) & \phi_2(t) \\ k_b \phi_1'(b) + h_b \phi_1(b) & 0 \end{vmatrix} dt$$

$$= \int_a^b \frac{-f(t)\phi_2(t)}{p(t)W(t)} [k_b \phi_1'(b) + h_b \phi_1(b)] \, dt$$

and thus

$$c_1 = \int_a^b \frac{-f(t)\phi_2(t)}{p(t)W(t)} \, dt \qquad (3.11.9)$$

Now that both c_1 and c_2 are known, the complete solution (3.11.4) is

$$y(x) = c_1 \phi_1(x) + c_2 \phi_2(x) + y_p(x)$$

$$= \phi_1(x) \int_a^b \frac{-f(t)\phi_2(t)}{p(t)W(t)} \, dt + \int_a^x \frac{-f(t)}{p(t)} \frac{\tilde{W}(x, t)}{W(t)} \, dt$$

$$= \phi_1(x) \int_a^b \frac{-f(t)\phi_2(t)}{p(t)W(t)} \, dt + \int_a^x \frac{-f(t)}{p(t)W(t)} [\phi_1(t)\phi_2(x) - \phi_1(x)\phi_2(t)] \, dt$$

$$= \int_a^x \frac{-f(t)\phi_1(x)\phi_2(t)}{p(t)W(t)} \, dt + \int_x^b \frac{-f(t)\phi_1(x)\phi_2(t)}{p(t)W(t)} \, dt$$

$$+ \int_a^x \frac{-f(t)\phi_1(t)\phi_2(x)}{p(t)W(t)} \, dt - \int_a^x \frac{-f(t)\phi_1(x)\phi_2(t)}{p(t)W(t)} \, dt$$

$$= \int_a^x \frac{-f(t)}{p(t)W(t)} \phi_1(t)\phi_2(x) \, dt + \int_x^b \frac{-f(t)}{p(t)W(t)} \phi_1(x)\phi_2(t) \, dt \qquad (3.11.10)$$

We make a brief diversion here to show that $p(t)W(t)$ is a constant. Recall that ϕ_i satisfy $L[\phi_i] = 0$; thus

$$\frac{d}{dx} [p(x)\phi_1'] + q(x)\phi_1 = 0$$

and

$$\frac{d}{dx} [p(x)\phi_2'] + q(x)\phi_2 = 0$$

Multiplying the first by ϕ_2, the second by ϕ_1 and then subtracting, we obtain

$$\phi_1 \frac{d}{dx} [p(x)\phi_2'] - \phi_2 \frac{d}{dx} [p(x)\phi_1'] = 0$$

or

$$\phi_1[p'\phi_2' + p\phi_2''] - \phi_2[p'\phi_1' + p\phi_1''] = 0$$

which is equivalent to

$$\frac{d}{dx} [p(\phi_1\phi_2' - \phi_2\phi_1')] = 0$$

and thus

$$p(\phi_1\phi_2' - \phi_2\phi_1') = p(x)W(x) = c, \text{ a constant}$$

Recall that $\phi_1(x)$ and $\phi_2(x)$ had one arbitrary constant left in each; thus without any loss of generality we can choose these constants such that

$$p(x)W(x) = 1 \tag{3.11.11}$$

and then the solution (3.11.10) becomes

$$y(x) = \int_a^x [-f(t)]\phi_1(t)\phi_2(x) \, dt + \int_x^b [-f(t)]\phi_1(x)\phi_2(t) \, dt$$

This may be rewritten compactly as

$$y(t) = \int_a^b [-f(t)]G(x, t) \, dt \tag{3.11.12}$$

where

$$G(x, t) = \begin{cases} \phi_1(t)\phi_2(x), & a \le t \le x \\ \phi_1(x)\phi_2(t), & x \le t \le b \end{cases} \tag{3.11.13}$$

is called the *Green's function* for the boundary value problem. It is apparent that the Green's function is *symmetric*, since $G(x, t) = G(t, x)$ for all x and t such that $x \in [a, b]$ and $t \in [a, b]$.

Note that in similarity with initial value problems [cf. eq. (3.5.17)], the Green's function is solely a function of the differential operator and the BCs, and *not* of the nonhomogeneous term $f(x)$. Given a differential operator $L[\ ]$ and *homogeneous* BCs, the Green's function $G(x, t)$ is determined once only, for all nonhomogeneities $f(x)$ from eq. (3.11.13), where recall that

$$\phi_1(x) \text{ and } \phi_2(x) \text{ are two linearly independent solutions of } L[y] = 0$$

such that

$$\phi_1(x) \text{ satisfies the left end (i.e., } x = a) \text{ BC}$$

$$\phi_2(x) \text{ satisfies the right end (i.e., } x = b) \text{ BC}$$

and

$$p(x)W(x) = 1$$

3.11.2 Properties of the Green's Function

The Green's function satisfies certain properties which are now examined. They are important in that $G(x, t)$ can be found either by the method indicated in the previous section or, alternatively, by finding a function of the variables x and t which satisfies the properties enumerated below.

Property 1. $G(x, t)$ is a continuous function in *both* variables x and t; that is,

$$G(x, t)|_{t=x_-} = G(x, t)|_{t=x_+} \text{ and } G(x, t)|_{x=t_-} = G(x, t)|_{x=t_+}$$

From eq. (3.11.13) we obtain

$$G(x, t) = \begin{cases} \phi_1(t)\phi_2(x), & a \le t \le x \\ \phi_1(x)\phi_2(t), & x \le t \le b \end{cases}$$

so we have

$$G(x, t)|_{t=x_-} = \phi_1(x_-)\phi_2(x)$$

$$G(x, t)|_{t=x_+} = \phi_1(x)\phi_2(x_+)$$

and since the ϕ_j are continuous functions we obtain

$$G(x, t)|_{t=x_-} = G(x, t)|_{t=x_+} \qquad (3.11.14a)$$

Similarly,

$$G(x, t)|_{x=t_-} = \phi_1(t_-)\phi_2(t)$$

$$G(x, t)|_{x=t_+} = \phi_1(t)\phi_2(t_+)$$

and thus

$$G(x, t)|_{x=t_-} = G(x, t)|_{x=t_+} \qquad (3.11.14b)$$

Property 2. $G(x, t)$ satisfies the same homogeneous BCs as y [i.e., eq. (3.11.2)] with respect to *both* the variables x and t. This follows directly from the definition of G, where recall that

$$\phi_1(x) \text{ satisfies the left end (i.e., } x = a) \text{ BC}$$

$$\phi_2(x) \text{ satisfies the right end (i.e., } x = b) \text{ BC.}$$

With respect to the variable x, we have

$$G(a, t) = \phi_1(a)\phi_2(t) \quad \text{and} \quad G(b, t) = \phi_1(t)\phi_2(b)$$

Since $\phi_1(x)$ satisfies the condition at $x = a$, so does $G(x, t)$; similarly, $G(x, t)$ satisfies the condition at $x = b$ because $\phi_2(b)$ does.

The result follows identically for the variable t, since

$$G(x, a) = \phi_1(a)\phi_2(x) \quad \text{and} \quad G(x, b) = \phi_1(x)\phi_2(b)$$

Property 3. $G(x, t)$ satisfies the *homogeneous* ODE for y, with respect to *both* the variables x and t. Thus

$$L_x[G(x,\ t)] = \frac{d}{dx}\left(p(x)\frac{dG}{dx}\right) + q(x)G = 0, \qquad x \neq t \qquad (3.11.15a)$$

and

$$L_t[G(x,\ t)] = \frac{d}{dt}\left(p(t)\frac{dG}{dt}\right) + q(t)G = 0, \qquad x \neq t \qquad (3.11.15b)$$

Let us show that the second assertion is true; the first follows similarly. By definition,

$$G(x,\ t) = \begin{cases} \phi_1(t)\phi_2(x), & a \leq t \leq x \\ \phi_1(x)\phi_2(t), & x \leq t \leq b \end{cases}$$

Differentiating with respect to t, we obtain

$$\frac{dG}{dt} = \begin{cases} \phi_1'(t)\phi_2(x), & a \leq t \leq x \\ \phi_1(x)\phi_2'(t), & x \leq t \leq b \end{cases}$$

$$\frac{d^2G}{dt^2} = \begin{cases} \phi_1''(t)\phi_2(x), & a \leq t \leq x \\ \phi_1(x)\phi_2''(t), & x \leq t \leq b \end{cases}$$

Thus from eq. (3.11.15b) we have

$$L_t[G] = \frac{d}{dt}\left(p(t)\frac{dG}{dt}\right) + q(t)G = \begin{cases} \phi_2(x)L_t[\phi_1], & a \leq t \leq x \\ \phi_1(x)L_t[\phi_2], & x \leq t \leq b \end{cases}$$

$$= \begin{cases} 0, & a \leq t \leq x \\ 0, & x \leq t \leq b \end{cases}$$

since the $\phi_i(t)$ satisfy $L_t[\phi_i] = 0$.

Property 4. Although $G(x, t)$ is continuous in its variables, its derivatives exhibit a jump discontinuity:

$$\left.\frac{dG}{dx}\right|_{x=t_+} - \left.\frac{dG}{dx}\right|_{x=t_-} = \frac{1}{p(t)} \qquad (3.11.16a)$$

$$\left.\frac{dG}{dt}\right|_{t=x_+} - \left.\frac{dG}{dt}\right|_{t=x_-} = \frac{1}{p(x)} \qquad (3.11.16b)$$

Considering the derivatives with respect to x first, we have

$$\left.\frac{dG}{dx}\right|_{x=t_+} = \phi_1(t)\phi_2'(t)$$

$$\left.\frac{dG}{dx}\right|_{x=t_-} = \phi_1'(t)\phi_2(t)$$

and thus

$$
\frac{dG}{dx}\bigg|_{x=t_+} - \frac{dG}{dx}\bigg|_{x=t_-} = \phi_1(t)\phi_2'(t) - \phi_1'(t)\phi_2(t)
$$

$$
= \begin{vmatrix} \phi_1(t) & \phi_2(t) \\ \phi'_1(t) & \phi_2'(t) \end{vmatrix}
$$

$$
= W(t)
$$

$$
= \frac{1}{p(t)}
$$

since $pW = 1$ from eq. (3.11.11). Similarly,

$$
\frac{dG}{dt}\bigg|_{t=x_+} = \phi_1(x)\phi_2'(x)
$$

$$
\frac{dG}{dt}\bigg|_{t=x_-} = \phi_1'(x)\phi_2(x)
$$

and thus

$$
\frac{dG}{dt}\bigg|_{t=x_+} - \frac{dG}{dt}\bigg|_{t=x_-} = \phi_1(x)\phi_2'(x) - \phi_1'(x)\phi_2(x)
$$

$$
= W(x)
$$

$$
= \frac{1}{p(x)}
$$

3.11.3 Some Examples

In this section we consider two examples to illustrate the procedures used to find the Green's function, $G(x, t)$. As noted earlier, we have a choice in this regard. We can find $G(x, t)$ either by its definition, which involves the determination of suitable ϕ_1 and ϕ_2 as discussed in section 3.11.1, or by utilizing its properties identified in section 3.11.2. Both of these methods are illustrated by the first example.

Example 1. Determine the Green's function corresponding to the operator

$$
L[y] = \frac{d^2y}{dx^2}, \qquad x \in (0, 1)
$$

with the homogeneous BCs

$$
-k\frac{dy}{dx} + hy = 0, \qquad x = 0
$$

$$
k\frac{dy}{dx} + hy = 0, \qquad x = 1
$$

METHOD 1. The functions ϕ_i satisfy

$$L[\phi] = \frac{d^2\phi}{dx^2} = 0$$

which has the general solution

$$\phi(x) = c_1 x + c_2$$

where c_1 and c_2 are arbitrary constants. Thus let us denote

$$\phi_1(x) = c_{11} x + c_{12}$$

and

$$\phi_2(x) = c_{21} x + c_{22}$$

$\phi_1(x)$ satisfies the condition at $x = 0$:

$$-k \frac{d\phi_1}{dx} + h\phi_1 = 0, \qquad x = 0$$

which gives

$$c_{12} = \frac{k}{h} c_{11}$$

so that

$$\phi_1(x) = \left(x + \frac{k}{h} \right) c_{11}$$

$\phi_2(x)$ satisfies the condition at $x = 1$:

$$k \frac{d\phi_2}{dx} + h\phi_2 = 0, \qquad x = 1$$

leading to

$$c_{22} = -\left(1 + \frac{k}{h} \right) c_{21}$$

and thus

$$\phi_2(x) = \left[x - \left(1 + \frac{k}{h} \right) \right] c_{21}$$

By definition,

$$G(x, t) = \begin{cases} \phi_1(t)\phi_2(x), & 0 \le t \le x \\ \phi_1(x)\phi_2(t), & x \le t \le 1 \end{cases}$$

$$= \begin{cases} \left(t + \dfrac{k}{h} \right)\left[x - \left(1 + \dfrac{k}{h} \right) \right] c_{11} c_{21}, & 0 \le t \le x \\[3mm] \left(x + \dfrac{k}{h} \right)\left[t - \left(1 + \dfrac{k}{h} \right) \right] c_{11} c_{21}, & x \le t \le 1 \end{cases}$$

The remaining constant, the product $c_{11} c_{21}$, is to be chosen such that

$$p(x)W(x) = 1$$

For the operator at hand, $p(x) = 1$. Also, by definition,

$$W(x) = \begin{vmatrix} \phi_1(x) & \phi_2(x) \\ \phi_1'(x) & \phi_2'(x) \end{vmatrix}$$

$$= \begin{vmatrix} \left(x + \dfrac{k}{h}\right)c_{11} & \left[x - \left(1 + \dfrac{k}{h}\right)\right]c_{21} \\ c_{11} & c_{21} \end{vmatrix}$$

$$= \left(1 + \dfrac{2k}{h}\right)c_{11}c_{21}$$

Since $p(x)W(x) = 1$, and $p(x) = 1$, we have

$$c_{11}c_{21} = \frac{1}{1 + (2k/h)}$$

and it follows that

$$G(x, t) = \begin{cases} \left(t + \dfrac{k}{h}\right)\left[x - \left(1 + \dfrac{k}{h}\right)\right] \Big/ \left(1 + \dfrac{2k}{h}\right), & 0 \leq t \leq x \\ \left(x + \dfrac{k}{h}\right)\left[t - \left(1 + \dfrac{k}{h}\right)\right] \Big/ \left(1 + \dfrac{2k}{h}\right), & x \leq t \leq 1 \end{cases}$$

METHOD 2. In the second method, either x or t may be used as the independent variable. If we use x, from property 3 we obtain

$$\frac{d^2G}{dx^2} = 0, \qquad x \neq t$$

Thus

$$G(x, t) = c_1(t)x + c_2(t)$$

and we can write

$$G(x, t) = \begin{cases} c_{11}(t)x + c_{12}(t), & 0 \leq x \leq t \\ c_{21}(t)x + c_{22}(t), & t \leq x \leq 1 \end{cases}$$

Property 2 states that $G(x, t)$ satisfies the same homogeneous BCs as y. Thus at $x = 0$ we have

$$-k[c_{11}(t)] + h[c_{12}(t)] = 0$$

leading to

$$c_{12}(t) = \frac{k}{h}c_{11}(t)$$

Similarly, at $x = 1$ we have

$$k[c_{21}(t)] + h[c_{21}(t) + c_{22}(t)] = 0$$

which implies that

$$c_{22}(t) = -\left(1 + \frac{k}{h}\right)c_{21}(t)$$

Hence

$$G(x, t) = \begin{cases} \left(x + \dfrac{k}{h} \right) c_{11}(t), & 0 \le x \le t \\[2ex] \left[x - \left(1 + \dfrac{k}{h} \right) \right] c_{21}(t), & t \le x \le 1 \end{cases}$$

Property 1 ensures the continuity of $G(x, t)$:

$$G(x, t)\big|_{x=t_+} = G(x, t)\big|_{x=t_-}$$

This means that

$$\left[t + \frac{k}{h} \right] c_{11}(t) = \left[t - \left(1 + \frac{k}{h} \right) \right] c_{21}(t)$$

so that

$$c_{21}(t) = \frac{t + (k/h)}{t - [1 + (k/h)]} \, c_{11}(t)$$

and hence

$$G(x, t) = \begin{cases} \left[x + \dfrac{k}{h} \right] c_{11}(t), & 0 \le x \le t \\[2ex] \dfrac{t + (k/h)}{t - [1 + (k/h)]} \left[x - \left(1 + \dfrac{k}{h} \right) \right] c_{11}(t), & t \le x \le 1 \end{cases}$$

Finally, employing the jump discontinuity relationship for the derivative of G, given by property 4, we obtain

$$\frac{dG}{dx}\bigg|_{x=t_+} - \frac{dG}{dx}\bigg|_{x=t_-} = \frac{1}{p(t)} = 1$$

which leads to

$$\left[\frac{t + (k/h)}{t - [1 + (k/h)]} - 1 \right] c_{11}(t) = 1$$

and then

$$c_{11}(t) = \frac{t - [1 + (k/h)]}{1 + (2k/h)}$$

Thus we have

$$G(x, t) = \begin{cases} \left(x + \dfrac{k}{h} \right) \left[t - \left(1 + \dfrac{k}{h} \right) \right] \Big/ \left(1 + \dfrac{2k}{h} \right), & 0 \le x \le t \\[2ex] \left(t + \dfrac{k}{h} \right) \left[x - \left(1 + \dfrac{k}{h} \right) \right] \Big/ \left(1 + \dfrac{2k}{h} \right), & t \le x \le 1 \end{cases}$$

which is the same expression as obtained by the first method.

Example 2. Consider one-dimensional steady-state heat conduction in a rod, with internal heat generation and heat loss to the surroundings (Figure 3.3). Let us define the following:

A	area of cross section (m^2)
g	rate of heat generation per unit volume (cal/m^3-sec)
h	heat transfer coefficient (cal/sec-m^2-K)
k	thermal conductivity of rod (cal/sec-m-K)
L	length of rod (m)
p	perimeter of rod (m)
T	temperature of rod (K)
T_a	temperature of surroundings (K)
x	distance along rod (m)

Then an energy balance over the differential element gives

$$\text{Input} = -kA \left.\frac{dT}{dx}\right|_x + g(x)A\Delta x \qquad (\text{cal/sec})$$

$$\text{Output} = -kA \left.\frac{dT}{dx}\right|_{x+\Delta x} + hp\Delta x(T - T_a) \qquad (\text{cal/sec})$$

since Input = Output at steady state,

$$-kA \left.\frac{dT}{dx}\right|_x + g(x)A\Delta x = -kA \left.\frac{dT}{dx}\right|_{x+\Delta x} + hp\Delta x(T - T_a)$$

Assuming constant properties, dividing through by $kA\Delta x$, and taking the limit as $\Delta x \to 0$ yields

$$\frac{d^2T}{dx^2} - \frac{hp}{kA}(T - T_a) = -\frac{g(x)}{k}, \qquad x \in (0, L)$$

Let us further assume for convenience that the ends of the rod are also maintained at the temperature of surroundings; thus the BCs are

$$T(0) = T_a = T(L)$$

Introducing the quantities

$$y = T - T_a, \quad \mu^2 = \frac{hp}{kA} > 0 \quad \text{and} \quad f(x) = \frac{g(x)}{k} \tag{3.11.17}$$

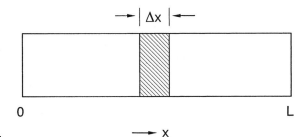

Figure 3.3 Heat conduction in a rod.

leads to

$$L[y] = \frac{d^2y}{dx^2} - \mu^2 y = -f(x), \qquad x \in (0, L) \tag{3.11.18}$$

along with the BCs

$$y(0) = 0 = y(L) \tag{3.11.19}$$

Following eq. (3.11.12), the solution is given by

$$y(x) = \int_0^L [-f(t)] G(x, t) \, dt \tag{3.11.20}$$

In order to find G, we use method 2—that is, the one that uses the properties of G.
From property 3 we obtain

$$\frac{d^2G}{dx^2} - \mu^2 G = 0, \qquad x \neq t$$

which has solution

$$G(x, t) = a_1(t) \sinh \mu x + a_2(t) \cosh \mu x$$

so that

$$G(x, t) = \begin{cases} a_{11}(t) \sinh \mu x + a_{12}(t) \cosh \mu x, & 0 \leq x \leq t \\ a_{21}(t) \sinh \mu x + a_{22}(t) \cosh \mu x, & t \leq x \leq L \end{cases}$$

From property 2 we obtain

$$G(0, t) = 0$$

which gives

$$a_{12}(t) = 0$$

Similarly,

$$G(L, t) = 0$$

and hence

$$a_{21}(t) \sinh \mu L + a_{22}(t) \cosh \mu L = 0$$

that is,

$$a_{22}(t) = -a_{21}(t) \tanh \mu L$$

Thus we have

$$G(x, t) = \begin{cases} a_{11}(t) \sinh \mu x, & 0 \leq x \leq t \\ \dfrac{a_{21}(t)}{\cosh \mu L} [\sinh \mu x \cdot \cosh \mu L - \cosh \mu x \cdot \sinh \mu L], & t \leq x \leq L \end{cases}$$

$$= \begin{cases} a_{11}(t) \sinh \mu x, & 0 \leq x \leq t \\ -a_{21}(t) \dfrac{\sinh \mu(L - x)}{\cosh \mu L}, & t \leq x \leq L \end{cases}$$

From the continuity of $G(x, t)$, ensured by property 1, we obtain

$$G(x, t)\big|_{x=t_-} = G(x, t)\big|_{x=t_+}$$

which gives

$$a_{11}(t) \sinh \mu t = -a_{21}(t) \frac{\sinh \mu(L - t)}{\cosh \mu L}$$

so that

$$G(x, t) = \begin{cases} a_{11}(t) \sinh \mu x, & 0 \le x \le t \\ a_{11}(t) \dfrac{\sinh \mu t \cdot \sinh \mu(L - x)}{\sinh \mu(L - t)}, & t \le x \le L \end{cases}$$

Finally, using the jump discontinuity for the derivative of G, from property 4 we obtain

$$\frac{dG}{dx}\bigg|_{x=t_+} - \frac{dG}{dx}\bigg|_{x=t_-} = \frac{1}{p(t)} = 1$$

that is,

$$\left[\sinh \mu t \left\{ -\frac{\mu \cosh \mu(L - t)}{\sinh \mu(L - t)} \right\} - \mu \cosh \mu t \right] a_{11}(t) = 1$$

or

$$a_{11}(t) = -\frac{\sinh \mu(L - t)}{\mu \sinh \mu L}$$

and thus we have

$$G(x, t) = \begin{cases} -\dfrac{\sinh \mu(L - t) \sinh \mu x}{\mu \sinh \mu L}, & 0 \le x \le t \\ -\dfrac{\sinh \mu(L - x) \sinh \mu t}{\mu \sinh \mu L}, & t \le x \le L \end{cases} \tag{3.11.21}$$

which can be substituted in eq. (3.11.20) to give the temperature distribution, $y(x)$, for any heat generation function, $f(x)$.

3.12 NONHOMOGENEOUS BOUNDARY CONDITIONS

In the previous section we considered problems in which the nonhomogeneous term occurs *only* in the ODE, but not in the BCs. Boundary value problems with nonhomogeneous BCs can also be handled readily. The strategy is to first convert the problem with nonhomogeneous BCs into one with homogeneous BCs and then utilize the Green's function approach to solve the reformulated problem.

Thus suppose that we wish to solve

$$L[y] = \frac{d}{dx}\left(p(x)\frac{dy}{dx}\right) + q(x)y = -f(x), \qquad x \in (a, b) \qquad (3.12.1)$$

$$-k_a \frac{dy}{dx} + h_a y = \alpha_1, \qquad\qquad x = a \qquad\qquad (3.12.2a)$$

$$k_b \frac{dy}{dx} + h_b y = \alpha_2, \qquad\qquad x = b \qquad\qquad (3.12.2b)$$

a problem in which nonhomogeneous terms appear *both* in the ODE and in the BCs. Make the simple change of dependent variables:

$$y(x) = z(x) + c_1 x + c_2 \qquad (3.12.3)$$

where c_1 and c_2 are as yet undetermined constants, to be chosen such that the problem above in y, when converted to z, has homogeneous BCs. Noting that

$$y' = z' + c_1 \quad \text{and} \quad y'' = z''$$

the ODE (3.12.1) takes the form

$$\frac{d}{dx}\left(p(x)\left\{\frac{dz}{dx} + c_1\right\}\right) + q(x)\{z + c_1 x + c_2\} = -f(x)$$

or

$$\frac{d}{dx}\left(p(x)\frac{dz}{dx}\right) + q(x)z = -\left[f(x) + q(x)\{c_1 x + c_2\} + c_1 \frac{dp}{dx}\right]$$
$$= -g(x) \qquad (3.12.4)$$

The BCs (3.12.2) become

$$-k_a z' + h_a z = \alpha_1 + k_a c_1 - h_a(c_1 a + c_2), \qquad x = a \qquad (3.12.5a)$$

$$k_b z' + h_b z = \alpha_2 - k_b c_1 - h_b(c_1 b + c_2), \qquad x = b \qquad (3.12.5b)$$

Now if the BCs for z are to be homogeneous, c_1 and c_2 should be chosen such that

$$(k_a - h_a a)c_1 - h_a c_2 = -\alpha_1 \qquad (3.12.6a)$$

$$(k_b + h_b b)c_1 + h_b c_2 = \alpha_2 \qquad (3.12.6b)$$

This system has a unique solution if and only if

$$\Delta = \begin{vmatrix} k_a - h_a a & -h_a \\ k_b + h_b b & h_b \end{vmatrix} \neq 0$$

which is ensured since

$$\Delta = (k_a - h_a a)h_b + (k_b + h_b b)h_a$$
$$= (k_a h_b + k_b h_a) + h_a h_b(b - a)$$
$$> 0$$

Thus using Cramer's rule, we obtain

$$c_1 = \frac{1}{\Delta} \begin{vmatrix} -\alpha_1 & -h_a \\ \alpha_2 & h_b \end{vmatrix} = \frac{\alpha_2 h_a - \alpha_1 h_b}{\Delta} \qquad (3.12.7a)$$

and

$$c_2 = \frac{1}{\Delta} \begin{vmatrix} k_a - h_a a & -\alpha_1 \\ k_b + h_b b & \alpha_2 \end{vmatrix} = \frac{\alpha_2(k_a - h_a a) + \alpha_1(k_b + h_b b)}{\Delta} \qquad (3.12.7b)$$

With the constants c_1 and c_2 known, the problem in z becomes

$$L[z] = \frac{d}{dx}\left(p(x)\frac{dz}{dx}\right) + q(x)z = -g(x), \qquad x \in (a, b) \qquad (3.12.8)$$

$$-k_a \frac{dz}{dx} + h_a z = 0, \qquad\qquad\qquad x = a \qquad (3.12.9a)$$

$$k_b \frac{dz}{dx} + h_b z = 0, \qquad\qquad\qquad x = b \qquad (3.12.9b)$$

which has homogeneous BCs and where

$$g(x) = f(x) + \{c_1 x + c_2\}q(x) + c_1 \frac{dp}{dx} \qquad (3.12.10)$$

is fully known.

3.13 PHYSICAL INTERPRETATION OF THE GREEN'S FUNCTION

In the previous sections, we have seen that the Green's function is useful for solving nonhomogeneous boundary value problems, where the nonhomogeneity may occur in either the ODE or the BCs. We now provide a physical interpretation of the Green's function.

Recall from section 3.11 that the solution of the problem

$$L[y] = -f(x), \qquad x \in (a, b) \qquad (3.13.1)$$

$$B[y] = 0 \qquad (3.13.2)$$

where $L[\ ]$ is a self-adjoint second-order linear differential operator and $B[\ ]$ represents self-adjoint BCs, is given by

$$y(x) = \int_a^b [-f(t)]G(x, t)\, dt \qquad (3.13.3)$$

Let the nonhomogeneous term $-f(x)$ be the Dirac delta function, $\delta(x - \bar{x})$ located at $x = \bar{x}$, as shown in Figure 3.4; that is,

$$-f(x) = 0, \qquad x \neq \bar{x} \qquad (3.13.4a)$$

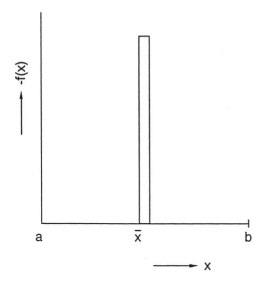

Figure 3.4 The Dirac delta function located at $x = \bar{x}$.

and

$$\int_a^b [-f(x)] \, dx = 1 \tag{3.13.4b}$$

Then from eq. (3.13.3) we have

$$y(x) = G(x, \bar{x}) \tag{3.13.5}$$

This means that the Green's function $G(x, t)$ is the solution of eqs. (3.13.1) and (3.13.2) when the nonhomogeneous term $-f(x)$ is a Dirac delta function located at $x = t$. Hence $G(x, t)$ represents the effect at position x of a forcing function of unit strength located at position t. In this context then, eq. (3.13.3) can be viewed as the cumulative effect of a large (infinite) number of individual localized forcing functions of strength $[-f(t)]$ dt.

III. EIGENVALUE PROBLEMS

3.14 ORIGIN OF EIGENVALUE PROBLEMS

Consider the transient heating or cooling of a slab of material. Making a differential energy balance, as outlined in section 3.9, the relevant *partial* differential equation (PDE) for the temperature distribution $T(x, t)$ is

$$\frac{\partial}{\partial x} \left(k \frac{\partial T}{\partial x} \right) = \frac{\partial}{\partial t} (C_p \rho T), \qquad x \in (0, L), \quad t > 0$$

If the properties are constant, then

$$\frac{\partial^2 T}{\partial x^2} = \frac{1}{\alpha}\frac{\partial T}{\partial t}, \qquad x \in (0, L), \quad t > 0 \tag{3.14.1}$$

where $\alpha = k/C_p\rho$ is the thermal diffusivity.

Let the BCs be

$$T(0, t) = T_a = T(L, t), \qquad t > 0 \tag{3.14.2}$$

which imply rapid heat transfer between the slab and the surroundings at the ends.

We also need an initial condition (IC) describing the temperature distribution at $t = 0$:

$$T(x, 0) = T_0(x), \qquad x \in (0, L) \tag{3.14.3}$$

Note that the BCs (3.14.2) are *nonhomogeneous*. These can be made homogeneous by the change of variable:

$$y = T - T_a \tag{3.14.4}$$

so that eqs. (3.14.1) to (3.14.3) reduce to

$$\frac{\partial^2 y}{\partial x^2} = \frac{1}{\alpha}\frac{\partial y}{\partial t}, \qquad x \in (0, L), \quad t > 0 \tag{3.14.5}$$

$$y(0, t) = 0 = y(L, t), \quad t > 0 \tag{3.14.6}$$

$$y(x, 0) = T_0(x) - T_a$$
$$= y_0(x), \qquad x \in (0, L) \tag{3.14.7}$$

which is an *initial-boundary value problem*. One method for solving the problem is by *separation of variables*, where we *assume* that the solution can be represented as

$$y(x, t) = X(x)\Theta(t) \tag{3.14.8}$$

that is, a product of two functions, each of which depends on *only* one independent variable. For this form of y,

$$\frac{\partial^2 y}{\partial x^2} = \frac{\partial^2}{\partial x^2}(X\Theta) = \Theta\frac{d^2 X}{dx^2}$$

and

$$\frac{\partial y}{\partial t} = \frac{\partial}{\partial t}(X\Theta) = X\frac{d\Theta}{dt}$$

so that eq. (3.14.5) becomes

$$\Theta\frac{d^2 X}{dx^2} = \frac{1}{\alpha}X\frac{d\Theta}{dt}$$

and dividing through by $X\Theta$ gives

$$\frac{1}{X}\frac{d^2 X}{dx^2} = \frac{1}{\alpha}\frac{1}{\Theta}\frac{d\Theta}{dt}$$

In this equation, since the lhs is a function of x only, and the rhs is a function only of t, *and x and t are independent variables*, both sides must be constant. Let us denote this constant by $-\lambda$; then

$$\frac{1}{X}\frac{d^2X}{dx^2} = \frac{1}{\alpha}\frac{1}{\Theta}\frac{d\Theta}{dt} = -\lambda$$

so that the original PDE (3.14.5) has been transformed into *two* ODEs

$$\frac{d^2X}{dx^2} = -\lambda X, \qquad x \in (0, L) \tag{3.14.9}$$

$$\frac{d\Theta}{dt} = -\alpha\lambda\Theta, \qquad t > 0 \tag{3.14.10}$$

Substituting the form (3.14.8) for y in the BCs (3.14.6) gives

$$X(0)\Theta(t) = 0 = X(L)\Theta(t), \qquad t > 0$$

Since $\Theta(t) \neq 0$ for all $t > 0$, because otherwise $y(x, t) = 0$ from eq. (3.14.8), which cannot hold for arbitrary $y_0(x)$ in the IC (3.14.7), we must have

$$X(0) = 0 = X(L) \tag{3.14.11}$$

Also, the IC (3.14.7) implies that

$$X(x)\Theta(0) = y_0(x) \tag{3.14.12}$$

Note that so far nothing has been said about λ, except the recognition that it is a constant. Furthermore, we have found that X satisfies the boundary value problem:

$$L[X] = \frac{d^2X}{dx^2} = -\lambda X, \qquad x \in (0, L) \tag{3.14.13}$$

$$X(0) = 0 = X(L) \tag{3.14.14}$$

This problem certainly admits the trivial solution $X(x) = 0$. However, this leads to $y(x, t) = 0$ from eq. (3.14.8), which is unacceptable for the reasons discussed above. So we must seek nontrivial solutions $X(x)$.

At this stage we ask the question, Are there any special values of the constant λ which lead to nontrivial solutions $X(x)$? If so, what are these values? (Note the similarity with the matrix eigenvalue problem: $\mathbf{Ax} = \lambda\mathbf{x}$, discussed in section 1.11.) Such values of λ, which give nontrivial X, are called *eigenvalues* of the operator $L[\ ]$ along with the homogeneous BCs, and the corresponding solutions X are called the *eigenfunctions*. The entire problem (3.14.13) and (3.14.14) is called an *eigenvalue problem*.

These considerations fall under the scope of Sturm–Liouville theory (cf. Weinberger, 1965, chapter 7). As will become clear, problems which are relatively easy to solve are those that involve a self-adjoint differential operator $L[\ ]$, along with self-adjoint homogeneous BCs. This is the reason for our stress on self-adjointness.

3.15   AN INTRODUCTORY EXAMPLE

Consider the eigenvalue problem posed in the previous section:

$$L[y] = \frac{d^2y}{dx^2} = -\lambda y, \qquad x \in (0, L) \tag{3.15.1}$$

$$y(0) = 0 = y(L) \tag{3.15.2}$$

where we seek values of the constant λ such that the solution $y(x)$ is nontrivial.

The general solution of eq. (3.15.1) is

$$y = c_1 \sin \sqrt{\lambda}\, x + c_2 \cos \sqrt{\lambda}\, x \tag{3.15.3}$$

where c_1 and c_2 are constants to be determined by satisfying the BCs (3.15.2). Using the condition at $x = 0$ gives

$$c_2 = 0$$

so that

$$y = c_1 \sin \sqrt{\lambda}\, x \tag{3.15.4}$$

The condition $y(L) = 0$ implies that

$$c_1 \sin \sqrt{\lambda}\, L = 0$$

which leads either to

$$c_1 = 0$$

or to

$$\sin \sqrt{\lambda}\, L = 0$$

If $c_1 = 0$, then from eq. (3.15.4), y would be the trivial solution. Since we seek nontrivial solutions, we must have

$$\sin \sqrt{\lambda}\, L = 0 \tag{3.15.5}$$

which yields

$$\sqrt{\lambda}\, L = \pm n\pi, \qquad n = 0, 1, 2, \ldots$$

that is,

$$\lambda = \frac{n^2\pi^2}{L^2}, \qquad n = 0, 1, 2, \ldots$$

Now $n \neq 0$, because otherwise $y = 0$ again from eq. (3.15.4). Thus we conclude that nontrivial solutions of the given eigenvalue problem occur with

$$\lambda = \lambda_n = \frac{n^2\pi^2}{L^2}, \qquad n = 1, 2, \ldots \tag{3.15.6}$$

which are an *infinite* number of eigenvalues. The corresponding solutions

$$y_n(x) = \sin \sqrt{\lambda_n}\, x = \sin\left(\frac{n\pi x}{L}\right) \tag{3.15.7}$$

are the eigenfunctions which belong to λ_n. In similarity with matrices, the eigenfunction y_n is determined only up to an arbitrary multiplicative constant c_1.

In writing solutions for $\sin\sqrt{\lambda}\, L = 0$, we concluded that

$$\sqrt{\lambda}\, L = n\pi, \qquad n = 0, 1, 2, \ldots$$

thus assuming real λ. However, can there be complex values of λ which satisfy $\sin\sqrt{\lambda}\, L = 0$? The answer is that there are *none*, which is shown below by contradiction.

Assume that $\sin z = 0$ has complex solutions; then if we let

$$z = a + ib, \qquad \text{with } b \neq 0$$

we have

$$\sin(a + ib) = 0$$

which on expansion gives

$$\sin a \cos(ib) + \cos a \sin(ib) = 0$$

However,

$$\cos(ib) = \cosh b \text{ and } \sin(ib) = i \sinh b$$

so that

$$\sin a \cosh b + i \cos a \sinh b = 0$$

Since the lhs is a complex number, we must have that a and b satisfy

$$\sin a \cosh b = 0$$

and

$$\cos a \sinh b = 0$$

For $b \neq 0$, $\sinh b$ and $\cosh b$ are never zero (in fact, they are both >0). Thus we conclude that a is a real number, which satisfies

$$\sin a = 0 \quad \text{and} \quad \cos a = 0$$

However, this is a contradiction, seen readily from geometric arguments. Thus, there are only real solutions of $\sin z = 0$, and the eigenvalues λ_n are therefore real.

The eigenfunctions $y_n(x)$ given by eq. (3.15.7) satisfy the important property of *orthogonality*. To see what this means, let λ_n and λ_m be distinct eigenvalues and let y_n and y_m be their corresponding eigenfunctions. Since the operator $L[\ ]$ along with the BCs is self-adjoint, we obtain

$$\int_0^L \{y_n L[y_m] - y_m L[y_n]\}\, dx = 0$$

Noting that

$$L[y_n] = -\lambda_n y_n \quad \text{and} \quad L[y_m] = -\lambda_m y_m$$

we have

$$\int_0^L \{-\lambda_m y_n y_m + \lambda_n y_n y_m\} \, dx = 0$$

or

$$(\lambda_n - \lambda_m) \int_0^L y_n y_m \, dx = 0$$

Since by assumption $\lambda_n \neq \lambda_m$, we conclude that

$$\int_0^l y_n y_m \, dx = 0 \tag{3.15.8}$$

which defines orthogonality of eigenfunctions y_n and y_m. It is interesting to observe that, without using any trigonometry, we have thus shown that

$$\int_0^L \sin\left(\frac{n\pi x}{L}\right) \sin\left(\frac{m\pi x}{L}\right) \, dx = 0$$

where $m \neq n$ are integers.

3.16   GENERAL SELF-ADJOINT SECOND-ORDER OPERATOR

In the previous section, we saw that the eigenvalues and the corresponding eigenfunctions of a specific operator plus its BCs possess some important properties; that is, the eigenvalues are real and the eigenfunctions corresponding to distinct eigenvalues are orthogonal. These properties hold for *all* self-adjoint second-order differential operators along with self-adjoint BCs.

Consider the eigenvalue problem

$$L[y] = \frac{d}{dx}\left[p(x) \frac{dy}{dx} \right] - q(x)y = -\lambda \rho(x)y, \qquad x \in (a, b) \tag{3.16.1}$$

$$-k_1 \frac{dy}{dx} + h_1 y = 0, \qquad x = a \tag{3.16.2a}$$

$$k_2 \frac{dy}{dx} + h_2 y = 0, \qquad x = b \tag{3.16.2b}$$

where $\rho(x)$ and $q(x)$ are real continuous functions, and $p(x)$ is continuously differentiable. Also, $p(x)$ and $\rho(x)$ are both real and positive for $x \in [a, b]$. The constants h_i and k_i are real and ≥ 0, but both together are not equal to zero for each i. We know from section 3.10 that $L[\ ]$ as well as the BCs are self-adjoint.

The above eigenvalue problem is sometimes called a *Sturm–Liouville problem*. As is always the case with eigenvalue problems, specific values of λ which lead to nontrivial solutions y are of interest.

We will state the theorem which guarantees existence of the eigenvalues later, but for now let us assume that they exist and proceed to derive some important properties.

Property 1. If λ_1 and λ_2 are *distinct* eigenvalues, then the corresponding eigenfunctions $\phi_1(x)$ and $\phi_2(x)$ are orthogonal on $x \in [a, b]$ with the weight function $\rho(x)$.

Proof
Since the operator $L[\ ]$ in eq. (3.16.1) along with the BCs (3.16.2) is self-adjoint, we obtain

$$\int_a^b \{\phi_1 L[\phi_2] - \phi_2 L[\phi_1]\}\ dx = 0$$

However,

$$L[\phi_i] = -\lambda_i \rho \phi_i$$

so that

$$\int_a^b \{-\lambda_2 \rho \phi_1 \phi_2 + \lambda_1 \rho \phi_1 \phi_2\}\ dx = 0$$

or

$$(\lambda_1 - \lambda_2) \int_a^b \rho(x)\phi_1(x)\phi_2(x)\ dx = 0$$

Since $\lambda_1 \neq \lambda_2$, this means that

$$\int_a^b \rho(x)\phi_1(x)\phi_2(x)\ dx = 0 \tag{3.16.3}$$

That is, the eigenfunctions corresponding to distinct eigenvalues are orthogonal on the interval $x \in [a, b]$, with the weight function $\rho(x)$.

Property 2. The eigenvalues λ_i are real.

Proof
Let us assume that they are not, so there exists a complex eigenvalue μ. The corresponding eigenfunction $\phi(x)$ must also be complex, since p, q, and ρ are all real; thus we have

$$L[\phi] = -\mu\rho\phi$$

along with BCs

$$B[\phi] = 0$$

where the BCs (3.16.2) have been written compactly in operator form.
    Taking complex conjugates gives

$$L[\overline{\phi}] = -\overline{\rho\mu\phi} = -\rho(x)\ \overline{\mu\phi}$$
$$B[\overline{\phi}] = 0$$

Thus $\bar{\mu}$, the complex conjugate of μ, has eigenfunction $\bar{\phi}$, the complex conjugate of ϕ. Since μ and $\bar{\mu}$ are distinct eigenvalues, from eq. (3.16.3) we obtain

$$\int_a^b \rho(x)\phi(x)\bar{\phi}(x)\ dx = 0$$

that is,

$$\int_a^b \rho(x)|\phi(x)|^2\ dx = 0$$

where $|\phi(x)|$ is the modulus of $\phi(x)$. Since $\rho(x) > 0$, the above equality requires $\phi(x) = 0$ for $x \in [a, b]$. However, being an eigenfunction, $\phi(x)$ does not equal zero. Thus we have a contradiction, and hence the eigenvalues λ_i are *real*.

Property 3. If $q(x) \geq 0$ for $x \in [a, b]$, then the eigenvalues λ_i are positive.

Proof
Let λ be an eigenvalue, and let y be the corresponding eigenfunction. From eq. (3.16.1) we have

$$\int_a^b yL[y]\ dx = -\lambda \int_a^b \rho(x)y^2(x)\ dx \tag{3.16.4}$$

and

$$\int_a^b yL[y]\ dx = \int_a^b \left\{ y\ \frac{d}{dx}\left[p(x)\ \frac{dy}{dx} \right] - q(x)y^2 \right\} dx$$

Integrating once by parts,

$$= yp(x)\ \frac{dy}{dx} \Bigg|_a^b - \int_a^b \left\{ p(x)\left(\frac{dy}{dx} \right)^2 + q(x)y^2 \right\} dx$$

where, using the BCs (3.16.2), the first term can be written as

$$yp(x)\ \frac{dy}{dx} \Bigg|_a^b = y(b)p(b)\ \frac{dy}{dx}\ (b) - y(a)p(a)\ \frac{dy}{dx}\ (a)$$

$$= -\frac{h_2}{k_2}\ p(b)y^2(b) - \frac{h_1}{k_1}\ p(a)y^2(a)$$

Thus

$$\int_a^b yL[y]\ dx = -\left[\frac{h_2}{k_2}\ p(b)y^2(b) + \frac{h_1}{k_1}\ p(a)y^2(a) \right.$$
$$\left. + \int_a^b \left\{ p(x)\left(\frac{dy}{dx} \right)^2 + q(x)y^2 \right\} dx \right] \tag{3.16.5}$$

and from eqs. (3.16.4) and 3.16.5), we have

$$\lambda = \frac{\dfrac{h_2}{k_2}\ p(b)y^2(b) + \dfrac{h_1}{k_1}\ p(a)y^2(a) + \displaystyle\int_a^b \left\{ p(x)\left(\frac{dy}{dx} \right)^2 + q(x)y^2 \right\} dx}{\displaystyle\int_a^b \rho(x)y^2\ dx} \tag{3.16.6}$$

which is positive if $q(x) \geq 0$ for $x \in [a, b]$. Note that this representation is used for proving the existence of the eigenvalues and is called the *Rayleigh quotient*.

Finally, note that if the BCs are of the first kind: $y(a) = 0 = y(b)$, following similar arguments, we obtain

$$\lambda = \frac{\int_a^b \left\{ p(x)\left(\frac{dy}{dx}\right)^2 + q(x)y^2 \right\} dx}{\int_a^b \rho(x)y^2 \, dx} \qquad (3.16.7)$$

The properties discussed above, along with some additional results, can be summarized as follows (for proof, see chapter 7 of Weinberger, 1965).

● **THEOREM**

For the eigenvalue problem (3.16.1) and (3.16.2), there exists a denumerable set of *real* values of λ, say

$$\lambda_1, \lambda_2, \ldots, \lambda_n, \ldots$$

and a set of eigenfunctions,

$$y_1, y_2, \ldots, y_n, \ldots$$

where $y_i(x)$ belong to λ_i The eigenvalues λ_i form an *infinite* sequence and can be ordered according to increasing magnitude:

$$\lambda_1 < \lambda_2 < \cdots < \lambda_{n-1} < \lambda_n < \lambda_{n+1} < \cdots$$

with $\lim_{i \to \infty} \lambda_i = \infty$. If $q(x) \geq 0$, then all λ_i are *positive*.

The eigenfunctions y_j and y_k corresponding to distinct eigenvalues λ_j and λ_k are *orthogonal* on the interval $[a, b]$ with the weight function $\rho(x)$. Finally, in the open interval $x \in (a, b)$, the eigenfunction $y_n(x)$ has $(n - 1)$ zeroes; this last property is called *Sturm's oscillation theorem*.

For the example considered in section 3.14, namely,

$$L[y] = \frac{d^2y}{dx^2} = -\lambda y, \qquad x \in (0, L)$$

$$y(0) = 0 = y(L)$$

we have

$$\lambda_n = \frac{n^2\pi^2}{L^2}, \qquad n = 1, 2, \ldots$$

and

$$y_n = \sin\left(\frac{n\pi x}{L}\right)$$

For this example, all the points of the theorem above, except that involving zeroes of the eigenfunctions, have been discussed previously. However, this is true as well, as shown in Figure 3.5 for $n = 1$ to 3.

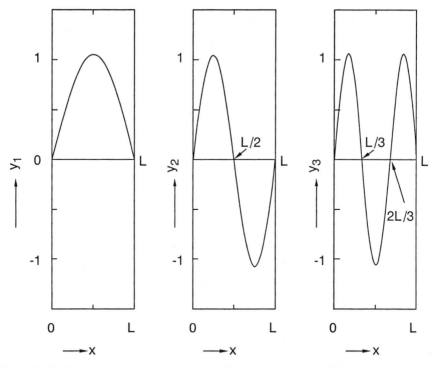

Figure 3.5 The first few eigenfunctions, $y_n = \sin(n\pi x/L)$, of the eigenvalue problem given by eqs. (3.15.1) and (3.15.2).

Orthonormal Set of Eigenfunctions.   We showed in property 1 that the set of eigenfunctions $\{\phi_j(x)\}$ is orthogonal on $[a, b]$ with the weight function $\rho(x)$; that is,

$$\int_a^b \rho(x)\phi_m(x)\phi_n(x)\, dx \begin{cases} = 0, & m \neq n \\ \neq 0, & m = n \end{cases}$$

From the set $\{\phi_j(x)\}$, we can generate a new set of eigenfunctions $\{\psi_j(x)\}$ such that

$$\| \psi_j(x) \| = 1$$

where $\| y \|$ denotes the *norm* of y, defined by

$$\| y \|^2 = \int_a^b \rho(x) y^2(x)\, dx \tag{3.16.8}$$

The relationship between ψ_j and ϕ_j is then simply

$$\psi_j(x) = \frac{\phi_j(x)}{\| \phi_j(x) \|} \tag{3.16.9}$$

For the eigenvalue problem given by eqs. (3.15.1) and (3.15.2), we obtain

$$\phi_j(x) = \sin\left(\frac{j\pi x}{L}\right), \qquad j = 1, 2, \ldots$$

and

$$\rho(x) = 1$$

Thus

$$\| \phi_j(x) \|^2 = \int_0^L \sin^2\left(\frac{j\pi x}{L}\right) dx = \frac{L}{2}$$

and hence the eigenfunctions with unity norm are

$$\psi_j(x) = \frac{\phi_j}{\| \phi_j \|} = \sqrt{\frac{2}{L}} \sin\left(\frac{j\pi x}{L}\right), \qquad j = 1, 2, \ldots \qquad (3.16.10)$$

The set of eigenfunctions $\{\psi_j(x)\}$, in which the members are orthogonal and have unit norm, is called an *orthonormal* set.

3.17 SOME FURTHER EXAMPLES

In this section we consider some eigenvalue problems involving BCs or differential operators which are different from the introductory example of section 3.15. The aim is to determine the eigenvalues and their corresponding eigenfunctions.

Example 1. Solve the eigenvalue problem

$$L[y] = \frac{d^2y}{dx^2} = -\lambda y, \qquad x \in (0, L) \qquad (3.17.1)$$

$$-\frac{dy}{dx} + \alpha_1 y = 0, \qquad x = 0 \qquad (3.17.2a)$$

$$\frac{dy}{dx} + \alpha_2 y = 0, \qquad x = L \qquad (3.17.2b)$$

where $\alpha_i \geq 0$. The general solution of eq. (3.17.1) is

$$y = c_1 \cos \sqrt{\lambda}\, x + c_2 \sin \sqrt{\lambda}\, x \qquad (3.17.3)$$

where c_1 and c_2 are constants to be determined by the BCs (3.17.2), which give at $x = 0$

$$-\sqrt{\lambda}\, c_2 + \alpha_1 c_1 = 0$$

and at $x = L$

$$\sqrt{\lambda}\, [-c_1 \sin \sqrt{\lambda}\, L + c_2 \cos \sqrt{\lambda}\, L] + \alpha_2 [c_1 \cos \sqrt{\lambda}\, L + c_2 \sin \sqrt{\lambda}\, L] = 0$$

These may be rewritten in the form

$$\alpha_1 c_1 - \sqrt{\lambda}\, c_2 = 0 \qquad (3.17.4a)$$

$$[-\sqrt{\lambda}\, \sin \sqrt{\lambda}\, L + \alpha_2 \cos \sqrt{\lambda}\, L]c_1 + [\sqrt{\lambda}\, \cos \sqrt{\lambda}\, L$$
$$+ \alpha_2 \sin \sqrt{\lambda}\, L]c_2 = 0 \qquad (3.17.4b)$$

which represent a set of two homogeneous linear algebraic equations in the two unknowns c_1 and c_2. The condition for nontrivial solutions y is

$$\begin{vmatrix} \alpha_1 & -\sqrt{\lambda} \\ -\sqrt{\lambda}\, \sin \sqrt{\lambda}\, L + \alpha_2 \cos \sqrt{\lambda}\, L & \sqrt{\lambda}\, \cos \sqrt{\lambda}\, L + \alpha_2 \sin \sqrt{\lambda}\, L \end{vmatrix} = 0$$

which simplifies to

$$\cot \sqrt{\lambda}\, L = \frac{\lambda - \alpha_1 \alpha_2}{(\alpha_1 + \alpha_2)\sqrt{\lambda}} \qquad (3.17.5)$$

Since eq. (3.17.5) is a transcendental equation, we investigate its solution by plotting the left- and the right-hand sides and noting the intersections. For this, let

$$\sqrt{\lambda}\, L = s \qquad (3.17.6)$$

so that eq. (3.17.5) becomes

$$F(s) = \cot s = \frac{1}{\alpha_1 + \alpha_2}\left[\frac{s}{L} - \frac{\alpha_1 \alpha_2 L}{s}\right] = G(s) \qquad (3.17.7)$$

Note the following properties of $G(s)$:

$$G(0_+) = -\infty$$
$$G = 0 \qquad \text{for } s = L\sqrt{\alpha_1 \alpha_2}$$

and

$$G \to \frac{s}{(\alpha_1 + \alpha_2)L} \qquad \text{as } s \to \infty$$

In Figure 3.6, the functions F and G are shown for positive s values only since λ is real. Observe that

$$(j - 1)\pi < s_j < j\pi, \qquad j = 1, 2, \ldots$$

and for j large,

$$s_j \to (j - 1)\pi + \varepsilon$$

where $0 < \varepsilon \ll 1$. Since

$$s_j = \sqrt{\lambda_j}\, L$$

we have

$$\lambda_j = \frac{s_j^2}{L^2}$$

and thus

$$\frac{(j - 1)^2 \pi^2}{L^2} < \lambda_j < \frac{j^2 \pi^2}{L^2}, \qquad j = 1, 2, \ldots \qquad (3.17.8)$$

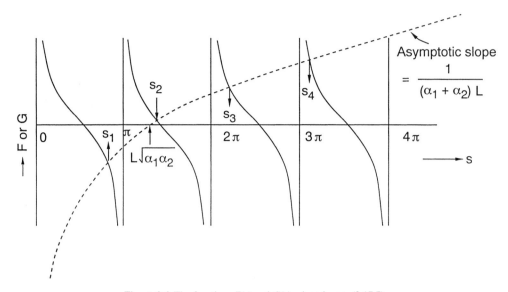

Figure 3.6 The functions $F(s)$ and $G(s)$, given by eq. (3.17.7).

and

$$\lim_{j \to \infty} \lambda_j = \frac{(j-1)^2 \pi^2}{L^2} \tag{3.17.9}$$

It is worth noting that the actual values of λ_j are obtained only by a numerical root-finding technique, such as the Newton–Raphson method. However, the bounds and limits on λ_j prescribed by eqs. (3.17.8) and (3.17.9) aid in their numerical search. Finally, since $\partial G / \partial \alpha_i < 0$, the roots s_j increase with α_1 or α_2. Also, if both $\alpha_i \to \infty$, then $G(s) \to -\infty$ for all s, hence $\lambda_j \to j^2 \pi^2 / L^2$, in conformity with the example of section 3.15.

The eigenfunction corresponding to λ_j can be found from eq. (3.17.3), by using either one among eqs. (3.17.4) to relate c_1 and c_2. Using the first of these,

$$c_2 = \frac{\alpha_1}{\sqrt{\lambda}} c_1$$

so that

$$y_j = [\sqrt{\lambda_j} \cos \sqrt{\lambda_j}\, x + \alpha_1 \sin \sqrt{\lambda_j}\, x] \frac{c_1}{\sqrt{\lambda_j}}, \qquad j = 1, 2, \ldots$$

Since the eigenfunctions are determined only up to an arbitrary multiplicative constant, we have

$$y_j = \sqrt{\lambda_j} \cos \sqrt{\lambda_j}\, x + \alpha_1 \sin \sqrt{\lambda_j}\, x, \qquad j = 1, 2, \ldots \tag{3.17.10}$$

Example 2. Solve the eigenvalue problem for the one-dimensional spherical Laplacian operator

$$L[y] = \frac{1}{x^2} \frac{d}{dx}\left[x^2 \frac{dy}{dx}\right] = -\lambda y, \qquad x \in (a, b) \tag{3.17.11}$$

with BCs

$$y(a) = 0 = y(b) \tag{3.17.12}$$

The operator $L[\ ]$ is not self-adjoint. We convert it to the self-adjoint operator $L_1[\ ]$ upon multiplying both sides of eq. (3.17.11) by x^2, to give

$$L_1[y] = \frac{d}{dx}\left(x^2 \frac{dy}{dx}\right) = -\lambda x^2 y, \qquad x \in (a, b) \tag{3.17.13}$$

where, by comparison with the standard form (3.16.1), we see that

$$p(x) = x^2 \quad \text{and} \quad \rho(x) = x^2 \tag{3.17.14}$$

The ODE (3.17.13) has variable coefficients:

$$x^2 \frac{d^2 y}{dx^2} + 2x \frac{dy}{dx} = -\lambda x^2 y \tag{3.17.15}$$

However, the coefficients can be made constant by the simple change of variables:

$$y = \frac{v}{x} \tag{3.17.16}$$

so that

$$\frac{dy}{dx} = \frac{1}{x}\frac{dv}{dx} - \frac{v}{x^2}$$

and

$$\frac{d^2 y}{dx^2} = \frac{1}{x}\frac{d^2 v}{dx^2} - \frac{2}{x^2}\frac{dv}{dx} + \frac{2v}{x^3}$$

Equation (3.17.15) then takes the form

$$L_2[v] = \frac{d^2 v}{dx^2} = -\lambda v, \qquad x \in (a, b) \tag{3.17.17}$$

and the BCs become

$$v(a) = 0 = v(b)$$

The problem in variable v is now the same as the example of section 3.15, except that the domain is (a, b) rather than $(0, L)$. Proceeding similarly, we get

$$\lambda_n = \frac{n^2 \pi^2}{(b - a)^2}, \qquad n = 1, 2, \ldots \tag{3.17.18}$$

and up to an arbitrary multiplicative constant, the corresponding eigenfunctions are

$$v_n = \sin\left(n\pi \frac{x - a}{b - a}\right), \qquad n = 1, 2, \ldots$$

The eigenfunctions of the original operator $L[\ ]$ are thus

$$y_n = \frac{v_n}{x} = \frac{1}{x}\sin\left(n\pi \frac{x - a}{b - a}\right), \qquad n = 1, 2, \ldots \tag{3.17.19}$$

3.18 EIGENFUNCTION EXPANSIONS AND THE FINITE FOURIER TRANSFORM

Consider a self-adjoint second-order linear differential operator $L[\ ]$, along with some self-adjoint *homogeneous* BCs, $B[\ ] = 0$. Further, consider the eigenvalue problem associated with $L[\ ]$ and $B[\ ] = 0$, with $\rho(x)$ that depends on the system

$$L[y] = -\lambda\rho(x)y, \qquad x \in (a, b) \tag{3.18.1}$$

$$B[y] = 0 \tag{3.18.2}$$

Since $L[\ ]$ along with $B[\ ]$ is self-adjoint, from the Theorem of section 3.16 we are ensured that there exist

1. an infinite set of real eigenvalues λ_n and
2. corresponding to each λ_n, an eigenfunction $\phi_n(x)$, where the set $\{\phi_j(x)\}$ is orthogonal.

Further, let the set $\{\phi_n\}$ be orthonormal—that is,

$$\int_a^b \rho(x)\phi_m(x)\phi_n(x)\,dx = \delta_{mn}$$

where δ_{mn} is the Kronecker delta:

$$\delta_{mn} = \begin{cases} 0, & m \neq n \\ 1, & m = n \end{cases}$$

Given a function $f(x)$, we now consider its expansion in terms of an infinite series of the eigenfunctions $\{\phi_n(x)\}$. Thus we are interested in representing a function by other functions.

Assume that $f(x)$ can be represented by the convergent series

$$f(x) = \sum_{j=1}^{\infty} c_j\phi_j(x) \tag{3.18.3}$$

where c_j are unknown constants. Multiplying both sides by $\rho(x)\phi_n(x)$ and integrating from a to b, we get

$$\int_a^b \rho(x)f(x)\phi_n(x)\,dx = \int_a^b \left[\sum_{j=1}^{\infty} c_j\phi_j(x)\right]\rho(x)\phi_n(x)\,dx$$

Exchanging the sum with integration, which is possible owing to the assumed convergence of the series, we have

$$\begin{aligned} \text{rhs} &= \sum_{j=1}^{\infty} c_j \int_a^b \rho(x)\phi_j(x)\phi_n(x)\,dx \\ &= c_n \end{aligned} \tag{3.18.4}$$

because, due to orthonormality, all other terms (except when $j = n$) are zero and

$$\int_a^b \rho(x)\phi_n^2(x)\ dx = 1$$

Since the unknown constants c_j are now determined, the representation of $f(x)$ by eq. (3.18.3) takes the form

$$f(x) = \sum_{j=1}^{\infty} \left[\int_a^b \rho(t)f(t)\phi_j(t)\ dt \right] \phi_j(x) \tag{3.18.5}$$

and is called a *generalized Fourier series*.

Note that although (3.18.3) was an assumed representation, it is indeed valid. This follows from the fact that $\{\phi_j(x)\}$ form a *complete set* (for a proof, see Weinberger, 1965, chapter 7). The completeness of $\{\phi_j(x)\}$ means that with c_n determined as in eq. (3.18.4), the series (3.18.3) converges absolutely and uniformly to $\phi(x)$, for all functions $f(x)$ for which $\int_a^b \rho(x)f^2(x)\ dx$ is finite.

From above, it is then clear that given a complete set $\{\phi_j(x)\}$ and given a function $f(x)$, we can find its corresponding $\{c_j\}$. Conversely, given $\{c_j\}$, we can identify the specific $f(x)$ they belong to. There is thus a *one-to-one correspondence* between $f(x)$ and $\{c_j\}$, which can be exploited to solve boundary value problems and PDEs as will be seen later. First we note that

$$c_j = \int_a^b \rho(x)f(x)\phi_j(x)\ dx = \mathcal{F}[f(x)] \tag{3.18.6}$$

is called the *finite Fourier transform* (FFT) of $f(x)$, where

$$\mathcal{F}[\ ] = \int_a^b \rho(x)\phi_j(x)[\ ]\ dx$$

is a linear integral operator [cf. Sneddon, 1951, chapter 3]. From eq. (3.18.6),

$$f(x) = \mathcal{F}^{-1}[c_j] \tag{3.18.7}$$

where $\mathcal{F}^{-1}[\ ]$ is the *inverse* FFT. The inverse FFT is a summation operation, since by definition

$$f(x) = \sum_{j=1}^{\infty} c_j\phi_j(x) = \mathcal{F}^{-1}[c_j] \tag{3.18.8}$$

From eqs. (3.18.6) and (3.18.7) we obtain

$$\mathcal{F}[f(x)] = \mathcal{F}\mathcal{F}^{-1}[c_j] = c_j$$

and thus

$$\mathcal{F}\mathcal{F}^{-1}[\ ] = I[\ ]$$

the identity operator. Similarly,

$$\mathscr{F}^{-1}[c_j] = \mathscr{F}^{-1}\,\mathscr{F}[f(x)] = f(x)$$

and thus

$$\mathscr{F}\mathscr{F}^{-1}[\ ] = I[\ ] = \mathscr{F}^{-1}\,\mathscr{F}[\ ] \tag{3.18.9}$$

so that $\mathscr{F}[\ ]$ is a linear integral operator which commutes with its inverse.

Finally, it is important to remember that $\mathscr{F}[\ ]$ operates on a *function* and generates a *number* c_j, while $\mathscr{F}^{-1}[\ ]$ operates on a *number* c_j and produces a *function* $f(x)$.

Exercise

Show that for $f(x) = 1$ and the set of orthonormal eigenfunctions given by eq. (3.16.10) with $L = 1$, we have the expansion

$$1 = \sum_{j=1}^{\infty} \frac{2}{j\pi} [1 - (-1)^j] \sin(j\pi x) \tag{3.18.10}$$

The sum of the first N terms of the series, s_N, is shown in Figure 3.7 for various values of N. Note that the finite sum exhibits oscillations around the expected value. The amplitude of the oscillation grows near the ends of the interval, owing to the BCs satisfied by the eigenfunctions. This feature is called the *Gibbs phenomenon*.

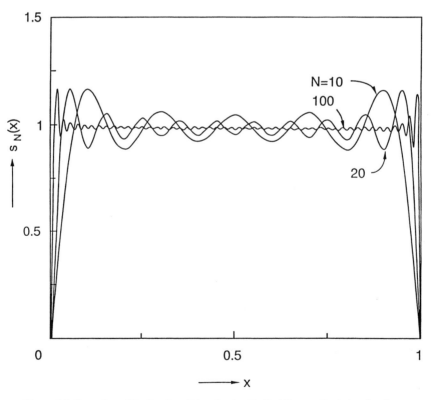

Figure 3.7 Expansion of the function $f(x) = 1$ using the first N normalized eigenfunctions.

3.19   SOLUTION OF BOUNDARY VALUE PROBLEMS BY THE METHOD OF FINITE FOURIER TRANSFORMS

In this section we see how boundary value problems can be solved, rather conveniently and directly, by the method of FFT. This will also be the central approach for solving PDEs in chapter 7.

   The method can be illustrated clearly by an example, and for this we consider a problem solved previously (example 2 of section 3.11.3). The problem involves finding the temperature distribution in a cooled one-dimensional rod with internal heat generation:

$$\frac{d^2y}{dx^2} - U^2y = -f(x), \qquad x \in (0, L) \tag{3.19.1}$$

with BCs

$$y(0) = 0 = y(L) \tag{3.19.2}$$

where $U^2 = hp/kA > 0$ is a constant. This problem was solved in section 3.11.3 by use of the Green's function. We solve it here by using the method of FFT, which turns out to be a shrewd approach.

   There are *two* ways to solve the problem by this method, and both lead to the same result. The reason is that we can define the differential operator in two different ways:

METHOD 1

$$L_1[y] = \frac{d^2y}{dx^2}$$

along with the BCs

$$y(0) = 0 = y(L)$$

METHOD 2

$$L_2[y] = \frac{d^2y}{dx^2} - U^2y$$

along with the BCs

$$y(0) = 0 = y(L)$$

   Following method 1, we define the eigenvalue problem corresponding to $L_1[\ ]$ and the given homogeneous BCs:

$$L_1[\phi] = \frac{d^2\phi}{dx^2} = -\lambda\phi, \qquad x \in (0, L)$$
$$\phi(0) = \quad 0 \quad = \phi(L)$$

From eqs. (3.15.6) and (3.16.9), the eigenvalues and the corresponding *orthonormal* eigenfunctions are

$$\lambda_n = \frac{n^2\pi^2}{L^2}, \qquad n = 1, 2, \ldots$$

and

$$\phi_n(x) = \sqrt{\frac{2}{L}} \sin\left(\frac{n\pi x}{L}\right), \qquad n = 1, 2, \ldots$$

Now, the original ODE (3.19.1) can be rewritten as

$$L_1[y] - U^2 y = -f(x), \qquad x \in (0, L) \tag{3.19.3}$$

along with

$$y(0) = 0 = y(L) \tag{3.19.2}$$

where observe that by construction, the eigenfunctions $\phi_n(x)$ satisfy the specified BCs.

Multiplying eq. (3.19.3) by $\phi_n(x)$ and integrating from 0 to L gives

$$\int_0^L \phi_n L_1[y] \, dx - U^2 \int_0^L y\phi_n \, dx = -\int_0^L f(x)\phi_n(x) \, dx$$

By definition,

$$\int_0^L y(x)\phi_n(x) \, dx = \mathcal{F}_1[y(x)] = y_n$$

and

$$\int_0^L f(x)\phi_n(x) \, dx = \mathcal{F}_1[f(x)] = f_n$$

where y_n and f_n are the FFTs of $y(x)$ and $f(x)$, respectively. Thus we have

$$\int_0^L \phi_n L_1[y] \, dx - U^2 y_n = -f_n \tag{3.19.4}$$

However, due to self-adjointness of $L_1[\ ]$ and the BCs, we obtain

$$\int_0^L \phi_n L_1[y] \, dx = \int_0^L y L_1[\phi_n] \, dx$$

$$= \int_0^L y(-\lambda_n \phi_n) \, dx$$

$$= -\lambda_n y_n$$

Thus eq. (3.19.4) takes the form

$$-(\lambda_n + U^2)y_n = -f_n$$

so that

$$y_n = \mathcal{F}_1[y(x)] = \frac{f_n}{\lambda_n + U^2}$$

Finally, by definition,

$$y(x) = \mathcal{F}_1^{-1}[y_n] = \sum_{n=1}^{\infty} y_n \phi_n(x)$$

so that the desired temperature distribution is

$$y(x) = \sum_{n=1}^{\infty} \frac{f_n}{\lambda_n + U^2} \phi_n(x)$$

Writing this more fully gives

$$y(x) = \frac{2}{L} \sum_{n=1}^{\infty} \frac{1}{U^2 + n^2\pi^2/L^2} \left[\int_0^L f(t) \sin\left(\frac{n\pi t}{L}\right) dt \right] \sin\left(\frac{n\pi x}{L}\right) \quad (3.19.5)$$

which is an infinite series representation of the solution (3.11.20) obtained earlier in terms of the Green's function. The series converges rapidly because of the n^2 term in the denominator.

Let us now follow the second alternative in defining the operator. In method 2 the eigenvalue problem corresponding to $L_2[\ ]$ is given by

$$L_2[\psi] = \frac{d^2\psi}{dx^2} - U^2\psi = -\eta\psi, \qquad x \in (0, L)$$

$$\psi(0) = 0 = \psi(L)$$

where η are the eigenvalues and ψ are the associated eigenfunctions. Note that since the eigenvalue problem may be rewritten as

$$\frac{d^2\psi}{dx^2} = -(\eta - U^2)\,\psi = -\sigma\psi$$

the values σ_n are the *same* as λ_n. Thus

$$\sigma_n = \eta_n - U^2 = \frac{n^2\pi^2}{L^2}$$

so that the eigenvalues are

$$\eta_n = U^2 + \frac{n^2\pi^2}{L^2}, \qquad n = 1, 2, \ldots$$

and the corresponding orthonormal eigenfunctions are

$$\psi_n(x) = \sqrt{\frac{2}{L}} \sin\left(\frac{n\pi x}{L}\right)$$

The ODE to be solved—that is, eq. (3.19.1)—is

$$L_2[y] = \frac{d^2T}{dx^2} - U^2 y = -f(x), \qquad x \in (0, L)$$

As in method 1, multiplying both sides by $\psi_n(x)$, integrating from 0 to L, and realizing that

$$\int_0^L \psi_n L_2[y] \, dx = \int_0^L y L_2[\psi_n] \, dx$$

$$= \int_0^L y(-\eta_n \psi_n) \, dx$$

$$= -\eta_n y_n$$

where

$$y_n = \mathscr{F}_2[y(x)] = \int_0^L y(x) \, \psi_n(x) \, dx$$

is the FFT of $y(x)$, we get

$$-\eta_n y_n = -f_n$$

where

$$f_n = \mathscr{F}_2[f(x)]$$

is the FFT of $f(x)$. Thus we have simply

$$y_n = \frac{f_n}{\eta_n} = \frac{f_n}{U^2 + (n^2\pi^2/L^2)}, \qquad n = 1, 2, \ldots$$

Since

$$y(x) = \mathscr{F}_2^{-1}[y_n] = \sum_{n=1}^{\infty} y_n \psi_n(x)$$

the solution is given by

$$y(x) = \frac{2}{L} \sum_{n=1}^{\infty} \frac{1}{U^2 + (n^2\pi^2/L^2)} \left[\int_0^L f(t) \sin\left(\frac{n\pi t}{L}\right) dt \right] \sin\left(\frac{n\pi x}{L}\right) \quad (3.19.6)$$

which is the *same* as for method 1. Thus the two different ways of defining the differential operator actually give the same result in the end.

Now note that by the Green's function approach, we have from eq. (3.11.20)

$$y(x) = \int_0^L [-f(t)]G(x, t) \, dt \quad (3.19.7)$$

where, from eq. (3.11.21), we obtain

$$G(x, t) = \begin{cases} -\dfrac{\sinh U(L - t) \sinh Ux}{U \sinh UL}, & 0 \le x \le t \\[2ex] -\dfrac{\sinh U(L - x) \sinh Ut}{U \sinh UL}, & t \le x \le L \end{cases} \quad (3.19.8)$$

Comparing the two forms of the solution $y(x)$, given by eqs. (3.19.6) and (3.19.7), we obtain a series representation for the Green's function $G(x, t)$:

$$G(x, t) = -\frac{2}{L} \sum_{n=1}^{\infty} \frac{1}{U^2 + (n^2\pi^2/L^2)} \sin\left(\frac{n\pi x}{L}\right) \sin\left(\frac{n\pi t}{L}\right) \qquad (3.19.9)$$

Although we have seen the application of the method of FFT to a specific problem only, it should be apparent that it is applicable to *all* nonhomogeneous boundary value problems in which

1. the BCs are homogeneous and
2. the differential operator and the BCs are self-adjoint.

Using the method of FFT, the solution of the boundary value problem is *always* in the form of an infinite series (i.e., the generalized Fourier series), which is usually rapidly convergent.

Finally, note that if the BCs in a given problem are nonhomogeneous, then by following the approach outlined in section 3.12 we can change the problem to one in which the BCs are indeed homogeneous. Thus the *only* restriction in applying the method of finite Fourier transform is that the differential operator and the corresponding homogeneous BCs be self-adjoint.

The problems become much more complicated when the differential operator and/ or the BCs are *not* self-adjoint. In such cases the theorem of section 3.16 does not apply. Thus we have the possibility of real or complex, and discrete or continuous set of eigenvalues. The corresponding eigenvectors may not be orthogonal, and the completeness of the eigenvectors is also not ensured. However, in a few special cases some analytical results can be obtained. For a discussion of these, the interested reader is referred to chapter 11 of Ramkrishna and Amundson (1985).

References

M. D. Greenberg, *Application of Green's Functions in Science and Engineering*, Prentice-Hall, Englewood Cliffs, NJ, 1971.

E. L. Ince, *Ordinary Differential Equations*, Dover, New York, 1956.

D. Ramkrishna and N. R. Amundson, *Linear Operator Methods in Chemical Engineering*, Prentice-Hall, Englewood Cliffs, NJ, 1985.

G. F. Roach, *Green's Functions*, 2nd ed., Cambridge University Press, Cambridge, 1982.

I. N. Sneddon, *Fourier Transforms*, McGraw-Hill, New York, 1951.

H. F. Weinberger, *A First Course in Partial Differential Equations*, Blaisdell, Waltham, MA, 1965.

Additional Reading

R. V. Churchill, *Operational Mathematics*, 3rd ed., McGraw-Hill, New York, 1972.

R. V. Churchill and J. W. Brown, *Fourier Series and Boundary Value Problems*, 4th ed., McGraw-Hill, New York, 1987.

E. A. Coddington and N. Levinson, *Theory of Ordinary Differential Equations*, McGraw-Hill, New York, 1955.

R. Courant and D. Hilbert, *Methods of Mathematical Physics*, volume 1, Interscience, New York, 1966.

H. Sagan, *Boundary and Eigenvalue Problems in Mathematical Physics*, Wiley, New York, 1961.

Problems

3.1 Determine whether the following differential operators are linear or nonlinear.

(a) $L[y] = x^2 \dfrac{d^2y}{dx^2} + x \dfrac{dy}{dx} + 4y$

(b) $L[y] = x \dfrac{d^2y}{dx^2} + 2\left(\dfrac{dy}{dx}\right)^2 + 3x^2y$

(c) $L[y] = a_0(x) \dfrac{d^3y}{dx^3} + a_1(x) \dfrac{d^2y}{dx^2} + a_2(x)y \dfrac{dy}{dx} + a_3(x)y$

(d) $L[y] = \dfrac{dy}{dx} + \exp(x)y$

3.2 Find a fundamental set of solutions for the linear ODEs shown below. In each case, also evaluate the Wronskian determinant.

(a) $y'' + 4y' + 3y = 0$
(b) $y''' - y'' - y' + y = 0$
(c) $y''' + 6y'' + 11y' + 6y = 0$
(d) $y'^v - y = 0$
(e) $y'' + 2y' + 2y = 0$
(f) $y''' + 3y'' + 3y' + y = 0$

3.3 In each case of problem 3.2, find the fundamental set of solutions; $x_0 = 0$.

3.4 The Euler equation of nth order is

$$L[y] = x^n y^{(n)} + a_1 x^{n-1} y^{(n-1)} + \cdots + a_{n-1} x y^{(1)} + a_n y = 0$$

where a_j are real constants, and $x > 0$. With the change of variables

$$x = e^z$$

it can be reduced to an nth order ODE with constant coefficients.

(a) Show that the above assertion is correct for $n = 3$.
(b) Following this technique, find the general solution of

$$x^3 y^{(3)} + x y^{(1)} - y = 0$$

3.5 Consider the Euler equation of second order:

$$x^2 y'' + axy' + by = 0, \qquad x > 0$$

where a and b are constants. Following the change of variables given in problem 3.4, write down the two linearly independent solutions for various combinations of a and b.

3.6   Consider the differential equation

$$\left\{ \frac{d}{dx} + a \right\}^{n} y = 0$$

where a is constant. Show that the functions $\{e^{-ax}x^{k}\}$, $k = 0, 1, 2, \ldots, n-1$ form a set of linearly independent solutions.

3.7   For eq. (3.5.27), find a particular solution by the method of undetermined coefficients. Show that, in accord with Theorem 2 of section 3.4, the particular solution thus obtained differs from eq. (3.5.29) by at most a linear combination of a fundamental set of solutions of the homogeneous equation.

3.8   Using the one-sided Green's function, find particular solutions as well the general solutions of the following ODEs.

(a) $y'' - y = x,$ $x_0 = x_0$
(b) $y'' + y = \sin x,$ $x_0 = 0$
(c) $y''' - 2y'' - y' + 2y = e^{4x},$ $x_0 = 0$
(d) $y'^{v} - y = g(x),$ $x_0 = 0$
(e) $x^3 y''' + x^2 y'' - 2xy' + 2y = x^3 \ln x,$ $x_0 > 0$

3.9   Find the adjoint operator corresponding to the following operators.

(a) $L[y] = e^{-x}y'' - e^{-x}y' + (\cos x)y$
(b) $L[y] = a_0(x)y^{(3)} + a_1(x)y^{(2)} + a_2(x)y^{(1)} + a_3(x)y$

3.10   Consider the general fourth-order linear differential operator:

$$L[y] = a_0(x)y^{(4)} + a_1(x)y^{(3)} + a_2(x)y^{(2)} + a_3(x)y^{(1)} + a_4(x)y$$

Determine the conditions which the coefficients $a_j(x)$ should satisfy in order to ensure that $L[\ ]$ is self-adjoint.

3.11   Verify the conclusion given as the last sentence of section 3.10 for the following cases of BCs: (1, 3) and (3, 2), where (i, j) means that the BC is i kind at the left end $(x = a)$ and j kind at the right end $(x = b)$.

3.12   The dimensionless reactant concentration in an axial dispersion model tubular reactor, in which an isothermal first-order reaction occurs, is described by the ODE

$$\frac{d^2 u}{ds^2} - \text{Pe} \frac{du}{ds} - \text{Da}\, u = 0, \qquad 0 < s < 1 \qquad g(x) = \frac{1}{a_0(x)} e^{\int}$$

along with the BCs

$$\frac{du}{ds} = \text{Pe}(u - 1), \qquad s = 0$$

$$\frac{du}{ds} = 0, \qquad s = 1$$

where Pe (Peclet number) and Da (Damköhler number) are constants.

(a) Is the differential operator self-adjoint?
(b) If it is not, then is there a transformation which will make it self-adjoint?
(c) Is the transformed problem [i.e., ODE along with homogeneous BCs] self-adjoint?

3.13 Including a recycle stream in the reactor does not change the ODE, but it alters the BCs as

$$\frac{du}{dx} = \text{Pe}\left[u - \frac{1 + Ru(1)}{1 + R}\right], \qquad s = 0$$

$$\frac{du}{dx} = 0, \qquad\qquad\qquad s = 1$$

where R is recycle ratio. Does the same transformation of problem 3.12 make this problem self-adjoint also?

3.14 Given the differential operator

$$L[u] = \frac{d^2u}{dx^2}, \qquad 0 < x < 1$$

and a set of so-called *antiperiodic* BCs:

$$u(0) + u(1) = 0$$

$$u'(0) + u'(1) = 0$$

(a) Find the set of adjoint BCs.
(b) Is this a self-adjoint problem?

3.15 Consider the differential operator

$$L[y] = \frac{d^2y}{dx^2}, \qquad a < x < b$$

along with homogeneous BCs of type (i, j), which means i kind at $x = a$ and j kind at $x = b$, defined as follows:

BC at $x = a$	Type	BC at $x = b$
$y = 0$	1	$y = 0$
$y' = 0$	2	$y' = 0$
$-y' + \alpha_1 y = 0$	3	$y' + \alpha_2 y = 0$

where $\alpha_i > 0$. Find the Green's function corresponding to all combinations of (i, j) except $(2, 2)$. Why is the combination $(2, 2)$ not considered?

3.16 Repeat problem 3.15 for the differential operator

$$L[y] = \frac{d^2y}{dx^2} + K^2y, \qquad a < x < b$$

3.17 Using the Green's functions developed in problem 3.15 for each BC, determine explicitly the solution of

(a) $L[y] = x$
(b) $L[y] = e^x$

3.18 Find the Green's function for the operator

$$L[y] = \frac{d}{dx}\left[x\frac{dy}{dx}\right], \qquad 0 < a < x < b$$

along with the BCs

$$y(a) = 0 \quad \text{and} \quad y(b) = 0$$

3.19 For the operator

$$L[y] = \frac{d}{dx}\left[x^2\frac{dy}{dx}\right], \qquad 0 < a < x < b$$

along with the BCs

$$y' \qquad = 0 \qquad x = a$$

$$y' + hy = 0, \qquad x = b$$

(a) Find the Green's function.
(b) Using the Green's function, find the solution of $L[y] = x^2$, along the same homogeneous BCs. Carry out all integrations fully.

3.20 Consider the differential operator

$$L[y] = \frac{d}{dx}\left[p(x)\frac{dy}{dx}\right] + q(x)y$$

and the ODE

$$L[y] = -f(x), \qquad a < x < b$$

along with nonhomogeneous BCs

$$-k_1\frac{dy}{dx} + h_1 y = \alpha_1, \qquad x = a$$

$$k_2\frac{dy}{dx} + h_2 y = \alpha_2, \qquad x = b$$

Show that the solution for k_1 and $k_2 \neq 0$ is given by

$$y(x) = \int_a^b G(x, t)[-f(t)]\, dt - \frac{p(a)}{k_1}\alpha_1 G(x, a) - \frac{p(b)}{k_2}\alpha_2 G(x, b)$$

where $G(x, t)$ is Green's function corresponding to $L[\ ]$ and the associated homogeneous BCs.
Also show that

(a) If $k_1 = 0$, then $\dfrac{G(x, a)}{k_1}$ is replaced by $\dfrac{1}{h_1}\dfrac{\partial G}{\partial t}(x, a)$

(b) If $k_2 = 0$, then $\dfrac{G(x, b)}{k_2}$ is replaced by $-\dfrac{1}{h_2}\dfrac{\partial G}{\partial t}(x, b)$

3.21 For the eigenvalue problem

$$L[y] = \frac{d^2y}{dx^2} = -\lambda y, \qquad 0 < x < 1$$

along with type (i, j) homogeneous BCs (i.e., i kind at $x = 0$ and j kind at $x = 1$), determine the eigenvalues and the normalized eigenfunctions. As in problem 3.15, consider all (i, j) combinations, except $(2, 2)$.

3.22 Find the eigenvalues and normalized eigenfunctions for

$$L[u] = \frac{d^2u}{dx^2} - \alpha^2 u = -\lambda u, \qquad 0 < x < 1$$

$$-u' + \beta u = 0, \qquad x = 0$$
$$u' + \beta u = 0, \qquad x = 1$$

where $\alpha > \beta > 0$ are real constants.

3.23 In eq. (3.17.5), numerically determine λ_i, $i = 1 \to 5$, for the following parameter values:

(a) $\alpha_1 = \alpha_2 = 2$
(b) $\alpha_1 = 2, \alpha_2 = 10$

3.24 Consider the differential operator

$$L[y] = \frac{d^2y}{dx^2}, \qquad 0 < x < L$$

along with homogeneous third-kind BCs:

$$-k\frac{dy}{dx} + h_1 y = 0, \qquad x = 0$$

$$k_2 \frac{dy}{dx} + h_2 y = 0, \qquad x = L$$

Give a formal representation of the Green's function in terms of the normalized eigenfunctions.

3.25 Consider steady-state heat conduction in a one-dimensional slab of thickness L, with internal heat consumption at rate $A \sin \omega x$. The left (i.e., $x = 0$) end of the slab is maintained at constant temperature T_0, while the heat flux into the slab at the right end (i.e., $x = L$) is fixed as α.

(a) Set up the relevant boundary value problem for the steady-state temperature distribution in the slab.
(b) Solve by using the method of finite Fourier transform.

3.26 For the diffusion-reaction problem involving an isothermal first-order reaction in a catalyst pellet of slab geometry, with no external mass transfer resistance, the dimensionless reactant concentration u satisfies

$$\frac{d^2u}{dx^2} = \phi^2 u, \qquad 0 < x < 1$$

$$\frac{du}{dx} = 0, \qquad x = 0$$

$$u = 1, \qquad x = 1$$

where ϕ is a constant and represents the Thiele modulus. The analytic solution can be obtained readily for u as well as for the effectiveness factor, defined as

$$\eta = \int_0^1 u(x)\, dx$$

Solve by the method of finite Fourier transform, and determine the number of terms required in the infinite series so that the absolute error between the analytic solution and the infinite series solution for

(a) η
(b) $u(0)$

is less thatn 10^{-5}. Compute for three values of ϕ: 0.1, 1, and 5.

Chapter 4

SERIES SOLUTIONS AND SPECIAL FUNCTIONS

In chapter 2, various theorems were discussed which guaranteed the existence and uniqueness of solutions to a single or a system of ordinary differential equations (ODEs). The method used was a constructive one, because it proved the existence of the solution by exhibiting the solution itself. This method of obtaining solutions is not particularly useful in general, and it is our purpose here to discuss the solution of differential equations in series form. We will limit the discussion to linear differential equations of the second order. The methods for higher-order differential equations are similar, but the analysis becomes tedious and complicated. In this chapter we will consider the solution of linear differential equations in which the coefficients are analytic or have at most a few singularities. Most of the special functions which occur in the solution of engineering problems arise as solutions of differential equations. In this way we are led to the functions of Bessel, Legendre, Laguerre, Hermite, and to the hypergeometric function.

4.1 SOME DEFINITIONS

A *power series* is an infinite series of the form

$$\sum_{n=0}^{\infty} c_n (x - a)^n = c_0 + c_1(x - a) + c_2(x - a)^2 + \cdots \qquad (4.1.1)$$

where a is the *center* and c_j are the *coefficients* of the power series.

$$s_n = \sum_{j=0}^{n} c_j (x - a)^j = c_0 + c_1(x - a) + \cdots + c_n(x - a)^n \qquad (4.1.2)$$

is called the *nth partial sum* of the power series. If the *n*th partial sum is subtracted from the power series, the resulting expression is called the *remainder*,

$$R_n = \sum_{j=n+1}^{\infty} c_j(x-a)^j = c_{n+1}(x-a)^{n+1} + c_{n+2}(x-a)^{n+2} + \cdots \quad (4.1.3)$$

If at a point $x = x_0$, $\lim_{n\to\infty} s_n(x_0)$ exists and is finite—say, equal to $s(x_0)$—then the power series is said to *converge* at $x = x_0$. A series may or may not converge for all x. If it does not converge at a specific x, it is said to *diverge* for that value of x. It should be evident that if a series converges at x_0, then $|R_n(x_0)|$ can be made as small as desired by taking n sufficiently large.

It is clear that the power series $\sum_{j=0}^{\infty} c_j(x-a)^j$ certainly converges for $x = a$. If there are other values of x around the center a for which convergence occurs, then these values form the *convergence interval*, having the midpoint $x = a$. This interval may be finite or infinite; in the latter case the series converges for all x. The size of the convergence interval is denoted by the *radius of convergence*, R; hence convergence occurs for $|x - a| < R$.

The radius of convergence can be determined from the coefficients of the series by one of the following formulae:

$$\frac{1}{R} = \lim_{n\to\infty} [|c_n|]^{1/n} \quad (4.1.4a)$$

$$\frac{1}{R} = \lim_{n\to\infty} \left| \frac{c_{n+1}}{c_n} \right| \quad (4.1.4b)$$

which are called the *root method* and the *ratio method*, respectively. Their use is illustrated by the examples below.

Example 1. Consider the geometric series

$$\sum_{j=0}^{\infty} x^j = 1 + x + x^2 + \cdots$$

where the center is $a = 0$. Using eqs. (4.1.4), we obtain

$$\frac{1}{R} = \lim_{n\to\infty} [|1|]^{1/n} = 1 \quad (4.1.5a)$$

$$\frac{1}{R} = \lim_{n\to\infty} \left| \frac{1}{1} \right| = 1 \quad (4.1.5b)$$

so that both formulae give $R = 1$. The geometric series $\sum_{j=0}^{\infty} x^j$ thus converges for $|x| < 1$, which can be readily understood since

$$\sum_{j=0}^{\infty} x^j = \frac{1}{1-x}$$

Example 2. Consider the exponential series

$$e^x = \sum_{j=0}^{\infty} \frac{x^j}{j!} = 1 + x + \frac{x^2}{2!} + \frac{x^3}{3!} + \cdots$$

which again has center $a = 0$. Using the ratio method (4.1.4b), we have

$$\frac{1}{R} = \lim_{n\to\infty} \left| \frac{1/(n+1)!}{1/n!} \right| = \lim_{n\to\infty} \left| \frac{1}{n+1} \right| = 0$$

Thus $R = \infty$, and so the series converges for all x.

If we use the root method (4.1.4a), we get the same answer, although with some difficulty:

$$\frac{1}{R} = \lim_{n\to\infty} \left[\left| \frac{1}{n!} \right| \right]^{1/n} \tag{4.1.6}$$

To evaluate the limit, we need a representation of $n!$ for large n. From the Stirling formula,

$$n! \sim \sqrt{2\pi n} \left(\frac{n}{e} \right)^n \qquad \text{for } n \gg 1$$

so that from eq. (4.1.6) we get

$$\frac{1}{R} = \lim_{n\to\infty} \left| \frac{1}{\sqrt{2\pi n}} \left(\frac{e}{n} \right)^n \right|^{1/n}$$

$$= \lim_{n\to\infty} \left| \frac{e}{n} \frac{1}{(2\pi n)^{1/2n}} \right| \tag{4.1.7}$$

Since

$$\lim_{n\to\infty} (2\pi n)^{1/2n} = 1$$

eq. (4.1.7) yields

$$\frac{1}{R} = \lim_{n\to\infty} \left| \frac{e}{n} \right| = 0$$

and hence $R = \infty$ as before.

4.2 ELEMENTARY OPERATIONS

In developing solutions of ODEs in the form of power series, we need some elementary operations, which are now considered. For concreteness, let the two power series $\sum_{j=0}^{\infty} a_j(x - x_0)^j$ and $\sum_{j=0}^{\infty} b_j(x - x_0)^j$ converge to the functions $f(x)$ and $g(x)$, respectively, in the common convergence interval $|x - x_0| < \rho$. In this interval then, the following results hold.

1. The two series can be added or subtracted term-by-term, and

$$f(x) \pm g(x) = \sum_{j=0}^{\infty} (a_j \pm b_j)(x - x_0)^j \tag{4.2.1}$$

2. The two series may be multiplied, and

$$f(x) \cdot g(x) = \left[\sum_{j=0}^{\infty} a_j(x - x_0)^j \right] \left[\sum_{j=0}^{\infty} b_j(x - x_0)^j \right]$$

$$= \sum_{j=0}^{\infty} c_j(x - x_0)^j \tag{4.2.2}$$

where $c_m = a_0 b_m + a_1 b_{m-1} + \cdots + a_m b_0$.

3. The function $f(x)$ is continuous and has derivatives of all orders. These derivatives $f^{(1)}, f^{(2)}, \ldots$ can be computed by differentiating the series term-by-term. Thus

$$f^{(1)}(x) = a_1 + 2a_2(x - x_0) + 3a_3(x - x_0)^2 + \cdots$$

$$= \sum_{j=1}^{\infty} ja_j(x - x_0)^{j-1} \tag{4.2.3}$$

and so on.

4. When the coefficients a_j are given by

$$a_j = \frac{f^{(j)}(x_0)}{j!} \tag{4.2.4}$$

the series $\sum_{j=0}^{\infty} a_j(x - x_0)^j$ is called the *Taylor series* for the function $f(x)$ about $x = x_0$. A function $f(x)$ that has a Taylor series expansion about $x = x_0$,

$$f(x) = \sum_{j=0}^{\infty} \frac{f^{(j)}(x_0)}{j!}(x - x_0)^j \tag{4.2.5}$$

with a radius of convergence $\rho > 0$, is said to be *analytic* at $x = x_0$. Note that for the specific case $x_0 = 0$, the Taylor series is called the *Maclaurin series*.

4.3 ODEs WITH ANALYTIC COEFFICIENTS

We consider the ODE

$$a_0(x) \frac{d^2y}{dx^2} + a_1(x) \frac{dy}{dx} + a_2(x)y = 0 \tag{4.3.1}$$

where the leading coefficient $a_0(x) \neq 0$ in some interval of interest $x \in [a, b]$, so that the equation may be written in the form

$$\frac{d^2y}{dx^2} + p(x) \frac{dy}{dx} + q(x)y = 0 \tag{4.3.2}$$

It is assumed that the coefficients $p(x)$ and $q(x)$ are analytic functions at $x = x_0$; the point x_0 is then called the *ordinary point* of the ODE. We seek solutions of eq. (4.3.2) of the form

$$y = \sum_{n=0}^{\infty} c_n(x - x_0)^n \tag{4.3.3}$$

by the power series method. Two questions arise about such solutions. The first is whether we can actually determine the coefficients c_n, so that eq. (4.3.3) satisfies eq. (4.3.2). The second is whether the obtained series solution converges, and, if so, what is the convergence interval. The question of convergence is important, because only if the series converges—for say, $|x - x_0| < R$—can we be certain that the termwise differentiation, addition of series, and other operations employed in obtaining the solution are in fact valid.

We will state the general theorem concerning the existence of the power series solution and its radius of convergence later in this section. It is instructive to first consider an example which illustrates how the coefficients c_n are determined. The basic technique of applying the power series method is simply to represent *all* functions in the ODE by power series in $(x - x_0)$ and then to determine c_n so that the ODE is satisfied.

Example. The ODE

$$\frac{d^2y}{dx^2} + y = 0, \qquad x \in (-\infty, \infty) \tag{4.3.4}$$

has two linearly independent solutions: $\cos x$ and $\sin x$. We now derive this result using the power series method.

For the equation at hand,

$$p(x) = 0, \qquad q(x) = 1$$

Hence $x_0 = 0$ is an ordinary point of the equation. We seek solutions of the form

$$y = \sum_{n=0}^{\infty} c_n x^n = c_0 + c_1 x + c_2 x^2 + c_3 x^3 + \cdots \tag{4.3.5}$$

so that

$$\frac{dy}{dx} = c_1 + 2c_2 x + 3c_3 x^2 + \cdots = \sum_{n=1}^{\infty} nc_n x^{n-1}$$

$$\frac{d^2y}{dx^2} = 2c_2 + 3 \cdot 2c_3 x + \cdots = \sum_{n=2}^{\infty} n(n - 1)c_n x^{n-2}$$

Substituting these series in eq. (4.3.4) gives

$$\sum_{n=2}^{\infty} n(n - 1)c_n x^{n-2} + \sum_{n=0}^{\infty} c_n x^n = 0$$

In the first sum, we shift the index of summation by replacing n by $m + 2$, so that

$$\sum_{m=0}^{\infty} (m + 2)(m + 1)c_{m+2} x^m + \sum_{n=0}^{\infty} c_n x^n = 0$$

However, because the index of summation is a dummy variable, we have

$$\sum_{n=0}^{\infty} [(n + 2)(n + 1)c_{n+2} + c_n]x^n = 0$$

Since this equation is to be satisfied for *all* x, the coefficient of *each* power of x must be zero; that is,

$$(n + 2)(n + 1)c_{n+2} + c_n = 0, \qquad n = 0, 1, 2, \ldots \qquad (4.3.6)$$

Using this *recursion formula*, we have

$$n = 0: \qquad 2c_2 + c_0 = 0, \quad \text{thus } c_2 = -\frac{c_0}{2} = -\frac{c_0}{2!}$$

$$n = 1: \qquad 3 \cdot 2c_3 + c_1 = 0, \quad \text{thus } c_3 = -\frac{c_1}{6} = -\frac{c_1}{3!}$$

$$n = 2: \qquad 4 \cdot 3c_4 + c_2 = 0, \quad \text{thus } c_4 = -\frac{c_2}{12} = \frac{c_0}{4!}$$

$$n = 3: \qquad 5 \cdot 4c_5 + c_3 = 0, \quad \text{thus } c_5 = -\frac{c_3}{20} = \frac{c_1}{5!}$$

and in general,

$$c_{2k} = \frac{(-1)^k}{(2k)!} c_0, \quad c_{2k+1} = \frac{(-1)^k}{(2k + 1)!} c_1, \qquad k = 0, 1, 2, \ldots$$

so that the even coefficients are found in terms of c_0, and the odd ones are found in terms of c_1. The series (4.3.5) thus takes the form

$$y = c_0 \left[1 - \frac{x^2}{2!} + \frac{x^4}{4!} + - \cdots + \frac{(-1)^n}{(2n)!} x^{2n} + - \cdots \right]$$

$$+ c_1 \left[x - \frac{x^3}{3!} + \frac{x^5}{5!} + - \cdots + \frac{(-1)^n}{(2n + 1)!} x^{2n+1} + - \cdots \right]$$

$$= c_0 \cos x + c_1 \sin x \qquad (4.3.7)$$

where c_0 and c_1 are arbitrary constants. This is what was expected.

We can now state without proof the general theorem which ensures the existence and convergence of power series solutions of ODEs with analytic coefficients. The proof requires complex analysis and may be found in texts such as Ince (1956, section 12.22).

● **THEOREM**

If x_0 is an ordinary point of the ODE

$$\frac{d^2y}{dx^2} + p(x) \frac{dy}{dx} + q(x)y = 0 \qquad (4.3.8)$$

that is, if functions $p(x)$ and $q(x)$ are analytic at x_0, then the general solution of eq. (4.3.8) is also analytic at x_0 and may be written in the form

$$y = \sum_{n=0}^{\infty} a_n(x - x_0)^n = c_0 y_0(x) + c_1 y_1(x) \qquad (4.3.9)$$

where c_0 and c_1 are arbitrary constants, and $y_0(x)$ and $y_1(x)$ are linearly independent series solutions. Further, the radii of convergence of each of the series solutions $y_0(x)$ and $y_1(x)$ are at least as large as the minimum of those for the series for p and q

4.4 LEGENDRE EQUATION AND LEGENDRE POLYNOMIALS

The ODE

$$(1 - x^2) \frac{d^2y}{dx^2} - 2x \frac{dy}{dx} + \alpha(\alpha + 1)y = 0 \qquad (4.4.1)$$

where α is a real constant, arises in the solution of PDEs in spherical geometry. Dividing through by $(1 - x^2)$ gives the standard form (4.3.8). The resulting coefficients are analytic at $x = 0$, and they remain so in the region $|x| < 1$. However, to find the solution by the power series method, it is more convenient to leave the equation in the form (4.4.1). It is not difficult to show that two linearly independent solutions are

$$y_1 = c_0\phi_1(x) = c_0\left[1 - \frac{(\alpha + 1)\alpha}{2!} x^2 + \frac{(\alpha + 1)(\alpha + 3)\alpha(\alpha - 2)}{4!} x^4 \right.$$

$$\left. - \frac{(\alpha + 1)(\alpha + 3)(\alpha + 5)\alpha(\alpha - 2)(\alpha - 4)}{6!} x^6 + \cdots \right] \qquad (4.4.2)$$

$$y_2 = c_1\phi_2(x) = c_1\left[x - \frac{(\alpha + 2)(\alpha - 1)}{3!} x^3 + \frac{(\alpha + 2)(\alpha + 4)(\alpha - 1)(\alpha - 3)}{5!} x^5 \right.$$

$$\left. - \frac{(\alpha + 2)(\alpha + 4)(\alpha + 6)(\alpha - 1)(\alpha - 3)(\alpha - 5)}{7!} x^7 + \cdots \right]$$

$$(4.4.3)$$

If α is not zero and also not a positive or negative integer, then both series for $\phi_1(x)$ and $\phi_2(x)$ will converge for $|x| < 1$ and diverge otherwise. On the other hand, if $\alpha = 0$ or an even positive integer or an odd negative integer, the series for $\phi_1(x)$ will terminate after a finite number of terms and therefore converges for all x. Similarly, if α is an odd positive integer or an even negative integer, $\phi_2(x)$ will terminate after a finite number of terms. Hence for positive or negative integral values of α, one solution is a polynomial and the other is a power series with a finite interval of convergence, $|x| < 1$. The differential equation (4.4.1) is called *Legendre's differential equation* and its solutions are called *Legendre's functions*. The polynomials obtained for $\alpha = n = 0, 1, 2, 3, \ldots$ are called *Legendre polynomials*, $P_n(x)$. For these, it is convenient to choose c_0 and c_1 so that the coefficient of x^n in $P_n(x)$ is

$$A_n = \frac{(2n)!}{2^n(n!)^2} \qquad (4.4.4)$$

The Legendre polynomials $P_n(x)$ may then be written in the form

$$P_n(x) = \sum_{m=0}^{M} (-1)^m \frac{(2n - 2m)!}{2^n m!(n - m)!(n - 2m)!} x^{n-2m} \qquad (4.4.5)$$

where $M = n/2$ or $(n - 1)/2$, whichever is an integer. The first few Legendre polynomials are

$$P_0(x) = 1, \qquad P_1(x) = x, \qquad P_2(x) = \tfrac{1}{2}(3x^2 - 1)$$

$$P_3(x) = \tfrac{1}{2}(5x^3 - 3x), \qquad P_4(x) = \tfrac{1}{8}(35x^4 - 30x^2 + 3)$$

$$P_5(x) = \tfrac{1}{8}(63x^5 - 70x^3 + 15x), \quad \text{etc.}$$

and are shown in Figure 4.1. Note that as a result of the choices of c_0 and c_1 above, we have

$$P_n(1) = 1$$

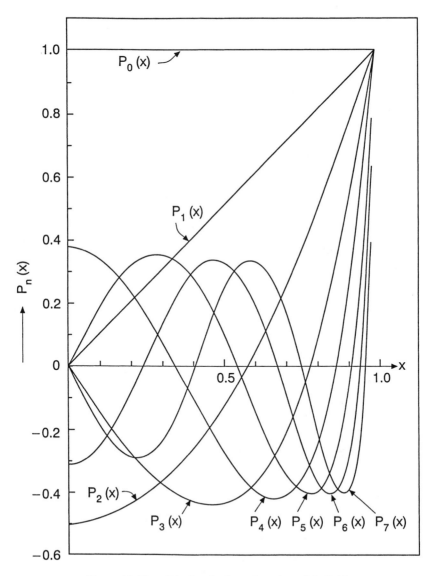

Figure 4.1 Plots of the first few Legendre polynomials, $P_n(x)$.

By using eq. (4.4.5), it is easily shown that Legendre polynomials satisfy the recurrence relation

$$P_{n+1}(x) = x \frac{2n + 1}{n + 1} P_n(x) - \frac{n}{n + 1} P_{n-1}(x), \qquad n \geq 1 \qquad (4.4.6)$$

which proves useful in finding them for large n.

One of the reasons for the importance of Legendre polynomials is that they are *orthogonal* over the interval $x \in [-1, 1]$, as shown next. Consider eq. (4.4.1) written for two different integral values of α:

$$(1 - x^2) \frac{d^2 P_n}{dx^2} - 2x \frac{dP_n}{dx} + n(n + 1)P_n = 0$$

$$(1 - x^2) \frac{d^2 P_m}{dx^2} - 2x \frac{dP_m}{dx} + m(m + 1)P_m = 0$$

which may be rewritten in the form

$$\frac{d}{dx} \left[(1 - x^2) \frac{dP_n}{dx} \right] + n(n + 1)P_n = 0$$

$$\frac{d}{dx} \left[(1 - x^2) \frac{dP_m}{dx} \right] + m(m + 1)P_m = 0$$

Multiply the first by $P_m(x)$ and the second by $P_n(x)$, and subtract to give

$$P_m(x) \frac{d}{dx} \left[(1 - x^2) \frac{dP_n}{dx} \right] - P_n(x) \frac{d}{dx} \left[(1 - x^2) \frac{dP_m}{dx} \right]$$
$$= [(m(m + 1) - n(n + 1)]P_m(x)P_n(x)$$

which may be rearranged as

$$\frac{d}{dx} \left[(1 - x^2) \left\{ P_m \frac{dP_n}{dx} - P_n \frac{dP_m}{dx} \right\} \right] = (m - n)(m + n + 1)P_m(x)P_n(x)$$

Integration from -1 to $+1$ gives

$$\left[(1 - x^2) \left\{ P_m \frac{dP_n}{dx} - P_n \frac{dP_m}{dx} \right\} \right]_{-1}^{1} = (m - n)(m + n + 1) \int_{-1}^{1} P_m(x)P_n(x) \, dx$$

and since the quantity on the lhs is zero, we have

$$\int_{-1}^{1} P_m(x)P_n(x) \, dx = 0, \qquad m \neq n \qquad (4.4.7)$$

which is a statement of orthogonality on the interval $[-1, 1]$.

The square of the *norm*

$$\| P_n \|^2 = \int_{-1}^{1} [P_n(x)]^2 \, dx \qquad (4.4.8)$$

is obtained with some difficulty. For this, consider now the sequence of powers of x:

$$1, x, x^2, x^3, \ldots, x^n$$

Then

$$1 = P_0(x) \quad \text{and} \quad x = P_1(x)$$

$P_2(x)$ contains terms in x^2 and constants only, so that x^2 may be expressed in terms of $P_0(x)$ and $P_2(x)$:

$$x^2 = \tfrac{2}{3}P_2(x) + \tfrac{1}{3}P_0(x)$$

$P_3(x)$ contains terms in x^3 and x only, so that x^3 may be expressed as a linear combination of $P_3(x)$ and $P_1(x)$:

$$x^3 = \tfrac{2}{5}P_3(x) + \tfrac{3}{5}P_1(x)$$

and similarly

$$x^4 = \tfrac{1}{35}[8P_4(x) + 20P_2(x) + 7P_0(x)]$$

It thus follows that any polynomial in x of degree n can be expressed as a linear combination of Legendre polynomials of order $n, n - 1, n - 2, \ldots, 1, 0$. Therefore $P_n(x)$ is orthogonal to *any* polynomial $p_j(x)$ of degree *less* than n—that is,

$$\int_{-1}^{1} P_n(x)p_j(x)\, dx = 0, \qquad j = 0, 1, 2, \ldots, (n - 1) \qquad (4.4.9)$$

We are now in a position to compute the value of $\| P_n \|$. Let the coefficient of x^n in $P_n(x)$ be A_n and that of x^{n-1} in $P_{n-1}(x)$ be A_{n-1}; then

$$q(x) = P_n(x) - \frac{A_n}{A_{n-1}} xP_{n-1}(x)$$

is a polynomial of degree $n - 2$. Hence

$$P_n(x) = \frac{A_n}{A_{n-1}} xP_{n-1}(x) + q(x)$$

and

$$\| P_n \|^2 = \int_{-1}^{1} P_n(x) \left[\frac{A_n}{A_{n-1}} xP_{n-1}(x) + q(x) \right] dx$$

$$= \frac{A_n}{A_{n-1}} \int_{-1}^{1} xP_n(x)P_{n-1}(x)\, dx$$

since $P_n(x)$ is orthogonal to $q(x)$. Using the recurrence relation (4.4.6), we have

$$\| P_n \|^2 = \frac{A_n}{A_{n-1}} \int_{-1}^{1} P_{n-1}(x) \left[\frac{n + 1}{2n + 1} P_{n+1}(x) + \frac{n}{2n + 1} P_{n-1}(x) \right] dx$$

$$= \frac{A_n}{A_{n-1}} \frac{n}{2n + 1} \int_{-1}^{1} [P_{n-1}(x)]^2\, dx = \frac{A_n}{A_{n-1}} \frac{n}{2n + 1} \| P_{n-1} \|^2 \qquad (4.4.10)$$

From eq. (4.4.4) we have

$$\frac{A_n}{A_{n-1}} = \frac{(2n)!}{2^n(n!)^2} \frac{2^{n-1}[(n-1)!]^2}{(2n-2)!}$$

$$= \frac{2n-1}{n}$$

and thus from eq. (4.4.10) we obtain

$$\| P_n \|^2 = \frac{2n-1}{2n+1} \| P_{n-1} \|^2 \tag{4.4.11}$$

Since $\| P_0 \|^2 = \int_{-1}^{1} 1^2 \, dx = 2$, from successive applications of eq. (4.4.11) we get

$$\| P_1 \|^2 = \tfrac{2}{3}$$

$$\| P_2 \|^2 = \tfrac{2}{5}$$

and in general

$$\| P_n \|^2 = \frac{2}{2n+1}, \qquad n = 0, 1, 2, \ldots \tag{4.4.12}$$

Since as shown above, any polynomial of degree n can be expanded as a linear combination of Legendre polynomials of order n and less, the question arises as to whether it is possible to expand an arbitrary function $f(x)$ in an infinite series of Legendre polynomials as

$$f(x) = \sum_{n=0}^{\infty} a_n P_n(x) \tag{4.4.13}$$

Multiplying each side by $P_m(x)$ and integrating from -1 to 1 gives

$$\int_{-1}^{1} f(x)P_m(x) \, dx = \sum_{n=0}^{\infty} a_n \int_{-1}^{1} P_m(x)P_n(x) \, dx$$

Since all but the term where $n = m$ on the rhs are zero, it follows that

$$a_m = \frac{2m+1}{2} \int_{-1}^{1} f(x)P_m(x) \, dx \tag{4.4.14}$$

which provides the coefficients a_m in exact similarity with eigenfunction expansions of section 3.18.

The condition of orthogonality of the Legendre polynomials enables us to prove a fact about the zeroes of $P_n(x)$ in the interval $(-1, 1)$. For greater generality, consider a set of orthogonal polynomials $\{p_n(x); n = 0, 1, 2, 3, \ldots\}$ which are orthogonal on an interval (a, b) with weight function $\rho(x)$—that is,

$$\int_{a}^{b} \rho(x)p_n(x)p_m(x) \, dx = 0, \qquad n \neq m \tag{4.4.15}$$

where $\rho(x) > 0$ for $x \in (a, b)$. Then, as before, since any polynomial $\pi(x)$ may be expressed in terms of this set of polynomials, it follows that if $\pi(x)$ is of degree *less than n*, then

$$\int_{a}^{b} \rho(x)p_n(x)\pi(x) \, dx = 0$$

Suppose $p_n(x)$ changes sign m times between a and b at $x_1, x_2, x_3, \ldots, x_m$, where $m < n$. If we let

$$\pi(x) = (x - x_1)(x - x_2) \cdots (x - x_m)$$

then $p_n(x)\pi(x)$ does not change sign and hence

$$\int_a^b \rho(x)p_n(x)\pi(x) \, dx$$

cannot be zero, leading to a contradiction for $m < n$. Since m cannot be larger than n, we must have $m = n$. Thus the orthogonal polynomial $p_n(x)$ vanishes *precisely* n times between a and b.

As a special case, the Legendre polynomial $P_n(x)$ must have n real zeroes for $x \in (-1, 1)$ since it does not vanish at $x = 1$ or $x = -1$. The zeroes of the first few Legendre polynomials are given in Table 4.1.

TABLE 4.1 Zeroes of the Legendre Polynomials on Interval $(-1, 1)$

Polynomial	Zeroes
$P_1(x)$	0
$P_2(x)$	± 0.57735027
$P_3(x)$	0
	± 0.77459667
$P_4(x)$	± 0.33998104
	± 0.86113631
$P_5(x)$	0
	± 0.53846931
	± 0.90617985
$P_6(x)$	± 0.23861919
	± 0.66120939
	± 0.93246951
$P_7(x)$	0
	± 0.40584515
	± 0.74153119
	± 0.94910791
$P_8(x)$	± 0.18343464
	± 0.52553241
	± 0.79666648
	± 0.96028986
$P_9(x)$	0
	± 0.32425342
	± 0.61337143
	± 0.83603111
	± 0.96816024
$P_{10}(x)$	± 0.14887434
	± 0.43339539
	± 0.67940957
	± 0.86506337
	± 0.97390653

Remark. While we have considered the Legendre equation and Legendre polynomials in some detail here, other well-known polynomials such as *Chebyshev polynomials* and *Hermite polynomials* also arise from series solutions of some specific ODEs around ordinary points. For their development, see problems 4.8 and 4.9 at the end of the chapter.

4.5 ODEs WITH REGULAR SINGULAR POINTS

In ODEs of the form

$$\frac{d^2y}{dx^2} + p(x)\frac{dy}{dx} + q(x)y = 0$$

the character of the solution depends critically on the nature of the functions $p(x)$ and $q(x)$. In section 4.3 we have seen how to obtain series solutions of equations where $p(x)$ and $q(x)$ are analytic at x_0, and hence x_0 is an ordinary point. However, many applications in science and engineering lead to ODEs where the functions $p(x)$ and $q(x)$ are not analytic. For example, Bessel's equation, which arises primarily in problems involving cylindrical geometry, has the form

$$\frac{d^2y}{dx^2} + \frac{1}{x}\frac{dy}{dx} + \left(1 - \frac{\alpha^2}{x^2}\right)y = 0$$

and thus $p(x)$ and $q(x)$ have singularities at $x = 0$. The Legendre equation considered in the last section,

$$\frac{d^2y}{dx^2} - \frac{2x}{1 - x^2}\frac{dy}{dx} + \frac{\alpha(\alpha + 1)}{1 - x^2}y = 0$$

has singularities at $x = \pm 1$ in its coefficients.

In many applications, we are interested in the behavior of the solution in the neighborhood of a singular point of the differential equation. The kinds and number of such points and their effect on the solution have been studied in detail, but we shall limit ourselves to second-order linear ODEs with regular singular points.

The point $x = x_0$ is said to be a *regular singular point* of the differential equation

$$\frac{d^2y}{dx^2} + \frac{p(x)}{x - x_0}\frac{dy}{dx} + \frac{q(x)}{(x - x_0)^2}y = 0 \tag{4.5.1}$$

if $p(x)$ and $q(x)$ are analytic at $x = x_0$; that is, they have series expansions

$$p(x) = \sum_{n=0}^{\infty} p_n(x - x_0)^n \tag{4.5.2a}$$

and

$$q(x) = \sum_{n=0}^{\infty} q_n(x - x_0)^n \tag{4.5.2b}$$

If $p_0 = 0$ and $q_0 = q_1 = 0$, then $x = x_0$ is an ordinary point of the equation, because then the coefficients are analytic. In general, for the nth-order ODE

$$\frac{d^n y}{dx^n} + \frac{p_1(x)}{x - x_0} \frac{d^{n-1} y}{dx^{n-1}} + \frac{p_2(x)}{(x - x_0)^2} \frac{d^{n-2} y}{dx^{n-2}} + \cdots$$

$$+ \frac{p_{n-1}(x)}{(x - x_0)^{n-1}} \frac{dy}{dx} + \frac{p_n(x)}{(x - x_0)^n} y = 0 \quad (4.5.3)$$

the point $x = x_0$ is a regular singular point if each of the $p_j(x)$ has a power series expansion about $x = x_0$, provided that x_0 is not an ordinary point. The point $x = \infty$ is a regular singular point if after the transformation $x = 1/t$, the point $t = 0$ is a regular singular point of the transformed equation. If a point is not ordinary or regular singular, then it is an *irregular singular point*. For example, the equation

$$\frac{d^2 y}{dx^2} + \frac{2}{x} \frac{dy}{dx} + \frac{a^2}{x^4} y = 0$$

has an irregular singular point at $x = 0$, and its solution is

$$y = A \sin \frac{a}{x} + B \cos \frac{a}{x}$$

Note that there is no series representation of the solution about $x = 0$.

4.6   THE EXTENDED POWER SERIES METHOD OF FROBENIUS

By changing the independent variable, eq. (4.5.1) may be rewritten in the form

$$L[y] = x^2 \frac{d^2 y}{dx^2} + xp(x) \frac{dy}{dx} + q(x)y = 0 \quad (4.6.1)$$

where $x = 0$ is a regular singular point; that is, $p(x)$ and $q(x)$ are analytic at $x = 0$. We seek a power series representation of the solution in the neighborhood of $x = 0$. The basic result here is due to Frobenius (for proof, see Ince, 1956, chapter 16):

● THEOREM

Let $x = 0$ be a regular singular point of eq. (4.6.1); that is, $p(x)$ and $q(x)$ are analytic at $x = 0$ with series representations

$$p(x) = \sum_{n=0}^{\infty} p_n x^n, \qquad q(x) = \sum_{n=0}^{\infty} q_n x^n$$

where the series have a common region of convergence $|x| < \rho$. Then eq. (4.6.1) has two linearly independent solutions; of these, *at least* one can be represented as

$$y(x) = x^r \sum_{n=0}^{\infty} c_n x^n \quad (4.6.2)$$

where r (real or complex) is chosen such that $c_0 \neq 0$. The series solution (4.6.2) converges for at least $|x| < \rho$. The second solution depends on the value of r, and we will develop it later.

To obtain the first solution, we can proceed by substituting eq. (4.6.2) and the series for $p(x)$ and $q(x)$ in eq. (4.6.1). First note that

$$y = x^r \sum_{n=0}^{\infty} c_n x^n = \sum_{n=0}^{\infty} c_n x^{n+r}$$

Thus

$$\frac{dy}{dx} = \sum_{n=0}^{\infty} (n+r)c_n x^{n+r-1} = x^{r-1} \sum_{n=0}^{\infty} (n+r)c_n x^n$$

and

$$\frac{d^2y}{dx^2} = \sum_{n=0}^{\infty} (n+r)(n+r-1)c_n x^{n+r-2} = x^{r-2} \sum_{n=0}^{\infty} (n+r)(n+r-1)c_n x^n$$

Now substituting all these in eq. (4.6.1), we have

$$x^r \left[\sum_{n=0}^{\infty} (n+r)(n+r-1)c_n x^n \right] + x^r \left[\sum_{n=0}^{\infty} p_n x^n \right]\left[\sum_{n=0}^{\infty} (n+r)c_n x^n \right]$$

$$+ x^r \left[\sum_{n=0}^{\infty} q_n x^n \right]\left[\sum_{n=0}^{\infty} c_n x^n \right] = 0 \qquad (4.6.3)$$

Comparing terms of equal powers of x on both sides gives the following for the lowest power of x (i.e., x^r):

$$[r(r-1) + p_0 r + q_0]c_0 = 0$$

Since $c_0 \neq 0$, we must have that r satisfies

$$r(r-1) + p_0 r + q_0 = 0 \qquad (4.6.4)$$

which is called the *indicial equation*. It has two roots

$$r_{1,2} = \frac{(1-p_0) \pm \sqrt{(p_0-1)^2 - 4q_0}}{2} \qquad (4.6.5)$$

and the various possibilities are as follows:

1. The roots are distinct, and do not differ by an integer.
2. The roots are equal.
3. The roots differ by an integer.

Note that complex roots always fall under case 1. We will now consider these cases one by one, and we also develop the second linearly independent solution.

Case 1. Unequal Roots Which Do Not Differ by an Integer
This is the simplest case. The strategy is to first set $r = r_1$ and then determine the coefficients $\{c_j\}$ by comparing terms of equal powers of x in eq. (4.6.3). With these determined, the first solution is

$$y_1(x) = x^{r_1} \sum_{n=0}^{\infty} c_n x^n \tag{4.6.6a}$$

The second solution is obtained similarly by putting $r = r_2$ and developing the coefficients, say $\{d_j\}$, to give

$$y_2(x) = x^{r_2} \sum_{n=0}^{\infty} d_n x^n \tag{4.6.6b}$$

Let us now see what the coefficients $\{c_j\}$ actually are; the procedure for finding them will also give an insight as to why cases 2 and 3 have to be treated separately. For this, rewrite eq. (4.6.3) as

$$x^r \Bigg[\sum_{n=0}^{\infty} (n + r)(n + r - 1)c_n x^n$$
$$+ \sum_{n=0}^{\infty} p_0(n + r)c_n x^n + \sum_{n=0}^{\infty} p_1(n + r)c_n x^{n-1} + \sum_{n=0}^{\infty} p_2(n + r)c_n x^{n+2} + \cdots$$
$$+ \sum_{n=0}^{\infty} q_0 c_n x^n + \sum_{n=0}^{\infty} q_1 c_n x^{n+1} + \sum_{n=0}^{\infty} q_2 c_n x^{n+2} + \cdots \Bigg] = 0$$

which is further rearranged to give

$$x^r \Bigg[\sum_{n=0}^{\infty} \{(n + r)(n + r - 1) + (n + r)p_0 + q_0\}c_n x^n$$
$$+ \sum_{n=1}^{\infty} \Bigg\{ \sum_{j=0}^{n-1} \{(r + j)p_{n-j} + q_{n-j}\}c_j \Bigg\} x^n \Bigg] = 0 \quad (4.6.7)$$

Setting the coefficients of various powers of x equal to zero gives the indicial equation (4.6.4) for x^r. For x^{r+n}, we get

$$[(n + r)(n + r - 1) + (n + r)p_0 + q_0]c_n$$
$$+ \sum_{j=0}^{n-1} \{(r + j)p_{n-j} + q_{n-j}\}c_j = 0, \qquad n \geq 1 \quad (4.6.8)$$

For notational convenience, define

$$I(r) = r(r - 1) + p_0 r + q_0$$
$$W_j(r) = p_j r + q_j, \qquad j = 1, 2, 3, \ldots$$

so that the indicial equation is $I(r) = 0$. For $n \geq 1$, eq. (4.6.8) may be written as

$$c_1 I(1 + r) + c_0 W_1(r) \qquad\qquad = 0$$
$$c_2 I(2 + r) + c_1 W_1(1 + r) + c_0 W_2(r) = 0$$
$$\vdots$$
$$c_n I(n + r) + \sum_{j=0}^{n-1} c_j W_{n-j}(j + r) \qquad = 0$$

which successively give

$$c_1 = -\frac{W_1(r)}{I(1 + r)} c_0$$

$$c_2 = -\frac{c_1 W_1(1 + r) + c_0 W_2(r)}{I(2 + r)}$$

$$\vdots$$

$$c_n = \frac{\sum_{j=0}^{n-1} c_j W_{n-j}(j + r)}{I(n + r)}, \qquad n \geq 1$$

(4.6.9)

and thus each c_j for $j = 1, 2, \ldots$ can be computed in terms of c_0 and the sets $\{p_j\}$ and $\{q_j\}$.

It should be evident that the above procedure will surely work for *one* of the roots of the indicial equation. It should also be clear that if the two roots r_1 and r_2 are equal (case 2), the *same* coefficients c_j will develop for both roots, and hence the two resulting series solutions will be identical. We will therefore not arrive at two linearly independent solutions of eq. (4.6.1), and thus this case will have to be dealt with separately.

Let us now consider what happens when the two roots r_1 and r_2 differ by an integer (case 3). Let $r_1 > r_2$ and denote $r_1 = r_2 + k$, where k is a positive integer. For the *larger* root r_1, we can find all the c_n from eq. (4.6.9); the difficulty arises when we want to find them for the *smaller* root, r_2. To see this, first note that the indicial equation is

$$I(r) = 0, \qquad \text{for } r = r_1, r_2$$

and since $r_1 = r_2 + k$ we have

$$I(r_1) = I(r_2 + k) = 0$$

Equation (4.6.9) is developed from

$$c_n I(n + r) + \sum_{j=0}^{n-1} c_j W_{n-j}(j + r) = 0$$

For $r = r_2$, the coefficients c_j for $j = 1, 2, \ldots, k - 1$ can be obtained in terms of c_0, $\{p_j\}$ and $\{q_j\}$. However, for $n = k$ we get

$$c_k I(k + r_2) + \sum_{j=0}^{k-1} c_j W_{k-j}(j + r_2) = 0$$

Since $I(k + r_2) = 0$, c_k is undefined; thus c_j for $j \geq k$ will remain undefined for the smaller root r_2.

Summarizing, in case 1 where the roots of the indicial equation are unequal and do not differ by an integer, the coefficients of the two linearly independent series solutions can be obtained from eq. (4.6.9) by successively taking $r = r_1$ and $r = r_2$. When the two roots are equal or differ by an integer, one series solution can be readily developed; the other will require special manipulation.

Case 2. Equal Roots

From eq. (4.6.5), we observe that the two roots of the indicial equation (4.6.4) are equal if

$$(p_0 - 1)^2 = 4q_0$$

and then $r_{1,2} = (1 - p_0)/2 = r$. One series solution

$$y_1(x) = x^r \sum_{n=0}^{\infty} c_n x^n$$

is determined as before, with the various c_j given in terms of c_0, $\{p_j\}$, and $\{q_j\}$ by eq. (4.6.9). The second linearly independent solution can be obtained by the method of *variation of parameters*. Thus let

$$y_2(x) = u(x)y_1(x) \tag{4.6.10}$$

where $u(x)$ is as yet an unknown function. This gives

$$\frac{dy_2}{dx} = y_1 \frac{du}{dx} + u \frac{dy_1}{dx}$$

and

$$\frac{d^2 y_2}{dx^2} = y_1 \frac{d^2 u}{dx^2} + 2 \frac{du}{dx}\frac{dy_1}{dx} + u \frac{d^2 y_1}{dx^2}$$

The ODE to be solved is

$$x^2 \frac{d^2 y}{dx^2} + xp(x) \frac{dy}{dx} + q(x)y = 0 \tag{4.6.1}$$

Substituting $y = y_2$ in eq. (4.6.1) gives

$$x^2 \left[y_1 \frac{d^2 u}{dx^2} + 2 \frac{du}{dx}\frac{dy_1}{dx} + u \frac{d^2 y_1}{dx^2} \right] + xp \left[y_1 \frac{du}{dx} + u \frac{dy_1}{dx} \right] + quy_1 = 0$$

or

$$x^2 \left[y_1 \frac{d^2 u}{dx^2} + 2 \frac{du}{dx}\frac{dy_1}{dx} \right] + xp \frac{du}{dx} y_1 + \left[x^2 \frac{d^2 y_1}{dx^2} + xp \frac{dy_1}{dx} + qy_1 \right] u = 0$$

The last term—that is, the coefficient of u above—is zero, since y_1 is a solution of eq. (4.6.1). Thus

$$x^2 \left[y_1 \frac{d^2 u}{dx^2} + 2 \frac{du}{dx}\frac{dy_1}{dx} \right] + xp \frac{du}{dx} y_1 = 0$$

and dividing through by $x^2 y_1$, we get

$$\frac{d^2 u}{dx^2} + \left[\frac{2}{y_1}\frac{dy_1}{dx} + \frac{p}{x} \right]\frac{du}{dx} = 0 \tag{4.6.11}$$

Now since

$$y_1 = x^r \sum_{n=0}^{\infty} c_n x^n, \qquad \frac{dy_1}{dx} = x^{r-1} \sum_{n=0}^{\infty} (n + r)c_n x^n, \qquad \text{and } p = \sum_{n=0}^{\infty} p_n x^n$$

eq. (4.6.11) takes the form

$$\frac{d^2u}{dx^2} + \left[\frac{2\{rc_0 + (r + 1)c_1x + (r + 2)c_2x^2 + \cdots\}}{x\{c_0 + c_1x + c_2x^2 + \cdots\}} \right.$$

$$\left. + \frac{p_0 + p_1x + p_2x^2 + \cdots}{x} \right] \frac{du}{dx} = 0$$

or

$$\frac{d^2u}{dx^2} + \left[\frac{2r + p_0}{x} + \alpha_0 + \alpha_1x + \alpha_2x^2 + \cdots \right] \frac{du}{dx} = 0 \qquad (4.6.12)$$

where α_j are constants. Also, since

$$r = \frac{1 - p_0}{2}$$

we have

$$2r + p_0 = 1$$

and eq. (4.6.12) becomes

$$\frac{d^2u}{dx^2} + \left[\frac{1}{x} + \sum_{n=0}^{\infty} \alpha_n x^n \right] \frac{du}{dx} = 0$$

This is a separable equation

$$\frac{u''}{u'} = -\left[\frac{1}{x} + \sum_{n=0}^{\infty} \alpha_n x^n \right]$$

which upon integration gives

$$\ln\left[\frac{du}{dx} \right] = -\ln x - \left[\alpha_0 x + \alpha_1 \frac{x^2}{2} + \cdots \right]$$

where the integration constant may be assumed to be zero, without any loss of generality. A rearrangement leads to

$$\frac{du}{dx} = \frac{1}{x} \exp\left[-\left\{ \alpha_0 x + \alpha_1 \frac{x^2}{2} + \cdots \right\} \right]$$

$$= \frac{1}{x} [1 + \beta_1 x + \beta_2 x^2 + \cdots], \qquad \text{on expanding the exponential}$$

$$= \frac{1}{x} + \beta_1 + \beta_2 x + \cdots$$

A further integration gives

$$u = \ln x + \beta_1 x + \beta_2 \frac{x^2}{2} + \cdots + \beta \qquad (4.6.13)$$

where β is the integration constant. The form of the desired function $u(x)$ is thus known. From eq. (4.6.10), the second linearly independent solution is

$$y_2(x) = u(x)y_1(x)$$

$$= [\ln x + \beta_1 x + \beta_2 x^2 + \cdots + \beta]y_1(x)$$

$$= y_1(x) \ln x + x^r \sum_{n=1}^{\infty} d_n x^n \qquad (4.6.14)$$

where the coefficients d_n are uniquely found in terms of d_1, $\{p_j\}$, and $\{q_j\}$, when eq. (4.6.14) is substituted in the ODE (4.6.1); terms of equal powers of x on both sides are compared.

Note that the term involving $\ln x$ does not appear when eq. (4.6.14) is substituted in eq. (4.6.1). To see this, let us call

$$y_{21}(x) = y_1(x) \ln x$$

then

$$\frac{dy_{21}}{dx} = \frac{dy_1}{dx} \ln x + \frac{y_1}{x}$$

and

$$\frac{d^2 y_{21}}{dx^2} = \frac{d^2 y_1}{dx^2} \ln x + \frac{2}{x} \frac{dy_1}{dx} - \frac{y_1}{x^2}$$

The y_{21} contributions to eq. (4.6.1) are

$$\left[x^2 \ln x \frac{d^2 y_1}{dx^2} + 2x \frac{dy_1}{dx} - y_1 \right] + p \left[x \ln x \frac{dy_1}{dx} + y_1 \right] + q y_1 \ln x$$

$$= \left[x^2 \frac{d^2 y_1}{dx^2} + xp \frac{dy_1}{dx} + q y_1 \right] \ln x + 2x \frac{dy_1}{dx} + (p - 1) y_1$$

$$= 2x \frac{dy_1}{dx} + (p - 1) y_1$$

since y_1 is a solution of eq. (4.6.1).

There is an alternative, and more direct, way of arriving at the conclusion that eq. (4.6.14) is indeed the second linearly independent solution. For this, if we denote

$$y_1(x) = \phi(x) = x^r \sum_{n=0}^{\infty} c_n x^n$$

with c_n determined from eq. (4.6.9), eq. (4.6.7) tells us that

$$L[\phi](r, x) = c_0 x^r I(r)$$

and indeed, with r determined by setting $I(r) = 0$, $y_1(x)$ is a solution—as we already know. In the case of repeated roots, $r = r_1$ is a double root of $I(r) = 0$, so that

$$I(r) = (r - r_1)^2$$

and

$$L[\phi](r, x) = c_0 x^r (r - r_1)^2 \tag{4.6.15}$$

Let us consider r to be a continuous variable. Then

$$\frac{\partial}{\partial r} [L[\phi](r, x)] = L\left[\frac{\partial \phi}{\partial r} \right](r, x) = -c_0 [x^r \ln x (r - r_1)^2 + 2x^r (r - r_1)]$$

and

$$L\left[\frac{\partial \phi}{\partial r} \right](r_1, x) = 0$$

so that $\partial\phi(r, x)/\partial r|_{r=r_1}$ is also a solution of the ODE (4.6.1). Since

$$\phi(r, x) = y_1(x) = x^r \left[c_0 + \sum_{n=1}^{\infty} c_n(r)x^n \right]$$

where, observe that for $n \geq 1$, the c_n determined from eq. (4.6.9) are functions of r,

$$y_2(x) = \left. \frac{\partial\phi(r, x)}{\partial r} \right|_{r=r_1} = x^{r_1}\ln x \left[c_0 + \sum_{n=1}^{\infty} c_n(r_1)x^n \right] + x^{r_1}\left[\left. \sum_{n=0}^{\infty} \frac{dc_n}{dr}\right|_{r_1} x^n \right]$$

$$= y_1(x) \ln x + x^{r_1} \sum_{n=1}^{\infty} \left. \frac{dc_n}{dr}\right|_{r_1} x^n \qquad (4.6.16)$$

is the second linearly independent solution, which has the *same* form as eq. (4.6.14). Note that in practice, it is more convenient to proceed through eq. (4.6.14) than by eq. (4.6.16) because it is generally cumbersome to evaluate dc_n/dr.

Case 3. Roots Differ by an Integer

When the two roots of the indicial equation differ by an integer—that is,

$$r_1 = r, \quad r_2 = r - k \qquad (4.6.17)$$

where k is a positive integer—we have seen that for the larger root r_1 the coefficients of the power series

$$y_1(x) = x^{r_1} \sum_{n=0}^{\infty} c_n x^n, \qquad c_0 \neq 0$$

can be determined by eq. (4.6.9). However, a difficulty arises in evaluating them for the smaller root r_2, to provide the second linearly independent solution. In this case as well, we could adopt the second method discussed in case 2, but it would be tedious. Following the method of variation of parameters, set

$$y_2(x) = u(x)y_1(x) \qquad (4.6.18)$$

where the unknown function $u(x)$ is to be determined. Proceeding exactly as in case 2 for equal roots, we arrive at eq. (4.6.12):

$$\frac{d^2u}{dx^2} + \left[\frac{2r + p_0}{x} + \alpha_0 + \alpha_1 x + \alpha_2 x^2 + \cdots \right]\frac{du}{dx} = 0$$

where α_j are constants. Now from eqs. (4.6.5) and (4.6.17)

$$r_1 + r_2 = 2r - k = 1 - p_0$$

so that

$$2r + p_0 = 1 + k$$

and we have the separable equation

$$\frac{u''}{u'} = -\left[\frac{1 + k}{x} + \sum_{n=0}^{\infty} \alpha_n x^n \right]$$

Integrating once gives

$$\ln\left[\frac{du}{dx}\right] = -\left[(1 + k) \ln x + \alpha_0 x + \alpha_1 \frac{x^2}{2} + \cdots \right]$$

and following the same steps as before leads to

$$\frac{du}{dx} = \frac{1}{x^{1+k}} [1 + \beta_1 x + \beta_2 x^2 + \cdots]$$

$$= \frac{1}{x^{1+k}} + \frac{\beta_1}{x^k} + \frac{\beta_2}{x^{k-1}} + \cdots + \frac{\beta_k}{x} + \beta_{k+1} + \beta_{k+2} x + \cdots$$

where β_j are constants. Integrating once again leads to

$$u = -\frac{1}{kx^k} - \frac{\beta_1}{(k-1)x^{k-1}} + \cdots + \beta_k \ln x + \beta_{k+1} x + \cdots + \beta$$

where β is the integration constant. The form of the second linearly independent solution is then

$$y_2(x) = u(x)y_1(x)$$

$$= Ay_1(x) \ln x + x^{r-k} \sum_{n=0}^{\infty} d_n x^n$$

$$= Ay_1(x) \ln x + x^{r_2} \sum_{n=0}^{\infty} d_n x^n, \qquad d_0 \neq 0 \qquad (4.6.19)$$

where the constants A and d_n are uniquely determined in terms of d_0 when eq. (4.6.19) is substituted in the ODE (4.6.1) and terms of equal powers of x are compared. Note that the constant A in some cases may turn out to be zero.

Remark. From section 4.8 onward, we will apply the method of Frobenius to Bessel's equation, which will lead to Bessel functions. The same method, applied to other specific ODEs, leads to other well-known functions such as the *hypergeometric function* and *Laguerre polynomials* (see problems 4.12 and 4.13 at the end of the chapter).

4.7 SOME SPECIAL FUNCTIONS

In this section we consider three special functions, defined as integrals. They prove useful in a variety of applications.

4.7.1 The Error Function

A function which occurs frequently in the solution of partial differential equations describing heat conduction and mass diffusion is related to the integral:

$$f(x) = \int_0^x \exp(-s^2)\, ds$$

The value of $f(x)$ as $x \to \infty$ may be computed by considering the product of two such integrals:

$$\int_0^\infty \exp(-s^2)\, ds \int_0^\infty \exp(-t^2)\, dt$$

considered as a double integral:

$$\int_0^\infty \int_0^\infty \exp[-(s^2 + t^2)]\, ds dt$$

This may be interpreted as the integration of the function $\exp[-(s^2 + t^2)]$ over the entire first quadrant of the (s, t) plane. We may convert to polar coordinates by

$$s = \rho \cos \theta \quad \text{and} \quad t = \rho \sin \theta$$

where the element of area in polar coordinates is $\rho \, d\rho \, d\theta$, so that the double integral becomes

$$\int_0^{\pi/2} \int_0^\infty \exp(-\rho^2) \, \rho \, d\rho d\theta = \frac{\pi}{4}$$

Thus we have

$$\int_0^\infty \exp(-s^2) \, ds = \frac{\sqrt{\pi}}{2}$$

It is then convenient to define

$$\text{erf}(x) = \frac{2}{\sqrt{\pi}} \int_0^x \exp(-s^2) \, ds \tag{4.7.1}$$

as the *error function*. A plot of this function is shown in Figure 4.2. The *complementary error function* is defined as

$$\text{erfc}(x) = \frac{2}{\sqrt{\pi}} \int_x^\infty \exp(-s^2) \tag{4.7.2}$$

so that

$$\text{erf}(x) + \text{erfc}(x) = 1 \tag{4.7.3}$$

One can obtain useful asymptotic formulae for eqs. (4.7.1) and (4.7.2) by integrating eq. (4.7.2) by parts. Writing in the form

$$\text{erfc}(x) = \frac{2}{\sqrt{\pi}} \int_x^\infty \frac{1}{s} \exp(-s^2) s \, ds$$

successive integration by parts gives

$$\begin{aligned}
\text{erfc}(x) &= \frac{2}{\sqrt{\pi}} \left[\frac{\exp(-x)^2}{2x} \left(1 - \frac{1}{2x^2} + \frac{1 \cdot 3}{(2x^2)^2} - \frac{1 \cdot 3 \cdot 5}{(2x^2)^3} \right) \right. \\
&\quad \left. + \frac{1 \cdot 3 \cdot 5 \cdot 7}{2^4} \int_x^\infty \frac{\exp(-s)^2}{s^8} \, ds \right] \\
&= \frac{2}{\sqrt{\pi}} \left[\frac{\exp(-x)^2}{2x} \sum_{k=0}^n (-1)^k \frac{1 \cdot 3 \cdot 5 \cdots (2k-1)}{(2x^2)^k} \right. \\
&\quad \left. + (-1)^{n+1} \frac{1 \cdot 3 \cdot 5 \cdots (2n+1)}{2^{n+1}} \int_x^\infty \frac{\exp(-s^2)}{s^{2n+2}} \, ds \right] \\
&= \frac{2}{\sqrt{\pi}} [S_n(x) + R_n(x)] \tag{4.7.4}
\end{aligned}$$

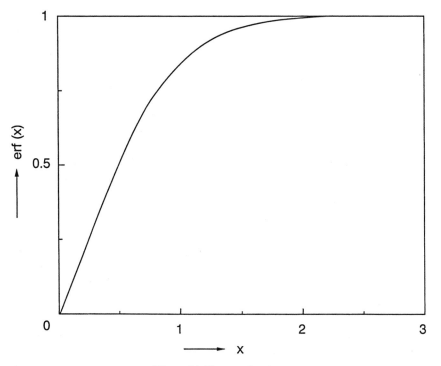

Figure 4.2 The error function.

The term

$$R_n = (-1)^{n+1} \frac{1 \cdot 3 \cdot 5 \cdots (2n+1)}{2^{n+1}} \int_x^{\infty} \frac{\exp(-s^2)}{s^{2n+2}} \, ds$$

may be written as

$$|R_n| = \frac{1 \cdot 3 \cdot 5 \cdots (2n+1)}{2^{n+1}} \int_x^{\infty} s \frac{\exp(-s^2)}{s^{2n+3}} \, ds$$

$$\leq \frac{1 \cdot 3 \cdot 5 \cdots (2n+1)}{2^{n+1}} \frac{1}{x^{2n+3}} \int_x^{\infty} s \exp(-s^2) \, ds$$

$$\leq \frac{1 \cdot 3 \cdot 5 \cdots (2n+1)}{2^{n+1}} \frac{1}{x^{2n+3}} \left(\frac{1}{2} \exp(-x^2) \right)$$

so that

$$\frac{2}{\sqrt{\pi}} |R_n| \leq \frac{2}{\sqrt{\pi}} \frac{\exp(-x^2)}{2x} \frac{1 \cdot 3 \cdot 5 \cdots (2n+1)}{(2x^2)^{n+1}} \tag{4.7.5}$$

is the first neglected term in the expression for the asymptotic development of erfc(x).

To give some idea of the accuracy of the asymptotic development of erfc(x), consider its value for $x = 3$. The error in truncating the development at $n = 3$ is

$$\frac{2}{\sqrt{\pi}} |R_3| \le \frac{2}{\sqrt{\pi}} \frac{\exp(-x^2)}{2x} \frac{1 \cdot 3 \cdot 5 \cdot 7}{(2x^2)^4}$$

which for $x = 3$ is less than 0.25×10^{-12}.

4.7.2   The Gamma Function and the Beta Function

In the remainder of this chapter, we will work extensively with Bessel functions. In their definition, as well as in other applications, it is useful to consider a function defined by the integral

$$\Gamma(\alpha) = \int_0^\infty t^{\alpha-1} \exp(-t)\, dt, \qquad \alpha > 0 \tag{4.7.6}$$

which is called the *Gamma function*. It has several properties of interest, and among them the most important is

$$\Gamma(\alpha + 1) = \alpha \Gamma(\alpha) \tag{4.7.7}$$

which is readily shown. From eq. (4.7.6) we have

$$\Gamma(\alpha + 1) = \int_0^\infty t^\alpha \exp(-t)\, dt$$

which, by integrating once by parts, gives

$$\Gamma(\alpha + 1) = -t^\alpha \exp(-t)\big|_0^\infty + \int_0^\infty \alpha t^{\alpha-1} \exp(-t)\, dt$$
$$= \alpha \Gamma(\alpha)$$

Also, by definition, we have

$$\Gamma(1) = \int_0^\infty \exp(-t)\, dt = -\exp(-t)\big|_0^\infty = 1$$

so that eq. (4.7.7) leads to

$$\Gamma(2) = 1\Gamma(1) = 1!$$
$$\Gamma(3) = 2\Gamma(2) = 2!$$
$$\Gamma(4) = 3\Gamma(3) = 3! \tag{4.7.8}$$
$$\vdots$$
$$\Gamma(n + 1) = n\Gamma(n) = n!$$

Thus the Gamma function extends the idea of the factorial function for noninteger positive values of α and is related to the factorial for integer values of the argument. For example, $(5/2)!$ has no meaning, but $\Gamma(7/2)$ may be computed readily since

$$\Gamma(\tfrac{7}{2}) = \tfrac{5}{2}\Gamma(\tfrac{5}{2}) = \tfrac{5}{2} \cdot \tfrac{3}{2}\Gamma(\tfrac{3}{2}) = \tfrac{5}{2} \cdot \tfrac{3}{2} \cdot \tfrac{1}{2}\Gamma(\tfrac{1}{2})$$

and, by definition, we have

$$\Gamma(\tfrac{1}{2}) = \int_0^\infty t^{-1/2} \exp(-t)\, dt$$

If one makes the change of variables

$$t = u^2$$

then

$$\Gamma(\tfrac{1}{2}) = 2\int_0^\infty \exp(-u^2)\, du$$
$$= \sqrt{\pi} \qquad\qquad\qquad (4.7.9)$$

using eq. (4.7.1). This gives

$$\Gamma(\tfrac{7}{2}) = \tfrac{5}{2} \cdot \tfrac{3}{2} \cdot \tfrac{1}{2} \sqrt{\pi}$$

Note that formula (4.7.6) does not hold for α equal to zero or a negative number. However, eq. (4.7.7) may be used to extend the definition to negative *noninteger* values of α. For example, with $\alpha = -\tfrac{1}{2}$ we have

$$\Gamma(\tfrac{1}{2}) = -\tfrac{1}{2}\, \Gamma(-\tfrac{1}{2})$$

and so

$$\Gamma(-\tfrac{1}{2}) = -2\Gamma(\tfrac{1}{2})$$

Proceeding similarly, in general,

$$\Gamma(\alpha) = \frac{\Gamma(\alpha + k + 1)}{\alpha(\alpha + 1) \cdots (\alpha + k)}, \qquad \alpha \neq 0, -1, -2, \ldots \qquad (4.7.10)$$

may be used to define the Gamma function for negative noninteger α, choosing for k the *smallest* positive integer such that $\alpha + k + 1 > 0$. Together with eq. (4.7.6), this provides a definition of $\Gamma(\alpha)$ for all α not equal to zero or a negative integer. When α tends to zero or a negative integer, $|\Gamma(\alpha)| \rightarrow \infty$.

A plot of the Gamma function is shown in Figure 4.3, where the above noted limits may be observed. Numerical values of the Gamma function (cf. Abramowitz and Stegun, 1965, section 6), however, are usually given only for α in the range $1 \leq \alpha \leq 2$, since it can be computed for all other values using eq. (4.7.7).

Also note that from eq. (4.7.8) we have

$$1! = \Gamma(2) = 1\Gamma(1) = 0!$$

so that 0! in the extended definition equals one.

As α increases, $\Gamma(\alpha)$ increases very rapidly, and a useful approximation for large positive α is the *Stirling formula*:

$$\Gamma(\alpha) \sim \sqrt{\frac{2\pi}{\alpha}} \left(\frac{\alpha}{e}\right)^\alpha \qquad\qquad (4.7.11)$$

where e is the base of the natural logarithm.

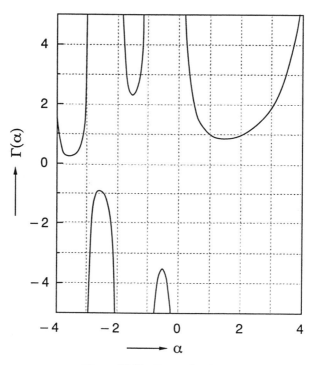

Figure 4.3 The Gamma function.

The *Beta function*, $\beta(m, n)$, is related to the Gamma function and is defined by

$$\beta(m, n) = \int_0^1 t^{m-1}(1-t)^{n-1} \, dt, \qquad m, n > 0 \qquad (4.7.12)$$

The substitution $t = 1 - s$ shows that it is symmetric; that is,

$$\beta(m, n) = \beta(n, m)$$

Consider now

$$\Gamma(m)\Gamma(n) = \int_0^\infty x^{m-1} \exp(-x) \, dx \int_0^\infty y^{n-1} \exp(-y) \, dy$$

$$= \int_0^\infty \int_0^\infty x^{m-1} y^{n-1} \exp[-(x+y)] \, dx dy$$

and let $x = s^2$ and $y = t^2$; then

$$\Gamma(m)\Gamma(n) = 4 \int_0^\infty \int_0^\infty s^{2m-1} t^{2n-1} \exp[-(s^2 + t^2)] \, ds dt$$

Changing further to polar coordinates

$$s = r \cos \phi, \qquad t = r \sin \phi$$

leads to

$$\Gamma(m)\Gamma(n) = 4\int_0^\infty \int_0^{\pi/2} r^{2(m+n)-1}(\cos\phi)^{2m-1}(\sin\phi)^{2n-1}\exp(-r^2)\,d\phi dr$$

$$= 4\int_0^\infty r^{2(m+n)-1}\exp(-r^2)\,dr\int_0^{\pi/2}(\cos\phi)^{2m-1}(\sin\phi)^{2n-1}\,d\phi$$

In the first integral let $r^2 = z$, and in the second let $u = \cos^2\phi$; then

$$\Gamma(m)\Gamma(n) = \int_0^\infty z^{m+n-1}\exp(-z)\,dz\int_0^1 u^{m-1}(1-u)^{n-1}\,du$$

$$= \Gamma(m+n)\beta(m, n)$$

so

$$\beta(m, n) = \frac{\Gamma(m)\Gamma(n)}{\Gamma(m+n)} \qquad (4.7.13)$$

which provides a relationship between the Beta and Gamma functions.

4.8 BESSEL'S EQUATION: BESSEL FUNCTIONS OF THE FIRST KIND

The ODE

$$x^2\frac{d^2y}{dx^2} + x\frac{dy}{dx} + (x^2 - \alpha^2)y = 0 \qquad (4.8.1)$$

where α is a real non-negative constant, arises in many applications and is called *Bessel's equation*. A great deal is known about the solutions of this equation and is available in the treatise of Watson (1966).

Comparing eq. (4.8.1) with the standard form

$$\frac{d^2y}{dx^2} + \frac{p(x)}{x}\frac{dy}{dx} + \frac{q(x)}{x^2}y = 0$$

provides

$$p(x) = 1, \qquad q(x) = -\alpha^2 + x^2$$

so that $p(x)$ and $q(x)$ are analytic at $x = 0$, with coefficients

$$p_0 = 1; \qquad p_j = 0, \quad j \ge 1$$
$$q_0 = \alpha^2, \quad q_1 = 0, \quad q_2 = 1; \qquad q_j = 0, \quad j \ge 3$$

and their series converge for all x. The theorem of Frobenius thus assures us that there is at least one solution of the form

$$y = x^r\sum_{n=0}^\infty c_n x^n \qquad (4.8.2)$$

where r is chosen such that $c_0 \neq 0$. The indicial equation (4.6.4) is

$$I(r) = r(r - 1) + r - \alpha^2 = 0$$

which simplifies to

$$r^2 = \alpha^2$$

so that

$$r_1 = +\alpha, \qquad r_2 = -\alpha \tag{4.8.3}$$

If $\alpha \neq 0$ or $2\alpha \neq k$, where k is an integer, then the two linearly independent solutions of Bessel's ODE are given by eq. (4.8.2) with r equal to r_1 and r_2, respectively. Furthermore, even if $\alpha = 0$ or $2\alpha = k$, one solution is still given by eq. (4.8.2) for $r = r_1 = \alpha$, where c_j satisfy the recurrence relationship eq. (4.6.8):

$$c_n I(n + r) + \sum_{j=0}^{n-1} c_j W_{n-j}(j + r) = 0, \qquad n \geq 1 \tag{4.8.4}$$

where

$$I(r) = r^2 - \alpha^2$$
$$W_j(r) = p_j r + q_j, \qquad j \geq 1$$

A straightforward computation shows that

$$I(n + \alpha) = (n + \alpha)^2 - \alpha^2$$
$$= n(n + 2\alpha) \tag{4.8.5}$$

and

$$W_j(\alpha) = q_j$$

so that

$$W_1 = 0; \qquad W_2 = 1; \qquad W_j = 0, \ j \geq 3 \tag{4.8.6}$$

From the recurrence relation (4.8.4), we have

$$c_1 = 0$$

and

$$n(n + 2\alpha)c_n + c_{n-2} = 0, \qquad n \geq 2$$

Thus all odd c_n equal 0. For even c_n let $n = 2m$; then

$$2^2 m(m + \alpha)c_{2m} + c_{2m-2} = 0$$

Hence

$$c_{2m} = -\frac{1}{2^2 m(m + \alpha)} c_{2m-2}, \qquad m \geq 1 \tag{4.8.7}$$

whereby the even c_n are determined in terms of an arbitrary c_0.

It is customary to put

$$c_0 = \frac{1}{2^\alpha \Gamma(\alpha + 1)} \tag{4.8.8}$$

where $\Gamma(\alpha)$ is the Gamma function. With this choice and the relationship (4.7.7)

$$\Gamma(\alpha + 1) = \alpha \Gamma(\alpha)$$

the various even c_j are

$$c_2 = -\frac{1}{2^2 \cdot 1 \cdot (1 + \alpha)}, \qquad c_0 = \frac{1}{2^{\alpha+2} \cdot 1! \cdot \Gamma(\alpha + 2)}$$

$$c_4 = -\frac{1}{2^2 \cdot 2 \cdot (2 + \alpha)}, \qquad c_2 = \frac{(-1)^2}{2^{\alpha+4} \cdot 2! \cdot \Gamma(\alpha + 3)}$$

and in general,

$$c_{2m} = \frac{(-1)^m}{2^{\alpha+2m} \cdot m! \cdot \Gamma(m + \alpha + 1)}, \qquad m \geq 1 \tag{4.8.9}$$

Thus one solution of Bessel's ODE is

$$y_1(x) = J_\alpha(x) = x^\alpha \sum_{m=0}^{\infty} \frac{(-1)^m x^{2m}}{2^{\alpha+2m} m! \Gamma(m + \alpha + 1)}$$

$$= \sum_{m=0}^{\infty} \frac{(-1)^m (x/2)^{2m+\alpha}}{m! \Gamma(m + \alpha + 1)} \tag{4.8.10}$$

and is called *Bessel function of the first kind of order* α. The ratio test shows that the series converges for all x.

If α is an integer n, then from eq. (4.7.8) we obtain

$$\Gamma(m + n + 1) = (m + n)!$$

and so we have

$$J_n(x) = \sum_{m=0}^{\infty} \frac{(-1)^m (x/2)^{2m+n}}{m!(m + n)!} \tag{4.8.11}$$

The first few Bessel functions of the first kind are

$$J_0(x) = 1 - \frac{x^2}{2^2} + \frac{x^4}{2^4(2!)^2} - \frac{x^6}{2^6(3!)^2} + \cdots$$

$$J_1(x) = \frac{x}{2} - \frac{x^3}{2^3 2!} + \frac{x^5}{2^5 2!3!} - \frac{x^7}{2^7 3!4!} + \cdots$$

$$J_2(x) = \frac{x^2}{2^2 2!} - \frac{x^4}{2^4 3!} + \frac{x^6}{2^6 2!4!} - \frac{x^8}{2^8 3!5!} + \cdots$$

and plots of these as well as $J_3(x)$ are shown in Figure 4.4. From the various curves, it appears that Bessel functions have many zeroes and asymptotically approach sinusoidal

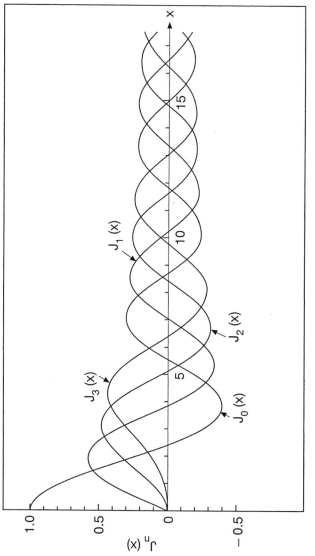

Figure 4.4 Bessel functions of the first kind of order n, $J_n(x)$, for $n = 0, 1, 2, 3$.

behavior for large values of the argument. This aspect is discussed further in section 4.12. Also, note that $J_0(0) = 1$, while $J_n(0) = 0$ for all integer values of n.

If $\alpha \neq 0$ or $2\alpha \neq k$, a positive integer, then the second linearly independent solution is simply given by

$$y_2(x) = J_{-\alpha}(x) = \sum_{m=0}^{\infty} \frac{(-1)^m (x/2)^{2m-\alpha}}{m!\, \Gamma(m - \alpha + 1)} \tag{4.8.12}$$

and so the general solution of Bessel's ODE in such cases is

$$y(x) = a_1 J_\alpha(x) + a_2 J_{-\alpha}(x) \tag{4.8.13}$$

where a_1 and a_2 are arbitrary constants. We will address the issue of $y_2(x)$ for the cases $\alpha = 0$ and $2\alpha = k$ in section 4.10.

4.9   SOME PROPERTIES OF BESSEL FUNCTIONS OF THE FIRST KIND

In working with Bessel functions, it is frequently required to perform various operations on them. These operations are greatly facilitated by the use of various relationships, with each other and with elementary trigonometric functions. To develop some of these, note that

$$J_\alpha(x) = x^\alpha \sum_{m=0}^{\infty} \frac{(-1)^m x^{2m}}{2^{\alpha+2m} m! \Gamma(m + \alpha + 1)}$$

which leads to

$$x^\alpha J_\alpha(x) = \sum_{m=0}^{\infty} \frac{(-1)^m x^{2m+2\alpha}}{2^{\alpha+2m} m! \Gamma(m + \alpha + 1)}$$

Differentiating once gives

$$\frac{d}{dx}[x^\alpha J_\alpha(x)] = \sum_{m=0}^{\infty} \frac{(-1)^m 2(m + \alpha) x^{2m+2\alpha-1}}{2^{\alpha+2m} m! \Gamma(m + \alpha + 1)}$$

$$= x^\alpha x^{\alpha-1} \sum_{m=0}^{\infty} \frac{(-1)^m x^{2m}}{2^{2m+\alpha-1} m! \Gamma(m + \alpha)}$$

$$= x^\alpha J_{\alpha-1}(x) \tag{4.9.1}$$

Similarly, we have

$$\frac{d}{dx}[x^{-\alpha} J_\alpha(x)] = -x^{-\alpha} J_{\alpha+1}(x) \tag{4.9.2}$$

and both of these are useful in evaluating integrals involving $J_\alpha(x)$.

A recurrence relation for Bessel functions of the first kind can be developed from the above two relations for the derivatives. Expanding the left-hand sides, we have

$$\alpha x^{\alpha-1} J_\alpha(x) + x^\alpha \frac{d}{dx} J_\alpha(x) = x^\alpha J_{\alpha-1}(x) \tag{4.9.3a}$$

and

$$-\alpha x^{-\alpha-1}J_\alpha(x) + x^{-\alpha}\frac{d}{dx}J_\alpha(x) = -x^{-\alpha}J_{\alpha+1}(x) \qquad (4.9.3b)$$

so that

$$\frac{d}{dx}J_\alpha(x) = J_{\alpha-1}(x) - \frac{\alpha}{x}J_\alpha(x) = \frac{\alpha}{x}J_\alpha(x) - J_{\alpha+1}(x)$$

which leads to the recurrence relation

$$J_{\alpha-1}(x) + J_{\alpha+1}(x) = \frac{2\alpha}{x}J_\alpha(x) \qquad (4.9.4)$$

Similarly, from eqs. (4.9.3) we obtain

$$\frac{\alpha}{x}J_\alpha(x) = J_{\alpha-1}(x) - \frac{d}{dx}J_\alpha(x) = \frac{d}{dx}J_\alpha(x) + J_{\alpha+1}(x)$$

and so we have

$$J_{\alpha-1}(x) - J_{\alpha+1}(x) = 2\frac{d}{dx}J_\alpha(x) \qquad (4.9.5)$$

These relationships are useful in evaluating Bessel functions of higher order in terms of those of lower order. For example, using eq. (4.9.4) for $\alpha = 1$ and 2 gives

$$J_0(x) + J_2(x) = \frac{2}{x}J_1(x)$$

$$J_1(x) + J_3(x) = \frac{4}{x}J_2(x)$$

respectively, so that

$$J_2(x) = \frac{2}{x}J_1(x) + J_0(x)$$

and

$$J_3(x) = \left[\frac{8}{x^2} - 1\right]J_1(x) - \frac{4}{x}J_0(x)$$

Another important set of relationships is with trigonometric functions. For example,

$$J_{1/2}(x) = \sqrt{\frac{2}{\pi x}}\sin x \qquad (4.9.6)$$

and to see this, note that the definition (4.8.10) leads to

$$
\begin{aligned}
J_{1/2}(x) &= \left(\frac{x}{2}\right)^{1/2} \sum_{m=0}^{\infty} \frac{(-1)^m x^{2m}}{2^{2m} m! \, \Gamma(m + \frac{3}{2})} \\
&= \left(\frac{2}{x}\right)^{1/2} \sum_{m=0}^{\infty} \frac{(-1)^m x^{2m+1}}{2^{2m+1} m! \, \Gamma(m + \frac{3}{2})} \\
&= \left(\frac{2}{x}\right)^{1/2} \left[\frac{x}{2\Gamma(\frac{3}{2})} - \frac{x^3}{2^3 \Gamma(\frac{5}{2})} + \frac{x^5}{2^5 2! \, \Gamma(\frac{7}{2})} - + \cdots \right]
\end{aligned}
$$

Using the relation (4.7.7), this reduces to

$$
\begin{aligned}
J_{1/2}(x) &= \left(\frac{2}{x}\right)^{1/2} \frac{1}{\Gamma(\frac{1}{2})} \left[\frac{x}{2 \cdot \frac{1}{2}} - \frac{x^3}{2^3 \cdot \frac{3}{2} \cdot \frac{1}{2}} + \frac{x^5}{2^5 \cdot 2! \cdot \frac{5}{2} \cdot \frac{3}{2} \cdot \frac{1}{2}} - + \cdots \right] \\
&= \left(\frac{2}{x}\right)^{1/2} \frac{1}{\Gamma(\frac{1}{2})} \left[x - \frac{x^3}{3!} + \frac{x^5}{5!} - + \cdots \right] \\
&= \sqrt{\frac{2}{\pi x}} \sin x
\end{aligned}
$$

since $\Gamma(\frac{1}{2}) = \sqrt{\pi}$ from eq. (4.7.9).
Similarly, we have

$$
J_{-1/2}(x) = \sqrt{\frac{2}{\pi x}} \cos x \tag{4.9.7}
$$

and using the recurrence relationship (4.9.4), it is evident that *all* $J_{+n/2}(x)$, where n is an integer, are related to the trigonometric functions.

4.10   BESSEL FUNCTIONS OF THE SECOND KIND

From the development of the method of Frobenius in section 4.6, it is apparent that if $\alpha = 0$ or $2\alpha = k$, a positive integer, the second linearly independent solution of Bessel's ODE will be given in a slightly different manner, because the roots of the indicial equation then are either equal ($\alpha = 0$) or differ by an integer ($2\alpha = k$). Three distinct cases arise, depending on whether α is zero, a half-integer, or an integer, and are considered separately below.

Case 1. $\alpha = 0$
In this case, the roots of the indicial equation are both equal to zero and Bessel's ODE (4.8.1) is equivalent to

$$
x \frac{d^2 y}{dx^2} + \frac{dy}{dx} + xy = 0 \tag{4.10.1}
$$

As determined in section 4.8, $J_0(x)$ is one solution of eq. (4.10.1). From eq. (4.6.14) the second linearly independent solution is given by

$$y_2(x) = J_0(x) \ln x + \sum_{m=1}^{\infty} d_m x^m \tag{4.10.2}$$

where the coefficients d_m are found by substituting eq. (4.10.2) in eq. (4.10.1) and comparing terms of equal powers of x.

Differentiating eq. (4.10.2) gives

$$\frac{dy_2}{dx} = \frac{dJ_0}{dx} \ln x + \frac{J_0(x)}{x} + \sum_{m=1}^{\infty} m d_m x^{m-1}$$

$$\frac{d^2 y_2}{dx^2} = \frac{d^2 J_0}{dx^2} \ln x + \frac{2}{x} \frac{dJ_0}{dx} - \frac{J_0(x)}{x^2} + \sum_{m=1}^{\infty} m(m-1) d_m x^{m-2}$$

and substituting in eq. (4.10.1) gives

$$x \left[\frac{d^2 J_0}{dx^2} \ln x + \frac{2}{x} \frac{dJ_0}{dx} - \frac{J_0(x)}{x^2} + \sum_{m=1}^{\infty} m(m-1) d_m x^{m-2} \right]$$

$$+ \left[\frac{dJ_0}{dx} \ln x + \frac{J_0(x)}{x} + \sum_{m=1}^{\infty} m d_m x^{m-1} \right] + x \left[J_0(x) \ln x + \sum_{m=1}^{\infty} d_m x^m \right] = 0$$

This simplifies to

$$2 \frac{dJ_0}{dx} + \sum_{m=1}^{\infty} m^2 d_m x^{m-1} + \sum_{m=1}^{\infty} d_m x^{m+1} = 0 \tag{4.10.3}$$

since $J_0(x)$ is a solution of eq. (4.10.1). Also, by definition (4.8.11)

$$J_0(x) = \sum_{m=0}^{\infty} \frac{(-1)^m x^{2m}}{2^{2m} m! m!}$$

so that

$$\frac{dJ_0}{dx} = \sum_{m=1}^{\infty} \frac{(-1)^m 2m x^{2m-1}}{2^{2m} m! m!}$$

$$= \sum_{m=1}^{\infty} \frac{(-1)^m x^{2m-1}}{2^{2m-1}(m-1)! m!}$$

and eq. (4.10.3) becomes

$$\sum_{m=1}^{\infty} \frac{(-1)^m x^{2m-1}}{2^{2m-2}(m-1)! m!} + \sum_{m=1}^{\infty} m^2 d_m x^{m-1} + \sum_{m=1}^{\infty} d_m x^{m+1} = 0 \tag{4.10.4}$$

Now, setting the coefficients of each power of x equal to zero gives

$$x^0: \qquad d_1 = 0$$
$$x^{2s}: \qquad (2s+1)^2 d_{2s+1} + d_{2s-1} = 0, \qquad s = 1, 2, \ldots$$

Hence *all* odd d_j are zero. Similarly,

$$x^1: \qquad -1 + 2^2 d_2 = 0, \qquad \text{so } d_2 = \frac{1}{4}$$

$$x^{2s+1}: \qquad \frac{(-1)^{s+1}}{2^{2s} s!(s+1)!} + (2s+2)^2 d_{2s+2} + d_{2s} = 0, \qquad s = 1, 2, \ldots$$

The last relation for $s = 1$ gives

$$\frac{1}{2^2 1! 2!} + 4^2 d_4 + d_2 = 0$$

or

$$d_4 = \frac{1}{4^2}\left[-\frac{1}{4} - \frac{1}{8}\right] = -\frac{3}{128}$$

and, in general,

$$d_{2m} = \frac{(-1)^{m+1}}{2^{2m}(m!)^2}\, h_m, \qquad m = 1, 2, \ldots \qquad (4.10.5)$$

where

$$h_m = 1 + \frac{1}{2} + \frac{1}{3} + \cdots + \frac{1}{m}$$

is the mth partial sum of the harmonic series $\sum_{n=1}^{\infty} 1/n$. From eq. (4.10.2), the second solution is then given by

$$y_2(x) = J_0(x)\ln x + \sum_{m=1}^{\infty} \frac{(-1)^{m+1} h_m}{2^{2m}(m!)^2} x^{2m} \qquad (4.10.6)$$

Thus the two linearly independent solutions of Bessel's ODE for $\alpha = 0$ are $y_1(x) = J_0(x)$ and $y_2(x)$, given by eq. (4.10.6).

Now since other solutions can be defined by taking linear combinations of these two solutions, we can take

$$a(y_2 + bJ_0)$$

where $a \neq 0$ and b are constants, as the second linearly independent solution. It is customary to choose

$$a = \frac{2}{\pi}, \qquad b = \gamma - \ln 2$$

where

$$\gamma = \lim_{m \to \infty} [h_m - \ln m] = 0.577215 \cdots \qquad (4.10.7)$$

is the *Euler constant*. The second solution is then

$$Y_0(x) = \frac{2}{\pi}\left[\left\{\ln\left(\frac{x}{2}\right) + \gamma\right\} J_0(x) + \sum_{m=1}^{\infty} \frac{(-1)^{m+1} h_m}{2^{2m}(m!)^2} x^{2m}\right] \qquad (4.10.8)$$

and is called *Bessel function of the second kind of order zero*.
For $\alpha = 0$, the general solution of Bessel's ODE

$$x\frac{d^2 y}{dx^2} + \frac{dy}{dx} + xy = 0$$

is then given by

$$y(x) = c_1 J_0(x) + c_2 Y_0(x) \qquad (4.10.9)$$

where c_1 and c_2 are arbitrary constants. Note that as $x \to 0$, $J_0(x) \to 1$, but $Y_0(x) \to -\infty$ due to the logarithmic term. Thus if *finite* solutions of Bessel's ODE for $\alpha = 0$ are sought in a region that includes $x = 0$, the solution will simply be

$$y(x) = c_1 J_0(x)$$

where c_1 is an arbitrary constant.

Case 2. α Is a Half-Integer $(\frac{1}{2}, \frac{3}{2}, \frac{5}{2}, \ldots)$

If α is a half-integer, $\alpha = n/2$, the roots of the indicial equation are

$$r_1 = n/2, \qquad r_2 = -n/2$$

and for the larger root r_1 the solution of Bessel's ODE is $y_1(x) = J_{n/2}(x)$. From eq. (4.6.19) the second linearly independent solution is given by

$$y_2(x) = A J_{n/2}(x) \ln x + x^{-n/2} \sum_{m=0}^{\infty} d_m x^m, \qquad d_0 \neq 0 \qquad (4.10.10)$$

where A and d_m are uniquely determined by substituting eq. (4.10.10) in eq. (4.8.1) and comparing terms of equal powers of x. It is, however, more instructive to not follow this approach, but rather return to eq. (4.6.8) which defines the coefficient c_m:

$$c_m I(m + r) + \sum_{j=0}^{m-1} c_j W_{m-j}(j + r) = 0, \qquad m \geq 1 \qquad (4.10.11)$$

where, as in eqs. (4.8.5) and (4.8.6),

$$I(m + r) = (m + r)^2 - \alpha^2$$
$$= (m + r + \alpha)(m + r - \alpha), \qquad m \geq 1$$

and

$$W_j(r) = q_j, \qquad j \geq 1$$

so that

$$W_1 = 0, \quad W_2 = 1, \quad W_j = 0, \qquad j \geq 3$$

For $\alpha = n/2$, $r = r_2 = -n/2$,

$$I(m + r) = m(m - n), \qquad m \geq 1$$

and eq. (4.10.11) states that

$$(1 - n)c_1 = 0 \qquad (4.10.12)$$

$$m(m - n)c_m + c_{m-2} = 0, \qquad m \geq 2 \qquad (4.10.13)$$

If $n = 1$ (i.e., $\alpha = 1/2$), eq. (4.10.12) implies that c_1 is arbitrary. For this case, from eq. (4.10.13)

$$c_m = \frac{1}{m(m - 1)} c_{m-2}, \qquad m \geq 2$$

so that all the even c_j are determined in terms of c_0, and the odd c_j in terms of c_1. These lead to

$$c_{2m} = \frac{(-1)^m}{(2m)!}\, c_0, \qquad m \geq 1$$

$$c_{2m+1} = \frac{(-1)^m}{(2m+1)!}\, c_1 \qquad m \geq 1$$

and thus

$$
\begin{aligned}
y_2(x) &= x^{r_2} \sum_{m=0}^{\infty} c_m x^m \\
&= x^{-1/2}\left[c_0 \sum_{m=0}^{\infty} \frac{(-1)^m}{(2m)!} x^{2m} + c_1 \sum_{m=0}^{\infty} \frac{(-1)^m}{(2m+1)!} x^{2m+1} \right] \\
&= x^{-1/2}[c_0 \cos x + c_1 \sin x]
\end{aligned}
$$

However, we showed in eq. (4.9.6) that

$$y_1(x) = J_{1/2}(x) = \sqrt{\frac{2}{\pi x}} \sin x$$

so the second part of $y_2(x)$ is simply a constant times $y_1(x)$. Thus the linearly independent portion is

$$y_2(x) = c_0 x^{-1/2} \cos x$$

If we take $c_0 = \sqrt{2/\pi}$, then

$$y_2(x) = \sqrt{\frac{2}{\pi x}} \cos x \tag{4.10.14}$$

is the second linearly independent solution of Bessel's ODE for $\alpha = 1/2$. Note that as observed before in eq. (4.9.7), this is precisely what we would get if we put $\alpha = -1/2$ in the general expression (4.8.10) for $J_\alpha(x)$.

For $n = 3, 5, 7, \ldots$, $c_1 = 0$ from eq. (4.10.12). Since n is an odd integer, all the even c_j are determined in terms of an arbitrary c_0 from eq. (4.10.13). From the same recurrence relationship, all the odd c_j up to $j = n - 2$ are zero, while c_n is arbitrary and all the later c_j can be determined in terms of c_n. The *even* coefficients satisfy

$$
\begin{aligned}
c_{2m} &= -\frac{1}{2m(2m-n)} c_{2m-2} \\
&= -\frac{1}{2^2 m(m-\alpha)} c_{2m-2}, \qquad m \geq 1
\end{aligned}
$$

which is exactly the same relationship as eq. (4.8.7) with α replaced by $-\alpha$. The even coefficients thus lead to the solution $J_{-n/2}(x)$. In analogy with the case of $\alpha = 1/2$, the odd coefficients lead to a multiple of $J_{n/2}(x)$, which is already the first solution.

It may thus be summarized that when α is a half-integer, the two linearly independent solutions of Bessel's ODE (4.8.1) are $J_\alpha(x)$ and $J_{-\alpha}(x)$, where $J_{-\alpha}(x)$ is given by the series (4.8.10) with α replaced by $-\alpha$.

Note that the same conclusion is reached if we start with the expression (4.10.10); in that case, A turns out to be zero.

Case 3. $\alpha = n$, an Integer

When $\alpha = n$, an integer, one solution is $y_1(x) = J_n(x)$, and for the second we must start with the form (4.6.19). Proceeding as in case 1 for $\alpha = 0$, the second linearly independent solution is eventually obtained as

$$y_2(x) = Y_n(x) = \frac{2}{\pi} \left[J_n(x) \ln\left(\frac{x}{2}\right) - \frac{1}{2} \sum_{m=0}^{n-1} \frac{(n-m-1)!}{m!} \left(\frac{x}{2}\right)^{2m-n} \right.$$

$$- \frac{1}{2} \sum_{m=0}^{\infty} (-1)^{m+1} \{\phi(m+1) + \phi(m+n+1)\}$$

$$\left. \frac{(x/2)^{2m+n}}{m!(m+n)!} \right] \tag{4.10.15}$$

where

$$\phi(l) = -\gamma$$

$$\phi(m) = -\gamma + \sum_{s=1}^{m-1} \frac{1}{s}, \qquad m \geq 2$$

and γ is the Euler constant. $Y_n(x)$ is called the *Bessel function of the second kind of order n*, and plots of a few of these as well as of $Y_0(x)$ are shown in Figure 4.5. Note that as $x \to 0$, $Y_n(x) \to -\infty$ for $n \geq 1$ as $-(2n/\pi)(n-1)!x^{-n}$, since $J_n(x) \ln (x/2) \to 0$ as $x \to 0$.

A Normalization. From all we have discussed in this section, we conclude that when α is not zero or an integer, the second linearly independent solution of Bessel's ODE is given by $J_{-\alpha}(x)$, while when α is zero or an integer, it is given by $Y_\alpha(x)$. To provide uniformity of notation, in the former cases, rather than using $J_{-\alpha}(x)$, the second solution is usually defined as

$$y_2(x) = Y_\alpha(x) = \frac{J_\alpha(x) \cos \alpha\pi - J_{-\alpha}(x)}{\sin \alpha\pi} \tag{4.10.16}$$

so that the general solution of Bessel's ODE

$$x^2 \frac{d^2y}{dx^2} + x \frac{dy}{dx} + (x^2 - \alpha^2)y = 0$$

is *always* given by

$$y(x) = c_1 J_\alpha(x) + c_2 Y_\alpha(x) \tag{4.10.17}$$

where c_1 and c_2 are arbitrary constants. Since *all* $Y_\alpha(x) \to -\infty$ as $x \to 0$, the finite general solution in a region that includes $x = 0$ is simply

$$y(x) = c_1 J_\alpha(x)$$

It is worth noting that eq. (4.10.16) does reduce to eq. (4.10.15) in the limit $\alpha \to n$.

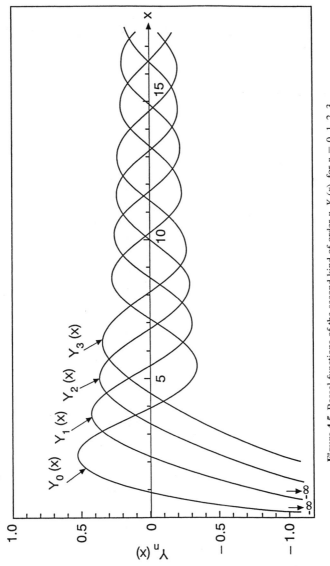

Figure 4.5 Bessel functions of the second kind of order n, $Y_n(x)$, for $n = 0, 1, 2, 3$.

323

4.11   THE MODIFIED BESSEL FUNCTIONS

The ODE

$$x^2 \frac{d^2y}{dx^2} + x \frac{dy}{dx} - (x^2 + \alpha^2)y = 0 \qquad (4.11.1)$$

where $\alpha \geq 0$, also arises in applications involving the cylindrical geometry. With the change of variables

$$t = ix$$

it reduces to the Bessel's ODE of order α

$$t^2 \frac{d^2y}{dt^2} + t \frac{dy}{dt} + (t^2 - \alpha^2)y = 0 \qquad (4.11.2)$$

Since one solution of eq. (4.11.2) is $J_\alpha(t)$, that of eq. (4.11.1) will be $J_\alpha(ix)$. Recall from eq. (4.8.10) that

$$J_\alpha(x) = \sum_{m=0}^{\infty} \frac{(-1)^m (x/2)^{2m+\alpha}}{m!\,\Gamma(m + \alpha + 1)}$$

thus

$$J_\alpha(ix) = i^\alpha \sum_{m=0}^{\infty} \frac{(x/2)^{2m+\alpha}}{m!\,\Gamma(m + \alpha + 1)}$$

The quantity

$$I_\alpha(x) = \sum_{m=0}^{\infty} \frac{(x/2)^{2m+\alpha}}{m!\,\Gamma(m + \alpha + 1)} \qquad (4.11.3)$$

is called the *modified Bessel function of the first kind of order* α. As before, if α is not zero or an integer, then

$$I_{-\alpha}(x) = \sum_{m=0}^{\infty} \frac{(x/2)^{2m-\alpha}}{m!\,\Gamma(m - \alpha + 1)} \qquad (4.11.4)$$

is a second linearly independent solution. For $\alpha = 0$ or n, the second solution takes the form

$$K_0(x) = -\left[\ln\left(\frac{x}{2}\right) + \gamma\right]I_0(x) + \sum_{m=1}^{\infty} \frac{h_m(x/2)^{2m}}{(m!)^2} \qquad (4.11.5)$$

$$K_n(x) = (-1)^{n+1}\ln\left(\frac{x}{2}\right)I_n(x) + \frac{1}{2}\sum_{m=0}^{n-1}(-1)^m \frac{(n - m - 1)!}{m!}\left(\frac{x}{2}\right)^{2m-n}$$

$$+ \frac{1}{2}(-1)^n \sum_{m=0}^{\infty} \{\phi(m + 1) + \phi(m + n + 1)\} \frac{(x/2)^{2m+n}}{m!(m + n)!} \qquad (4.11.6)$$

where h_m and $\phi(m)$ have been defined in eqs. (4.10.5) and (4.10.15), respectively. $K_n(x)$ is called the *modified Bessel function of the second kind of order* n.

Finally, in analogy with Bessel functions of the first kind, rather than $I_{-\alpha}(x)$ when α is not zero or an integer, the second solution is defined as

$$K_\alpha(x) = \frac{\pi}{2} \frac{I_{-\alpha}(x) - I_\alpha(x)}{\sin \alpha\pi} \tag{4.11.7}$$

which is the *modified Bessel function of the second kind of order* α. A tedious manipulation shows that as $\alpha \to n$, the form (4.11.7) reduces to (4.11.6).

The general solution of eq. (4.11.1) is then always given by

$$y(x) = c_1 I_\alpha(x) + c_2 K_\alpha(x) \tag{4.11.8}$$

where c_1 and c_2 are arbitrary constants.

For $n = 0$ and 1, eq. (4.11.3) leads to

$$I_0(x) = 1 + \frac{x^2}{2^2} + \frac{x^4}{2^4(2!)^2} + \frac{x^6}{2^6(3!)^2} + \frac{x^8}{2^8(4!)^2} + \cdots$$

$$I_1(x) = \frac{x}{2} + \frac{x^3}{2^2 2!} + \frac{x^5}{2^5(2!3!)} + \frac{x^7}{2^7(3!4!)} + \cdots$$

which are shown in Figure 4.6, along with $K_0(x)$ and $K_1(x)$. It is clear from eqs. (4.11.3) to (4.11.7) that except for I_0, all $I_\alpha(0) = 0$, while all the $K_\alpha(x)$ become infinite as $x \to 0$.

Further, it may also be shown that

$$I_{1/2}(x) = \sqrt{\frac{2}{\pi x}} \sinh x, \qquad I_{-1/2}(x) = \sqrt{\frac{2}{\pi x}} \cosh x \tag{4.11.9}$$

formulae analogous to those for $J_{1/2}(x)$ and $J_{-1/2}(x)$.

4.12 ASYMPTOTIC BEHAVIOR OF BESSEL FUNCTIONS

In many applications, asymptotic behavior of Bessel functions for small or large values of the argument is required. It may be shown with ease that the power series portions of all the Bessel functions converge for all values of x; however, there are terms of the form $x^{-\alpha}$ or $\ln x$ as multiplying factors, and these diverge as x approaches zero.

For *small* values of x it is easy to determine the behavior of the various Bessel functions from their definitions, and these are

$$J_\alpha(x) \sim \frac{x^\alpha}{2^\alpha \Gamma(\alpha + 1)}$$

$$Y_\alpha(x) \sim -\frac{2^\alpha \Gamma(\alpha)}{\pi} x^{-\alpha}, \qquad \alpha \neq 0$$

$$Y_0(x) \sim \frac{2}{\pi} \ln x$$

$$I_\alpha(x) \sim \frac{x^\alpha}{2^\alpha \Gamma(\alpha + 1)}$$

$$K_\alpha(x) \sim 2^{\alpha-1} \Gamma(\alpha) x^{-\alpha}, \qquad \alpha \neq 0$$

$$K_0(x) \sim -\ln x$$

The behavior of Bessel functions for *large* values of x is more difficult to establish, but an idea of the general character may be obtained by making the substitution

$$y = x^{-1/2}u$$

in the ODE

$$x^2 \frac{d^2y}{dx^2} + x \frac{dy}{dx} + (x^2 - \alpha^2)y = 0$$

which gives

$$\frac{d^2u}{dx^2} + \left[1 - \frac{\alpha^2 - \frac{1}{4}}{x^2} \right]u = 0$$

For large values of x, this equation is approximately

$$\frac{d^2u}{dx^2} + u = 0$$

which has solutions $\sin x$ and $\cos x$. The solution in terms of y may therefore be written in the form

$$y = \frac{A}{\sqrt{x}} \sin(x + a) + \frac{B}{\sqrt{x}} \cos(x + b)$$

where A, B, a, and b are constants. A considerable amount of analysis is required to compute the constants, but it can be shown that

$$J_\alpha(x) \sim \sqrt{\frac{2}{\pi x}} \cos\left(x - \frac{\pi\alpha}{2} - \frac{\pi}{4} \right)$$

$$Y_\alpha(x) \sim \sqrt{\frac{2}{\pi x}} \sin\left(x - \frac{\pi\alpha}{2} - \frac{\pi}{4} \right)$$

for large x.

It may also be shown that for large x, the modified Bessel functions have asymptotic behavior

$$I_\alpha(x) \sim \frac{\exp(x)}{\sqrt{2\pi x}}$$

$$K_\alpha(x) \sim \sqrt{\frac{\pi}{2x}} \exp(-x)$$

so that for large argument, $I_\alpha \to \infty$ and $K_\alpha \to 0$ as shown in Figure 4.6. Also observe that both $I_\alpha(x)$ and $K_\alpha(x)$ become independent of α for large x.

These formulae indicate that for large values of the argument, the zeroes of the Bessel functions of the first and second kind approach those of the trigonometric functions and are therefore infinite in number. Tables 4.2 and 4.3 list the first few zeroes of

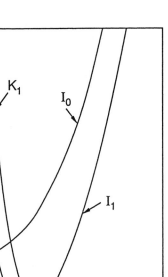

Figure 4.6 Modified Bessel functions of the first kind, $I_n(x)$, and second kind $K_n(x)$, for $n = 0, 1$.

TABLE 4.2 Zeroes of the Bessel Functions of the First Kind

	$J_0(x_i) = 0$	$J_1(x_i) = 0$	$J_2(x_i) = 0$	$J_3(x_i) = 0$	$J_4(x_i) = 0$
x_1	2.4048	3.8317	5.1356	6.3802	7.5883
x_2	5.5201	7.0156	8.4172	9.7610	11.0647
x_3	8.6537	10.1735	11.6199	13.0152	14.3725
x_4	11.7915	13.3237	14.7960	16.2235	17.6160
x_5	14.9309	16.4706	17.9598	19.4094	20.8269
x_6	18.0711	19.6159	21.1170	22.5827	24.0190
x_7	21.2116	22.7601	24.2701	25.7482	27.1991
x_8	24.3525	25.9037	27.4206	28.9084	30.3710
x_9	27.4935	29.0468	30.5692	32.0649	33.5371
x_{10}	30.6346	32.1897	33.7165	35.2187	36.6990

$X > 10$ use asymptotic

TABLE 4.3 Zeroes of the Bessel Functions of the Second Kind

	$Y_0(x_i) = 0$	$Y_1(x_i) = 0$	$Y_2(x_i) = 0$	$Y_3(x_i) = 0$	$Y_4(x_i) = 0$
x_1	0.8936	2.1971	3.3842	4.5270	5.6451
x_2	3.9577	5.4297	6.7938	8.0976	9.3616
x_3	7.0861	8.5960	10.0235	11.3965	12.7301
x_4	10.2223	11.7492	13.2100	14.6231	15.9996
x_5	13.3611	14.8974	16.3790	17.8185	19.2244
x_6	16.5009	18.0434	19.5390	20.9973	22.4248
x_7	19.6413	21.1881	22.6940	24.1662	25.6103
x_8	22.7820	24.3319	25.8456	27.3288	28.7859
x_9	25.9230	27.4753	28.9951	30.4870	31.9547
x_{10}	49.0640	30.6183	32.1430	33.6420	35.1185

the Bessel functions of the first and second kind, and we can see that for large x adjacent zeroes tend to differ by a value approximately equal to π. In solving partial differential equations involving cylindrical geometry, the zeroes of the Bessel functions play an important role as eigenvalues. In section 4.16, it is shown that they are all real. The zeroes of the modified Bessel functions are purely imaginary and their values are related to those of Bessel functions, since the differential equations which lead to them are related as shown in section 4.11.

4.13 RELATIONSHIPS AMONG BESSEL FUNCTIONS

In the same manner as relationships among various $J_\alpha(x)$ and $J'_\alpha(x)$ were obtained in section 4.9 starting with series representations, similar relationships for the other Bessel functions can also be derived; their proof is left as an exercise for the reader. If E stands for either J, Y, I, or K, then

$$E_{\alpha-1}(x) + E_{\alpha+1}(x) = \frac{2\alpha}{x} E_\alpha(x) \qquad (J, Y)$$

$$E_{\alpha-1}(x) - E_{\alpha+1}(x) = 2E'_\alpha(x) \qquad (J, Y)$$

$$E'_\alpha(x) = E_{\alpha-1}(x) - \frac{\alpha}{x} E_\alpha(x) \qquad (J, Y, I)$$

$$E'_\alpha(x) = -E_{\alpha+1}(x) + \frac{\alpha}{x} E_\alpha(x) \qquad (J, Y, K)$$

$$\frac{d}{dx}[x^\alpha E_\alpha(x)] = x^\alpha E_{\alpha-1}(x) \qquad (J, Y, I)$$

$$\frac{d}{dx}[x^{-\alpha} E_\alpha(x)] = -x^{-\alpha} E_{\alpha+1}(x) \qquad (J, Y, K)$$

where the formulae above hold for those Bessel functions given on the right. The analogs of these formulae for the cases not covered above are

$$I_{\alpha-1}(x) - I_{\alpha+1}(x) = \frac{2\alpha}{x} I_\alpha(x)$$

$$K_{\alpha-1}(x) - K_{\alpha+1}(x) = -\frac{2\alpha}{x} K_\alpha(x)$$

$$I_{\alpha-1}(x) + I_{\alpha+1}(x) = 2I'_\alpha(x) \quad 2I_\alpha$$

$$-K_{\alpha-1}(x) - K_{\alpha+1}(x) = 2K'_\alpha(x)$$

$$K'_\alpha(x) = -K_{\alpha-1}(x) - \frac{\alpha}{x} K_\alpha(x)$$

$$I'_\alpha(x) = I_{\alpha+1}(x) + \frac{\alpha}{x} I_\alpha(x)$$

$$\frac{d}{dx}[x^\alpha K_\alpha(x)] = -x^\alpha K_{\alpha-1}(x)$$

$$\frac{d}{dx}[x^{-\alpha} I_\alpha(x)] = x^{-\alpha} I_{\alpha+1}(x)$$

4.14 DIFFERENTIAL EQUATIONS LEADING TO BESSEL FUNCTIONS

One of the reasons why Bessel functions are so important in applied mathematics is that a large number of differential equations, seemingly unrelated to Bessel's equation, may be reduced to it. For example, the ODE

$$\frac{d^2y}{dx^2} + ax^2 y = 0$$

where $a > 0$, has the general solution

$$y = \sqrt{x}\left[c_1 J_{1/4}\left(\frac{\sqrt{a}}{2} x^2\right) + c_2 Y_{1/4}\left(\frac{\sqrt{a}}{2} x^2\right)\right]$$

where c_1 and c_2 are arbitrary constants. This may be verified by making the substitutions $u = y/\sqrt{x}$ and $t = (\sqrt{a}/2)x^2$.

A large family of differential equations whose solutions can be expressed in terms of Bessel functions is represented by the result below.

● **THEOREM**

If $(1 - a)^2 \geq 4c$ and if neither d, p, nor q is zero, then except in the special case when it reduces to Euler's equation, the ODE

$$x^2 \frac{d^2y}{dx^2} + x(a + 2bx^p) \frac{dy}{dx} + [c + dx^{2q} + b(a + p - 1)x^p + b^2x^{2p}]y = 0 \quad (4.14.1)$$

has the general solution

$$y(x) = x^\alpha \exp(-\beta x^p)[c_1 J_\nu(\lambda x^q) + c_2 Y_\nu(\lambda x^q)] \quad (4.14.2)$$

where

$$\alpha = \frac{1-a}{2}, \ \beta = \frac{b}{p}, \ \lambda = \frac{\sqrt{|d|}}{q}, \ \nu = \frac{[(1-a)^2 - 4c]^{1/2}}{2q} \qquad (4.14.3)$$

If $d < 0$, J_ν and Y_ν should be replaced by I_ν and K_ν, respectively. If ν is not zero or an integer, Y_ν and K_ν can be replaced by $J_{-\nu}$ and $I_{-\nu}$, if desired.

The proof of the theorem is straightforward and is therefore not presented in full. It consists of transforming the ODE by the change of variables

$$y = x^{(1-a)/2} \exp\left(-\frac{b}{p} x^p\right) u, \qquad t = \frac{\sqrt{|d|}}{q} x^q$$

and verifying that the resulting ODE in u and t is precisely Bessel's equation of order ν.

The special case of Euler's equation is

$$a_0 x^2 \frac{d^2y}{dx^2} + a_1 x \frac{dy}{dx} + a_2 y = 0 \qquad (4.14.4)$$

where a_j are constants. With the change of independent variable

$$x = e^z \quad \text{or} \quad z = \ln x$$

it reduces to an ODE with constant coefficients:

$$a_0 \frac{d^2y}{dz^2} + (a_1 - a_0) \frac{dy}{dz} + a_2 y = 0$$

which may be solved by standard methods.

4.15 SOME APPLICATIONS INVOLVING BESSEL FUNCTIONS

In this section we consider two engineering problems which can be solved by the use of either Bessel or modified Bessel functions.

Example 1. Determine the steady-state temperature distribution in an infinitely long solid cylinder of radius R, in which heat is generated per unit volume at rate $(a + bT)$, where a and b are positive constants and T is temperature. The external surface is maintained at constant temperature T_0.

The relevant ODE (see Figure 4.7) is

$$\frac{k}{r} \frac{d}{dr}\left[r \frac{dT}{dr}\right] = -(a + bT), \qquad r \in (0, R) \qquad (4.15.1)$$

where k is thermal conductivity of the solid and r is radial distance from the center. The BCs are given by

$$T = T_0, \qquad r = R \qquad (4.15.2)$$

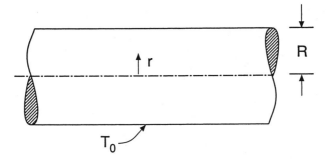

Figure 4.7 An infinitely long cylinder.

and the associated condition that T is finite. Note that the finiteness of T is equivalent to the "symmetry" condition at the center

$$\frac{dT}{dr} = 0, \qquad r = 0 \tag{4.15.3}$$

as will become clear later.

Note that the ODE (4.15.1) can be rewritten as

$$r \frac{d^2T}{dr^2} + \frac{dT}{dr} + (a + bT)\frac{r}{k} = 0$$

or

$$r^2 \frac{d^2T}{dr^2} + r \frac{dT}{dr} + (a + bT)\frac{r^2}{k} = 0 \tag{4.15.4}$$

while the standard form of Bessel's ODE is (4.8.1):

$$x^2 \frac{d^2y}{dx^2} + x \frac{dy}{dx} + (x^2 - \alpha^2)y = 0, \qquad \alpha \geq 0 \tag{4.15.5}$$

The first point to notice in comparing eqs. (4.15.4) and (4.15.5) is that the last term is somewhat different. If we let

$$u = a + bT$$

then

$$\frac{dT}{dr} = \frac{1}{b} \frac{du}{dr}, \qquad \frac{d^2T}{dr^2} = \frac{1}{b} \frac{d^2u}{dr^2}$$

and eq. (4.15.4) becomes

$$r^2 \frac{d^2u}{dr^2} + r \frac{du}{dr} + \frac{b}{k} r^2 u = 0 \tag{4.15.6}$$

which is almost in the standard form (4.15.5), except for the coefficient of the last term. There are two ways to proceed now; the first is to use the theorem of section 4.14, and the second is to make another change of variables.

Following the *first* method, we recognize in comparing eqs. (4.15.6) and (4.14.1) that

$$a = 1, \qquad b = 0, \qquad c = 0, \qquad d = \frac{b}{k} > 0, \qquad q = 1$$

so that

$$\alpha = 0, \qquad \beta = 0, \qquad \lambda = \sqrt{(b/k)}, \qquad \nu = 0$$

From eq. (4.14.2) the general solution is

$$u(r) = c_1 J_0(r\sqrt{b/k}) + c_2 Y_0(r\sqrt{b/k}) \qquad (4.15.7)$$

where c_1 and c_2 are arbitrary constants.

In the *second* method, if we change the independent variable

$$r = cx$$

where c is as yet an undetermined constant, eq. (4.15.6) takes the form

$$x^2 \frac{d^2u}{dx^2} + x \frac{du}{dx} + \frac{bc^2}{k} x^2 u = 0$$

Hence with the choice

$$c = \sqrt{k/b}$$

we have

$$x^2 \frac{d^2u}{dx^2} + x \frac{du}{dx} + x^2 u = 0$$

which is the standard form of Bessel's ODE, with $\alpha = 0$. This has the general solution

$$u(x) = c_1 J_0(x) + c_2 Y_0(x)$$

which is the same as eq. (4.15.7)

Thus, recalling that

$$u = a + bT$$

the general solution is

$$T(r) = \frac{1}{b} [c_1 J_0(r\sqrt{b/k}) + c_2 Y_0(r\sqrt{b/k}) - a] \qquad (4.15.8)$$

where the arbitrary constants c_1 and c_2 must now be determined.

Since $Y_0(x) \to -\infty$ as $x \to 0$, finiteness implies that $c_2 = 0$. We can, alternatively, replace the finiteness condition by the symmetry condition (4.15.3) as follows. From eq. (4.15.8) and the derivative formula

$$\frac{d}{dx} [x^{-\alpha} E_\alpha(x)] = -x^{-\alpha} E_{\alpha+1}(x) \qquad (J, Y, K)$$

we have

$$\frac{dT}{dr} = -[c_1 J_1(r\sqrt{b/k}) + c_2 Y_1(r\sqrt{b/k})]/\sqrt{b/k}$$

so that unless $c_2 = 0$, $(dT/dr)(0) \neq 0$.

   With $c_2 = 0$, the solution (4.15.8) reduces to

$$T(r) = \frac{1}{b}[c_1 J_0(r\sqrt{b/k}) - a]$$

Applying the BC (4.15.2) gives

$$T_0 = \frac{1}{b}[c_1 J_0(R\sqrt{b/k}) - a]$$

so that

$$c_1 = \frac{a + bT_0}{J_0(R\sqrt{b/k})}$$

and hence

$$T(r) = \frac{1}{b}\left[(a + bT_0)\frac{J_0(r\sqrt{b/k})}{J_0(R\sqrt{b/k})} - a\right] \qquad (4.15.9)$$

is the desired temperature distribution.

Example 2. Determine the steady-state concentration distribution $C(r)$ and the effectiveness factor for an isothermal first-order reaction occurring in an infinitely long cylindrical porous catalyst pellet of radius R. The external surface is maintained at constant reactant concentration, C_0.

The relevant ODE is

$$\frac{D}{r}\frac{d}{dr}\left[r\frac{dC}{dr}\right] = kC, \qquad r \in (0, R)$$

where D is the diffusion coefficient of reactant in the pellet, and k is the reaction rate constant. The associated BCs are

$$C = C_0, \qquad r = R$$

and C is finite.

   The problem can be made dimensionless by introducing the variables

$$s = \frac{r}{R}, \qquad u = \frac{C}{C_0}, \qquad \phi^2 = \frac{kR^2}{D}$$

where ϕ is called *Thiele modulus*, to give

$$\frac{1}{s}\frac{d}{ds}\left[s\frac{du}{ds}\right] = \phi^2 u, \qquad s \in (0, 1) \tag{4.15.10}$$

and the BCs

$$u(1) = 1 \tag{4.15.11a}$$

$$u(s) = \text{finite} \tag{4.15.11b}$$

The ODE (4.15.10) can be rearranged as

$$s^2\frac{d^2u}{ds^2} + s\frac{du}{ds} - \phi^2 s^2 u = 0$$

Comparing with the standard form (4.14.1) in the theorem of section 4.14, we have

$$a = 1, \qquad b = 0, \qquad c = 0, \qquad d = -\phi^2 < 0, \qquad q = 1$$

so that

$$\alpha = 0, \qquad \beta = 0, \qquad \lambda = \phi, \qquad \nu = 0$$

and the general solution is

$$u(s) = c_1 I_0(\phi s) + c_2 K_0(\phi s) \tag{4.15.12}$$

where c_1 and c_2 are arbitrary constants.

Since $K_0(x) \to +\infty$ as $x \to 0$, finiteness of u implies that $c_2 = 0$, leading to

$$u(s) = c_1 I_0(\phi s)$$

Now applying the external surface condition (4.15.11a) yields

$$c_1 = \frac{1}{I_0(\phi)}$$

so that the dimensionless concentration distribution is

$$u(s) = \frac{I_0(\phi s)}{I_0(\phi)} \tag{4.15.13}$$

which has the form shown in Figure 4.8.

The *effectiveness factor* of the catalyst pellet is defined as the ratio of the actual rate of reaction to that which would prevail in the absence of any transport limitation. Thus

$$\eta = \frac{\displaystyle\int_0^R rkC(r)\,dr}{\displaystyle\int_0^R rkC_0\,dr} = \frac{\displaystyle\int_0^R r[C(r)/C_0]\,dr}{\displaystyle\int_0^R r\,dr}$$

$$= 2\int_0^1 su(s)\,ds \tag{4.15.14}$$

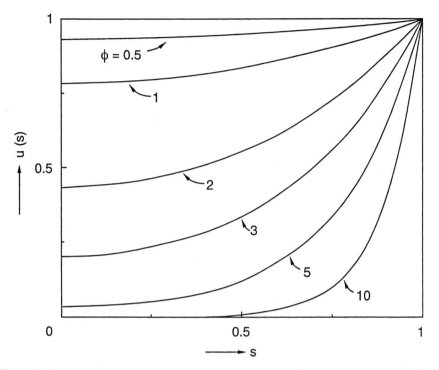

Figure 4.8 Dimensionless concentration profiles, $u(s)$ given by eq. (4.15.13), for various values of the Thiele modulus, ϕ.

There are two ways of evaluating the integral. The *first* is to use the expression (4.15.13) for $u(s)$ and to carry out the integration

$$\eta = 2 \int_0^1 s \, \frac{I_0(\phi s)}{I_0(\phi)} \, ds$$

Letting $x = \phi s$, the integral becomes

$$\eta = \frac{2}{\phi^2 I_0(\phi)} \int_0^\phi x I_0(x) \, dx$$

However, utilizing the fifth relationship in section 4.13 for I_α and $\alpha = 1$, namely,

$$\frac{d}{dx} [x I_1(x)] = x I_0(x)$$

leads to

$$\eta = \frac{2}{\phi^2 I_0(\phi)} \int_0^\phi \frac{d}{dx} [x I_1(x)] \, dx$$

$$= \frac{2}{\phi} \frac{I_1(\phi)}{I_0(\phi)} \qquad\qquad (4.15.15)$$

In the *second* method, we utilize the ODE (4.15.10) directly in eq. (4.15.14):

$$\eta = 2 \int_0^1 s u(s) \, ds = \frac{2}{\phi^2} \int_0^1 \frac{d}{ds}\left[s \frac{du}{ds} \right] ds = \frac{2}{\phi^2} \frac{du}{ds} \quad (1)$$

Now, from the expression (4.15.13) for u we obtain

$$\frac{du}{ds} = \frac{1}{I_0(\phi)} \frac{d}{ds}[I_0(\phi s)] = \frac{\phi}{I_0(\phi)} \frac{d}{d(\phi s)}[I_0(\phi s)]$$

and the last relationship in section 4.13 is

$$\frac{d}{dx}[x^{-\alpha} I_\alpha(x)] = x^{-\alpha} I_{\alpha+1}(x)$$

so that

$$\frac{d}{dx}[I_0(x)] = I_1(x)$$

and

$$\frac{du}{ds} = \frac{\phi I_1(\phi s)}{I_0(\phi)}$$

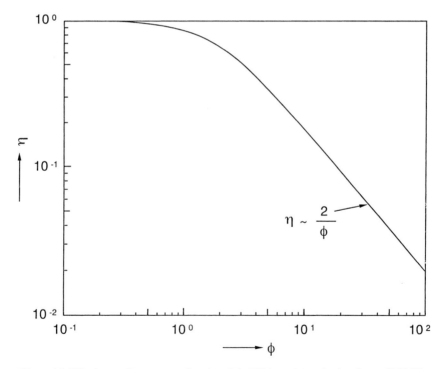

Figure 4.9 Effectiveness factor, η, as a function of the Thiele modulus, ϕ, given by eq. (4.15.15).

Thus we have

$$\eta = \frac{2}{\phi^2} \frac{\phi I_1(\phi)}{I_0(\phi)} = \frac{2}{\phi} \frac{I_1(\phi)}{I_0(\phi)}$$

as before. A plot of the effectiveness factor (η) versus the Thiele modulus (ϕ) is shown in Figure 4.9, where we observe that as a result of the asymptotic behavior discussed in section 4.12:

$$I_\alpha(x) \sim \frac{x^\alpha}{2^\alpha \Gamma(\alpha + 1)} \qquad \text{as } x \to 0$$

$$I_\alpha(x) \sim \frac{\exp(x)}{\sqrt{2\pi x}} \qquad \text{as } x \to \infty$$

the expression (4.15.15) leads to

$$\eta = 1 \qquad \text{as } \phi \to 0$$

and

$$\eta \to \frac{2}{\phi} \qquad \text{as } \phi \to \infty$$

4.16 EIGENVALUE PROBLEMS INVOLVING EQUATIONS WITH SINGULAR END-POINTS

In section 3.16, while considering the eigenvalue problem for a general second-order self-adjoint operator

$$L[y] = \frac{d}{dx}\left[p(x) \frac{dy}{dx} \right] - q(x)y = -\lambda \rho(x)y, \qquad x \in (a, b) \qquad (4.16.1)$$

we restricted attention to the case where $p(x)$ and $\rho(x)$ are both continuous and positive for all $x \in [a, b]$. However, in many cases of physical interest, particularly when solving partial differential equations involving cylindrical or spherical geometries, we encounter problems where $p(x)$ and $\rho(x)$ are positive in the *open* interval $x \in (a, b)$ but vanish at one or both ends. The Sturm–Liouville theory applies in these cases as well, as we shall see in this section.

Let us consider in detail the eigenvalue problem for a full cylinder:

$$L[\phi] = \frac{d}{dx}\left[x \frac{d\phi}{dx} \right] = -\lambda x \phi, \qquad x \in (0, a) \qquad (4.16.2)$$

$$\phi = 0, \qquad x = a \qquad (4.16.3a)$$

$$\phi = \text{finite} \qquad (4.16.3b)$$

where, as discussed in the previous section, finiteness of the solution serves in lieu of a boundary condition. Comparison of eq. (4.16.2) with the standard form (4.16.1) yields

$$p(x) = x, \qquad q(x) = 0, \qquad \rho(x) = x \qquad (4.16.4)$$

so that both $p(x)$ and $\rho(x)$ vanish at one end of the domain.

Equation (4.16.2) can be rearranged as

$$x\phi'' + \phi' + \lambda x\phi = 0, \qquad x \in (0, a)$$

or

$$x^2\phi'' + x\phi' + \lambda x^2\phi = 0$$

A comparison of this with the standard form (4.14.1) implies that the general solution of eq. (4.16.2) is

$$\phi(x) = c_1 J_0(\sqrt{\lambda}x) + c_2 Y_0(\sqrt{\lambda}x) \qquad (4.16.5)$$

where c_1 and c_2 are arbitrary constants. Since the solution is valid over the closed domain $x \in [0, a]$, and $Y_0(x) \to -\infty$ as $x \to 0$, the finiteness condition (4.16.3b) implies that $c_2 = 0$. Thus

$$\phi(x) = c_1 J_0(\sqrt{\lambda}x) \qquad (4.16.6)$$

Applying the surface BC given by eq. (4.16.3a) then gives

$$c_1 J_0(\sqrt{\lambda}a) = 0 \qquad (4.16.7)$$

Now $c_1 \neq 0$, because otherwise $\phi(x)$ from eq. (4.16.6) would be identically zero, and this cannot hold since $\phi(x)$ is an eigenfunction. Thus eq. (4.16.7) can be satisfied only by requiring that the eigenvalues λ satisfy

$$J_0(\sqrt{\lambda}a) = 0 \qquad (4.16.8)$$

From sections 4.8 and 4.12, we know that there are an infinite number of roots of $J_0(x) = 0$; hence there are an infinite number of eigenvalues, λ_n, satisfying the equation

$$J_0(\sqrt{\lambda_n}a) = 0, \qquad n = 1, 2, \ldots \qquad (4.16.9)$$

and up to an arbitrary multiplying constant, corresponding to each λ_n is the eigenfunction

$$\phi_n(x) = J_0(\sqrt{\lambda_n}x), \qquad n = 1, 2, \ldots \qquad (4.16.10)$$

As noted earlier, since the operator of eq. (4.16.2), namely,

$$L[\phi] = \frac{d}{dx}\left[x\frac{d\phi}{dx}\right]$$

is in the standard form (4.16.1) with $p(x) = x$, it is indeed self-adjoint. The BCs given by eq. (4.16.3) are also self-adjoint, as is verified next.

From the definition of $L[\ ]$ and integrating by parts, we have

$$\int_0^a uL[v]\ dx = \int_0^a u\frac{d}{dx}\left[x\frac{dv}{dx}\right]dx$$

$$= uxv'\Big|_0^a - \int_0^a u'xv'\ dx$$

$$= uxv'\Big|_0^a - (u'x)v\Big|_0^a + \int_0^a v(u'x)'\ dx$$

$$= x(uv' - u'v)\Big|_0^a + \int_0^a vL[u]\ dx$$

so that

$$\int_0^a \{uL[v] - vL[u]\}\ dx = a[u(a)v'(a) - u'(a)v(a)] \qquad (4.16.11)$$

From eq. (4.16.11), it is easy to see that if v satisfies the condition

$$v(a) = 0 \qquad (4.16.12)$$

then

$$\int_0^a \{uL[v] - vL[u]\}\ dx = 0 \qquad (4.16.13)$$

with the choice

$$u(a) = 0 \qquad (4.16.14)$$

which is the same as eq. (4.16.12). From the discussion of section 3.10, it follows that the operator $L[\ ]$ of eq. (4.16.2), along with the BCs (4.16.3), constitutes a self-adjoint problem. Thus all conclusions of the Sturm–Liouville theory discussed in section 3.16 apply. In particular, the eigenvalues λ_n which satisfy eq. (4.16.9) are all real, and we have the following important results.

Orthogonality of Distinct Eigenfunctions.   Let λ_m and λ_n be two distinct eigenvalues and let $\phi_m(x)$ and $\phi_n(x)$ be their respective eigenfunctions; that is,

$$\phi_m(x) = J_0(\sqrt{\lambda_m}x), \qquad \phi_n(x) = J_0(\sqrt{\lambda_n}x) \qquad (4.16.15)$$

From eq. (4.6.13) we have

$$\int_0^a \{\phi_m L[\phi_n] - \phi_n L[\phi_m]\}\ dx = 0$$

which, by using eq. (4.16.2), reduces to

$$\int_0^a \{\phi_m(-\lambda_n x\phi_n) - \phi_n(-\lambda_m x\phi_m)\}\ dx = 0$$

or

$$(\lambda_m - \lambda_n)\int_0^a x\phi_m\phi_n\ dx = 0$$

Since $\lambda_m \neq \lambda_n$, this means that

$$\int_0^a x\phi_m\phi_n \, dx = 0 \tag{4.16.16}$$

That is, eigenfunctions corresponding to distinct eigenvalues are orthogonal with respect to the weight function $\rho(x) = x$.

Expansion of an Arbitrary Function. Similar to the discussion in section 3.18, we can consider the expansion of an arbitrary function $f(x)$ in terms of an infinite series of eigenfunctions $\{\phi_n(x)\}$ given by eq. (4.16.10). Thus we write

$$f(x) = \sum_{n=1}^{\infty} c_n \phi_n(x) \tag{4.16.17}$$

where c_n are as yet unknown constants. In order to determine these, we multiply both sides of eq. (4.16.17) by $x\phi_m(x)$ and integrate from 0 to a:

$$\int_0^a f(x)x\phi_m(x) \, dx = \int_0^a \left[\sum_{n=1}^{\infty} c_n \phi_n(x) \right] x\phi_m(x) \, dx$$

Exchanging the sum with integration, we obtain

$$\text{rhs} = \sum_{n=1}^{\infty} c_n \int_0^a x\phi_m(x)\phi_n(x) \, dx$$

$$= c_m \int_0^a x\phi_m^2(x) \, dx$$

since all other terms in the sum are zero, thanks to the orthogonality condition (4.16.16). Thus

$$c_m = \frac{\displaystyle\int_0^a f(x)x\phi_m(x) \, dx}{\displaystyle\int_0^a x\phi_m^2(x) \, dx} \tag{4.16.18}$$

which provides the constants c_n required for the expansion (4.16.17). Now, given a function $f(x)$, we can compute the constants explicitly if we can calculate the quantity

$$\| \phi_m \|^2 = \int_0^a x\phi_m^2(x) \, dx \tag{4.16.19}$$

where $\| \phi_m \|$ is the norm of $\phi_m(x)$. The evaluation of the norm requires some effort, and the details are given next.

Let λ_n be an eigenvalue, and let λ be an arbitrary constant. Thus λ_n satisfies eq. (4.16.9). Also let

$$u = J_0(\sqrt{\lambda}x), \qquad v = J_0(\sqrt{\lambda_n}x) \tag{4.16.20}$$

so that *both* u and v satisfy

$$L[\phi] = -\lambda x \phi$$

but *only* v satisfies the BC $v(a) = 0$.
  From eq. (4.16.11) we have

$$\int_0^a \{uL[v] - vL[u]\}\ dx = a[u(a)v'(a) - u'(a)v(a)]$$

which by substituting eq. (4.16.20) becomes

$$\int_0^a [J_0(\sqrt{\lambda}x)\{-\lambda_n x J_0(\sqrt{\lambda_n}x)\} - J_0(\sqrt{\lambda_n}x)\{-\lambda x J_0(\sqrt{\lambda}x)\}]\ dx$$
$$= a[J_0(\sqrt{\lambda}a) \frac{d}{dx}\{J_0(\sqrt{\lambda_n}x)\}|_{x=a}]$$

and then

$$(\lambda - \lambda_n) \int_0^a x J_0(\sqrt{\lambda}x)J_0(\sqrt{\lambda_n}x)\ dx = aJ_0(\sqrt{\lambda}a)[-\sqrt{\lambda_n}J_1(\sqrt{\lambda_n}a)]$$

where in the rhs we have used the derivative formula from section 4.13:

$$\frac{d}{dx}[x^{-\alpha}J_\alpha(x)] = -x^{-\alpha}J_{\alpha+1}(x) \qquad (4.16.21)$$

Thus

$$\int_0^a x J_0(\sqrt{\lambda}x)J_0(\sqrt{\lambda_n}x)\ dx = -a\sqrt{\lambda_n}J_1(\sqrt{\lambda_n}a)\frac{J_0(\sqrt{\lambda}a)}{\lambda - \lambda_n} \qquad (4.16.22)$$

Now, if we take the limit $\lambda \to \lambda_n$, then we obtain

$$\text{lhs} = \|J_0(\sqrt{\lambda_n}x)\|^2$$

while

$$\text{rhs} \to \frac{0}{0}$$

and is thus indeterminate. Applying L'Hôpital's rule gives

$$\|J_0(\sqrt{\lambda_n}x)\|^2 = \frac{-a\sqrt{\lambda_n}J_1(\sqrt{\lambda_n}a)}{2\sqrt{\lambda_n}} \frac{d}{d\sqrt{\lambda}}[J_0(\sqrt{\lambda}a)]|_{\lambda \to \lambda_n}$$
$$= \frac{a^2}{2} J_1^2(\sqrt{\lambda_n}a) \qquad (4.16.23)$$

where the derivative formula (4.16.21) has been used again.

Finally, from eq. (4.16.18) we have the relation

$$c_m = \frac{\int_0^a x f(x) \phi_m(x)\, dx}{(a^2/2)J_1^2(\sqrt{\lambda_n}\, a)} \tag{4.16.24}$$

which completes the expansion formula (4.16.17).

Note that with the norm determined, the *orthonormal* eigenfunctions are

$$X_n(x) = \frac{\phi_n(x)}{\| \phi_n(x) \|} = \frac{J_0(\sqrt{\lambda_n}\, x)}{(a/\sqrt{2})J_1(\sqrt{\lambda_n}\, a)} \tag{4.16.25}$$

and the orthogonality condition (4.16.16) implies that

$$\int_0^a x X_m(x) X_n(x)\, dx = \delta_{mn}$$

the Kronecker delta. If we use the normalized eigenfunctions for expanding $f(x)$, then in place of eq. (4.16.17) we have

$$f(x) = \sum_{n=1}^{\infty} d_n X_n(x) \tag{4.16.26a}$$

where

$$d_n = \mathcal{H}[f(x)] = \int_0^a x f(x) X_n(x)\, dx \tag{4.16.26b}$$

is a generalized finite Fourier transform of $f(x)$, called the *finite Hankel transform*.

Exercise 1
Show that for $f(x) = 1$, eqs. (4.16.26) give the expansion

$$1 = \sum_{n=1}^{\infty} \frac{2}{a} \frac{J_0(\sqrt{\lambda_n}\, x)}{\sqrt{\lambda_n}\, J_1(\sqrt{\lambda_n}\, a)} \tag{4.16.27}$$

The partial sum, s_N of the series on the rhs, for various values of N, is shown in Figure 4.10. It exhibits the Gibbs phenomenon, discussed previously in section 3.18.

Exercise 2
If the BC (4.16.3a) is of the third kind:

$$\frac{d\phi}{dx} + \alpha\phi = 0; \qquad x = a,\, \alpha > 0 \tag{4.16.28}$$

then in place of eq. (4.16.9) the eigenvalue equation is

$$\alpha J_0(\sqrt{\lambda_n}\, a) = \sqrt{\lambda_n}\, J_1(\sqrt{\lambda_n}\, a); \qquad n = 1, 2, \ldots \tag{4.16.29}$$

and the normalized eigenfunctions are

$$\phi_n(x) = \frac{J_0(\sqrt{\lambda_n}\, x)}{a J_0(\sqrt{\lambda_n}\, a)\left[\dfrac{\alpha^2 + \lambda_n}{2\lambda_n}\right]^{1/2}} \tag{4.16.30}$$

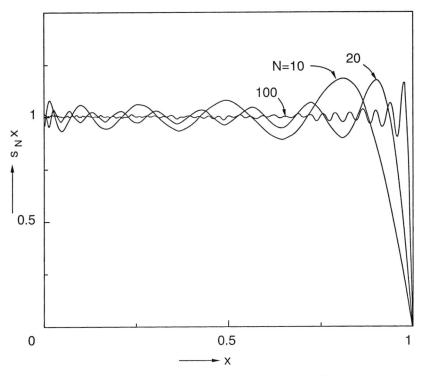

Figure 4.10 Expansion of the function $f(x) = 1$ using the first N normalized eigenfunctions.

Remark. We have discussed the eigenvalue problem for only the cylindrical geometry in this section. The case of the spherical geometry, where $p(x) = \rho(x) = x^2$ in eq. (4.16.1), is simpler because it can be reduced to the case of a slab geometry (see Example 2, section 3.17). Note that even when $x = 0$ is a part of the domain, the conclusions of the Sturm–Liouville theory continue to apply since the singularity occurs only at the ends.

References

M. Abramowitz and I. A. Stegun, *Handbook of Mathematical Functions*, Dover, New York, 1965.

L. C. Andrews, *Special Functions for Engineers and Applied Mathematicians*, Macmillan, New York, 1985.

E. L. Ince, *Ordinary Differential Equations*, Dover, New York, 1956.

G. N. Watson, *A Treatise on the Theory of Bessel Functions*, 2nd ed., Cambridge University Press, Cambridge, 1966.

Additional Reading

A. Erdélyi, W. Magnus, F. Oberhettinger, and F. Tricomi, *Higher Transcendental Functions*, volumes 1, 2, and 3, McGraw-Hill, New York, 1953.

N. N. Lebedev, *Special Functions and Their Applications*, Dover, New York, 1972.

T. M. MacRobert, *Spherical Harmonics: An Elementary Treatise on Harmonic Functions with Applications*, Dover, New York, 1948.

Problems

4.1   Find the radius of convergence of the following series:

(a) $\displaystyle\sum_{m=0}^{\infty} \frac{x^m}{2^m}$

(b) $\displaystyle\sum_{m=0}^{\infty} \frac{(2m)!}{(m!)^2} x^m$

(c) $\displaystyle\sum_{m=0}^{\infty} \frac{x^{m^2}}{2^m}$

(d) $\displaystyle\sum_{m=1}^{\infty} \frac{(m+1)}{m \cdot 3^m} (x-2)^m$

(e) $\displaystyle\sum_{m=1}^{\infty} \frac{(2x+1)^m}{m^2}$

4.2   Solve by using the power series method:

$$(x - 3)y' - xy = 0, \qquad x_0 = 0$$

Show that the solution is the same as that obtained by separation of variables.

4.3   Solve by using the power series method:

$$xy' - y = x^2 e^x, \qquad x_0 = 0$$

Show that the solution is the same as that obtained by the integrating factor approach.

4.4   Apply the power series method to solve *Airy's equation*:

$$y'' = xy, \qquad x_0 = 0$$

In finding the solutions, determine the general terms and also test for convergence of the series solutions.

4.5   In the context of Legendre equation,
   (a) Confirm that the two linearly independent solutions, y_1 and y_2, are given by eqs. (4.4.2) and (4.4.3), respectively.
   (b) By using the ratio test, show that the series for y_1 and y_2 both converge for $|x| < 1$.
   (c) Confirm that eq. (4.4.5) is correct.
   (d) By using eq. (4.4.5), develop the recurrence relation (4.4.6).

4.6   Legendre polynomials can be represented by the *Rodrigues formula*:

$$P_n(x) = \frac{1}{2^n n!} \frac{d^n}{dx^n} [(x^2 - 1)^n]$$

   (a) Verify the formula for $n = 0, 1, 2,$ and 3.

(b) For an arbitrary n, integrating the formula n times by parts, show that

$$\| P_n \|^2 = \int_{-1}^{1} P_n^2(x) \, dx = \frac{2}{2n + 1}$$

4.7 In eq. (4.4.13), determine the coefficients a_n for the following functions:

(a)
$$f(x) = \begin{cases} 0 & \text{for } -1 \leq x < 0 \\ 1 & \text{for } 0 < x \leq 1 \end{cases}$$

(b)
$$f(x) = \begin{cases} 1 & \text{for } -1 \leq x < 0 \\ 0 & \text{for } 0 < x \leq 1 \end{cases}$$

Also, in the interval $-1 \leq x \leq 1$, plot $s_N(x) = \sum_{n=0}^{N} a_n P_n(x)$ for $N = 5$, 15, and 25 in both cases.

4.8 Consider the Chebyshev differential equation:

$$(1 - x^2)y'' - xy' + \alpha^2 y = 0$$

where α is a non-negative constant.

(a) Obtain series solutions for $x_0 = 0$. What is the assured radius of convergence of the series solutions? Report the general term in the series.

(b) Show that if $\alpha = n$, an integer, then there is a polynomial solution of degree n. When properly normalized, these polynomials are called *Chebyshev polynomials*, $T_n(x)$ (cf. Andrews, 1985, section 5.4.2).

(c) Find a polynomial solution for $\alpha = n = 0$, 1, 2, and 3.

4.9 The ODE

$$y'' - 2xy' + \lambda y = 0$$

where λ is a constant, is called the Hermite equation.

(a) Obtain series solutions for $x_0 = 0$.

(b) Observe that if λ is a non-negative even integer, say $\lambda = 2n$, then one or the other of the infinite series terminates as a polynomial of degree n. When normalized such that the coefficient of x^n equals 2^n, these polynomials are called *Hermite polynomials*, $H_n(x)$ (cf. Andrews, 1985, section 5.2).

(c) Determine the Hermite polynomials for $n = 0$, 1, 2, and 3.

4.10 Classify all the singular points (finite and infinite) of the following ODEs:

(a) $y'' = xy$ (Airy equation)

(b) $x^2 y'' + xy' + (x^2 - \alpha^2)y = 0$ (Bessel equation)

(c) $x(1 - x)y'' + [c - (a + b + 1)x]y' - aby = 0$ (Gauss hypergeometric equation)

(d) $2(x - 2)^2\, xy'' + 3xy' + (x - 2)y = 0$

4.11 Using the Frobenius method, find a fundamental set of solutions of the following ODEs. If possible, identify the obtained series as expansions of known functions.

(a) $xy'' + 2y' + xy$ $= 0$

(b) $x^2 y'' + 6xy' + (6 - x^2)y$ $= 0$

(c) $xy'' + (1 - 2x)y' + (x - 1)y$ $= 0$

(d) $x^2(1 + x)y'' - x(1 + 2x)y' + (1 + 2x)y = 0$

4.12 The ODE

$$x(1 - x)y'' + [c - (a + b + 1)x]y' - aby = 0$$

where a, b, and c are constants, is called the *Gauss hypergeometric equation*.

(a) Show that the roots of the indicial equation are $r_1 = 0$ and $r_2 = 1 - c$.

(b) Show that for $r_1 = 0$, the Frobenius method yields the solution

$$y_1(x) = 1 + \frac{ab}{c} x + \frac{a(a + 1)b(b + 1)}{c(c + 1)} \frac{x^2}{2!}$$

$$+ \frac{a(a + 1)(a + 2)b(b + 1)(b + 2)}{c(c + 1)(c + 2)} \frac{x^3}{3!} + \cdots$$

for $c \neq 0, -1, -2, \ldots$. This series is called the *hypergeometric series*, and $y_1(x)$ is commonly denoted by $F(a, b, c; x)$, the *hypergeometric function* (cf. Andrews, 1985, section 8.3).

(c) Using the expression for $y_1(x)$ above, show that the series converges for $|x| < 1$.

(d) Many elementary functions are special cases of $F(a, b, c; x)$. Show that

    (i) $F(1, b, b; x) = 1 + x + x^2 + \cdots$ (the geometric series)

    (ii) $xF(1, 1, 2; -x) = \ln(1 + x)$

    (iii) $F(-n, b, b; -x) = (1 + x)^n$

4.13 The ODE

$$xy'' + (1 - x)y'' + \lambda y = 0$$

where λ is a constant, is called the *Laguerre equation*.

(a) Show that $x = 0$ is a regular singular point.

(b) Determine the roots of the indicial equation and one solution $(x > 0)$ using the Frobenius method.

(c) Show that for $\lambda = n$, a positive integer, this solution reduces to a polynomial. When suitably normalized, these polynomials are called the *Laguerre polynomials*, $L_n(x)$ (cf. Andrews, 1985, section 5.3).

4.14 Evaluate the functions $\text{erf}(x)$ and $\text{erfc}(x)$ at $x = 0$ and ∞.

4.15 The error function is useful in solving problems involving the normal (also called *Gaussian*) distribution in probability theory. A normal random variable x is described by the probability density function

$$p(x) = \frac{1}{\sqrt{2\pi}\sigma} \exp\left[\frac{-(x - \bar{x})^2}{2\sigma^2}\right]$$

where $\bar{x}$ is the mean value of x and σ^2 is the variance. The probability that $x \leq X$ is defined to be

$$P(x \leq X) = \int_{-\infty}^{X} p(x)\, dx$$

Show that for a normal distribution we obtain

$$P(x \leq X) = \frac{1}{2}\left[1 + \text{erf}\left(\frac{X - \bar{x}}{\sqrt{2}\sigma}\right)\right]$$

4.16 Evaluate the accuracy of the Stirling formula (4.7.11), for $\alpha = 10$ and 15.

4.17 Using the definition in eq. (4.8.10), show that the series for $J_\alpha(x)$ converges for all x. This series in fact converges quite rapidly (why?). Determine the number of terms required in the sum to represent $J_0(1)$ accurately, with an error less than 10^{-5}.

4.18 Verify that eq. (4.10.16) indeed yields eq. (4.10.8) for $Y_0(x)$, in the limit as $\alpha \to 0$. Start with eq. (4.8.10), which defines $J_\alpha(x)$.

4.19 Verify that eq. (4.11.9) is correct.

4.20 By making the indicated substitutions, verify that eq. (4.14.2) indeed represents the general solution of eq. (4.14.1).

4.21 By making the indicated substitutions, reduce the following ODEs to Bessel's equation and report the general solution in terms of Bessel functions.
   (a) $xy'' + y' + \frac{1}{4}y = 0,$ $\sqrt{x} = z$
   (b) $xy'' + (1 + 2n)y' + xy = 0,$ $y = x^{-n}u$
   (c) $y'' - xy = 0,$ $y = u\sqrt{x},$ $\frac{2}{3}x^{3/2} = z$

4.22 For the temperature distribution (4.15.9), derive a formula for the average temperature in the cylinder.

4.23 Solve eq. (4.15.1) for the case of a hollow cylinder, of inside and outside radii R_1 and R_2, respectively. The inner and outer surfaces are maintained at T_1 and T_2, respectively. Also, derive a formula for the average temperature.

4.24 Repeat problem 4.23 for the case where the inner surface is insulated.

4.25 Repeat Example 2 of section 4.15 for the case of finite external mass transfer resistance at the pellet surface. The BC at $r = R$ is then replaced by

$$D \frac{dC}{dr} = k_g(C_f - C), \qquad r = R$$

where C_f is reactant concentration in the bulk fluid, and k_g is the mass transfer coefficient. In making the problem dimensionless, an additional dimensionless parameter called the Biot number, $\text{Bi} = k_g R/D$, arises. Also report asymptotic values of the effectiveness factor, η, for small and large values of the Thiele modulus, ϕ.

4.26 Consider, as shown in Figure 4.11, a circular fin of thickness b attached to a heated pipe with outside wall temperature T_0. Heat loss from the end area $(2\pi R_1 b)$ may be neglected. The ambient air is maintained at temperature T_a.

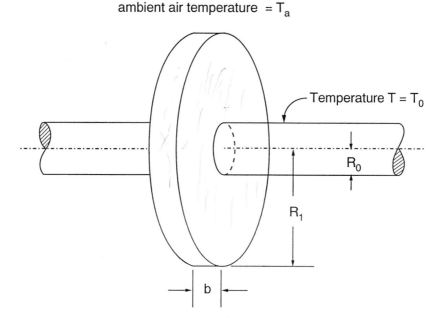

Figure 4.11 A circular fin on a heated pipe.

(a) Derive an expression for the temperature profile within the fin.

(b) Derive an expression for the total heat loss from the fin.

(c) Derive an expression for the fin efficiency, η, which is defined as the ratio of heat actually dissipated by the fin, to that which would be lost if the entire fin surface were held at T_0.

4.27 Following the remark at the end of section 4.16, consider the eigenvalue problem for a sphere:

$$L[\phi] = \frac{d}{dr}\left[r^2 \frac{d\phi}{dr}\right] = -\lambda r^2 \phi, \qquad r \in (0, a)$$
$$\phi = 0, \qquad r = a$$
$$\phi = \text{finite}$$

Derive expressions for the eigenvalues, normalized eigenfunctions, orthogonality relationships, and coefficients c_n for the expansion (4.16.17) of an arbitrary function.

Chapter 5

FUNDAMENTALS OF PARTIAL DIFFERENTIAL EQUATIONS

With this chapter we begin a systematic discussion of partial differential equations (PDEs). Equations of this type arise when several independent variables are involved in a differential equation. In typical physical problems the independent variables are one or more spatial variables and time. Examples of PDEs include (a) the heat equation, which describes evolution of spatial temperature distribution in a conducting body, and (b) the kinematic wave equation, which describes processes such as adsorption of a component from a flowing fluid.

Some fundamental aspects of PDEs are discussed in this chapter. These equations are first classified with regard to whether they are linear, quasilinear, or nonlinear. Following this, characteristic curves of first-order PDEs, which provide a parametric description of the solutions, are discussed in detail. Finally, second-order PDEs are treated and a general classification of commonly encountered two-dimensional equations is presented.

These fundamental aspects are utilized in chapters 6 and 7, which also deal with PDEs. Chapter 6 is devoted to first-order equations, while generalized Fourier transform methods for solving second-order equations are discussed in chapter 7.

5.1 ORDINARY AND PARTIAL DIFFERENTIAL EQUATIONS

In general, an m-dimensional nth-order PDE can be represented as a function of m independent variables $(x_1, x_2, \ldots, x_m)$, one dependent variable (the unknown z), and all the partial derivatives of z with respect to the independent variables up to order n:

$$F\left(z, x_1, x_2, \ldots, x_m, \frac{\partial z}{\partial x_1}, \frac{\partial z}{\partial x_2}, \ldots, \frac{\partial z}{\partial x_m}, \ldots, \frac{\partial^n z}{\partial x_1^n}, \frac{\partial^n z}{\partial x_1^{n-1} \partial x_2}, \ldots, \frac{\partial^n z}{\partial x_m^n}\right) = 0$$

$$(5.1.1)$$

where F is an arbitrary function and the order n is defined as the highest order of the derivatives present in the equation.

The solution $z(x_1, x_2, \ldots, x_m)$ is a continuous function whose derivatives up to the nth order are also continuous; when it is substituted in the lhs of eq. (5.1.1), the latter becomes identically equal to zero. Such a solution is a *general solution* and it contains n arbitrary functions, which can be determined by enforcing suitable conditions usually applied at the boundaries of the domain of integration. The resulting solution is the one normally sought in applications.

Note the strict analogy with ordinary differential equations (ODEs) of order n whose general solution contains n arbitrary constants (rather than functions), which have to be determined from suitable initial or boundary conditions.

To illustrate this point, let us consider the ODE describing an object moving along the z axis with constant acceleration, a (Figure 5.1a):

$$\frac{d^2 z}{dt^2} = a, \quad t > 0$$

The general solution is given by

$$z(t) = c_1 + c_2 t + \tfrac{1}{2} a t^2$$

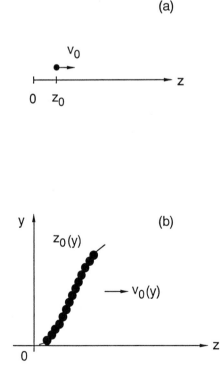

Figure 5.1 One-dimensional motion: (a) Single object. (b) String of objects.

where c_1 and c_2 are arbitrary constants whose values are determined by satisfying two conditions, such as for example the initial position (z_0) and velocity (v_0) of the object:

$$z(t) = z_0 + v_0 t + \tfrac{1}{2}at^2$$

On the other hand, let us consider the same movement along the z axis for a string of objects (Figure 5.1b), but now assuming that their acceleration is proportional to the distance y of each object from the z axis. In this case, by treating the string of objects as a continuum, the movement can be described by the following PDE:

$$\frac{\partial^2 z}{\partial t^2} = ky, \qquad t > 0$$

whose general solution is given by

$$z(t, y) = c_1(y) + c_2(y)t + \tfrac{1}{2}kyt^2$$

where $c_1(y)$ and $c_2(y)$ are arbitrary functions. Again, these can be determined by providing the initial position, $z_0(y)$, and velocity, $v_0(y)$, of the string of objects, both assumed to be continuous functions of y. The solution is then given by

$$z(t, y) = z_0(y) + v_0(y)t + \tfrac{1}{2}kyt^2$$

Note that in the case of the ODE, c_1 and c_2 are arbitrary *constants*, while in the case of the PDE, $c_1(y)$ and $c_2(y)$ are arbitrary *functions*.

The comparison between ordinary and partial differential equations can be pursued further by considering the behavior of first-order equations from a geometrical point of view. In the case of an ODE, which in general can be represented in the implicit form

$$F\left(z, x, \frac{dz}{dx}\right) = 0 \qquad (5.1.2)$$

the general solution is given by a family of lines $z(x)$ lying in the two-dimensional (z, x) space. The ODE (5.1.2) provides for any point in the (z, x) plane, the value of the derivative of the solution at that point. Thus, the ODE constructs the solution curve by prescribing at each point (z, x) the tangent to the solution curve passing through the point.

The so-called *isocline method* for solving ODEs is in fact based on the above observation. The isocline is a curve in the (z, x) plane which represents the locus of points where the slope (or derivative value) of the solution curve is equal to a constant, q. The function describing such an isocline is readily obtained by solving eq. (5.1.2) for z, with a fixed derivative value $dz/dx = q$; that is,

$$z_{\text{iso}} = z(x, q)$$

Using the above equation, various isoclines corresponding to different q values can be drawn in the (z, x) plane. This provides a network of lines which enables us to immediately evaluate the slope of the solution curve at every point in the (z, x) plane. Thus, taking an arbitrary initial point, it is possible to draw the solution curve originating

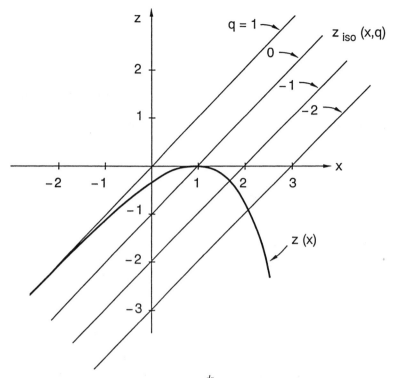

Figure 5.2 Application of the isocline method. ODE: $\dfrac{dz}{dx} = z - x + 1$ with IC $z(1) = 0$. Isocline: $z_{iso} = x - 1 + q$.

from it by moving ahead each time by a "small step" in the direction indicated by the isocline field, as shown in Figure 5.2 for an illustrative example.

The geometrical interpretation above cannot be applied to a first-order PDE, which in the case of two independent variables (x and y) can be represented as follows:

$$F\left(z, x, y, \frac{\partial z}{\partial x}, \frac{\partial z}{\partial y}\right) = 0 \qquad (5.1.3)$$

In this case the general solution is given by a family of surfaces $z(x, y)$ lying in the three-dimensional (z, x, y) space. For each point in such a space, eq. (5.1.3) provides only a relationship between the two derivatives $\partial z/\partial x$ and $\partial z/\partial y$, but not their specific values. Thus the PDE does not prescribe for each point in the three-dimensional (z, x, y) space the directions of the plane tangent to the solution surface at that point. Consequently, the idea of the isocline method cannot be applied for solving PDEs.

A closer insight into this aspect of PDE behavior can be gained by considering the particular case where eq. (5.1.3) can be written in the form

$$P(z, x, y) \frac{\partial z}{\partial x} + Q(z, x, y) \frac{\partial z}{\partial y} = R(z, x, y) \qquad (5.1.4)$$

This equation may be regarded as expressing the condition that the scalar product of the two vectors $[P, Q, R]$ and $[\partial z/\partial x, \partial z/\partial y, -1]$ is equal to zero, which geometrically means that the two vectors are orthogonal at any given point on the solution surface $z(x, y)$. Recalling that for the solution surface:

$$dz = \frac{\partial z}{\partial x}\,dx + \frac{\partial z}{\partial y}\,dy \qquad (5.1.5)$$

it follows that the vector $[\partial z/\partial x, \partial z/\partial y, -1]$ is orthogonal to the surface. Thus it can be concluded that the PDE (5.1.4), from a geometrical point of view, represents the condition that at each point in the (z, x, y) space the solution surface is tangent to the vector $[P, Q, R]$, as shown in Figure 5.3—a condition which is obviously less stringent than prescribing the tangent plane. In particular, the PDE (5.1.4) identifies a pencil of planes whose axis is given by the vector $[P, Q, R]$, among which is the one tangent to the solution surface.

5.2 INITIAL AND BOUNDARY CONDITIONS

A typical problem involving differential equations is the so-called *Cauchy problem*. It consists of finding a solution of the differential equation in a given domain which attains some specified values at the boundaries. The specific geometric form of such a domain

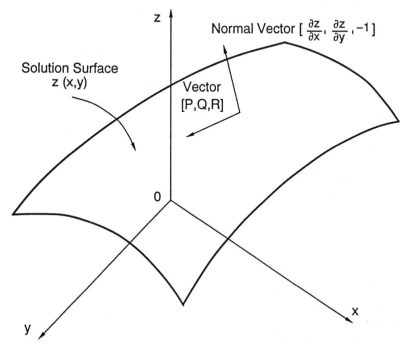

Figure 5.3 Geometrical interpretation of the PDE (5.1.4).

and of its boundaries depends upon the number of independent variables involved in the problem. In the following, this issue is addressed for a number of different cases, in order to introduce the fundamental question of properly assessing the boundary conditions of a partial differential equation.

In the case of one independent variable, such as the ODE (5.1.2), the solution is a one-dimensional curve $z(x)$ lying in the two-dimensional space (z, x), which takes on a prescribed value at some specific point (zero-dimensional)—for example, $z = z_0$ at $x = x_0$. Usually, in this case the domain of integration of the differential equation is some interval along the x axis (one-dimensional). The location where the boundary or initial conditions are provided is one or more points at the interval extrema (zero-dimensional).

In the case of two independent variables, such as x and y in the PDE (5.1.3), the geometric picture described above increases by one dimension. The solution is a two-dimensional surface $z(x, y)$ lying in the three-dimensional (z, x, y) space, whose values along some one-dimensional curve $y(x)$ are specified. Usually, in applications the domain of integration of the PDE is some finite or semi-infinite portion of the (x, y) plane and some specific values for the solution are prescribed at the boundaries of such a region.

The picture above can be generalized to the case of a PDE involving m independent variables, such as that given by eq. (5.1.1). The solution for the Cauchy problem $z(x_1, x_2, \ldots, x_m)$ is an m-dimensional surface, lying in an $(m + 1)$-dimensional $(z, x_1, x_2, \ldots, x_m)$ space, which takes on some prescribed values along an $(m - 1)$-dimensional boundary, I.

The specific form of initial and/or boundary conditions for a PDE strongly depends upon the problem at hand and the physical meaning of the equation considered. This is a fundamental aspect in developing the model of a physical phenomenon in terms of PDEs, which challenges the skill of the engineer. In order to illustrate this point, we next discuss a few models with increasing dimensions which describe the same physical picture: a homogeneous isothermal tubular reactor in which a single first-order reaction occurs. The focus is on the development of the initial and boundary conditions and on their physical meaning in the context of the model.

The first model (the *axial dispersion* model) accounts only for one independent variable (the reactor length, x) and describes the steady-state changes of the reactant concentration along the reactor axis, including the effect of axial dispersion. Then, the reactant mass balance reduces to

$$\frac{1}{\text{Pe}_a} \frac{d^2u}{dx^2} - \frac{du}{dx} = Bu, \qquad 0 < x < 1 \qquad (5.2.1)$$

B = Damköhler number, lk/v
C = reactant concentration, mol/liter
C_r = reference concentration, mol/liter
D_a = axial dispersion coefficient, m^2/sec
k = first-order reaction rate constant, l/sec
l = reactor length, m
Pe_a = axial Peclet number, lv/D_a
u = dimensionless reactant concentration, C/C_r

v = fluid velocity, m/sec

x = dimensionless reactor axial coordinate, z/l

z = reactor axial coordinate, m

Since eq. (5.2.1) is a second-order ODE, we need to provide two conditions at one or two points in the domain of integration—that is, the reactor axis $0 \le x \le 1$. This problem has a long and interesting history (cf. Varma, 1982). Langmuir first in 1908 and subsequently Danckwerts in 1953 developed such conditions by considering the mass balances at the reactor inlet and outlet, leading to

$$u = u^i + \frac{1}{\mathrm{Pe}_a} \frac{du}{dx} \qquad \text{at } x = 0 \qquad (5.2.2a)$$

$$\frac{du}{dx} = 0 \qquad \text{at } x = 1 \qquad (5.2.2b)$$

In the case of very large values of the Peclet number, $\mathrm{Pe}_a \to \infty$, eq. (5.2.1) reduces to a first-order ODE (the *plug-flow* model), which needs only one condition, taken as the initial condition (5.2.2a) at the reactor inlet, which in this case reduces to $u = u^i$.

As a second example, let us consider the unsteady-state version of the previous model. A second independent variable (time) has to be taken into account now and the mass balance of the reactant leads to the following two-dimensional PDE:

$$-\frac{\partial u}{\partial t} + \frac{1}{\mathrm{Pe}_a} \frac{\partial^2 u}{\partial x^2} - \frac{\partial u}{\partial x} = Bu, \qquad 0 < x < 1 \text{ and } t > 0 \qquad (5.2.3)$$

where t represents the dimensionless time defined as the real time $\bar{t}$ divided by l/v. In this case the conditions to the solution surface $u(x, t)$ need to be provided along some line in the (x, t) plane. On physical grounds it is convenient to select such conditions as follows:

One initial condition for the independent variable t (since eq. 5.2.3 is first-order with respect to t):

$$u = u^0(x) \qquad \text{at } 0 \le x \le 1 \text{ and } t = 0 \qquad (5.2.4a)$$

which describes the initial state of the reactor.

Two boundary conditions for the independent variable x (since eq. 5.2.3 is second order with respect to x):

$$u = u^i(t) + \frac{1}{\mathrm{Pe}_a} \frac{\partial u}{\partial x} \qquad \text{at } x = 0 \text{ and } t > 0 \qquad (5.2.4b)$$

$$\frac{\partial u}{\partial x} = 0 \qquad \text{at } x = 1 \text{ and } t > 0 \qquad (5.2.4c)$$

which describe the mass balances at the reactor inlet and outlet according to the axial dispersion model as in eqs. (5.2.2).

It is worth noting that the curve I on the (x, t) plane where the initial and boundary conditions are provided is given by a set of three straight lines with discontinuities in the derivatives at the connection points, as shown in Figure 5.4*a*.

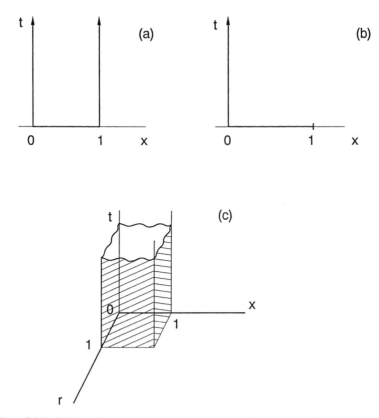

Figure 5.4 Definitions of the regions I, where the initial and boundary conditions are provided.

Again, in the case of plug flow ($Pe_a \to \infty$), the model reduces to a first-order PDE in x and t, the second boundary condition (5.2.4c) drops out, and the curve I above reduces to that shown in Figure 5.4b.

The last example involves the spatially two-dimensional form of the previous model, where the radial coordinate of the tubular reactor is introduced as the third independent variable. The transient mass balance of the reactant leads to the following three-dimensional PDE:

$$-\frac{\partial u}{\partial t} + \frac{p}{Pe_r}\left(\frac{\partial^2 u}{\partial r^2} + \frac{1}{r}\frac{\partial u}{\partial r}\right) + \frac{1}{Pe_a}\frac{\partial^2 u}{\partial x^2} - \frac{\partial u}{\partial x} = Bu,$$

(5.2.5)

$$0 < x < 1, \, 0 < r < 1, \, t > 0$$

where D_r = radial dispersion coefficient, m²/sec

$\quad\quad p$ = aspect ratio, l/R_t

$\quad Pe_r$ = radial Peclet number, $\upsilon R_t/D_r$

$\quad\quad r$ = dimensionless reactor radial coordinate, s/R_t

$\quad\, R_t$ = radius of reactor tube, m

$\quad\quad s$ = reactor radial coordinate, m

In this case the general solution $u(x, t, r)$ is a family of three-dimensional surfaces in the four-dimensional (u, x, t, r) space. To obtain the desired solution, we need to provide suitable conditions on some surface in the (x, t, r) space. Based upon physical arguments, these conditions can be specified as described below.

One initial condition which defines the initial state of the reactor

$$u = u^0(x, r) \qquad \text{at } 0 \leq x \leq 1, 0 \leq r \leq 1, \text{ and } t = 0 \qquad (5.2.6a)$$

Two boundary conditions for the independent variable x, which represent the mass balances at the reactor inlet and outlet:

$$u = u^i(t, r) + \frac{1}{\text{Pe}_a} \frac{\partial u}{\partial x} \qquad \text{at } x = 0, 0 \leq r \leq 1, \text{ and } t > 0 \qquad (5.2.6b)$$

$$\frac{\partial u}{\partial x} = 0 \qquad \text{at } x = 1, 0 \leq r \leq 1, \text{ and } t > 0 \qquad (5.2.6c)$$

Two boundary conditions for the independent variable r, which establish that no mass flux in the radial direction is possible either at the reactor centerline (symmetry) or at the reactor wall (impermeability):

$$\frac{\partial u}{\partial r} = 0 \qquad \text{at } 0 < x < 1, r = 0, \text{ and } t > 0 \qquad (5.2.6d)$$

$$\frac{\partial u}{\partial r} = 0 \qquad \text{at } 0 < x < 1, r = 1, \text{ and } t > 0 \qquad (5.2.6e)$$

The selected surface I where the initial and boundary conditions are provided is then given by the five faces of the semi-infinite right parallelopiped with square base lying on the $t = 0$ plane, as shown in Figure 5.4c.

5.3 CLASSIFICATION OF PARTIAL DIFFERENTIAL EQUATIONS

Partial differential equations exhibit the general form indicated in eq. (5.1.1). Depending upon the specific form of the function F in eq. (5.1.1), they can be classified as linear, quasilinear, or nonlinear. In particular, when the function F is a polynomial of degree one with respect to the unknown z and to each of its partial derivatives of any order, then the PDE is called *linear*. When F is a polynomial of degree one only with respect to the partial derivatives of z of the highest order, then the PDE is called *quasilinear*. In all other cases the PDE is called *nonlinear*.

To illustrate this, let us consider first-order PDEs with two independent variables, x and y. The general form is

$$F\left(z, x, y, \frac{\partial z}{\partial x}, \frac{\partial z}{\partial y}\right) = 0 \qquad (5.3.1)$$

where F is an arbitrary function. For specific forms of F, we have

Linear PDEs

$$P(x, y) \frac{\partial z}{\partial x} + Q(x, y) \frac{\partial z}{\partial y} + S(x, y)z = R(x, y) \tag{5.3.2}$$

Quasilinear PDEs

$$P(z, x, y) \frac{\partial z}{\partial x} + Q(z, x, y) \frac{\partial z}{\partial y} = R(z, x, y) \tag{5.3.3}$$

while for any other form of F, eq. (5.3.1) is nonlinear.

In the case of second-order two-dimensional PDEs having the general form

$$F\left(z, x, y, \frac{\partial z}{\partial x}, \frac{\partial z}{\partial y}, \frac{\partial^2 z}{\partial x^2}, \frac{\partial^2 z}{\partial x \partial y}, \frac{\partial^2 z}{\partial y^2}\right) = 0 \tag{5.3.4}$$

we define

Linear PDEs

$$A(x, y) \frac{\partial^2 z}{\partial x^2} + B(x, y) \frac{\partial^2 z}{\partial x \partial y} + C(x, y) \frac{\partial^2 z}{\partial y^2} + D(x, y) \frac{\partial z}{\partial x}$$
$$+ E(x, y) \frac{\partial z}{\partial y} + F(x, y)z = R(x, y) \tag{5.3.5}$$

Quasilinear PDEs

$$A\left(z, x, y, \frac{\partial z}{\partial x}, \frac{\partial z}{\partial y}\right) \frac{\partial^2 z}{\partial x^2} + B\left(z, x, y, \frac{\partial z}{\partial x}, \frac{\partial z}{\partial y}\right) \frac{\partial^2 z}{\partial x \partial y}$$
$$+ C\left(z, x, y, \frac{\partial z}{\partial x}, \frac{\partial z}{\partial y}\right) \frac{\partial^2 z}{\partial y^2} = R\left(z, x, y, \frac{\partial z}{\partial x}, \frac{\partial z}{\partial y}\right) \tag{5.3.6}$$

and any other equation of the type (5.3.4) is a nonlinear PDE.

In the theory of PDEs, first-order quasilinear equations play a particularly important role. This is because it is possible, under very general conditions, to reduce a system of nonlinear PDEs to an equivalent system of quasilinear first-order PDEs. Let us consider as an example the nonlinear first-order equation in the general form (5.3.1) together with the initial condition

$$z = z^0(y) \qquad \text{at } x = 0 \tag{5.3.7}$$

Substituting the new variables defined as

$$p = \frac{\partial z}{\partial x} \quad \text{and} \quad q = \frac{\partial z}{\partial y} \tag{5.3.8}$$

eq. (5.3.1) reduces to

$$F(z, x, y, p, q) = 0 \qquad (5.3.9)$$

which upon differentiating both sides with respect to x leads to

$$\frac{\partial F}{\partial x} + \frac{\partial F}{\partial z}\frac{\partial z}{\partial x} + \frac{\partial F}{\partial p}\frac{\partial p}{\partial x} + \frac{\partial F}{\partial q}\frac{\partial q}{\partial x} = 0 \qquad (5.3.10)$$

Thus, combining eqs. (5.3.8) and (5.3.10), the following system of three first-order PDEs in the three unknowns z, p, and q is obtained:

$$\frac{\partial F}{\partial x} + \frac{\partial F}{\partial z}p + \frac{\partial F}{\partial p}\frac{\partial p}{\partial x} + \frac{\partial F}{\partial q}\frac{\partial q}{\partial x} = 0 \qquad (5.3.11a)$$

$$\frac{\partial z}{\partial x} = p \qquad (5.3.11b)$$

$$\frac{\partial z}{\partial y} = q \qquad (5.3.11c)$$

The above equations are all quasilinear, since they are linear in the partial derivatives of z, p, and q; that is, they are of the form (5.3.3) involving three rather than only one independent variable. The three initial conditions needed to complete the system (5.3.11) are obtained directly from the original eq. (5.3.7) and by substituting it in the second of eq. (5.3.8) and in eq. (5.3.9), both considered at $x = 0$; that is,

$$z = z^0(y) \qquad \text{at } x = 0 \qquad (5.3.12a)$$

$$q = \frac{dz^0}{dy} \qquad \text{at } x = 0 \qquad (5.3.12b)$$

$$F\left(z^0, x, y, p, \frac{dz^0}{dy}\right) = 0 \qquad \text{at } x = 0 \qquad (5.3.12c)$$

where eq. (5.3.12c) implicitly provides the value of p at $x = 0$. The proof that the obtained system of quasilinear PDEs is equivalent to the original nonlinear problem (5.3.1) and (5.3.7) is straightforward. Specifically, eq. (5.3.11a) states that $dF/dx = 0$, which coupled with $F = 0$ at $x = 0$ imposed by eq. (5.3.12c) leads to the conclusion that $F = 0$ in the entire domain of interest, thus satisfying eq. (5.3.9) or, alternatively, eq. (5.3.1).

The result above can be extended to higher-order PDEs, to larger number of independent variables, and to systems of equations. In general, it can be concluded that the initial value problem for a system of nonlinear PDEs of any order and any dimension can be reduced to an equivalent initial value problem for a system of quasilinear first-order PDEs, with the only limitation that the initial data is not given along a characteristic curve (defined in the next section).

For example, let us consider the following initial value problem for the two-dimensional, nonlinear, second-order PDE:

$$\frac{\partial z}{\partial x} + \left(\frac{\partial z}{\partial y}\right)^2 + \sqrt{\frac{\partial^2 z}{\partial y^2}} = z^2 \tag{5.3.13}$$

with the ICs provided along the curve $y = y_0(x)$:

$$z = z_0(x) \qquad \text{at } y = y_0(x) \tag{5.3.14a}$$

$$\frac{\partial z}{\partial y} = z_1(x) \qquad \text{at } y = y_0(x) \tag{5.3.14b}$$

By introducing the new variables

$$p = \frac{\partial z}{\partial x}, \qquad q = \frac{\partial z}{\partial y}, \qquad r = \frac{\partial q}{\partial y} \tag{5.3.15}$$

eq. (5.3.13) reduces to

$$p + q^2 + \sqrt{r} - z^2 = 0 \tag{5.3.16}$$

Differentiating both sides with respect to x and substituting eq. (5.3.15) leads to the following system of first-order quasilinear equations:

$$\frac{\partial p}{\partial x} + 2q \frac{\partial q}{\partial x} + \frac{1}{2\sqrt{r}} \frac{\partial r}{\partial x} = 2zp \tag{5.3.17a}$$

$$\frac{\partial z}{\partial x} = p \tag{5.3.17b}$$

$$\frac{\partial z}{\partial y} = q \tag{5.3.17c}$$

$$\frac{\partial q}{\partial y} = r \tag{5.3.17d}$$

The ICs are given by

$$z = z_0(x) \qquad\qquad\qquad \text{at } y = y_0(x) \tag{5.3.18a}$$

$$p = \frac{dz_0}{dx} \qquad\qquad\qquad \text{at } y = y_0(x) \tag{5.3.18b}$$

$$q = z_1(x) \qquad\qquad\qquad \text{at } y = y_0(x) \tag{5.3.18c}$$

$$r = \left(z_0^2 - \frac{dz_0}{dx} - z_1^2\right)^2 \qquad \text{at } y = y_0(x) \tag{5.3.18d}$$

where eqs. (5.3.18a)–(5.3.18c) derive from the original ICs (5.3.14) along with the definitions (5.3.15), while eq. (5.3.18d) arises from eq. (5.3.16).

5.4   FIRST-ORDER PARTIAL DIFFERENTIAL EQUATIONS

First-order PDEs arise often when modeling physicochemical processes. In particular, the most commonly encountered operator is

$$\overline{L}[\ ] = \frac{\partial[\ ]}{\partial \overline{t}} + v\,\frac{\partial[\ ]}{\partial z} \tag{5.4.1a}$$

where $\overline{t}$ represents time and z a spatial coordinate, while v is the velocity of convective transport along z. The operator can also be put in dimensionless form by introducing the variables $t = \overline{t}v/l$ and $x = z/l$, where l is a characteristic length, to give

$$L[\ ] = \frac{\partial[\ ]}{\partial t} + \frac{\partial[\ ]}{\partial x} \tag{5.4.1b}$$

The two partial derivatives in the operator above represent the accumulation term and the convective transport term, respectively, in a conservation equation which may relate to either mass, energy, or momentum. The first-order PDE representing the conservation equation is then obtained by equating the above terms to a function representing the generation or consumption of the quantity being balanced. This usually represents an elementary model for the physicochemical phenomenon or process unit, since acceleration terms (involving second-order derivatives with respect to time) and dispersion terms (involving second-order derivatives with respect to space) are ignored. Nevertheless, models of this type provide useful insight into physical processes, and it is therefore important to understand their behavior. In the following, examples of such models are given by considering the conservation of mass, energy, and momentum (cf. Rhee et al., 1986).

As a first example, consider the homogeneous, tubular *plug-flow reactor* where a single isothermal reaction occurs. In the case of a first-order reaction, the mass balance of the reactant is given by

$$\frac{\partial u}{\partial t} + \frac{\partial u}{\partial x} = -Bu, \qquad 0 < x < 1 \text{ and } t > 0 \tag{5.4.2}$$

with ICs

$$u = u^0(x) \qquad \text{at } 0 \le x \le 1 \text{ and } t = 0 \tag{5.4.3a}$$

$$u = u^i(t) \qquad \text{at } x = 0 \text{ and } t > 0 \tag{5.4.3b}$$

where all the symbols have been introduced in the previous section in the context of eq. (5.2.3), and the 0 and i superscripts indicate the initial and inlet conditions, respectively, for the reactor.

For a nonisothermal reaction the above equation has to be coupled with the energy balance, where in addition to heat accumulation and convective transport, terms involving heat exchange with the reactor jacket (for nonadiabatic units) and heat of reaction need to be included. Such a balance gives

$$\frac{\partial \theta}{\partial t} + \frac{\partial \theta}{\partial x} = A(\theta_j - \theta) + Gu \, \exp\left[\gamma\left(1 - \frac{1}{\theta}\right)\right], \qquad 0 < x < 1 \text{ and } t > 0 \tag{5.4.4}$$

with ICs

$$\theta = \theta^0(x) \qquad \text{at } 0 \le x \le 1 \text{ and } t = 0 \qquad\qquad (5.4.5a)$$

$$\theta = \theta^i(t) \qquad \text{at } x = 0 \text{ and } t > 0 \qquad\qquad (5.4.5b)$$

where, in addition to the quantities defined above, the following have been used:

A = dimensionless parameter, $2lU/\upsilon\rho C_p R_t$

C_p = specific heat of fluid, cal/g K

E = activation energy, cal/mol

G = dimensionless parameter, $l(-\Delta H)kC_r/\upsilon\rho C_p T_r$

ΔH = enthalpy of reaction, cal/mol

R_t = radius of reactor tube, m

R = universal gas constant, cal/mol K

T = reactor temperature, K

T_j = jacket temperature, K

T_r = reference temperature, K

U = overall heat transfer coefficient at reactor wall, cal/sec K m^2

γ = dimensionless activation energy, E/RT_r

θ = dimensionless temperature, T/T_r

ρ = density of fluid, g/m^3

It should be noted that the first-order partial differential operator (5.4.1b) appears in the left-hand sides of both eqs. (5.4.2) and (5.4.4).

Another example is offered by *chromatography* separation processes. These are based on the different selectivities exhibited by the various components of a fluid mixture for adsorption on a solid phase (e.g., activated carbon, molecular sieve, ion exchange resin, etc.). The simplest model of these separation units is one which neglects transport resistances as well as dispersion phenomena and which further assumes that the two phases in contact are always at equilibrium within the unit. These models constitute the basis for the *equilibrium theory* for chromatography, which is widely used.

Two configurations are most often adopted for contacting the fluid (containing the mixture to be separated) and the adsorbent phases. In the *fixed bed*, the adsorbent phase is immobilized (and constitutes the fixed bed), while the fluid phase flows through it from the bed inlet to the outlet. In the *moving bed*, both phases are mobile and they flow along the unit in either the same (cocurrent) or opposite (countercurrent) directions.

In the case of an isothermal fixed bed with only one adsorbable component in the fluid mixture (as, for example, when removing one pollutant from an inert stream), the model of the unit reduces to the following mass balance for the adsorbable component (see Figure 5.5a)

$$\upsilon \frac{\partial C}{\partial z} + \varepsilon \frac{\partial C}{\partial t} + (1 - \varepsilon) \frac{\partial \Gamma}{\partial t} = 0, \qquad 0 < z < l \text{ and } t > 0 \qquad (5.4.6)$$

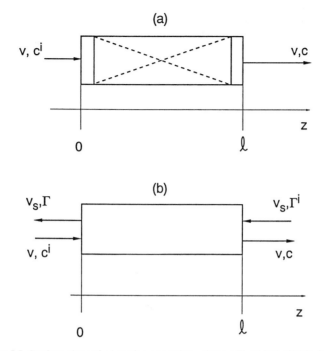

Figure 5.5 Configurations of adsorption separation units: (*a*) Fixed bed. (*b*) Moving bed.

where the first term represents convective transport, v being the superficial velocity of the fluid through the bed, while the other two terms refer to accumulation in the fluid and the adsorbent phase, respectively. In particular, C represents the concentration of the adsorbable component in the fluid phase, Γ that in the absorbent phase, and ε is the void fraction of the bed.

In the context of equilibrium theory, as noted above, local adsorption equilibrium is assumed between the two phases, which implies an algebraic relationship between the two concentrations, C and Γ, such as

$$\Gamma = F(C) \tag{5.4.7}$$

This relationship is usually referred to as the adsorption *equilibrium isotherm*, which, for example, in the particular case of the Langmuir adsorption isotherm, takes the form

$$\Gamma = \frac{\Gamma^\infty KC}{1 + KC} \tag{5.4.8}$$

where Γ^∞ represents the saturation concentration for the adsorbent, while K is the adsorption equilibrium constant.

Substituting the equilibrium relationship (5.4.7) in eq. (5.4.6), we obtain

$$V\frac{\partial C}{\partial z} + \frac{\partial C}{\partial t} = 0, \qquad 0 < z < l \text{ and } t > 0 \tag{5.4.9}$$

which again contains the partial differential operator (5.4.1a), and where the parameter V has dimensions of length/time and is defined as

$$V = v \Big/ \left[\varepsilon + (1 - \varepsilon) \frac{dF}{dC} \right] \qquad (5.4.10)$$

In the case of Langmuir isotherm, eq. (5.4.10) reduces to

$$V = v \Big/ \left[\varepsilon + (1 - \varepsilon) \frac{\Gamma^\infty K}{(1 + KC)^2} \right] \qquad (5.4.11)$$

The ICs, as in the case of the plug-flow reactor, define the initial concentration profile of the adsorbable component within the bed, $C^0(z)$, and the inlet concentration variation with time, $C^i(t)$:

$$C = C^0(x) \qquad \text{at } 0 \le z \le l \text{ and } \bar{t} = 0 \qquad (5.4.12a)$$

$$C = C^i(\bar{t}) \qquad \text{at } z = 0 \text{ and } \bar{t} > 0 \qquad (5.4.12b)$$

The corresponding values for the adsorbed phase concentration, Γ, are readily obtained by substituting $C^0(z)$ and $C^i(\bar{t})$ in the equilibrium relationship (5.4.7).

In the case of the moving-bed adsorber, since both phases move, the corresponding mass balance of the adsorbable component should include two convective transport terms. Thus, with reference to the countercurrent operation shown in Figure 5.5b, the model of the adsorber in the framework of the equilibrium theory reduces to

$$v \frac{\partial C}{\partial z} - v_s \frac{\partial \Gamma}{\partial z} + \varepsilon \frac{\partial C}{\partial \bar{t}} + (1 - \varepsilon) \frac{\partial \Gamma}{\partial \bar{t}} = 0, \qquad 0 < z < l \text{ and } \bar{t} > 0 \quad (5.4.13)$$

where v_s is the superficial velocity of the adsorbent. Substituting the equilibrium relationship (5.4.7), eq. (5.4.13) leads to the first-order PDE:

$$V \frac{\partial C}{\partial z} + \frac{\partial C}{\partial \bar{t}} = 0, \qquad 0 < z < l \text{ and } \bar{t} > 0 \qquad (5.4.14)$$

which is identical to the corresponding eq. (5.4.9) for the fixed-bed configuration, but the parameter V is now defined as

$$V = \frac{v - v_s(dF/dC)}{\varepsilon + (1 - \varepsilon)(dF/dC)} \qquad (5.4.15)$$

The difference with the fixed-bed mode arises in the ICs, which in this case are given by:

$$C = C^0(z) \qquad \text{at } 0 \le z \le l \text{ and } \bar{t} = 0 \qquad (5.4.16a)$$

$$C = C^i(\bar{t}) \qquad \text{at } z = 0 \text{ and } \bar{t} > 0 \qquad (5.4.16b)$$

$$\Gamma = \Gamma^i(\bar{t}) \qquad \text{at } z = l \text{ and } \bar{t} > 0 \qquad (5.4.16c)$$

The last example involves an application of conservation equations of mass, energy and momentum for describing the one-dimensional motion of a compressible, nonviscous fluid. Since generation terms are not present, the conservation equations reduce to an application of the partial differential operator (5.4.1a) to

Mass

$$\frac{\partial \rho}{\partial t} + \frac{\partial (\rho v)}{\partial z} = 0 \qquad\qquad (5.4.17)$$

Energy

$$\frac{\partial}{\partial t}\left(\rho U + \frac{\rho v^2}{2}\right) + \frac{\partial}{\partial z}\left[\rho v\left(U + \frac{v^2}{2} + \frac{P}{\rho}\right)\right] = 0 \qquad\qquad (5.4.18)$$

Momentum

$$\frac{\partial (\rho v)}{\partial t} + \frac{\partial}{\partial z}(P + \rho v^2) = 0 \qquad\qquad (5.4.19)$$

These equations, coupled with a suitable equation of state for the fluid—that is, $P = P(\rho, U)$—lead to the evaluation of the four variables describing the fluid motion in time and space: density (ρ), velocity (v), pressure (P), and internal energy (U). The ICs are again provided by the initial state of the fluid and by the possible presence of physical constraints in space.

5.4.1   The Characteristic Curves

The concept of a characteristic curve is of fundamental importance in the theory of PDEs. It is introduced next with reference to quasilinear, two-dimensional, first-order PDEs of the form

$$P(z, x, y)\frac{\partial z}{\partial x} + Q(z, x, y)\frac{\partial z}{\partial y} = R(z, x, y), \qquad 0 < x < 1 \text{ and } 0 < y < 1 \quad (5.4.20)$$

with IC

$$z = z^0(x, y) \text{ along the curve } I^0: \qquad y = y^0(x) \qquad\qquad (5.4.21)$$

As we have seen in section 5.2, the solution of the PDE (5.4.20) is given by a surface $z = z(x, y)$ in the three-dimensional (z, x, y) space which goes through the curve I defined by the initial condition (5.4.21), which in parametric form can be written as follows:

$$I: \qquad x = \xi, y = y^0(\xi), z = z^0(\xi) \text{ with } a < \xi < b \qquad\qquad (5.4.22)$$

and whose projection on the (x, y) plane is I^0. The solution surface can also be represented in parametric form as a function of the parameter $s: z = z(x(s), y(s))$, so that by applying the chain differentiation rule we obtain

$$\frac{dz}{ds} = \frac{\partial z}{\partial x}\frac{dx}{ds} + \frac{\partial z}{\partial y}\frac{dy}{ds} \qquad\qquad (5.4.23)$$

By comparing term by term eq. (5.4.20) with eq. (5.4.23) leads to the system of ODEs:

$$\frac{dx}{ds} = P(z, x, y) \qquad (5.4.24a)$$

$$\frac{dy}{ds} = Q(z, x, y) \qquad (5.4.24b)$$

$$\frac{dz}{ds} = R(z, x, y) \qquad (5.4.24c)$$

whose solution provides the parametric description of a family of curves in the (z, x, y) space, which are called the *characteristic curves* of the PDE (5.4.20). It is readily seen by substituting eqs. (5.4.24) into eq. (5.4.23) that each curve of the family satisfies eq. (5.4.20). Among all such curves we need to select those which satisfy also the IC (5.4.21)—that is, those going through the curve I (5.4.22). This is done by taking eq. (5.4.22) as the ICs (at $s = 0$) for integrating the system of characteristic equations (5.4.24). Thus we can see, with reference to Figure 5.6, that to each $\xi \varepsilon [a, b]$ corresponds a point $\bar{P}(\xi, y^0(\xi), z^0(\xi))$ on the initial condition curve I (5.4.22), from which originates the characteristic curve that is obtained by integrating the system (5.4.24) along the parameter s. Accordingly, by considering each point on the curve I (i.e., various values of $\xi \varepsilon [a, b]$) we can generate a one-parameter (ξ) family of characteristic curves, each

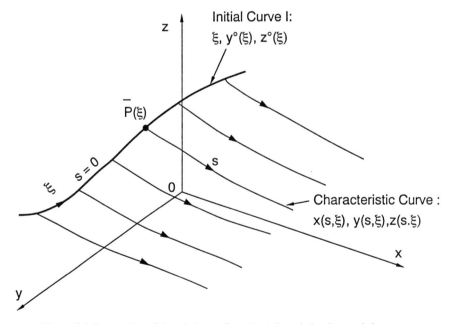

Figure 5.6 Construction of the solution surface $z(x, y)$ through the characteristic curves.

described by the parameter s, which fully describes the solution surface. This surface can be represented in the parametric form:

$$x = x(s, \xi) \tag{5.4.25a}$$

$$y = y(s, \xi) \tag{5.4.25b}$$

$$z = z(s, \xi) \tag{5.4.25c}$$

Note that since the family of characteristic curves has been obtained as the solution of an initial value problem for a system of ODEs, we are guaranteed that they do not intersect (see section 2.2). Moreover, it should be noted that the domain of integration covered by the solution is determined by the form of the IC and the characteristic curves. Thus, in the case of eqs. (5.4.20) and (5.4.21) we are not guaranteed that as ξ and s vary in $\xi \in [a, b]$ and $s \geq 0$, the values of x and y given by eqs. (5.4.25a) and (5.4.25b), respectively, cover the entire domain $x \in [0, 1]$ and $y \in [0, 1]$. For the values of x and y not covered, the solution does not exist.

In order to recover the solution in the form $z = z(x, y)$ from the above equations, we need to solve eqs. (5.4.25a) and (5.4.25b) for s and ξ as functions of x and y and then to substitute these in eq. (5.4.25c). This can be done, at least in a small neighborhood of the initial condition curve I, if and only if the Jacobian of the algebraic system of eqs. (5.4.25a) and (5.4.25b) does not vanish at any point of I; that is,

$$J = \begin{vmatrix} \dfrac{\partial x}{\partial s} & \dfrac{\partial x}{\partial \xi} \\[2mm] \dfrac{\partial y}{\partial s} & \dfrac{\partial y}{\partial \xi} \end{vmatrix}_{s=0} \neq 0 \tag{5.4.26}$$

In this case the algebraic system can be uniquely solved for s and ξ, leading to

$$s = s(x, y) \quad \text{and} \quad \xi = \xi(x, y) \tag{5.4.27}$$

which, when substituted in eq. (5.4.25c), lead to the solution in the explicit form, $z = z(x, y)$.

An interesting geometrical interpretation of condition (5.4.26) can be obtained by rewriting it in the form

$$\frac{\partial y/\partial s}{\partial x/\partial s} \neq \frac{\partial y/\partial \xi}{\partial x/\partial \xi} \quad \text{at } s = 0 \text{ for any } \xi \in [a, b] \tag{5.4.28}$$

The left-hand side of the above equation represents, for any point $\overline{P}$ on the curve I (corresponding to a given value of $\xi \in [a, b]$ and $s = 0$), the slope of the tangent to the projection onto the (x, y) plane of the characteristic curve passing through the point $\overline{P}$. The right-hand side represents the slope of the tangent to the projection onto the (x, y) plane of the initial condition curve I passing through the same point $\overline{P}$. Thus, condition (5.4.28) requires that the two projections should not be tangent at any point.

Let us consider the case where the Jacobian, J, is equal to zero along the entire curve I—that is, for all values of $\xi \in [a, b]$. Thus, the projections of curve I and of the characteristic curve are tangents at all points, which means that they are coincident. Accordingly, we can represent the projection of curve I on the (x, y) plane in parametric form as follows:

$$\frac{dx}{d\xi} = P(z, x, y) \qquad (5.4.29a)$$

$$\frac{dy}{d\xi} = Q(z, x, y) \qquad (5.4.29b)$$

By representing the curve I in the parametric form $z = z(x(\xi), y(\xi))$ and applying the chain differentiation rule together with eqs. (5.4.29), we obtain

$$\frac{dz}{d\xi} = \frac{\partial z}{\partial x}\frac{dx}{d\xi} + \frac{\partial z}{\partial y}\frac{dy}{d\xi} = P(z, x, y)\frac{\partial z}{\partial x} + Q(z, x, y)\frac{\partial z}{\partial y} \qquad (5.4.30)$$

Since the solution surface must pass through curve I, by comparing eqs. (5.4.20) and (5.4.30) it is seen that

$$\frac{dz}{d\xi} = R(z, x, y) \qquad (5.4.31)$$

Integrating eqs. (5.4.29) and (5.4.31) yields the parametric description of the initial condition curve, I. Since this system of ODEs is identical to that defined by eqs. (5.4.24), we see that in this case the curve I is itself a characteristic curve.

We can then conclude that when the Jacobian vanishes, the PDE (5.4.20) with IC (5.4.21) has a solution only in the case where the initial condition curve, I is also a characteristic curve. Otherwise the problem has no solution.

It is worth noting that when the curve I is a characteristic curve, the procedure for constructing the solution surface outlined above fails. In particular, when integrating the system (5.4.24) with any point $\overline{P}$ on the curve I taken as initial condition, we always obtain as a result the same curve I, since the curve I is itself a solution of the system. Thus the procedure above does not lead to a surface but rather only to a curve in the (z, x, y) space which coincides with the initial condition curve I. The problem in this case does not have a unique solution, since there is an infinite number of surfaces passing through curve I which are all solutions to the problem since, in addition to satisfying the initial condition, they also satisfy the characteristic system of equations (5.4.24) and hence the PDE (5.4.20).

As an illustrative example let us consider the plug-flow reactor model given by eq. (5.4.2), with IC (5.4.3). The system of characteristic equations is given by

$$\frac{dx}{ds} = 1 \qquad (5.4.32a)$$

$$\frac{dt}{ds} = 1 \qquad (5.4.32b)$$

$$\frac{du}{ds} = -Bu \qquad (5.4.32c)$$

which upon integration lead to

$$x = s + c_1 \tag{5.4.33a}$$

$$t = s + c_2 \tag{5.4.33b}$$

$$u = c_3 \exp(-Bs) \tag{5.4.33c}$$

The three integration constants c_1, c_2, and c_3 are evaluated using the ICs (5.4.3), which describe the initial condition curve I. It is readily seen that in this case, curve I has a discontinuity at the origin, unless $u^0(0) = u^i(0)$. Therefore, in solving this problem, we have to consider separately the two branches of curve I: first the one given by eq. (5.4.3a) whose projection onto the (x, t) plane is the segment $[0, 1]$ on the x axis, and then that given by eq. (5.4.3b) whose projection is the positive portion of the t axis. For each of these taken as initial condition, the integration of the characteristic system (5.4.32) leads to a family of characteristic curves describing a surface in the (u, x, t) space. The combination of the two surfaces thus obtained is the solution of the initial value problem.

The first branch of the initial condition curve (5.4.3a) can be written in parametric form as follows:

$$x = \xi, \quad t = 0, \quad u = u^0(\xi) \qquad \text{with } 0 \le \xi \le 1 \tag{5.4.34}$$

Using this condition at $s = 0$, the integration constants appearing in eq. (5.4.33) can be evaluated, leading to the one-parameter family of characteristic curves

$$x = s + \xi \tag{5.4.35a}$$

$$t = s \tag{5.4.35b}$$

$$u = u^0(\xi) \exp(-Bs) \tag{5.4.35c}$$

By solving eqs. (5.4.35a) and (5.4.35b) for x and t and substituting these in eq. (5.4.35c), the solution in explicit form is obtained:

$$u = u^0(x - t) \exp(-Bt) \tag{5.4.36}$$

In order to establish the region of validity for the solution (5.4.36), it is convenient to consider the projection of the family of characteristic curves (5.4.35) on the (x, t) plane as shown in Figure 5.7. Using eqs. (5.4.35a) and (5.4.35b), these are described by the one-parameter family

$$t = x - \xi \tag{5.4.37}$$

which, since ξ lies between 0 and 1 [see eq. (5.4.34)], represents a set of parallel straight lines with slope unity covering a portion of the domain bounded by

$$t \le x \tag{5.4.38}$$

This then represents the domain of validity of the solution (5.4.36).

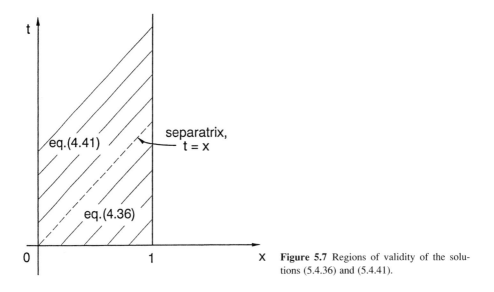

Figure 5.7 Regions of validity of the solutions (5.4.36) and (5.4.41).

Let us now turn to the second branch of the initial condition curve, given by eq. (5.4.3b), which in parametric form becomes

$$x = 0, \quad t = \xi, \quad u = u^i(\xi) \qquad \text{with } \xi > 0 \tag{5.4.39}$$

Using eq. (5.4.39) for evaluating the integration constants in eq. (5.4.33), the following one-parameter family of characteristic curves is obtained:

$$x = s \tag{5.4.40a}$$

$$t = s + \xi \tag{5.4.40b}$$

$$u = u^i(\xi) \exp(-Bs) \tag{5.4.40c}$$

from which the solution is explicitly given as

$$u = u^i(t - x) \exp(-Bx) \tag{5.4.41}$$

To determine the range of validity of this solution, we again consider the projection of the family of characteristic curves on the (x, t) plane, which is given by eqs. (5.4.40a) and (5.4.40b) as follows:

$$t = x + \xi \tag{5.4.42}$$

Recalling from eq. (5.4.39) that ξ varies from zero to infinity, eq. (5.4.42) describes a set of parallel straight lines (see Figure 5.7) with slope unity, covering the portion of the domain bounded by

$$t > x \tag{5.4.43}$$

Thus summarizing, the solution of eqs. (5.4.2) and (5.4.3) in the domain of interest (i.e., $t \geq 0$ and $0 \leq x \leq 1$) is given by

$$u = u^0(x - t) \exp(-Bt) \qquad \text{for } t \leq x \qquad \text{(5.4.44a)}$$

$$u = u^i(t - x) \exp(-Bx) \qquad \text{for } t > x \qquad \text{(5.4.44b)}$$

It is worth noting that the solution surface in the (u, x, t)-space exhibits a discontinuity along the curve whose projection onto the (x, t)-plane is given by the straight line $t = x$, which is the separatrix between the domains of validity of the two solutions given by eq. (5.4.44). This is a consequence of the discontinuity present in the initial condition curve I at the origin, which propagates on the solution surface following the characteristic curve passing through the origin, whose projection on the (x, t) plane is in fact $t = x$. This propagation is a general feature of characteristic curves. Note that when $u^0(0) = u^i(0)$, i.e. when there is no discontinuity on the IC curve, then there is also no discontinuity on the solution surface, as it can be seen by comparing eqs. (5.4.44a) and (5.4.44b) for $t = x$. In this case, discontinuities in the first or higher derivatives can exist, depending on the nature of functions $u^0(x)$ and $u^i(t)$.

In the above example, we had no difficulty in recovering the solution in explicit form from the parametric one since condition (5.4.26) holds on both branches of the initial condition curve. In particular, we have $J = -1$ for the system (5.4.35) and $J = +1$ for the system (5.4.40). This is not true in general, as illustrated by the next example where the same PDE (5.4.2) is considered, but is now coupled with a different initial condition curve I (without discontinuity points) given in parametric form by

$$x = \xi, \quad t = \xi, \quad u = \bar{u} \qquad \text{with } \xi \geq 0 \qquad \text{(5.4.45)}$$

Note that in this case the IC is provided on the line $x = t$, as opposed to the x and t axes in the previous example. This is an unusual region from a physical point of view, but it is selected in order to clarify a specific point. The one-parameter family of characteristic curves is obtained by integrating the characteristic system (5.4.32) with the ICs (5.4.45), leading to

$$x = s + \xi \qquad \text{(5.4.46a)}$$

$$t = s + \xi \qquad \text{(5.4.46b)}$$

$$u = \bar{u} \exp(-Bs) \qquad \text{(5.4.46c)}$$

In order to recover the solution in explicit form, we need to solve the system of algebraic equations (5.4.46a) and (5.4.46b), whose Jacobian according to eq. (5.4.26) is $J = 1 - 1 = 0$. It is evident that s and ξ cannot be expressed in terms of x and t, and hence the solution cannot be obtained.

From a geometrical point of view, it may be noted that in this case the projections of the characteristic curve [as given by eqs. (5.4.46a) and (5.4.46b)] and of the initial condition curve [as given by the first two of eqs. (5.4.45)] on the (x, t) plane are tangent. In fact, being straight lines, they coincide. However, the two curves in the (u, x, t) space are not coincident, as is clear by comparing eq. (5.4.46c) and the third of eq. (5.4.45). Accordingly, as mentioned above, in this case the problem has no solution.

A different situation arises when the following initial condition curve I is coupled to the PDE (5.4.2):

$$t = \xi, \quad x = \xi, \quad u = \bar{u} \exp(-B\xi) \qquad \text{with } \xi \geq 0 \qquad (5.4.47)$$

In this case the one-parameter family of characteristic curves is given by

$$x = s + \xi \qquad (5.4.48a)$$

$$t = s + \xi \qquad (5.4.48b)$$

$$u = \bar{u} \exp[-B(s + \xi)] \qquad (5.4.48c)$$

and again condition (5.4.26) is not satisfied since $J = 1 - 1 = 0$. However, by comparing eqs. (5.4.47) and (5.4.48) it follows that the initial condition curve I and the characteristic curve in the (u, x, t) space are identical.

In this case the initial value problem admits infinite solutions, which can be represented by the expression

$$u = \bar{u} f(w) \exp(-Bt) \qquad (5.4.49)$$

where $w = t - x$ and f is any arbitrary function such that $f(0) = 1$. The latter condition is required to ensure that eq. (5.4.49) satisfies the IC (5.4.47). By direct substitution of eq. (5.4.49) it can be shown that the original PDE (5.4.2) is indeed satisfied for any function $f(w)$.

5.4.2 First-Order PDEs with m Independent Variables

The method for solving first-order PDEs, based upon integration of the system of characteristic equations, described above in the case of two independent variables, is now extended to the general case of m independent variables. Let us illustrate this point with reference to the m-dimensional, first-order, quasilinear PDE:

$$\sum_{i=1}^{m} P_i(z, \mathbf{x}) \frac{\partial z}{\partial x_i} = R(z, \mathbf{x}) \qquad (5.4.50)$$

where $\mathbf{x}$ represents the vector of m independent variables. As initial condition we assume that the solution takes a prescribed set of values, z^0, along an $(m - 1)$-dimensional curve in the $(m + 1)$-dimensional $(z, \mathbf{x})$ space; that is,

$$z = z^0 \text{ for } x_m = f(x_1, x_2, \ldots, x_{m-1}) \qquad (5.4.51)$$

The solution $z = z(\mathbf{x})$ can be represented in the parametric form $z = z(\mathbf{x}(s))$, which upon differentiation with respect to s leads to

$$\frac{dz}{ds} = \sum_{i=1}^{m} \frac{\partial z}{\partial x_i} \frac{dx_i}{ds} \qquad (5.4.52)$$

By equating eqs. (5.4.50) and (5.4.52) term by term, the following system of ODEs is obtained:

$$\frac{dx_1}{ds} = P_1(z, \mathbf{x}) \tag{5.4.53a}$$

$$\frac{dx_2}{ds} = P_2(z, \mathbf{x}) \tag{5.4.53b}$$

$$\vdots$$

$$\frac{dx_m}{ds} = P_m(z, \mathbf{x}) \tag{5.4.53c}$$

$$\frac{dz}{ds} = R(z, \mathbf{x}) \tag{5.4.53d}$$

which constitutes the characteristic system associated with the PDE (5.4.50). Its general solution provides a family of characteristic curves $z = z(\mathbf{x})$ in the $(m + 1)$-dimensional $(z, \mathbf{x})$ space, which satisfy the PDE (5.4.50). The particular solution is then constructed by selecting those curves which originate from the initial condition curve I (5.4.51), which in parametric form is written as

$$x_1 = \xi_1, \quad x_2 = \xi_2, \ldots, x_{m-1} = \xi_{m-1}, \quad x_m = f(\xi_1, \xi_2, \ldots, \xi_{m-1}),$$
$$z = z^0(\xi_1, \xi_2, \ldots, \xi_{m-1}) \tag{5.4.54}$$

These curves, which constitute an $(m - 1)$-parameter family, are obtained by integrating the characteristic system (5.4.53) with ICs given by eqs. (5.4.54), leading to the solution in parametric form:

$$x_1 = x_1(s, \xi_1, \xi_2, \ldots, \xi_{m-1}) \tag{5.4.55a}$$

$$x_2 = x_2(s, \xi_1, \xi_2, \ldots, \xi_{m-1}) \tag{5.4.55b}$$

$$\vdots$$

$$x_m = x_m(s, \xi_1, \xi_2, \ldots, \xi_{m-1}) \tag{5.4.55c}$$

$$z = z(s, \xi_1, \xi_2, \ldots, \xi_{m-1}) \tag{5.4.55d}$$

The first m equations above represent an algebraic system which can be solved to yield the parameters $s, \xi_1, \ldots, \xi_{m-1}$ as functions of the m independent variables $x_1, x_2, \ldots, x_m$, which, when substituted in eq. (5.4.55d), produce the solution in the explicit form $z = z(\mathbf{x})$.

5.5   SECOND-ORDER PARTIAL DIFFERENTIAL EQUATIONS

Second-order derivatives arise in partial differential equations when modeling physical phenomena where acceleration (second-order derivative with respect to time) or diffusion of mass, energy, or momentum (second-order derivative with respect to space) are important.

5.5.1 Examples of Second-Order PDEs

In the previous section we have seen that the first-order operator (5.4.1) arises, for example, in transient processes where the sole mechanism for transport is convection. A typical second-order PDE describes processes where the mechanism for transport includes diffusion. This is the case, for example, of heat transport through the solid slab shown in Figure 5.8, where the energy balance equation reduces to

$$k_t \frac{\partial^2 T}{\partial x^2} - \rho C_p \frac{\partial T}{\partial t} = -f(x, t), \qquad 0 < x < L \text{ and } t > 0 \tag{5.5.1}$$

where C_p = specific heat, cal/g K
$\quad f(x, t)$ = rate of local heat generation, cal/sec m^3
$\quad k_t$ = thermal conductivity, cal/sec m K
$\quad L$ = slab thickness, m
$\quad T$ = temperature, K
$\quad t$ = time, sec
$\quad x$ = axial coordinate, m
$\quad \rho$ = density, g/m^3

The solution of the PDE (5.5.1) requires supplementary information. This includes:

one IC, representing the initial temperature profile within the slab

$$T = T^0(x) \qquad \text{at } 0 \le x \le L \text{ and } t = 0 \tag{5.5.2a}$$

two BCs, representing either the temperature values at the two sides of the slab

$$T = T_1(t) \qquad \text{at } x = 0 \text{ and } t > 0 \tag{5.5.2b}$$

$$T = T_2(t) \qquad \text{at } x = L \text{ and } t > 0 \tag{5.5.2c}$$

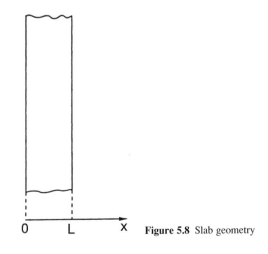

0   L   X   **Figure 5.8** Slab geometry

or the heat fluxes

$$\frac{\partial T}{\partial x} = \phi_1(t) \qquad \text{at } x = 0 \text{ and } t > 0 \tag{5.5.2d}$$

$$\frac{\partial T}{\partial x} = \phi_2(t) \qquad \text{at } x = L \text{ and } t > 0 \tag{5.5.2e}$$

or linear combinations of the temperature and heat flux values. As discussed in section 3.9, these boundary conditions arise commonly in modeling and are referred to as first-, second-, and third-kind conditions, respectively.

A second example is provided by the analogous problem of mass transport, where a chemical species diffuses in a porous solid (say a catalyst pellet) and undergoes a chemical reaction. The mass balance, again for a slab geometry, leads to

$$D\frac{\partial^2 C}{\partial x^2} - \frac{\partial C}{\partial t} = r(C), \qquad 0 < x < L \text{ and } t > 0 \tag{5.5.3}$$

where C is the molar concentration of the reactant, D is its diffusion coefficient in the porous medium, and r is the reaction rate. The ICs and BCs are of the same form as noted above in eq. (5.5.2), where temperature and heat flux are now replaced by concentration and mass flux, respectively.

In general, PDEs such as (5.5.1) and (5.5.3) are referred to as the *heat equation* (or sometimes the *diffusion equation*, independently of whether energy or mass is actually balanced), and constitute a widely studied family of equations. Another class is the *Laplace equation*, which typically arises when modeling steady-state transport processes in two- or three-dimensional bodies. An example is the steady-state version of the previous mass transport problem in the case of a two-dimensional porous catalyst:

$$D\left(\frac{\partial^2 C}{\partial x^2} + \frac{\partial^2 C}{\partial y^2}\right) = r(C) \tag{5.5.4}$$

where x and y are two independent space variables, and the pellet boundary is given by the curve I: $y = f(x)$, as shown in Figure 5.9. Similarly to the previous case of the heat equation, the BCs associated with the PDE (5.5.4) are usually given in the form of specified concentration values on curve I:

$$C = C^0(x) \qquad \text{for } y = f(x) \text{ and } a < x < b \tag{5.5.a}$$

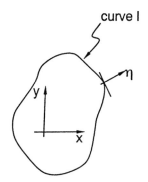

Figure 5.9 Two-dimensional catalyst particle.

or specified mass fluxes:

$$\frac{\partial C}{\partial n} = \phi(x) \qquad \text{for } y = f(x) \text{ and } a < x < b \tag{5.5.5b}$$

where n is the normal to curve I, or a linear combination of concentration and mass flux values.

The last class of second-order PDEs worth noting is the *wave equation*, which is characterized by a second-order derivative with respect to time. These equations arise in the application of the second law of mechanics; that is, acceleration times mass equals the applied force. An example is the classical problem of a one-dimensional vibrating string, referred to as the *D'Alembert problem*.

Consider a homogeneous elastic string of length L, which vibrates as a result of an external force or initial perturbation. When the movement of the generic point $P(x)$ on the string is small, then the displacement $z(x, t)$ normal to the x axis shown in Figure 5.10 is described by the PDE:

$$\frac{\partial^2 z}{\partial t^2} + p\,\frac{\partial z}{\partial t} = c^2\,\frac{\partial^2 z}{\partial x^2} + f(x, t), \qquad 0 < x < L \text{ and } t > 0 \tag{5.5.6}$$

where $p \geq 0$ and $c \geq 0$ are constants. The lhs terms z_{tt} and pz_t represent acceleration and viscous resistance from the surrounding medium, respectively. The rhs terms $c^2 z_{xx}$ and f arise from an internal elastic force and an external force, respectively.

The conditions coupled with eq. (5.5.6) are usually in the form of ICs, describing the initial state of the string in terms of its displacement and velocity:

$$z = \phi(x) \qquad \text{at } 0 \leq x \leq L \text{ and } t = 0 \tag{5.5.7a}$$

$$\frac{\partial z}{\partial t} = \psi(x) \qquad \text{at } 0 \leq x \leq L \text{ and } t = 0 \tag{5.5.7b}$$

and BCs, specifying the displacement of the two string extrema as a function of time:

$$z = z_1(t) \qquad \text{at } x = 0 \text{ and } t > 0 \tag{5.5.8a}$$

$$z = z_2(t) \qquad \text{at } x = L \text{ and } t > 0 \tag{5.5.8b}$$

The examples discussed above represent three classes of equations based upon their mathematical structure related to their physical meaning (cf. Sneddon, 1957). A

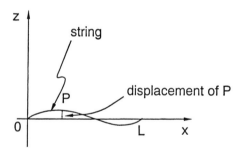

Figure 5.10 Coordinate system for the vibrating string problem.

rigorous classification of second-order PDEs is possible and is presented in section 5.5.4. However, it is first necessary to introduce some concepts about the characteristic behavior of second-order PDEs.

5.5.2 The One-Dimensional Wave Equation

Let us now consider, in some detail, the vibrating string problem given by eq. (5.5.6) in the case of negligible resistance of the surrounding medium and in the absence of external forces:

$$\frac{\partial^2 z}{\partial t^2} - c^2 \frac{\partial^2 z}{\partial x^2} = 0, \qquad 0 < x < L \text{ and } t > 0 \tag{5.5.9}$$

with ICs (5.5.7) and BCs (5.5.8). This is commonly referred to as the *one-dimensional wave* equation and also applies to other physical processes such as one-dimensional motion of an elastic solid and propagation of sound waves in a gas.

In order to obtain the general solution of eq. (5.5.9), it is helpful to introduce the new variables:

$$\xi = x + ct \tag{5.5.10a}$$

$$\eta = x - ct \tag{5.5.10b}$$

Using the chain rule for differentiation we have

$$\frac{\partial^2 z}{\partial t^2} = \frac{\partial}{\partial t}\left[\frac{\partial z}{\partial \xi}\frac{\partial \xi}{\partial t} + \frac{\partial z}{\partial \eta}\frac{\partial \eta}{\partial t} \right] = \frac{\partial^2 z}{\partial \xi^2}\left(\frac{\partial \xi}{\partial t} \right)^2 + 2\frac{\partial^2 z}{\partial \eta \partial \xi}\frac{\partial \eta}{\partial t}\frac{\partial \xi}{\partial t} + \frac{\partial^2 z}{\partial \eta^2}\left(\frac{\partial \eta}{\partial t} \right)^2$$
$$= c^2 \frac{\partial^2 z}{\partial \xi^2} - 2c^2 \frac{\partial^2 z}{\partial \eta \partial \xi} + c^2 \frac{\partial^2 z}{\partial \eta^2} \tag{5.5.11}$$

$$\frac{\partial^2 z}{\partial x^2} = \frac{\partial^2 z}{\partial \xi^2} + 2\frac{\partial^2 z}{\partial \eta \partial \xi} + \frac{\partial^2 z}{\partial \eta^2} \tag{5.5.12}$$

which substituted in eq. (5.5.9) lead to

$$\frac{\partial^2 z}{\partial \eta \partial \xi} = 0 \tag{5.5.13}$$

Integrating twice, once with respect to ξ and then to η, the following solution is obtained:

$$z = F(\xi) + G(\eta) = F(x + ct) + G(x - ct) \tag{5.5.14}$$

where F and G are arbitrary functions, each of one variable alone.

From eq. (5.5.14) it is seen that the general solution is given by the superposition of two waves: $F(x + ct)$ and $G(x - ct)$, traveling along the x axis without changing shape with the *same* constant velocity c but in *opposite* directions. In particular, since F and G are both functions of one variable only, it follows that the same value of F at time $t = 0$ and location $x = x_0$ is exhibited again after time t at the location

$$x = x_0 - ct \tag{5.5.15}$$

since $F(x_0) = F(x + ct)$ if $x_0 = x + ct$. Similarly, since $G(x_0) = G(x - ct)$ if $x_0 = x - ct$, the same value of G at $t = 0$ and $x = x_0$ is exhibited again after time t at the location

$$x = x_0 + ct \qquad (5.5.16)$$

As an illustration, examples of two waves traveling without changing shape in opposite directions are shown in Figures 5.11a and 5.11b. Their sum, shown in Figure 5.11c, is also a traveling wave, but one whose shape does change with time.

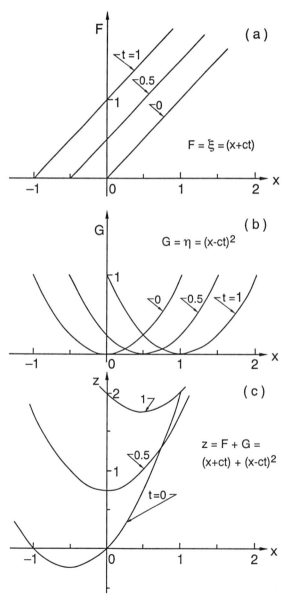

Figure 5.11 Geometrical representation of the solution (5.5.14) of the wave equation. $c = 1, L = 1$.

The ICs (5.5.7) impose some constraints on the arbitrary functions F and G. Substituting eq. (5.5.14) in eqs. (5.5.7a) and (5.5.7b) for $t = 0$, we obtain

$$F(x) + G(x) = \phi(x) \qquad 0 \leq x \leq L \text{ and } t = 0 \qquad (5.5.17a)$$

$$cF'(x) - cG'(x) = \psi(x) \qquad 0 \leq x \leq L \text{ and } t = 0 \qquad (5.5.17b)$$

Integrating eq. (5.5.17b) leads to

$$F(x) - G(x) = K + \frac{1}{c} \int_0^x \psi(\tau) \, d\tau \qquad (5.5.18)$$

where K is an arbitrary constant. Combining eqs. (5.5.17a) and (5.5.18) and solving for $F(x)$ and $G(x)$ gives

$$F(\xi) = \frac{\phi(\xi)}{2} + \frac{K}{2} + \frac{1}{2c} \int_0^\xi \psi(\tau) \, d\tau, \qquad 0 \leq \xi \leq L \qquad (5.5.19a)$$

$$G(\eta) = \frac{\phi(\eta)}{2} - \frac{K}{2} - \frac{1}{2c} \int_0^\eta \psi(\tau) \, d\tau, \qquad 0 \leq \eta \leq L \qquad (5.5.19b)$$

where the argument x has been replaced by the variables ξ and η in F and G, respectively. This is possible because eqs. (5.5.19) have been derived using the IC at $t = 0$, where $\xi = x$ and $\eta = x$, with $0 \leq x \leq L$. Since F and G are functions of one variable alone, eq. (5.5.19) holds as long as their arguments ξ or η remain in the interval $[0, L]$.

The solution of the initial value problem is then given by eqs. (5.5.14) and (5.5.19) as follows:

$$z(x, t) = \frac{1}{2} \left[\phi(x + ct) + \phi(x - ct) + \frac{1}{2c} \int_{x-ct}^{x+ct} \psi(\tau) \, d\tau \right] \qquad (5.5.20)$$

which is valid only in the region of validity of both eqs. (5.5.19a) and (5.5.19b); that is,

$$0 \leq (x + ct) \leq L \quad \text{and} \quad 0 \leq (x - ct) \leq L \qquad (5.5.21)$$

When plotted in the (x, t) plane, these inequalities identify the crosshatched quadrilateral region shown in Figure 5.12a. Since we are interested only in the domain $0 \leq x \leq L$ and $t \geq 0$, the solution given by eq. (5.5.20) is valid *only* in the triangular region defined by

$$t \geq 0, \qquad t \leq \frac{x}{c} \quad \text{and} \quad t \leq \frac{L - x}{c} \qquad (5.5.22)$$

From eq. (5.5.20) it is seen that the value of $z(\bar{x}, \bar{t})$ at any point $Q(\bar{x}, \bar{t})$ within the region (5.5.22) depends upon the initial displacement $\phi(x)$ at two specific locations, $x_1 = \bar{x} - c\bar{t}$ and $x_2 = \bar{x} + c\bar{t}$, as well as upon the initial velocity $\psi(x)$ in the entire interval $[x_1, x_2]$. The points x_1 and x_2, and the interval enclosed therein on the x axis, can be identified geometrically by drawing two straight lines with slope $1/c$ and $-1/c$ through the point $Q(\bar{x}, \bar{t})$, as shown in Figure 5.12b. On the line with slope $1/c$, $x - ct$ is constant, hence the contribution of the term $\phi(x - ct)$ in eq. (5.5.20) is also constant. Similarly, $\phi(x + ct)$ remains constant on the line with slope $-1/c$. Thus we can say that the initial displacement ϕ propagates into the (x, t) plane with two constant velocity values: c and

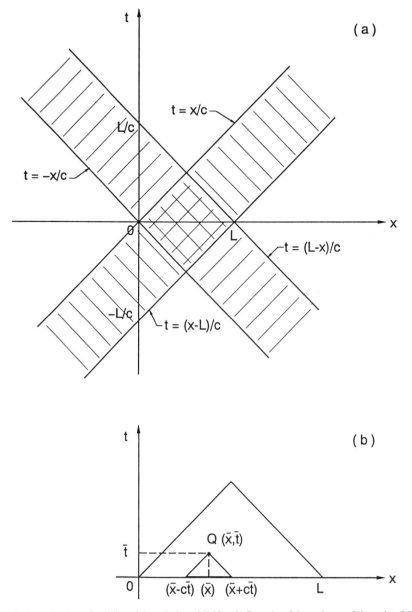

Figure 5.12 (*a*) Region of validity of the solution (5.5.20). (*b*) Domain of dependence of the point $Q(\bar{x}, \bar{t})$.

$-c$. On the other hand, because of the integral in eq. (5.5.20), the solution $z(\bar{x}, \bar{t})$ at point $Q(\bar{x}, \bar{t})$ is affected by *all* the initial velocity values $\psi(x)$ in the interval $[x_1, x_2]$, which then propagate in the (x, t) plane at all speeds bounded by $-c$ and c.

It is evident from above that the straight lines of slope $1/c$ and $-1/c$ play a fundamental role in the analysis of PDEs of this type. They are called *characteristic curves* of the wave equation (5.5.9). Furthermore, the interval $[(\bar{x} - c\bar{t}), (\bar{x} + c\bar{t})]$, obtained by intersecting the x axis with the two characteristics passing through a given point $Q(\bar{x}, \bar{t})$, is called the *domain of dependence* of Q.

As discussed above, the obtained solution (5.5.20) is valid only in the triangular region defined by eq. (5.5.22). It is remarkable that in this region the solution is not affected by the BCs (5.5.8). Since we seek the solution in the entire domain $0 \leq x \leq L$ and $t \geq 0$, we need to analyze the problem further. In particular, in order to evaluate the solution $z(x, t)$ in this domain using eq. (5.5.14), we need to determine the functions $F(\xi)$ and $G(\eta)$ in the ranges indicated:

$$F(\xi) \quad \text{with } \xi = x + ct, \quad \text{for } 0 \leq \xi < \infty \quad\quad (5.5.23a)$$

$$G(\eta) \quad \text{with } \eta = x - ct, \quad \text{for } -\infty < \eta \leq L \quad\quad (5.5.23b)$$

Since eq. (5.5.19a) provides the values of $F(\xi)$ in the interval $0 \leq \xi \leq L$ and eq. (5.5.19b) provides those of $G(\eta)$ in the interval $0 \leq \eta \leq L$, we now need to extend these intervals in order to fully cover the ranges indicated by eqs. (5.5.23a) and (5.5.23b), respectively. This can be done by using the BCs (5.5.8), where for simplicity it is assumed that the two extrema are fixed—that is, $z_1(t) = z_2(t) = 0$. Thus, substituting eq. (5.5.14) in eqs. (5.5.8a) and (5.5.8b) we obtain

$$z(0, t) = F(ct) + G(-ct) = 0, \quad\quad t \geq 0 \quad\quad (5.5.24a)$$

$$z(L, t) = F(L + ct) + G(L - ct) = 0, \quad\quad t \geq 0 \quad\quad (5.5.24b)$$

which, substituting $\eta = -ct$ in the first equation and $\xi = L + ct$ in the second one, lead to

$$G(\eta) = -F(-\eta), \quad\quad -\infty < \eta \leq 0 \qu\quad (5.5.25a)$$

$$F(\xi) = -G(2L - \xi), \quad\quad L \leq \xi < +\infty \quad\quad (5.5.25b)$$

The ranges for η and ξ noted in eqs. (5.5.25) are established as follows. Since $\eta = -ct$ and $t \geq 0$ in eq. (5.5.24a), we have $-\infty < \eta \leq 0$. Similarly, $\xi = L + ct$ and $t \geq 0$ in eq. (5.5.24b) yields $L \leq \xi < +\infty$. Together with the ranges in eqs. (5.5.19), these cover the entire domain of interest indicated in eqs. (5.5.23).

These equations can be used as recursive formulae, starting from eq. (5.5.19a) valid in $0 \leq \xi \leq L$ and eq. (5.5.19b) valid in $0 \leq \eta \leq L$. In particular, by substituting the value of $F(\eta)$ in $0 \leq \eta \leq L$ given by eq. (5.5.19a) in eq. (5.5.25a), the value of $G(\eta)$ in the interval $-L \leq \eta \leq 0$ is obtained:

$$G(\eta) = -\frac{\phi(-\eta)}{2} - \frac{K}{2} - \frac{1}{2c} \int_0^{-\eta} \psi(\tau) \, d\tau, \quad\quad -L \leq \eta \leq 0 \quad\quad (5.5.26a)$$

Similarly, the value of $F(\xi)$ in the interval $L \leq \xi \leq 2L$ is obtained from eq. (5.5.25b) by substituting the value of $G(\eta)$ with $0 \leq \eta \leq L$ given by eq. (5.5.19b):

$$F(\xi) = -\frac{\phi(2L - \xi)}{2} + \frac{K}{2} + \frac{1}{2c} \int_0^{2L-\xi} \psi(\tau) \, d\tau, \quad\quad L \leq \xi \leq 2L \quad\quad (5.5.26b)$$

This procedure is then iterated using the F and G values obtained above to evaluate $G(\eta)$ over $-2L \leq \eta \leq -L$ [substituting $F(\xi)$ with $L \leq \xi \leq 2L$ in eq. (5.5.25a)] and

$F(\xi)$ over $2L \leq \xi \leq 3L$ [substituting $G(\eta)$ with $-L \leq \eta \leq 0$ in eq. (5.5.25b)], and so on. The values of F and G thus obtained can be combined according to eq. (5.5.14) to give the solution $z(x, t)$ in the entire domain of interest. It is worth noting in this connection that the constant K in the expressions (5.5.19) and (5.5.26) of F and G does not need to be evaluated since it does not affect the value of $z(x, t)$. This is because the coefficients of K in F and G are identical but opposite in sign. Then, when summing up F and G to get $z(x, t)$, K drops out.

A concept which may be viewed as the dual of the domain of dependence discussed above is the *domain of influence* of a point $(\bar{x}, \bar{t})$. This is defined as the locus of points in the (x, t) plane where the solution is affected by a variation of z at $(\bar{x}, \bar{t})$. Let us take, for example, the point $(\bar{x}, 0)$ on the x axis and consider a nonzero value for the first IC—that is, $\phi(\bar{x}) \neq 0$ in eq. (5.5.7a). The domain of influence of $(\bar{x}, 0)$ includes all those points in the (x, t) plane where the solution is affected by the value of $\phi(\bar{x})$. Since the argument of ϕ in the solution (5.5.20) is either $x + ct$ or $x - ct$, we can conclude that the domain of influence of $(\bar{x}, 0)$ includes all points satisfying the two equations

$$x + ct = \bar{x} \tag{5.5.27a}$$

$$x - ct = \bar{x} \tag{5.5.27b}$$

These correspond to the left (α) and right (β) characteristics originating from the point $(\bar{x}, 0)$, as shown in Figure 5.13. For larger time values the solution is obtained from the recursive formulae (5.5.25), which after the first iteration lead to the G and F expressions given by eqs. (5.5.26a) and (5.5.26b), respectively. Considering again the arguments of ϕ in these expressions, it may be seen that the value $\phi(\bar{x})$ propagates along the straight lines γ and δ:

$$-\eta = \bar{x}, \quad \text{i.e., } t = \frac{x + \bar{x}}{c} \tag{5.5.28a}$$

$$2L - \xi = \bar{x}, \quad \text{i.e., } t = \frac{2L - x - \bar{x}}{c} \tag{5.5.28b}$$

By further iterating this procedure, we get other portions of the domain of influence of the point $(\bar{x}, 0)$, which are always in the form of straight lines. The final picture, as shown in Figure 5.13, consists of two sets of straight lines. Each set is obtained by following one of the two characteristics originating from the point $(\bar{x}, 0)$ up to one of the two boundaries of the domain—that is, $x = 0$ or $x = L$. When this is reached the characteristic is *reflected*; that is, it continues with the same slope but with opposite sign, so that the left characteristic becomes a right characteristic and vice versa. All these lines belong to the domain of influence of the point $(\bar{x}, 0)$.

It is worth noting, with reference to Figure 5.13, that while the contribution of $\phi(\bar{x})$ along the characteristics α and β is positive [see eq. (5.5.19)], it becomes negative along the characteristics γ and δ, as seen from eqs. (5.5.26). This is a general behavior; after each reflection, the sign of the contribution to the general solution changes.

In order to complete the domain of influence of the point $(\bar{x}, 0)$, we need to consider those points which are affected by changes in the second IC at $(\bar{x}, 0)$, i.e.

$\psi(\bar{x}) \neq 0$ in eq. (5.5.7b). Let us first consider the triangular region in the (x, t) plane defined by eq. (5.5.22), where the solution is given by eq. (5.5.20). From the integral in the rhs of the latter equation it is seen that the value of $\psi(\bar{x})$ affects the solution at all those points $S(x_1, t_1)$ which satisfy

$$(x_1 - ct_1) < \bar{x} < (x_1 + ct_1) \tag{5.5.29}$$

By rearranging the inequality above in the form

$$(\bar{x} - ct_1) < x_1 < (\bar{x} + ct_1) \tag{5.5.30}$$

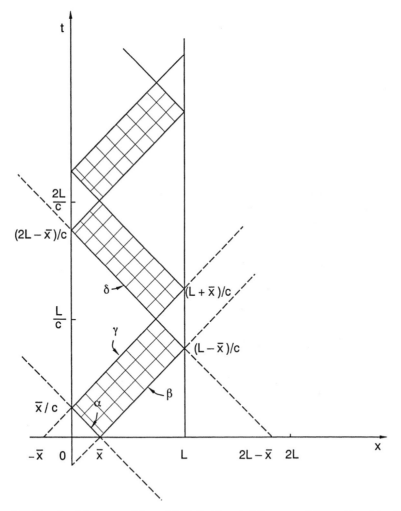

Figure 5.13 Domain of influence of the point $(\bar{x}, 0)$. Characteristic curves: $(\alpha)\ t = (\bar{x} - x)/c;\ (\beta)\ t = (x - \bar{x})/c;\ (\gamma)\ t = (\bar{x} + x)/c;\ (\delta)\ t = (2L - x - \bar{x})/c.$

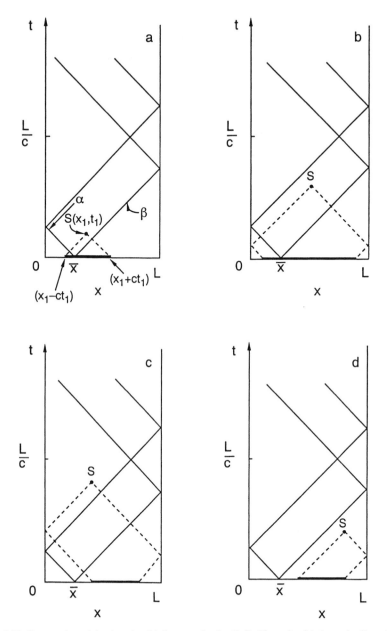

Figure 5.14 Construction of the domain of influence of point $(\bar{x}, 0)$. The point S belongs (a, b) or does not belong (c, d) to the domain.

it is readily seen that the points $S(x_1, t_1)$ are included between the two characteristics α and β, as shown in Figure 5.14a. In the region outside the triangle of validity of eq. (5.5.20) a similar conclusion applies, even though the proof becomes cumbersome because of the reflections of the characteristics against the boundaries of the domain. Thus, all the points inside the parallelograms formed by the characteristic lines are influenced by a change in the initial velocity at $\bar{x}$—that is, $\psi(\bar{x})$. A few examples to illustrate this statement, for which no rigorous proof is reported here (for details see Weinberger, 1965, page 20), are shown in Figure 5.14. It is seen that if S lies inside the parallelograms (Figure 5.14b), then $\bar{x}$ belongs to its domain of dependence, while if S lies outside the parallelograms (Figures 5.14c and 5.14d), then $\bar{x}$ does not belong to its domain of dependence.

Thus summarizing, the domain of influence of the point $(\bar{x}, 0)$ consists of all the points lying either inside or on the boundaries of the parallelograms formed by the characteristic curves, as shown by the crosshatched regions in Figure 5.13.

Another important aspect concerning the one-dimensional wave equation is related to the occurrence of discontinuities in the first-order derivative of the solution. We have seen that using the characteristic curve as coordinates, this equation can be reduced to the form (5.5.13), which does not involve second-order derivatives with respect to a single independent variable. Thus, discontinuities in the derivatives, of first or higher order, can be present and their propagation is determined by the characteristic curves.

Let us consider, for example, the directional derivative of $z(x, t)$ along the left characteristic (i.e., $x + ct = $ constant):

$$\left(\frac{dz}{dx}\right)_{x+ct=\text{const}} = \frac{\partial z}{\partial x} - \frac{1}{c}\frac{\partial z}{\partial t} \tag{5.5.31}$$

and assume that it exhibits a discontinuity at a point $(\bar{x}, \bar{t})$. By limiting our analysis to the triangular region (5.5.22) of the (x, t) plane, where the solution is given by eq. (5.5.20), it is found that

$$\frac{\partial z}{\partial x} = \frac{1}{2}\left[\phi'(x + ct) + \phi'(x - ct)\right] + \frac{1}{2c}\left[\psi(x + ct) - \psi(x - ct)\right] \tag{5.5.32a}$$

$$\frac{\partial z}{\partial t} = \frac{1}{2}\left[c\phi'(x + ct) - c\phi'(x - ct)\right] + \frac{1}{2c}\left[c\psi(x + ct) + c\psi(x - ct)\right] \tag{5.5.32b}$$

Substituting eqs. (5.5.32), the directional derivative (5.5.31) reduces to

$$\left(\frac{dz}{dx}\right)_{x+ct=\text{const}} = \phi'(\eta) - \frac{1}{c}\psi(\eta) \tag{5.5.33}$$

where $\eta = x - ct$. Thus if the lhs of eq. (5.5.33) is discontinuous at $(\bar{x}, \bar{t})$, then also the rhs is discontinuous at $\bar{\eta} = \bar{x} - c\bar{t}$. Since this equation is valid for all x and t values in the domain of interest, it follows that the directional derivative exhibits the same discontinuity for all pairs (x, t) which satisfy the equation $x - ct = \bar{\eta}$. This means that the discontinuity propagates in the (x, t) plane along straight line $x - ct = $ constant, which is the right characteristic.

In general, a discontinuity in the derivative in the direction of the left characteristic propagates along the right characteristic. Similarly, a discontinuity in the derivative in the direction of the right characteristic propagates along the left one.

Note that this conclusion holds in the entire domain of interest and not only in the triangular region (5.5.22) considered in the proof above. This can be shown using the extended expressions for F and G given by the recursive formulae (5.5.25), in the solution (5.5.14). Moreover, it should be noted that the discontinuities are reflected by the boundaries of the domain in the same manner as the corresponding characteristic curves.

Following similar arguments it is possible to investigate the propagation of a discontinuity in a derivative in any other direction. For example, we can consider a discontinuity at $(\bar{x}, \bar{t})$ of the partial derivative $\partial z / \partial x$ given by eq. (5.5.32a), which can be rearranged to yield

$$\frac{\partial z}{\partial x} = \frac{1}{2}\left[\phi'(\xi) + \frac{1}{c}\psi(\xi)\right] + \frac{1}{2}\left[\phi'(\eta) + \frac{1}{c}\psi(\eta)\right] \qquad (5.5.34)$$

where ξ and η are defined by eqs. (5.5.10). This implies that at least one of the two terms in brackets should exhibit a discontinuity—that is, either

$$\left[\phi'(\bar{\xi}) + \frac{1}{c}\psi(\bar{\xi})\right] \qquad \text{at } \bar{\xi} = \bar{x} + c\bar{t} \qquad (5.5.35a)$$

or

$$\left[\phi'(\bar{\eta}) + \frac{1}{c}\psi(\bar{\eta})\right] \qquad \text{at } \bar{\eta} = \bar{x} - c\bar{t} \qquad (5.5.35b)$$

In the first case the discontinuity propagates along the left characteristic (i.e., $x + ct = $ constant), while in the second case it propagates along the right characteristic (i.e., $x - ct = $ constant). If both terms simultaneously undergo a discontinuity at the same point $(\bar{x}, \bar{t})$, then this propagates along *both* the characteristics. Some examples of this behavior are discussed in the next section.

5.5.3 Application to the Vibrating String Problem

In the previous section we derived the analytical solution of the one-dimensional wave equation

$$\frac{\partial^2 z}{\partial t^2} - c^2 \frac{\partial^2 z}{\partial x^2} = 0, \qquad 0 < x < L \text{ and } t > 0 \qquad (5.5.36)$$

with ICs:

$$z = \phi(x) \qquad \text{at } 0 \le x \le L \text{ and } t = 0 \qquad (5.5.37a)$$

$$\frac{\partial z}{\partial t} = \psi(x) \qquad \text{at } 0 \le x \le L \text{ and } t = 0 \qquad (5.5.37b)$$

and BCs:

$$z = 0 \qquad \text{at } x = 0 \text{ and } t > 0 \qquad\qquad (5.5.38\text{a})$$

$$z = 0 \qquad \text{at } x = L \text{ and } t > 0 \qquad\qquad (5.5.38\text{b})$$

As mentioned above, this corresponds to a string which is free to vibrate in the absence of external forces and viscous resistance from the surrounding medium, while maintaining the ends at $x = 0$ and $x = L$ fixed. In order to better illustrate the properties of the solution, we consider the particular form of IC (5.5.37) where

$$\phi(x) = \begin{cases} 2x & \text{for } 0 < x < \dfrac{L}{4} \\[2mm] \dfrac{2}{3}(L - x) & \text{for } \dfrac{L}{4} < x < L \end{cases} \qquad\qquad (5.5.39\text{a})$$

$$\psi(x) = 0 \qquad\qquad (5.5.39\text{b})$$

This means that initially the string is pulled from a point located at one quarter of its length (i.e., $x = L/4$), so as to attain the initial position shown in Figure 5.15a, and then it is suddenly released with zero initial velocity.

On physical grounds, due to the absence of either forcing or damping terms in the wave equation, we expect the string to undergo a periodic vibrating movement which continues unchanged in time.

The solution is given by eq. (5.5.14), where the functions $F(\xi)$ and $G(\eta)$ can be computed through eqs. (5.5.19) or eqs. (5.5.25) depending upon the values of ξ and η. Let us compute, for example, the position of the string—that is, $z = z(x)$—at time $t = L/8c$.

From the definitions (5.5.10), it is seen that as x varies in the interval of interest (i.e., $0 \le x \le L$), ξ and η vary in the intervals

$$\frac{L}{8} \le \xi \le \frac{9L}{8} \quad \text{and} \quad -\frac{L}{8} \le \eta \le \frac{7L}{8} \qquad\qquad (5.5.40)$$

Thus, in order to compute the functions F and G, it is convenient to divide the interval of x into three portions.

1. $0 \le x \le L/8$. From eq. (5.5.10): $L/8 \le \xi \le L/4$, $-L/8 \le \eta \le 0$.
F and G can be computed from eqs. (5.5.19a) and (5.5.26a), respectively, leading to

$$z = F(\xi) + G(\eta) = \tfrac{1}{2}\phi(\xi) - \tfrac{1}{2}\phi(-\eta) \qquad\qquad (5.5.41)$$

From eq. (5.5.39a), considering the intervals of interest for ξ and η above, it is seen that $\phi(\xi) = 2\xi$ and $\phi(-\eta) = -2\eta$. Substituting in eq. (5.5.41), together with eq. (5.5.10), leads to

$$z = \xi + \eta = 2x \qquad\qquad (5.5.42)$$

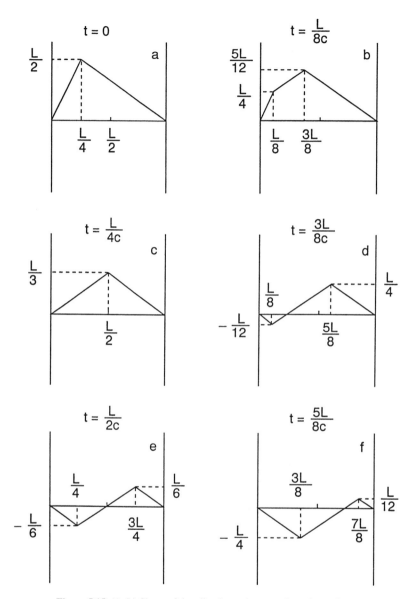

Figure 5.15 (A–L) Shape of the vibrating string at various time values.

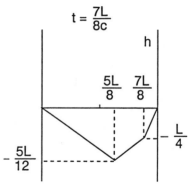

$$t = \frac{3L}{4c}$$

g

$$\frac{L}{2}$$

$$-\frac{L}{3}$$

$$t = \frac{7L}{8c}$$

h

$$\frac{5L}{8} \quad \frac{7L}{8}$$

$$-\frac{L}{4}$$

$$-\frac{5L}{12}$$

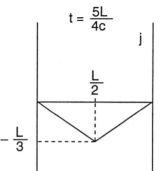

$$t = \frac{L}{c}$$

i

$$\frac{3L}{4}$$

$$\frac{L}{2}$$

$$-\frac{L}{2}$$

$$t = \frac{5L}{4c}$$

j

$$\frac{L}{2}$$

$$-\frac{L}{3}$$

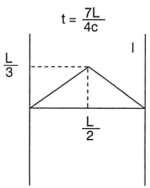

$$t = \frac{3L}{2c}$$

k

$$\frac{L}{4}$$

$$\frac{L}{6}$$

$$-\frac{L}{6}$$

$$\frac{3L}{4}$$

$$t = \frac{7L}{4c}$$

l

$$\frac{L}{3}$$

$$\frac{L}{2}$$

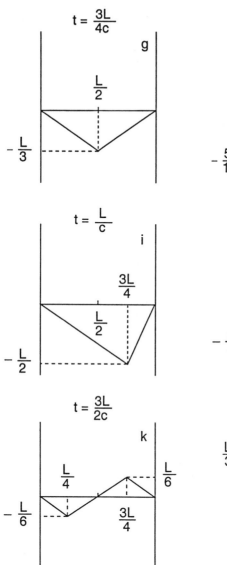

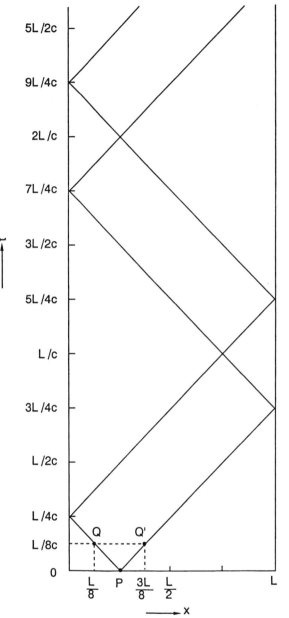

Figure 5.16 Characteristic curves on the (x, t) plane for the vibrating string problem.

2. $L/8 \le x \le 3L/8$. From eq. (5.5.10): $L/4 \le \xi \le L/2, 0 \le \eta \le L/4$.
In this case, F and G are given by eqs. (5.5.19), and then

$$z = \tfrac{1}{2}\phi(\xi) + \tfrac{1}{2}\phi(\eta) \tag{5.5.43}$$

which, using eq. (5.5.39a), gives

$$z = \tfrac{1}{3}(L - \xi) + \eta = \tfrac{1}{6}(4x + L) \tag{5.5.44}$$

3. $3L/8 \le x \le 7L/8$. From eq. (5.5.10): $L/2 \le \xi \le L, L/4 \le \eta \le 3L/4$.
F and G are again given by eq. (5.5.43). However, due to the change of intervals for ξ and η, when substituting eq. (5.5.39a) the solution becomes

$$z = \tfrac{1}{3}(L - \xi) + \tfrac{1}{3}(L - \eta) = \tfrac{2}{3}(L - x) \tag{5.5.45}$$

4. $7L/8 \le x \le L$. From eq. (5.5.10): $L \le \xi \le 9L/8, 3L/4 \le \eta \le 7L/8$.
F and G are given by eq. (5.5.26b) and eq. (5.5.19b), respectively. Accordingly,

$$z = -\tfrac{1}{2}\phi(2L - \xi) + \tfrac{1}{2}\phi(\eta) \tag{5.5.46}$$

which, using eq. (5.5.39a), reduces to

$$z = -\tfrac{1}{3}(L - 2L + \xi) + \tfrac{1}{3}(L - \eta) = \tfrac{2}{3}(L - x) \tag{5.5.47}$$

By plotting the expressions for z derived above for each interval of x, the position of the string at $t = L/8c$ is obtained as shown in Figure 5.15b. It is seen that the slope discontinuity initially located at $x = L/4$ (see Figure 5.15a), has evolved into *two* slope discontinuities located at $x = L/8$ and $x = 3L/8$, respectively.

This behavior can be understood by referring to the characteristic curves of eq. (5.5.36) in the (x, t) plane. In particular, those emanating from the point $P(x = L/4, t = 0)$, which is the initial location of the slope discontinuity, are shown in Figure 5.16. These curves constitute the domain of influence of point P, and hence are the locus of points where the solution exhibits a slope discontinuity. Thus, for example, in Figure 5.16 it is seen that at $t = L/8c$ the two slope discontinuities are located at $x = L/8$ and $x = 3L/8$, respectively—as computed earlier from the analytical solution shown in Figure 5.15b.

The characteristic curves on the (x, t) plane can be used effectively for computing the solution, $z(x, t)$. First, we need to identify the domain of dependence for any given point on the plane. This is particularly simple in this case since, due to the condition of zero initial velocity given by eq. (5.5.39b), the integral terms in eqs. (5.5.19) and (5.5.26) vanish. Thus considering, for example, the point $Q(x = L/8, t = L/8c)$ in Figure 5.17, the domain of dependence is given by the two characteristic lines originating from the x axis at $x = 0$ and $x = L/4$, respectively. Recalling eqs. (5.5.14) and (5.5.19), the solution is obtained by combining the values of the initial displacement in $x = 0$ and $x = L/4$, which propagate along each characteristic curve, as follows:

$$z = \frac{1}{2}\left[\phi(0) + \phi\left(\frac{L}{4}\right)\right] = \frac{L}{4} \tag{5.5.48}$$

The value above is the same as that given by the analytical solution (5.5.44) for $x = L/8$ (see also Figure 5.15b). Similarly, for the point $R(x = L/4, t = L/2c)$ in Figure

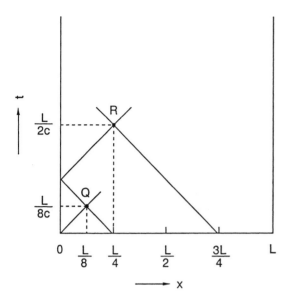

Figure 5.17 Domains of dependence.

5.17, the domain of dependence is given by the two characteristic curves originating from the x axis at $x = L/4$ and $x = 3L/4$. Thus, the value of the solution at R is given by

$$z = \frac{1}{2}\left[-\phi\left(\frac{L}{4}\right) + \phi\left(\frac{3L}{4}\right)\right] = -\frac{L}{6} \qquad (5.5.49)$$

where the sign of the first contribution has been changed because the corresponding characteristic line has been reflected once by the boundary of the domain before reaching R. Again, this value is identical to that computed from the analytical solution, which is shown in Figure 5.15e for $t = L/2c$.

As discussed in section 5.5.2, the characteristic lines can also be used to study the propagation of discontinuities. Let us consider the space partial derivative at $t = 0$, which from eq. (5.5.39a) is seen to exhibit a discontinuity at $x = L/4$—that is, point P in Figure 5.16. In order to see in which direction such a discontinuity propagates, we need to consider the two terms defined by eqs. (5.5.35a) and (5.5.35b), with $\bar{t} = 0$ and $\bar{x} = L/4$. Since $\bar{\xi} = \bar{\eta} = L/4$, from eq. (5.5.39a) we see that each term exhibits a discontinuity. Thus we conclude that at point P the discontinuity propagates along *both* the characteristics.

A different result is obtained for the direction of propagation of the discontinuity at point $Q(\bar{x} = L/8, \bar{t} = L/8c)$ in the same figure. In this case $\bar{\xi} = L/4$ and $\bar{\eta} = 0$, and only the first term (5.5.35a) exhibits a discontinuity. Thus, the discontinuity at point Q propagates along only the left characteristic. On the other hand, the discontinuity at point $Q'(\bar{x} = 3L/8, \bar{t} = L/8c)$ propagates along only the right characteristic, as it is readily seen noting that in this case $\bar{\xi} = L/2$ and $\bar{\eta} = L/4$, so that only the second term (5.5.35b) is discontinuous.

Finally, an overview of the complete movement of the vibrating string is provided by Figures 5.15a to 5.15l, where its position at various time instants is shown. At earlier times (i.e., $0 < t < L/4c$), the shape of the string is characterized by two slope discontinuities (cf. Figure 5.15b) which move along the x axis in opposite directions, as is also apparent from the characteristic lines in Figure 5.16. At $t = L/4c$, one of the two discontinuities reaches the string end at $x = 0$, leading to the singular shape shown in Figure 5.15c. At larger times, since the first discontinuity is reflected by the boundary at $x = 0$, the shape of the string is again characterized by two slope-discontinuity points, but now the value of the displacement is negative for the first one and positive for the second (cf. Figure 5.15d). The string maintains this shape for the time interval $L/4c < t < 3L/4c$, where the two discontinuities move along the x axis in the same direction (cf. Figure 5.15d–5.15f), as is also evident from the characteristic lines in Figure 5.16. In the same figure it may be seen that at $t = 3L/4c$, the second discontinuity reaches the boundary at $x = L$. At this point the shape of the string, shown in Figure 5.15g, is again singular—similar to that in Figure 5.15c—but is now characterized by all negative displacement values. Subsequently, the shape of the string is characterized by two negative slope-discontinuity points (cf. Figure 5.15h), which travel again in opposite directions along the x axis. From Figure 5.16 it is seen that at $t = L/c$ the two characteristic lines meet, and hence also the two slope discontinuities, at $x = 3L/4$. The resulting shape of the string is again of the singular type, as shown in Figure 5.15i. The string has now reached a position symmetric to the initial one (cf. Figure 5.15a). Then, in the following time interval $L/c < t < 2L/c$, the string repeats its movement, as illustrated by Figures 5.15j–5.15l. From Figure 5.16 it is seen that this movement is characterized by a time interval, $L/c < t < 5L/4c$, where the two slope discontinuities move along the x axis in opposite directions, followed by a time interval, $5L/4c < t < 7L/4c$, where they move in the same direction and finally by another one, $7L/4c < t < 2L/c$, where the directions are opposite. At time $t = 2L/c$ the string is back to the initial condition—that is, with the shape shown in Figure 5.15a and zero velocity. This completes this first period of vibration, which is repeated subsequently without changes.

5.5.4 Classification of Two-Dimensional Second-Order PDEs

The classification of two-dimensional, second-order PDEs can be best illustrated with reference to the operator:

$$L[z] = A(x, y)\,\frac{\partial^2 z}{\partial x^2} + B(x, y)\,\frac{\partial^2 z}{\partial x \partial y} + C(x, y)\,\frac{\partial^2 z}{\partial y^2} \qquad (5.5.50)$$

As in the case of the wave equation, we seek a nonlinear change of variables:

$$\xi = \xi(x, y) \qquad (5.5.51a)$$

$$\eta = \eta(x, y) \qquad (5.5.51b)$$

which transforms the original operator into a simpler form. Using the chain rule of differentiation and substituting eqs. (5.5.51) in eq. (5.5.50), we obtain

$$
\begin{aligned}
L[z] = & \left[A(x, y)\left(\frac{\partial \xi}{\partial x}\right)^2 + B(x, y)\left(\frac{\partial \xi}{\partial x}\right)\left(\frac{\partial \xi}{\partial y}\right) + C(x, y)\left(\frac{\partial \xi}{\partial y}\right)^2 \right] \frac{\partial^2 z}{\partial \xi^2} \\
& + \left[2A(x, y)\left(\frac{\partial \eta}{\partial x}\right)\left(\frac{\partial \xi}{\partial x}\right) + B(x, y)\left(\frac{\partial \xi}{\partial x}\right)\left(\frac{\partial \eta}{\partial y}\right) \right. \\
& \left. + B(x, y)\left(\frac{\partial \xi}{\partial y}\right)\left(\frac{\partial \eta}{\partial x}\right) + 2C(x, y)\left(\frac{\partial \xi}{\partial y}\right)\left(\frac{\partial \eta}{\partial y}\right) \right] \frac{\partial^2 z}{\partial \xi \partial \eta} \\
& + \left[A(x, y)\left(\frac{\partial \eta}{\partial x}\right)^2 + B(x, y)\left(\frac{\partial \eta}{\partial x}\right)\left(\frac{\partial \eta}{\partial y}\right) + C(x, y)\left(\frac{\partial \eta}{\partial y}\right)^2 \right] \frac{\partial^2 z}{\partial \eta^2}
\end{aligned}
\tag{5.5.52}
$$

In order to have the simpler form

$$
L[z] = D(x, y)\frac{\partial^2 z}{\partial \xi \partial \eta}
\tag{5.5.53}
$$

we require that the coefficients of $\partial^2 z/\partial \xi^2$ and $\partial^2 z/\partial \eta^2$ should vanish. For this we need the following conditions to be satisfied:

$$
A\left(\frac{\partial \xi}{\partial x}\right)^2 + B\left(\frac{\partial \xi}{\partial x}\right)\left(\frac{\partial \xi}{\partial y}\right) + C\left(\frac{\partial \xi}{\partial y}\right)^2 = 0
\tag{5.5.54a}
$$

$$
A\left(\frac{\partial \eta}{\partial x}\right)^2 + B\left(\frac{\partial \eta}{\partial x}\right)\left(\frac{\partial \eta}{\partial y}\right) + C\left(\frac{\partial \eta}{\partial y}\right)^2 = 0
\tag{5.5.54b}
$$

Assuming both ξ and η to be functions of y(i.e., $\partial \xi/\partial y \neq 0$ and $\partial \eta/\partial y \neq 0$), we can take as unknowns in the equations above the two ratios $(\partial \xi/\partial x)/(\partial \xi/\partial y)$ and $(\partial \eta/\partial x)/(\partial \eta/\partial y)$. From eq. (5.5.51a) we have

$$
d\xi = \left(\frac{\partial \xi}{\partial x}\right) dx + \left(\frac{\partial \xi}{\partial y}\right) dy
\tag{5.5.55}
$$

and so ξ is constant (i.e., $d\xi = 0$) along the curve

$$
\left(\frac{dy}{dx}\right)_\xi = -\left(\frac{\partial \xi}{\partial x}\right) \Big/ \left(\frac{\partial \xi}{\partial y}\right)
\tag{5.5.56}
$$

Similarly, using eq. (5.5.51b) we can see that η is constant along the curve

$$
\left(\frac{dy}{dx}\right)_\eta = -\left(\frac{\partial \eta}{\partial x}\right) \Big/ \left(\frac{\partial \eta}{\partial y}\right)
\tag{5.5.57}
$$

Thus the algebraic system (5.5.54) reduces to

$$A\left(\frac{dy}{dx}\right)_\xi^2 - B\left(\frac{dy}{dx}\right)_\xi + C = 0 \qquad (5.5.58a)$$

$$A\left(\frac{dy}{dx}\right)_\eta^2 - B\left(\frac{dy}{dx}\right)_\eta + C = 0 \qquad (5.5.58b)$$

which are two identical but independent equations in the unknowns $(dy/dx)_\xi$ and $(dy/dx)_\eta$, whose solutions have to be different in order to avoid a singular transformation. Thus we have

$$\left(\frac{dy}{dx}\right)_\xi = \frac{B - \sqrt{B^2 - 4AC}}{2A} \qquad (5.5.59a)$$

$$\left(\frac{dy}{dx}\right)_\eta = \frac{B + \sqrt{B^2 - 4AC}}{2A} \qquad (5.5.59b)$$

The solution of the ODEs (5.5.59) provides *two* distinct families of curves, called the *characteristic curves* of the operator $L[\ ]$, if the following condition is satisfied:

$$B^2 - 4AC > 0 \qquad (5.5.60)$$

Second-order PDEs which satisfy this condition are called *hyperbolic* and their operator has the canonical form

$$L[z] = \frac{\partial^2 z}{\partial \xi \partial \eta} \qquad (5.5.61)$$

The wave equation (5.5.9) discussed above can be readily seen to be hyperbolic. In particular, for $A = -c^2$, $B = 0$, and $C = 1$ the operator (5.5.50) reduces to that involved in eq. (5.5.9), where the independent variable t is replaced by y. In this case, the expressions (5.5.59) for the characteristic curves become

$$\left(\frac{dy}{dx}\right)_\xi = -\frac{1}{c} \qquad (5.5.62a)$$

$$\left(\frac{dy}{dx}\right)_\eta = \frac{1}{c} \qquad (5.5.62b)$$

in agreement with the eqs. (5.5.10) derived previously.

The treatment of the wave equation reported in section 5.5.2 can be applied in general to all hyperbolic equations. Specifically, there are two families of characteristic curves which determine the domains of influence and dependence as well as the propagation of discontinuities.

Second-order PDEs for which

$$B^2 - 4AC = 0 \qquad (5.5.63)$$

are called *parabolic*. In this case, since the discriminant in eq. (5.5.59) is zero, there exists only *one* family of characteristic curves, which can be taken as those where one parameter, say ξ, is constant:

$$\left(\frac{dy}{dx}\right)_\xi = \frac{B}{2A} \tag{5.5.64}$$

Since eq. (5.5.59a) arises from (5.5.54a), it is clear that the coefficient of $\partial^2 z/\partial\xi^2$ in eq. (5.5.52) is equal to zero. However, the coefficient of $\partial^2 z/\partial\eta^2$ cannot be zero, because otherwise ξ and η would not be independent variables. The coefficient $\partial^2 z/\partial\xi\partial\eta$, on the other hand, in fact vanishes. This can be seen by dividing the coefficient of $\partial^2 z/\partial\xi\partial\eta$ by $(\partial\xi/\partial y)(\partial\eta/\partial y)$ and using eqs. (5.5.56) and (5.5.57) along with eqs. (5.5.63) and (5.5.64). Thus the canonical form of the second-order partial differential operator for parabolic PDEs is

$$L[z] = \frac{\partial^2 z}{\partial\eta^2} \tag{5.5.65}$$

Note that the conclusion above was reached independently of the specific form of the second transformation (5.5.51b), which for simplicity can be taken as $\eta = y$. Thus if the characteristic curve $\xi = $ constant is taken as one of the coordinates, the only second-order derivative with respect to η is present in the operator $L[z]$. Also in this case, discontinuities in derivatives propagate along the characteristics, which again determine the domains of dependence.

Using condition (5.5.63) it is readily seen that the heat equation described in section 5.5.1 is indeed a parabolic PDE. A particularly simple parabolic equation is

$$\frac{\partial^2 z}{\partial\eta^2} = 0 \tag{5.5.66}$$

The general solution is obtained by integrating eq. (5.5.66) twice, leading to

$$z = F(\xi) + \eta G(\xi) \tag{5.5.67}$$

where F and G are two arbitrary functions of ξ alone. It appears that $z(x, y)$, similar to the case of the wave equation, is given by the superposition of two waves: $F(\xi)$ and $\eta G(\xi)$. The first wave travels with velocity given by eq. (5.5.64) without changing its shape; that is, $F(\xi)$ is constant along the $y = y(x)$ curve defined by eq. (5.5.64), where ξ is constant. The second wave travels with the same velocity, but it grows linearly along the coordinate y, since we have selected $\eta = y$ as the second transformation (5.5.51b).

The final case arises when

$$B^2 - 4AC < 0 \tag{5.5.68}$$

In this case there are no real solutions for eqs. (5.5.59), hence no transformations are available for eliminating the terms containing $\partial^2 z/\partial\xi^2$ and $\partial^2 z/\partial\eta^2$. Second-order PDEs which satisfy this condition are called *elliptic*.

The canonical form of elliptic PDEs is obtained by introducing a change of variables which makes the coefficient of $\partial^2 z/\partial\xi\partial\eta$ in eq. (5.5.52) equal to zero. This transformation, $\xi = \xi(x, y)$, is given by

$$\left(\frac{dy}{dx}\right)_\xi = -\frac{(\partial\xi/\partial x)}{(\partial\xi/\partial y)} = \frac{B(\partial\eta/\partial x) + 2C(\partial\eta/\partial y)}{2A(\partial\eta/\partial x) + B(\partial\eta/\partial y)} \tag{5.5.69}$$

which defines the curves in the (y, x) plane along which the value of ξ is constant. The second change of variables (5.5.51b) remains arbitrary. This leads to the canonical form of the second-order partial differential operator for elliptic PDEs:

$$L[z] = \frac{\partial^2 z}{\partial\xi^2} + \frac{\partial^2 z}{\partial\eta^2} \tag{5.5.70}$$

It can be easily seen, using condition (5.5.68), that the Laplace equation described in section 5.5.1 belongs to this class. Since in the elliptic operator (5.5.70) second-order derivatives with respect to both ξ and η are involved, the partial derivatives of the solution of $L[z] = 0$ can have no discontinuities. In elliptic PDEs no characteristic curve exists and the domain of dependence of each point is the entire domain of integration.

A last observation concerns the dependence of the coefficients A, B, and C of the operator (5.5.50) upon the two independent variables x and y. For a PDE to be called hyperbolic, parabolic, or elliptic, we require that the coefficients A, B, and C satisfy the conditions (5.5.60), (5.5.63), or (5.5.68), respectively for *all* (x, y) values in the entire domain. In some cases the coefficients do not satisfy the same condition in the entire domain and then the equation is classified as of *mixed* type. An example is the Tricomi equation:

$$y\frac{\partial^2 z}{\partial x^2} + \frac{\partial^2 z}{\partial y^2} = 0 \tag{5.5.71}$$

which is hyperbolic for $y < 0$ and elliptic for $y > 0$, as can be readily seen by applying the conditions (5.5.60) and (5.5.68).

We will return to second-order PDEs in chapters 7 and 8, where methods for deriving analytical solutions are discussed.

References

H.-K. Rhee, R. Aris, and N. R. Amundson, *First-Order Partial Differential Equations*, volumes 1 and 2, Prentice-Hall, Englewood Cliffs, NJ, 1986.

I. N. Sneddon, *Elements of Partial Differential Equations*, McGraw-Hill, New York, 1957.

A. Varma, "Some Historical Notes on the Use of Mathematics in Chemical Engineering," in *A Century of Chemical Engineering*, W. F. Furter, ed., Plenum Press, New York, 1982.

H. F. Weinberger, *A First Course in Partial Differential Equations*, Blaisdell, Waltham, MA, 1965.

Additional Reading

R. Aris, *Mathematical Modelling Techniques*, Pitman, London, 1978.

P. R. Garabedian, *Partial Differential Equations*, Wiley, New York, 1964.

R. Haberman, *Elementary Applied Partial Differential Equations*, 2nd ed., Prentice-Hall, Englewood Cliffs, NJ, 1987.

F. John, *Partial Differential Equations*, 3rd ed., Springer-Verlag, New York, 1978.

T. Myint-U and L. Debnath, *Partial Differential Equations for Scientists and Engineers*, 3rd ed., North-Holland, New York, 1987.

E. C. Zachmanoglou and D. W. Thoe, *Introduction to Partial Differential Equations with Applications*, Dover, New York, 1986.

Problems

5.1   Use the isocline method to solve the first-order ODE:

$$z\left(\frac{dz}{dx}\right)^2 = x + 1$$

with the IC

$$z = 0 \qquad \text{at } x = -1$$

5.2   Several transport problems are considered in this book. They include the problems defined by eqs. (5.5.4), (7.1.1), (7.3.2), and (7.9.1). In each case, identify the dimension of the space for both the initial condition and the solution.

5.3   Write the heat balance equations for the three tubular reactor models whose mass balances are given by eqs. (5.2.1), (5.2.3), and (5.2.5), respectively. The new variables to be used, in addition to those listed after eq. (5.4.5), are:

$$E_{ah} \quad \text{axial heat dispersion coefficient, cal/K sec m}$$
$$E_{rh} \quad \text{radial heat dispersion coefficient, cal/K sec m}$$
$$k_f \quad \text{fluid thermal conductivity, cal/K sec m}$$
$$h_w \quad \text{wall heat transfer coefficient, cal/K sec m}^2$$

Provide appropriate initial conditions for these equations.

5.4   Classify the following PDEs in terms of linear, quasilinear, and nonlinear:

$$\frac{\partial z}{\partial x}\frac{\partial^3 z}{\partial y^3} + \frac{\partial^2 z}{\partial y^2}\frac{\partial^2 z}{\partial x^2} = x$$

$$\frac{\partial^3 z}{\partial x^3}\frac{\partial^3 z}{\partial y^3} = xy$$

$$\frac{\partial^3 z}{\partial x^3} + x\frac{\partial^2 z}{\partial y^2} + y\frac{\partial z}{\partial x} = xy$$

$$\frac{\partial^2 z}{\partial t^2} + \sqrt{z}\left(\frac{\partial^2 z}{\partial x \partial y}\right)^3 - \frac{\partial z}{\partial x}\frac{\partial z}{\partial y} = \frac{1}{z}$$

5.5   In section 5.2 the mass balances corresponding to various models of a tubular reactor are presented: steady-state axial dispersion model (5.2.1), transient axial dispersion model

(5.2.3), and a transient model including both axial and radial dispersion (5.2.5). Classify these models in terms of linear, quasilinear, and nonlinear PDEs. Does this classification change when considering the corresponding heat balance equations? Also classify the transport models considered in problem 5.2. Is it possible to draw some general conclusions?

5.6 Transform the last PDE in problem 5.4, along with the following ICs:

$$z = z_0(x, y) \qquad \text{at } t = t_0(x, y)$$

$$\frac{\partial z}{\partial t} = z_1(x, y) \qquad \text{at } t = t_0(x, y)$$

into a system of quasilinear PDEs.

5.7 Consider the first-order PDE

$$z \frac{\partial z}{\partial x} + \frac{\partial z}{\partial y} = 0, \qquad -\infty < x < +\infty, y > 0$$

along with the IC

$$x = \xi, y = 0, z = \phi(\xi) \qquad \text{with } -\infty < \xi < +\infty$$

Plot the characteristic curves on the (x, y) plane and find the solution $z(x, y)$.

5.8 Consider a fluid flowing with velocity v through a pipe immersed in a bath of constant temperature T_a. The fluid temperature changes according to the heat balance:

$$\frac{\partial T}{\partial t} + v \frac{\partial T}{\partial x} = U(t)(T_a - T), \qquad 0 < x < L \text{ and } t > 0$$

where $U(t) = U_0 e^{-\alpha t}$ is the overall heat transfer coefficient which decreases in time due to fouling. Determine the evolution of the fluid temperature assuming that it is initially constant and equal to T_i, while the fluid enters the pipe at $T = T_f$.

5.9 Consider a tubular reactor where a zeroth-order reaction is taking place. In this case the heat balance (5.4.4) reduces to

$$\frac{\partial \theta}{\partial t} + \frac{\partial \theta}{\partial x} = A(\theta_j - \theta) + G \exp\left[\gamma\left(1 - \frac{1}{\theta}\right)\right], \qquad 0 < x < 1, t > 0$$

Develop the characteristic curves of the PDE. The evaluation of one of these requires numerical quadrature. Following this procedure, compute the transient temperature profile using $\theta^0 = \theta^i = 1$ for the initial conditions (5.4.5) and using $\theta_j = 1$, $A = 0.5$, and $G = 2$ for the dimensionless parameters.

5.10 Determine the regions where the solution of the following PDE exists

$$-x \frac{\partial z}{\partial x} + \frac{\partial z}{\partial y} = f(z, x, y), \qquad 0 < x < 1, y > 0$$

for each of the three ICs:

IC$_1$: $x = \xi$, $y = 0$, $z = z_0(x, y)$,   and $0 < \xi < 1$
IC$_2$: $x = 1$, $y = \xi$, $z = z_0(x, y)$,   and $\xi > 0$
IC$_3$: $x = \xi$, $y = \xi$, $z = z_0(x, y)$,   and $0 < \xi < 1$

5.11 Use the method of characteristics to solve the first-order PDE

$$\frac{\partial u}{\partial x} + x \frac{\partial u}{\partial y} + u \frac{\partial u}{\partial z} = 1$$

with the ICs

$$x = \xi_1, \qquad y = \xi_2, \qquad z = \xi_1\xi_2, \qquad u = 0, \quad \text{and} \quad 0 < \xi_1, \xi_2 < 1$$

5.12 Solve the vibrating string problem (5.5.36) to (5.5.38) with

$$\phi(x) = \begin{cases} x & \text{for } 0 < x < L/2 \\ (L - x) & \text{for } L/2 < x < L \end{cases}$$

and

$$\psi(x) = 0$$

Identify the propagation of the slope discontinuity in time and the solution at $t = L/4c$, $L/2c$, $3L/4c$, L/c, $3L/2c$, and $2L/c$, so as to explore the entire period of the oscillation.

5.13 Solve the vibrating string problem (5.5.36) to (5.5.38) with

$$\phi(x) = 0$$

and

$$\psi(x) = \begin{cases} 0 & \text{for } 0 \le x < L/4 \\ 1 & \text{for } L/4 \le x \le 3L/4 \\ 0 & \text{for } 3L/4 < x \le L \end{cases}$$

5.14 Determine the oscillations of the free vibrating string (5.5.36) to (5.5.38) in the case where no vertical force is applied at the end $x = 0$, which is then free to vibrate. In this case the BC (5.5.38a) is replaced by

$$\frac{\partial u}{\partial x} = 0, \qquad x = 0, \qquad t > 0$$

5.15 Show that if $\partial^2 u/\partial x^2$ has a discontinuity at $(\bar{x}, \bar{t})$, then this discontinuity is propagated along at least one of the characteristics through $(\bar{x}, \bar{t})$.

5.16 Find the oscillations of a vibrating string with free ends, in the presence of an external force, given by the PDE

$$\frac{\partial^2 u}{\partial t^2} - \frac{\partial^2 u}{\partial x^2} = e^x, \qquad 0 < x < 1, t > 0$$

with the BCs

$$\frac{\partial u}{\partial x} = 0, \qquad x = 0, t > 0$$

$$\frac{\partial u}{\partial x} = 0, \qquad x = 1, t > 0$$

and the ICs

$$u = 0, \qquad 0 \le x \le 1, t = 0$$

$$\frac{\partial u}{\partial t} = 0, \qquad 0 \le x \le 1, t = 0$$

Chapter 6

FIRST-ORDER PARTIAL DIFFERENTIAL EQUATIONS

First-order partial differential equations (PDEs) arise in a variety of problems in science and engineering. In this chapter we cover some basic concepts related to the solution of these equations, which are helpful in the analysis of problems of practical interest. These problems are briefly reviewed in the first section of the chapter, where they are cast as special cases of the kinematic wave equation.

We then proceed to discuss the solution of first-order PDEs using the method of characteristics introduced in chapter 5. This leads to solutions in the form of simple waves, including linear, expansion, compression, and centered waves. The occurrence of shocks (i.e., jump discontinuities) in the solution is then examined. This is a peculiar feature of first-order PDEs, and such solutions are sometimes referred to as *weak solutions*. These various aspects are illustrated in the context of a practical example, involving the dynamic behavior of a countercurrent separation unit.

Finally, in the last section the occurrence of constant-pattern solutions for PDEs is discussed. Although these equations are of order greater than one, their analysis is relevant to this chapter since it provides further support to the existence of weak solutions for first-order PDEs.

6.1   THE KINEMATIC WAVE EQUATION

In section 5.4 we discussed some examples indicating the importance of first-order PDEs in modeling various physical processes. These equations arise typically when applying a conservation law (e.g., mass) to a flowing system under transient conditions. Let us consider the case of a one-dimensional process which involves two independent variables: space (x) and time (t). The system is described by two dependent variables:

one representing the amount per *unit volume*, γ (e.g., concentration), and the other representing the amount per *unit area and per unit time*, ϕ (e.g., mass flux), which is transported by the flowing stream. The conservation law equates the accumulation rate at any time and position to the difference between the flows entering and leaving that position—that is,

$$\frac{\partial \gamma}{\partial t} + \frac{\partial \phi}{\partial x} = 0 \qquad (6.11.1)$$

The above equation is homogeneous and applies when internal generation and exchange with the external environment do not occur. These models are completed by coupling the variables γ and ϕ, most often by algebraic relationships such as

$$\phi = \phi(\gamma) \qquad (6.1.2)$$

Before discussing some examples to illustrate this relationship, let us first substitute eq. (6.1.2) in eq. (6.1.1), leading to

$$\frac{\partial \gamma}{\partial t} + c \frac{\partial \gamma}{\partial x} = 0 \qquad (6.1.3)$$

where $c = d\phi/d\gamma$ represents the velocity with which a constant value of γ propagates along the x direction. This can be understood by considering a coordinate moving along the x-axis according to $x = f(t)$, so that

$$\frac{d\gamma}{dt} = \frac{\partial \gamma}{\partial t} + \frac{\partial \gamma}{\partial x} \frac{dx}{dt} \qquad (6.1.4)$$

Comparison with eq. (6.1.3) indicates that γ remains constant for $dx/dt = f'(t) = c$. This means that the value of γ is constant along a straight line with slope $1/c$ in the (x, t) plane; that is, a given value of γ propagates unchanged along the x coordinate with speed c. Accordingly, PDEs of type (6.1.3) are referred to as *kinematic wave equations* and the parameter c is defined as the *wave velocity*. Note from eq. (6.1.2) that a constant value of ϕ also propagates along the wave together with a constant value of γ.

A variety of examples of kinematic wave equations arise in chemical engineering, and their behavior has been studied in detail (cf. Rhee et al., 1986). Perhaps the simplest one is provided by the flow of a homogeneous fluid through a pipe, where eq. (6.1.1) describes the conservation of mass in the case where dispersion in the axial direction can be neglected. Here, γ represents the concentration ($\gamma = C$) and ϕ represents the convective flux ($\phi = vC$, where v is the space velocity of the fluid). The relation (6.1.2) is then simply $\phi = vC = v\gamma$, thus indicating that $c = d\phi/d\gamma = v$; that is, the wave velocity is identical to the space velocity of the fluid.

For a heterogeneous system, with one mobile (concentration C) and one stationary (concentration Γ) phase, as in the case of the chromatography process described by eq. (5.4.9), the mass conservation law may again be represented by eq. (6.1.1). In this case the definition of concentration γ involves both phases; that is,

$$\gamma = \varepsilon C + (1 - \varepsilon)\Gamma = \varepsilon C + (1 - \varepsilon)F(C) \qquad (6.1.5)$$

where the local equilibrium condition (5.4.7) between the mobile and the stationary phases has been used. The flux ϕ is related only to the mobile phase concentration C through the superficial velocity v as follows:

$$\phi = vC \tag{6.1.6}$$

Substituting eqs. (6.1.5) and (6.1.6), it follows that eq. (6.1.1) reduces to the mass balance (5.4.9), which is equivalent to eq. (6.1.3) with the wave velocity given by

$$c = V = \frac{v}{\varepsilon + (1 - \varepsilon)(dF/dC)} \tag{6.1.7}$$

The last example involves an adiabatic packed-bed heated by direct contact with a hot fluid stream. In this case, for constant physical properties, the energy conservation law gives

$$\left[\rho_b C_{pb}(1 - \varepsilon) \frac{\partial T_b}{\partial t} + \rho_f C_{pf} \varepsilon \frac{\partial T_f}{\partial t} \right] + v\rho_f C_{pf} \frac{\partial T_f}{\partial x} = 0 \tag{6.1.8}$$

where the subscript b refers to the packed bed while f refers to the fluid, and the other symbols are defined as follows:

C_p specific heat, cal/g K
T temperature, K
ε bed void fraction
ρ density, g/m^3

Assuming that the two phases are in thermal equilibrium (i.e., the interphase heat transfer resistances are negligible), we have

$$T_b = T_f \tag{6.1.9}$$

Then, comparing with eq. (6.1.1), the variable γ represents the "energy concentration":

$$\gamma = [\rho_b C_{pb}(1 - \varepsilon) + \rho_f C_{pf}\varepsilon]T_f \tag{6.1.10}$$

where the "energy flux" ϕ is given by

$$\phi = (v\rho_f C_{pf}T_f) \tag{6.1.11}$$

From eqs. (6.1.10) and (6.1.11) we obtain

$$\phi = \frac{v\rho_f C_{pf}}{[\rho_b C_{pb}(1 - \varepsilon) + \rho_f C_{pf}\varepsilon]} \gamma \tag{6.1.12}$$

which provides the ϕ–γ relationship equivalent to (6.1.2). Thus the wave velocity is given by

$$c = \frac{d\phi}{d\gamma} = \frac{v\rho_f C_{pf}}{[\rho_b C_{pb}(1 - \varepsilon) + \rho_f C_{pf}\varepsilon]} \tag{6.1.13}$$

In physical terms the preceding equation represents the velocity of propagation of the temperature wave along the fixed bed. By introducing the parameter α, defined as the ratio between the heat capacities of the bed and the fluid, that is,

$$\alpha = \frac{\rho_b C_{pb}(1 - \varepsilon)}{\rho_f C_{pf}\varepsilon} \tag{6.1.14}$$

eq. (6.1.13) reduces to

$$c = \frac{v}{\varepsilon}\frac{1}{\alpha + 1} \tag{6.1.15}$$

We observe that if the bed heat capacity is large, then the temperature wave velocity is small; that is, $c \to 0$ as $\alpha \to \infty$. As the bed heat capacity decreases, the wave velocity increases. In the limit where the bed heat capacity is negligible relative to that of the fluid, the temperature wave velocity becomes equal to the interstitial velocity of the fluid stream; that is, $c \to v/\varepsilon$ as $\alpha \to 0$.

In addition to their intrinsic value in the simulation of systems of practical relevance, kinematic wave equations also occur as limiting behavior of higher-order models. In particular, such equations arise when purely convective transport (either of mass or heat) is considered, while other transport mechanisms involving dispersion or diffusion which introduce second-order space derivatives are neglected. A typical example is the model describing axial dispersion of mass in a one-dimensional flow system, given by eq. (5.2.3) with $B = 0$. As the axial dispersion coefficient vanishes (i.e., Pe $\to \infty$), this model approaches the kinematic wave equation (6.1.3).

As we have seen in the last two examples, kinematic wave equations also arise in heterogeneous systems when the two phases are assumed to be at equilibrium. These *equilibrium models* can be regarded as the limiting behavior of *nonequilibrium models* for vanishing transport resistances. Thus they can be used to derive approximate solutions, either analytic or numerical, of the nonequilibrium models by treating them as perturbations of the corresponding equilibrium models.

6.2 SIMPLE WAVES

From the classification of PDEs, it is seen that the general form of a homogeneous, quasilinear first-order PDE is given by [cf. eq. (5.3.3)]:

$$P(z, x, y)\frac{\partial z}{\partial x} + Q(z, x, y)\frac{\partial z}{\partial y} = 0 \tag{6.2.1}$$

together with an IC given as follows:

$$z = z^0(x) \text{ along the curve } I: y = y^0(x) \tag{6.2.2}$$

The main feature of these homogeneous equations arises from the third in the set of equations describing the characteristic curve (5.4.24), that is,

$$\frac{dz}{ds} = 0 \tag{6.2.3}$$

which requires z to be constant along the entire characteristic curve. As a consequence, the characteristic curves of eq. (6.2.1) are always parallel to the (x, y) plane. Furthermore, since the value of z is constant along a characteristic curve, we can claim that the state (i.e., the value of z) *propagates* along the characteristics. Accordingly, we introduce the concept of *rate of propagation of the state*, which makes it easier to construct the solution.

In the following, two specific cases which have some interesting applications in engineering problems are examined in detail. The first refers to homogeneous equations with *constant coefficients* (i.e., P and Q are constants), while the second refers to *reducible equations* (i.e., P and Q are functions of z alone).

6.2.1 Homogeneous Quasilinear Equations with Constant Coefficients: Linear Waves

Let us consider, for example, the chromatography model (5.4.9) in the case of linear equilibrium between the mobile and the stationary phases. In this case the local equilibrium condition (5.4.7) reduces to

$$\Gamma = F(C) = KC \tag{6.2.4}$$

The chromatography model (5.4.9) is then given by

$$V \frac{\partial C}{\partial x} + \frac{\partial C}{\partial t} = 0, \qquad 0 < x < L \text{ and } t > 0 \tag{6.2.5}$$

where

$$V = \frac{v}{[\varepsilon + (1 - \varepsilon)K]} \tag{6.2.5a}$$

along with the following ICs, which define the initial concentration profile in the column and the inlet concentration value as a function of time:

$$C = C^0(x) \text{ along the curve } I_1: 0 \le x \le L, t = 0 \tag{6.2.6a}$$

$$C = C^i(t) \text{ along the curve } I_2: x = 0, t > 0 \tag{6.2.6b}$$

The solution procedure, based on the characteristic curves, is the same as that applied previously for solving the nonhomogeneous, constant coefficient, first-order PDE (5.4.2). We repeat this procedure here, with reference to eq. (6.2.5), in order to better emphasize some concepts. Following eqs. (5.4.20) and (5.4.24), the characteristic curves associated with eq. (6.2.5) are given by

$$\frac{dx}{ds} = V \tag{6.2.7a}$$

$$\frac{dt}{ds} = 1 \tag{6.2.7b}$$

$$\frac{dC}{ds} = 0 \tag{6.2.7c}$$

whose IC for $s = 0$ is given along the *discontinuous* curve $(I_1 \cup I_2)$ according to eqs. (6.2.6). This requires analyzing the problem using each branch of the initial condition curve separately.

Let us first consider the IC on curve I_1, which in parametric form is

$$x = \xi, \quad t = 0, \quad C = C^0(\xi) \qquad \text{with } 0 \leq \xi \leq L \qquad (6.2.8)$$

Integrating eqs. (6.2.7) with the IC (6.2.8) for $s = 0$ leads to

$$x = Vs + \xi \qquad (6.2.9a)$$

$$t = s \qquad (6.2.9b)$$

$$C = C^0(\xi) \qquad (6.2.9c)$$

with $0 \leq \xi \leq L$ and $s \geq 0$. Eliminating the parameters s and ξ from the eqs. (6.29a) and (6.29b) and substituting in eq. (6.29c) yields the solution

$$C = C^0(x - Vt) \qquad (6.2.10)$$

with $0 \leq (x - Vt) \leq L$. Since we are considering only $t \geq 0$ and $0 \leq x \leq L$, this means that $0 \leq t \leq x/V$. Equation (6.2.10) indicates that the solution is given by the initial concentration profile along the column, but with the independent variable replaced by $(x - Vt)$.

Using the second branch of the initial condition curve I_2

$$x = 0, \quad t = \xi, \quad C = C^i(\xi) \qquad \text{with } \xi > 0 \qquad (6.2.11)$$

as IC at $s = 0$ to integrate eqs. (6.2.7), we obtain

$$x = Vs \qquad (6.2.12a)$$

$$t = s + \xi \qquad (6.2.12b)$$

$$C = C^i(\xi) \qquad (6.2.12c)$$

with $\xi > 0$ and $s \geq 0$. Eliminating s and ξ yields

$$C = C^i\left(t - \frac{x}{V}\right) \qquad (6.2.13)$$

which applies for $t > x/V$. Thus in this region of the (x, t) plane, the solution is given by the time evolution of the inlet concentration, but with the independent variable now replaced by $(t - x/V)$. In summary, the full solution of the problem is

$$C = C^0(x - Vt) \qquad \text{for } 0 \leq t \leq x/V \qquad (6.2.14a)$$

$$C = C^i\left(t - \frac{x}{V}\right) \qquad \text{for } t > x/V \qquad (6.2.14b)$$

The same solution can also be obtained through a different procedure based upon geometrical considerations. From eqs. (6.2.7a) and (6.2.7b) it is seen that the projections of the characteristic curves on the (x, t) plane are straight lines with slope

$$\frac{dt}{dx} = \frac{1}{V} \tag{6.2.15}$$

As shown in Figure 6.1, these constitute a family of parallel straight lines originating from different points of the initial condition curve I, on which the value of the solution C is prescribed by eq. (6.2.6). From eq. (6.2.7c) it is seen that C is constant along the characteristics; that is, the value of the solution remains unchanged along the straight lines shown in Figure 6.1. Thus it is clear that the solution C propagates from the IC curve into the (x, t) plane with a speed given by the reciprocal of the slope of the characteristic lines in Figure 6.1, i.e., V.

In the region below the curve $t = x/V$, as given by eq. (6.2.14a), the solution is determined by the *initial* condition $C^0(x)$. For a given x, at any time $t < x/V$, the value of C is equal to the initial value at $\bar{x} = x - Vt$, i.e. $C(x, t) = C^0(\bar{x})$. For $t = x/V$, the influence of the fluid initially preceding location x is washed out. For $t > x/V$, the solution loses all memory of the initial condition and depends only on the *inlet* condition, $C^i(t)$.

It is worth reiterating that since the chromatography model (6.2.5) belongs to the family of kinematic wave equations, the rate of propagation of the solution V coincides with the wave velocity c defined by eq. (6.1.7). In fact, using eqs. (6.2.4) and (6.2.5a) yields

$$c = \frac{\upsilon}{\varepsilon + (1 - \varepsilon)K} = V \tag{6.2.16}$$

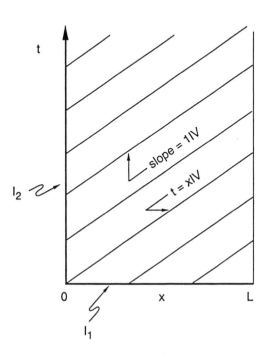

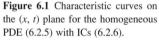

Figure 6.1 Characteristic curves on the (x, t) plane for the homogeneous PDE (6.2.5) with ICs (6.2.6).

From eq. (6.2.16), it can be seen that the rate of propagation of an adsorbable component is always smaller than v/ε, which is the velocity of an inert component flowing in the same packed bed without being adsorbed (i.e., $K = 0$). This is so because the adsorption process retards the movement of the adsorbable component, thus increasing the time required for it to reach the end of the column. This leads us to the concept of *retention time*, which can be defined as the time required by a given concentration value at the column inlet to reach the column outlet. In this case, since the propagation rate is constant along x, the retention time is simply

$$t_R = \frac{L}{V} \tag{6.2.17}$$

and then using eq. (6.2.16) we obtain

$$t_R = \frac{\varepsilon L}{v} + \frac{(1 - \varepsilon)KL}{v} \tag{6.2.18}$$

From the above equation, it is clear that the retention time is given by the sum of two contributions. The first, due to the convective transport through the packed bed, is proportional to the reciprocal of the interstitial velocity, v/ε. The second, due to the retardation effect caused by the adsorption process, is proportional to the adsorption constant K. In this connection, it is worth noting that eq. (6.2.18) can be used in practice to estimate the adsorption constant of a component from two measurements of retention time: one obtained using the adsorbable component and the other using an inert tracer.

As an example, let us consider the case of a column initially filled with an inert, which is fed for a time interval equal to Δt with a stream containing an adsorbable component at a fixed concentration and thereafter with pure inert. Thus the ICs (6.2.6) are

$$C^0 = 0, \qquad 0 \leq x \leq L \text{ and } t = 0 \tag{6.2.19a}$$

and

$$C^i = \begin{cases} \overline{C}, & x = 0 \text{ and } 0 < t < \Delta t \\ 0, & x = 0 \text{ and } t > \Delta t \end{cases} \tag{6.2.19b}$$

Since the inlet concentration is given by a step function, the ICs exhibit two discontinuity points at $x = 0$, $t = 0$ and at $x = 0$, $t = \Delta t$, which propagate into the (x, t) plane like any other constant value of the solution.

As discussed above, the solution is given by the values of C on the IC curves (6.2.19) which propagate unchanged in the (x, t) plane along the characteristics. These are shown in Figure 6.2 with a different notation depending upon the specific constant value of C. In particular, the characteristics corresponding to the value $C = 0$ are represented by a broken line, while those corresponding to $C = \overline{C}$ are indicated by a continuous line. The discontinuities of the solution propagate along the bolder lines.

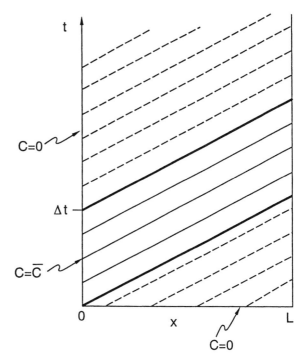

Figure 6.2 Characteristic curves on the (x, t) plane for the homogeneous PDE (6.2.5) with ICs (6.2.19).

The solution can be obtained from eq. (6.2.14) using eqs. (6.2.19) as follows:

$$C = C^i\left(t - \frac{x}{V}\right) = \begin{cases} \overline{C} & \text{for } \dfrac{x}{V} < t < \dfrac{x}{V} + \Delta t \\[2mm] 0 & \text{for } t > \dfrac{x}{V} + \Delta t \end{cases} \tag{6.2.20a}$$

$$C = C^0(x - Vt) = 0 \qquad \text{for } 0 \le t \le \frac{x}{V} \tag{6.2.20b}$$

From Figure 6.3, it is clear that the solution has the form of a step function, identical to that representing the inlet concentration, which travels with constant shape through the column with a constant velocity, V. This is the typical behavior of *linear waves*.

Let us now consider the case where the duration of the inlet concentration step, Δt, is varied, while the overall amount of the adsorbable component, Q, is kept fixed; that is,

$$\int_0^\infty FC^i(t)\, dt = Q \tag{6.2.21}$$

where F is the constant inlet volumetric flowrate. Substituting eq. (6.2.19b) in eq. (6.2.21), it follows that

$$\overline{C} = \frac{Q}{F\Delta t} \tag{6.2.22}$$

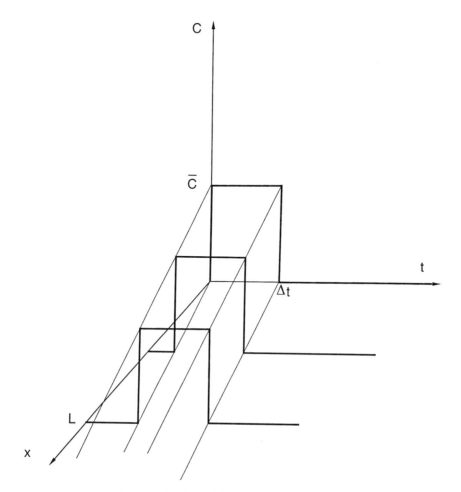

Figure 6.3 Qualitative behavior of the solution (6.2.20).

As $\Delta t \rightarrow 0$, the step function (6.2.19b) approaches the Dirac delta function

$$C^i(t) = \frac{Q}{F}\,\delta(t) \tag{6.2.23}$$

where, by definition of the Dirac delta function, $\delta(t) = 0$ for all $t \neq 0$ and $\int_0^\infty \delta(t)\,dt = 1$. In this case the solution (6.2.20) reduces to

$$C = \frac{Q}{F}\,\delta\!\left(t - \frac{x}{V}\right) \qquad \text{for } 0 \leq x \leq L \text{ and } t > 0 \tag{6.2.24}$$

which represents a peak of nonzero concentration traveling through the column with constant speed, V. The time required by this peak to reach the end of the column is the retention time, t_R, given by eqs. (6.2.17) and (6.2.18). This latter equation is the basis for estimating the linear adsorption constant, K, from measurements of the retention time in pulse-chromatography experiments.

6.2.2 Reducible Equations: Expansion, Compression, and Centered Waves

A first-order homogeneous, quasilinear PDE is called *reducible* when the coefficients P and Q in the general form (6.2.1) are functions only of the dependent variable, z; that is,

$$P(z) \frac{\partial z}{\partial x} + Q(z) \frac{\partial z}{\partial y} = 0 \qquad (6.2.25)$$

From eqs. (5.4.24), the characteristics are given by

$$\frac{dx}{ds} = P(z) \qquad (6.2.26a)$$

$$\frac{dy}{ds} = Q(z) \qquad (6.2.26b)$$

$$\frac{dz}{ds} = 0 \qquad (6.2.26c)$$

Thus the projection of the characteristic curve on the (x, y) plane is described by

$$\frac{dy}{dx} = \frac{Q(z)}{P(z)} = \psi(z) \qquad (6.2.27)$$

From eq. (6.2.26c) it is clear that the value of z remains constant along a characteristic curve and hence the slope, $\psi(z)$, is also constant. This allows us to conclude not only that the characteristic curves are parallel to the (x, y) plane, as for all homogeneous PDEs, but also that they are straight lines. We are again in the simple situation where the solution can be readily constructed by propagating the value of z given on the IC curve I into the (x, y) plane following the straight characteristic lines. The difference relative to the previous case of constant coefficient PDEs (section 6.2.1) is that the slope of the characteristics $\psi(z)$ is now a function of the value of z at the particular point on the initial curve from which the characteristic originates. Accordingly, it is not possible for reducible PDEs to plot the characteristics on the (x, y) plane independently of the particular initial condition, as was possible in the previous case.

Integrating eqs. (6.2.26) along a characteristic, using at $s = 0$ the IC curve in parametric form

$$I: \qquad x = x^0(\xi), \quad y = y^0(\xi), \quad z = z^0(\xi) \qquad (6.2.28)$$

yields

$$x = x^0(\xi) + P(z^0(\xi))s, \quad y = y^0(\xi) + Q(z^0(\xi))s, \quad z = z^0(\xi) \qquad (6.2.29)$$

The x and y relations above give

$$s = \frac{x - x^0(\xi)}{P(z^0(\xi))} = \frac{y - y^0(\xi)}{Q(z^0(\xi))} \tag{6.2.30}$$

so that

$$y - y^0(\xi) = \psi(z^0(\xi))[x - x^0(\xi)] \tag{6.2.31}$$

The parameter ξ can be eliminated by the third relation of (6.2.29) to give the solution in implicit form:

$$F(x, y, z) = 0 \tag{6.2.32}$$

Note that eq. (6.2.31) is the projection of the characteristic curves on the (x, y) plane. It represents a one-parameter (ξ) family of straight lines, with slope $\psi(z^0(\xi))$, originating from the projection of curve I on the (x, y) plane; that is, I^0: $x = x^0(\xi)$, $y = y^0(\xi)$.

To analyze the structure of this family, let us consider the straight lines obtained for increasing values of the parameter ξ. As ξ increases from say a to b, we traverse the curve I^0 and the variation in the slope of the straight lines is given by

$$\frac{d\psi}{d\xi} = \frac{d\psi}{dz^0}\frac{dz^0}{d\xi} \tag{6.2.33}$$

Equation (6.2.33) involves the product of two terms. The first, $d\psi/dz^0$, depends upon the coefficients of the PDE [cf. eq. (6.2.26)], while the second, $dz^0/d\xi$, is a function *only* of the ICs. Two different situations arise.

Case 1. $d\psi/d\xi < 0$
In this case, as ξ increases moving along I^0, the slopes of the characteristic lines decrease, as shown in the upper part of Figure 6.4. For concreteness, let us consider an IC $z^0(\xi)$, which decreases monotonically as ξ increases from a to b, as shown in Figure 6.5. The z versus x curve for a constant value of $y = y_1$ is shown in the lower part of Figure 6.4. Each point is obtained by projecting in the lower part of Figure 6.4 the intersections of the $y = y_1$ line with the characteristics in the upper part of the figure, where recall that each characteristic carries a specific value of z given by its origin on curve I^0. For a larger value of $y = y_2$, the new z versus x curve is obtained similarly. As a consequence of the decrease in slope of the characteristics as ξ increases from a to b, the gap Δx between the two curves increases monotonically; that is, the wave becomes less steep. This gives rise to wave solutions which expand while moving forward; hence they are referred to as *expansion waves*.

Case 2. $d\psi/d\xi > 0$
In this case the slopes of the characteristic curves emanating from the curve I^0 increase as ξ increases from a to b, as shown in the upper part of Figure 6.6. Following arguments similar to those described above, the wave solutions compress (i.e., become steeper) as they move forward, as illustrated in the lower part of the same figure. These are referred to as *compression waves*.

It is clear from the above discussion that if $d\psi/d\xi = 0$, then the solution waves do not expand or compress. They propagate without change in shape. This is the case of linear waves arising from constant coefficient PDEs discussed in section 6.2.1. These, together with expansion and compression waves, are referred to as *simple waves*.

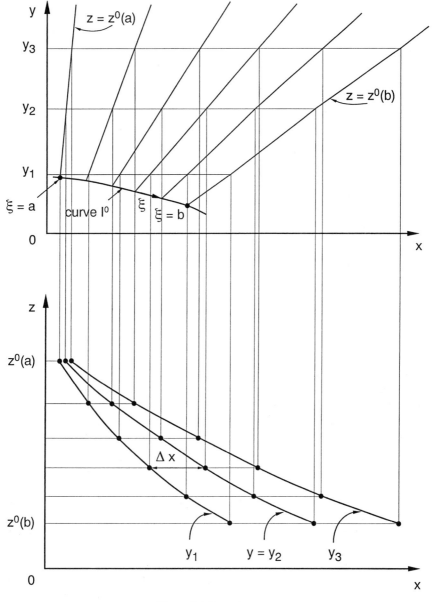

Figure 6.4 Expansion wave.

It should be noted that the conclusion reached above about the definition of expansion and compression waves depends upon the convention adopted in defining the parameter ξ. Thus if we take ξ increasing in the opposite direction along I^0 in Figures 6.4 and 6.6, the sign of $d\psi/d\xi$ which identifies expansion or compression waves would be reversed. The convention that we have adopted is that an observer moving along the

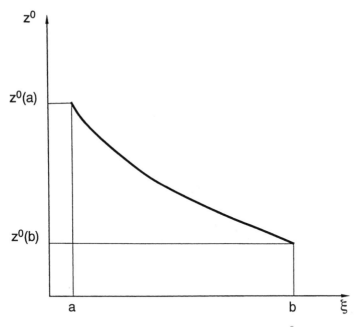

Figure 6.5 Qualitative behavior of the initial condition $z^0(\xi)$.

IC curve I^0 as ξ increases would see the characteristics propagating in the lhs region. For example, to study the propagation of characteristics in the first quadrant if we take I^0 to be the x axis, then $\xi = x$, while if we take I^0 to be the t axis, then $\xi = -t$.

The classification of simple waves reported above cannot be applied when $z^0(\xi)$ is not a continuous function, since $d\psi/d\xi$ in eq. (6.2.33) cannot be defined. This occurs, for example, when z^0 undergoes a step change at a point $c \in [a, b]$, as illustrated in Figure 6.7. This is indeed the case of the IC (6.2.19b) considered previously in connection with the chromatography model.

Let us consider the initial condition $z^0(\xi)$ considered above in Figure 6.5, where z^0 decreases continuously from $z^0(a)$ to $z^0(b)$ as ξ goes from a to b. If we move a and b toward an intermediate point c, while keeping unchanged $z^0(a)$ and $z^0(b)$, the continuous curve $z^0(\xi)$ becomes steeper and approaches in the limit the step function shown in Figure 6.7. Correspondingly, the shape of the characteristic straight lines in Figure 6.4 also changes, leading to the situation illustrated in Figure 6.8a. In this case the characteristics originating from the portions of curve I^0 given by $a \leq \xi < c$ and $c < \xi \leq b$ are parallel straight lines with slopes $\psi(z^0(a))$ and $\psi(z^0(b))$, respectively. On the other hand, all the straight lines whose slopes range from $\psi(z^0(a))$ to $\psi(z^0(b))$ originate from the point $\xi = c$. This can be understood considering that at $\xi = c$ the step function z^0 takes on all values between $z^0(a)$ and $z^0(b)$. In this case the wave is called a *centered simple wave*.

Before considering an example of simple wave solution, it is worth pointing out a difficulty which arises in the case of compression waves as illustrated in Figure 6.6. In the upper part of the figure it is seen that since the characteristic straight lines are

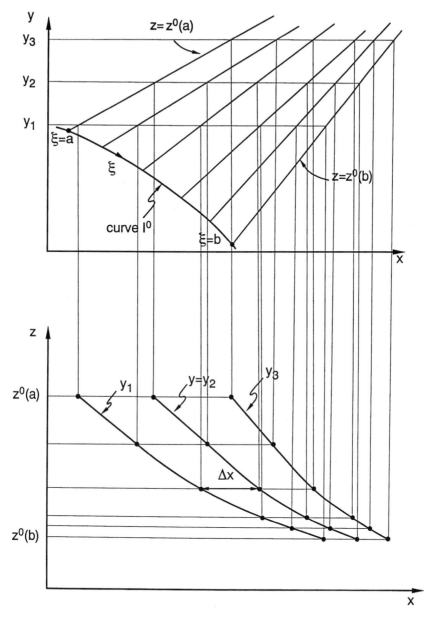

Figure 6.6 Compression wave.

converging, they will intersect at some points away from the initial condition curve, I^0. Recalling that on each of such characteristics a different constant value of the solution propagates, it follows that a conflict arises at each point where two such characteristics intersect. In particular, at this point (i.e., for a given pair of values for x and y) the solution z should take simultaneously two different values. This is physically meaning-

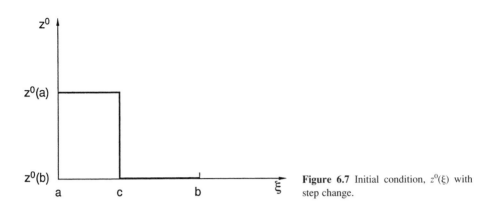

Figure 6.7 Initial condition, $z^0(\xi)$ with step change.

less, and indeed we need to formulate some new argument in order to resolve this conflict and develop a satisfactory solution.

The same difficulty arises in the case of the centered wave illustrated in Figure 6.8b, which is derived from the compression wave shown in the upper part of Figure 6.6 using the same procedure described above—that is, by letting a and b approach an intermediate point c on the I^0 curve while keeping both $z^0(a)$ and $z^0(b)$ constant. In this case the intersections among the characteristic straight lines occur immediately from the point $\xi = c$ on the curve I^0.

Example. A Simple Wave Solution. Let us now consider the model of a physical process where the difficulties mentioned above do not arise. In particular, we refer again to the chromatography model (6.5.9) examined in the previous section, but now assuming that the local equilibrium between the mobile and the stationary phases is not linear but rather described by a nonlinear Langmuir isotherm. Thus eq. (6.2.4) is replaced by

$$\Gamma = \Gamma^\infty \frac{KC}{1 + KC} \tag{6.2.34}$$

and the chromatography model is then given by

$$V(C)\frac{\partial C}{\partial x} + \frac{\partial C}{\partial t} = 0, \qquad 0 < x < L \text{ and } t > 0 \tag{6.2.35}$$

Due to the nonlinearity in the equilibrium condition (6.2.34), the parameter $V(C)$ is now a function of the concentration in the mobile phase, that is,

$$V(C) = \frac{v}{\varepsilon + (1 - \varepsilon)\Gamma^\infty K/(1 + KC)^2} \tag{6.2.35a}$$

Let us consider a desorption or regeneration operation, where the chromatographic column, initially saturated with a concentration $C = \bar{C}$ of an adsorbable component, is fed with an inert stream—that is, $C = 0$. This is represented by the following IC:

$$I_1: \qquad C = C^0(\xi) = \bar{C}, \quad x = \xi, t = 0, \quad \text{and} \quad 0 \le \xi \le L \tag{6.2.36a}$$

$$I_2: \qquad C = C^i(\xi) = 0, \quad x = 0, t = \xi, \quad \text{and} \quad \xi > 0 \tag{6.2.36b}$$

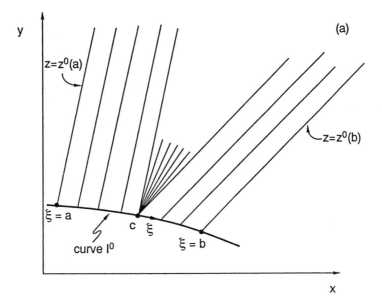

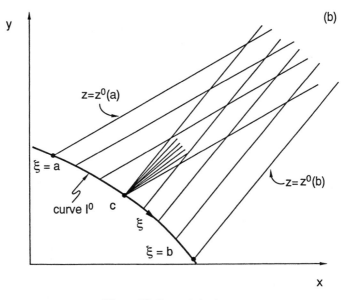

Figure 6.8 Centered simple wave.

The slope of the projections of the characteristic straight lines on the (x, t) plane is given by

$$\frac{dt}{dx} = \frac{1}{V(C)} \tag{6.2.37}$$

Thus by combining eqs. (6.2.36) and (6.2.37) it is possible to determine the characteristics on the (x, t) plane and then to derive the solution.

Along the characteristics originating from the IC curve I_1, as indicated by eq. (6.2.36a), the solution is constant, $C = \overline{C}$. Therefore, these are parallel straight lines whose slope is given by eqs. (6.2.35a) and (6.2.37) as follows:

$$\frac{1}{V(\overline{C})} = \frac{1}{v}\left[\varepsilon + (1 - \varepsilon)\frac{\Gamma^\infty K}{(1 + K\overline{C})^2}\right] \tag{6.2.38}$$

Similarly, the characteristics originating from the IC curve I_2, as given by eq. (6.2.36b), constitute a family of parallel straight lines where the solution is constant (i.e., $C = 0$), and their slope is given by

$$\frac{1}{V(0)} = \frac{1}{v}[\varepsilon + (1 - \varepsilon)\Gamma^\infty K] \tag{6.2.39}$$

where note that $1/V(0) > 1/V(\overline{C})$.

At the point $(x = 0, t = 0)$, where the initial condition curves I_1 and I_2 meet, the solution exhibits a discontinuity going from the value $C = \overline{C}$, given by $C^0(0)$ in eq. (6.2.36a), to the value $C = 0$, given by $C^i(0)$ in eq. (6.2.36b). This leads to a centered wave, constituted by a family of straight lines originating from the point $(x = 0, t = 0)$, whose slopes vary from $1/V(\overline{C})$ to $1/V(0)$; that is,

$$t = \frac{x}{V(C)} \qquad \text{with } 0 \le C \le \overline{C} \tag{6.2.40}$$

A complete picture of the characteristic lines on the (x, t) plane is shown in Figure 6.9a. The two sets of parallel characteristics derived above correspond to two linear waves, while the centered wave originates at $(x = 0, t = 0)$. Note that the characteristics of the centered wave never intersect those of the two linear waves. This guarantees that the solution is unique at any point on the (x, t) plane.

An analytical representation of the solution can be obtained using the parametric expressions of the characteristics. Thus integrating eq. (6.2.37) with the IC given by eq. (6.2.36) yields

$$C = \overline{C} \qquad \text{at } t = \frac{1}{V(\overline{C})}(x - \xi) \quad \text{for } 0 \le \xi \le L \tag{6.2.41a}$$

$$C = 0 \qquad \text{at } t = \xi + \frac{x}{V(0)} \qquad \text{for } \xi > 0 \tag{6.2.41b}$$

Eliminating the parameter ξ, an explicit expression for the solution is obtained:

$$C = \overline{C} \qquad \text{for } 0 \le t \le \frac{x}{V(\overline{C})} \tag{6.2.42a}$$

$$C = 0 \qquad \text{for } t > \frac{x}{V(0)} \tag{6.2.42b}$$

In the remaining time interval, the solution is given by the centered wave (6.2.40); that is,

$$t = \frac{x}{V(C)} \qquad \text{with } 0 < C < \overline{C} \quad \text{for } \frac{x}{V(\overline{C})} < t < \frac{x}{V(0)} \tag{6.2.42c}$$

Thus the solution has the form of a concentration wave moving in time through the column. This is illustrated in Figure 6.9b, where the concentration profile $C(x)$ is shown for various values of t. It is seen that, while moving along the column, the concentration wave, which

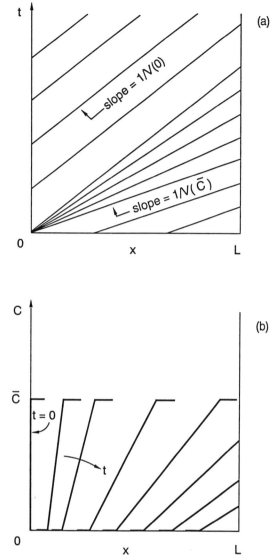

Figure 6.9 Solution of eq. (6.2.35) with ICs (6.2.36). (*a*) Complete picture of the characteristic lines on the (*x, t*) plane. (*b*) Concentration profile $C(x)$ for various values of *t*.

is initially very steep (i.e., a step function), becomes wider. This is the typical feature of expansion waves that we have discussed earlier.

6.3   SHOCKS AND WEAK SOLUTIONS

In the previous section we noted that a problem arises when the characteristic curves on the (*x, y*) plane intersect at some point. In particular, the adopted procedure leads to a multivalued solution z—that is, one which attains more than one value at the same point (*x, y*). This is not physically permitted, and so the obtained solution is not acceptable. This problem can be overcome by allowing the solution to exhibit a discontinuity, referred to as a *shock*, which propagates in the domain of integration.

In this section we discuss the occurrence of discontinuities in the solution of first-order PDEs. We will see that discontinuities may either propagate from the initial conditions or originate somewhere within the domain of integration. This is a rather unique feature of this type of equation. For example, in the case of second-order PDEs considered in section 5.5, discontinuities in the solution occur only if introduced in the boundary conditions, from which they propagate into the domain of integration.

The treatment is developed for homogeneous quasilinear PDEs, and in particular with reference to reducible equations. To illustrate the physical implications of the development of discontinuities, we first introduce this subject through an example.

6.3.1 The Occurrence of Shocks: An Introductory Example

Let us consider again the chromatography model (6.2.35) but now providing new ICs:

$$I_1: \qquad C = C^0(\xi_1) = 0, \qquad x = \xi_1, t = 0, 0 \le \xi_1 \le L \qquad (6.3.1a)$$

$$I_2: \qquad C = \begin{cases} C^i(\xi_2) = \dfrac{\overline{C}t}{\Delta t}, & x = 0, t = \xi_2, 0 < \xi_2 < \Delta t \\ C^i(\xi_3) = \overline{C}, & x = 0, t = \xi_3, \xi_3 > \Delta t \end{cases} \qquad (6.3.1b)$$

This corresponds to the adsorption mode of operation of a chromatographic column initially filled with an inert. The concentration of the adsorbable component in the feed stream increases linearly from $C = 0$ to $C = \overline{C}$ as time goes from $t = 0$ to $t = \Delta t$ and then remains constant at $C = \overline{C}$ for $t > \Delta t$. In this situation, which is not frequent in practice, there is no discontinuity in the value of C in the IC. Note that such discontinuities were in fact present [see eq. (6.2.36)] in the desorption problem discussed in section 6.2.2. This choice allows us to better illustrate the origin of discontinuities in the solution of first-order PDEs.

Let us introduce the following positive dimensionless quantities:

$$z = \frac{x}{L}, \quad \tau = \frac{t}{\Delta t}, \quad u = \frac{C}{\overline{C}}, \quad \alpha = \frac{L\varepsilon}{v \Delta t}, \quad \beta = \frac{(1 - \varepsilon)\Gamma^\infty K}{\varepsilon}, \quad \sigma = K\overline{C} \quad (6.3.2)$$

Thus eqs. (6.2.35) and (6.3.1) reduce to

$$\frac{\partial u}{\partial z} + \psi(u) \frac{\partial u}{\partial \tau} = 0, \qquad 0 < z < 1, \tau > 0 \qquad (6.3.3)$$

where $\psi(u)$ is the reciprocal of the dimensionless rate of wave propagation defined as follows:

$$\psi(u) = \frac{L}{\Delta t V(C)} = \alpha \left[1 + \frac{\beta}{(1 + \sigma u)^2} \right] \qquad (6.3.3a)$$

with ICs in parametric form

$$I_1: \quad u = 0 \quad z = \xi_1, \tau = 0, 0 \le \xi_1 \le 1 \qquad (6.3.4a)$$

$$I_2: \quad u = \begin{cases} \xi_2 & z = 0, \tau = \xi_2, 0 < \xi_2 < 1 \\ 1 & z = 0, \tau = \xi_3, \xi_3 > 1 \end{cases} \qquad (6.3.4b)$$

The characteristic straight lines on the (z, τ) plane are obtained by integrating the equation

$$\frac{d\tau}{dz} = \psi(u) \qquad (6.3.5)$$

with appropriate ICs. In particular, the IC (6.3.4a) yields the following family of parallel straight lines originating from curve I_1, on which $u = 0$:

$$\tau = \psi(0)(z - \xi_1) \quad \text{for } 0 \le \xi_1 \le 1 \qquad (6.3.6a)$$

where

$$\psi(0) = \alpha(1 + \beta) \qquad (6.3.6b)$$

For the characteristics emanating from the IC I_2, two different expressions are obtained depending upon whether the portion with $0 < \xi_2 < 1$ or with $\xi_3 > 1$ is considered. Thus using the IC (6.3.4b) we get

$$\tau = \xi_2 + \psi(\xi_2)z \quad \text{for } 0 < \xi_2 < 1 \qquad (6.3.7a)$$

and

$$\tau = \xi_3 + \psi(1)z \quad \text{for } \xi_3 > 1 \qquad (6.3.7b)$$

where

$$\psi(1) = \alpha\left[1 + \frac{\beta}{(1 + \sigma)^2}\right] < \psi(0) \qquad (6.3.7c)$$

A complete picture of these three one-parameter families of straight lines is shown in Figure 6.10. It is seen that eqs. (6.3.6a) and (6.3.7b) represent two families of parallel straight lines, which lead to linear waves. They enclose the third family of straight lines (6.3.7a), whose slopes decrease while moving along the τ axis, and lead to a compression wave.

Indeed all members of the third family of straight lines intersect, leading to the difficulty mentioned earlier. Let us now identify where such intersection occurs for the first time—that is, for the smallest value of z (or τ). For this, we need to consider the intersection of each member of the family (6.3.7a) with the closest member of the other two—that is, the one given by $\xi_1 = 0$ in eq. (6.3.6a) and that given by $\xi_3 = 1$ in eq. (6.3.7b).

In the first case, the location $z = z_1$ where the intersection occurs is obtained by

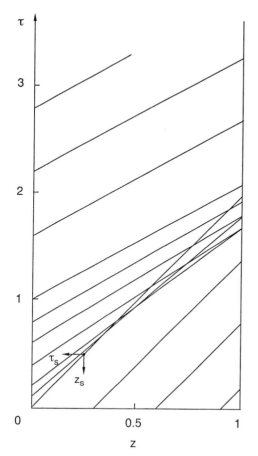

Figure 6.10 Characteristic curves on the (z, τ) plane for the chromatography model (6.3.3) with ICs (6.3.4); $\alpha = 1$, $\beta = 1$, $\sigma = 2$.

combining eq. (6.3.6a) with $\xi_1 = 0$ and eq. (6.3.7a)—that is,

$$\psi(0)z_1 = \psi(\xi_2)z_1 + \xi_2 \qquad \text{with } 0 < \xi_2 < 1 \tag{6.3.8}$$

Solving for z_1, using eqs. (6.3.3a) and (6.3.6b), leads to

$$z_1 = \frac{(1 + \sigma\xi_2)^2}{\alpha\beta\sigma(2 + \sigma\xi_2)} \qquad \text{with } 0 < \xi_2 < 1 \tag{6.3.9}$$

We are interested in the member of the family (6.3.7a) which first intersects the straight line $\tau = \psi(0)z$—that is, in the ξ_2 value which makes z_1 minimum in eq. (6.3.9). Such a value is readily seen to be $\xi_2 = 0$, noting from eq. (6.3.9) that since $\sigma > 0$, $dz_1/d\xi_2 > 0$ for $0 < \xi_2 < 1$. Accordingly, the coordinates of the point (z_s, τ_s) where the first intersection occurs are given by

$$z_s = \frac{1}{2\alpha\beta\sigma}, \qquad \tau_s = \frac{(1 + \beta)}{2\beta\sigma} \tag{6.3.10}$$

Similarly, we can determine the point $z = z_2$ where the generic member of the family (6.3.7a) intersects the straight line $\tau = 1 + \psi(1)z$. By combining eq. (6.3.7b) with $\xi_3 = 1$ and eq. (6.3.7a), we get

$$z_2 = \frac{(1 + \sigma\xi_2)^2(1 + \sigma)^2}{\alpha\beta\sigma[2 + \sigma(1 + \xi_2)]} \quad \text{with } 0 < \xi_2 < 1 \quad (6.3.11)$$

Comparing with eq. (6.3.9), the equation above may be rewritten as follows:

$$z_2 = z_1 \frac{(2 + \sigma\xi_2)(1 + \sigma)^2}{[2 + \sigma(1 + \xi_2)]} = z_1 g(\sigma, \xi_2) \quad (6.3.12)$$

Noting that $g(0, \xi_2) = 1$ and $\partial g/\partial\sigma > 0$ for all $\sigma > 0$ and $0 < \xi_2 < 1$, we see that $z_2 > z_1$. Thus we only need to consider the intersections with the straight line $\tau = \psi(0)z$ discussed earlier, since they occur at smaller values of z.

As a result of the foregoing analysis, we can conclude that the smaller value of z at which the characteristic curves on the (z, τ) plane intersect is $z = z_s$, as given by eq. (6.3.10). Therefore, when the parameters are such that $z_s > 1$, that is,

$$\alpha\beta\sigma < \tfrac{1}{2} \quad (6.3.13)$$

the characteristic curves do not intersect in the domain of interest, and the solution is obtained through the procedure developed previously in section 6.2.2. The difficulties arise only when condition (6.3.13) is violated.

Let us now derive the analytic solution of eq. (6.3.3) to investigate its behavior. By eliminating ξ_1 from eq. (6.3.6a) and ξ_3 from eq. (6.3.7b), we obtain

$$u = 0 \quad \text{for } 0 < \tau < \psi(0)z \quad (6.3.14a)$$

$$u = 1 \quad \text{for } \tau > 1 + \psi(1)z \quad (6.3.14b)$$

In the intermediate region of τ values the characteristics on the (z, τ) plane are given by eq. (6.3.7a). Thus recalling that along these the IC $u = \xi_2$ propagates, the solution can be written in the following implicit form:

$$\tau = u + \psi(u)z \quad \text{with } 0 < u < 1 \quad (6.3.14c)$$

Using eq. (6.3.3a), eq. (6.3.14c) can be made explicit with respect to z as follows:

$$z = \frac{(\tau - u)(1 + \sigma u)^2}{\alpha[\beta + (1 + \sigma u)^2]} \quad \text{for } 0 < u < 1 \quad (6.3.15)$$

This form of the solution can be used for computing the concentration profile along the column for a given time value, τ since it provides for each $u \in [0, 1]$ the corresponding value of the axial coordinate z. For z values not covered by this equation, the concentration profile can be completed using eqs. (6.3.14a) and (6.3.14b).

An example is shown in Figure 6.11. It is seen that the solution has the form of a wave moving through the column, which represents the concentration of the adsorbable component entering the initially empty column. Moreover, as is characteristic of compression waves, the profiles become steeper while moving along the column.

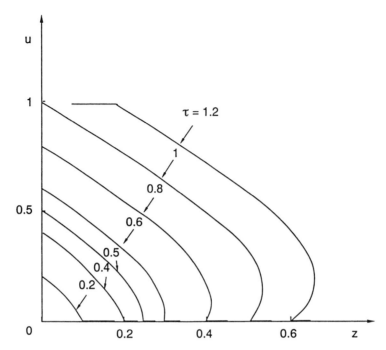

Figure 6.11 Concentration profiles given by eq. (6.3.14) for various time values; $\alpha = 1$, $\beta = 1$, $\sigma = 2$.

From a physical point of view, this can be understood noting that the velocity of propagation of the solution in the (z, τ) plane, which from eq. (6.3.3a) is given by

$$\frac{1}{\psi(u)} = \frac{\Delta t V(C)}{L} = \frac{(1 + \sigma u)^2}{\alpha[\beta + (1 + \sigma u)^2]} \tag{6.3.16}$$

is a monotonically increasing function of u. In other words, the larger concentration values move along the column faster than smaller concentration values. As a consequence, in the adsorption mode where the inlet concentration is larger than that present in the column initially, the concentration wave becomes steeper while moving along the column; that is, it is a compression wave. Note that in the desorption mode, where the inlet concentration is smaller than that present in the column initially, the wave becomes wider while moving along the column (i.e., it is an expansion wave), in agreement with the results obtained in section 6.2.2.

This behavior is evident in Figure 6.11, where it is seen that the larger u values tend to reach and overtake the lower ones. This leads to concentration profiles whose slope, in the region of small u, increases continuously in time up to the point where the profile becomes multivalued. Let us now determine the point in the (z, τ) plane where the transition from a one-valued to a two-valued profile occurs. For this, it is convenient to regard z as the dependent and u as the independent variable. From eq. (6.3.15) we get

$$\frac{\partial z}{\partial u} = \frac{(1 + \sigma u)}{\alpha} \frac{2\sigma\beta(\tau - u) - (1 + \sigma u)[\beta + (1 + \sigma u)^2]}{[\beta + (1 + \sigma u)^2]^2} \tag{6.3.17}$$

and so the curve $z = z(u)$ exhibits an extremum at $u = u_M$, given by

$$F(u_M) = 2\sigma\beta(\tau - u_M) - (1 + \sigma u_M)[\beta + (1 + \sigma u_M)^2] = 0 \qquad (6.3.18)$$

The second-order derivative of $z(u)$ is obtained from eq. (6.3.17) after some algebraic manipulations as follows:

$$\frac{\partial^2 z}{\partial u^2} = \frac{2\beta\sigma}{\alpha[\beta + (1 + \sigma u)^2]^3} \times \{\sigma(\tau - u)[\beta - 3(1 + \sigma u)^2]$$

$$- 2(1 + \sigma u)[\beta + (1 + \sigma u)^2]\} \qquad (6.3.19)$$

and using eq. (6.3.18), we have:

$$\left(\frac{\partial^2 z}{\partial u^2}\right)_{u=u_M} = -\frac{3\sigma(1 + \sigma u_M)}{[\beta + (1 + \sigma u_M)^2]} < 0 \qquad (6.3.20)$$

This allows us to conclude that the extreme point at $u = u_M$ is a maximum. Furthermore, to investigate the effect of τ on u_M, we can differentiate both sides of eq. (6.3.18), to give

$$\frac{dF}{d\tau} = \frac{\partial F}{\partial \tau} + \frac{\partial F}{\partial u_M} \frac{du_M}{d\tau} = 0 \qquad (6.3.21)$$

thus leading to

$$\frac{du_M}{d\tau} = \frac{2\beta}{3[\beta + (1 + \sigma u_M)^2]} > 0 \qquad (6.3.22)$$

Thus the maximum of the $z(u)$ curve occurs at larger values of u_M as time increases. This indicates that the shape of the $z(u)$ curve changes in time as illustrated schematically in Figure 6.12. In particular, it is seen that the transition from a one-valued to a two-valued $u(z)$ profile in Figure 6.11 corresponds to the point where the maximum in the $z(u)$ curve in Figure 6.12 first crosses the z axis—that is, when $u_M = 0$. Substituting this in eq. (6.3.18), the critical time value at which this transition occurs is obtained:

$$\tau_c = \frac{(1 + \beta)}{2\beta\sigma} \qquad (6.3.23)$$

The corresponding value of z is obtained by substituting eq. (6.3.23) in eq. (6.3.15), to give

$$z_c = \frac{\tau_c}{\alpha(1 + \beta)} = \frac{1}{2\alpha\beta\sigma} \qquad (6.3.24)$$

By comparing with eq. (6.3.10) it is readily seen that these are in fact the coordinates of the point (z_s, τ_s) where the characteristics first intersect.

We have now seen that the point (z_s, τ_s) represents the first location where the solution becomes multivalued. Clearly, although mathematically correct, this is physically not permitted since we cannot have two distinct values of concentration simultaneously at the same point in space. This inconsistent result arises because the first-order model (6.3.3) which we are considering fails and loses its reliability in this region.

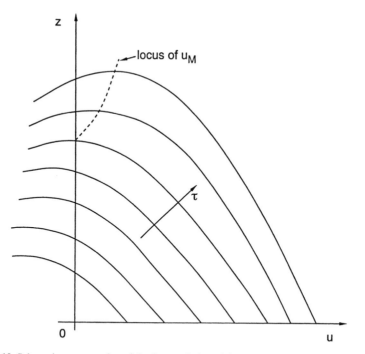

Figure 6.12 Schematic representation of the time evolution of the $z = z(u)$ curve given by eq. (6.3.15).

The reason is that here the concentration profiles are steep and exhibit large second-order derivative values. Thus the effect of axial dispersion is no longer negligible, and we should revise the model by taking it into account. We will return to this issue in section 6.5.1, where a revised model is discussed.

Since the model (6.3.3) provides a physically invalid solution, we *replace* it in this region with a new one based on the following statement: The larger values of concentration do not overtake the smaller ones but rather they superimpose, thus creating a *discontinuity*, which is referred to as a *shock*. The shape of the concentration profile takes the form shown in Figure 6.13, where C_+ and C_- represent the values of the concentration which immediately follow and precede the discontinuity, respectively. Note that the discontinuity originates from (z_s, τ_s) and at this point $C_+ = C_- = 0$.

To complete the solution of the problem, we need to characterize the discontinuity. In particular, we now determine its propagation velocity in the (z, τ) plane, V_s. For this we require that the movement of the shock satisfy the mass balance. The difference relative to the original model (6.3.3), which also states the conservation of mass, is that in this case we do not represent the terms involved in the balance in differential form but rather in algebraic form, because of the discontinuity.

The steady-state mass balance at the shock interface requires that no mass be accumulated—that is, that the flows of adsorbable component entering and leaving the interface be identical. Both flows can be regarded as given by the sum of the two contributions. The first is associated with the convective flow of the mobile phase, whose velocity with respect to the shock is $(v/\varepsilon - V_s)$. The second arises from the stationary phase which, to maintain local equilibrium with the mobile phase, releases or subtracts finite amounts of the adsorbable component. Thus, with reference to Figure 6.13, the mass balance at the shock interface is given by

$$\varepsilon C_+ \left(\frac{v}{\varepsilon} - V_s \right) + (1 - \varepsilon)\Gamma_- V_s = \varepsilon C_- \left(\frac{v}{\varepsilon} - V_s \right) + (1 - \varepsilon)\Gamma_+ V_s \quad (6.3.25)$$

which can be rearranged as follows:

$$\frac{v}{V_s} = \varepsilon + (1 - \varepsilon) \frac{\Gamma_+ - \Gamma_-}{C_+ - C_-} \quad (6.3.26)$$

Using the Langmuir equilibrium isotherm (6.2.34) and the dimensionless quantities (6.3.2), eq. (6.3.26) yields

$$\frac{v}{\varepsilon V_s} = 1 + \frac{\beta}{(u_+ - u_-)} \left[\frac{u_+}{(1 + \sigma u_+)} - \frac{u_-}{(1 + \sigma u_-)} \right] \quad (6.3.27)$$

To determine the trajectory of the shock on the (z, τ) plane, we derive from eq. (6.3.27) the slope ψ_s as follows:

$$\psi_s = \left(\frac{d\tau}{dz} \right)_{shock} = \frac{L}{\Delta t V_s} = \alpha \left[1 + \frac{\beta}{(1 + \sigma u_+)(1 + \sigma u_-)} \right] \quad (6.3.28)$$

Thus the shock originates at the point (z_s, τ_s) and then propagates in the (z, τ) plane along the trajectory defined by eq. (6.3.28). In particular, at the origin (z_s, τ_s), since as seen above, $u_+ = u_- = 0$, the slope of the trajectory is given by

$$\psi_s = \alpha(1 + \beta) \quad (6.3.29)$$

Since from eq. (6.3.3a), $\psi_s = \psi(0)$, the shock trajectory at its origin is tangent to the characteristic (6.3.6a) with $\xi_1 = 0$. Then, as shown in Figure 6.14, this trajectory continues in the (z, τ) plane by separating the two families of characteristics originating from the IC curves I_1 and I_2, respectively. Each one of the characteristics impinging

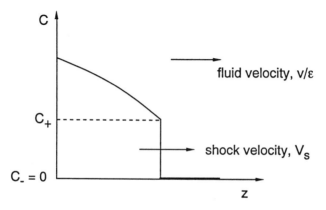

Figure 6.13 Concentration profile exhibiting a shock.

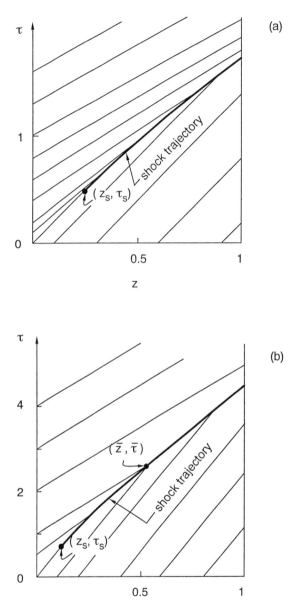

Figure 6.14 Shock trajectories on the (z, τ) plane: (a) $\alpha = 1$, $\beta = 1$, $\sigma = 2$; (b) $\alpha = 2$, $\beta = 2$, $\sigma = 1$.

into the shock trajectory carries a constant value of u. In particular, u_+ and u_- are the concentration values which propagate along the characteristics impinging on the shock trajectory from the left- and right-hand side, respectively. Thus, to compute the shock trajectory we need to integrate eq. (6.3.28) together with the relations for u_+ and u_-, arising from the definitions of the characteristics on the (z, τ) plane.

Since all the characteristics on the right-hand side of the shock trajectory originate from the IC curve I_1, they propagate the same concentration value $u = 0$, so we have

$$u_- = 0 \tag{6.3.30}$$

For the value of u_+, two different situations may arise. If the characteristic impinging the shock trajectory from the left-hand side originates from the first portion of the IC curve I_2, as given by eq. (6.3.7a), then the concentration value is given by eq. (6.3.14c) and accordingly

$$\tau = u_+ + \alpha \left[1 + \frac{\beta}{(1 + \sigma u_+)^2} \right] z \tag{6.3.31a}$$

If the impinging characteristic originates instead from the second portion of the IC curve I_2, as given by eq. (6.3.7b), then the propagated value of the solution is given by eq. (6.3.14b)—that is,

$$u_+ = 1 \tag{6.3.31b}$$

Using eq. (6.3.30), eq. (6.3.28) reduces to

$$\frac{d\tau}{dz} = \alpha \left[1 + \frac{\beta}{(1 + \sigma u_+)} \right] \tag{6.3.32}$$

which, integrated together with eqs. (6.3.31) and the IC $\tau = \tau_s$ at $z = z_s$, where $u_+ = 0$, leads to the shock trajectory shown in Figure 6.14.

To derive an analytical expression for the shock trajectory, it is convenient to represent it in parametric form by taking $u_+ \in [0, 1]$ as a parameter; that is, $z = z(u_+)$ and $\tau = \tau(u_+)$. Thus, differentiating both sides of eq. (6.3.31a) with respect to u_+ yields

$$\frac{d\tau}{du_+} = 1 - \frac{2\alpha\beta\sigma z}{(1 + \sigma u_+)^3} + \alpha \left[1 + \frac{\beta}{(1 + \sigma u_+)^2} \right] \frac{dz}{du_+} \tag{6.3.33}$$

By rewriting eq. (6.3.32) in the equivalent form

$$\frac{d\tau}{du_+} = \alpha \left[1 + \frac{\beta}{(1 + \sigma u_+)} \right] \frac{dz}{du_+} \tag{6.3.34}$$

and substituting in eq. (6.3.33), we obtain

$$\frac{\alpha\beta\sigma u_+}{(1 + \sigma u_+)^2} \frac{dz}{du_+} = 1 - \frac{2\alpha\beta\sigma z}{(1 + \sigma u_+)^3} \tag{6.3.35}$$

Since by definition

$$\frac{d}{du_+} \left[\frac{z u_+^2}{(1 + \sigma u_+)^2} \right] = \frac{u_+^2}{(1 + \sigma u_+)^2} \frac{dz}{du_+} + \frac{2 u_+ z}{(1 + \sigma u_+)^3} \tag{6.3.36}$$

comparing eqs. (6.3.35) and (6.3.36) gives

$$\frac{d}{du_+} \left[\frac{z u_+^2}{(1 + \sigma u_+)^2} \right] = \frac{u_+}{\alpha\beta\sigma} \tag{6.3.37}$$

Integrating with the IC $u_+ = 0$ at $z = z_s$, given by eq. (6.3.10), leads to

$$z = \frac{(1 + \sigma u_+)^2}{2\alpha\beta\sigma} \tag{6.3.38}$$

Solving for u_+ gives

$$u_+ = \frac{1}{\sigma} [\sqrt{2\alpha\beta\sigma z} - 1] \tag{6.3.39}$$

and substituting in eq. (6.3.31a) leads to the expression for the shock trajectory:

$$\tau = \alpha z - \frac{1}{2\sigma} + \sqrt{\frac{2\alpha\beta z}{\sigma}} \tag{6.3.40}$$

The expression above is valid as long as the shock trajectory does not intersect the first straight-line characteristic of the family (6.3.7b)—that is, $\tau = 1 + \psi(1)z$. Using eq. (6.3.40), it is readily seen that this intersection occurs at the point $(\bar{z}, \bar{\tau})$, where

$$\bar{z} = \frac{(1 + \sigma)^2}{2\alpha\beta\sigma}, \qquad \bar{\tau} = \frac{(1 + \sigma)^2 + \beta(1 + 2\sigma)}{2\beta\sigma} \tag{6.3.41}$$

and from eq. (6.3.39), $u_+ = 1$. Note that in the case where $\bar{z} > 1$, that is,

$$\alpha < \frac{(1 + \sigma)^2}{2\beta\sigma} \tag{6.3.42}$$

eq. (6.3.40) provides the shock trajectory in the entire domain of interest. This is the situation illustrated in Figure 6.14a.

When condition (6.3.42) is violated, as shown in Figure 6.14b, the portion of the trajectory beyond the point $(\bar{z}, \bar{\tau})$ is obtained by integrating eq. (6.3.32) with the value of u_+ given by eq. (6.3.31b) and the IC (6.3.41), as follows:

$$\tau = \bar{\tau} + \alpha \left[1 + \frac{\beta}{(1 + \sigma)} \right] (z - \bar{z}) \tag{6.3.43}$$

Thus beyond the point $(\bar{z}, \bar{\tau})$, the shock trajectory becomes a straight line whose slope is larger than $\psi(1)$, and then it intersects only the characteristics (6.3.7b) originating from the portion of the curve I_2 defined by $\xi_3 > 1$.

In summary, the shock trajectory originates, as shown in Figure 6.14, at the point (z_s, τ_s) where it is tangent to the characteristic $\tau = \psi(0)z$, emanating from the origin. Then it moves into the (z, τ) plane with a continuously decreasing slope (as u_+ increases from 0 to 1) given by eq. (6.3.32), up to the point $(\bar{z}, \bar{\tau})$ where $u_+ = 1$. Beyond this point, which is in fact reached only if condition (6.3.42) is violated, the shock trajectory becomes the straight line (6.3.43).

We can now report the complete solution of the problem, illustrated by the concentration waves moving through the column as shown in Figure 6.15. Note that the multivalued regions in Figure 6.11 have now been replaced by discontinuities. The form of the solution depends on the position of such discontinuities. Thus before the occurrence of the shock, that is,

$$z < z_s \quad \text{or} \quad \tau < \tau_s \tag{6.3.44}$$

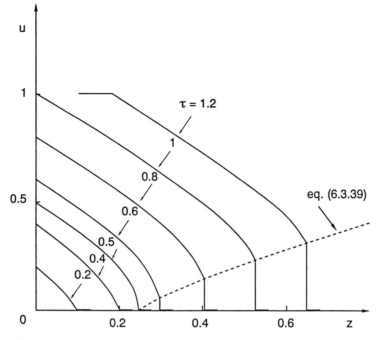

Figure 6.15 Concentration profiles for various time values; $\alpha = 1$, $\beta = 1$, $\sigma = 2$.

the solution is given by the relations (6.3.14) developed earlier. After the shock has been formed—that is, when both conditions (6.3.44) are not satisfied—the solution is $u = 0$ in the region below the shock trajectory (see Figure 6.14), that is,

$$\tau < \alpha z - \frac{1}{2\sigma} + \sqrt{\frac{2\alpha\beta z}{\sigma}} \qquad \text{for } z_s < z < \bar{z} \qquad (6.3.45a)$$

and

$$\tau < \bar{\tau} + \alpha\left[1 + \frac{\beta}{(1 + \sigma)}\right](z - \bar{z}) \qquad \text{for } z > \bar{z} \qquad (6.3.45b)$$

In the region above the shock trajectory, the solution is given by eq. (6.3.15) for

$$\alpha z - \frac{1}{2\sigma} + \sqrt{\frac{2\alpha\beta z}{\sigma}} < \tau < 1 + \psi(1)z \quad \text{and} \quad z_s < z < \bar{z} \qquad (6.3.46a)$$

while it is $u = 1$ for larger time values; that is,

$$\tau > 1 + \psi(1)z \quad \text{for } z_s < z < \bar{z} \qquad (6.3.46b)$$

and

$$\tau > \bar{\tau} + \alpha\left(1 + \frac{\beta}{1 + \sigma}\right)(z - \bar{z}) \qquad \text{for } z > \bar{z} \qquad (6.3.46c)$$

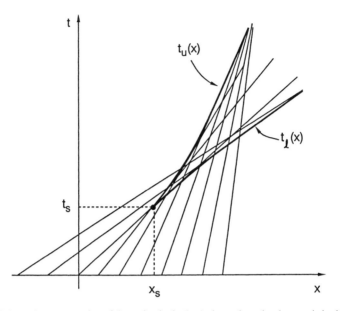

Figure 6.16 Schematic representation of the region in the (x, t) plane where the characteristics intersect each other.

Using the same parameter values as in Figure 6.14a, we can compute the solution profiles shown in Figure 6.15. From these, and particularly from that corresponding to $\tau = 1.2$, we can recognize the various portions of the solution corresponding to the three simple waves and the shock shown in Figure 6.14a. The two extreme constant portions—that is, $u = 1$ and $u = 0$—arise from the two linear waves: The vertical segment is the shock, while the remaining portion is the compression wave.

6.3.2 Formation and Propagation of Shocks

In the previous section, in order to obtain a physically admissible solution, we had to permit the occurrence of a discontinuity. Solutions of this kind, typical of first-order PDEs, are referred to as *weak solutions*, which indicates that they are not continuously differentiable.

It is worth stressing that this discontinuity does not necessarily arise from discontinuities in the ICs. We have seen several instances where a discontinuity, originally introduced in the ICs, propagates into the domain of integration. These include the first-order linear chromatography model (6.2.5) whose solution is shown in Figure 6.3, as well as the second-order vibrating string model (5.5.36) illustrated in Figure 5.15. In both cases the discontinuity propagates along a characteristic line, with a velocity determined by the slope of the characteristic itself. This is not the case for the discontinuity of interest here. As shown in the example of section 6.3.1, this originates at some point *within* the domain of integration and then propagates along a shock trajectory which is not a characteristic curve.

In the following we discuss the general procedure used to derive weak solutions, and in particular to characterize the propagation of discontinuities. The two basic steps are as follows:

1. Determine the location from where the shock originates in the (x, t) plane.

2. Evaluate the propagation velocity of the shock:

$$\left(\frac{dx}{dt}\right)_{\text{shock}} = V_s \tag{6.3.47}$$

The trajectory of the shock on the (x, t) plane is obtained by integrating the propagation velocity (6.3.47), taking as IC the origin of the shock. As seen in section 6.3.1, the expression of the shock velocity involves the values of the solution which immediately follow and precede the shock. Thus we have to couple eq. (6.3.47) with the equations of the simple waves originating from the IC curves and impinging on the two sides of the shock trajectory.

We consider the evaluation of the shock origin and propagation velocity in the general context of the kinematic wave equation. This is a reducible equation given by eq. (6.1.3):

$$\frac{\partial \gamma}{\partial x} + \psi(\gamma) \frac{\partial \gamma}{\partial t} = 0 \tag{6.3.48}$$

where

$$\frac{1}{\psi(\gamma)} = c(\gamma) = \frac{d\phi}{d\gamma} \tag{6.3.48a}$$

is the wave propagation velocity related to the equilibrium condition $\phi = \phi(\gamma)$, as given by eq. (6.1.2).

Origin of the Shock.   Let us first address the case where the wave velocity is positive [i.e., $c(\gamma) > 0$], and the initial condition is given along the x axis:

$$\gamma = \gamma^0(\xi), \qquad x = \xi, \qquad t = 0, \quad \text{and} \quad 0 \leq \xi \leq 1 \tag{6.3.49}$$

The characteristic curves of eq. (6.3.48) on the (x, t) plane are obtained by integrating the equation

$$\frac{dt}{dx} = \psi(\gamma) \tag{6.3.50}$$

using curve (6.3.49) as the IC. Since we are dealing with a reducible PDE, the characteristic curves are straight lines along which the solution γ is constant. In parametric form these are given by

$$t = \psi(\gamma^0(\xi))(x - \xi), \qquad 0 \leq \xi \leq 1 \tag{6.3.51}$$

which for any given value of ξ represents a straight line originating from the point

$(x = \xi, t = 0)$ with slope $\psi(\gamma^0(\xi))$. We have seen in section 6.3.1 that the shock originates at the point (x_s, t_s) where the first intersection among these characteristics occurs.

In the case where eq. (6.3.51) represents a compression wave, we can in fact expect that there exists a region in the (x, t) plane where the characteristic curves intersect each other, as shown schematically in Figure 6.16. Such a region is enveloped by two curves, $t_u(x)$ and $t_l(x)$, which for any given x value provide the upper and lower bounds for the values of t where such intersections occur. The origin (x_s, t_s) of the shock is then given by the point where these two curves originate—that is, $t_s = t_u(x_s) = t_l(x_s)$. This is the location where the first intersection among the characteristics occurs, which, as seen in the previous section, is also the point where the solution starts to be multivalued and hence physically not admissible.

To determine the point (x_s, t_s), we first need to define the upper and lower bounds of the region of characteristic intersections—that is, $t_u(x)$ and $t_l(x)$.

The characteristic straight line passing through a given point (x_0, t_0) on the (x, t) plane can be determined by solving the following equation for $0 \leq \xi \leq 1$:

$$t_0 = \psi(\gamma^0(\xi))(x_0 - \xi) = H(\xi) \tag{6.3.52}$$

and substituting the obtained value of ξ in eq. (6.3.51). If eq. (6.3.52) exhibits multiple solutions—that is, it is satisfied by two or more values of $\xi \in [0, 1]$—then there are two or more characteristic straight lines passing through the point (x_0, t_0). Thus, for any given value of x_0, the upper and lower bounds of the region where the characteristics intersect—that is, $t_u(x_0)$ and $t_l(x_0)$—are in fact the upper and lower bounds of the region where eq. (6.3.52) exhibits multiple solutions.

To determine the multiplicity region of eq. (6.3.52), we investigate the behavior of the function $H(\xi)$. Since we are interested only in positive time values and we have assumed $\psi = 1/c > 0$, eq. (6.3.52) implies that the domain of interest is $\xi \in [0, x_0]$, where $x_0 \in [0, 1]$. The values of $H(\xi)$ at the two extrema of this interval are given by

$$H(0) = \psi(\gamma^0(0))x_0 > 0 \tag{6.3.53a}$$

$$H(x_0) = 0 \tag{6.3.53b}$$

Further,

$$\frac{dH}{d\xi} = \frac{d\psi}{d\xi}(x_0 - \xi) - \psi \tag{6.3.54}$$

and two cases typically arise; they are discussed separately next.

Case 1. $d\psi/d\xi < 0$ for $0 \leq \xi \leq 1$

From eq. (6.3.54) it follows that, for any given value x_0, $dH/d\xi < 0$ for $0 < \xi < x_0$. Thus, as shown by curve a in Figure 6.17, $H(\xi)$ decreases monotonically and hence a unique solution of eq. (6.3.52) exists for all $t_0 \leq H(0)$. This means that for any point on the vertical axis $x = x_0$ in the (x, t) plane, only one characteristic straight line passes through. Since this argument holds for any $x_0 \in [0, 1]$, it follows that the characteristics never intersect on the (x, t) plane, and thus discontinuities (or shocks) cannot arise. This conclusion is in agreement with the results reported in section 6.2.2, which indicate that the condition $d\psi/d\xi < 0$ corresponds to a solution in the form of an expansion wave.

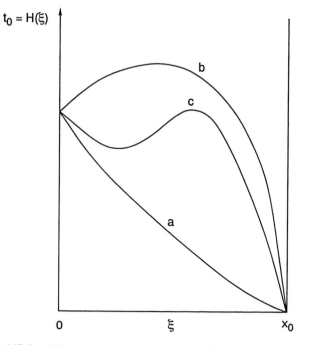

Figure 6.17 Qualitative behavior of the function $t_0 = H(\xi)$ defined by eq. (6.3.52).

Case 2. $d\psi/d\xi > 0$ for $0 < \xi < 1$

From eq. (6.3.54) it is seen that, for any given value of x_0, $dH/d\xi$ may be either negative or positive, and hence we can have intervals of t_0 values where eq. (6.3.52) exhibits multiple solutions. For example, if $dH/d\xi > 0$ at $\xi = 0$, then as shown by curve b in Figure 6.17, there is a multiplicity region which is bounded by $t_l = H(0)$ and $t_u = H(\xi_M)$ where ξ_M is the location at which $H(\xi)$ exhibits its maximum value—that is, $dH/d\xi = 0$ at $\xi = \xi_M$.

Another possible shape is that shown by curve c in Figure 6.17, where $H(\xi)$ exhibits a minimum and a maximum at $\xi = \xi_m$ and $\xi = \xi_M$, respectively. Also in this case we can determine the upper and lower bounds of the interval of t_0 where eq. (6.3.52) exhibits multiple solutions; that is, $t_l = H(\xi_m) < t_0 < t_u = H(\xi_M)$.

In general, we can conclude that the lower and upper bounds of the region of multiplicity for eq. (6.3.52) are given by

$$t_l = \text{minimum of } H(\xi_m) \text{ and } H(0) \qquad (6.3.55a)$$

where $dH/d\xi = 0$ and $d^2H/d\xi^2 > 0$ at $\xi = \xi_m$, and

$$t_u = H(\xi_M) \qquad (6.3.55b)$$

where $dH/d\xi = 0$ and $d^2H/d\xi^2 < 0$ at $\xi = \xi_M$.

Note that in eqs. (6.3.55), ξ_m is the location of the smallest of the local minima of $H(\xi)$, while ξ_M is the location of the largest of the local maxima.

The curves $t_l(x_0)$ and $t_u(x_0)$ on the (x, t) plane are obtained from eqs. (6.3.52) and (6.3.55) as follows:

$$t_0 = \psi(\xi)(x_0 - \xi) = H(\xi) \tag{6.3.56a}$$

$$\frac{dH}{d\xi} = \frac{d\psi}{d\xi}(x_0 - \xi) - \psi(\xi) = 0 \tag{6.3.56b}$$

For any given value of $x_0 \in [0, 1]$, we solve eq. (6.3.56b) for $\xi = \xi_m$ (minimum of H) and $\xi = \xi_M$ (maximum of H) and substitute them in eq. (6.3.56a) to yield $t_0 = t_l(x_0)$ and $t_0 = t_u(x_0)$, respectively. In the case where $H(\xi_m) > H(0)$, eq. (6.3.56b) for the lower bound is replaced by $\xi = 0$ and then the curve $t_l(x_0)$ is given by

$$t_l = \psi(0)x_0 \tag{6.3.57}$$

Now that we have determined the upper and lower bounds, $t_l(x_0)$ and $t_u(x_0)$, for any given value of $x_0 \in [0, 1]$, we next proceed to locate (x_s, t_s). This is done by finding the common point between $t_l(x_0)$ and $t_u(x_0)$, as seen in Figure 6.16.

Since both t_0 and ξ in eqs. (6.3.56a) and (6.3.56b) are functions of x_0—that is, $t_0 = t_0(x_0)$ and $\xi = \xi(x_0)$—differentiating both sides of eq. (6.3.56a) we obtain

$$\frac{dt_0}{dx_0} = \left[\frac{d\psi}{d\xi}(x_0 - \xi) - \psi(\xi) \right] \frac{d\xi}{dx_0} + \psi(\xi) \tag{6.3.58}$$

Noting from eq. (6.3.56b) that the term in square brackets vanishes, eq. (6.3.58) reduces to

$$\frac{dt_0}{dx_0} = \psi(\xi) > 0 \tag{6.3.59a}$$

where t_0 is either t_l or t_u, depending on $\xi = \xi_m$ or ξ_M, respectively. A similar conclusion is reached in the case where eq. (6.3.57) applies—that is,

$$\frac{dt_l}{dx_0} = \psi(0) > 0 \tag{6.3.59b}$$

Therefore we can conclude that the curves $t_l(x)$ and $t_u(x)$, which envelop the region of the (x, t) plane where the characteristics intersect, always have positive slope. Thus the point where they meet—that is, the origin of the shock (x_s, t_s)—represents the minimum value of both x and t at which an intersection between two characteristics occurs.

To obtain the value of $x_0 = x_s$ where $t_s = t_l(x_s) = t_u(x_s)$, let us consider eqs. (6.3.55). The condition $t_l = t_u$ implies that either

$$\xi_M = \xi_m \tag{6.3.60a}$$

or

$$\xi_M = 0 \tag{6.3.60b}$$

depending upon whether $H(\xi_m) < H(0)$ or $H(\xi_m) > H(0)$, respectively. In the first case, eq. (6.3.60a) indicates that the maximum and minimum of $H(\xi)$ occur at the same

ξ value, which means that $H(\xi)$ exhibits a horizontal inflection point; that is,

$$\frac{d^2 H}{d\xi^2} = 0 \qquad \text{at } \xi = \xi_M = \xi_m \qquad (6.3.61)$$

Accordingly, the values of the shock origin coordinates t_s and x_s, and of the corresponding ξ, are obtained solving eqs. (6.3.56) together with the condition

$$\frac{d^2 H}{d\xi^2} = \frac{d^2 \psi}{d\xi^2} (x_0 - \xi) - 2 \frac{d\psi}{d\xi} = 0 \qquad (6.3.62)$$

In the case where $H(\xi_m) > H(0)$, substituting $\xi = 0$ in eqs. (6.3.56) yields

$$x_s = \frac{\psi(0)}{(d\psi/d\xi)_{\xi=0}} \qquad (6.3.63a)$$

$$t_s = \psi(0) x_s \qquad (6.3.63b)$$

The relations above for the shock origin were derived for the case of IC (6.3.49), given along the x axis. We now repeat the analysis for the case where the IC is given along the t axis. Thus eq. (6.3.49) is replaced by

$$\gamma = \gamma^0(\xi) \qquad x = 0, t = \xi, \text{ and } 0 \le \xi \le 1 \qquad (6.3.64)$$

The characteristics of eq. (6.3.48) in this case become

$$t = \psi(\gamma^0(\xi))x + \xi \qquad 0 \le \xi \le 1 \qquad (6.3.65)$$

where again $\psi(\gamma) > 0$ since we consider positive wave propagation velocity, that is, $c(\gamma) > 0$.

Before proceeding it is worth noting that since in eq. (6.3.64) we have selected the parameter ξ increasing along the t axis, we are looking for the solution in the half-plane lying to the right, rather than to the left, of the curve along which the parameter ξ increases. Thus, as discussed in section 6.2.2, compression waves arise for $d\psi/d\xi < 0$, while expansion waves arise for $d\psi/d\xi > 0$—that is, the opposite of what we observed in the previous case where ξ increased along the x axis.

The determination of upper, $t_u(x)$, and lower, $t_l(x)$, bounds of the region where the characteristics intersect each other follows the same lines described earlier. Similarly to eq. (6.3.52), we introduce the new function $H(\xi)$ defined as

$$x_0 = \frac{t_0 - \xi}{\psi(\gamma_0(\xi))} = H(\xi) \qquad \text{for } 0 \le \xi \le 1 \qquad (6.3.66)$$

This equation provides one or more values of ξ which identify the characteristic straight lines passing through the point (x_0, t_0). It is readily seen that the boundaries of the multiplicity region of eq. (6.3.66) again coincide with the curves $t_l(x)$ and $t_u(x)$.

Since we are interested only in positive values of x_0, let us investigate the behavior of $H(\xi)$ in the interval $0 \le \xi \le t_0$. Moreover, since at the extrema of this interval, $H(0) = t_0/\psi(\gamma^0(0)) > 0$ and $H(t_0) = 0$, we conclude that the shape of $H(\xi)$ is similar to that of the function (6.3.52) introduced in the previous case and illustrated in Figure 6.17.

Accordingly, determination of the boundaries of the multiplicity region proceeds along similar arguments, leading to the following relations for the shock origin:

$$x_s = \frac{1}{\psi(\xi)} (t_s - \xi) \tag{6.3.67a}$$

$$\frac{dH}{d\xi} = -\frac{1}{\psi(\xi)} \left[1 + \frac{d\psi}{d\xi} \frac{(t_s - \xi)}{\psi(\xi)} \right] = 0 \tag{6.3.67b}$$

$$\frac{d^2H}{d\xi^2} = \frac{1}{\psi(\xi)^2} \left[2\frac{d\psi}{d\xi} - \frac{d^2\psi}{d\xi^2}(t_s - \xi) + 2\left(\frac{d\psi}{d\xi}\right)^2 \frac{(t_s - \xi)}{\psi(\xi)} \right] = 0 \tag{6.3.67c}$$

for the case where $H(\xi_m) < H(0)$. Otherwise, the last equation is replaced by $\xi = 0$, thus leading to

$$t_s = -\frac{\psi(0)}{(d\psi/d\xi)_{\xi=0}} \tag{6.3.68a}$$

$$x_s = \frac{t_s}{\psi(0)} \tag{6.3.68b}$$

To illustrate the results obtained above, let us consider the example of adsorption process in a chromatographic column discussed in section 6.3.1, and recast it in the form of eq. (6.3.48) with IC (6.3.64) that we have examined here. By comparing eqs. (6.3.3) and (6.3.48), we see that the variables u, z, and τ are now replaced by γ, x, and t, respectively. Thus we have from eq. (6.3.3a)

$$\psi(\gamma) = \alpha\left[1 + \frac{\beta}{(1 + \sigma\gamma)^2} \right] \tag{6.3.69}$$

and from eq. (6.3.4b)

$$\gamma^0(\xi) = \xi \qquad \text{for } 0 < \xi < 1 \tag{6.3.70}$$

Next we introduce a new function $G(\xi)$ defined by

$$\frac{dH}{d\xi} = -\frac{G(\xi)}{\psi(\xi)} \tag{6.3.71}$$

which, using eqs. (6.3.66), (6.3.69), and (6.3.70), leads to

$$G(\xi) = \frac{(1 + \sigma\xi)^3 + \beta(1 + \sigma\xi) - 2\beta\sigma(t_0 - \xi)}{(1 + \sigma\xi)^3 + \beta(1 + \sigma\xi)} \tag{6.3.72}$$

Thus the stationary points of $H(\xi)$ are given by the zeroes of $G(\xi)$, which correspond to the roots of the cubic polynomial in the numerator of eq. (6.3.72). The latter is a monotonically increasing function of ξ because its derivative is positive—that is,

$$3\sigma(1 + \sigma\xi)^2 + 3\beta\sigma > 0 \tag{6.3.73}$$

Recalling that $G(\xi)$ is continuous in $0 \le \xi \le t_0$, since the denominator in eq. (6.3.72) does not vanish, it follows that $G(\xi)$ has at most one zero. Moreover, since $G(t_0) = 1 > 0$, in order for this zero to be in the interval of interest (i.e., $0 \le \xi \le t_0$), we need

$G(0) < 0$. In this case, $dH/d\xi > 0$ at $\xi = 0$ and hence $H(\xi)$ exhibits one maximum in the interval $0 < \xi < t_0$. Accordingly, $H(\xi)$ has the shape of curve b in Figure 6.17, and a shock occurs in the solution. In particular, its origin is given by eqs. (6.3.68), which, when using eqs. (6.3.69) and (6.3.70), yield

$$t_s = \frac{1 + \beta}{2\beta\sigma} \tag{6.3.74a}$$

$$x_s = \frac{1}{2\alpha\beta\sigma} \tag{6.3.74b}$$

These shock origin coordinates are the same as those determined earlier in eqs. (6.3.10).

In the case where $G(0) > 0$, since $G(\xi)$ is continuous and monotonically increasing, we have $G(\xi) > 0$ in the entire interval of interest. Accordingly, from eq. (6.3.71), $H(\xi)$ decreases monotonically in $0 \le \xi \le t_0$, with the shape of curve a in Figure 6.17. Thus, the characteristic straight lines do not intersect and no shock develops in the solution.

Propagation Velocity of the Shock.   In section 6.3.1 we have seen that in some region of the domain of integration, the solution of a first-order PDE may become multivalued and therefore not physically admissible. This occurs because in these regions the representation of the physical reality provided by the first-order PDE is insufficient. In particular, the contribution of terms involving second-order derivatives becomes significant. Typically, this is the case where the solution exhibits steep profiles, leading to large second-order derivatives. To overcome this problem we replace the first-order model, in the portion of the domain of integration where its solution is not physically admissible, with a new one. Specifically, we develop a weak solution which exhibits a discontinuity in the region where the original model leads to a multivalued solution. The propagation velocity of this discontinuity can be determined using the same physical principles on which the original model is based. In the example discussed in section 6.3.1, both the original model (6.3.3) as well as the equation for the shock velocity (6.3.25) represent mass balances. The only difference is in the formulation of the balance, which in the latter case, due to the presence of a discontinuity, involves algebraic rather than differential equations.

In this section we derive the propagation velocity of a shock in the kinematic wave equation (6.3.48). As discussed in section 6.1, this equation represents the balance in a one-dimensional system of a generic quantity, for which γ and ϕ represent the amount per unit volume and the amount transported by convective flow per unit area and unit time, respectively. Similarly, the velocity of the shock is obtained by balancing the same generic quantity across the discontinuity. Thus, with reference to Figure 6.18, we require that the amount of this quantity entering the discontinuity equals that leaving it, so that no accumulation occurs at the discontinuity interface. This implies that the amount per unit time which is accumulated in the new volume made available by the movement of the shock equals the net flow entering the shock associated with the convective flow—that is,

$$V_s(\gamma_+ - \gamma_-) = \phi(\gamma_+) - \phi(\gamma_-) \tag{6.3.75}$$

The shock velocity is then obtained as follows:

$$V_s = \frac{\phi(\gamma_+) - \phi(\gamma_-)}{\gamma_+ - \gamma_-} = \frac{\Delta\phi}{\Delta\gamma} \qquad (6.3.76)$$

which is usually referred to as the *jump condition*.

Let us apply, as an example, the equation above to the chromatography model (6.2.35). In this case the definitions of γ and ϕ are given by eqs. (6.1.5) and (6.1.6), respectively, so that

$$\Delta\gamma = \gamma_+ - \gamma_- = \varepsilon(C_+ - C_-) + (1 - \varepsilon)[F(C_+) - F(C_-)] \qquad (6.3.77a)$$

$$\Delta\phi = \phi_+ - \phi_- = v(C_+ - C_-) \qquad (6.3.77b)$$

Substituting in eq. (6.3.76), we obtain the expression of the shock velocity:

$$V_s = \frac{v(C_+ - C_-)}{\varepsilon(C_+ - C_-) + (1 - \varepsilon)[F(C_+) - F(C_-)]} \qquad (6.3.78)$$

which, recalling that $\Gamma = F(C)$ from the equilibrium condition (5.4.7), is identical to eq. (6.3.26) derived earlier.

From eq. (6.3.76) we see that for given values of the state (i.e., solution) on one side of the shock and of the shock velocity, we can compute the state on the other side of the shock. However, it remains to be determined which of the two states is on the left (subscript $+$ as in Figure 6.18) and which on the right (subscript $-$) of the shock. This ambiguity arises because in eq. (6.3.76) we can exchange the values of γ_+ and

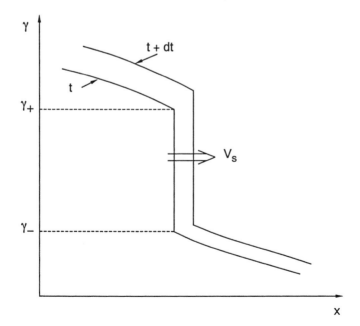

Figure 6.18 Schematic representation of the propagation of a shock.

γ_- without changing the value of V_s. Thus we need a condition to determine the position of the two states relative to the shock.

In section 6.3.1 we saw that a shock arises when the characteristics of the PDE intersect each other. These impinge on the shock trajectory as shown in Figure 6.19. Since the characteristics of eq. (6.3.48) are straight lines with slope given by $\psi(\gamma)$, we conclude that

$$\psi(\gamma_+) < \psi(\gamma_-) \tag{6.3.79}$$

That is, the slope of the characteristic impinging from the left is lower than that impinging from the right. From eq. (6.3.48a) and recalling that we consider only positive values of $\psi(\gamma)$, we obtain

$$c(\gamma_+) = \left(\frac{d\phi}{d\gamma}\right)_{\gamma=\gamma_+} > \left(\frac{d\phi}{d\gamma}\right)_{\gamma=\gamma_-} = c(\gamma_-) \tag{6.3.80}$$

This condition allows us to distinguish between the states on the right- and on the left-hand side of the shock. In particular, two typical cases may arise.

Case 1. $dc/d\gamma = d^2\phi/d\gamma^2 < 0$
As shown in Figure 6.20*a*, condition (6.3.80) leads to

$$\gamma_+ < \gamma_- \tag{6.3.81a}$$

which means that the value of the state increases across the shock.

Case 2. $dc/d\gamma = d^2\phi/d\gamma^2 > 0$
From Figure 6.20*b* we see that condition (6.3.80) leads to

$$\gamma_+ > \gamma_- \tag{6.3.81b}$$

That is, the state decreases across the shock.

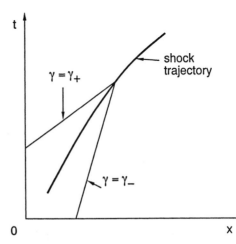

Figure 6.19 Characteristic lines impinging on the shock trajectory.

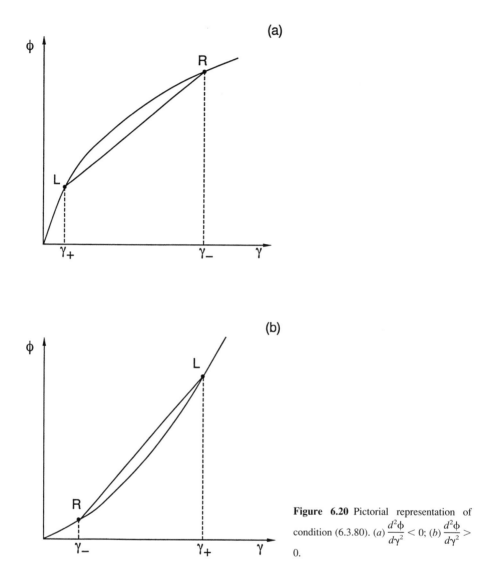

Figure 6.20 Pictorial representation of condition (6.3.80). (a) $\dfrac{d^2\phi}{d\gamma^2} < 0$; ($b$) $\dfrac{d^2\phi}{d\gamma^2} > 0$.

In both cases above, the shock velocity, V_s as given by eq. (6.3.76) is equal to the slope of the straight line connecting points L and R in Figures 6.20a and 6.20b, so that

$$\frac{1}{\psi(\gamma_+)} = \left(\frac{d\phi}{d\gamma}\right)_{\gamma=\gamma_+} > V_s > \left(\frac{d\phi}{d\gamma}\right)_{\gamma=\gamma_-} = \frac{1}{\psi(\gamma_-)} \qquad (6.3.82)$$

This is the so-called *shock condition*, which allows to remove the ambiguity in eq. (6.3.76) about the definition of the states propagating on the lhs and on the rhs of the shock.

The shock condition implies that the propagation velocity of the state impinging from the lhs of the shock $c(\gamma_+) = 1/\psi(\gamma_+)$ is larger than that of the state impinging from the right, $c(\gamma_-) = 1/\psi(\gamma_-)$. Since none of these can cross the shock trajectory— that is, the states cannot overtake or be overtaken by the shock—it follows that the shock remains sharp during its propagation. This is a peculiar feature of these discontinuities which is usually referred to as the *self-sharpening tendency* of a shock.

In the case of the chromatography model, we have from eq. (6.1.7)

$$\frac{1}{\psi(\gamma)} = \frac{d\phi}{d\gamma} = \frac{v}{\varepsilon + (1 - \varepsilon)(dF/dC)} \tag{6.3.83}$$

Noting that

$$\frac{d^2\phi}{d\gamma^2} = \left[\frac{d}{dC}\left(\frac{d\phi}{d\gamma}\right)\right]\frac{dC}{d\gamma} \tag{6.3.84}$$

from eqs. (6.3.83) and (6.1.5) we obtain

$$\frac{d^2\phi}{d\gamma^2} = -\frac{v(1 - \varepsilon)\dfrac{d^2F}{dC^2}}{\left[\varepsilon + (1 - \varepsilon)\dfrac{dF}{dC}\right]^3} \tag{6.3.85}$$

Since in general the equilibrium isotherm is a monotonically increasing curve (i.e., $dF/dC > 0$), we can conclude that in the case of convex equilibrium isotherms, for which $d^2F/dC^2 < 0$ [as, for example, the Langmuir isotherm (6.2.34)], we are in case 2 above. Thus, for a shock to exist, the concentration must decrease across the discontinuity; that is $C_+ > C_-$. This indicates that the column is operating in the adsorption mode; that is, the inlet concentration of the adsorbable component is larger than that present initially in the column. On the other hand, concave equilibrium isotherms, for which $d^2F/dC^2 > 0$, fall under case 1 above. For these, the concentration must increase across the shock; that is, $C_+ < C_-$. Thus the column operates in the desorption mode; that is, the inlet concentration is smaller than the initial one. It is worth mentioning that the same conclusion about the relation between the type of concavity of the equilibrium isotherm and the column operation mode can also be reached by considering the occurrence of compression or expansion waves (see problem 6.5).

It should be noted that the shock condition (6.3.82) is not valid in general. Specifically, in the case where the equilibrium curve $\phi = \phi(\gamma)$ exhibits one or more inflection points, condition (6.3.82) may not be correct. The proper condition of general validity has been developed by Oleinik (1959) and can be formulated as follows:

$$\frac{\phi(\gamma_+) - \phi(\gamma_-)}{(\gamma_+ - \gamma_-)} \leq \frac{\phi(\gamma_+) - \phi(\gamma)}{(\gamma_+ - \gamma)} \quad \text{for all } \gamma \text{ values between } \gamma_+ \text{ and } \gamma_- \tag{6.3.86}$$

This is usually referred to as the *entropy condition*. It can be proven that a weak solution of eq. (6.3.48), which at the discontinuity point satisfies both this and the jump condition (6.3.76), is unique. We do not prove the entropy condition here; the interested reader may refer to the original papers (Oleinik, 1963; Lax, 1971).

Note that although limited to equilibrium curves $\phi(\gamma)$ which do not have inflection points, the shock condition (6.3.82) has the same role as condition (6.3.86) in removing the ambiguity of eq. (6.3.76), thus making the solution of eq. (6.3.48) unique. The limitation noted above is not serious because most equilibrium curves of practical interest do not exhibit inflection points.

6.3.3 An Example

As an application of the results obtained in the previous section, consider a binary mixture where both components can be adsorbed. In this case the equilibrium model (6.2.34) has to be modified in order to account for the competition between the two components in the adsorption process. This can be done using the multicomponent Langmuir isotherm:

$$\Gamma_i = \Gamma^\infty \frac{K_i C_i}{1 + K_1 C_1 + K_2 C_2}, \qquad i = 1, 2 \qquad (6.3.87)$$

where the indexes 1 and 2 refer to the two components.

Since the molar density of the mixture,

$$C^0 = C_1 + C_2 \qquad (6.3.88)$$

can be assumed constant, we can represent the binary Langmuir isotherm above in terms of the concentration of a single component:

$$\Gamma_1 = \Gamma^\infty \frac{K_1 C_1}{1 + K_2 C^0 + (K_1 - K_2) C_1} \qquad (6.3.89)$$

Using this equilibrium model, it is possible to describe the binary adsorption process using the single component chromatography model given by eq. (6.2.35). Differentiating eq. (6.3.89) twice yields

$$\frac{d^2\Gamma_1}{dC_1^2} = -2\Gamma^\infty K_1 (1 + K_2 C^0) \frac{K_1 - K_2}{[1 + K_2 C^0 + (K_1 - K_2) C_1]^3}$$

which indicates that the equilibrium curve is convex (i.e., $d^2\Gamma_1/dC_1^2 < 0$) for $K_1 > K_2$ and concave (i.e., $d^2\Gamma_1/dC_1^2 > 0$) for $K_1 < K_2$. Accordingly, we expect that the more adsorbable component, which has a larger equilibrium constant K_i, exhibits a shock when it is adsorbed during the process, while the less adsorbable component exhibits a shock when it is desorbed.

This feature may be seen in the two experimental runs shown in Figure 6.21 (Morbidelli et al., 1984). The data represent the concentration values in the outlet stream of a chromatography column, packed with zeolite Y, operated with a binary mixture of p-chlorotoluene (1) and toluene (2). Note that in this case p-chlorotoluene is more adsorbable than toluene ($K_1/K_2 = 2.4$). In run I, pure p-chlorotoluene is fed to the column initially filled with toluene. Thus the strong component is adsorbed while the weak one is desorbed. Under these conditions we expect the occurrence of a shock for *both* components. In run II, pure toluene is fed to the column initially filled with p-chlorotoluene. In this case the strong component is desorbed while the weak one is

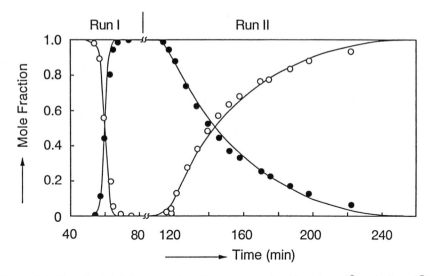

Figure 6.21 Experimental elution curves for a binary mixture of *p*-chlorotoluene (●) and toluene (○).

adsorbed. Thus no shock should occur and the concentration profiles should have the form of expansion waves. These conclusions are in agreement with the shape of the elution curves shown in Figure 6.21. In the case where the shock occurs (run I), we may note that the profile, although quite steep, is not exactly a step function. This is due to the effects of dissipation mechanisms, such as axial dispersion and mass transfer resistance, which are discussed in section 6.5.

6.4   APPLICATION TO A COUNTERCURRENT SEPARATION

Let us consider a unit where two phases α and β (e.g., a liquid and a gas, two liquids, or a fluid and a solid) move countercurrently with velocities v^α and v^β, respectively. In the framework of equilibrium models, where axial dispersion in the two phases is neglected and local equilibrium between the two phases is assumed everywhere along the unit, the mass balance yields

$$\varepsilon \frac{\partial C^\alpha}{\partial t} + (1 - \varepsilon) \frac{\partial C^\beta}{\partial t} + v^\alpha \frac{\partial C^\alpha}{\partial x} - v^\beta \frac{\partial C^\beta}{\partial x} = 0, \qquad 0 < x < L, t > 0 \quad (6.4.1a)$$

where C^α and C^β indicate the concentrations of a specific component in each of the two phases, t is time, x is the axial coordinate, and ε is the volume fraction of the unit occupied by the phase α. The local equilibrium condition may be represented as follows:

$$C^\beta = F(C^\alpha) \qquad (6.4.1b)$$

which, when substituted in eq. (6.4.1a), leads to

$$V \frac{\partial C^\alpha}{\partial x} + \frac{\partial C^\alpha}{\partial t} = 0, \qquad 0 < x < L, t > 0 \qquad (6.4.2)$$

where

$$V = \frac{v^\alpha - v^\beta (dF/dC^\alpha)}{\varepsilon + (1 - \varepsilon)(dF/dC^\alpha)} \tag{6.4.2a}$$

The above equation has the typical form of the one-dimensional kinematic wave equation (6.3.48), where V is the wave propagation velocity, $c = 1/\psi$. It is possible to derive the mass balance (6.4.1a) from the general form of the kinematic wave equation (6.1.1) through a proper definition of the variables γ and ϕ. In particular, γ involves the concentrations in both phases; that is,

$$\gamma = \varepsilon C^\alpha + (1 - \varepsilon)C^\beta \tag{6.4.3}$$

which is similar to eq. (6.1.5) derived in the context of the fixed-bed chromatography model. About the flux ϕ, we consider that the two phases move in opposite directions, so that

$$\phi = v^\alpha C^\alpha - v^\beta C^\beta \tag{6.4.4}$$

Substituting eqs. (6.4.3) and (6.4.4), it is readily seen that eq. (6.1.1) reduces to the mass balance (6.4.1a). Let us introduce the following dimensionless quantities:

$$u = \frac{C^\alpha}{C_r}, \qquad w = \frac{C^\beta}{C_r} \tag{6.4.5a}$$

$$z = \frac{x}{L}, \qquad \tau = \frac{tv^\alpha}{\varepsilon L} \tag{6.4.5b}$$

$$f(u) = \frac{F(C^\alpha)}{C_r}, \qquad \mu = \frac{v^\beta}{v^\alpha}, \qquad \nu = \frac{1 - \varepsilon}{\varepsilon} \tag{6.4.5c}$$

where τ is the dimensionless time, z is the dimensionless axial coordinate, C_r is a reference concentration value, and L is the length of the unit. Equation (6.4.2) can then be rewritten in dimensionless form:

$$\frac{\partial u}{\partial z} + \psi(u)\frac{\partial u}{\partial \tau} = 0, \qquad 0 < z < 1, \tau > 0 \tag{6.4.6}$$

where

$$\psi(u) = \frac{v^\alpha}{\varepsilon V} = \frac{1 + \nu f'(u)}{1 - \mu f'(u)} \tag{6.4.6a}$$

is the reciprocal of the dimensionless rate of wave propagation (6.4.2a) and the equilibrium condition (6.4.1b) now has the dimensionless form:

$$w = f(u) \tag{6.4.7}$$

Note that ψ is an increasing function of $f'(u)$. This can be seen from eq. (6.4.6a), which, upon differentiation, leads to

$$\frac{d\psi}{d\rho} = \frac{\nu + \mu}{(1 - \mu\rho)^2} > 0, \qquad \text{where } \rho = f'(u) \tag{6.4.8}$$

The initial conditions of eq. (6.4.6) are given typically along the three straight lines shown in Figure 6.22. In general, they have the following dimensionless form:

$$I_1: \qquad u = u^0 \qquad \text{for } 0 \le z \le 1 \text{ and } \tau = 0 \tag{6.4.9a}$$

$$I_2: \qquad u = u_{in} \qquad \text{for } z = 0 \text{ and } \tau > 0 \tag{6.4.9b}$$

$$I_3: \qquad w = w_{in} \qquad \text{for } z = 1 \text{ and } \tau > 0 \tag{6.4.9c}$$

The first one provides the initial concentration value in the α phase throughout the column, while according to the local equilibrium condition, the initial concentration in the β phase is given by eq. (6.4.7); that is, $w^0 = f(u^0)$. The last two conditions provide the concentration values in the two phases entering from the two opposite ends of the unit, as shown in Figure 6.23.

An important feature of the solution can be seen by realizing that the local equilibrium condition (6.4.7) applies *only* inside the unit. The inlet and outlet streams at both ends of the column are not necessarily at equilibrium; that is, in general $w_{out} \ne f(u_{in})$ and $w_{in} \ne f(u_{out})$. The consequence is that the concentration profiles *exhibit a discontinuity at least at one end of the unit*.

This can be understood by considering the unit operation at steady-state conditions. The overall mass balance, in dimensionless form, is given by

$$\mu w_{out} + u_{out} = \mu w_{in} + u_{in} \tag{6.4.10}$$

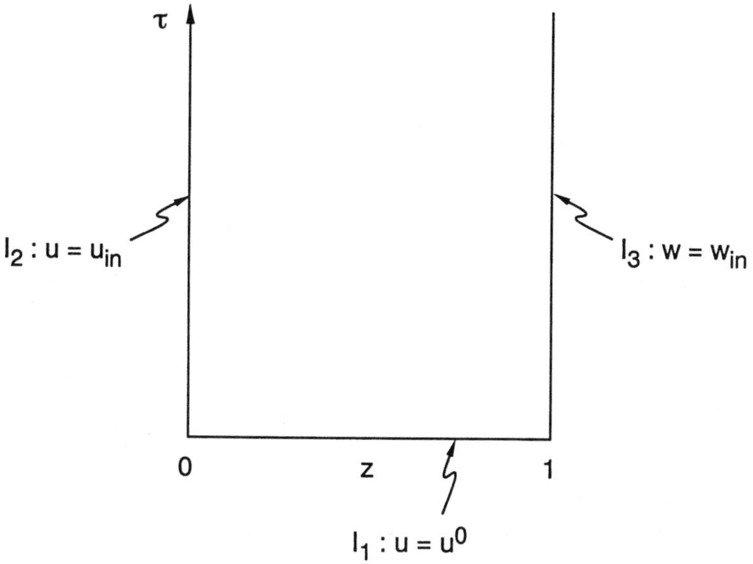

Figure 6.22 Domain of integration and initial conditions for the countercurrent equilibrium model (6.4.6).

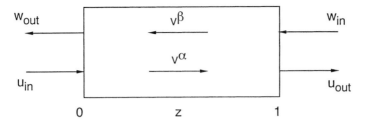

Figure 6.23 Schematic representation of a continuous countercurrent separation unit.

Let us assume that the concentration profiles are continuous at the end $z = 0$. Thus we have

$$\lim_{z \to 0^+} u(z) = u(0^+) = u_{in} \qquad (6.4.11a)$$

$$\lim_{z \to 0^+} w(z) = w(0^+) = w_{out} \qquad (6.4.11b)$$

Since the streams inside the unit are at equilibrium [i.e., $w(0^+) = f(u(0^+))$], from eqs. (6.4.11) it follows that also the streams outside the end $z = 0$ are at equilibrium [i.e., $w_{out} = f(u_{in})$]. Similarly, continuity at the end $z = 1$ implies $w_{in} = f(u_{out})$. Now, since the inlet concentrations u_{in} and w_{in} are given arbitrarily, it follows that the values of w_{out} and u_{out} given by the equilibrium relationships will, in general, not satisfy the steady-state overall mass balance (6.4.10).

Therefore we should expect a discontinuity to occur at either one or both ends of the unit. Note that this discontinuity is not related to the occurrence of a shock or to the propagation of a discontinuity present originally in the initial conditions. It is located at the *boundary* of the unit, (i.e., $z = 0$ or $z = 1$), and it does not propagate in the domain of integration. Therefore the PDE (6.4.6), which applies only for $0 < z < 1$, does not have to deal with this discontinuity.

The concentrations in the outlet streams can be computed by requiring that no mass accumulates at the two boundaries of the unit. Thus at $z = 0$ the mass balance gives

$$u_{in} - u(0^+) - \mu w_{out} + \mu w(0^+) = 0 \qquad (6.4.12)$$

which, using eq. (6.4.7), leads to

$$w_{out} = f(u(0^+)) + \frac{1}{\mu}[u_{in} - u(0^+)] \qquad (6.4.13)$$

Similarly, at the $z = 1$ end we obtain for the outlet stream concentration

$$u_{out} = u(1^-) + \mu[w_{in} - f(u(1^-))] \qquad (6.4.14)$$

Thus, using eqs. (6.4.13) and (6.4.14), from a knowledge of the inlet concentrations, u_{in} and w_{in}, and of the internal concentration profile $u(z)$, we can compute the outlet concentration values, w_{out} and u_{out}.

We now proceed to calculate the internal concentration profile $u(z)$ by solving eq. (6.4.6). In particular, for the ICs (6.4.9) let us select the initial concentration value u^0 to be equal to the inlet value:

$$u^0 = u_{in} \qquad (6.4.15)$$

Moreover, we treat the case where the component to be separated is transferred from the β-phase to the α-phase, so that $u_{in} < u_{out}$ and $w_{in} > w_{out}$. This is the case, for example, where a pollutant is removed from a gas stream (β phase, $w_{in} > w_{out}$) by contacting it countercurrently with a liquid stream (α phase, $u_{out} > u_{in}$).

Because of the countercurrent flow arrangement, the concentration in each outlet stream can at most reach the value in equilibrium with the concentration in the corresponding inlet stream. This leads to the constraints

$$u_{out} \leq u^*_{out} \qquad (6.4.16a)$$

and

$$w_{out} \geq w^*_{out} \qquad (6.4.16b)$$

where the equilibrium concentrations denoted by the asterisk are given by

$$w_{in} = f(u^*_{out}) \qquad (6.4.17a)$$

and

$$w^*_{out} = f(u_{in}) \qquad (6.4.17b)$$

Thus we can conclude that for the separation problem at hand, the inlet and outlet stream concentrations satisfy the conditions

$$u_{in} < u_{out} \leq u^*_{out} \qquad (6.4.18a)$$

and

$$w_{in} > w_{out} \geq w^*_{out} \qquad (6.4.18b)$$

Since eq. (6.4.6) is reducible, the characteristic lines on the (z, τ) plane are straight lines with slope given by

$$\frac{d\tau}{dz} = \psi(u) \qquad (6.4.19)$$

Using the ICs (6.4.9) and eq. (6.4.15), three sets of parallel characteristic straight lines are obtained. Two of these, both with slope $\psi(u_{in})$, originate from I_1 and I_2, while the third one, with slope $\psi(u_{out})$, originates from I_3. Moreover, at the point $(z = 1, \tau = 0)$ the ICs exhibit a discontinuity, since the concentration value jumps from u_{in} to u_{out}.

Before proceeding with the analysis, it is worth noting that the sign of the slope of each characteristic determines whether this affects the solution in the domain of interest—that is, $0 < z < 1$ and $\tau > 0$. For example, if $\psi(u_{in}) < 0$, then the characteristics originating from the τ axis (i.e., I_2 in Figure 6.22) propagate, for increasing time values, in the second quadrant of the (z, τ) plane and therefore do not affect the solution in the domain of interest. Similarly, in the case where $\psi(u_{out}) > 0$, the characteristics originating from curve I_3 propagate, for increasing time values, in the portion of the first quadrant where $z > 1$, again a region of no concern to us. Thus in the two examples above we can conclude that the conditions at the unit end $z = 0$ or $z = 1$, respectively, do not affect the solution.

Let us now consider two separate cases depending upon the shape of the equilibrium curve (6.4.7).

6.4.1   Convex Equilibrium Curve

The Langmuir adsorption isotherm (6.2.34) is a typical example of a convex equilibrium curve [i.e., $f''(u) < 0$], as shown in Figure 6.24. In this case, from eq. (6.4.18a) we note that

$$f'(u_{in}) > f'(u_{out}) \geq f'(u_{out}^*) \tag{6.4.20}$$

and three different types of solutions are obtained depending upon the signs of the slopes of the characteristics.

Case 1. $\mu > 1/f'(u_{out}^*)$

From eqs. (6.4.6a), (6.4.8), and (6.4.20) it follows that $\psi(u_{out}^*) \leq \psi(u_{out}) < \psi(u_{in}) < 0$; that is, all the characteristics have negative slope. Thus, as discussed, the linear wave originating from curve I_2 propagates for increasing time in a region away from the domain of interest. The relevant characteristics are shown in Figure 6.25, where the two linear waves originating from curves I_1 and I_3 are shown along with the centered wave located at ($z = 1, \tau = 0$).

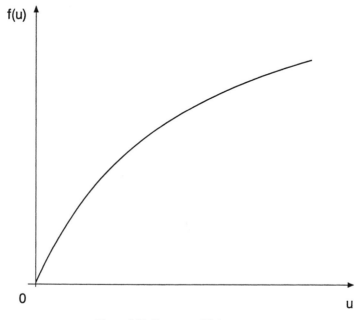

Figure 6.24 Convex equilibrium curve.

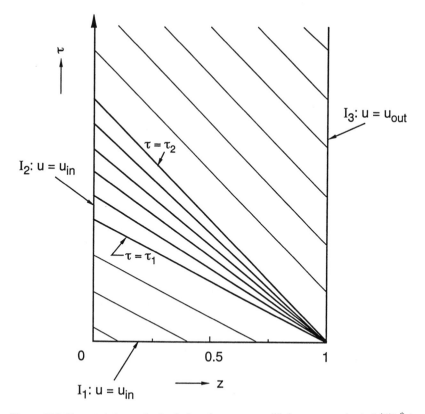

Figure 6.25 Characteristics on the (z, τ) plane for convex equilibrium curve and $\mu > 1/f'(u^*_{out})$.

It should be noted that in Figure 6.25 we have represented the characteristics originating from I_3, where $u = u_{out}$, as parallel straight lines, which implies that u_{out} is constant
in time. This is an assumption at this stage; however, it will soon be shown to be true.
    From the characteristics shown in Figure 6.25, whose expressions are obtained by
integrating eq. (6.4.19) using ICs (6.4.9a), with u^0 given by eqs. (6.4.15), (6.4.9b) and
$u = u_{out}$ at $z = 1$ and $\tau > 0$ in place of eq. (6.4.9c), we obtain the solution

$$u = u_{in} \qquad\qquad\qquad\qquad\qquad\qquad \text{for } \tau \leq \tau_1 \qquad (6.4.21\text{a})$$

$$\tau = \psi(u)(z - 1) \text{ with } u_{in} \leq u \leq u_{out} \qquad \text{for } \tau_1 < \tau < \tau_2 \qquad (6.4.21\text{b})$$

$$u = u_{out} \qquad\qquad\qquad\qquad\qquad\qquad \text{for } \tau \geq \tau_2 \qquad (6.4.21\text{c})$$

where $\tau_1 = \psi(u_{in})(z - 1)$ and $\tau_2 = \psi(u_{out})(z - 1)$. Thus the solution $u(z, \tau)$ is explicit for
$\tau \leq \tau_1$ and $\tau \geq \tau_2$, but is implicit for the interval $\tau_1 < \tau < \tau_2$.
    Having developed an expression for the internal concentration profile, we can now
determine the concentrations in the outlet streams using eqs. (6.4.13) and (6.4.14). From
eqs. (6.4.21), since at $z = 1$ we have $\tau_1 = \tau_2 = 0$, then $u(1^-) = u_{out}$ for all $\tau > 0$ (see
also Figure 6.25). Thus, eq. (6.4.14) gives

$$w_{in} = f(u(1^-)) = f(u_{out}) \qquad\qquad\qquad\qquad\qquad\qquad (6.4.22)$$

and then, since $w_{in} = f(u^*_{out})$ by definition, we obtain

$$u_{out} = u^*_{out} \tag{6.4.23}$$

Thus the solution at the right end of the unit in Figure 6.23 (i.e., $z = 1$), is continuous, since $u(1^-) = u_{out}$ and $w(1^-) = f(u(1^-)) = w_{in}$. The two streams, one entering the other leaving $z = 1$, are at equilibrium as indicated by eq. (6.4.23).

In summary, the solution of eq. (6.4.6) is now completely determined. From the inlet concentration values we can compute the outlet concentration, u_{out}, using eq. (6.4.23) and then through eqs. (6.4.21) the internal concentration profile, $u(z, \tau)$. Finally, the concentration in the remaining outlet stream, w_{out}, is obtained from eq. (6.4.13).

Using this solution, we can describe the dynamic behavior of the unit, starting from the uniform initial profile $u = u_{in}$. As time increases the concentration profiles take the form of an expansion wave as shown in Figure 6.26. It is seen that the state at the right end of the unit, u_{out}, enters at $z = 1$, propagates upstream and eventually makes the concentration in the entire column uniformly equal to u_{out}. On the other hand, the state at the left end, u_{in}, does not propagate within the unit. This is a consequence of the negative slope of the characteristics originating from curve I_2, as discussed above.

The concentration profiles shown in Figure 6.26 are continuous for all $\tau > 0$ at $z = 1$ but not at $z = 0$. The outlet concentration w_{out} at $z = 0$ is given by eq. (6.4.13), which, using eqs. (6.4.21) to compute $u(0^+)$ as a function of time, leads to

$$w_{out} = f(u_{in}) \qquad \text{for } \tau < \tau_1 \tag{6.4.24a}$$

$$w_{out} = f(u(0^+)) + (u_{in} - u(0^+)) \qquad \text{for } \tau_1 < \tau < \tau_2 \tag{6.4.24b}$$

$$w_{out} = f(u_{out}) + \frac{1}{\mu}(u_{in} - u_{out}) \qquad \text{for } \tau > \tau_2 \tag{6.4.24c}$$

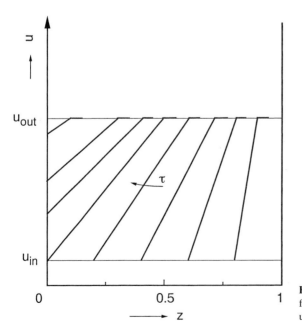

Figure 6.26 Concentration waves as a function of time corresponding to Figure 6.25.

where the value of $u(0^+)$ in eq. (6.4.24b) is given by the implicit equation

$$\tau + \psi(u(0^+)) = 0 \tag{6.4.25}$$

The steady-state solution, as $\tau \to \infty$, corresponds to a uniform concentration profile $u = u_{out}$, with the outlet concentrations u_{out} and w_{out} given by eqs. (6.4.23) and (6.4.24c), respectively. On physical grounds, we see that the outlet concentration of the α-phase, u_{out}, is determined by the equilibrium condition with the inlet β-phase. This occurs because we are dealing with sufficiently large μ (i.e., v^β/v^α) values, so that the inlet β-phase imposes the concentration value (or state) which propagates inside the unit. Moreover, at least in the framework of the equilibrium model, we see that under these conditions the β-phase drives the α-phase to equilibrium right from the first point of contact—that is, $z = 0$. For example this would be the case where we attempt to desorb a gas stream (β-phase) using a small flowrate of a liquid (α-phase). Then, at steady state, the gas stream would saturate the liquid in the entire unit right from the inlet section at $z = 0$. Here a step change in the liquid concentration would occur, from u_{in} to u_{out}.

Case 2. $1/f'(u_{in}) < \mu < 1/f'(u^*_{out})$
Let us first determine the characteristics on the (z, τ) plane. From eq. (6.4.6a) we see that $\psi(u^*_{out}) > 0$ and $\psi(u_{in}) < 0$. Moreover, note that in this case $\psi(u_{out}) > 0$. In order to prove this, let us assume that $\psi(u_{out}) < 0$. Then, all characteristics have negative slope, as in the previous case illustrated in Figure 6.25, and the solution is given by eq. (6.4.21). This implies, through eq. (6.4.23), that $\psi(u_{out}) = \psi(u^*_{out})$, which, since $\psi(u^*_{out}) > 0$, contradicts the assumption.

Thus, in addition to the characteristics originating from curve I_2, also those originating from I_3 propagate outside the domain of interest. The characteristic straight lines originating from curve I_1, as shown in Figure 6.27, constitute a family of parallel straight lines with negative slope given by $\psi(u_{in})$. Similarly to the previous case, at $z = 1$ the IC jumps from u_{in} to u_{out}, so that at ($z = 1, \tau = 0$) is located a centered wave consisting of straight lines whose slopes vary from a negative to a positive value—that is, from $\psi(u_{in})$ to $\psi(u_{out})$. From eq. (6.4.6a) we see that $\psi(u)$ changes sign at $u = u^\infty$, given by the implicit equation

$$f'(u^\infty) = \frac{1}{\mu} \tag{6.4.26}$$

where $\psi(u^\infty) = \pm \infty$. Since we are interested only in those characteristics which propagate in the region $0 < z < 1$ and $\tau > 0$, we consider in Figure 6.27 only that portion of the centered wave which is formed by straight lines with negative slope—that is, ranging from $\psi(u_{in})$ to $\psi(u^\infty) = -\infty$.

The solution of eq. (6.4.6) is obtained directly from Figure 6.27 using the expressions of the characteristics as follows:

$$u = u_{in} \qquad \text{for } \tau < \tau_1 \tag{6.4.27a}$$

$$\tau = \psi(u)(z - 1) \text{ with } u_{in} < u < u^\infty \qquad \text{for } \tau > \tau_1 \tag{6.4.27b}$$

where $\tau_1 = \psi(u_{in})(z - 1)$.

The concentrations in the outlet streams are obtained from eqs. (6.4.13) and (6.4.14), using the solution for the internal concentration profile derived above. The value of $u(0^+)$ is given by eqs. (6.4.27) as

$$u(0^+) = u_{in} \qquad \text{for } \tau < -\psi(u_{in}) \tag{6.4.28a}$$

$$\tau + \psi(u(0^+)) = 0 \qquad \text{for } \tau > -\psi(u_{in}) \tag{6.4.28b}$$

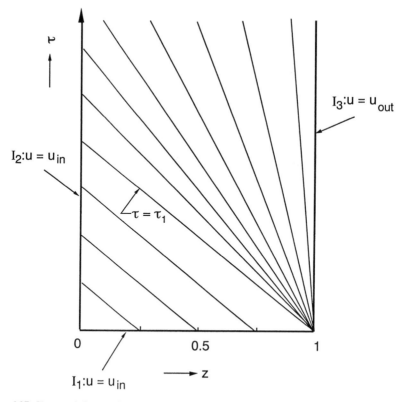

Figure 6.27 Characteristics on the (z, τ) plane for convex equilibrium curve and $1/f'(u_{in}) < \mu < 1/f'(u^*_{out})$.

Substituting in eq. (6.4.13) we obtain

$$w_{out} = f(u_{in}) \qquad \qquad \text{for } \tau < -\psi(u_{in}) \qquad (6.4.29a)$$

$$w_{out} = f(u(0^+)) + \frac{1}{\mu}[u_{in} - u(0^+)] \qquad \text{for } \tau > -\psi(u_{in}) \qquad (6.4.29b)$$

where $u(0^+)$ is given as a function of τ by eq. (6.4.28b). Similarly, from eqs. (6.4.27) we see that

$$u(1^-) = u^\infty \qquad \text{for } \tau > 0 \qquad (6.4.30)$$

where u^∞ is given by eq. (6.4.26). Substituting in eq. (6.4.14) leads to

$$u_{out} = u^\infty + \mu[w_{in} - f(u^\infty)] \qquad \text{for } \tau > 0 \qquad (6.4.31)$$

Using eqs. (6.4.27) we can represent the concentration profiles inside the unit as functions of time, as shown in Figure 6.28. These have the form of an expansion wave which enters the unit from the rhs, where the value is fixed at $u = u^\infty$, and propagates within the unit. As $\tau \to \infty$, the asymptotic solution is $u = u^\infty$, a constant through the entire unit. A comparison with the solution obtained in the previous case, shown in Figure 6.26, indicates that in both cases we have expansion waves entering from the rhs of the unit. However,

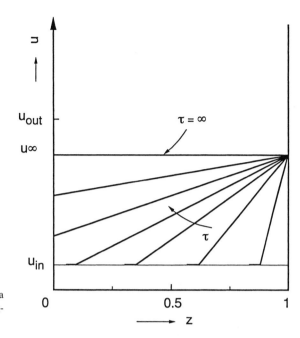

Figure 6.28 Concentration waves as a function of time corresponding to Figure 6.27.

here the concentration at $z = 1$ is less than u_{out}—that is, $u(1^-) = u^\infty < u_{out}$. Thus, in contrast with eq. (6.4.23) obtained previously, now the two outlet streams at the rhs of the unit are not at equilibrium, i.e., $u_{out} < u^*_{out}$. Since a discontinuity develops also at this end of the unit, the solution is characterized by two discontinuities, one at each end. Physically, this occurs because neither of the two stream flow rates dominates, their ratio μ being bounded.

Case 3. $\mu < 1/f'(u_{in})$
Proceeding as in the previous two cases, from eqs. (6.4.6a), (6.4.8), and (6.4.20) we see that $\psi(u_{in}) > \psi(u_{out}) > 0$. Thus all characteristic lines have positive slope, and as shown in Figure 6.29, only those originating from curves I_1 and I_2 propagate in the domain of interest. The solution in this case is particularly simple—that is, a constant concentration value through the entire column for all time values:

$$u = u_{in} \qquad \text{for } 0 < z < 1 \text{ and } \tau > 0 \qquad (6.4.32)$$

Thus the unit has *no* transient behavior since the initial concentration profile is also the steady-state solution.

The concentration values in the outlet streams are again obtained from eqs. (6.4.13) and (6.4.14), which, substituting $u(0^+) = u_{in}$ and $u(1^-) = u_{in}$, yield

$$w_{out} = f(u_{in}) \qquad (6.4.33a)$$

$$u_{out} = u_{in} + \mu[w_{in} - f(u_{in})] \qquad (6.4.33b)$$

It is seen that at the unit end $z = 0$ the two outlet streams are now at equilibrium, and no discontinuity occurs—that is, $w_{out} = w^*_{out}$. At the $z = 1$ end, the discontinuity remains since $u(1^-) = u_{in} \neq u_{out}$.

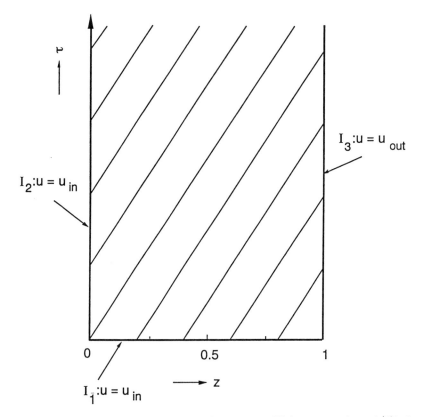

Figure 6.29 Characteristics on the (z, τ) plane for convex equilibrium curve and $\mu < 1/f'(u_{in})$.

On physical grounds, this solution represents the dual of that obtained earlier for large μ values—that is, $\mu > 1/f'(u^*_{out})$. Here, it is the flowrate of the α-phase which is so large that the outlet β-phase is forced to be at equilibrium with the inlet α-phase.

Let us return to the same example considered previously where a pollutant is removed from a given gas stream (β-phase) by absorbing it in a liquid stream (α-phase). The results obtained here provide some practical guidelines for selecting the operating conditions. The value $\mu = 1/f'(u_{in})$ represents the minimum liquid flowrate which allows the outlet gas stream to reach the minimum attainable pollutant concentration— that is, that in equilibrium with the inlet liquid stream, $w_{out} = f(u_{in})$. It is worth stressing that this conclusion holds *only* in the framework of equilibrium models, where mass dispersion and transport resistances are neglected.

6.4.2 Concave Equilibrium Curve

In this case, $f''(u) > 0$ as shown in Figure 6.30, and a typical example is the Freundlich isotherm:

$$f(u) = Ku^n, \qquad n > 1 \tag{6.4.34}$$

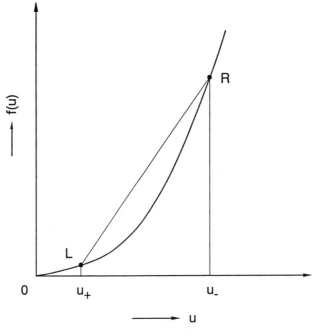

Figure 6.30 Concave equilibrium curve.

From eq. (6.4.18a) we see that

$$f'(u_{in}) < f'(u_{out}) \le f'(u^*_{out}) \tag{6.4.35}$$

which, compared with eq. (6.4.20) for the case of a convex equilibrium curve, indicates that the inequalities are reversed. The form of the solution again depends upon the value of the flowrate ratio parameter, μ; in particular we consider three intervals.

Case 1. $\mu > 1/f'(u_{in})$
From eqs. (6.4.6a), (6.4.8), and (6.4.35) we see that $\psi(u_{in}) < \psi(u_{out}) < 0$. All characteristics have negative slope, so that only those originating from curves I_1 and I_3 propagate in the domain of interest. However, in contrast to the previous cases, the straight lines now intersect as shown in Figure 6.31, and hence a shock develops in the solution.

In order to determine the shock trajectory, we need to identify its origin (z_s, τ_s) in the (z, τ) plane—that is, the point where the characteristics *first* intersect. Let us represent the characteristics originating from curves I_1 and I_3, respectively, in parametric form:

$$\tau = \psi(u_{in})(z - \xi_1), \qquad 0 \le \xi_1 \le 1 \tag{6.4.36a}$$

$$\tau = \psi(u_{out})(z - 1) + \xi_2, \qquad \xi_2 \ge 0 \tag{6.4.36b}$$

Eliminating z between the above equations, we obtain the value of τ where two straight lines, each identified by a value of the corresponding parameter $0 \le \xi_1 \le 1$ and $\xi_2 \ge 0$, intersect as follows:

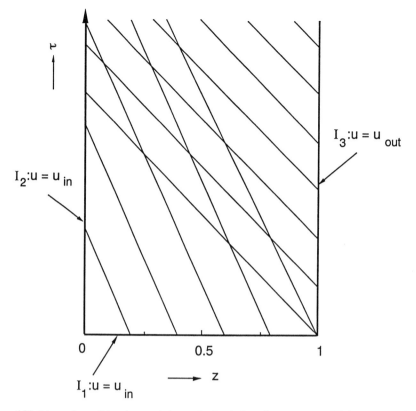

Figure 6.31 Intersections of the characteristics on the (z, τ) plane for concave equilibrium curve and $\mu > 1/f'(u_{in})$.

$$\tau = \psi(u_{in}) \frac{\psi(u_{out})(\xi_1 - 1) + \xi_2}{\psi(u_{in}) - \psi(u_{out})} \tag{6.4.37}$$

It is apparent that the smallest non-negative value of τ, with ξ_1 and ξ_2 in the respective admissible intervals, is obtained for $\xi_1 = 1$ and $\xi_2 = 0$. These lead to $\tau = 0$, which through eqs. (6.4.36) identify the origin of the shock as

$$z_s = 1, \qquad \tau_s = 0 \tag{6.4.38}$$

The propagation velocity of the shock, V_s, can be obtained from the jump condition (6.3.76) using the definitions of γ and ϕ given by eqs. (6.4.3) and (6.4.4), respectively:

$$V_s = \frac{\Delta \phi}{\Delta \gamma} = \frac{v^\alpha (C_+^\alpha - C_-^\alpha) - v^\beta [F(C_+^\alpha) - F(C_-^\alpha)]}{\varepsilon (C_+^\alpha - C_-^\alpha) + (1 - \varepsilon)[F(C_+^\alpha) - F(C_-^\alpha)]} \tag{6.4.39}$$

Using the dimensionless quantities (6.4.5), this provides the slope of the shock trajectory on the (z, τ) plane:

$$\psi_s = \frac{v^\alpha}{\varepsilon V_s} = \frac{1 + v[f(u_+) - f(u_-)]/(u_+ - u_-)}{1 - \mu[f(u_+) - f(u_-)]/(u_+ - u_-)} \tag{6.4.40}$$

From Figure 6.31 we see that the values of the solution propagating along the characteristics impinging to the left (i.e., u_+), and to the right (i.e., u_-), of the shock trajectory are given by

$$u_+ = u_{\text{in}} \text{ and } u_- = u_{\text{out}} \tag{6.4.41}$$

The shock trajectory is obtained by integrating eq. (6.4.40), which, using eqs. (6.4.41), reduces to

$$\left(\frac{d\tau}{dz}\right)_{\text{shock}} = \psi_s = \frac{1 + \nu[f(u_{\text{in}}) - f(u_{\text{out}})]/(u_{\text{in}} - u_{\text{out}})}{1 - \mu[f(u_{\text{in}}) - f(u_{\text{out}})]/(u_{\text{in}} - u_{\text{out}})} \tag{6.4.42}$$

with IC (6.4.38). Since the value of ψ_s is constant, the shock trajectory is the straight line originating at $(z_s = 1, \tau_s = 0)$:

$$\tau = \psi_s(z - 1) \tag{6.4.43}$$

By comparing the expressions for the slope of the characteristics (6.4.6a) and the slope of the shock trajectory (6.4.40), we observe that they can both be represented by

$$\psi(\rho) = \frac{1 + \nu\rho}{1 - \mu\rho} \tag{6.4.44}$$

where $\rho = f'(u)$ in eq. (6.4.6a), while $\rho = \rho_s = [f(u_+) - f(u_-)]/(u_+ - u_-)$ in eq. (6.4.40). By differentiating eq. (6.4.44) it is readily seen that $\psi(\rho)$ is an increasing function of ρ; that is,

$$\frac{d\psi}{d\rho} = \frac{\mu + \mu}{(1 - \mu\rho)^2} > 0 \tag{6.4.45}$$

From Figure 6.30 we see that $f'(u_+) < \rho_s < f'(u_-)$, since ρ_s is equal to the slope of the straight line passing through the points L and R. Thus from eq. (6.4.45) we can conclude that

$$\psi(u_+) < \psi_s < \psi(u_-) \tag{6.4.46}$$

Hence the slope of the shock trajectory is intermediate between those of the two characteristics impinging from the lhs and the rhs, leading to the configuration on the (z, τ) plane shown in Figure 6.32. This implies that the solution is given by

$$u = u_{\text{in}} \qquad \text{for } \tau < \psi_s(z - 1) \tag{6.4.47a}$$

$$u = u_{\text{out}} \qquad \text{for } \tau > \psi_s(z - 1) \tag{6.4.47b}$$

Note that since the shock separates two linear waves, the strength of the discontinuity remains constant and equal to $(u_{\text{out}} - u_{\text{in}})$ while moving along the unit. The solution profiles are shown in Figure 6.33 for various time values, τ. They are waves, in the form of step functions, which enter the unit from the boundary at $z = 1$ and propagate toward the opposite end with constant velocity, $1/\psi_s$. From eq. (6.4.43) we see that at time $\tau = -\psi_s > 0$ the shock reaches the end $z = 0$ and thereafter the solution remains constant and equal to u_{out} in the entire unit. Thus, at steady state the solution is given by

$$u = u_{\text{out}} \qquad \text{for } 0 < z < 1 \tag{6.4.48}$$

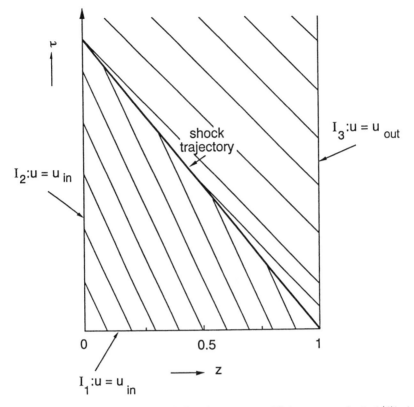

Figure 6.32 Shock trajectory on the (z, τ) plane for concave equilibrium curve and $\mu > 1/f'(u_{in})$.

The corresponding outlet stream concentrations are given by eqs. (6.4.13) and (6.4.14) as follows:

$$w_{out} = f(u_{out}) + \frac{1}{\mu}(u_{in} - u_{out}) \qquad (6.4.49)$$

$$w_{in} = f(u_{out}) \qquad (6.4.50)$$

where the latter equation along with eq. (6.4.17a) leads to

$$u_{out} = u^*_{out} \qquad (6.4.51)$$

Thus the solution exhibits only one discontinuity at $z = 0$, while it is continuous at $z = 1$ where the two streams, one entering and the other leaving, are at equilibrium.

It is worth noting that the steady-state solution above is the same as that obtained earlier for a convex equilibrium curve and large values of μ—that is, $\mu > 1/f'(u^*_{out})$—as given by eqs. (6.4.21c), (6.4.23), and (6.4.24c). The difference is in the transient behavior, which involves linear waves or shocks, depending upon the shape of the equilibrium curve, $f(u)$.

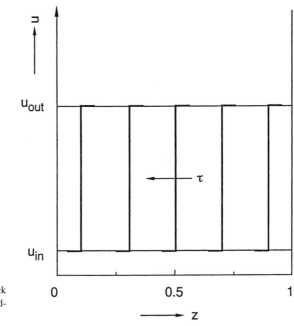

Figure 6.33 Concentration shock waves as a function of time corresponding to Figure 6.32.

Case 2. $1/f'(u_{in}) > \mu > 1/f'(u^*_{out})$

The definition (6.4.6a) and eq. (6.4.35) indicate that in this case we have $\psi(u^*_{out}) < 0 < \psi(u_{in})$. We further assume that $\psi(u_{out}) < 0$ as in case 1 above, which is indeed true for values of μ sufficiently close to $1/f'(u_{in})$.

The characteristic straight lines on the (z, τ) plane are shown in Figure 6.34. It is seen that in this case all of them, whether originating from I_1, I_2, or I_3, propagate inside the domain of integration. Since the characteristics intersect, the solution exhibits a shock. We first proceed to characterize the shock trajectory and then to develop the solution of the problem.

Since the derivation of the shock trajectory is identical to that illustrated in case 1 above, we do not repeat it here in detail. It is found that the origin of the shock is $z_s = 1$, $\tau_s = 0$, and its propagation velocity is given by eq. (6.4.42). Thus the shock trajectory is again represented by eq. (6.4.43); that is,

$$\tau = \psi_s(z - 1) \tag{6.4.52}$$

where

$$\psi_s = \frac{1 + \nu\rho}{1 - \mu\rho} \quad \text{and} \quad \rho = \frac{f(u_{in}) - f(u_{out})}{u_{in} - u_{out}} \tag{6.4.52a}$$

The solution, which can be obtained graphically from Figure 6.35, is then given by eq. (6.4.47) and has the form of a step function propagating from the $z = 1$ end of the unit to the opposite one. Thus it is similar to the solution derived earlier for case 1 above and shown in Figure 6.33. At steady state the constant value $u = u_{out}$ prevails inside the unit,

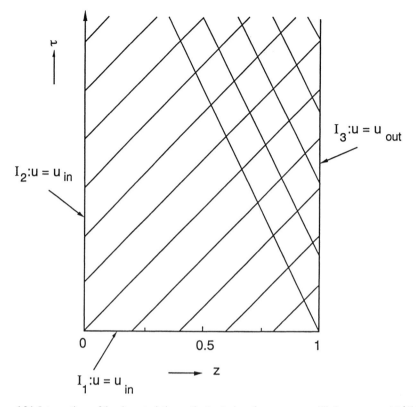

Figure 6.34 Intersections of the characteristics on the (z, τ) plane for concave equilibrium curve and $1/f'(u_{in})$ $> \mu > \mu_c$.

and the concentrations in the outlet streams u_{out} and w_{out} are given by eqs. (6.4.49) and (6.4.51)—that is,

$$w_{out} = f(u_{out}) + \frac{1}{\mu}(u_{in} - u_{out}) \tag{6.4.53}$$

$$u_{out} = u_{out}^* \tag{6.4.54}$$

Note that from eq. (6.4.54), recalling that we are considering values of μ such that $\psi(u_{out}^*) < 0$, we can not confirm that $\psi(u_{out}) < 0$, as assumed at the beginning of this analysis.

In order to better understand the behavior of the solution, let us consider the equilibrium curve $w = f(u)$ illustrated in Figure 6.36a. The coordinates of point $A(u = u_{out}^*$ and $w = w_{in})$ represent, according to eq. (6.4.54), the concentrations in the stream leaving and entering the unit at the end $z = 1$, respectively. Equation (6.4.53) represents a straight line on the (u, w) plane of slope $1/\mu$ passing through point A. In particular, the point on this straight line corresponding to $u = u_{in}$ provides the concentration value in the stream leaving the unit from the end $z = 0$—that is, $w = w_{out}$.

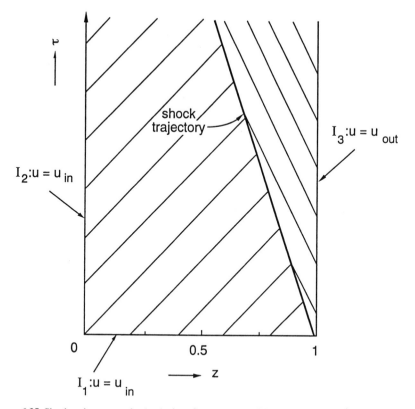

Figure 6.35 Shock trajectory on the (z, τ) plane for concave equilibrium curve and $1/f'(u_{in}) > \mu > \mu_c$.

Figure 6.36a shows the specific straight line obtained for $1/\mu = f'(u_{in})$, leading to $w_{out} = \overline{w}_{out}$, which separates the present case 2 from case 1 above. Thus for $1/\mu < f'(u_{in})$ we have $\overline{w}_{out} < w_{out} < w_{in}$, and this corresponds to the solution determined in the previous case and illustrated in Figure 6.32. As $1/\mu$ increases above $f'(u_{in})$, w_{out} becomes smaller than $\overline{w}_{out}$ and approaches the value $w_{out} = w^*_{out} = f(u_{in})$ as $1/\mu$ tends to the critical value $1/\mu_c$ given by

$$\frac{1}{\mu_c} = \frac{f(u^*_{out}) - f(u_{in})}{u^*_{out} - u_{in}} = \frac{w_{in} - w^*_{out}}{u^*_{out} - u_{in}} \qquad (6.4.55)$$

This represents the singular operating condition for which equilibrium conditions, and hence continuity, prevail at *both* ends of the unit. For $1/\mu_c > 1/\mu > f'(u_{in})$ we have $w^*_{out} < w_{out} < \overline{w}_{out}$, and the solution has the form illustrated in Figure 6.35. At $1/\mu = 1/\mu_c$ the form of the solution changes since the slope of the shock trajectory, as given by eq. (6.4.52a), becomes unbounded (i.e., $\psi_s \rightarrow -\infty$).

For $1/\mu > 1/\mu_c$, the slope ψ_s becomes positive. Thus the characteristics originating from I_3 in Figure 6.35 cannot propagate inside the domain of integration, which is now fully covered by the characteristics originating from I_1 and I_2, as shown in Figure 6.37.

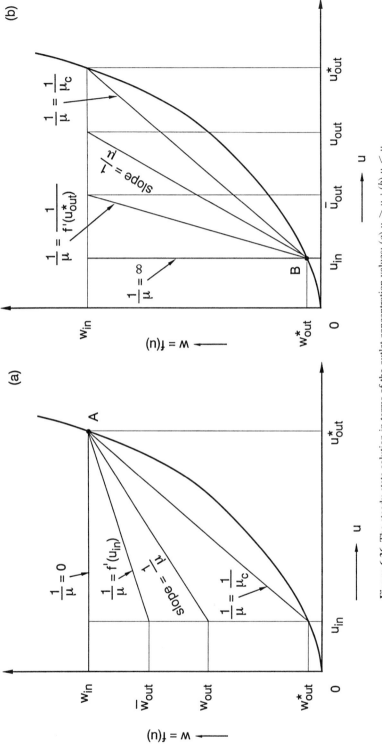

Figure 6.36 The steady-state solution in terms of the outlet concentration values: (a) $\mu > \mu_c$; (b) $\mu < \mu_c$.

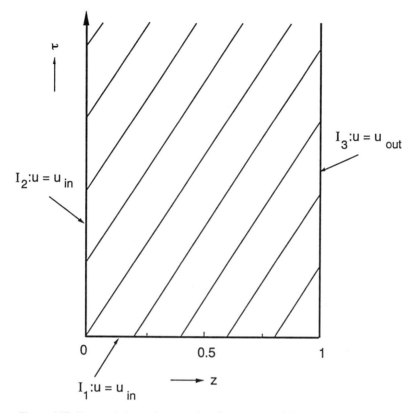

Figure 6.37 Characteristics on the (z, τ) plane for concave equilibrium curve and $\mu < \mu_c$.

The solution is then given by the initial condition which remains unchanged in time; that is,

$$u = u_{\text{in}} \qquad \text{for } 0 < z < 1 \text{ and } \tau > 0 \tag{6.4.56}$$

The concentrations in the outlet streams are given as usual by eqs. (6.4.13) and (6.4.14), which, using the above equation, lead to

$$w_{\text{out}} = f(u_{\text{in}}) = w^*_{\text{out}} \tag{6.4.57}$$

and

$$u_{\text{out}} = u_{\text{in}} + \mu[w_{\text{in}} - f(u_{\text{in}})] \tag{6.4.58}$$

Thus for $1/\mu > 1/\mu_c$ equilibrium conditions prevail at the $z = 0$ end of the unit, while at the opposite end a discontinuity in the solution occurs, with $u_{\text{out}} < u^*_{\text{out}}$. This is illustrated in Figure 6.36b, where according to eq. (6.4.57) point B, with coordinates $u = u_{\text{in}}$ and $w = w^*_{\text{out}}$, represents the $z = 0$ end of the unit.

Equation (6.4.58) describes a straight line, of slope $1/\mu$ and passing through point B. For $w = w_{\text{in}}$, it provides the concentration value in the outlet stream at $z = 1$—that is,

u_{out}. We see that as $1/\mu$ becomes larger than $1/\mu_c$, the value of u_{out} becomes smaller than u_{out}^*. For $1/\mu = f'(u_{out}^*)$, which represents the upper bound of the $1/\mu$ interval at hand, the concentration in the outlet stream at $z = 1$ is given by $u_{out} = \bar{u}_{out}$, as indicated by the straight line shown in Figure 6.36b.

Note that the sign of $\psi(u_{out})$ turns from negative to positive at some μ value in the interval $1/f'(u_{out}^*) \leq \mu \leq \mu_c$. This can be verified since for $\mu = \mu_c$ we have $u_{out} = u_{out}^*$ and hence $\psi(u_{out}) < 0$. On the other hand, for $\mu = 1/f'(u_{out}^*)$, $u_{out} < u_{out}^*$ which implies $1/\mu = f'(u_{out}^*) > f'(u_{out})$, so that $\psi(u_{out}) > 0$.

Case 3. $\mu < 1/f'(u_{out}^*)$
From eqs. (6.4.6a) and (6.4.35) it is readily seen that in this case all the characteristic straight lines on the (z, τ) plane have positive slope—that is, $\psi(u_{out}^*) > \psi(u_{in}) > 0$. Thus, only the characteristics originating from I_1 and I_2 propagate in the domain of interest, leading to the same situation discussed above and shown in Figure 6.37. Accordingly, the solution is again given by the initial condition which remains unchanged in time—that is, by eq. (6.4.56).

The outlet stream concentrations are given by eqs. (6.4.57) and (6.4.58), and from Figure 6.36b it is readily seen that for values of $1/\mu > 1/f'(u_{out}^*)$ we have $u_{out} < \bar{u}_{out}$. In particular, as $1/\mu \to \infty$, $u_{out} \to u_{in}$. In physical terms this means that as the flowrate of the α-phase becomes much larger than that of the β phase [i.e., $v^\alpha \gg v^\beta$, which from eq. (6.4.5c) implies $\mu \ll 1$], the value of u_{out} becomes close to u_{in}. In other words, since the α phase is present in large excess relative to the β phase, the concentration change in this phase across the unit is rather small.

It is worth noting that this solution is identical to that obtained in the case of a convex equilibrium curve for $\mu < 1/f'(u_{in})$, as given by eqs. (6.4.32) and (6.4.33).

We now mention some practical implications of this analysis by considering the example of a pollutant in a gaseous stream (β phase) which is removed by absorption in a liquid stream (α-phase). Our aim is to select the minimum liquid flowrate (i.e., the maximum μ) which provides the lowest possible pollutant concentration in the outlet gaseous stream, w_{out}. Following equilibrium theory, the lowest possible value for w_{out} is $w_{out} = w_{out}^*$, which, as seen above, is obtained for all values of the liquid flowrate as long as $\mu \leq \mu_c$. Hence the minimum liquid flowrate is given by $\mu = \mu_c$, and it can be readily evaluated from eqs. (6.4.5c) and (6.4.55).

6.5 CONSTANT PATTERN ANALYSIS

Some PDEs of interest in applications have solutions which take the form of a wave propagating in space with constant velocity without changing its shape. This is referred to as a *constant pattern*. In order to obtain the constant-pattern solution, we typically have to remove the original BCs so as to avoid end effects which disturb the shape of the propagating wave. This is done by assuming that the wave propagates in an infinite space domain. In this case constant-pattern solutions can be viewed as approximations of the true solutions, since the original BCs have been altered. In this section we illustrate the techniques for developing constant-pattern solutions using two examples.

Constant-pattern solutions arise from PDEs of order larger than one and are continuous; hence they differ from the weak solutions developed earlier in this chapter for first-order PDEs, which exhibit jump discontinuities.

A typical example is the kinematic wave equation modified so as to include a dissipation mechanism. This is the first example discussed in this section—that is, the chromatography model along with axial dispersion. We will see that the dispersion process has the effect of smoothening the discontinuity of the shock. The model then admits a constant-pattern solution, which approaches a shock as axial dispersion vanishes. The constant-pattern solutions, which are also referred to as *shock layers*, provide additional physical support to the weak solutions of first-order PDEs, discussed earlier in this chapter. This serves as the motivation for discussing PDEs of order greater than one.

6.5.1 Chromatography Model with Axial Dispersion

Let us consider the chromatography model (5.4.9) including the effect of axial dispersion. The mass balance in the fixed-bed adsorber is given by

$$-\varepsilon E_a \frac{\partial^2 C}{\partial x^2} + v \frac{\partial C}{\partial x} + \varepsilon \frac{\partial C}{\partial t} + (1 - \varepsilon) \frac{\partial \Gamma}{\partial t} = 0 \tag{6.5.1}$$

where the stationary phase is at equilibrium with the mobile phase everywhere along the bed; that is,

$$\Gamma = F(C) \tag{6.5.2}$$

We introduce the dimensionless quantities

$$u = \frac{C}{C_r}, \qquad f(u) = \frac{F(C)}{C_r}, \qquad z = \frac{x}{L}$$
$$=$$
$$\tau = \frac{tv}{\varepsilon L}, \qquad \text{Pe} = \frac{vL}{\varepsilon E_a}, \qquad v = \frac{1 - \varepsilon}{\varepsilon} \tag{6.5.3}$$

where C_r is a reference concentration, L is a characteristic length, and Pe is the Peclet number describing the effect of axial dispersion in the bed. Substituting eqs. (6.5.2) and (6.5.3), eq. (6.5.1) becomes

$$-\frac{1}{\text{Pe}} \frac{\partial^2 u}{\partial z^2} + \frac{\partial u}{\partial z} + \frac{\partial u}{\partial \tau} + v \frac{\partial f}{\partial \tau} = 0 \tag{6.5.4}$$

Note that in the case where axial dispersion is negligible (i.e., Pe $\rightarrow \infty$), the above model reduces to the kinematic wave equation (6.3.48); that is,

$$\frac{\partial u}{\partial z} + \psi(u) \frac{\partial u}{\partial \tau} = 0 \tag{6.5.5}$$

where

$$\frac{1}{\psi(u)} = c(u) = \frac{1}{1 + vf'(u)} \tag{6.5.5a}$$

is the wave propagation velocity.

In order to avoid end effects, we consider an infinitely long column. Accordingly, the BCs are given by

$$u = u_+ \qquad \text{as } z \to -\infty \tag{6.5.6a}$$

$$u = u_- \qquad \text{as } z \to +\infty \tag{6.5.6b}$$

where u_+ and u_- are the asymptotic values prevailing at the two ends of the unit. Let us now investigate whether the PDE (6.5.4) with BCs (6.5.6) admits a constant pattern (or shock layer) solution. For this, we assume that there exists a moving coordinate ξ, defined by

$$\xi = z - \lambda\tau \tag{6.5.7}$$

such that the solution can be represented as a one-variable function of ξ, that is,

$$u(z, \tau) = u(\xi) \tag{6.5.8}$$

and further satisfies the conditions

$$u = u_+, \quad \frac{du}{d\xi} = 0 \qquad \text{as } \xi \to -\infty \tag{6.5.9a}$$

$$u = u_-, \quad \frac{du}{d\xi} = 0 \qquad \text{as } \xi \to +\infty \tag{6.5.9b}$$

If such a moving coordinate exists, then $u(\xi)$ is the shock-layer solution and λ is its constant speed of propagation along the space coordinate, z.

In order to introduce the new variable ξ, we use the chain differentiation rule to yield

$$\frac{\partial u}{\partial z} = \frac{du}{d\xi}\frac{\partial \xi}{\partial z} = \frac{du}{d\xi}, \qquad \frac{\partial^2 u}{\partial z^2} = \frac{d^2 u}{d\xi^2} \tag{6.5.10a}$$

$$\frac{\partial u}{\partial \tau} = \frac{du}{d\xi}\frac{\partial \xi}{\partial \tau} = -\lambda\frac{du}{d\xi}, \qquad \frac{\partial f}{\partial \tau} = -\lambda\frac{df}{d\xi} \tag{6.5.10b}$$

so that eq. (6.5.4) reduces to

$$\frac{1}{\text{Pe}}\frac{d^2 u}{d\xi^2} - \frac{du}{d\xi} + \lambda\frac{du}{d\xi} + \lambda v\frac{df}{d\xi} = 0, \qquad -\infty < \xi < +\infty \tag{6.5.11}$$

Integrating both sides of eq. (6.5.11) from $\xi = -\infty$ to the generic value of ξ and using conditions (6.5.9a), we obtain

$$\frac{1}{\text{Pe}}\frac{du}{d\xi} - (1 - \lambda)(u - u_+) + \lambda v[f(u) - f(u_+)] = 0 \tag{6.5.12}$$

Since this equation applies for all values of ξ, as $\xi \to +\infty$ it must satisfy the condition (6.5.9b). For this, we select the value of λ, which is as yet unknown, as follows:

$$\lambda = \frac{(u_- - u_+)}{(u_- - u_+) + v[f(u_-) - f(u_+)]} = \frac{\Delta u}{\Delta u + v\Delta f(u)} \tag{6.5.13}$$

where Δ indicates the difference between the values attained at the two ends of the space domain—that is, as $\xi \rightarrow -\infty$ and $\xi \rightarrow +\infty$, respectively.

Note that in general the equilibrium curve is monotonically increasing; that is, $f'(u) > 0$ for $0 \leq u \leq 1$. Thus the ratio $\Delta f/\Delta u$ is greater than zero and hence the wave propagation velocity λ is greater than zero as well.

Substituting the value of λ in eq. (6.5.12) yields

$$\frac{1}{\text{Pe}} \frac{du}{d\xi} = \frac{v\Delta f}{\Delta u + v\Delta f} (u - u_+) - v \frac{\Delta u}{\Delta u + v\Delta f} [f(u) - f(u_+)] \quad (6.5.14)$$

We now investigate the existence of a solution $u(\xi)$ of the above ODE which also satisfies the two conditions:

$$u = u_+ \qquad \text{as } \xi \rightarrow -\infty \qquad\qquad (6.5.15a)$$

and

$$u = u_- \qquad \text{as } \xi \rightarrow +\infty \qquad\qquad (6.5.15b)$$

Note that eq. (6.5.14) is a first-order ODE which requires only one IC. Thus a solution which satisfies both conditions may not exist. However, if it does, it is the shock-layer solution.

Let us rewrite eq. (6.5.14) in the equivalent form:

$$\frac{1}{\text{Pe}} \frac{du}{d\xi} = v\lambda G(u) \qquad\qquad (6.5.16)$$

where

$$G(u) = \frac{\Delta f}{\Delta u} (u - u_+) - [f(u) - f(u_+)] \qquad\qquad (6.5.16a)$$

Note that both u_+ and u_- are stationary points of the ODE (6.5.16), since $G(u_+) = G(u_-) = 0$. Moreover, from eq. (6.5.16a) we have

$$G''(u) = -f''(u) \qquad\qquad (6.5.17)$$

Thus if we assume that the equilibrium curve is either always convex or concave—that is, $f''(u) < 0$ or $f''(u) > 0$ for $0 \leq u \leq 1$, respectively—then we can conclude that u_+ and u_- are the *only* zeroes of $G(u)$.

To investigate the existence of a solution of the ODE (6.5.16) which connects the two stationary points u_+ and u_- as ξ goes from $-\infty$ to $+\infty$, let us now apply the stability theory discussed in chapter 2. Since these are the only stationary points for the ODE, such a solution exists if the stationary point u_+ is unstable, while u_- is stable.

To investigate the stability character of each steady state, let us linearize locally the function $G(u)$ in eq. (6.5.16). The stability character is determined by the sign of the first-order coefficient, which, using eq. (6.5.16a), is given by

$$G'(u) = \Delta f/\Delta u - f'(u) \qquad \text{at } u = u_\pm \qquad\qquad (6.5.18)$$

Two cases arise depending upon the shape of the equilibrium curve.

Case 1. Convex Equilibrium Curve, f″(u) < 0

From Figure 6.38 we see that $\Delta f/\Delta u$ is the slope of the straight line connecting the points $u = u_+$ and $u = u_-$ on the equilibrium curve. To analyze the sign of $G'(u)$ at each steady state, let us first consider the case where $u_- > u_+$. Since $f''(u) < 0$, we have

$$f'(u_+) > \frac{\Delta f}{\Delta u} > f'(u_-) \tag{6.5.19a}$$

and from eq. (6.5.18) we obtain

$$G'(u_+) < 0 \quad \text{and} \quad G'(u_-) > 0 \tag{6.5.19b}$$

These conditions indicate that u_+ is stable while u_- is unstable. Since this is the opposite of what is required, the shock-layer solution does not exist in this case.

Let us now assume that $u_- < u_+$. Thus

$$f'(u_-) > \frac{\Delta f}{\Delta u} > f'(u_+) \tag{6.5.20a}$$

while

$$G'(u_-) < 0, \qquad G'(u_+) > 0 \tag{6.5.20b}$$

which indicate that u_- is stable while u_+ is unstable. Thus in this case we have a shock-layer solution.

Case 2. Concave Equilibrium Curve, f″(u) > 0

In the case where $u_- > u_+$, since $f''(u) > 0$, we have

$$f'(u_-) > \frac{\Delta f}{\Delta u} > f'(u_+) \tag{6.5.21a}$$

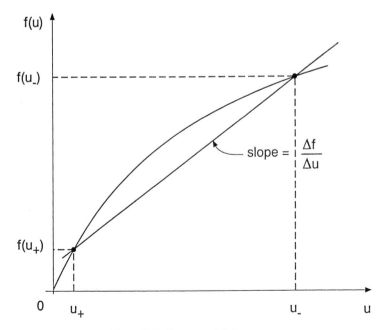

Figure 6.38 Convex equilibrium curve.

and then from eq. (6.5.18) we obtain

$$G'(u_-) < 0 \quad \text{and} \quad G'(u_+) > 0 \tag{6.5.21b}$$

These indicate that the steady-state u_- is stable, while u_+ is unstable, as required for the existence of the shock-layer solution. Along similar arguments, we see that for $u_- < u_+$ we have

$$G'(u_-) > 0, \qquad G'(u_+) < 0 \tag{6.5.22}$$

and then the solution does not exist.

From the above analysis we conclude that for a convex equilibrium curve, the shock-layer solution exists only for $u_- < u_+$. This corresponds to the case where the column, which initially contains a fluid with concentration of the adsorbable component given by u_-, is fed with a flowstream where the concentration of the same component is u_+. Since $u_- < u_+$, the column is operated in the adsorption mode. On the other hand, for a concave equilibrium curve, this solution exists only when the column is operated in the desorption mode (i.e., $u_- > u_+$).

Let us now investigate the *shape* of the shock-layer solution. For a convex equilibrium curve and $u_- < u_+$—that is, for a column operated in the adsorption mode— we see from eq. (6.5.17) that $G''(u) > 0$ for $0 \le u \le 1$. Since $G(u_-) = 0$ and $G(u_+) = 0$, it follows that $G(u) < 0$ for $u_- < u < u_+$, as shown in Figure 6.39. Substituting in eq. (6.5.16) and recalling that $\lambda > 0$, we see that $u(\xi)$ decreases monotonically from $u = u_+$ at $\xi = -\infty$ to $u = u_-$ at $\xi = +\infty$. The shock-layer solution has the form of an adsorption wave, as illustrated in Figure 6.40a. Its location relative to the ξ-axis has

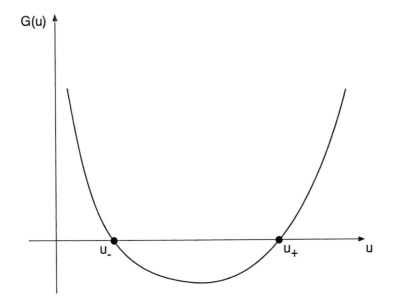

Figure 6.39 Schematic representation of the function $G(u)$ defined by eq. (6.5.16a) in the case of a convex equilibrium curve and $u_- < u_+$.

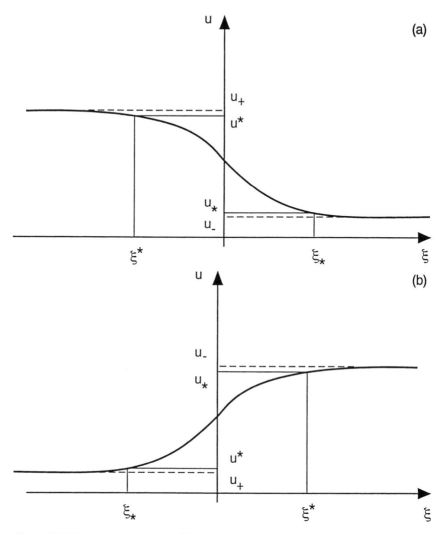

Figure 6.40 Schematic representation of the shock-layer solution: (*a*) Convex equilibrium curve and $u_- <$ u_+ (*b*) Concave equilibrium curve and $u_- > u_+$.

been selected arbitrarily so that $u = (u_- + u_+)/2$ at $\xi = 0$. The ICs (6.5.9) in fact allow for any translation of $u(\xi)$ along the ξ axis.

In the case of a concave equilibrium curve—that is, $f''(u) > 0$, with $u_- > u_+$— following similar arguments we conclude that $G(u) > 0$ in $u_+ < u < u_-$. Thus from eq. (6.5.16) we see that the shock-layer solution is a monotonically increasing function from $u = u_+$ at $\xi = -\infty$ to $u = u_-$ at $\xi = +\infty$. The shape of $u(\xi)$, illustrated in Figure 6.40*b*, corresponds to a desorption wave. This is the case where the column, initially containing a fluid with concentration u_- of the adsorbable component, is fed with a stream with a lower concentration u_+ of the same component.

In order to quantify the steepness of the waves, let us introduce the wave-width parameter, δ, defined as follows:

$$\delta = \xi_* - \xi^* \tag{6.5.23}$$

where ξ^* and ξ_* represent the extrema of the ξ interval where most of the concentration change occurs. Thus for the case of a convex equilibrium curve, illustrated in Figure 6.40a, we have

$$u = u^* = u_+ - \varepsilon \qquad \text{at } \xi = \xi^* \tag{6.5.24a}$$

and

$$u = u_* = u_- - \varepsilon \qquad \text{at } \xi = \xi_* \tag{6.5.24b}$$

where $\varepsilon > 0$ is a small parameter. For example, if we take $\varepsilon = 0.005 \, (u_+ - u_-)$, the parameter δ represents the width of the interval $(\xi_* - \xi^*)$ over which 99% of the overall change occurs; that is, $(u^* - u_*) = 0.99(u_+ - u_-)$.

Similarly, the width of the wave in the case of the concave equilibrium curve, shown in Figure 6.40b, is defined using eq. (6.5.23), where the extrema of the interval on the ξ axis are now given by

$$u = u^* = u_+ + \varepsilon \qquad \text{at } \xi = \xi^* \tag{6.5.25a}$$

and

$$u = u_* = u_- - \varepsilon \qquad \text{at } \xi = \xi_* \tag{6.25b}$$

In both cases, by integrating eq. (6.5.16) from $\xi = \xi^*$ to $\xi = \xi_*$ we obtain the same expression for the wave-width parameter:

$$\delta = \frac{1}{\text{Pe } \nu \lambda} \int_{u^*}^{u_*} \frac{1}{G(u)} \, du \tag{6.5.26}$$

An interesting application of eq. (6.5.26) is in establishing the wave behavior as axial dispersion vanishes. By taking the limit of both sides as Pe $\to \infty$, since the integral on the rhs has a finite value which does not depend on Pe, we have that for any given $\varepsilon > 0$:

$$\lim_{\text{Pe} \to \infty} \delta = 0 \tag{6.5.27}$$

This indicates that as Pe $\to \infty$, the wave-width tends to zero, and the shock-layer solution approaches a step function—that is, a jump discontinuity. We have seen earlier that as Pe $\to \infty$, the original PDE (6.5.4) reduces to the kinematic wave equation (6.5.5). Thus we conclude that the solution of the kinematic wave equation has the form of a step function. This confirms the results obtained in section 6.3.2 for the kinematic wave equation in general form (6.3.48). We found there that the solution is characterized by a shock—that is, a jump discontinuity—which moves along the z axis with a velocity given by the jump condition (6.3.76). In the case of the chromatography model, this provides the shock propagation velocity V_s given by eq. (6.3.78), which, using the dimensionless quantities (6.5.3), reduces to

$$\frac{\varepsilon V_s}{v} = \frac{(u_- - u_+)}{(u_- - u_+) + v[f(u_-) - f(u_+)]} \tag{6.5.28}$$

This is identical to the dimensionless propagation velocity of the shock layer, λ, given by eq. (6.5.13). Moreover, the conditions derived here for the existence of the shock-layer solution, that is,

$$u_+ < u_- \qquad \text{for } f''(u) > 0 \tag{6.5.29a}$$

and

$$u_+ > u_- \qquad \text{for } f''(u) < 0 \tag{6.5.29b}$$

are equivalent to the conditions for the existence of the shock derived previously in eqs. (6.3.81). By comparing eqs. (6.5.5) and (6.3.48), we observe that $\gamma = u$ and $c(\gamma = 1/\psi(u)$. From eq. (6.5.5a) it follows that

$$\frac{dc}{d\gamma} = \frac{d}{du}\frac{1}{\psi(u)} = -\frac{vf''(u)}{[1 + vf'(u)]^2} \tag{6.5.30}$$

Thus, $f''(u) > 0$ in eq. (6.5.5) corresponds to $dc/d\gamma < 0$ in eq. (6.3.48), and then the shock-layer condition (6.5.29a) coincides with the shock condition (6.3.81a). Conversely, for $f''(u) < 0$ we have $dc/d\gamma > 0$, and then conditions (6.5.29b) and (6.3.81b) are also identical.

The results of this analysis indicate that the axial dispersion model (6.5.4) can be regarded as the revised version of the chromatography model given by the kinematic wave equation (6.5.5). In particular, the model (6.5.4) accounts for second-order derivatives, which are neglected in the first-order PDE (6.5.5). As discussed in section 6.3.1, when second-order derivatives become large, the first-order model provides a multi-valued solution, which is physically unrealistic. In such cases we developed a weak solution, based on the formation of a shock. We have seen here that the second-order model provides a meaningful continuous solution in all cases, which approaches the weak solution as $1/\text{Pe}$ vanishes. This provides a sound basis to the procedure for deriving weak solutions, involving the formation and propagation of shocks, discussed in section 6.3.2.

It is worth noting that the propagation velocity of the shock-layer solution, λ given by eq. (6.5.13), does not depend on the coefficient of the second-order derivative, $1/\text{Pe}$. On the other hand, the width of the wave, δ in eq. (6.5.26), is proportional to $1/\text{Pe}$. Thus axial dispersion acts purely as a dissipation mechanism. This result is confirmed in more general terms in section 8.6.3, with reference to a chromatography column fed by a pulse of an adsorbable component. In this case, Laplace transforms are used to analyze the effects of two dissipation mechanisms, namely, axial dispersion and interphase mass transport. It is shown there that the wave propagation velocity is independent of the dissipation phenomena, which affect only the width of the wave. The effect of interphase mass transport can also be investigated using the shock-layer analysis (see problem 6.15).

6.5.2   Combustion Waves in One-Dimensional Systems

An interesting example of constant pattern solution is the propagation of a temperature wave in a medium where a combustion reaction occurs. A thermal wave is formed in a homogeneous, one-dimensional system when the temperature at one end is increased above the ignition temperature to initiate the reaction. In this case the reaction propagates along the space coordinate in the form of a moving reaction front, accompanied by a temperature wave which, under appropriate conditions, propagates as a constant pattern wave.

The mechanism of propagation of the reaction front is based on the interaction between the production of heat by chemical reaction and its dispersion by thermal conduction. Thus at the location where the system is heated above its ignition temperature, the reaction starts and liberates heat, which leads to a further temperature increase. Simultaneously, due to thermal conduction, heat is dispersed in the neighborhood regions which are as yet unreacted. When these also reach ignition temperature, reaction starts, generating additional heat. This leads to the propagation of a thermal wave along the space coordinate, sustained by the heat of reaction.

In this section we describe the movement of the combustion wave using the constant pattern approximation for a homogeneous, one-dimensional system. This is the case of a flame propagating in a homogeneous stagnant system consisting of a stoichiometric mixture of oxygen and, say, a hydrocarbon. Another example is the highly exothermic solid–solid reaction between titanium and carbon to produce titanium carbide. Here we consider a cylindrical pellet consisting of titanium and carbon uniformly mixed in the form of fine powders. Similar to the case of a flame, this system can be ignited at one end by initiating a reaction front which propagates through the pellet. This process is referred to as *combustion synthesis* and can be applied to produce a variety of advanced materials, such as ceramics and intermetallics.

Let us consider a one-dimensional, homogeneous, adiabatic stagnant system where a combustion reaction takes place. Indicating the space coordinate with x and time with t, the mass and energy balances are given by

$$\frac{\partial C}{\partial t} = -r(C, T) \tag{6.5.31}$$

$$\rho C_p \frac{\partial T}{\partial t} = (-\Delta H) r(C, T) + k_e \frac{\partial^2 T}{\partial x^2} \tag{6.5.32}$$

where C is the concentration of the limiting reactant, T is temperature, $(-\Delta H) > 0$ is the heat of reaction, ρC_p is the heat capacity per unit volume, k_e is the thermal conductivity, and $r(C, T)$ is the reaction rate. Assuming first-order kinetics and using the Arrhenius temperature dependence, we have

$$r(C, T) = AC\, e^{-E/RT} = k(T)C \tag{6.5.33}$$

where A is the preexponential factor, E is the activation energy, and R is the ideal gas constant. Note that the two terms on the rhs of the energy balance (6.5.32) represent the contributions of the heat generated by reaction and of heat conduction. On the other hand, only the contribution of the reaction is accounted for in the mass balance (6.5.31), while diffusion of mass is neglected. This is indeed correct in the case of a reaction between two solids, while it should be regarded as an approximation for a flame propagating in a gaseous system.

In order to study the propagation of the combustion wave in the form of a constant pattern, we consider an infinitely long system. Let us assume that at one end ($x = -\infty$) the reaction is complete. Since we have neglected heat losses, the temperature here reaches its adiabatic value. At the other extreme ($x = +\infty$) the system is unreacted, and so the initial conditions prevail for both concentration and temperature. Thus summarizing,

$$C = 0, \quad T = T_{ad} \qquad \text{as } x \to -\infty \tag{6.5.34a}$$

$$C = C^0, \quad T = T^0 \qquad \text{as } x \to +\infty \tag{6.5.34b}$$

Note that the adiabatic temperature of the system is given by

$$T_{ad} = T^0 + \frac{(-\Delta H)C^0}{\rho C_p} \tag{6.5.35}$$

Introducing the dimensionless quantities

$$u = \frac{C}{C^0}, \qquad \theta = \frac{T}{T_{ad}}, \qquad z = \frac{x}{L}, \qquad \tau = tk(T_{ad})$$

$$\beta = \frac{(-\Delta H)C^0}{\rho C_p T_{ad}}, \qquad \gamma = \frac{E}{RT_{ad}}, \qquad \phi = \frac{k_e}{\rho C_p L^2 k(T_{ad})} \tag{6.5.36}$$

where L is a characteristic length, eqs. (6.5.31), (6.5.32), and (6.5.34) reduce to

$$\frac{\partial u}{\partial \tau} = -u \exp\left[\gamma\left(1 - \frac{1}{\theta}\right)\right] \tag{6.5.37}$$

$$\frac{\partial \theta}{\partial \tau} = \beta u \exp\left[\gamma\left(1 - \frac{1}{\theta}\right)\right] + \phi \frac{\partial^2 \theta}{\partial z^2} \tag{6.5.38}$$

with BCs

$$u = 0, \quad \theta = 1 \qquad \text{as } z \to -\infty \tag{6.5.39a}$$

$$u = 1, \quad \theta = 1 - \beta \qquad \text{as } z \to +\infty \tag{6.5.39b}$$

In order to derive the constant pattern solution, we introduce the moving coordinate

$$\xi = z - \lambda\tau \tag{6.5.40}$$

where λ is the constant propagation velocity. Substituting eq. (6.5.40) and using the chain differentiation rule, eqs. (6.5.37) and (6.5.38) lead to

$$\lambda \frac{du}{d\xi} = u \exp\left[\gamma\left(1 - \frac{1}{\theta}\right)\right] \tag{6.5.41}$$

$$\lambda \frac{d\theta}{d\xi} = -\beta u \exp\left[\gamma\left(1 - \frac{1}{\theta}\right)\right] - \phi \frac{d^2\theta}{d\xi^2} \tag{6.5.42}$$

If it exists, the constant pattern solution must satisfy the above ODEs, together with the BCs derived from eqs. (6.5.39):

$$u = 0, \quad \theta = 1, \quad \frac{d\theta}{d\xi} = 0 \quad \text{as } \xi \to -\infty \tag{6.5.43a}$$

$$u = 1, \quad \theta = 1 - \beta, \quad \frac{d\theta}{d\xi} = 0 \quad \text{as } \xi \to +\infty \tag{6.5.43b}$$

By combining eqs. (6.5.41) and (6.5.42) to eliminate the nonlinear term, we obtain

$$\lambda \frac{d\theta}{d\xi} = -\lambda\beta \frac{du}{d\xi} - \phi \frac{d^2\theta}{d\xi^2} \tag{6.5.44}$$

which, integrated from ξ to $+\infty$ using eq. (6.5.43b), leads to

$$\phi \frac{d\theta}{d\xi} = \lambda[(1 - \beta - \theta) + \beta(1 - u)] \tag{6.5.45}$$

Note that the above equation satisfies the remaining IC (6.5.43a) for *any* value of the propagation velocity λ. Thus, contrary to the case of the chromatography model discussed in the previous section, here we cannot derive an explicit expression for λ from the ICs. Thus we need to derive an alternative procedure.

Let us divide eq. (6.5.41) by eq. (6.5.45) to obtain

$$\lambda^2 \frac{du}{d\theta} = \frac{\phi u \, \exp[\gamma(1 - 1/\theta)]}{(1 - \beta - \theta) + \beta(1 - u)} \tag{6.5.46}$$

The value of λ, if it exists, is that which forces the solution $u(\theta)$ of eq. (6.5.46) to satisfy *both* conditions (6.5.43)—that is, to connect the two points on the (u, θ) plane given by

$$y = 0, \quad \theta = 1 \quad \text{and} \quad u = 1, \quad \theta = 1 - \beta \tag{6.5.47}$$

Since eq. (6.5.46) cannot be integrated analytically, we introduce two approximations particularly suited for combustion systems which are characterized by large values of the heat of reaction and activation energy—that is, $\beta\gamma > 1$.

We note that the rhs of eq. (6.5.46) is significantly different from zero only in the region where the reaction occurs, because otherwise $\theta \simeq 1 - \beta$ and for large $\beta\gamma$, the exponential in the numerator is small. Since we expect the reaction to be completed in a narrow region (the so-called *thin-zone approximation*), we can approximate the value of θ in the denominator of the rhs of eq. (6.5.46) with the adiabatic value $\theta = 1$, leading to

$$(1 - \beta - \theta) + \beta(1 - u) = -\beta u \tag{6.5.48}$$

Note that the error introduced by this approximation is expected to be large only for those θ values (i.e., $\theta \simeq 1 - \beta$) where the contribution to the integral of $du/d\theta$ is small and hence of little consequence. Substituting eq. (6.5.48) in eq. (6.5.46) yields the separable equation

$$\lambda^2 \frac{du}{d\theta} = -\frac{\phi}{\beta} \exp\left[\gamma\left(1 - \frac{1}{\theta}\right)\right] \tag{6.5.49}$$

Along the same lines, we can expand the argument of the exponential in eq. (6.5.49) in the neighborhood of $\theta = 1$, taking only the leading term, to give

$$\lambda^2 \frac{du}{d\theta} = -\frac{\phi}{\beta} \exp[\gamma(\theta - 1)] \tag{6.5.50}$$

This approximation of a negative exponential of the temperature reciprocal by a positive exponential of a temperature difference is generally referred to as the *Frank–Kamenetskii approximation*. Even though its accuracy is not always satisfactory, it has been widely used in the analysis of nonisothermal reacting systems since it often permits an analytical treatment (for a more detailed discussion see Aris, 1975, section 4.3).

Using the limits given by eq. (6.5.47), eq. (6.5.50) may be integrated through a separation of variables to yield

$$\lambda^2 = \frac{\phi}{\beta\gamma} (1 - e^{-\gamma\beta}) \tag{6.5.51}$$

Assuming that $\gamma\beta \gg 1$, this reduces to

$$\lambda = \sqrt{\frac{\phi}{\beta\gamma}} \tag{6.5.52}$$

We can now obtain the wave propagation velocity in dimensional form, v_f, using eqs. (6.5.36) as follows:

$$v_f = \frac{dx}{dt} = Lk(T_{ad}) \frac{dz}{d\tau} = Lk(T_{ad})\lambda \tag{6.5.53}$$

which, after substituting eq. (6.5.52), gives

$$v_f = \left[\frac{k_e k(T_{ad}) R T_{ad}^2}{(-\Delta H) C^0 E} \right]^{1/2} \tag{6.5.54}$$

Note that the propagation velocity of the combustion front increases for increasing values of the thermal conductivity k_e and of the reaction rate constant $k(T_{ad})$. In this case, contrary to the chromatography example discussed in the previous section, the coefficient of the second-order derivative k_e affects the propagation velocity. This is because, as discussed earlier, heat conduction is a part of the mechanism responsible for the movement of the front, while in chromatography axial dispersion is only a dissipative phenomenon, which does not contribute toward advancement of the shock-layer.

The existence of a constant-pattern solution has been verified experimentally for many combustion systems. An example is the gasless combustion synthesis of titanium carbide from the elements titanium and carbon, where the temperature of the propagating combustion front reaches values as high as 3000 K. The location of the front, monitored as a function of time using a video camera, is shown in Figure 6.41 for two

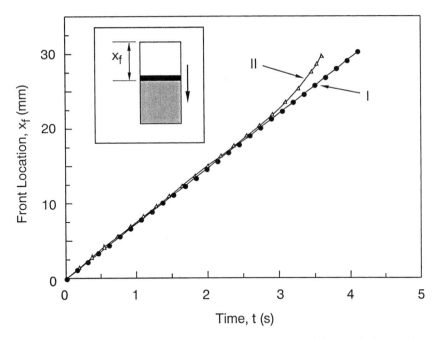

Figure 6.41 Location of the propagating combustion front as a function of time for titanium carbide synthesis.

cylindrical samples (Lebrat and Varma, 1993). It may be observed that the propagation velocity is constant, in agreement with a constant-pattern solution. For sample II, where the end was thermally insulated, the velocity increases somewhat as the front approaches the end. This is due to end effects which, as discussed earlier, are neglected in the constant-pattern analysis.

References

R. Aris, *The Mathematical Theory of Diffusion and Reaction in Permeable Catalysts*, volume 1, Clarendon Press, Oxford, 1975.

P. D. Lax, "Shock Waves and Entropy" in *Contributions to Nonlinear Functional Analysis*, E. H. Zarantonello, ed., Academic Press, New York, 1971, pp. 603–634.

J.-P. Lebrat and A. Varma, "Self-Propagating Reactions in Finite Pellets: Synthesis of Titanium Carbide," *AIChE Journal*, **39**, 1732–1734 (1993).

M. Morbidelli, G. Storti, S. Carra, G. Neiderjaufner, and A. Pontoglio, "Study of a Separation Process Through Adsorption on Molecular Sieves," *Chemical Engineering Science*, **39**, 383–393 (1984).

O. A. Oleinik, "Uniqueness and Stability of the Generalized Solution of the Cauchy Problem for a Quasi-Linear Equation," English translation in *American Mathematical Society Translation Services 2*, **33**, 285–290 (1963).

H. K. Rhee, R. Aris, and N. Amundson, *First Order Partial Differential Equations*, volumes 1 and 2, Prentice-Hall, Englewood Cliffs, NJ, 1986.

Additional Reading

R. Courant and K. O. Friedrichs, *Supersonic Flow and Shock Waves*, Wiley-Interscience, New York, 1948.

R. Courant and D. Hilbert, *Methods of Mathematical Physics: Partial Differential Equations*, volume 2, Interscience, New York, 1962.

A. Jeffrey, *Quasilinear Hyperbolic Systems and Waves*, Pitman, London, 1976.

R. L. Streeter, *The Analysis and Solution of Partial Differential Equations*, Brooks Cole, Monterey, CA, 1973.

G. B. Whitham, *Linear and Nonlinear Waves*, Wiley, New York, 1974.

Problems

6.1 Consider the traffic flow on a road, where γ represents the number of cars per mile. The velocity v (miles per hour) of the cars tends to be a function of their concentration γ, and can be approximated by the following linear relation:

$$v = v_{max}\left(1 - \frac{\gamma}{\gamma_j}\right)$$

The velocity is maximum at $\gamma = 0$, while it becomes zero at the jamming concentration γ_j, where the traffic flow ceases. Show that the kinematic wave equation which describes the concentration of cars along the road is given by

$$\frac{\partial \gamma}{\partial t} + v_{max}\left(1 - \frac{2\gamma}{\gamma_j}\right)\frac{\partial \gamma}{\partial x} = 0$$

where x and t represent position and time, respectively. Also, compare the wave velocity with the car velocity.

6.2 The sedimentation of solid particles in a column of fluid with constant cross section can be described by a kinematic wave equation. The terminal velocity of a particle falling in a fluid is determined by Stokes' law. However, when the concentration of particles γ (number of particles per unit volume) increases, their interactions become significant and the settling velocity v is then a function of γ. A possible analytical expression is

$$v(\gamma) = v_0\left(1 - \frac{\gamma}{\gamma_m}\right)\left(1 - \frac{\gamma/\gamma_m}{3\alpha - 1}\right)$$

where γ_m is the maximum particle concentration corresponding to a settled suspension [i.e., $v(\gamma_m) = 0$], α is an adjustable parameter, and v_0 is the settling velocity of a single particle. Show that the kinematic wave equation governing sedimentation is given by

$$\frac{\partial \gamma}{\partial t} + v_0\left[\left(1 - \frac{2\gamma}{\gamma_m}\right)\left(1 - \frac{\gamma/\gamma_m}{3\alpha - 1}\right) - \left(1 - \frac{\gamma}{\gamma_m}\right)\frac{\gamma/\gamma_m}{3\alpha - 1}\right]\frac{\partial \gamma}{\partial x} = 0$$

and compare the wave propagation and settling velocities.

6.3 Consider the kinematic wave equation

$$\frac{\partial \gamma}{\partial t} + \gamma \frac{\partial \gamma}{\partial x}, \qquad x \in (-\infty, +\infty), t > 0$$

with two possible initial conditions

IC$_1$: $\gamma = \gamma_1^0 = \dfrac{1}{1 + \exp(x)}$, $x \in (-\infty, +\infty)$, $t = 0$

IC$_2$: $\gamma = \gamma_2^0 = \dfrac{1}{1 + \exp(-x)}$ $x \in (-\infty, +\infty)$, $t = 0$

Define the type of wave solution (i.e., expansion or compression) in the two cases. In addition, using the pattern of the characteristics on the (x, t) plane, illustrate the evolution of the shape of the solution in time. Do not consider time values where some characteristics on the (x, t) plane intersect.

6.4 Repeat the previous problem by considering the initial conditions as given along the t axis rather than along the x axis—that is,

IC$_1$: $\gamma = \gamma_1^0 = \dfrac{1}{1 + \exp(t)}$ at $x = 0$, $t > 0$

IC$_2$: $\gamma = \gamma_2^0 = \dfrac{1}{1 + \exp(-t)}$ at $x = 0$, $t > 0$

6.5 Consider the chromatography model (6.2.35) with the IC:

$C = C^0(\xi)$, $t = 0$, $x = \xi$ with $0 \le \xi \le 1$

Show that when the equilibrium curve is convex [e.g., the Langmuir isotherm (6.2.34)], the solution has the form of a compression wave in the adsorption mode of operation (i.e., C^0 decreases along the column, $dC^0/d\xi < 0$), while it is an expansion wave in the desorption mode (i.e., $dC^0/d\xi > 0$). On the other hand, for a concave equilibrium curve such as the Freundlich isotherm

$\Gamma = KC^n$, $n > 1$

show that the solution becomes an expansion wave in adsorption and a compression wave in desorption.

6.6 Solve the chromatography model (6.2.35) and (6.2.35a) in the desorption mode—that is, with the IC:

I_1: $C = \overline{C}e^{\alpha\xi}$, $x = \xi$, $t = 0$, $0 \le \xi \le L$
I_2: $C = \overline{C}$, $x = 0$, $t = \xi$, $\xi > 0$

6.7 Repeat the problem above for the discontinuous IC:

I_1: $C = \overline{C}$, $x = \xi$, $t = 0$, $0 \le \xi \le L$
I_2: $C = 0$, $x = 0$, $t = \xi$, $\xi > 0$

6.8 Using the Freundlich isotherm $\Gamma = KC^n$ (with $n > 1$) in the chromatography model (6.2.35), determine the parameter $V(C)$. Then solve with the IC:

I_1: $C = \overline{C}e^{-\alpha\xi}$, $x = \xi$, $t = 0$, $0 \le \xi \le L$
I_2: $C = \overline{C}$, $x = 0$, $t = \xi$, $\xi > 0$

6.9 Repeat the problem above with the IC:

I_1: $C = 0$, $x = \xi$, $t = 0$, $0 \le \xi \le L$
I_2: $C = \overline{C}$, $x = 0$, $t = \xi$, $\xi > 0$

6.10 As discussed in problem 6.1, the traffic flow on a road can be described by the following dimensionless kinematic wave equation

$$\frac{\partial u}{\partial \tau} + (1 - 2u)\frac{\partial u}{\partial z} = 0, \qquad -\infty < z < +\infty, \tau > 0$$

where $u = \gamma/\gamma_j$ is the dimensionless car concentration, $\tau = tv_{max}/L$ is dimensionless time, and $z = x/L$ the dimensionless space coordinate where L is a reference length. Let us assume that a traffic light located at $z = 0$ turns green when an infinitely long queue of cars is waiting in front of it; that is, the IC is given by

I_1: $u = 1$, $z < 0$, $\tau = 0$

I_2: $u = 0$, $z > 0$, $\tau = 0$

Determine the concentration of cars, $u(z, \tau)$. Also, calculate how long it takes for a car, initially located at a distance L from the light (i.e., $z = -1$), to reach it.

6.11 Solve the chromatography model (6.3.3) and (6.3.3a) in the adsorption mode—that is, with the IC given by

I_1: $u = 0$, $\tau = 0$, $z = \xi$, $0 \le \xi \le 1$

I_2: $u = 1$, $\tau = \xi$, $z = 0$, $\xi > 1$

6.12 Repeat problem 6.8 with the IC corresponding to the desorption mode:

I_1: $C = \overline{C}$, $x = \xi$, $t = 0$, $0 \le \xi \le L$

I_2: $C = 0$, $x = 0$, $t = \xi$, $\xi \le 0$

6.13 Consider the traffic problem 6.10 in the case where two traffic lights are located on the x axis at a distance equal to L; that is, the first light is at $z = 0$ and the second at $z = -1$. Assume that this interval of road is packed with cars when the first light turns green, so that

I_1: $u = 0$, $z > 0$, $\tau = 0$

I_2: $u = 1$, $-1 \le z < 0$, $\tau = 0$

I_3: $u = 0$, $z < -1$, $\tau = 0$

How long should the green light last in order to completely empty the road? (Assume that the second light is always red.)

6.14 According to the discussion in problem 6.2, the sedimentation of solid particles in a column of fluid with constant cross section and height H can be described by the following kinematic wave equation:

$$\frac{\partial u}{\partial \tau} + \left[(1 - 2u)\left\{ 1 - \frac{u}{3\alpha - 1} \right\} - \left\{ \frac{u(1 - u)}{3\alpha - 1} \right\} \right] \frac{\partial u}{\partial z} = 0$$

where $u = \gamma/\gamma_m$ is the dimensionless particle concentration, $z = x/H$ is the dimensionless position, and $\tau = tv_0/H$ is the dimensionless time.

Determine the sedimentation dynamics in the case where the concentration of particles in the fluid is initially uniform:

$u = u_0$, $\tau = 0$, $0 \le z \le 1$

$u = 0$, $\tau > 0$, $z = 0$

$u = 1$, $\tau > 0$, $z = 1$

Note that the last two conditions correspond to zero flux of particles at the top and bottom of the column respectively.

6.15 Analyze the behavior of a column of cars reached from behind by cars traveling at higher speed. This is given by the solution of the traffic equation shown in problem 6.10 with the IC:

$$I_1: \quad u = 0, \qquad \tau > 0, \quad x = 0$$

$$I_2: \quad u = \frac{x}{x_0} u_0, \quad \tau = 0, \quad 0 \le x \le x_0$$

$$I_3: \quad u = u_0, \qquad \tau = 0, \quad x > x_0$$

Develop the solution for u_0 larger and smaller than 0.5.

6.16 Derive the shock velocity (6.4.39) through a mass balance at the shock interface.

6.17 Solve the countercurrent separation model (6.4.6) and (6.4.6a) for the case where the phase β initially inside the bed is the same as that entering the unit:

$$I_1: \quad w = w_{in}, \quad \tau = 0, \quad 0 \le z \le 1$$
$$I_2: \quad u = u_{in}, \quad \tau > 0, \quad z = 0$$
$$I_3: \quad w = w_{in}, \quad \tau > 0, \quad z = 1$$

Assume that the component to be separated is transferred from the phase β to the phase α, so that $u_{in} < u_{out}$ and $w_{in} > w_{out}$.

6.18 A fixed bed has been packed with two different adsorbent layers which follow the Langmuir adsorption isotherm (6.2.34) with the same loading capacity Γ^∞. Using the chromatography model (6.3.3) in the adsorption mode—that is, with the IC of problem 6.11—determine the trajectory of the shock which originates at the column inlet. Consider the two cases where the adsorbent in the first half of the column has an equilibrium constant larger than that of the adsorbent in the second half, as well as the opposite one. All remaining parameters for the two adsorbents are identical.

6.19 Consider the chromatography model for the case where the intraparticle mass transport resistance is significant. Then the mass balance in the fluid phase (6.5.4) must be replaced by two mass balances in the fluid and the solid phases

$$\frac{\partial u}{\partial z} + \frac{\partial u}{\partial \tau} + v \frac{\partial \gamma}{\partial \tau} = 0$$

$$\frac{\partial \gamma}{\partial \tau} = St[f(u) - \gamma]$$

where axial dispersion has been neglected (i.e., Pe $\to \infty$), $\gamma = \Gamma/C_r$ is the dimensionless concentration in the solid phase, St $= k \varepsilon a_p L/v$ is the Stanton number, a_p is the particle surface area per unit volume, and k is the mass transfer coefficient. Determine the shock-layer propagation velocity and width, using the end-effect free BCs given by

$$u = u_+ \quad \text{as } z \to -\infty$$
$$u = u_- \quad \text{as } z \to +\infty$$

Compare the effect of intraparticle mass transport resistance (i.e., St) with that of axial dispersion (i.e., Pe) discussed in section 6.5.1.

6.20 Repeat the previous problem by considering both axial dispersion and intraparticle mass transport resistance:

$$-\frac{1}{\text{Pe}} \frac{\partial^2 u}{\partial z^2} + \frac{\partial u}{\partial z} + \frac{\partial u}{\partial \tau} + v \frac{\partial \gamma}{\partial \tau} = 0$$

$$\frac{\partial \gamma}{\partial \tau} = St[f(u) - \gamma]$$

with BCs

$$u = u_+ \qquad \text{as } z \to -\infty$$
$$u = u_- \qquad \text{as } z \to +\infty$$

6.21 Consider a tube containing a mixture of monomer (M) and initiator (I) at the initial temperature $T = T_0$. At time $t = 0$, the temperature at one end of the tube is increased, leading to decomposition of the initiator and production of radical species (R) which attack the monomer molecules and initiate the radical chain polymerization process. This generates heat and consequently a local temperature increase which, due to thermal conduction, propagates along the tube. This process leads to a temperature (and conversion) wave which propagates as a shock layer with a mechanism similar to that discussed in section 6.5.2 for combustion waves. To evaluate the propagation velocity of the shock layer, consider the following mass balances for the reacting species M and R (assuming constant initiator concentration I for simplicity) and the heat balance:

$$\frac{\partial M}{\partial t} = -k_p(T)MR$$

$$\frac{\partial R}{\partial t} = 2k_d(T)I - k_t(T)R^2$$

$$\rho C_p \frac{\partial T}{\partial t} = k_e \frac{\partial^2 T}{\partial x^2} + (-\Delta H)k_p(T)MR$$

where k_p, k_d, and k_t are the rate constants of the monomer propagation, initiator decomposition, and radical termination reactions, $-\Delta H$ is the heat produced by the propagation reaction (which is the dominant one in the process), ρ is the density, C_p is the heat capacity per unit volume, and k_e is the thermal conductivity. The rate constants follow Arrhenius law,

$$k = k(T_{ad}) \exp\left[\frac{E}{R_g T_{ad}}\left(1 - \frac{T_{ad}}{T}\right)\right]$$

Since the radical species are quite reactive, we can assume $\partial R/\partial t \simeq 0$ in the radical mass balance (i.e., pseudo-steady-state conditions), thus leading to $R = [2k_d(T)I/k_t(T)]^{1/2}$. Using this result and introducing the dimensionless quantities

$$u = \frac{M}{M_0}, \qquad \theta = \frac{T}{T_{ad}}, \qquad z = \frac{x}{L}, \qquad \tau = \frac{t}{t*}$$
$$t* = [k_t(T_{ad})/2Ik_d(T_{ad})k_p^2(T_{ad})]^{1/2}, \qquad \gamma = [(2E_p + E_d - E_t)/2R_g T_{ad}]$$
$$L = \left(\frac{k_e t*}{\rho C_p}\right)^{1/2}, \qquad \beta = \frac{(-\Delta H)M_0}{\rho C_p T_{ad}}$$

the model equations reduce to

$$\frac{\partial u}{\partial \tau} = -u \exp\left[\gamma\left(1 - \frac{1}{\theta}\right)\right]$$

$$\frac{\partial \theta}{\partial \tau} = \frac{\partial^2 \theta}{\partial z^2} + \beta u \exp\left[\gamma\left(1 - \frac{1}{\theta}\right)\right]$$

Note that above we have used the adiabatic temperature T_{ad} given by

$$T_{ad} = T_0 + \frac{(-\Delta H)M_0}{\rho C_p}$$

where the subscript 0 indicates initial conditions. In dimensionless form, this reduces to

$$1 = \theta_0 + \beta$$

The boundary conditions are given so as to avoid end effects as follows:

$$u = 0, \quad \theta = 1 \qquad \text{as } z \to -\infty$$
$$u = 1, \quad \theta = \theta_0 = 1 - \beta \qquad \text{as } z \to +\infty$$

Develop the eigenvalue problem corresponding to eq. (6.5.46). Numerically evaluate the propagation velocity of the shock layer for the parameter values: $\gamma = 50$ and $\beta = 0.5$. Finally, utilizing the thin-zone and Frank—Kamenetskii approximations, derive an analytical expression for the propagation velocity as well.

Chapter 7

**GENERALIZED
FOURIER
TRANSFORM
METHODS FOR
LINEAR PARTIAL
DIFFERENTIAL
EQUATIONS**

In section 3.19 we discussed how the solution of boundary value (BV) problems involving ordinary differential equations (ODEs) can be obtained, directly and very conveniently, by the method of finite Fourier transforms (FFTs). Methods based on the Fourier transforms are developed in the present chapter to solve linear partial differential equations (PDEs). These methods are quite powerful and can be used for problems involving two or more independent variables and finite, semi-infinite, or infinite domains. Also, these methods are of general utility and can tackle second-order PDEs of all types—that is, hyperbolic, parabolic, and elliptic (see chapter 5 for this classification). As in the case of BV problems involving ODEs, the only restriction in applying the generalized Fourier transform methods is that the differential operator and the associated boundary conditions (BCs) in each spatial dimension should be self-adjoint. The books by Sneddon (1951) and Churchill (1972) are excellent sources for additional details about these methods.

7.1 AN INTRODUCTORY EXAMPLE OF THE FINITE FOURIER TRANSFORM: TRANSIENT HEAT CONDUCTION

A class of problems which can be solved by the method of FFT is that involving time and one space variable. An example is the parabolic heat conduction equation. In this context, let us consider the problem of transient cooling of a one-dimensional rod with internal heat generation. The corresponding steady-state problem was treated previously in section 3.19. The relevant PDE is

$$\frac{\partial^2 T}{\partial x^2} - U^2 T + f(x) = \frac{1}{\kappa} \frac{\partial T}{\partial t}; \qquad x \in (0, L), \quad t > 0 \tag{7.1.1}$$

with BCs

$$T = 0; \qquad x = 0, \quad t > 0 \tag{7.1.2a}$$

$$T = 0; \qquad x = L, \quad t > 0 \tag{7.1.2b}$$

and IC

$$T = q(x); \qquad x \in (0, L), \quad t = 0 \tag{7.1.3}$$

where $U^2 = hp/kA$, $\kappa = k/C_p\rho$, and $f(x) = g(x)/k$, with $g(x)$ being the rate of heat generation per unit volume. The notation has been described in section 3.13.

The plane of independent variables (x and t) is shown in Figure 7.1. The boundary and initial conditions (7.1.2) and (7.1.3) provide the value of the temperature T on all the hatched surfaces. By solving eq. (7.1.1), we wish to find T at any generic point (x, t) in the plane.

In solving this PDE, just like the corresponding steady-state problem involving an ODE (see section 3.19), we have a choice in defining the operator. There are two such choices, and both lead to the same answer.

Consider the differential operator

$$L[\phi] = \frac{d^2\phi}{dx^2} - U^2\phi \tag{7.1.4}$$

and the associated eigenvalue problem

$$L[\phi] = -\lambda\phi, \qquad x \in (0, L) \tag{7.1.5}$$

$$\phi = 0, \qquad x = 0 \tag{7.1.6a}$$

$$\phi = 0, \qquad x = L \tag{7.1.6b}$$

where the BCs (7.1.6) arise from the original homogeneous BCs (7.1.2).

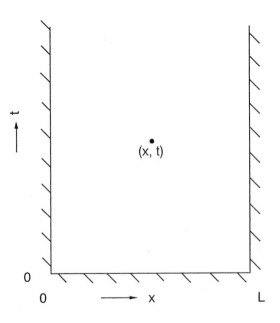

Figure 7.1 The (x, t) plane.

From section 3.19 the eigenvalues are

$$\lambda_n = U^2 + \frac{n^2\pi^2}{L^2}, \qquad n = 1, 2, \ldots \qquad (7.1.7)$$

and the corresponding orthonormal eigenfunctions are

$$\phi_n(x) = \sqrt{\frac{2}{L}} \sin\left(\frac{n\pi x}{L}\right), \qquad n = 1, 2, \ldots \qquad (7.1.8)$$

The PDE (7.1.1) can be rewritten as

$$L[T] + f(x) = \frac{1}{\kappa} \frac{\partial T}{\partial t}$$

Taking the FFT—that is, multiplying both sides by $\phi_n(x)$ and integrating from $x = 0$ to L—gives

$$\int_0^L \phi_n(x)L[T]\,dx + \int_0^L \phi_n(x)f(x)\,dx = \frac{1}{\kappa}\int_0^L \phi_n(x)\frac{\partial T}{\partial t}\,dx \qquad (7.1.9)$$

Now, due to self-adjointness of $L[\ ]$ and the associated BCs (7.1.6), we obtain

$$\int_0^L \phi_n(x)L[T]\,dx = \int_0^L T(x, t)L[\phi_n]\,dx$$

$$= \int_0^L T(x, t)[-\lambda_n\phi_n(x)]\,dx$$

$$= -\lambda_n T_n(t) \qquad (7.1.10)$$

where

$$T_n(t) = \int_0^L T(x, t)\phi_n(x)\,dx = \mathcal{F}[T(x, t)] \qquad (7.1.11)$$

is the FFT of $T(x, t)$. Note that T_n is a function of time, t. Also

$$\int_0^L \phi_n(x)f(x)\,dx = \mathcal{F}[f(x)] = f_n \qquad (7.1.12)$$

is the FFT of the heat generation function $f(x)$ and is a constant.

The rhs of eq. (7.1.9) can be rewritten as

$$\frac{1}{\kappa}\int_0^L \phi_n(x)\frac{\partial T}{\partial t}\,dx = \frac{1}{\kappa}\frac{\partial}{\partial t}\int_0^L \phi_n(x)T(x, t)\,dx$$

$$= \frac{1}{\kappa}\frac{dT_n}{dt} \qquad (7.1.13)$$

Note that the derivative with respect to t becomes an ordinary derivative since T_n is a function *only* of t. The first step above means that we can exchange the two operations: differentiation with respect to t and integration over x; that is, we can carry out either one of them first. This is assured if $\partial T/\partial t$ is a continuous function, which is expected from a physical viewpoint.

With this, eq. (7.1.9) takes the form

$$-\lambda_n T_n + f_n = \frac{1}{\kappa} \frac{dT_n}{dt}$$

which is a first-order ODE in T_n:

$$\frac{1}{\kappa} \frac{dT_n}{dt} + \lambda_n T_n = f_n \qquad (7.1.14)$$

The solution of eq. (7.1.14) is

$$T_n(t) = \frac{f_n}{\lambda_n} + \left[T_n(0) - \frac{f_n}{\lambda_n} \right] \exp(-\lambda_n \kappa t) \qquad (7.1.15)$$

where $T_n(0)$ is the value of $T_n(t)$ at $t = 0$. Now, by definition,

$$T_n(0) = \lim_{t \to 0} T_n(t)$$

$$= \lim_{t \to 0} \int_0^L T(x, t) \phi_n(x) \, dx$$

$$= \int_0^L \phi_n(x) \, [\lim_{t \to 0} T(x, t)] \, dx, \qquad \text{exchanging integration with limit}$$

$$= \int_0^L \phi_n(x) q(x) \, dx, \qquad \text{using the IC (7.1.3)}$$

$$= q_n = \mathcal{F}[q(x)] \qquad (7.1.16)$$

Note that $T_n(0) = q_n$ implies that the initial condition for the FFT of $T(x, t)$ equals the FFT of the initial condition for $T(x, t)$—that is, $q(x)$. Thus from eq. (7.1.15) we have

$$T_n(t) = \frac{f_n}{\lambda_n} + \left[q_n - \frac{f_n}{\lambda_n} \right] \exp(-\lambda_n \kappa t) \qquad (7.1.17)$$

Hence,

$$T(x, t) = \mathcal{F}^{-1}[T_n(t)]$$

$$= \sum_{n=1}^{\infty} T_n(t) \phi_n(x), \qquad \text{from eq. (3.18.8)}$$

$$= \sum_{n=1}^{\infty} \left[\frac{f_n}{\lambda_n} + \left(q_n - \frac{f_n}{\lambda_n} \right) \exp(-\lambda_n \kappa t) \right] \phi_n(x) \qquad (7.1.18)$$

represents the complete solution of eqs. (7.1.1) to (7.1.3), where

$$f_n = \mathcal{F}[f(x)] = \int_0^L f(x)\phi_n(x)\,dx$$

$$q_n = \mathcal{F}[q(x)] = \int_0^L q(x)\phi_n(x)\,dx$$

$$\lambda_n = U^2 + \frac{n^2\pi^2}{L^2}, \qquad \phi_n(x) = \sqrt{\frac{2}{L}}\sin\left(\frac{n\pi x}{L}\right); \qquad n = 1, 2, \ldots$$

Since λ_n grow rapidly with n, generally only the first few terms are required in eq. (7.1.18) to give an adequate representation of the solution. Also, from eq. (7.1.18), note that since κ and $\lambda_n > 0$, as $t \to \infty$, we obtain

$$T(x, t) \to \sum_{n=1}^{\infty} \frac{f_n}{\lambda_n}\phi_n(x) \tag{7.1.19}$$

which matches with the steady-state solution (3.19.6) developed earlier.

7.2 ANOTHER EXAMPLE: LAMINAR FLOW IN A RECTANGULAR DUCT

The previous example involved a parabolic equation, in time and one space variable. The same technique employed for obtaining the solution can also be applied to solve elliptic equations. An interesting example involves determining the velocity distribution for fully developed laminar flow in a rectangular duct. The geometry is shown in Figure 7.2. This problem involves *no* time but *two* space variables.

The Navier–Stokes equations (cf. Bird et al., 1960, Table 3.4-2) describing the velocity distribution give

$$\mu\left[\frac{\partial^2 u}{\partial x^2} + \frac{\partial^2 u}{\partial y^2}\right] = \frac{\partial p}{\partial z}; \qquad x \in (0, a), \quad y \in (0, b) \tag{7.2.1}$$

along with the BCs

$$
\begin{aligned}
u &= 0, & x &= 0, & y &\in (0, b) \\
u &= 0, & x &= a, & y &\in (0, b) \\
u &= 0, & y &= 0, & x &\in (0, a) \\
u &= 0, & y &= b, & x &\in (0, a)
\end{aligned}
\tag{7.2.2}
$$

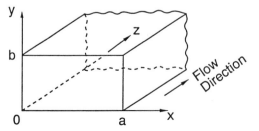

Figure 7.2 Flow in a rectangular duct.

For fully developed flow, u is a function *only* of x and y (and not z), and $\partial p/\partial z$ is a constant. Let $\partial p/\partial z = \Delta p/L$, then eq. (7.2.1) becomes

$$\frac{\partial^2 u}{\partial x^2} + \frac{\partial^2 u}{\partial y^2} = -\alpha \tag{7.2.3}$$

where $\alpha = -\Delta p/(L\mu)$ is a positive constant. In order to solve the problem by the method of FFT, we need to first separate the operators, so that either only the x derivative or only the y derivative appears on one side of the equation. If we choose the x derivative, then eq. (7.2.3) can be rewritten as

$$L_x[u] = \frac{\partial^2 u}{\partial x^2} = -\frac{\partial^2 u}{\partial y^2} - \alpha \tag{7.2.4}$$

The eigenvalue problem corresponding to $L_x[\ ]$, and its associated BCs in the variable x, is

$$L_x[\phi] = \frac{d^2\phi}{dx^2} = -\lambda\phi, \qquad x \in (0, a)$$

$$\phi(0) = 0 = \phi(a)$$

which can be solved readily to give the eigenvalues

$$\lambda_n = \frac{n^2\pi^2}{a^2}, \qquad n = 1, 2, \ldots \tag{7.2.5a}$$

and the corresponding orthonormal eigenfunctions

$$\phi_n(x) = \sqrt{\frac{2}{a}} \sin\left(\frac{n\pi x}{a}\right), \qquad n = 1, 2, \ldots \tag{7.2.5b}$$

Furthermore, given a function $f(x)$,

$$\int_0^a f(x)\phi_n(x)\, dx = \mathcal{F}_x[f(x)] = f_n \tag{7.2.6a}$$

defines the FFT of $f(x)$, and conversely

$$f(x) = \mathcal{F}_x^{-1}[f_n] = \sum_{n=1}^{\infty} f_n \phi_n(x) \tag{7.2.6b}$$

provides an infinite series representation for the function $f(x)$ in terms of the complete set $\{\phi_n(x)\}$.

To solve eq. (7.2.4), taking the FFT—that is, multiplying both sides by $\phi_n(x)$ and integrating from 0 to a—gives the following for the lhs:

$$\int_0^a \phi_n(x)L_x[u]\ dx = \int_0^a u(x, y)L_x[\phi_n]\ dx,$$

due to self-adjointness of $L_x[\ ]$

and its associated BCs

$$= \int_0^a u(x, y)[-\lambda_n\phi_n(x)]\ dx,$$

since $L_x[\phi_n] = -\lambda_n\phi_n$

$$= -\lambda_n u_n(y) \tag{7.2.7}$$

where

$$u_n(y) = \mathcal{F}_x[u(x, y)] = \int_0^a u(x, y)\phi_n(x)\ dx$$

is the FFT of $u(x, y)$ with respect to the x variable. On the other hand, for the rhs we have

$$-\int_0^a \frac{\partial^2 u}{\partial y^2}\ \phi_n(x)\ dx - \alpha \int_0^a \phi_n(x)\ dx = -\frac{d^2}{dy^2}\int_0^a u(x, y)\phi_n(x)\ dx$$

$$- \alpha\sqrt{\frac{2}{a}}\int_0^a \sin\left(\frac{n\pi x}{a}\right)\ dx$$

exchanging the operations involving derivative with respect to y and integral over x

$$= -\frac{d^2 u_n}{dy^2} + a_n \tag{7.2.8}$$

where

$$a_n = \sqrt{\frac{2}{a}}\ \alpha\ \frac{a}{n\pi}\ [(-1)^n - 1] \tag{7.2.8a}$$

is a constant. By equating the lhs from eq. (7.2.7) and the rhs from eq. (7.2.8), we have

$$\frac{d^2 u_n}{dy^2} - \lambda_n u_n = a_n \tag{7.2.9}$$

Note that, in essence, by using the FFT once, the PDE (7.2.3) in *two* independent space variables (x and y) has been reduced to an ODE for the transform involving only *one* space variable (y).

Let us now determine the BCs for $u_n(y)$. Note that

$$u_n(y) = \int_0^a u(x, y)\phi_n(x)\ dx$$

so that

$$u_n(0) = \lim_{y \to 0} \int_0^a u(x, y)\phi_n(x) \, dx$$

$$= \int_0^a u(x, 0)\phi_n(x) \, dx, \qquad \text{exchanging limit with integration}$$

$$= 0 \tag{7.2.10a}$$

since $u(x, 0) = 0$, $x \in (0, a)$ from the BC (7.2.2). Similarly,

$$u_n(b) = 0 \tag{7.2.10b}$$

To determine $u_n(y)$, we need to solve eq. (7.2.9) along with BCs given by eqs. (7.2.10). Once $u_n(y)$ has been determined, then from eq. (7.2.6b) we have

$$u(x, y) = \sum_{n=1}^{\infty} u_n(y)\phi_n(x) \tag{7.2.11}$$

There are two ways of solving eq. (7.2.9), and both are considered next.

METHOD 1. The first method involves solving the ODE for u_n *directly*, which itself can be done in two ways, by using

(a) the method of undetermined coefficients or

(b) the Green's function approach.

Following the former method we have

$$\text{homogeneous solution} = c_1 \cosh\sqrt{\lambda_n} \, y + c_2 \sinh\sqrt{\lambda_n} \, y$$

$$\text{particular solution} = -\frac{a_n}{\lambda_n}$$

so that

$$u_n(y) = c_1 \cosh\sqrt{\lambda_n} \, y + c_2 \sinh\sqrt{\lambda_n} \, y - \frac{a_n}{\lambda_n} \tag{7.2.12}$$

where c_1 and c_2 are constants, to be determined by applying the BCs (7.2.10). In particular, using eq. (7.2.10a) gives

$$0 = c_1 - \frac{a_n}{\lambda_n}$$

so that

$$c_1 = \frac{a_n}{\lambda_n}$$

and from eq. (7.2.12) we obtain

$$u_n(y) = \frac{a_n}{\lambda_n} [-1 + \cosh\sqrt{\lambda_n} \, y] + c_2 \sinh\sqrt{\lambda_n} \, y$$

Similarly, application of BC (7.2.10b) gives

$$0 = \frac{a_n}{\lambda_n}[-1 + \cosh\sqrt{\lambda_n}\, b] + c_2 \sinh\sqrt{\lambda_n}\, b$$

Thus

$$c_2 = \frac{a_n}{\lambda_n}\left[\frac{1 - \cosh\sqrt{\lambda_n}\, b}{\sinh\sqrt{\lambda_n}\, b}\right]$$

and

$$u_n(y) = \frac{a_n}{\lambda_n}\left[-1 + \frac{\sinh\sqrt{\lambda_n}\, y + \sinh\sqrt{\lambda_n}\, (b - y)}{\sinh\sqrt{\lambda_n}\, b}\right] \tag{7.2.13}$$

Finally, from eq. (7.2.11),

$$u(x, y) = \sum_{n=1}^{\infty} u_n(y)\phi_n(x)$$

represents the solution explicitly, where $u_n(y)$, λ_n and $\phi_n(x)$, and a_n are given by eqs. (7.2.13), (7.2.5), and (7.2.8a), respectively. This provides the desired velocity distribution for laminar flow in a rectangular duct and is a generalization of Poiseuille equation in circular tubes.

METHOD 2. The second method involves taking a FFT once again. The problem to be solved is

$$\frac{d^2u_n}{dy^2} - \lambda_n u_n = a_n, \qquad y \in (0, b) \tag{7.2.9}$$

along with homogeneous BCs

$$u_n(0) = 0, \qquad u_n(b) = 0 \tag{7.2.10}$$

We can again identify an operator:

$$L_y[\ ] = \frac{d^2[\ ]}{dy^2}$$

along with homogeneous BCs at $y = 0$ and b. The operator and the BCs are self-adjoint. The corresponding eigenvalue problem

$$L_y[\psi] = \frac{d^2\psi}{dy^2} = -\mu\psi, \qquad y \in (0, b)$$

$$\psi(0) = 0, \qquad \psi(b) = 0$$

is similar to that for $L_x[\ ]$ and gives the eigenvalues and orthonormal eigenfunctions as

$$\mu_m = \frac{m^2\pi^2}{b^2}, \qquad\qquad m = 1, 2, \dots \tag{7.2.14a}$$

$$\psi_m(y) = \sqrt{\frac{2}{b}}\sin\left(\frac{m\pi y}{b}\right), \qquad m = 1, 2, \dots \tag{7.2.14b}$$

In an analogous manner, given a function $g(y)$,

$$\int_0^y g(y)\psi_m(y)\, dy = \mathcal{F}_y[g(y)] = g_m \qquad (7.2.15a)$$

defines the FFT of $g(y)$, and its inverse

$$g(y) = \mathcal{F}_y^{-1}[g_m] = \sum_{m=1}^{\infty} g_m\psi_m(y) \qquad (7.2.15b)$$

provides an infinite series representation for the function $g(y)$.

Taking the FFT of eq. (7.2.9) gives

$$\int_0^b \psi_m(y)L_y[u_n]\, dy - \lambda_n \int_0^b u_n(y)\psi_m(y)\, dy = a_n \int_0^b \psi_m(y)\, dy \qquad (7.2.16)$$

Let us define

$$u_{nm} = \mathcal{F}_y[u_n(y)] = \int_0^b u_n(y)\psi_m(y)\, dy \qquad (7.2.17)$$

to be the FFT of $u_n(y)$. Then in eq. (7.2.16),

$$\text{the first term of lhs} = \int_0^b \psi_m(y)L_y[u_n]\, dy$$

$$= \int_0^b u_n(y)L_y[\psi_m]\, dy, \text{ since } L_y[\ ] \text{ and its associated BCs} \atop \text{are self-adjoint}$$

$$= \int_0^b u_n(y)[-\mu_m\psi_m]\, dy, \qquad \text{since } L_y[\psi_m] = -\mu_m\psi_m$$

$$= -\mu_m u_{nm} \qquad (7.2.18a)$$

$$\text{the second term of lhs} = -\lambda_n u_{nm} \qquad (7.2.18b)$$

$$\text{the rhs} = a_n \int_0^b \psi_m(y)\, dy$$

$$= a_n \sqrt{\frac{2}{b}} \int_0^b \sin\left(\frac{m\pi y}{b}\right) dy$$

$$= a_n b_m \qquad (7.2.18c)$$

where

$$b_m = \sqrt{\frac{2}{b}}\, \frac{b}{m\pi}\, [(-1)^m - 1]$$

Thus from eqs. (7.2.16) to (7.2.18) we have

$$-(\mu_m + \lambda_n)u_{nm} = a_n b_m$$

and hence

$$u_{nm} = -\frac{a_n b_m}{\mu_m + \lambda_m} \qquad (7.2.19)$$

which is an explicit expression for the FFT of $u_n(y)$. From (7.2.15b) the inverse is

$$u_n(y) = \sum_{m=1}^{\infty} u_{nm} \psi_m(y)$$

$$= \sum_{m=1}^{\infty} -\left\{\frac{a_n b_m}{\mu_m + \lambda_n}\right\} \psi_m(y) \qquad (7.2.20)$$

and finally, from eq. (7.2.6b)

$$u(x, y) = \sum_{n=1}^{\infty} u_n(y)\phi_n(x)$$

$$= \sum_{n=1}^{\infty} \left[\sum_{m=1}^{\infty} -\left\{\frac{a_n b_m}{\mu_m + \lambda_n}\right\} \psi_m(y)\right]\phi_n(x) \qquad (7.2.21)$$

provides the desired velocity distribution.

In summary, following the two methods, we have two different representations of the solution, given by eqs. (7.2.11) and (7.2.21). Mathematically, they are equivalent. However, from a computational viewpoint, the first method is simpler because it involves only one infinite series.

7.3 MULTIDIMENSIONAL PROBLEMS

In the two previous sections, we discussed the solution of linear PDEs involving two independent variables: either (a) time and one space variable or (b) two space variables. Both problems involved self-adjoint differential operators along with self-adjoint homogeneous BCs. We saw that the method of FFT provides a powerful procedure for solving such problems.

In applications, we also encounter problems involving more than two independent variables. A typical example is the heat equation describing transient heat conduction (or mass diffusion) in a three-dimensional body, Ω, along with internal heat generation:

$$k\left[\frac{\partial^2 T}{\partial x^2} + \frac{\partial^2 T}{\partial y^2} + \frac{\partial^2 T}{\partial z^2}\right] + g(x, y, z, t) = C_p\rho\,\frac{\partial T}{\partial t}, \qquad (x, y, z) \in \Omega, t > 0 \quad (7.3.1)$$

where the heat generation term $g(x, y, z, t)$ is in general a function of all the independent variables. It should be evident that as long as the BCs are self-adjoint and homogeneous, we can solve problems of this type following Method 2 of section 7.2. This will involve taking successive FFTs—for example, first in x, then in y, and finally in z. Each FFT eliminates one space variable, so the successive application of three FFTs will yield a first-order ODE (with t as the independent variable) in the triple transform. The solution $T(x, y, z, t)$ is then obtained by three successive inverse transforms (taken in the reverse order z, y, and x), leading to the final solution in the form of a triple infinite series.

As an example, let us determine the transient temperature distribution in a rectangular parallelopiped of dimensions shown in Figure 7.3, by the method of FFTs. The external surfaces are maintained at constant temperature T_s, and the initial temperature T_i is also constant.

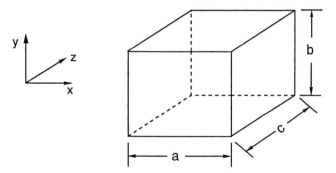

Figure 7.3 A rectangular parallelopiped.

The heat equation describing the temperature distribution is

$$k\left[\frac{\partial^2 T}{\partial x^2} + \frac{\partial^2 T}{\partial y^2} + \frac{\partial^2 T}{\partial z^2}\right]$$

$$= C_p \rho \frac{\partial T}{\partial t}, \qquad x \in (0, a), \quad y \in (0, b), \quad z \in (0, c), \quad t > 0 \quad (7.3.2)$$

along with the BC

$$T = T_0 \qquad\qquad\qquad (7.3.3a)$$

on the *six* surfaces

$$
\begin{array}{llll}
x = 0, & y \in (0, b), & z \in (0, c), & t > 0 \\
x = a, & y \in (0, b), & z \in (0, c), & t > 0 \\
x \in (0, a), & y = 0, & z \in (0, c), & t > 0 \\
x \in (0, a), & y = b, & z \in (0, c), & t > 0 \\
x \in (0, a), & y \in (0, b), & z = 0, & t > 0 \\
x \in (0, a), & y \in (0, b), & z = c, & t > 0
\end{array}
$$

and the IC

$$T = T_i \qquad \text{for } x \in (0, a), \quad y \in (0, b), \quad z \in (0, c), \quad t = 0 \quad (7.3.3b)$$

Let us introduce the variables

$$\theta = \frac{T - T_0}{T_i - T_0} \quad \text{and} \quad \tau = \frac{kt}{C_p \rho} \qquad\qquad (7.3.4)$$

so that eq. (7.3.2) takes the form

$$\frac{\partial^2 \theta}{\partial x^2} + \frac{\partial^2 \theta}{\partial y^2} + \frac{\partial^2 \theta}{\partial z^2} = \frac{\partial \theta}{\partial \tau} \qquad\qquad (7.3.5)$$

all the BCs become homogeneous, and the IC reduces to

$$\theta = 1; \quad x \in (0, a), \quad y \in (0, b), \quad z \in (0, c), \quad \tau = 0 \qquad (7.3.6)$$

Rearrange eq. (7.3.5) as

$$L_x[\theta] = \frac{\partial^2 \theta}{\partial x^2} = \frac{\partial \theta}{\partial \tau} - \frac{\partial^2 \theta}{\partial y^2} - \frac{\partial^2 \theta}{\partial z^2} \qquad (7.3.7)$$

The operator $L_x[\ ]$ and the associated homogeneous BCs define the eigenvalue problem

$$L_x[\phi] = \frac{d^2 \phi}{dx^2} = -\lambda \phi, \qquad x \in (0, a)$$

$$\phi(0) = 0, \qquad \phi(a) = 0$$

which gives the eigenvalues

$$\lambda_n = \frac{n^2 \pi^2}{a^2}, \qquad n = 1, 2, \ldots \qquad (7.3.8a)$$

and the orthonormal eigenfunctions

$$\phi_n(x) = \sqrt{\frac{2}{a}} \sin\left(\frac{n\pi x}{a}\right), \qquad n = 1, 2, \ldots \qquad (7.3.8b)$$

Taking the FFT of eq. (7.3.7) gives

$$\int_0^a L_x[\theta]\phi_n(x)\ dx = \int_0^a \left[\frac{\partial \theta}{\partial \tau} - \frac{\partial^2 \theta}{\partial y^2} - \frac{\partial^2 \theta}{\partial z^2}\right]\phi_n(x)\ dx \qquad (7.3.9)$$

As usual, we now simplify the lhs and rhs of this equation.

$$\text{lhs} = \int_0^a \theta(x, y, z, \tau) L_x[\phi_n]\ dx$$

$$= \int_0^a \theta(x, y, z, \tau)[-\lambda_n \phi_n(x)]\ dx$$

$$= -\lambda_n \theta_n(y, z, \tau)$$

where

$$\theta_n(y, z, \tau) = \mathcal{F}_x[\theta(x, y, z, \tau)] = \int_0^a \theta(x, y, z, \tau)\phi_n(x)\ dx$$

is the FFT of $\theta(x, y, z, \tau)$ with respect to the variable x. The rhs of eq. (7.3.9) takes the form

$$\text{rhs} = \frac{\partial \theta_n}{\partial \tau} - \frac{\partial^2 \theta_n}{\partial y^2} - \frac{\partial^2 \theta_n}{\partial z^2}$$

so that we have

$$-\lambda_n \theta_n = \frac{\partial \theta_n}{\partial \tau} - \frac{\partial^2 \theta_n}{\partial y^2} - \frac{\partial^2 \theta_n}{\partial z^2}$$

which may be rearranged to give

$$L_y[\theta_n] = \frac{\partial^2 \theta_n}{\partial y^2} = \frac{\partial \theta_n}{\partial \tau} - \frac{\partial^2 \theta_n}{\partial z^2} + \lambda_n \theta_n \qquad (7.3.10)$$

The eigenvalue problem corresponding to $L_y[\ ]$ and its associated homogeneous BCs is:

$$L_y[\psi] = \frac{d^2 \psi}{dy^2} = -\mu\psi, \qquad y \in (0, b)$$

$$\psi(0) = 0, \qquad \psi(b) = 0$$

The eigenvalues and orthonormal eigenfunctions are

$$\mu_m = \frac{m^2 \pi^2}{b^2}, \qquad\qquad m = 1, 2, \ldots \qquad (7.3.11a)$$

$$\psi_m(y) = \sqrt{\frac{2}{b}} \sin\left(\frac{m\pi y}{b}\right), \qquad m = 1, 2, \ldots \qquad (7.3.11b)$$

Taking the FFT of eq. (7.3.10) leads to

$$\int_0^b L_y[\theta_n]\psi_m(y)\, dy = \int_0^b \left[\frac{\partial \theta_n}{\partial \tau} - \frac{\partial^2 \theta_n}{\partial z^2} + \lambda_n \theta_n\right]\psi_m(y)\, dy \qquad (7.3.12)$$

We immediately recognize that

$$\text{lhs} = \int_0^b L_y[\theta_n]\psi_m(y)\, dy$$

$$= \int_0^b \theta_n(y, z, \tau) L_y[\psi_m]\, dy$$

$$= \int_0^b \theta_n(y, z, \tau)[-\mu_m\psi_m(y)]\, dy$$

$$= -\mu_m\theta_{nm}(z, \tau)$$

where

$$\theta_{nm}(z, \tau) = \mathcal{F}_y[\theta_n(y, z, \tau)] = \int_0^b \theta_n(y, z, \tau)\psi_m(y)\, dy$$

is the FFT of $\theta_n(y, z, \tau)$ with respect to the variable y. The rhs of eq. (7.3.12) simplifies to

$$\text{rhs} = \frac{\partial \theta_{nm}}{\partial \tau} - \frac{\partial^2 \theta_{nm}}{\partial z^2} + \lambda_n \theta_{nm}$$

Thus we have

$$-\mu_m \theta_{nm} = \frac{\partial \theta_{nm}}{\partial \tau} - \frac{\partial^2 \theta_{nm}}{\partial z^2} + \lambda_n \theta_{nm}$$

which rearranges as

$$L_z[\theta_{nm}] = \frac{\partial^2 \theta_{nm}}{\partial z^2} = \frac{\partial \theta_{nm}}{\partial \tau} + (\lambda_n + \mu_m)\theta_{nm} \qquad (7.3.13)$$

Repeating the procedure followed above, now for the z variable, gives rise to the eigenvalue problem:

$$L_z[\xi] = \frac{d^2\xi}{dz^2} = -\eta\xi, \qquad z \in (0, c)$$

$$\xi(0) = 0, \qquad \xi(c) = 0$$

which has eigenvalues and orthonormal eigenfunctions:

$$\eta_p = \frac{p^2\pi^2}{c^2}, \qquad \xi_p = \sqrt{\frac{2}{c}} \sin\left(\frac{p\pi z}{c}\right), \qquad p = 1, 2, \ldots \qquad (7.3.14)$$

Taking the FFT of eq. (7.3.13) gives, as before,

$$-\eta_p \theta_{nmp} = \frac{d\theta_{nmp}}{d\tau} + (\lambda_n + \mu_m)\theta_{nmp}$$

which is rearranged to give

$$\frac{d\theta_{nmp}}{d\tau} = -(\lambda_n + \mu_m + \eta_p)\theta_{nmp} \qquad (7.3.15)$$

where

$$\theta_{nmp}(\tau) = \mathcal{F}_z[\theta_{nm}(z, \tau)] = \int_0^c \theta_{nm}(z, \tau)\xi_p(z) \, dz$$

is the FFT of $\theta_{nm}(z, \tau)$ with respect to the variable z. Note that θ_{nmp} is the triple FFT of the original variable $\theta(x, y, z, \tau)$, obtained by taking three successive transforms of θ with respect to the variables x, y, and z, respectively. The solution of eq. (7.3.15) is

$$\theta_{nmp}(\tau) = \theta_{nmp}(0) \exp[-(\lambda_n + \mu_m + \eta_p)\tau] \qquad (7.3.16)$$

The only unknown in expression (7.3.16) is $\theta_{nmp}(0)$, which we determine now. For this, recall that eq. (7.3.6) provides the IC for θ:

$$\theta(x, y, z, 0) = 1, \qquad x \in (0, a), y \in (0, b), z \in (0, c)$$

Then,

$$\theta_n(y, z, 0) = \int_0^a 1 \, \phi_n(x) \, dx$$

$$= \sqrt{\frac{2}{a}} \int_0^a \sin\left(\frac{n\pi x}{a}\right) dx$$

$$= \frac{\sqrt{2a}}{n\pi} [1 - (-1)^n]$$

Similarly,

$$\theta_{nm}(z, 0) = \int_0^b \theta_n(y, z, 0)\psi_m(y) \, dy$$

$$= \frac{\sqrt{2a}}{n\pi} \frac{\sqrt{2b}}{m\pi} [1 - (-1)^n][1 - (-1)^m]$$

and finally,

$$\theta_{nmp}(0) = \int_0^c \theta_{nm}(z, 0)\xi_p(z) \, dz$$

$$= \frac{\sqrt{2a}}{n\pi} \frac{\sqrt{2b}}{m\pi} \frac{\sqrt{2c}}{p\pi} [1 - (-1)^n][1 - (-1)^m][1 - (-1)^p]$$

$$= 8 \frac{\sqrt{2a}}{n\pi} \frac{\sqrt{2b}}{m\pi} \frac{\sqrt{2c}}{p\pi}; \qquad n, m, \text{ and } p \text{ are } odd \text{ integers} \qquad (7.3.17)$$

so that $\theta_{nmp}(\tau)$ in eq. (7.3.16) is fully known.

We now need to take the inverse transforms. These are taken in the *reverse* order of variables. Thus

$$\theta_{nm}(z, \tau) = \mathcal{F}_z^{-1}[\theta_{nmp}(\tau)] = \sum_{p=1}^{\infty} \theta_{nmp}(\tau)\xi_p(z)$$

$$\theta_n(y, z, \tau) = \mathcal{F}_y^{-1}[\theta_{nm}(z, \tau)] = \sum_{m=1}^{\infty} \theta_{nm}(z, \tau)\psi_m(y)$$

and

$$\theta(x, y, z, \tau) = \mathcal{F}_x^{-1}[\theta_n(y, z, \tau)] = \sum_{n=1}^{\infty} \theta_n(y, z, \tau)\phi_n(x)$$

In explicit form,

$$\theta(x, y, z, \tau) = \sum_{n=1,2}^{\infty} \sum_{m=1,2}^{\infty} \sum_{p=1,2}^{\infty} \frac{64}{\pi^3 nmp} \sin\left(\frac{n\pi x}{a}\right) \sin\left(\frac{m\pi y}{b}\right) \sin\left(\frac{p\pi z}{c}\right)$$

$$\exp[-(\lambda_n + \mu_m + \eta_p)\tau] \qquad (7.3.18)$$

represents the temperature distribution as a function of position and time, where recall that

$$\theta = \frac{T - T_0}{T_i - T_0}, \qquad \tau = \frac{kt}{C_p \rho}, \qquad \lambda_n = \frac{n^2 \pi^2}{a^2}, \qquad \mu_m = \frac{m^2 \pi^2}{b^2}, \qquad \eta_p = \frac{p^2 \pi^2}{c^2}$$

The notation $\sum_{j=1,2}^{\infty}$ means summation from 1 to ∞, in steps of 2; that is, terms involving $j = 1, 3, 5, \ldots$ are summed.

Remark. The procedure discussed above in the context of eq. (7.3.2) can also be used for solving eq. (7.3.1), with a general heat generation term $g(x, y, z, t)$. Taking the FFTs with respect to the x, y, and z variables successively accounts for the contributions from portions of the g function depending on these variables, respectively. The resulting linear first-order ODE in the triple transform T_{nmp} is then slightly modified from eq. (7.3.15) and includes a nonhomogeneous term of the form $\alpha h(t)$, where α is a constant involving a product of the x, y, and z transforms of g. This equation can be solved using standard techniques. The remaining steps of the procedure, involving the successive inverse transforms, are the same as those followed above.

7.4 NONHOMOGENEOUS BOUNDARY CONDITIONS

The problems considered so far have all involved homogeneous BCs. In applications, we also encounter problems with nonhomogeneous BCs. An example is the determination of transient temperature distribution in a slab, with internal heat generation, and time-dependent BCs. For this, let us consider the PDE

$$\frac{\partial^2 u}{\partial x^2} + g_1(x) g_2(t) = \frac{\partial u}{\partial t}; \qquad x \in (0, 1), \quad t > 0 \qquad (7.4.1)$$

with BCs

$$u = f_1(t); \qquad x = 0, \qquad t > 0 \qquad (7.4.2a)$$

$$u = f_2(t); \qquad x = 1, \qquad t > 0 \qquad (7.4.2b)$$

and IC

$$u = h(x); \qquad x \in (0, 1), \quad t = 0 \qquad (7.4.3)$$

In order to solve this problem, we consider instead the following four problems.

Problem 1

$$\frac{\partial^2 u_1}{\partial x^2} + g_1(x) g_2(t) = \frac{\partial u_1}{\partial t}; \qquad x \in (0, 1), \quad t > 0$$

$$u_1 = 0; \qquad x = 0, \qquad t > 0$$

$$u_1 = 0; \qquad x = 1, \qquad t > 0$$

$$u_1 = 0; \qquad x \in (0, 1), \quad t = 0$$

Problem 2

$$\frac{\partial^2 u_2}{\partial x^2} = \frac{\partial u_2}{\partial t}; \qquad x \in (0, 1), \quad t > 0$$

$$u_2 = f_1(t); \qquad x = 0, \qquad t > 0$$

$$u_2 = 0; \qquad x = 1, \qquad t > 0$$

$$u_2 = 0; \qquad x \in (0, 1), \quad t = 0$$

Problem 3

$$\frac{\partial^2 u_3}{\partial x^2} = \frac{\partial u_3}{\partial t}; \qquad x \in (0, 1), \quad t > 0$$

$$u_3 = 0; \qquad x = 0, \qquad t > 0$$

$$u_3 = f_2(t); \qquad x = 1, \qquad t > 0$$

$$u_3 = 0; \qquad x \in (0, 1), \quad t = 0$$

Problem 4

$$\frac{\partial^2 u_4}{\partial x^2} = \frac{\partial u_4}{\partial t}; \qquad x \in (0, 1), \quad t > 0$$

$$u_4 = 0; \qquad x = 0, \qquad t > 0$$

$$u_4 = 0; \qquad x = 1, \qquad t > 0$$

$$u_4 = h(x); \qquad x \in (0, 1), \quad t = 0$$

If we define the function

$$y = u_1 + u_2 + u_3 + u_4 \tag{7.4.4}$$

then

$$\frac{\partial^2 y}{\partial x^2} = \frac{\partial^2}{\partial x^2} [u_1 + u_2 + u_3 + u_4]$$

Using the PDEs of Problems 1 to 4, we obtain

$$\frac{\partial^2 y}{\partial x^2} = \frac{\partial}{\partial t} [u_1 + u_2 + u_3 + u_4] - g_1(x)g_2(t)$$

$$= \frac{\partial y}{\partial t} - g_1(x)g_2(t)$$

that is,

$$\frac{\partial^2 y}{\partial x^2} + g_1(x)g_2(t) = \frac{\partial y}{\partial t}$$

Thus the function y satisfies the *same* PDE (7.4.1) as u does. As far as the BCs and IC for y are concerned, note that

$$\text{at } x = 0, t > 0: \qquad y = f_1(t)$$

$$\text{at } x = 1, t > 0: \qquad y = f_2(t)$$

$$\text{for } x \in (0, 1), t = 0: \qquad y = h(x)$$

which are also the *same* as given by eqs. (7.4.2) and (7.4.3) for u. Thus we can conclude that

$$u = y = u_1 + u_2 + u_3 + u_4 \qquad (7.4.5)$$

is indeed the solution of the original problem. This result that the solution of the original problem can be obtained by summing up the solutions of several individual problems (four in this case) is called the *principle of superposition*.

Since Problems 1 and 4 for u_1 and u_4, respectively, involve homogeneous BCs, we can solve them by the method of FFTs discussed in section 7.1. At this stage we do not know how to solve Problems 2 and 3, because they involve nonhomogeneous BCs. The method to solve these problems is similar, so we may focus on, say Problem 2:

$$\frac{\partial^2 u_2}{\partial x^2} = \frac{\partial u_2}{\partial t}; \qquad x \in (0, 1), \quad t > 0 \qquad (7.4.6)$$

$$u_2 = f_1(t); \qquad x = 0, \qquad t > 0 \qquad (7.4.7a)$$

$$u_2 = 0; \qquad x = 1, \qquad t > 0 \qquad (7.4.7b)$$

$$u_2 = 0; \qquad x \in (0, 1), \quad t = 0 \qquad (7.4.8)$$

where $f_1(t)$ is a given function of t. In order to solve this, let us *assume* that we can solve the problem where the nonhomogeneous term $f_1(t)$ is a square pulse, $f(t)$, of area unity located in the interval $\tau < t < \tau + a$, as shown in Figure 7.4. Let us denote the solution of this pulse problem as

$$v(x, t; \tau, a)$$

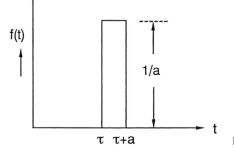

Figure 7.4 A rectangular pulse of unit strength.

where τ and a are parameters on which the solution v depends. Now, to determine v, we need to solve eqs. (7.4.6) to (7.4.8) with $f_1(t)$ replaced by $f(t)$. From the shape of $f(t)$, it is clear that for $t < \tau$, the problem is *fully* homogeneous, and so its solution will be just the trivial solution. For $t > \tau$, the time variable of importance in solving for v will in fact be $(t - \tau)$. Thus the proper form of the solution for the problem with $f_1(t)$ replaced by the pulse $f(t)$ will actually be

$$v(x, t - \tau; a), \qquad t > \tau + a$$

which simply indicates that the time variable will appear as $(t - \tau)$. If we now take the limit as $a \to 0$ and denote

$$W(x, t - \tau) = \lim_{a \to 0} v(x, t - \tau; a) \tag{7.4.9}$$

then W can be viewed as the *impulse response*, because the limiting pulse is an impulse of strength (i.e., area) unity, located at $t = \tau$—that is, the Dirac delta function $\delta(t - \tau)$.

For our problem, however, $f_1(t)$ is in general an arbitrary function as shown in Figure 7.5. The shaded area in the figure represents an impulse of strength $f_1(\tau)\, d\tau$. Since $W(x, t - \tau)$ is the solution (i.e., response) of the system for time $t > \tau$, to a unit impulse applied at time $t = \tau$, it follows that

$$W(x, t - \tau)f_1(\tau)\, d\tau$$

will be the response to an impulse of strength $f_1(\tau)\, d\tau$. Finally, from the principle of superposition, the response of the system to function $f_1(t)$ as the input will then simply be the sum of the responses resulting from all past impulses applied up to time t—that is,

$$u_2(x, t) = \int_0^t W(x, t - \tau)f_1(\tau)\, d\tau \tag{7.4.10}$$

This equation is a form of the so-called *Duhamel's formula*. It shows how to obtain the solution to a problem with time-dependent BCs from the solution of the same problem with constant BCs.

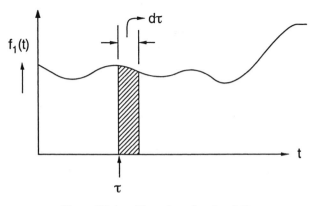

Figure 7.5 An arbitrary input function, $f_1(t)$.

The only remaining problem is to find the impulse response, W. For this, let us first determine the *step response* $w(x, t)$—that is, the solution $u_2(x, t)$ for $f_1(t) = 1$. Equation (7.4.10) gives

$$
\begin{aligned}
w(x, t) &= \int_0^t W(x, t - \tau) \cdot 1 \cdot d\tau \\
&= -\int_t^0 W(x, \eta) \, d\eta, \qquad \text{if we let } t - \tau = \eta \\
&= \int_0^t W(x, \eta) \, d\eta
\end{aligned}
$$

Differentiating both sides with respect to t gives

$$
\frac{\partial w}{\partial t} = W(x, t) \tag{7.4.11}
$$

Thus the impulse response, $W(x, t)$ is merely the first partial derivative of the step response, $w(x, t)$ with respect to t.

In summary, the procedure to determine the solution of eqs. (7.4.6) to (7.4.8) is as follows:

1. First, find the step response $w(x, t)$—that is, the solution of eqs. (7.4.6) to (7.4.8) with $f_1(t)$ replaced by 1.
2. Then, determine $W(x, t)$ from eq. (7.4.11).
3. Finally, determine $u_2(x, t)$ by substituting for $W(x, t)$ in eq. (7.4.10).

This procedure will also work for problems involving multiple space dimensions. However, the method for determining the step response becomes different, depending on whether one or more space variables are involved. For this reason, we next consider these two categories separately.

7.4.1 Single Space Dimension

To describe the strategy in this case, we can continue discussing the same problem. Let us now proceed to obtain the step response, $w(x, t)$. This is the solution of the problem.

$$
\frac{\partial^2 w}{\partial x^2} = \frac{\partial w}{\partial t}; \qquad x \in (0, 1), \quad t > 0 \tag{7.4.12}
$$

$$
w = 1; \qquad x = 0, \qquad t > 0 \tag{7.4.13a}
$$

$$
w = 0; \qquad x = 1, \qquad t > 0 \tag{7.4.13b}
$$

$$
w = 0; \qquad x \in (0, 1), \quad t = 0 \tag{7.4.14}
$$

To solve for w, we need to first make the BCs homogeneous. For this, change the dependent variable as

$$
z(x, t) = w(x, t) + c_1 x + c_2 \tag{7.4.15}
$$

where c_1 and c_2 are constants, to be chosen such that the problem in z has homogeneous BCs. Thus we require that

$$\text{at } x = 0: \qquad 1 + c_2 = 0$$
$$\text{at } x = 1: \qquad c_1 + c_2 = 0$$

so that

$$c_1 = 1, \qquad c_2 = -1$$

and from eq. (7.4.15)

$$z = w + (x - 1) \tag{7.4.16}$$

is the desired change of variable. From eqs. (7.4.12) to (7.4.14) the problem in z is therefore given by

$$\frac{\partial^2 z}{\partial x^2} = \frac{\partial z}{\partial t}; \qquad x \in (0, 1), \quad t > 0$$

$$z = 0; \qquad x = 0, \qquad t > 0$$

$$z = 0; \qquad x = 1, \qquad t > 0$$

$$z = (x - 1); \qquad x \in (0, 1), \quad t = 0$$

which is similar to the introductory example treated in section 7.1. From eq. (7.1.18) the solution is

$$z(x, t) = \sum_{n=1}^{\infty} q_n \exp(-n^2\pi^2 t) \sqrt{2} \sin(n\pi x) \tag{7.4.17}$$

where

$$q_n = \int_0^1 (x - 1) \sqrt{2} \sin(n \pi x) \, dx$$

$$= -\frac{\sqrt{2}}{n\pi} \tag{7.4.17a}$$

From eq. (7.4.16) the step response is thus given by

$$w(x, t) = 1 - x + z(x, t)$$

$$= 1 - x - \frac{2}{\pi} \sum_{n=1}^{\infty} \frac{1}{n} \sin(n\pi x) \exp(-n^2\pi^2 t) \tag{7.4.18}$$

Then, from eq. (7.4.11) the impulse response is

$$W(x, t) = \frac{\partial w}{\partial t}$$

$$= 2\pi \sum_{n=1}^{\infty} n \sin(n\pi x) \exp(-n^2\pi^2 t) \tag{7.4.19}$$

Finally, from eq. (7.4.10) the solution of eqs. (7.4.6)–(7.4.8) is given as

$$u_2(x, t) = 2\pi \sum_{n=1}^{\infty} n \sin(n\pi x) \left[\int_0^t f_1(\tau) \exp[-n^2\pi^2(t - \tau)] \, d\tau \right] \quad (7.4.20)$$

Remark. It should be clear from the above discussion that the procedure to solve Problem 3 for $u_3(x, t)$ will involve finding the corresponding step response, by replacing $f_2(t)$ by 1 in the BC at $x = 1$. The impulse response will then be the derivative defined by eq. (7.4.11), and the solution $u_3(x, t)$ will be obtained from eq. (7.4.10), with $f_1(\tau)$ replaced by $f_2(\tau)$. The full solution of the original problem given by eq. (7.4.4) can then be easily constructed (see problem 7.11 at the end of the chapter).

7.4.2 Multiple Space Dimensions

We now describe the procedure to solve problems with nonhomogeneous boundary conditions, which also involve multiple space dimensions. As noted earlier, the procedure is the same as when there is only one space dimension, with the only exception that the step response is determined differently. In order to illustrate this portion of the general procedure, we consider the two-dimensional analog of problem (7.4.12) to (7.4.14):

$$\frac{\partial^2 w}{\partial x^2} + \frac{\partial^2 w}{\partial y^2} = \frac{\partial w}{\partial t}; \qquad x \in (0, 1), \quad y \in (0, 1), \quad t > 0 \qquad (7.4.21)$$

$$w = 1; \qquad x = 0, \qquad y \in (0, 1), \quad t > 0 \qquad (7.4.22a)$$

$$w = 0; \qquad x = 1, \qquad y \in (0, 1), \quad t > 0 \qquad (7.4.22b)$$

$$w = 0; \qquad x \in (0, 1), \quad y = 0, \qquad t > 0 \qquad (7.4.22c)$$

$$w = 0; \qquad x \in (0, 1), \quad y = 1, \qquad t > 0 \qquad (7.4.22d)$$

$$w = 0; \qquad x \in (0, 1), \quad y \in (0, 1), \quad t = 0 \qquad (7.4.23)$$

It can be seen readily that in this case it is not possible to make *all* the BCs (7.4.22) homogeneous through a simple change of the dependent variables, as we did in the previous section for the problem with only one space dimension. Thus we follow a different procedure. Specifically, we take eq. (7.4.21) in the finite Fourier transformed space, by multiplying both sides of the equation by $[\sqrt{2} \sin(n\pi x)] [\sqrt{2} \sin(m\pi y)]$ and integrating twice over $x \in [0, 1]$ and $y \in [0, 1]$. The first term on the lhs, using conditions (7.4.22a) and (7.4.22b) and integrating twice by parts in x, leads to

$$2 \int_0^1 \int_0^1 \frac{\partial^2 w}{\partial x^2} \sin(n\pi x) \sin(m\pi y) \, dx \, dy = 2 \int_0^1 \left[n\pi - (n\pi)^2 \int_0^1 w \sin(n\pi x) \, dx \right]$$

$$\sin(m\pi y) \, dy$$

$$= \frac{2n}{m} [1 - \cos(m\pi)] - (n\pi)^2 \, w_{nm}(t)$$

$$= \frac{2n}{m} [1 - (-1)^m] - (n\pi)^2 \, w_{nm}(t)$$

where we have introduced the double FFT of $w(x, y, t)$ defined by

$$w_{nm}(t) = 2 \int_0^1 \int_0^1 w(x, y, t) \sin(n\pi x) \sin(m\pi y) \, dx \, dy$$

For the second term on the lhs, using BCs (7.4.22c) and (7.4.22d) and integrating twice by parts in y, we obtain

$$2 \int_0^1 \int_0^1 \frac{\partial^2 w}{\partial y^2} \sin(n\pi x) \sin(m\pi y) \, dx \, dy = -2(m\pi)^2 \int_0^1 \int_0^1 w \sin(n\pi x) \sin(m\pi y) \, dx \, dy$$

$$= -(m\pi)^2 \, w_{nm}(t)$$

Finally, since it is readily seen that the term on the rhs leads to dw_{nm}/dt, eq. (7.4.21) in the transformed space takes the form

$$\frac{dw_{nm}}{dt} = -\pi^2(n^2 + m^2)w_{nm} + \frac{2n}{m} [1 - (-1)^m] \tag{7.4.24}$$

along with the IC

$$w_{nm} = 0 \quad \text{at } t = 0 \tag{7.4.25}$$

which arises from condition (7.4.23). Integrating the first-order ODE yields

$$w_{nm}(t) = \frac{2n[1 - (-1)^m]}{m\pi^2(n^2 + m^2)} [1 - \exp\{-\pi^2(n^2 + m^2)t\}] \tag{7.4.26}$$

We now need to invert the above double FFT. For this, we can proceed along the same lines as in section 7.3, so that

$$w_n(y, t) = \sqrt{2} \sum_{m=1}^\infty w_{nm}(t) \sin(m\pi y)$$

$$w(x, y, t) = 2 \sum_{n=1}^\infty \sum_{m=1}^\infty w_{nm}(t) \sin(n\pi x) \sin(m\pi y)$$

Substituting eq. (7.4.26), we have

$$w(x, y, t) = \frac{8}{\pi^2} \sum_{n=1}^\infty \sum_{m=1,2}^\infty \frac{n}{m(n^2 + m^2)} \sin(n\pi x) \sin(m\pi y) \tag{7.4.27}$$

$$[1 - \exp\{-\pi^2(n^2 + m^2)t\}]$$

where note that the second sum involves only odd terms—that is, $m = 1, 3, 5. \ldots$

Equation (7.4.27) provides the desired step response—that is, the solution of eqs. (7.4.21) to (7.4.23). Finally, note that using eq. (7.4.11), the impulse response is given by

$$W(x, y, t) = \frac{\partial w}{\partial t} = 8 \sum_{n=1}^\infty \sum_{m=1,2}^\infty \frac{n}{m} \sin(n\pi x) \sin(m\pi y) \exp\{-\pi^2(n^2 + m^2)t\}$$

and problems with a general nonhomogeneous term $f(y, t)$, in lieu of 1 in eq. (7.4.22a), can then be solved using the Duhamel formula (7.4.10).

7.5　A PROBLEM INVOLVING CYLINDRICAL GEOMETRY: THE FINITE HANKEL TRANSFORM

In section 4.15 we saw that one-dimensional steady-state problems in cylindrical geometry can be solved in terms of Bessel functions. Following the approach developed in previous sections, transient problems involving cylindrical geometry can also be solved readily. As an example to illustrate the technique, let us consider transient diffusion of mass within an infinitely long cylinder (or equivalently a finite cylinder with sealed ends) shown in Figure 7.6. A shell balance gives the PDE

$$\frac{D}{r}\frac{\partial}{\partial r}\left[r\frac{\partial C}{\partial r}\right] = \frac{\partial C}{\partial t}; \qquad r \in (0, a), \quad t > 0 \tag{7.5.1}$$

with BCs

$$C \text{ finite} \tag{7.5.2a}$$

$$C = C_0; \qquad r = a, \quad t > 0 \tag{7.2.5b}$$

and IC

$$C = f(r); \qquad r \in (0, a), \quad t = 0 \tag{7.5.3}$$

where C is concentration of the diffusing species, D is its diffusion coefficient, r is radial position within cylinder of radius a, and t is the time variable. Introducing the dimensionless quantities

$$u = \frac{C}{C_0}, \qquad s = \frac{r}{a}, \qquad \tau = \frac{Dt}{a^2} \tag{7.5.4}$$

eqs. (7.5.1) to (7.5.3) take the form

$$\frac{1}{s}\frac{\partial}{\partial s}\left[s\frac{\partial u}{\partial s}\right] = \frac{\partial u}{\partial \tau}; \qquad s \in (0, 1), \quad \tau > 0 \tag{7.5.5}$$

$$u = 1, \qquad s = 1, \qquad \tau > 0 \tag{7.5.6}$$

$$u = g(s); \qquad s \in (0, 1), \quad \tau = 0 \tag{7.5.7}$$

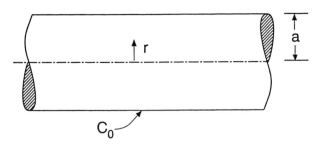

Figure 7.6 An infinitely long cylinder.

where

$$g(s) = \frac{f(as)}{C_0}$$

In order to solve the problem by the method of generalized FFTs, we need to identify a self-adjoint operator along with *homogeneous* self-adjoint BCs. For this, let us first make the BC at $s = 1$ homogeneous by introducing the variable

$$v = u - 1 \tag{7.5.8}$$

so that eqs. (7.5.5) to (7.5.7) take the form

$$\frac{1}{s}\frac{\partial}{\partial s}\left[s\frac{\partial v}{\partial s} \right] = \frac{\partial v}{\partial \tau}; \qquad s \in (0, 1), \quad \tau > 0 \tag{7.5.9}$$

$$v \text{ finite} \tag{7.5.10a}$$

$$v = 0; \qquad s = 1, \qquad \tau > 0 \tag{7.5.10b}$$

$$v = h(s); \qquad s \in (0, 1), \quad \tau = 0 \tag{7.5.11}$$

where

$$h(s) = g(s) - 1$$

Equation (7.5.9) can be rearranged as

$$L[v] = \frac{\partial}{\partial s}\left[s\frac{\partial v}{\partial s} \right] = s\frac{\partial v}{\partial \tau} \tag{7.5.12}$$

to yield the cylindrical diffusion operator $L[\ ]$ in the self-adjoint form. This along with the BCs given by eqs. (7.5.10) defines the eigenvalue problem:

$$L[\phi] = \frac{d}{ds}\left[s\frac{d\phi}{ds} \right] = -\lambda s\phi, \qquad s \in (0, 1) \tag{7.5.13}$$

$$\phi \text{ finite} \tag{7.5.14a}$$

$$\phi = 0, \qquad s = 1 \tag{7.5.14b}$$

which was analyzed in section 4.16. From eq. (4.16.9) the eigenvalues λ_n satisfy

$$J_0(\sqrt{\lambda_n}) = 0, \qquad n = 1, 2, \ldots \tag{7.5.15}$$

and from eq. (4.16.25) the corresponding orthonormal eigenfunctions are

$$\phi_n(s) = \sqrt{2}\frac{J_0(\sqrt{\lambda_n}s)}{J_1(\sqrt{\lambda_n})}, \qquad n = 1, 2, \ldots \tag{7.5.16}$$

Let us now take the generalized FFT, called the finite *Hankel* transform (FHT) in this case [see eq. (4.16.26)], of both sides of eq. (7.5.12):

$$\int_0^1 \phi_n(s)L[v]\, ds = \int_0^1 s\phi_n(s)\frac{\partial v}{\partial \tau}\, ds \tag{7.5.17}$$

Due to self-adjointness of $L[\ ]$ and the BCs (7.5.10), the lhs gives

$$\text{lhs} = \int_0^1 vL[\phi_n]\ ds = \int_0^1 v[-\lambda_n s\phi_n]\ ds = -\lambda_n v_n(\tau) \qquad (7.5.18)$$

where

$$v_n(\tau) = \int_0^1 s\phi_n(s)v(s,\tau)\ ds = \mathcal{H}[v(s,\tau)] \qquad (7.5.19)$$

is the FHT of $v(s,\tau)$.

The rhs of eq. (7.5.17) gives

$$\text{rhs} = \frac{\partial}{\partial \tau} \int_0^1 s\phi_n(s)v\ ds = \frac{dv_n}{d\tau} \qquad (7.5.20)$$

where the integration and derivative operations have been exchanged in the first step, and the partial derivative becomes an ordinary derivative since v_n is a function *only* of τ. Equations (7.5.17), (7.5.18), and (7.5.20) then lead to

$$\frac{dv_n}{d\tau} = -\lambda_n v_n$$

which has the solution

$$v_n(\tau) = v_n(0)\ \exp[-\lambda_n \tau] \qquad (7.5.21)$$

Now,

$$
\begin{aligned}
v_n(0) &= \lim_{\tau \to 0} v_n(\tau) \\
&= \lim_{\tau \to 0} \int_0^1 s\phi_n(s)v(s,\tau)\ ds \\
&= \int_0^1 s\phi_n(s)v(s,0)\ ds \\
&= \int_0^1 s\phi_n(s)h(s)\ ds \\
&= h_n \qquad (7.5.22)
\end{aligned}
$$

where

$$h_n = \mathcal{H}[h(s)] \qquad (7.5.23)$$

is the FHT of $h(s)$. Thus, as usual, the IC $v_n(0)$ for the FHT of $v(s,\tau)$ is given by the FHT of the IC for $v(s,\tau)$—that is, $h(s)$.

Finally, taking the inverse transform gives the complete solution:

$$
\begin{aligned}
v(s,\tau) = \mathcal{H}^{-1}[v_n(\tau)] &= \sum_{n=1}^{\infty} v_n(\tau)\phi_n(s) \\
&= \sum_{n=1}^{\infty} h_n\ \exp[-\lambda_n \tau]\ \frac{\sqrt{2}J_0(\sqrt{\lambda_n}s)}{J_1(\sqrt{\lambda_n})} \qquad (7.5.24)
\end{aligned}
$$

from which $u(s, \tau)$ and $C(r, t)$ can be calculated directly using the definitions given by eqs. (7.5.8) and (7.5.4), respectively.

One is sometimes interested in the *average concentration* within the cylinder. This is defined as

$$\overline{C}(t) = \frac{\displaystyle\int_0^a C(r, t)2\pi r \, dr}{\displaystyle\int_0^a 2\pi r \, dr} = \frac{C_0 \displaystyle\int_0^1 u(s, \tau)as \, ads}{\displaystyle\int_0^1 as \, ads}$$

$$= 2C_0 \int_0^1 u(s, \tau)s \, ds$$

$$= 2C_0 \int_0^1 [v(s, \tau) + 1] \, s \, ds$$

$$= C_0 + 2C_0 \int_0^1 s \sum_{n=1}^{\infty} h_n \, \exp[-\lambda_n \tau] \frac{\sqrt{2}J_0(\sqrt{\lambda_n} s)}{J_1(\sqrt{\lambda_n})} \, ds$$

$$= C_0 + 2\sqrt{2}C_0 \sum_{n=1}^{\infty} \frac{h_n \, \exp[-\lambda_n \tau]}{J_1(\sqrt{\lambda_n})} \int_0^1 sJ_0(\sqrt{\lambda_n} s) \, ds \qquad (7.5.25)$$

The last integral can be easily evaluated as follows. If we call

$$I = \int_0^1 sJ_0(\sqrt{\lambda_n} s) \, ds$$

then the change of variable

$$\sqrt{\lambda_n} s = z$$

gives

$$I = \frac{1}{\lambda_n} \int_0^{\sqrt{\lambda_n}} zJ_0(z) \, dz$$

Using a derivative formula from section 4.13:

$$\frac{d}{dx} [x^\alpha J_\alpha(x)] = x^\alpha J_{\alpha-1}(x)$$

when applied for $\alpha = 1$ implies that

$$\frac{d}{dz} [zJ_1(z)] = zJ_0(z)$$

Thus

$$I = \frac{1}{\lambda_n} [zJ_1(z)]_0^{\sqrt{\lambda_n}} = \frac{J_1(\sqrt{\lambda_n})}{\sqrt{\lambda_n}}$$

so that from eq. (7.5.25) we have an expression for the average concentration:

$$\overline{C}(t) = C_0 + 2\sqrt{2}C_0 \sum_{n=1}^{\infty} \frac{h_n \exp[-\lambda_n Dt/a^2]}{\sqrt{\lambda_n}} \qquad (7.5.26)$$

A special case involves the IC where the concentration distribution within the cylinder is initially *uniform*. In this case,

$$f(r) = C_i$$

a constant, so that

$$h(s) = \frac{C_i}{C_0} - 1$$

Then from eq. (7.5.23),

$$h_n = \int_0^1 \left[\frac{C_i}{C_0} - 1 \right] s \frac{\sqrt{2}J_0(\sqrt{\lambda_n} s)}{J_1(\sqrt{\lambda_n})} ds$$

$$= \left[\frac{C_i}{C_0} - 1 \right] \sqrt{\frac{2}{\lambda_n}} \qquad (7.5.27)$$

which can be substituted in eq. (7.5.24) to give the transient concentration distribution:

$$C(r, t) = C_0 + 2(C_i - C_0) \sum_{n=1}^{\infty} \frac{J_0(\sqrt{\lambda_n} r/a)}{\sqrt{\lambda_n}J_1(\sqrt{\lambda_n})} \exp[-\lambda_n Dt/a^2] \qquad (7.5.28)$$

For this special IC, the average concentration from eq. (7.5.26) is given by

$$\overline{C}(t) = C_0 + 4(C_i - C_0) \sum_{n=1}^{\infty} \frac{\exp[-\lambda_n Dt/a^2]}{\lambda_n} \qquad (7.5.29)$$

The expression (7.5.29) is sometimes used to obtain the diffusion coefficient D from experimental data. If a finite cylinder with sealed ends and volume V, initially with a uniform concentration C_i of a diffusing species, is exposed suddenly to a species concentration C_0, then its weight (e.g., measured in a gravimetric balance) will vary with time according to $V\overline{C}(t)$, where $\overline{C}(t)$ is given by eq. (7.5.29). Since $0 < \lambda_1 < \lambda_2 < \cdots$, the asymptotic behavior of $\overline{C}(t)$ for long time is

$$\overline{C}(t) \sim C_0 + \frac{4}{\lambda_1} (C_i - C_0) \exp\left[-\frac{\lambda_1 Dt}{a^2} \right] \qquad (7.5.30)$$

Thus as shown in Figure 7.7 the asymptotic slope of the function

$$X(t) = \ln\left[\frac{4(C_i - C_0)}{\lambda_1\{\overline{C}(t) - C_0\}} \right] \qquad (7.5.31)$$

when plotted versus t, is given by $\lambda_1 D/a^2$, from which the species diffusion coefficient D can be computed readily.

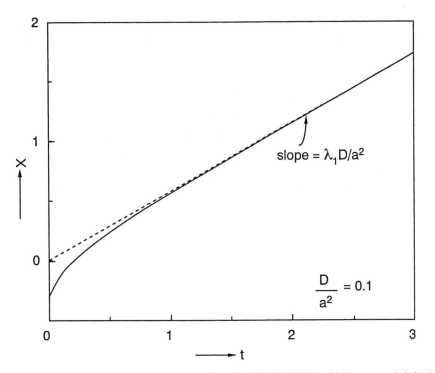

Figure 7.7 Comparison of the function $X(t)$ given by eq. (7.5.31) (solid line) with its asymptotic behavior for large t values (dashed line).

7.6 VIBRATION OF A CIRCULAR MEMBRANE

The hyperbolic PDE

$$\frac{\partial^2 u}{\partial t^2} = c^2 \left[\frac{\partial^2 u}{\partial s^2} + \frac{1}{s} \frac{\partial u}{\partial s} \right]; \qquad s \in (0, 1), \quad t > 0 \tag{7.6.1}$$

along with the BCs

$$u \text{ finite} \tag{7.6.2a}$$

$$u = 0; \qquad s = 1, \quad t > 0 \tag{7.6.2b}$$

and ICs

$$u = f(s); \qquad s \in (0, 1), \quad t = 0 \tag{7.6.3a}$$

$$\frac{\partial u}{\partial t} = 0; \qquad s \in (0, 1), \quad t = 0 \tag{7.6.3b}$$

describes the transverse motion of an elastic circular membrane, fixed at its boundary. The parameter $c^2 = T/\rho$, where tension T represents the force per unit length caused by stretching the membrane and ρ, is the membrane mass per unit area. The vibration of a membrane represents a two-dimensional analog of the vibrating string problem (see section 5.5.3).

We can rearrange eq. (7.6.1) as

$$L[u] = \frac{\partial}{\partial s}\left[s \frac{\partial u}{\partial s} \right] = \frac{s}{c^2} \frac{\partial^2 u}{\partial t^2} \tag{7.6.4}$$

The operator $L[\ ]$ and the BCs given by eqs. (7.6.2) again define the eigenvalue problem described by eqs. (7.5.13) to (7.5.14), with the corresponding eigenvalues λ_n and orthonormal eigenfunctions $\phi_n(s)$ given by eqs. (7.5.15) and (7.5.16), respectively. As usual, taking the finite Hankel transform of eq. (7.6.4) gives

$$-\lambda_n u_n(t) = \frac{1}{c^2} \frac{d^2 u_n}{dt^2} \tag{7.6.5}$$

where

$$u_n(t) = \int_0^1 s\phi_n(s)u(s, t)\, ds = \mathcal{H}[u(s, t)] \tag{7.6.6}$$

is the FHT of $u(s, t)$. The ICs for $u_n(t)$ can be obtained readily from the ICs for u, given by eqs. (7.6.3) as follows:

$$
\begin{aligned}
u_n(0) &= \lim_{t \to 0} u_n(t) = \lim_{t \to 0} \int_0^1 s\phi_n(s)u(s, t)\, ds \\
&= \int_0^1 s\phi_n(s)u(s, 0)\, ds \\
&= \int_0^1 s\phi_n(s)f(s)\, ds \\
&= f_n
\end{aligned}
\tag{7.6.7a}
$$

where

$$f_n = \mathcal{H}[f(s)]$$

Similarly, from eq. (7.6.3b) we have

$$\frac{du_n}{dt}(0) = 0 \tag{7.6.7b}$$

The general solution of eq. (7.6.5) is

$$u_n = d_1 \cos(c\sqrt{\lambda_n}t) + d_2 \sin(c\sqrt{\lambda_n}t)$$

where d_1 and d_2 are constants, to be determined by applying the ICs (7.6.7). The first of these, eq. (7.6.7a), gives

$$d_1 = f_n$$

while the second, eq. (7.6.7b), leads to

$$d_2 = 0$$

Thus we have

$$u_n(t) = f_n \cos(c\sqrt{\lambda_n} t) \qquad (7.6.8)$$

and inverting the transform gives the solution

$$u(s, t) = \sum_{n=1}^{\infty} f_n \cos(c\sqrt{\lambda_n} t) \frac{\sqrt{2} J_0(\sqrt{\lambda_n} s)}{J_1(\sqrt{\lambda_n})} \qquad (7.6.9)$$

Equation (7.6.9) implies that the transverse deflection of the membrane is a sum of sinusoidal motions with fixed space. The contribution from the nth term is called the nth *normal mode*. This has the *natural* or *characteristic frequency* $c\sqrt{\lambda_n}/2\pi$, determined by the nth eigenvalue λ_n. The shapes of the normal modes are determined by the corresponding eigenfunctions $\phi_n(s)$.

Note that in the ICs (7.6.3), if the initial position is zero but the initial velocity is finite, then the function $\cos(c\sqrt{\lambda_n} t)$ in eq. (7.6.9) is replaced by $\sin(c\sqrt{\lambda_n} t)$, so that the normal modes have the same frequency but a different phase. Obviously, if the initial position and velocity are both nonzero, then the normal modes will involve a linear combination of the corresponding sine and cosine terms.

We consider next the nonhomogeneous problem:

$$\frac{\partial^2 u}{\partial t^2} - c^2 \left[\frac{\partial^2 u}{\partial s^2} + \frac{1}{s} \frac{\partial u}{\partial s} \right] = g(s) \sin \omega t; \qquad s \in (0, 1), \quad t > 0 \qquad (7.6.10)$$

along with homogeneous BCs and ICs:

$$u \text{ finite} \qquad (7.6.11a)$$

$$u = 0; \qquad s = 1, \qquad t > 0 \qquad (7.6.11b)$$

$$u = 0; \qquad 0 < s < 1, \quad t = 0 \qquad (7.6.12a)$$

$$\frac{\partial u}{\partial t} = 0; \qquad s \in (0, 1), \quad t = 0 \qquad (7.6.12b)$$

which describes the transverse motion of the circular membrane driven purely by a distributed sinusoidal force $g(s) \sin \omega t$, of frequency $\omega/2\pi$.

Rearranging eq. (7.6.10) as

$$L[u] = \frac{\partial}{\partial s} \left[s \frac{\partial u}{\partial s} \right] = \frac{s}{c^2} \left[\frac{\partial^2 u}{\partial t^2} - g(s) \sin \omega t \right]$$

and taking the FHT gives

$$-\lambda_n u_n(t) = \frac{1}{c^2} \frac{d^2 u_n}{dt^2} - \frac{\sin \omega t}{c^2} g_n \qquad (7.6.13)$$

where $u_n(t)$ is the FHT of $u(s, t)$, defined by eq. (7.6.6) and similarly, $g_n = \mathcal{H}[g(s)]$. From eqs. (7.6.12) the ICs for $u_n(t)$ are

$$u_n = 0, \qquad t = 0 \tag{7.6.14a}$$

$$\frac{du_n}{dt} = 0, \qquad t = 0 \tag{7.6.14b}$$

Thus $u_n(t)$ satisfies the ODE

$$\frac{d^2 u_n}{dt^2} + c^2 \lambda_n u_n = g_n \sin \omega t \tag{7.6.15}$$

The solution of the corresponding homogeneous equation is

$$u_{n,h} = d_1 \cos(c\sqrt{\lambda_n}\, t) + d_2 \sin(c\sqrt{\lambda_n}\, t) \tag{7.6.16}$$

where d_1 and d_2 are constants. Assume that eq. (7.6.15) has a particular solution of the form

$$u_{n,p} = k_1 \cos \omega t + k_2 \sin \omega t \tag{7.6.17}$$

where k_1 and k_2 are also constants. Substituting $u_{n,p}$ in eq. (7.6.15) gives

$$(c^2 \lambda_n - \omega^2)[k_1 \cos \omega t + k_2 \sin \omega t] = g_n \sin \omega t$$

so that

$$k_1 = 0 \tag{7.6.18a}$$

and

$$k_2 = \frac{g_n}{c^2 \lambda_n - \omega^2} \tag{7.6.18b}$$

provided that $c^2 \lambda_n \neq \omega^2$.

Thus the general solution of eq. (7.6.15) is

$$u_n(t) = d_1 \cos(c\sqrt{\lambda_n}\, t) + d_2 \sin(c\sqrt{\lambda_n}\, t) + \frac{g_n}{c^2 \lambda_n - \omega^2} \sin \omega t$$

The constants d_1 and d_2 can be evaluated by applying the ICs (7.6.14), which give

$$u_n(t) = \frac{g_n}{c^2 \lambda_n - \omega^2} \left[\sin \omega t - \frac{\omega}{c\sqrt{\lambda_n}} \sin(c\sqrt{\lambda_n}\, t) \right] \tag{7.6.19}$$

By inverting the transform, we obtain

$$u(s, t) = \sum_{n=1}^{\infty} u_n(t) \frac{\sqrt{2} J_0(\sqrt{\lambda_n}\, s)}{J_1(\sqrt{\lambda_n})} \tag{7.6.20}$$

Thus we see that if $\omega \neq c\sqrt{\lambda_n}$ for all n, then $u(s, t)$ is a sum of the normal modes plus a mode corresponding to the frequency $\omega/2\pi$ of the driving force. In the limiting case where $\omega \to c\sqrt{\lambda_n}$ for some n, by applying L'Hôpital's rule to eq. (7.6.19) we get

$$u_n(t) = \lim_{\omega \to c\sqrt{\lambda_n}} g_n \frac{t\cos\omega t - [\sin(c\sqrt{\lambda_n}\,t)/(c\sqrt{\lambda_n})]}{-2\omega}$$

which becomes unbounded as $t \to \infty$. This phenomenon is called *resonance*, and it results in an oscillation of increasingly larger amplitude as time t increases. In this case we say that a sinusoidal driving force with frequency $\omega/2\pi$ excites a normal mode whose natural frequency $c\sqrt{\lambda_n}/2\pi$ is close to $\omega/2\pi$. Note that while the driving force is an externally applied entity, the normal modes are intrinsic to the system.

7.7 SEMI-INFINITE DOMAINS

All the problems treated so far have involved domains of finite size. In this section we consider the case where one or more space variables are semi-infinite—that is, extend from 0 to ∞.

7.7.1 Fourier Sine and Cosine Transforms

Consider a simpler version of the problem treated in Section 7.1, involving heat conduction in a finite one-dimensional rod with no cooling or internal heat generation:

$$L[T] \equiv \frac{\partial^2 T}{\partial x^2} = \frac{1}{\kappa}\frac{\partial T}{\partial t}; \qquad x \in (0, L), \quad t > 0 \tag{7.7.1}$$

with BCs

$$T = 0; \qquad x = 0, \quad t > 0 \tag{7.7.2a}$$

$$T = 0; \qquad x = L, \quad t > 0 \tag{7.7.2b}$$

and IC

$$T = q(x); \qquad x \in (0, L), \quad t = 0 \tag{7.7.3}$$

From section 7.1 the eigenvalues and normalized eigenfunctions corresponding to the operator $L[\ ]$ and the BCs (7.7.2) are

$$\lambda_n = \frac{n^2\pi^2}{L^2}, \qquad n = 1, 2, \ldots \tag{7.7.4}$$

$$\phi_n(x) = \sqrt{\frac{2}{L}}\sin\left(\frac{n\pi x}{L}\right), \qquad n = 1, 2, \ldots \tag{7.7.5}$$

Together, they define the FFT:

$$\mathcal{F}[f(x)] \equiv \int_0^L \phi_n(x)f(x)\,dx = f_n \tag{7.7.6}$$

and the inverse transform:

$$\mathcal{F}^{-1}[f_n] = f(x) = \sum_{n=1}^{\infty} f_n\phi_n(x) \tag{7.7.7}$$

From eq. (7.1.18), the solution of the problem given by eqs. (7.7.1) to (7.7.3) is

$$T(x, t) = \sum_{n=1}^{\infty} q_n \phi_n(x) \exp(-\lambda_n \kappa t) \tag{7.7.8}$$

where

$$q_n = \mathscr{F}[q(x)]$$

Let us now consider what happens as the length of the rod, L, becomes large. In this case the eigenvalues get very close together; in particular,

$$\lambda_{n+1} - \lambda_n = \frac{(n+1)\pi}{L} - \frac{n\pi}{L} = \frac{\pi}{L} \tag{7.7.9}$$

which tends to zero as $L \to \infty$. Let us call

$$\alpha_n = \frac{n\pi}{L} \tag{7.7.10a}$$

then

$$\Delta\alpha = \alpha_{n+1} - \alpha_n = \frac{\pi}{L} \tag{7.7.10b}$$

and

$$\frac{1}{L} = \frac{\Delta\alpha}{\pi} \tag{7.7.10c}$$

The fuller version of the expression (7.7.7) is

$$f(x) = \frac{2}{L} \sum_{n=1}^{\infty} \sin\left(\frac{n\pi x}{L}\right) \int_0^L f(s) \sin\left(\frac{n\pi s}{L}\right) ds \tag{7.7.11}$$

which now takes the form

$$f(x) = \frac{2}{\pi} \sum_{n=1}^{\infty} \sin(\alpha_n x)\Delta\alpha \int_0^L f(s) \sin(\alpha_n s) \, ds$$

If we let $L \to \infty$, then α_n becomes a continuous variable α, $\Delta\alpha$ becomes $d\alpha$, the summation is replaced by an integral, and we have

$$f(x) = \frac{2}{\pi} \int_0^{\infty} \sin(\alpha x) \, d\alpha \int_0^{\infty} f(s) \sin(\alpha s) \, ds$$

This can be split in two parts:

$$\mathscr{F}_s[f] = \sqrt{\frac{2}{\pi}} \int_0^{\infty} f(s) \sin(\alpha s) \, ds = F(\alpha) \tag{7.7.12a}$$

$$f(x) = \sqrt{\frac{2}{\pi}} \int_0^{\infty} F(\alpha) \sin(\alpha x) \, d\alpha \tag{7.7.12b}$$

which are called a pair of *Fourier sine formulae*. In particular, $\mathcal{F}_s[f]$ is called the *Fourier sine transform* of $f(x)$. These formulae are valid under the mild restriction that the function $f(x)$ is piecewise smooth and absolutely integrable—that is,

$$\int_0^\infty |f(s)|\ ds = M < \infty \tag{7.7.13}$$

Due to its origin, we can expect that the Fourier sine transform will be useful for solving PDEs in the semi-infinite domain $x \in (0, \infty)$, for the operator $L[\ ] = d^2[\ ]/dx^2$ with the homogeneous conditions $[\ ] = 0$ at $x = 0$ and $[\ ] \to 0$ as $x \to \infty$. Note that for finiteness of solution, the condition at the far end also implies that $d[\ ]/dx \to 0$ as $x \to \infty$.

If we repeat the same arguments as above, for the case where the BC (7.7.2a) is replaced by

$$\frac{dT}{dx} = 0, \qquad x = 0, \quad t > 0 \tag{7.7.14}$$

then we are led to the pair of *Fourier cosine formulae*:

$$\mathcal{F}_c[g] = \sqrt{\frac{2}{\pi}} \int_0^\infty g(s)\ \cos(\alpha s)\ ds = G(\alpha) \tag{7.7.15a}$$

$$g(x) = \sqrt{\frac{2}{\pi}} \int_0^\infty G(\alpha)\ \cos(\alpha x)\ d\alpha \tag{7.7.15b}$$

where $\mathcal{F}_c[g]$ is called the *Fourier cosine transform* of $g(x)$. The Fourier cosine transform will prove useful for solving PDEs in $x \in (0, \infty)$ for the operator $L[\ ] = d^2[\ ]/dx^2$, with the homogeneous condition $d[\ ]/dx = 0$ at $x = 0$. The condition at the far end remains $[\ ] \to 0$ and $d[\ ]/dx \to 0$ as $x \to \infty$.

The Fourier sine and cosine transforms of some elementary functions are shown in Tables 7.1 and 7.2, respectively. We next discuss some properties of these transforms which are useful for the solution of problems.

Linearity.   From eqs. (7.7.12a) and (7.7.15a), it is clear that both the sine and cosine transforms are linear integral operators—that is,

$$\mathcal{F}_s[c_1 f_1 + c_2 f_2] = c_1 \mathcal{F}_s[f_1] + c_2 \mathcal{F}_s[f_2] \tag{7.7.16a}$$

and

$$\mathcal{F}_c[c_1 g_1 + c_2 g_2] = c_1 \mathcal{F}_c[g_1] + c_2 \mathcal{F}_c[g_2] \tag{7.7.16b}$$

where c_1 and c_2 are constants.

Transforms of Derivatives of $f(x)$.   Let us first consider the Fourier cosine transform of $f^{(n)}(x)$:

$$\mathcal{F}_c[f^{(n)}] = \sqrt{\frac{2}{\pi}} \int_0^\infty f^{(n)}(s)\ \cos(\alpha s)\ ds$$

TABLE 7.1 Fourier Sine Transforms[a]

$f(x)$		$F(\alpha) = \sqrt{\dfrac{2}{\pi}} \displaystyle\int_0^\infty f(x)\,\sin(\alpha x)\,dx$
$\begin{cases} 1 & \text{for } 0 < x < a \\ 0 & \text{for } a < x < \infty \end{cases}$		$\sqrt{\dfrac{2}{\pi}} \dfrac{(1 - \cos \alpha a)}{\alpha}$
$f(ax)$	$a > 0$	$\dfrac{1}{a} F\left(\dfrac{x}{a}\right)$
$f(ax)\cos bx$	$a, b > 0$	$\dfrac{1}{2a}\left[F\left(\dfrac{\alpha + b}{a}\right) + F\left(\dfrac{\alpha - b}{a}\right) \right]$
$f(ax)\sin bx$	$a, b > 0$	$\dfrac{1}{2a}\displaystyle\int_0^\infty f(x)\left[\cos\left(\dfrac{\alpha - b}{a} x\right) - \cos\left(\dfrac{\alpha + b}{a} x\right) \right] dx$
$\dfrac{1}{x}$		$\sqrt{\dfrac{\pi}{2}}$
$\dfrac{1}{\sqrt{x}}$		$\dfrac{1}{\sqrt{\alpha}}$
x^{a-1}	$0 < a < 1$	$\sqrt{\dfrac{2}{\pi}} \dfrac{\Gamma(a)}{\alpha^a} \sin\left(\dfrac{a\pi}{2}\right)$
e^{-ax}	$a > 0$	$\sqrt{\dfrac{2}{\pi}} \dfrac{\alpha}{\alpha^2 + a^2}$
xe^{-ax^2}	$a > 0$	$\dfrac{\alpha}{2\sqrt{2}a^{3/2}} e^{-\alpha^2/4a}$
$\dfrac{\sin ax}{x}$	$a > 0$	$\dfrac{1}{\sqrt{2\pi}} \log\left(\left\|\dfrac{\alpha + a}{\alpha - a}\right\|\right)$
$\dfrac{\cos ax}{x}$	$a > 0$	$\begin{cases} 0 & \text{for } 0 < \alpha < a \\ \sqrt{\pi/8} & \text{for } \alpha = a \\ \sqrt{\pi/2} & \text{for } a < \alpha < \infty \end{cases}$
$\dfrac{\sin^2 ax}{x}$	$a > 0$	$\begin{cases} \sqrt{\pi/8} & \text{for } 0 < \alpha < 2a \\ \sqrt{\pi/32} & \text{for } \alpha = 2a \\ 0 & \text{for } 2a < \alpha < \infty \end{cases}$
$\sin\left(\dfrac{a^2}{x}\right)$	$a > 0$	$\sqrt{\dfrac{\pi}{2}} \dfrac{a}{\sqrt{\alpha}} J_1(2a\sqrt{\alpha})$

[a]For a more extensive list, see Erdélyi et al. (1954).

TABLE 7.2 Fourier Cosine Transforms[a]

$f(x)$		$F(\alpha) = \sqrt{\dfrac{2}{\pi}} \displaystyle\int_0^\infty f(x)\cos(\alpha x)\,dx$
$\begin{cases} 1 & \text{for } 0 < x < a \\ 0 & \text{for } a < x < \infty \end{cases}$		$\sqrt{\dfrac{2}{\pi}}\,\dfrac{\sin(\alpha a)}{\alpha}$
$f(ax)$	$a > 0$	$\dfrac{1}{a}F\left(\dfrac{\alpha}{a}\right)$
$f(ax)\cos bx$	$a, b > 0$	$\dfrac{1}{2a}\left[F\left(\dfrac{\alpha + b}{a}\right) + F\left(\dfrac{\alpha - b}{a}\right)\right]$
$f(ax)\sin bx$	$a, b > 0$	$\dfrac{1}{2a}\displaystyle\int_0^\infty f(x)\left[\sin\left(\dfrac{\alpha + b}{a}x\right) - \sin\left(\dfrac{\alpha - b}{a}x\right)\right]dx$
x^{a-1}	$0 < a < 1$	$\sqrt{\dfrac{2}{\pi}}\,\dfrac{\Gamma(a)}{\alpha^a}\sin\left(\dfrac{a\pi}{2}\right)$
$\dfrac{1}{x^2 + a^2}$	$a > 0$	$\sqrt{\dfrac{\pi}{2}}\,\dfrac{e^{-\alpha a}}{a}$
e^{-ax}	$a > 0$	$\sqrt{\dfrac{2}{\pi}}\,\dfrac{a}{\alpha^2 + a^2}$
e^{-ax^2}	$a > 0$	$\dfrac{1}{\sqrt{2a}}e^{-\alpha^2/4a}$
$\dfrac{1}{x}(e^{-bx} - e^{-ax})$	$a, b > 0$	$\dfrac{1}{\sqrt{2\pi}}\log\left(\dfrac{\alpha^2 + a^2}{\alpha^2 + b^2}\right)$
$\sin(ax^2)$	$a > 0$	$\dfrac{1}{2\sqrt{a}}\left[\cos\left(\dfrac{\alpha^2}{4a}\right) - \sin\left(\dfrac{\alpha^2}{4a}\right)\right]$
$\cos(ax^2)$	$a > 0$	$\dfrac{1}{2\sqrt{a}}\left[\cos\left(\dfrac{\alpha^2}{4a}\right) + \sin\left(\dfrac{\alpha^2}{4a}\right)\right]$
$e^{-bx}\sin ax$	$a, b > 0$	$\dfrac{1}{\sqrt{2\pi}}\left[\dfrac{(\alpha + a)}{b^2 + (\alpha + a)^2} + \dfrac{(a - \alpha)}{b^2 + (a - \alpha)^2}\right]$
$\dfrac{\sin ax}{x}$	$a > 0$	$\begin{cases} \sqrt{\pi/2} & \text{for } \alpha < a \\ \sqrt{\pi/8} & \text{for } \alpha = a \\ 0 & \text{for } \alpha > 0 \end{cases}$

[a]For a more extensive list, see Erdélyi et al. (1954).

Integrating once by parts, we get

$$\mathscr{F}_c[f^{(n)}] = \sqrt{\frac{2}{\pi}} \, [f^{(n-1)}(s) \, \cos(\alpha s)]_0^\infty + \alpha \sqrt{\frac{2}{\pi}} \int_0^\infty f^{(n-1)}(s) \, \sin(\alpha s) \, ds$$

which, assuming that

$$\lim_{s \to \infty} f^{(n-1)}(s) = 0$$

leads to

$$\mathscr{F}_c[f^{(n)}] = -\sqrt{\frac{2}{\pi}} \, f^{(n-1)}(0) + \alpha \mathscr{F}_s \, [f^{(n-1)}] \tag{7.7.17}$$

Similarly, for the Fourier sine transform, we have

$$\mathscr{F}_s[f^{(n)}] = -\alpha \mathscr{F}_c[f^{(n-1)}] \tag{7.7.18}$$

As an example, for $n = 2$, by successively applying eqs. (7.7.17) and (7.7.18), we get

$$\mathscr{F}_s[f^{(2)}] = -\alpha \mathscr{F}_c[f^{(1)}]$$

$$= -\alpha^2 \mathscr{F}_s[f] + \alpha \sqrt{\frac{2}{\pi}} \, f(0) \tag{7.7.19}$$

We now consider some examples illustrating the use of these transforms.

Example 1. Solve the one-dimensional heat conduction equation in the semi-infinite domain:

$$\frac{\partial^2 T}{\partial x^2} = \frac{1}{\kappa} \frac{\partial T}{\partial t}; \qquad x \in (0, \infty), \quad t > 0 \tag{7.7.20}$$

with the BCs

$$\frac{\partial T}{\partial x} = 0; \qquad x = 0, \qquad t > 0 \tag{7.7.21a}$$

$$T \to 0 \text{ and } \frac{\partial T}{\partial x} \to 0; \qquad \text{as } x \to \infty, \quad t > 0 \tag{7.7.21b}$$

and the IC

$$T = f(x); \qquad x \in (0, \infty), \quad t = 0 \tag{7.7.22}$$

It is worth noting that the compatibility between eqs. (7.7.21b) and (7.7.22) requires that $f(x) \to 0$ as $x \to \infty$.

From the previous discussion, it is clear that we should work with the Fourier sine transform. Thus, using eq. (7.7.12a), let us define

$$\mathscr{F}_s[T(x, t)] = \sqrt{\frac{2}{\pi}} \int_0^\infty T(s, t) \sin(\alpha s) \, ds = \bar{T}(\alpha, t) \tag{7.7.23}$$

as the sine transform of $T(x, t)$. Taking the sine transform of both sides of eq. (7.7.20) gives

$$\sqrt{\frac{2}{\pi}} \int_0^\infty \frac{\partial^2 T}{\partial x^2} \sin(\alpha x) \, dx = \frac{1}{\kappa} \sqrt{\frac{2}{\pi}} \int_0^\infty \frac{\partial T}{\partial t} \sin(\alpha x) \, dx \tag{7.7.24}$$

Using eqs. (7.7.19) and (7.7.21a) we have

$$\text{lhs} = -\alpha^2 \bar{T}(\alpha, t)$$

while

$$\text{rhs} = \frac{1}{\kappa} \frac{d\bar{T}}{dt}$$

exchanging the derivative and integration steps. This leads to

$$\frac{1}{\kappa} \frac{d\bar{T}}{dt} = -\alpha^2 \bar{T}$$

which has solution

$$\bar{T}(\alpha, t) = \bar{T}(\alpha, 0) \exp(-\alpha^2 \kappa t) \tag{7.7.25}$$

As usual,

$$\bar{T}(\alpha, 0) = \lim_{t \to 0} \bar{T}(\alpha, t)$$

$$= \lim_{t \to 0} \sqrt{\frac{2}{\pi}} \int_0^\infty T(s, t) \sin(\alpha s) \, ds$$

$$= \sqrt{\frac{2}{\pi}} \int_0^\infty f(s) \sin(\alpha s) \, ds \tag{7.7.26}$$

so that

$$\bar{T}(\alpha, t) = \sqrt{\frac{2}{\pi}} \left[\int_0^\infty f(s) \sin(\alpha s) \, ds \right] \exp(-\alpha^2 \kappa t) \tag{7.7.27}$$

is the complete expression for the sine transform of $T(x, t)$. The solution $T(x, t)$ is then given by the inverse transform relation (7.7.12b)—that is,

$$T(x, t) = \sqrt{\frac{2}{\pi}} \int_0^\infty \bar{T}(\alpha, t) \sin(\alpha x) \, d\alpha$$

$$= \frac{2}{\pi} \int_0^\infty \sin(\alpha x) \left[\int_0^\infty f(s) \sin(\alpha s) \, ds \right] \exp(-\alpha^2 \kappa t) \, d\alpha \tag{7.7.28}$$

This formal representation of the solution involves the sine transform of the initial condition $f(x)$. It can be simplified, as shown next.

First, we can reverse the order of integration; that is, first carry out the integration over α and then over s. This gives

$$T(x, t) = \frac{2}{\pi} \int_0^\infty f(s) \left[\int_0^\infty \sin(\alpha x) \sin(\alpha s) \exp(-\alpha^2 \kappa t) \, d\alpha \right] ds \tag{7.7.29}$$

The integral over α can be evaluated analytically. For this, note that

$$\sin(\alpha x) \sin(\alpha s) = \frac{1}{2} [\cos \alpha(s - x) - \cos \alpha(s + x)]$$

so the integral over α is

$$I_1 = \frac{1}{2} \int_0^\infty [\cos \alpha(s - x) - \cos \alpha(s + x)] \exp(-\alpha^2 \kappa t) \, d\alpha \tag{7.7.30}$$

Consider now the integral:

$$I_2 = \int_0^\infty \cos(\alpha\omega) \exp(-\alpha^2\beta) \, d\alpha \tag{7.7.31}$$

Noting that $\cos x$ is an even function and $\sin x$ is an odd function, we get

$$I_2 = \frac{1}{2} \int_{-\infty}^\infty \cos(\alpha\omega) \exp(-\alpha^2\beta) \, d\alpha$$

$$+ \frac{1}{2} i \int_{-\infty}^\infty \sin(\alpha\omega) \exp(-\alpha^2\beta) \, d\alpha$$

since the second integral is zero anyway. Combining the two integrals using the Euler formula gives

$$I_2 = \frac{1}{2} \int_{-\infty}^\infty \exp(i\alpha\omega) \exp(-\alpha^2\beta) \, d\alpha$$

$$= \frac{1}{2} \int_{-\infty}^\infty \exp\left[-\beta\left(\alpha^2 - \frac{i\omega\alpha}{\beta} \right) \right] d\alpha$$

$$= \frac{1}{2} \int_{-\infty}^\infty \exp\left[-\beta\left(\alpha - \frac{i\omega}{2\beta} \right)^2 - \frac{\omega^2}{4\beta} \right] d\alpha$$

$$= \frac{1}{2} \exp\left(-\frac{\omega^2}{4\beta} \right) \int_{-\infty}^\infty \exp\left[-\beta\left(\alpha - \frac{i\omega}{2\beta} \right)^2 \right] d\alpha$$

If we now change variables as

$$y = \sqrt{\beta} \left(\alpha - \frac{i\omega}{2\beta} \right)$$

so that

$$dy = \sqrt{\beta} \, d\alpha$$

then we have

$$I_2 = \frac{1}{2\sqrt{\beta}} \exp\left(-\frac{\omega^2}{4\beta} \right) \int_{-\infty}^\infty \exp(-y^2) \, dy$$

Further, since

$$\int_{-\infty}^\infty \exp(-y^2) \, dy = \sqrt{\pi} \tag{7.7.32}$$

as derived in section 4.7, we obtain

$$I_2 = \frac{1}{2} \sqrt{\frac{\pi}{\beta}} \exp\left(-\frac{\omega^2}{4\beta} \right) \tag{7.7.33}$$

From eq. (7.7.30) the integral I_1 then has the value

$$I_1 = \frac{1}{4} \sqrt{\frac{\pi}{\kappa t}} \left[\exp\left\{ -\frac{(s-x)^2}{4\kappa t} \right\} - \exp\left\{ -\frac{(s+x)^2}{4\kappa t} \right\} \right] \tag{7.7.34}$$

and from eq. (7.7.29),

$$T(x, t) = \frac{1}{2\sqrt{\pi\kappa t}} \int_0^\infty f(s) \left[\exp\left\{ -\frac{(s - x)^2}{4\kappa t} \right\} - \exp\left\{ -\frac{(s + x)^2}{4\kappa t} \right\} \right] ds \quad (7.7.35)$$

is the desired solution of eqs. (7.7.20) to (7.7.22).

A special case arises when the initial temperature distribution $f(x)$ is *uniform*—that is,

$$f(x) = T_i \quad (7.7.36)$$

In this case, changing the variables as

$$\frac{x - s}{2\sqrt{\kappa t}} = u \quad \text{and} \quad \frac{x + s}{2\sqrt{\kappa t}} = v \quad (7.7.37)$$

in the first and second integrals of eq. (7.7.35) gives

$$\begin{aligned}
T(x, t) &= \frac{T_i}{2\sqrt{\pi\kappa t}} \left[\int_{x/(2\sqrt{\kappa t})}^{-\infty} \exp(-u^2) \cdot 2\sqrt{\kappa t}(-du) \right. \\
&\quad \left. - \int_{x/(2\sqrt{\kappa t})}^{\infty} \exp(-v^2) \cdot 2\sqrt{\kappa t}\, dv \right] \\
&= \frac{T_i}{\sqrt{\pi}} \left[\int_{-\infty}^{x/(2\sqrt{\kappa t})} \exp(-u^2)\, du - \int_{x/(2\sqrt{\kappa t})}^{\infty} \exp(-u^2)\, du \right] \\
&= \frac{2T_i}{\sqrt{\pi}} \int_0^{x/(2\sqrt{\kappa t})} \exp(-u^2)\, du
\end{aligned}$$

because the contributions of the integrals, the first over $-\infty$ to 0 and the second over 0 to ∞, cancel out since $\exp(-u^2)$ is an even function. Thus we have

$$T(x, t) = T_i \, \text{erf}\left(\frac{x}{2\sqrt{\kappa t}} \right) \quad (7.7.38)$$

where erf(x) is the error function defined in section 4.7.1. An evolution of the temperature profile, as described by eq. (7.7.38), is shown in Figure 7.8.

7.7.2 Nonhomogeneous Boundary Conditions

From the result of Example 1 above, we can readily obtain the solutions of problems involving nonhomogeneous boundary conditions. This is illustrated in the present section by the use of several examples with increasingly complex BCs.

Example 2. Solve

$$\frac{\partial^2 T}{\partial x^2} = \frac{1}{\kappa} \frac{\partial T}{\partial t}; \qquad x \in (0, \infty), \quad t > 0 \quad (7.7.39)$$

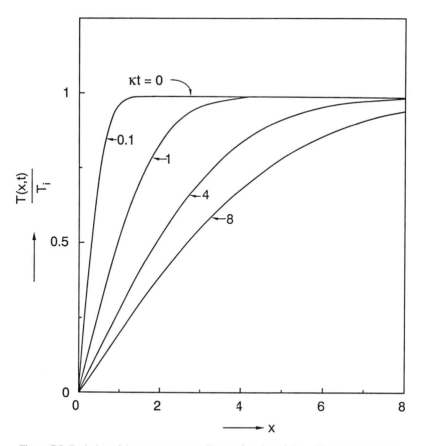

Figure 7.8 Evolution of the temperature profile as a function of time, given by eq. (7.7.38)

with the BCs

$$T = T_0; \qquad\qquad x = 0, \qquad t > 0 \qquad\qquad (7.7.40a)$$

$$T \to 0 \quad \text{and} \quad \frac{\partial T}{\partial x} \to 0; \qquad \text{as } x \to \infty, \quad t > 0 \qquad\qquad (7.7.40b)$$

and the IC

$$T = 0; \qquad\qquad x \in (0, \infty), \quad t = 0 \qquad\qquad (7.7.41)$$

This problem differs from the previous one in that the BC at $x = 0$ is nonhomogeneous (although only a constant), while the IC is homogeneous. Further, as in the previous example, the condition $T \to 0$ as $x \to \infty$, $t > 0$ is required for compatibility with the IC (7.7.41).

If we change the dependent variable as

$$w = T - T_0 \qquad\qquad (7.7.42)$$

then the problem in w becomes

$$\frac{\partial^2 w}{\partial x^2} = \frac{1}{\kappa}\frac{\partial w}{\partial t}; \qquad\qquad x \in (0, \infty), \quad t > 0 \qquad\qquad (7.7.43)$$

$$w = 0; \qquad\qquad x = 0, \qquad t > 0 \qquad\qquad (7.7.44a)$$

$$w \to -T_0 \quad\text{and}\quad \frac{\partial w}{\partial x} \to 0; \qquad \text{as } x \to \infty, \quad t > 0 \qquad\qquad (7.7.44b)$$

$$w = -T_0; \qquad\qquad x \in (0, \infty), \quad t = 0 \qquad\qquad (7.7.45)$$

which has a homogeneous BC at $x = 0$ and a nonhomogeneous constant IC. The solution from eq. (7.7.38) is

$$w(x, t) = -T_0 \operatorname{erf}\left(\frac{x}{2\sqrt{\kappa t}}\right) \qquad\qquad (7.7.46)$$

so that from eq. (7.7.42) we obtain

$$T(x, t) = T_0\left[1 - \operatorname{erf}\left(\frac{x}{2\sqrt{\kappa t}}\right)\right]$$

$$= T_0 \operatorname{erfc}\left(\frac{x}{2\sqrt{\kappa t}}\right) \qquad\qquad (7.7.47)$$

is the desired solution of eqs. (7.7.39) to (7.7.41).

The case where the BC at $x = 0$ is a function of t can now be handled, using the result of Example 2 and the ideas of step response and impulse response from section 7.4.

Example 3. Consider the problem

$$\frac{\partial^2 T}{\partial x^2} = \frac{1}{\kappa}\frac{\partial T}{\partial t}; \qquad\qquad x \in (0, \infty), \quad t > 0 \qquad\qquad (7.7.48)$$

with the BCs

$$T = g(t), \qquad\qquad x = 0, \qquad t > 0 \qquad\qquad (7.7.49a)$$

$$T \to 0 \quad\text{and}\quad \frac{\partial T}{\partial x} \to 0; \qquad \text{as } x \to \infty, \quad t > 0 \qquad\qquad (7.7.49b)$$

and the IC

$$T = 0; \qquad\qquad x \in (0, \infty), \quad t = 0 \qquad\qquad (7.7.50)$$

From Example 2, the solution when $g(t) = 1$ is the *step response* given by

$$w(x, t) = \operatorname{erfc}\left(\frac{x}{2\sqrt{\kappa t}}\right) = \frac{2}{\sqrt{\pi}}\int_{x/(2\sqrt{\kappa t})}^{\infty} \exp(-u^2)\, du \qquad\qquad (7.7.51)$$

Then from eq. (7.4.11), the *impulse response* is

$$
W(x, t) = \frac{\partial w}{\partial t}
$$

$$
= -\frac{2}{\sqrt{\pi}} \frac{x}{2\sqrt{\kappa}} \left(-\frac{1}{2t^{3/2}} \right) \exp\left(-\frac{x^2}{4\kappa t} \right)
$$

$$
= \frac{x}{2\sqrt{\pi \kappa t^3}} \exp\left(-\frac{x^2}{4\kappa t} \right) \tag{7.7.52}
$$

which is in fact the solution of eqs. (7.7.48) to (7.7.50), with $g(t)$ replace by $\delta(t)$, the Dirac delta function.

Finally, from eq. (7.4.10) the solution of eqs. (7.7.48) to (7.7.50) is

$$
T(x, t) = \int_0^t W(x, t - \tau)g(\tau)\, d\tau
$$

By the change of variable

$$
t - \tau = \eta
$$

we obtain the alternative convolution integral:

$$
T(x, t) = \int_0^t W(x, \eta)g(t - \eta)\, d\eta \tag{7.7.53}
$$

which provides the complete solution.

Finally, using the results of Examples 1 and 3, we can solve problems where the BC at $x = 0$ is a function of t, and the IC is also nonhomogeneous.

Example 4. Consider the problem

$$
\frac{\partial^2 T}{\partial x^2} = \frac{1}{\kappa}\frac{\partial T}{\partial t}; \qquad x \in (0, \infty), \quad t > 0 \tag{7.7.54}
$$

with the BCs

$$
T = g(t); \qquad x = 0, \qquad t > 0 \tag{7.7.55a}
$$

$$
T \to 0 \quad \text{and} \quad \frac{\partial T}{\partial x} \to 0; \qquad \text{as } x \to \infty, \quad t > 0 \tag{7.7.55b}
$$

and the IC

$$
T = f(x); \qquad x \in (0, \infty), \quad t = 0 \tag{7.7.56}
$$

Since the problems are linear, from the *principle of superposition*, the solution of eqs. (7.7.54) to (7.7.56) will be the sum of the solutions of Examples 1 and 3, given by eqs. (7.7.35) and (7.7.53), respectively.

7.7.3 Hankel Transforms

As with Fourier transforms, the *Hankel transform* is an extension of the finite Hankel transform (see sections 7.5 and 7.6) to include the case where the radial coordinate r goes to ∞. For this, recall that the eigenvalue problem of section 4.16,

$$L[\phi] = \frac{d}{dr}\left[r\frac{d\phi}{dr}\right] = -\lambda r\phi, \qquad r \in (0, a) \tag{7.7.57}$$

$$\phi = 0, \qquad r = a \tag{7.7.58a}$$

$$\phi = \text{finite} \tag{7.7.58b}$$

leads to the eigenvalue equation

$$J_0(\sqrt{\lambda_n}\,a) = 0, \quad n = 1, 2, \dots \tag{7.7.59}$$

where the orthonormal eigenfunctions are given by

$$\phi_n(r) = \frac{\sqrt{2}\,J_0(\sqrt{\lambda_n}\,r)}{a\,J_1(\sqrt{\lambda_n}\,a)}, \qquad n = 1, 2, \dots \tag{7.7.60}$$

Further, given a function $f(r)$, we can expand it as

$$f(r) = \sum_{n=1}^{\infty} c_n \phi_n(r) \tag{7.7.61}$$

where

$$c_n = \mathcal{H}[f(r)] = \int_0^a s f(s)\phi_n(s)\,ds \tag{7.7.62}$$

is the finite Hankel transform of $f(r)$.

Let us now consider the case where the radius a goes to ∞. From the asymptotic behavior of Bessel functions discussed in section 4.12, we have

$$J_\nu(x) \sim \sqrt{\frac{2}{\pi x}}\cos\left(x - \frac{\pi\nu}{2} - \frac{\pi}{4}\right) \qquad \text{as } x \to \infty$$

Thus for large a, the roots of eq. (7.7.59) are given by

$$\sqrt{\lambda_n}\,a \sim (2n + 1)\frac{\pi}{2} + \frac{\pi}{4} \tag{7.7.63}$$

so that

$$J_1(\sqrt{\lambda_n}\,a) \sim \sqrt{\frac{2}{\pi\sqrt{\lambda_n}\,a}}\cos\left[(2n + 1)\frac{\pi}{2} + \frac{\pi}{4} - \frac{\pi}{2} - \frac{\pi}{4}\right]$$

$$= \sqrt{\frac{2}{\pi\sqrt{\lambda_n}\,a}}(-1)^n$$

Substituting this in eqs. (7.7.61) and (7.7.62) yields

$$f(r) = \frac{2}{a^2}\sum_{n=1}^{\infty}\frac{J_0(\sqrt{\lambda_n}\,r)}{\dfrac{2}{\pi\sqrt{\lambda_n}\,a}}\int_0^a s f(s)J_0(\sqrt{\lambda_n}\,s)\,ds$$

$$= \frac{\pi}{a}\sum_{n=1}^{\infty}\sqrt{\lambda_n}\,J_0(\sqrt{\lambda_n}\,r)\int_0^a s f(s)J_0(\sqrt{\lambda_n}\,s)\,ds \tag{7.7.64}$$

From eq. (7.7.63) we have

$$\Delta \sqrt{\lambda_n} = \sqrt{\lambda_{n+1}} - \sqrt{\lambda_n}$$

$$= \frac{\pi}{a}$$

so that eq. (7.7.64) takes the form

$$f(r) = \sum_{n=1}^{\infty} \sqrt{\lambda_n} J_0(\sqrt{\lambda_n} r) \Delta \sqrt{\lambda_n} \int_0^a s f(s) J_0(\sqrt{\lambda_n} s) \, ds$$

As $a \to \infty$, $\lambda_n \to \lambda$, $\Delta \sqrt{\lambda_n} \to d\sqrt{\lambda}$, and the summation is replaced by an integral, yielding

$$f(r) = \int_0^{\infty} \alpha J_0(\alpha r) \, d\alpha \int_0^{\infty} s f(s) J_0(\alpha s) \, ds$$

This can be split in two parts:

$$\mathcal{H}[f] = \int_0^{\infty} s f(s) J_0(\alpha s) \, ds = F(\alpha) \tag{7.7.65a}$$

$$f(r) = \mathcal{H}^{-1}[F(\alpha)] = \int_0^{\infty} \alpha F(\alpha) J_0(\alpha r) \, d\alpha \tag{7.7.65b}$$

which define a pair of *Hankel transform formulae of order zero*. A generalization of this approach leads to the formulae of *order m* (cf. Sneddon, 1951, section 8):

$$\mathcal{H}_m[f] = \int_0^{\infty} s f(s) J_m(\alpha s) \, ds = F_m(\alpha) \tag{7.7.66a}$$

$$f(r) = \mathcal{H}_m^{-1}[F_m(\alpha)] = \int_0^{\infty} \alpha F_m(\alpha) J_m(\alpha r) \, d\alpha \tag{7.7.66b}$$

The function $F_m(\alpha)$ is called the order m Hankel transform of $f(r)$. The transforms of order 0 and 1 are widely used for solving problems involving the diffusion operator in a radially infinite cylindrical geometry. Note that, in particular, we have (see problem 7.19)

$$\mathcal{H}_m \left\{ \frac{1}{r} L[f] - \frac{m^2}{r^2} f \right\} = -\alpha^2 F_m(\alpha) \tag{7.7.67}$$

where $L[\ ]$ is defined in eq. (7.7.57), provided that both rf' and rf vanish as $r \to 0$ and $r \to \infty$. The Hankel transforms of some elementary functions are given in Table 7.3.

> **Example.** Obtain the solution for the free vibration of a large circular membrane.
> This is an extension of the problem treated in section 7.6, so that the relevant PDE describing the transverse motion is

$$L[u] = \frac{\partial}{\partial r} \left[r \frac{\partial u}{\partial r} \right] = \frac{r}{c^2} \frac{\partial^2 u}{\partial t^2}; \qquad 0 < r < \infty, \quad t > 0 \tag{7.7.68}$$

TABLE 7.3 Hankel Transforms[a]

$f(x)$	ν	$F(\alpha) = \int_0^\infty sf(s)J_\nu(\alpha s)\,ds$
$\begin{cases} x^\nu, & 0 < x < a \\ 0, & x > a \end{cases}$	> -1	$\dfrac{a^{\nu+1}}{\alpha} J_{\nu+1}(\alpha a)$
$\begin{cases} 1, & 0 < x < a \\ 0, & x > a \end{cases}$	0	$\dfrac{a}{\alpha} J_1(a\alpha)$
$\begin{cases} a^2 - x^2, & 0 < x < a \\ 0, & x > a \end{cases}$	0	$\dfrac{4a}{\alpha^3} J_1(\alpha a) - \dfrac{2a^2}{\alpha^2} J_0(\alpha a)$
$x^\nu e^{-px^2}$	> -1	$\dfrac{\alpha^\nu}{(2p)^{\nu+1}} e^{-\alpha^2/4p}$
$x^{\mu-1}$	> -1	$\dfrac{2\mu\Gamma(\frac{1}{2} + \frac{1}{2}\mu + \frac{1}{2}\alpha)}{\alpha^{\mu+1}\Gamma(\frac{1}{2} - \frac{1}{2}\mu + \frac{1}{2}\alpha)}$
$\dfrac{e^{-px}}{x}$	0	$(\alpha^2 + p^2)^{-1/2}$
e^{-px}	0	$p(\alpha^2 + p^2)^{-3/2}$
$x^{-2}e^{-px}$	1	$\dfrac{(\alpha^2 + p^2)^{1/2} - p}{\alpha}$
$\dfrac{e^{-px}}{x}$	1	$\dfrac{1}{\alpha} - \dfrac{p}{\alpha(\alpha^2 + p^2)^{1/2}}$
e^{-px}	1	$\alpha(\alpha^2 + p^2)^{-3/2}$
$\dfrac{a}{(a^2 + x^2)^{3/2}}$	0	$e^{-a\alpha}$
$\dfrac{\sin(ax)}{x}$	0	$\begin{cases} 0, & \alpha > a \\ (a^2 - \alpha^2)^{-1/2}, & 0 < \alpha < a \end{cases}$
$\dfrac{\sin(ax)}{x}$	1	$\begin{cases} \dfrac{a}{\alpha(\alpha^2 - a^2)^{1/2}}, & \alpha > a \\ 0, & \alpha < a \end{cases}$
$\dfrac{\sin(x)}{x^2}$	0	$\begin{cases} \sin^{-1}\left(\dfrac{1}{\alpha}\right), & \alpha > 1 \\ \dfrac{1}{2}\pi, & \alpha < 1 \end{cases}$

[a]For a more extensive list, see Erdélyi et al. (1954).

along with the BCs

$$ru = 0 \quad \text{and} \quad r\frac{\partial u}{\partial r} = 0; \qquad r = 0 \text{ and } r \to \infty, \quad t > 0 \qquad (7.7.69)$$

and the ICs

$$u = f(r) \quad \text{and} \quad \frac{\partial u}{\partial t} = g(r); \qquad 0 < r < \infty, \qquad t = 0 \qquad (7.7.70)$$

Taking the Hankel transform ($m = 0$) of eq. (7.7.68), using eq. (7.7.67), yields

$$-\alpha^2 U = \frac{1}{c^2} \frac{d^2 U}{dt^2} \tag{7.7.71}$$

where

$$U(\alpha, t) = \mathcal{H}_0[u] = \int_0^\infty r u(r, t) J_0(\alpha r) \, dr \tag{7.7.72}$$

is the Hankel transform of $u(r, t)$. The ICs (7.7.70) take the form

$$U = F(\alpha) \quad \text{and} \quad \frac{dU}{dt} = G(\alpha), \qquad t = 0 \tag{7.7.73}$$

where $F(\alpha)$ and $G(\alpha)$ are Hankel transforms of $f(r)$ and $g(r)$, respectively.
Equation (7.7.71) has the solution

$$U(\alpha, t) = A(\alpha) \cos(c\alpha t) + B(\alpha) \sin(c\alpha t)$$

Applying the ICs (7.7.73) yields

$$A(\alpha) = F(\alpha) \quad \text{and} \quad B(\alpha) = \frac{G(\alpha)}{c\alpha}$$

so that we have

$$U(\alpha, t) = F(\alpha) \cos(c\alpha t) + \frac{G(\alpha)}{c\alpha} \sin(c\alpha t) \tag{7.7.74}$$

The inversion formula (7.7.65b) then gives the desired solution

$$u(r, t) = \int_0^\infty \alpha F(\alpha) \cos(c\alpha t) J_0(\alpha r) \, d\alpha + \frac{1}{c} \int_0^\infty G(\alpha) \sin(c\alpha t) J_0(\alpha r) \, d\alpha \tag{7.7.75}$$

Note that it is, in general, difficult to evaluate the above integrals analytically. For the special case where the ICs are

$$f(r) = \frac{\gamma}{[1 + r^2/b^2]^{1/2}}, \qquad g(r) = 0$$

where γ and b are constants, we have $G(\alpha) = 0$ and

$$F(\alpha) = \gamma \int_0^\infty \frac{r J_0(\alpha r) \, dr}{[1 + r^2/b^2]^{1/2}}$$

$$= \gamma b^2 \int_0^\infty \frac{u J_0(\alpha b u) \, du}{[1 + u^2]^{1/2}}$$

It is possible to show that (cf. Sneddon, 1951, page 514)

$$\int_0^\infty e^{-wx} J_0(\rho x) \, dx = \frac{1}{[w^2 + \rho^2]^{1/2}} \tag{7.7.76}$$

so that by the inversion formula (7.7.65b) we obtain

$$\int_0^\infty \frac{\rho J_0(\rho x) \, d\rho}{[w^2 + \rho^2]^{1/2}} = \frac{e^{-wx}}{x}$$

and

$$F(\alpha) = \frac{\gamma b}{\alpha} e^{-b\alpha}$$

Substituting in eq. (7.7.75), this gives

$$u(r, t) = \gamma b \int_0^\infty e^{-b\alpha} \cos(c\alpha t) J_0(\alpha r) \, d\alpha$$

$$= \gamma b \, \text{Re}\left[\int_0^\infty e^{-\alpha(b+ict)} J_0(\alpha r) \, d\alpha\right]$$

where Re[z] denotes the real part of z. Using eq. (7.7.76) once again, we obtain

$$u(r, t) = \gamma b \, \text{Re}\left[\frac{1}{r^2 + (b + ict)^2}\right]^{1/2}$$

as the desired solution.

7.8   INFINITE DOMAINS

7.8.1   Fourier Series, Integrals, and Transforms

Before we can treat the solution of PDEs involving infinite spatial domains by transform methods, we need to be able to express periodic functions in a *Fourier series* (cf. Churchill and Brown, 1978, chapter 4; Kreyszig, 1993, chapter 10). For this, consider a function $f(x)$ of period $p = 2L$. Since the trigonometric functions $\{\sin nx\}$ and $\{\cos nx\}$, $n = 1, 2, \ldots$, have period 2π, the scaled trigonometric functions $\{\sin(n\pi x/L)\}$ and $\{\cos(n\pi x/L)\}$, $n = 1, 2, \ldots$, have period $2L$. The periodic function $f(x)$, of period $p = 2L$, has the Fourier series expansion

$$f(x) = \frac{a_0}{2} + \sum_{n=1}^\infty \left\{ a_n \cos \frac{n\pi x}{L} + b_n \sin \frac{n\pi x}{L} \right\} \tag{7.8.1}$$

where the Fourier coefficients a_n and b_n can be determined as follows. Integrating both sides from $-L$ to L gives

$$\int_{-L}^L f(x) \, dx = \int_{-L}^L \left[\frac{a_0}{2} + \sum_{n=1}^\infty \left\{ a_n \cos \frac{n\pi x}{L} + b_n \sin \frac{n\pi x}{L} \right\} \right] dx$$

Integrating the series on the rhs term by term, which is assured in the case of uniform convergent series (7.8.1), gives

$$\int_{-L}^L f(x) \, dx = a_0 L + \sum_{n=1}^\infty a_n \int_{-L}^L \cos \frac{n\pi x}{L} \, dx + \sum_{n=1}^\infty b_n \int_{-L}^L \sin \frac{n\pi x}{L} \, dx$$

$$= a_0 L$$

since all the integrals on the rhs are zero, as may be seen readily by direct integration.

Thus we have

$$a_0 = \frac{1}{L} \int_{-L}^{L} f(x)\, dx \qquad (7.8.2)$$

To determine the a_n, multiply both sides of eq. (7.8.1) by $\cos(m\pi x/L)$, $m = 1$, $2, \ldots$, and integrate from $-L$ to L, to give

$$\int_{-L}^{L} f(x) \cos \frac{m\pi x}{L}\, dx = \frac{a_0}{2} \int_{-L}^{L} \cos \frac{m\pi x}{L}\, dx + \sum_{n=1}^{\infty} a_n \int_{-L}^{L} \cos \frac{n\pi x}{L} \cos \frac{m\pi x}{L}\, dx$$

$$+ \sum_{n=1}^{\infty} b_n \int_{-L}^{L} \sin \frac{n\pi x}{L} \cos \frac{m\pi x}{L}\, dx \qquad (7.8.3)$$

The first integral on the rhs is zero. Using the trigonometric identities

$$\cos \frac{n\pi x}{L} \cos \frac{m\pi x}{L} = \frac{1}{2} \left[\cos(n + m) \frac{\pi x}{L} + \cos(n - m) \frac{\pi x}{L} \right]$$

and

$$\sin \frac{n\pi x}{L} \cos \frac{m\pi x}{L} = \frac{1}{2} \left[\sin(n + m) \frac{\pi x}{L} + \sin(n - m) \frac{\pi x}{L} \right]$$

we can evaluate the remaining integrals easily. This leads to

$$\int_{-L}^{L} \cos \frac{n\pi x}{L} \cos \frac{m\pi x}{L}\, dx = \frac{1}{2} \int_{-L}^{L} \left[\cos(n + m) \frac{\pi x}{L} + \cos(n - m) \frac{\pi x}{L} \right] dx$$

$$= \frac{1}{2} \left[\frac{\sin(n + m) \dfrac{\pi x}{L}}{(n + m) \dfrac{\pi}{L}} + \frac{\sin(n - m) \dfrac{\pi x}{L}}{(n - m) \dfrac{\pi}{L}} \right]_{-L}^{L}$$

$$= \begin{cases} 0, & n \neq m \\ L, & n = m \end{cases} \qquad \begin{matrix} (7.8.4a) \\ (7.8.4b) \end{matrix}$$

and similarly,

$$\int_{-L}^{L} \sin \frac{n\pi x}{L} \cos \frac{m\pi x}{L}\, dx = \frac{1}{2} \int_{-L}^{L} \left[\sin(n + m) \frac{\pi x}{L} + \sin(n - m) \frac{\pi x}{L} \right] dx$$

$$= \frac{1}{2} \left[-\frac{\cos(n + m) \dfrac{\pi x}{L}}{(n + m) \dfrac{\pi}{L}} - \frac{\cos(n - m) \dfrac{\pi x}{L}}{(n - m) \dfrac{\pi}{L}} \right]_{-L}^{L} \qquad (7.8.5)$$

$$= 0$$

Thus from eqs. (7.8.3) to (7.8.5) we have

$$a_n = \frac{1}{L} \int_{-L}^{L} f(x) \cos \frac{n\pi x}{L} \, dx, \qquad n = 1, 2, \ldots \tag{7.8.6}$$

Similarly, to determine the b_n, multiply both sides of eq. (7.8.1) by $\sin(m\pi x/L)$, $m = 1, 2, \ldots$, and integrate from $-L$ to L, to eventually give

$$b_n = \frac{1}{L} \int_{-L}^{L} f(x) \sin \frac{n\pi x}{L} \, dx, \qquad n = 1, 2, \ldots \tag{7.8.7}$$

In summary, then, eqs. (7.8.2), (7.8.6), and (7.8.7) provide expressions for the coefficients of the Fourier series expansion given by eq. (7.8.1). These expressions are sometimes called *Euler formulae*. Using these expressions, an alternative version of eq. (7.8.1) can be written as follows:

$$\begin{aligned}
f(x) &= \frac{1}{2L} \int_{-L}^{L} f(t) \, dt + \frac{1}{L} \sum_{n=1}^{\infty} \left[\cos \frac{n\pi x}{L} \int_{-L}^{L} f(t) \cos \frac{n\pi t}{L} \, dt \right.\\
&\qquad \left. + \sin \frac{n\pi x}{L} \int_{-L}^{L} f(t) \sin \frac{n\pi t}{L} \, dt \right]\\
&= \frac{1}{2L} \int_{-L}^{L} f(t) \, dt + \frac{1}{L} \sum_{n=1}^{\infty} \int_{-L}^{L} f(t) \cos \left\{ \frac{n\pi}{L} (x - t) \right\} dt
\end{aligned} \tag{7.8.8}$$

It is worth emphasizing once again that since both $\cos(n\pi x/L)$ and $\sin(n\pi x/L)$ are periodic functions of period $2L$, the expansion (7.8.1) is limited to periodic functions of the same period.

If we wish to extend the expansion (7.8.1) to *arbitrary* nonperiodic functions $f(x)$, then in essence we are seeking the expansion for periods $2L \to \infty$. Thus we want to consider the limit $L \to \infty$ in eq. (7.8.8).

First, let us suppose that the function $f(x)$ is absolutely integrable—that is, the integral

$$\int_{-\infty}^{\infty} |f(t)| \, dt$$

exists and is finite. Then the first term on the rhs of eq. (7.8.8) approaches zero as $L \to \infty$. Thus for a fixed x, as $L \to \infty$, we have

$$f(x) = \lim_{L \to \infty} \frac{1}{L} \sum_{n=1}^{\infty} \int_{-L}^{L} f(t) \cos \left\{ \frac{n\pi}{L} (x - t) \right\} dt \tag{7.8.9}$$

Now, let

$$\alpha_n = \frac{n\pi}{L}$$

so that

$$\Delta \alpha = \alpha_{n+1} - \alpha_n = (n + 1) \frac{\pi}{L} - \frac{n\pi}{L} = \frac{\pi}{L}$$

Hence

$$\frac{1}{L} = \frac{\Delta\alpha}{\pi} \tag{7.8.10}$$

and eq. (7.8.9) becomes

$$f(x) = \lim_{L\to\infty} \frac{1}{\pi} \sum_{n=1}^{\infty} \Delta\alpha \int_{-L}^{L} f(t) \cos\{\alpha_n(x-t)\} \, dt$$

$$= \lim_{L\to\infty} \sum_{n=1}^{\infty} G(\alpha_n) \, \Delta\alpha \tag{7.8.11}$$

where

$$G(\alpha) = \frac{1}{\pi} \int_{-L}^{L} f(t) \cos\{\alpha(x-t)\} \, dt \tag{7.8.12}$$

As $L \to \infty$, we have $\Delta\alpha \to 0$ and

$$\sum_{n=1}^{\infty} G(\alpha_n) \, \Delta\alpha \to \int_0^{\infty} G(\alpha) \, d\alpha \tag{7.8.13}$$

Thus we obtain

$$f(x) = \int_0^{\infty} G(\alpha) \, d\alpha$$

$$= \frac{1}{\pi} \int_0^{\infty} \left[\int_{-\infty}^{\infty} f(t) \cos\{\alpha(x-t)\} \, dt \right] d\alpha \tag{7.8.14}$$

which is called the representation of $f(x)$ by the *Fourier integral*.

Note that the convergence of the Fourier integral representation to $f(x)$ is not guaranteed but is merely suggested by the arguments given above. Indeed, it is clear that this representation is *not* valid for *all* functions. For example, it is invalid when $f(x)$ is a constant, for then the integral with respect to t is indeterminate. Conditions on $f(x)$ which ensure that eq. (7.8.14) is valid have been reported in various sources (cf. Sneddon, 1951, chapter 1).

In eq. (7.8.14), note that the inner integral over t is an *even* function of α, since $\cos\{\alpha(x-t)\}$ is an even function of α. Thus the outer integral over α equals one-half of the integral over $-\infty$ to ∞; that is,

$$f(x) = \frac{1}{2\pi} \int_{-\infty}^{\infty} \left[\int_{-\infty}^{\infty} f(t) \cos\{\alpha(x-t)\} \, dt \right] d\alpha \tag{7.8.15}$$

Also, since $\sin\{\alpha(x-t)\}$ is an *odd* function of α, we have

$$\frac{1}{2\pi} \int_{-\infty}^{\infty} \left[\int_{-\infty}^{\infty} f(t) \sin\{\alpha(x-t)\} \, dt \right] d\alpha = 0 \tag{7.8.16}$$

Now, if we multiply eq. (7.8.16) by i and add to eq. (7.8.15), we get

$$f(x) = \frac{1}{2\pi} \int_{-\infty}^{\infty} \left[\int_{-\infty}^{\infty} f(t) \exp\{i\alpha(x-t)\} \, dt \right] d\alpha \tag{7.8.17}$$

recalling that from the Euler formula we have

$$\exp(it) = \cos t + i \sin t \qquad (7.8.18)$$

Equation (7.8.17) is called the *complex form of the Fourier integral*. The exponential in eq. (7.8.17) can be written as the product of two exponentials, to give the following complementary relationships. If we define

$$F(\alpha) = \frac{1}{\sqrt{2\pi}} \int_{-\infty}^{\infty} f(t) \exp(-i\alpha t) \, dt \qquad (7.8.19a)$$

then

$$f(x) = \frac{1}{\sqrt{2\pi}} \int_{-\infty}^{\infty} F(\alpha) \exp(i\alpha x) \, d\alpha \qquad (7.8.19b)$$

The function $F(\alpha)$ is called the *Fourier transform* of $f(x)$ (i.e. $\mathcal{F}[f(x)]$), and $f(x)$ is called the *inverse Fourier transform* of $F(\alpha)$ (i.e. $\mathcal{F}^{-1}[F(\alpha)]$). Note that since the arguments of cosine and sine functions in eqs. (7.8.15) and (7.8.16) can alternatively be kept as $\alpha(t - x)$, eqs. (7.8.19) are sometimes defined with $\exp(i\alpha t)$ in (7.8.19a) and $\exp(-i\alpha x)$ in (7.8.19b).

In analogy with the *finite* Fourier transforms, it should be clear that we can go back and forth between the function $f(x)$ and its Fourier transform, $\mathcal{F}[f(x)] = F(\alpha)$. Given a function, $f(x)$ we can find its Fourier transform $F(\alpha)$ from eq. (7.8.19a). Alternatively, given the transform $F(\alpha)$, using eq. (7.8.19b) we can find the function, $f(x)$ to which it belongs. Fourier transforms of a few simple functions are shown in Table 7.4.

7.8.2 Properties of Fourier Transforms

Fourier transforms satisfy certain properties which are helpful in solving PDEs. Some of these are discussed next.

Linearity. The Fourier transform is a linear integral operator. If $f_1(x)$ and $f_2(x)$ are two functions, and c_1 and c_2 are constants, then the definition (7.8.19a) gives

$$\mathcal{F}[c_1 f_1(x) + c_2 f_2(x)] = \frac{1}{\sqrt{2\pi}} \int_{-\infty}^{\infty} [c_1 f_1(t) + c_2 f_2(t)] \exp(-i\alpha t) \, dt$$

$$= c_1 \mathcal{F}[f_1(x)] + c_2 \mathcal{F}[f_2(x)] \qquad (7.8.20)$$

Hence $\mathcal{F}[\ ]$ is a linear integral operator.

Shifting. If $F(\alpha) = \mathcal{F}[f(x)]$, then $\mathcal{F}[f(x - c)] = \exp(-i\alpha c) F(\alpha)$. Again, from the definition (7.8.19a), we have

$$\mathcal{F}[f(x - c)] = \frac{1}{\sqrt{2\pi}} \int_{-\infty}^{\infty} f(t - c) \exp(-i\alpha t) \, dt$$

$$= \frac{1}{\sqrt{2\pi}} \int_{-\infty}^{\infty} f(y) \exp[-i\alpha(c + y)] \, dy, \quad \text{where } y = t - c$$

$$= \exp(-i\alpha c) \, \mathcal{F}[f(x)] \qquad (7.8.21)$$

TABLE 7.4 Fourier Transforms[a]

$f(x)$	$F(\alpha) = \dfrac{1}{\sqrt{2\pi}} \displaystyle\int_{-\infty}^{\infty} f(t)\exp(-i\alpha t)\,dt$
$f\left(\dfrac{x}{a} + b\right)$　　$(a > 0)$	$a\exp(iab\alpha)G(a\alpha)$
$f\left(-\dfrac{x}{a} + b\right)$　　$(a > 0)$	$a\exp(-iab\alpha)F(-a\alpha)$
$f(ax)\exp(ibx)$　　$(a > 0)$	$\dfrac{1}{a}F\left(\dfrac{\alpha - b}{a}\right)$
$f(ax)\cos bx$　　$(a > 0)$	$\dfrac{1}{2a}\left[F\left(\dfrac{\alpha - b}{a}\right) + F\left(\dfrac{\alpha + b}{a}\right)\right]$
$f(ax)\sin bx$　　$(a > 0)$	$\dfrac{1}{2ai}\left[F\left(\dfrac{\alpha - b}{a}\right) - F\left(\dfrac{\alpha + b}{a}\right)\right]$
$\begin{cases} 1 & \text{for } \lvert x\rvert < a \\ 0 & \text{for } \lvert x\rvert > a \end{cases}$	$\sqrt{\dfrac{2}{\pi}}\,\dfrac{\sin(a\alpha)}{\alpha}$
$\begin{cases} 1 & \text{for } a < x < b \\ 0 & \text{for } x < a \text{ or } x > b \end{cases}$	$\dfrac{\exp(-ia\alpha) - \exp(ib\alpha)}{i\alpha\sqrt{2\pi}}$
$\dfrac{1}{\lvert x\rvert}$	$\dfrac{1}{\lvert \alpha\rvert}$
$\dfrac{1}{x^2 + a^2}$　　$(a > 0)$	$\sqrt{\dfrac{\pi}{2}}\,\dfrac{\exp(-a\lvert\alpha\rvert)}{a}$
e^{-ax^2}　　$(a > 0)$	$\dfrac{1}{\sqrt{2a}}\exp\left(-\dfrac{\alpha^2}{4a}\right)$
$\begin{cases} e^{ax} & \text{for } b < x < c \\ 0 & \text{for } x < b \text{ or } x > c \end{cases}$	$\dfrac{\exp[(a - i\alpha)c] - \exp[(a - i\alpha)b]}{\sqrt{2\pi}(a - i\alpha)}$
$\begin{cases} e^{-ax} & \text{for } x > 0 \\ 0 & \text{for } x < 0 \end{cases}$　$(a > 0)$	$\dfrac{1}{\sqrt{2\pi}(a + i\alpha)}$
$\begin{cases} e^{iax} & \text{for } \lvert x\rvert < b \\ 0 & \text{for } \lvert x\rvert > b \end{cases}$	$\sqrt{\dfrac{2}{\pi}}\,\dfrac{\sin[b(\alpha - a)]}{(\alpha - a)}$
$\begin{cases} e^{iax} & \text{for } b < x < c \\ 0 & \text{for } x < b \text{ or } x > c \end{cases}$	$\dfrac{i}{\sqrt{2\pi}}\,\dfrac{\exp[ib(a - \alpha)] - \exp[ic(a - \alpha)]}{(a - \alpha)}$
$\cos(ax^2)$	$\dfrac{1}{\sqrt{2a}}\cos\left(\dfrac{\alpha^2}{4a} - \dfrac{\pi}{4}\right)$
$\sin(ax^2)$	$\dfrac{1}{\sqrt{2a}}\sin\left(\dfrac{\alpha^2}{4a} + \dfrac{\pi}{4}\right)$
$\dfrac{\sin(ax)}{x}$　　$(a > 0)$	$\begin{cases} \sqrt{\dfrac{\pi}{2}} & \text{for } \lvert\alpha\rvert < a \\ 0 & \text{for } \lvert\alpha\rvert > a \end{cases}$

[a]For a more extensive list, see Erdélyi et al. (1954).

Scaling. If $F(\alpha) = \mathcal{F}[f(x)]$, then $\mathcal{F}[f(kx)] = (1/|k|)F\left(\dfrac{\alpha}{k}\right)$, where k is a nonzero constant.

$$\mathcal{F}[f(kx)] = \frac{1}{\sqrt{2\pi}} \int_{-\infty}^{\infty} f(kt) \exp(-i\alpha t)\, dt$$

$$= \frac{1}{|k|} \frac{1}{\sqrt{2\pi}} \int_{-\infty}^{\infty} f(y) \exp\left(-\frac{i\alpha y}{k}\right) dy, \qquad \text{where } y = kt$$

$$= \frac{1}{|k|} F\left(\frac{\alpha}{k}\right) \tag{7.8.22}$$

Fourier Transforms of Derivatives of $f(x)$. Let $f(x)$ be continuous on the x axis and $f(x) \to 0$ as $|x| \to \infty$. Then

$$\mathcal{F}[f'(x)] = i\alpha \mathcal{F}[f(x)] \tag{7.8.23}$$

In order to show this, note that from eq. (7.8.19a) we obtain

$$\mathcal{F}[f'(x)] = \frac{1}{\sqrt{2\pi}} \int_{-\infty}^{\infty} f'(t) \exp(-i\alpha t)\, dt$$

$$= \frac{1}{\sqrt{2\pi}} \left[f(t) \exp(-i\alpha t)\big|_{-\infty}^{\infty} - (-i\alpha) \int_{-\infty}^{\infty} f(t) \exp(-i\alpha t)\, dt \right]$$

Since $\exp(i\theta) = \cos\theta + i\sin\theta$, and $f(x) \to 0$ as $|x| \to \infty$, the first term above is zero, leading to

$$\mathcal{F}[f'(x)] = i\alpha \mathcal{F}[f(x)]$$

We can apply the relation (7.8.23) twice, to give

$$\mathcal{F}[f''(x)] = i\alpha \mathcal{F}[f'(x)] = (i\alpha)^2 \mathcal{F}[f(x)]$$

$$= -\alpha^2 \mathcal{F}[f(x)] \tag{7.8.24}$$

provided that $f'(x)$ is also continuous and tends to zero as $|x| \to \infty$.

A repeated application of eq. (7.8.23) gives the relationship

$$\mathcal{F}[f^{(n)}(x)] = (i\alpha)^n \mathcal{F}[f(x)], \qquad n = 0, 1, 2, \ldots \tag{7.8.25}$$

provided that f and its first $(n-1)$ derivatives are continuous and vanish as $|x| \to \infty$.

● **CONVOLUTION THEOREM**

The function

$$h(x) = \int_{-\infty}^{\infty} f(r)g(x-r)\, dr = \int_{-\infty}^{\infty} f(x-r)g(r)\, dr \tag{7.8.26}$$

is called the *convolution* (or *Faltung*) of the functions $f(x)$ and $g(x)$ over the interval $x \in (-\infty, \infty)$ and is generally denoted as $h = f * g$. The convolution theorem relates the Fourier transform of $f * g$ to the transforms of f and g, as follows:

$$\mathcal{F}[f * g] = \sqrt{2\pi}\, \mathcal{F}[f]\mathcal{F}[g] \tag{7.8.27}$$

To prove this, note that definitions (7.8.19a) and (7.8.26) lead to

$$\mathcal{F}[f * g] = \frac{1}{\sqrt{2\pi}} \int_{-\infty}^{\infty} \int_{-\infty}^{\infty} f(r)g(t - r) \exp(-i\alpha t) \, dr \, dt$$

$$= \frac{1}{\sqrt{2\pi}} \int_{-\infty}^{\infty} \int_{-\infty}^{\infty} f(r)g(t - r) \exp(-i\alpha t) \, dt \, dr$$

if we reverse the order of integration. Changing the variable as

$$t - r = y$$

leads to

$$\mathcal{F}[f * g] = \frac{1}{\sqrt{2\pi}} \int_{-\infty}^{\infty} \int_{-\infty}^{\infty} f(r)g(y) \exp[-i\alpha(r + y)] \, dy \, dr$$

$$= \frac{1}{\sqrt{2\pi}} \left[\int_{-\infty}^{\infty} f(r) \exp(-i\alpha r) \, dr \right] \left[\int_{-\infty}^{\infty} g(y) \exp(-i\alpha y) \, dy \right]$$

$$= \sqrt{2\pi}\mathcal{F}[f]\mathcal{F}[g]$$

which is eq. (7.8.27). Note that taking the inverse Fourier transform of both sides of eq. (7.8.27) gives, using the definition (7.8.19b),

$$\mathcal{F}^{-1}[F(\alpha)G(\alpha)] = \frac{1}{\sqrt{2\pi}} (f * g)(x) \qquad (7.8.28)$$

a formula which is useful in applications.

7.8.3 An Example

As an example of the use of Fourier transforms to solve PDEs in the infinite domain, let us return to the problem of one-dimensional transient heat conduction, this time in an infinite rod. Thus let us solve the problem

$$\frac{\partial^2 T}{\partial x^2} = \frac{1}{\kappa} \frac{\partial T}{\partial t}; \qquad x \in (-\infty, \infty), \quad t > 0 \qquad (7.8.29)$$

subject to the BCs

$$T \to 0 \text{ and } \frac{\partial T}{\partial x} \to 0 \qquad \text{as } x \to \pm\infty, \quad t > 0 \qquad (7.8.30)$$

and the IC

$$T = f(x); \qquad x \in (-\infty, \infty), \quad t = 0 \qquad (7.8.31)$$

In order to solve the problem, let us take the Fourier transform of both sides of eq. (7.8.29). This gives

$$\mathcal{F}\left[\frac{\partial^2 T}{\partial x^2} \right] = \frac{1}{\kappa} \mathcal{F}\left[\frac{\partial T}{\partial t} \right] \qquad (7.8.32)$$

Using the relation (7.8.24) gives for the lhs

$$-\alpha^2 \mathcal{F}[T(x, t)] = -\alpha^2 \overline{T}(\alpha, t) \tag{7.8.33}$$

where using the definition (7.8.19a)

$$\overline{T}(\alpha, t) = \frac{1}{\sqrt{2\pi}} \int_{-\infty}^{\infty} T(s, t) \exp(-i\alpha s) \, ds \tag{7.8.34}$$

is the Fourier transform of $T(x, t)$. The rhs of eq. (7.8.32) gives

$$\frac{1}{\kappa} \frac{1}{\sqrt{2\pi}} \int_{-\infty}^{\infty} \frac{\partial T(s, t)}{\partial t} \exp(-i\alpha s) \, ds$$

$$= \frac{1}{\kappa} \frac{\partial}{\partial t} \left[\frac{1}{\sqrt{2\pi}} \int_{-\infty}^{\infty} T(s, t) \exp(-i\alpha s) \, ds \right]$$

$$= \frac{1}{\kappa} \frac{d\overline{T}}{dt} \tag{7.8.35}$$

where we have assumed that the order of differentiation and integration can be exchanged.

From eqs. (7.8.32), (7.8.33), and (7.8.35) we have

$$\frac{1}{\kappa} \frac{d\overline{T}}{dt} = -\alpha^2 \overline{T}$$

which has solution

$$\overline{T}(\alpha, t) = \overline{T}(\alpha, 0) \exp(-\alpha^2 \kappa t) \tag{7.8.36}$$

Following arguments which have been used in previous sections, we have

$$\overline{T}(\alpha, 0) = \lim_{t \to 0} \overline{T}(\alpha, t)$$

$$= \lim_{t \to 0} \frac{1}{\sqrt{2\pi}} \int_{-\infty}^{\infty} T(s, t) \exp(-i\alpha s) \, ds$$

$$= \frac{1}{\sqrt{2\pi}} \int_{-\infty}^{\infty} f(s) \exp(-i\alpha s) \, ds \tag{7.8.37}$$

so that $\overline{T}(\alpha, t)$ in eq. (7.8.36) is fully known. Inverting the transform gives the solution

$$T(x, t) = \frac{1}{\sqrt{2\pi}} \int_{-\infty}^{\infty} \overline{T}(\alpha, t) \exp(i\alpha x) \, d\alpha$$

$$= \frac{1}{2\pi} \int_{-\infty}^{\infty} \left[\exp(-\alpha^2 \kappa t) \int_{-\infty}^{\infty} f(s) \exp(-i\alpha s) \, ds \right] \exp(i\alpha x) \, d\alpha$$

$$= \frac{1}{2\pi} \int_{-\infty}^{\infty} f(s) \left[\int_{-\infty}^{\infty} \exp\{i\alpha(x - s)\} \exp(-\alpha^2 \kappa t) \, d\alpha \right] ds \tag{7.8.38}$$

The inner integral I has been evaluated in section 7.7, and eq. (7.7.33) leads to

$$I = \sqrt{\frac{\pi}{\kappa t}} \exp\left\{-\frac{(x-s)^2}{4\kappa t}\right\}$$

Thus we have

$$T(x, t) = \frac{1}{2\sqrt{\pi \kappa t}} \int_{-\infty}^{\infty} f(s) \exp\left\{-\frac{(x-s)^2}{4\kappa t}\right\} ds \qquad (7.8.39)$$

which clearly shows how the initial temperature distribution $T(x, 0) = f(x)$ influences the solution. It is interesting to note that the temperature at any specific (x, t) value depends on the entire initial temperature distribution.

Note that the above result can alternatively be obtained by applying the convolution theorem. From eq. (7.8.37) we obtain

$$f(x) = \mathscr{F}^{-1}[\bar{T}(\alpha, 0)]$$

and from Table 7.1 we have

$$\frac{1}{\sqrt{2\kappa t}} \exp\left(-\frac{x^2}{4\kappa t}\right) = \mathscr{F}^{-1}[\exp(-\alpha^2 \kappa t)]$$

Then using eqs. (7.8.28) and (7.8.36) leads directly to eq. (7.8.39).

The function

$$G(x, t; s, 0) = \frac{1}{2\sqrt{\pi \kappa t}} \exp\left\{-\frac{(x-s)^2}{4\kappa t}\right\} \qquad (7.8.40)$$

is called either the *influence function* or a *fundamental solution* of the heat conduction equation. It is easy to verify by direct differentiation that it satisfies eq. (7.8.29)—that is,

$$\frac{\partial^2 G}{\partial x^2} = \frac{1}{\kappa}\frac{\partial G}{\partial t}; \qquad x \neq s, \quad t > 0 \qquad (7.8.41)$$

Also, the integral

$$\int_{-\infty}^{\infty} G(x, t; s, 0) \, ds = 1 \qquad (7.8.42)$$

may be verified by using the definition (7.8.40) and the relationship given by eq. (7.7.32).

Since eq. (7.8.39) may be rewritten as

$$T(x, t) = \int_{-\infty}^{\infty} f(s)G(x, t; s, 0) \, ds \qquad (7.8.43)$$

the influence function $G(x, t; s, 0)$ may be regarded as one that measures the effect of an initial (i.e., $t = 0$) temperature at position s on the temperature at time $t > 0$ and location x. A sketch of the influence function is shown in Figure 7.9. It is clear from

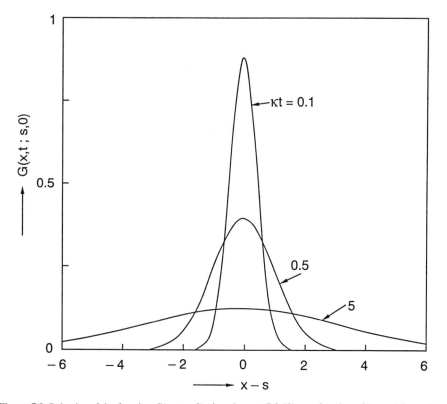

Figure 7.9 Behavior of the function $G(x, t; s, 0)$ given by eq. (7.8.40) as a function of $(x - s)$ for various values of κt.

the sketch that the width of the influence is small initially, but increases with time. Also, for any fixed t, the influence is largest at $x = s$ and decreases as x moves away from s. Note that $G(x, t; s, 0)$ is in fact the gaussian probability density function, with mean s and variance $2\kappa t$.

From eq. (7.8.43),

$$\lim_{t \to 0^+} T(x, t) = \lim_{t \to 0^+} \int_{-\infty}^{\infty} f(s) G(x, t; s, 0) \, ds$$

$$= \int_{-\infty}^{\infty} f(s) \lim_{t \to 0^+} G(x, t; s, 0) \, ds$$

However, since the lhs equals $f(x)$ from the IC (7.8.31), we can conclude that

$$\lim_{t \to 0^+} G(x, t; s, 0) = \delta(x - s) \qquad (7.8.44)$$

the Dirac delta function. This can also be understood from Figure 7.9 and eq. (7.8.42).

7.9 MULTIDIMENSIONAL SEMI-INFINITE AND INFINITE DOMAINS

Many types of problems can arise when more than one space dimension is involved. Thus we can have cases where each dimension is finite, semi-infinite, or infinite. Among these, problems where all dimensions are finite were treated in sections 7.3 and 7.4. The other cases can also be solved using the same technique—that is, taking successive transforms in each space dimension. It should be clear that for finite dimensions, the finite Fourier transform should be applied. For semi-infinite and infinite dimensions, the Fourier sine or cosine and Fourier transforms, respectively, would apply. For cylindrical geometries, the appropriate Hankel transforms should be used. Problems of this type have been treated in various sources, such as Sneddon, 1951, chapter 5; and Özisik, 1989, chapter 2. We illustrate the technique next by the use of an example.

Consider transient heat conduction in the semi-infinite rectangular strip shown in Figure 7.10, governed by the PDE

$$\kappa\left(\frac{\partial^2 T}{\partial x^2} + \frac{\partial^2 T}{\partial y^2}\right) = \frac{\partial T}{\partial t}; \qquad x \in (0, +\infty), \quad y \in (0, L), \quad t > 0 \qquad (7.9.1)$$

The initially cold strip is heated from the upper side as indicated by the following BCs:

$$T = 0; \qquad x = 0, \qquad\qquad y \in (0, L), \quad t > 0 \qquad\qquad (7.9.2a)$$

$$T = 0; \qquad x \in (0, +\infty), \quad y = 0, \qquad\quad t > 0 \qquad\qquad (7.9.2b)$$

$$T = 1; \qquad x \in (0, +\infty), \quad y = L, \qquad\quad t > 0 \qquad\qquad (7.9.2c)$$

and the IC

$$T = 0; \qquad x \in (0, +\infty), \quad y \in (0, L), \quad t = 0 \qquad\qquad (7.9.3)$$

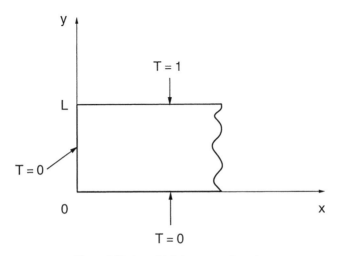

Figure 7.10 A semi-infinite rectangular strip.

Note that in eqs. (7.9.2) above, since there are two independent variables x and y, there is no simple change of variables [such as eq. (3.12.3)] which makes *all* the BCs homogeneous. Hence as in section 7.4.2 we proceed with the BCs as they stand.

Since the y space coordinate is finite while x is semi-infinite, we apply the finite Fourier transform in the y direction and the Fourier sine transform in the x direction, thus leading to the mixed transform

$$\bar{T}_n(\alpha, t) = \sqrt{\frac{2}{\pi}}\sqrt{\frac{2}{L}} \int_0^\infty \sin(\alpha x)\, dx \int_0^L T(x, y, t) \sin\left(\frac{n\pi y}{L}\right) dy \qquad (7.9.4)$$

Thus multiply both sides of eq. (7.9.1) by $\sqrt{2/\pi}\,\sin(\alpha x)$ and $\sqrt{2/L}\,\sin(n\pi y/L)$, and integrate over the whole strip. Integrating twice by parts in the variable x and using eq. (7.9.2a), the first term on the lhs leads to

$$\kappa \sqrt{\frac{2}{\pi}}\,\sqrt{\frac{2}{L}} \int_0^L \sin\left(\frac{n\pi y}{L}\right) dy \int_0^\infty \frac{\partial^2 T}{\partial x^2} \sin(\alpha x)\, dx$$

$$= -\kappa \sqrt{\frac{2}{\pi}}\sqrt{\frac{2}{L}}\,\alpha^2 \int_0^L \sin\left(\frac{n\pi y}{L}\right) dy \int_0^\infty T(x, y, t) \sin(\alpha x)\, dx$$

$$= -\kappa\alpha^2 \bar{T}_n(\alpha, t) \qquad (7.9.5)$$

In deriving this equation, we have assumed that

$$T(x, y, t) \rightarrow 0 \quad \text{and} \quad \frac{\partial T}{\partial x} \rightarrow 0 \qquad \text{as } x \rightarrow \infty \qquad (7.9.6)$$

Similarly, after integrating twice by parts in the variable y and using the BCs (7.9.2b) and (7.9.2c), the second term reduces to

$$\kappa \sqrt{\frac{2}{\pi}}\sqrt{\frac{2}{L}} \int_0^\infty \sin(\alpha x)\, dx \int_0^L \frac{\partial^2 T}{\partial y^2} \sin\left(\frac{n\pi y}{L}\right) dy$$

$$= \kappa \sqrt{\frac{2}{\pi}}\sqrt{\frac{2}{L}}\,(-1)^{n+1}\left(\frac{n\pi}{L}\right) \int_0^\infty \sin(\alpha x)\, dx$$

$$- \kappa \sqrt{\frac{2}{\pi}}\sqrt{\frac{2}{L}}\left(\frac{n\pi}{L}\right)^2 \int_0^\infty \sin(\alpha x)\, dx \int_0^L T(x, y, t) \sin\left(\frac{n\pi y}{L}\right) dy$$

$$= \kappa \sqrt{\frac{2}{\pi}}\sqrt{\frac{2}{L}}\,(-1)^{n+1}\left(\frac{n\pi}{L}\right) \int_0^\infty \sin(\alpha x)\, dx - \kappa\left(\frac{n\pi}{L}\right)^2 \bar{T}_n(\alpha, t) \qquad (7.9.7)$$

The integral remaining in the equation above can be evaluated by noting that

$$\int_0^\infty e^{-\varepsilon x} \sin(\alpha x)\, dx = \frac{\alpha}{\varepsilon^2 + \alpha^2} \qquad (7.9.8)$$

which, as $\varepsilon \rightarrow 0$, implies that

$$\int_0^\infty \sin(\alpha x)\, dx = \frac{1}{\alpha} \qquad (7.9.9)$$

With these results, eq. (7.9.1) in the transformed space becomes

$$\frac{1}{\kappa}\frac{d\bar{T}_n}{dt} = -\left[\alpha^2 + \left(\frac{n\pi}{L}\right)^2\right]\bar{T}_n + (-1)^{n+1}\sqrt{\frac{2}{\pi}}\sqrt{\frac{2}{L}}\left(\frac{n\pi}{L}\right)\frac{1}{\alpha} \qquad (7.9.10)$$

which integrated with the IC (7.9.3) leads to

$$\bar{T}_n(\alpha, t) = \sqrt{\frac{2}{\pi}}\sqrt{\frac{2}{L}}\frac{(-1)^{n+1}n\pi}{\alpha L\left[\alpha^2 + \left(\frac{n\pi}{L}\right)^2\right]}\left\{1 - \exp\left[-\left(\alpha^2 + \frac{n^2\pi^2}{L^2}\right)\kappa t\right]\right\} \qquad (7.9.11)$$

In order to obtain the solution in the real space, we need to invert the above mixed Fourier transform. For this, we use eqs. (7.2.6) and (7.7.12) to yield

$$T(x, y, t) = \sqrt{\frac{2}{\pi}}\sqrt{\frac{2}{L}}\sum_{n=1}^{\infty}\sin\left(\frac{n\pi y}{L}\right)\int_0^{\infty}\bar{T}_n(\alpha, t)\sin(\alpha x)\,d\alpha, \qquad (7.9.12)$$
$$n = 1, 2, \ldots$$

Recalling the value of the integral

$$\int_0^{\infty}\frac{\sin(\alpha x)}{\alpha\left[\alpha^2 + \left(\frac{n\pi}{L}\right)^2\right]}\,d\alpha = \frac{L^2}{2\pi n^2}\left[1 - \exp\left(-\frac{n\pi x}{L}\right)\right] \qquad (7.9.13)$$

and substituting eq. (7.9.11) in eq. (7.9.12), it follows that

$$T(x, y, t) = \frac{4}{\pi L}\sum_{n=1}^{\infty}\frac{n\pi}{L}(-1)^{n+1}\sin\left(\frac{n\pi y}{L}\right)\int_0^{\infty}\frac{\sin(\alpha x)}{\alpha\left[\alpha^2 + \left(\frac{n\pi}{L}\right)^2\right]}$$

$$\left\{1 - \exp\left[-\left(\alpha^2 + \frac{n^2\pi^2}{L^2}\right)\kappa t\right]\right\}\,d\alpha$$

$$= \frac{2}{\pi}\sum_{n=1}^{\infty}\frac{1}{n}(-1)^{n+1}\sin\left(\frac{n\pi y}{L}\right)\left[1 - \exp\left(-\frac{n\pi x}{L}\right)\right]$$

$$- \frac{4}{L^2}\sum_{n=1}^{\infty}n(-1)^{n+1}\sin\left(\frac{n\pi y}{L}\right)$$

$$\int_0^{\infty}\frac{\sin(\alpha x)}{\alpha\left[\alpha^2 + \left(\frac{n\pi}{L}\right)^2\right]}\exp\left[-\left(\alpha^2 + \frac{n^2\pi^2}{L^2}\right)\kappa t\right]\,d\alpha \qquad (7.9.14)$$

represents the solution of eqs. (7.9.1) to (7.9.3).

References

R. B. Bird, W. E. Stewart, and E. N. Lightfoot, *Transport Phenomena*, Wiley, New York, 1960.

R. V. Churchill, *Operational Mathematics*, 3rd ed., McGraw-Hill, New York, 1972.

R. V. Churchill and J. W. Brown, *Fourier Series and Boundary Value Problems*, 4th ed., Mc-Graw-Hill, New York, 1987.

A. Erdélyi, W. Magnus, F. Oberhettinger, and F. Tricomi, *Tables of Integral Transforms*, volumes 1 and 2, McGraw-Hill, New York, 1954.

E. Kreyszig, *Advanced Engineering Mathematics*, 7th ed., Wiley, New York, 1993.

M. N. Özisik, *Boundary Value Problems of Heat Conduction*, Dover, New York, 1989.

I. N. Sneddon, *Fourier Transforms*, McGraw-Hill, New York, 1951.

Additional Reading

H. Bateman, *Partial Differential Equations of Mathematical Physics*, Cambridge University Press, Cambridge, 1964.

H. S. Carslaw and J. C. Jaeger, *Conduction of Heat in Solids*, 2nd ed., Oxford University Press, Oxford, 1959.

J. Crank, *The Mathematics of Diffusion*, 2nd ed., Oxford University Press, Oxford, 1975.

E. C. Titchmarsh, *Eigenfunction Expansions*, Oxford University Press, Oxford, 1946.

E. C. Titchmarsh, *Introduction to the Theory of Fourier Integrals*, 2nd ed., Oxford University Press, Oxford, 1948.

H. F. Weinberger, *A First Course in Partial Differential Equations*, Blaisdell, Waltham, MA, 1965.

Problems

7.1 Repeat the transient heat conduction problem of section 7.1, for the BCs of type (i, j), where $i, j = 1 \rightarrow 3$.

BC at $x = 0$	Type	BC at $x = L$
$T = 0$	1	$T = 0$
$\dfrac{\partial T}{\partial x} = 0$	2	$\dfrac{\partial T}{\partial x} = 0$
$-\dfrac{\partial T}{\partial x} + \alpha T = 0$	3	$\dfrac{\partial T}{\partial x} + \alpha T = 0$

and α is a positive constant.

7.2 A solid spherical body of radius R with constant initial temperature T_i is dunked into a well-agitated bath whose temperature is maintained at T_b. Determine the following.

(a) the temperature distribution within the body, as a function of time and position.

(b) a formula for the average temperature, as a function of time.

7.3 Solve for the transient concentration distribution for an isothermal first-order reaction occurring in a slab catalyst pellet of half-thickness L with no external mass transfer limitations. The initial distribution of reactant concentration in the pellet is $C_0(x)$. Also compute a formula for the transient effectiveness factor $\eta(t)$. Finally, for $C_0(x) = 0$, plot $\eta(t)$ versus t for the Thiele modulus $\phi = 1, 2,$ and 5, showing the approach to the steady state.

7.4 Repeat the problem above for a spherical catalyst pellet of radius R.

7.5 Solve the problem associated with the free vibrations of a string of finite length, given by eqs. (5.5.36) to (5.5.38).

7.6 Solve the following Poisson equation in a rectangle:

$$\frac{\partial^2 u}{\partial x^2} + \frac{\partial^2 u}{\partial y^2} = y(1 - y)\sin^3 x; \qquad 0 < x < 1, \quad 0 < y < 1$$

along with the BCs

$$
\begin{aligned}
u(x, 0) &= 0; & 0 < x < 1 \\
u(x, 1) &= 0; & 0 < x < 1 \\
u(0, y) &= 0; & 0 < y < 1 \\
u(1, y) &= 0; & 0 < y < 1
\end{aligned}
$$

7.7 Solve the PDE

$$\frac{\partial^2 u}{\partial x^2} + \frac{\partial^2 u}{\partial y^2} = -f(x, y); \qquad 0 < x < a, \quad 0 < y < b$$

along with the BCs

$$
\begin{aligned}
u(x, 0) &= 0; & 0 < x < a \\
\frac{\partial u}{\partial y}(x, b) &= 0; & 0 < x < a \\
u(0, y) &= 0; & 0 < y < b \\
\frac{\partial u}{\partial x}(a, y) &= 0; & 0 < y < b
\end{aligned}
$$

7.8 Solve the PDE (7.3.1) along with the conditions given by eqs. (7.3.3).

7.9 An isothermal first-order reaction occurs at steady state in a rectangular parallelopiped (see Figure 7.3) catalyst pellet, with no external mass transfer resistance. Determine the reactant concentration distribution as a function of position, as well as the catalyst effectiveness factor.

7.10 The transverse displacement of a rectangular membrane during free vibrations is described by the PDE

$$\frac{1}{c^2}\frac{\partial^2 z}{\partial t^2} = \frac{\partial^2 z}{\partial x^2} + \frac{\partial^2 z}{\partial y^2}; \qquad 0 < x < a, \quad 0 < y < b, \quad t > 0$$

The edge of the membrane is fixed during the motion, while the ICs are given by

$$z = f(x, y), \qquad \frac{\partial z}{\partial t} = g(x, y), \qquad t = 0$$

7.11 Obtain the solution of the problem given by eqs. (7.4.1)–(7.4.3). See the remark at the end of section 7.4.

7.12 Solve the Laplace equation in a rectangle:

$$\frac{\partial^2 u}{\partial x^2} + \frac{\partial^2 u}{\partial y^2} = 0, \qquad 0 < x < 1, \quad 0 < y < 1$$

along with the BCs

$$
\begin{aligned}
u(0, y) &= 0, & 0 < y < 1 \\
u(1, y) &= 0, & 0 < y < 1 \\
u(x, 0) &= f(x), & 0 < x < 1 \\
u(x, 1) &= 0, & 0 < x < 1
\end{aligned}
$$

7.13 Repeat the problem given by eqs. (7.5.1)–(7.5.3) for the case of finite external mass transfer resistance. Thus the BC at $r = a$ is

$$D \frac{\partial C}{\partial r} = k_g(C_0 - C); \qquad r = a, \quad t > 0$$

and in making the problem dimensionless, an additional dimensionless parameter, the Biot number $\text{Bi} = k_g a/D$ arises.

7.14 Repeat problem 7.3 for an infinitely long cylindrical catalyst pellet of radius R.

7.15 Consider steady state diffusion of mass in a finite cylinder of radius a and length L, with no external mass transfer resistance. Determine the concentration distribution as a function of position.

7.16 Repeat the problem above for an isothermal first-order reaction. Also determine a formula for the catalyst effectiveness factor.

7.17 Repeat the problem involving vibrations of a circular membrane given by eqs. (7.6.1) and (7.6.2), with the ICs

$$u = 0; \qquad 0 < s < 1, \quad t = 0$$

$$\frac{\partial u}{\partial t} = g(s); \qquad 0 < s < 1, \quad t = 0$$

7.18 Repeat Examples 1–4 of section 7.7, with the only change that the BC at $x = 0$ is modified as follows.

Example 1: $\qquad \dfrac{\partial T}{\partial x} = 0; \qquad x = 0, \quad t > 0$

Example 2: $\qquad \dfrac{\partial T}{\partial x} = q < 0; \qquad x = 0, \quad t > 0$

Examples 3 & 4: $\qquad \dfrac{\partial T}{\partial x} = g(t) < 0; \qquad x = 0, \quad t > 0$

7.19 Show that eq. (7.7.67) is valid.

7.20 Determine the temperature distribution in a cylinder of large radius:

$$\frac{1}{\kappa} \frac{\partial T}{\partial t} = \frac{1}{r} \frac{\partial}{\partial r} \left[r \frac{\partial T}{\partial r} \right]; \qquad 0 < r < \infty, \quad t > 0$$

with the conditions

T finite
$T = f(r); \qquad 0 < r < \infty, \quad t = 0$

Hint: In solving this problem, Weber's formula

$$\int_0^\infty \alpha J_0(\alpha r) J_0(\alpha \eta) e^{-p\alpha^2} \, d\alpha = \frac{1}{2p} \exp\left[-\frac{r^2 + \eta^2}{4p} \right] I_0\left(\frac{r\eta}{2p} \right)$$

proves useful.

7.21 Repeat the problem of section 7.8.3, where in addition heat is generated within the rod at rate $g(x, t)$.

7.22 Solve the transient heat conduction problem (7.9.1) in an initially hot semi-infinite rectangular strip, which is being cooled down. The vertical boundary is adiabatic, hence heat can escape only from the horizontal boundaries:

$$T = 0; \qquad x \in (0, +\infty), \quad y = 0, \qquad t > 0$$
$$T = 0; \qquad x \in (0, +\infty), \quad y = L, \qquad t > 0$$
$$\frac{\partial T}{\partial x} = 0; \qquad x = 0, \qquad y \in (0, L), \quad t > 0$$

The IC is given by

$$T = e^{-x}; \qquad x \in (0, +\infty), \quad y \in (0, L), \quad t = 0$$

7.23 Solve the transient heat conduction problem (7.9.1)–(7.9.2) for the case where the BC (7.9.2c) is replaced by the more general condition

$$T = \phi(t); \qquad x \in (0, +\infty), \quad y = L, \quad t > 0$$

where $\phi(t)$ is an arbitrary function of time.

7.24 Determine the temperature distribution in a semi-infinite cylinder with large radius, where heat is generated at rate $g(r) = a/(r^2 + a^2)^{3/2}$, $a > 0$. The end at $z = 0$ is maintained at $T = 0$.

7.25 Solve the three-dimensional heat conduction equation in the infinite domain

$$\frac{1}{\kappa} \frac{\partial T}{\partial t} = \frac{\partial^2 T}{\partial x^2} + \frac{\partial^2 T}{\partial y^2} + \frac{\partial^2 T}{\partial z^2}; \qquad x \in (-\infty, \infty), \quad y \in (-\infty, \infty), \quad z \in (-\infty, \infty), \quad t > 0$$

with the IC

$$T = f(x, y, z); \qquad t = 0$$

7.26 Repeat the problem above for the case of heat generation within the medium, at rate $g(x, y, z, t)$.

Chapter 8

LAPLACE TRANSFORM

The Laplace transform provides a useful technique for solving *linear* differential equations. The basic idea is to first rewrite the equation in the transformed space, where the original unknown is replaced by its Laplace transform. Next, this equation is solved leading to the Laplace transform, and finally the solution is obtained through the inverse transform operation. This latter step is usually the most difficult and sometimes requires numerical techniques. The main advantage of the method of Laplace transform, as well as of any other integral transform such as the finite Fourier transform (see Section 3.19 and Chapter 7), is that the dimension of the differential equation decreases by one in the transformed space. Thus, ordinary differential equations (ODEs) reduce to algebraic equations, two-dimensional partial differential equations (PDEs) reduce to ODEs, and, in general, m-dimensional PDEs reduce to $(m - 1)$-dimensional PDEs.

In this chapter we first define the Laplace transform and introduce some of its most useful properties. We then proceed to discuss the application of these transforms to solve ODEs and PDEs.

8.1 DEFINITION AND EXISTENCE

In general, the integral transform $\mathcal{I}\{f(t)\}$ of a function $f(t)$ is defined by

$$\mathcal{I}\{f(t)\} = \int_a^b K(s, t)\, f(t)\, dt = F(s) \qquad (8.1.1)$$

where $K(s, t)$ is called the *kernel* of the transformation and we have assumed that the integral exists. The relation (8.1.1) establishes a correspondence between the function

$f(t)$ and its transform $F(s)$; that is, it transforms a function of the independent variable t into a function of the independent variable s (cf. Bellman and Roth, 1984).

Similarly, we can regard $f(t)$ as the *inverse transform* of $F(s)$; that is,

$$f(t) = \mathcal{S}^{-1}\{F(s)\} \tag{8.1.2}$$

In the case where $f(t)$ is defined for $t \geq 0$, by selecting $a = 0$, $b = \infty$, and the kernel $K(s, t) = e^{-st}$, the integral transform (8.1.1) becomes

$$\mathcal{L}\{f(t)\} = \int_0^\infty e^{-st} f(t) \, dt = F(s) \tag{8.1.3}$$

which is defined as the *Laplace transform*. Note that as a general convention we indicate the original function [e.g., $f(t)$] by using lowercase letters and indicate its transform by using the corresponding capital letters [e.g., $F(s)$].

The definition of Laplace transform (8.1.3) involves an improper integral. This is given by the limit of the corresponding proper (i.e., bounded) integral as follows:

$$\int_0^\infty e^{-st} f(t) \, dt = \lim_{T \to \infty} \int_0^T e^{-st} f(t) \, dt \tag{8.1.4}$$

This implies the existence of the proper integral for all $T > 0$, as well as the existence of its limit as $T \to \infty$.

For example, in the case where $f(t) = e^{at}$, the rhs of eq. (8.1.4) reduces to

$$\lim_{T \to \infty} \int_0^T e^{(a-s)t} \, dt = \frac{1}{(a - s)} \lim_{T \to \infty} [e^{(a-s)T} - 1] \tag{8.1.5}$$

for $s > a$, the limit exists and hence the improper integral *converges* (i.e., it exists) and is given by

$$\int_0^\infty e^{-st} e^{at} \, dt = \frac{1}{s - a} \tag{8.1.6}$$

On the other hand, for $s < a$, the limit (8.1.5) is unbounded and so the improper integral *diverges*; that is, it does not exist. This conclusion holds also for $s = a$, since in this case the proper integral in eq. (8.1.5) is equal to T, which diverges as $T \to \infty$.

The convergence of the improper integral in eq. (8.1.3) is the condition which limits the existence of the Laplace transform of a given function $f(t)$. For example, we have already seen that the Laplace transform of the function $f(t) = e^{at}$ exists only for $s > a$.

We now investigate *sufficient conditions* for the existence of the Laplace transform of the function $f(t)$. First, we require $f(t)$ to be *piecewise continuous* on any finite interval in $t \geq 0$, which ensures that the proper integral on the rhs of eq. (8.1.4) exists for all $T > 0$.

Let us consider an interval $a \leq t \leq b$, divided into N subintervals, $a_{i-1} \leq t \leq a_i$ with $i = 1, \ldots, N$ (where $a = a_0$, $b = a_N$ and N is arbitrarily large but finite). Then a function $f(t)$ is piecewise continuous on this interval if it is continuous within each subinterval and takes on finite values as t approaches either end of the subinterval; that is,

$$\lim_{t \to a_{i-1}^+} f(t) = \text{finite} \qquad \text{for } i = 1, \ldots, N$$

$$\lim_{t \to a_i^-} f(t) \;\; = \text{finite} \qquad \text{for } i = 1, \ldots, N \tag{8.1.7}$$

In other words, we require $f(t)$ to be continuous in $a \leq t \leq b$ with the exception of at most $(N - 1)$ points where it exhibits finite jump discontinuities; that is,

$$\lim_{t \to a_i^-} f(t) \neq \lim_{t \to a_i^+} f(t) \qquad \text{for } i = 1, \ldots, (N - 1) \tag{8.1.8}$$

Having ensured the existence of the proper integral in eq. (8.1.4), we now need to consider the existence of its limit as $T \to \infty$. For this, we have to put some limitation on the growth rate of the function $f(t)$ as t increases. This is done by the following theorem.

● **EXISTENCE THEOREM**

Consider a function $f(t)$ which is piecewise continuous on any finite interval in $t \geq 0$ and such that

$$|f(t)| \leq K e^{\alpha t} \qquad \text{for } t \geq 0 \tag{8.1.9}$$

where $K > 0$ and α are real constants. The corresponding Laplace transform $\mathscr{L}\{f(t)\}$ exists for $s > \alpha$.

Proof
The proof follows from the definition (8.1.3), using eq. (8.1.9). Thus we have

$$\mathscr{L}\{f(t)\} = \int_0^\infty e^{-st} f(t)\, dt \leq \int_0^\infty e^{-st} |f(t)|\, dt \leq K \int_0^\infty e^{-st} e^{\alpha t}\, dt \tag{8.1.10}$$

The last integral, as seen above, converges for $s > \alpha$, so that from eq. (8.1.6) we obtain

$$\mathscr{L}\{f(t)\} \leq \frac{K}{s - \alpha} \qquad \text{for } s > \alpha \tag{8.1.11}$$

This indicates that for $s > \alpha$, the improper integral in the definition of the Laplace transform is upper bounded and hence exists. This completes the proof.

Note that the existence theorem provides only a sufficient condition. Equation (8.1.10) in fact does not allow us to conclude that the Laplace transform of $f(t)$ does not exist for $s < \alpha$.

Let us further analyze the nature of correspondence between the function $f(t)$ and its Laplace transform, $F(s)$. From eq. (8.1.3) it is readily seen that, if the Laplace transform of $f(t)$ exists, it is unique. On the other hand, it can be shown that two different *continuous* functions, $f_1(t)$ and $f_2(t)$ cannot have the same transform, $F(s)$. We do not report the proof of this statement here; it is given in various places including Churchill (1972), section 69. However, in principle, two *piecewise continuous* functions may have the same Laplace transform. For this, the two functions may differ at most at those isolated points where they exhibit jump discontinuities. Consider, for example, the two functions $f_1(t)$ and $f_2(t)$ shown in Figure 8.1. They are identical for all $t \geq 0$, with the exception of the points corresponding to jump discontinuities; that is, $f_1(a_1) \neq f_2(a_1)$ and $f_1(a_2) \neq f_2(a_2)$. Since these differences do not affect the integral of $f(t)$ over any finite interval in $t \geq 0$, the Laplace transforms of $f_1(t)$ and $f_2(t)$ are identical.

It should be noted that since these are rather minor exceptions, we can assume in applications that the inverse of a given Laplace transform—that is, $\mathcal{L}^{-1}\{F(s)\}$—is unique.

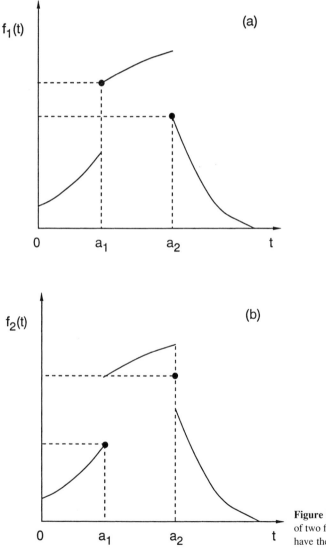

Figure 8.1 Schematic representation of two functions $f_1(t)$ and $f_2(t)$ which have the same Laplace transform.

8.2 ELEMENTARY PROPERTIES

We introduce in the following some properties of Laplace transforms which are useful in applications. It is the collection of these properties which makes the Laplace transform a useful tool for solving problems involving linear ODEs and PDEs.

8.2.1 Linearity

The Laplace transform is a linear operator. This means that the Laplace transform of a linear combination of functions is equal to the linear combination of the corresponding Laplace transforms:

$$\mathcal{L}\left\{\sum_{i=1}^{n} c_i f_i(t)\right\} = \sum_{i=1}^{n} c_i \mathcal{L}\{f_i(t)\} = \sum_{i=1}^{n} c_i F_i(s) \qquad (8.2.1)$$

The proof follows directly from the definition (8.1.3):

$$\mathcal{L}\left\{\sum_{i=1}^{n} c_i f_i(t)\right\} = \int_0^\infty e^{-st}\left[\sum_{i=1}^{n} c_i f_i(t)\right] dt$$

$$= \sum_{i=1}^{n} c_i \int_0^\infty e^{-st} f_i(t)\, dt = \sum_{i=1}^{n} c_i \mathcal{L}\{f_i(t)\} \qquad (8.2.2)$$

8.2.2 Shifting of the Independent Variable

As seen above, the Laplace transform of a function of the independent variable t [i.e., $f(t)$] is a function of the independent variable s [i.e., $F(s)$]. We now analyze the effect of a shift of $f(t)$ along the t axis on the Laplace transform $F(s)$ and, conversely, the effect of a shift of $F(s)$ along the s axis on the original function $f(t)$.

Consider a function $f(t)$ defined for $t \geq 0$, and the corresponding function $\bar{f}(t)$ obtained by shifting $f(t)$ along the t axis by quantity a, as shown in Figure 8.2. Note that since $\bar{f}(t)$ is not defined for $0 \leq t < a$, we arbitrarily set $\bar{f}(t) = 0$ in this interval. Accordingly, we can represent $\bar{f}(t)$ as follows:

$$\bar{f}(t) = u(t - a)f(t - a) \qquad (8.2.3)$$

where $u(t - a)$ is the unit step function defined by

$$u(t - a) = \begin{cases} 0 & \text{for } t < a \\ 1 & \text{for } t > a \end{cases} \qquad (8.2.4)$$

Thus we can evaluate the Laplace transform of $f(t)$, using eqs. (8.2.3) and (8.2.4) to give

$$\mathcal{L}\{\bar{f}(t)\} = \int_0^\infty e^{-st} u(t - a) f(t - a)\, dt = \int_a^\infty e^{-st} f(t - a)\, dt \qquad (8.2.5)$$

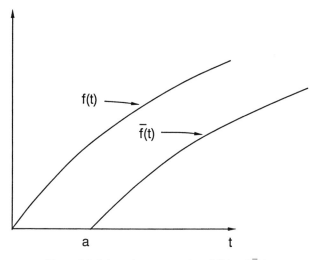

Figure 8.2 Schematic representation of $f(t)$ and $\bar{f}(t)$.

Substituting $\tau = t - a$ leads to

$$\mathscr{L}\{\bar{f}(t)\} = \int_0^\infty e^{-s(\tau+a)} f(\tau)\ d\tau = e^{-sa}\mathscr{L}\{f(t)\} \tag{8.2.6}$$

This indicates that when shifting the function $f(t)$ along the t axis by a, the corresponding Laplace transform is multiplied by e^{-sa}; that is,

$$\mathscr{L}\{\bar{f}(t)\} = e^{-sa}F(s) \tag{8.2.7}$$

Conversely, let us now consider a shift of the transform $F(s)$ along the s axis by quantity a. Thus from the definition

$$F(s) = \int_0^\infty e^{-st} f(t)\ dt \tag{8.2.8}$$

we have

$$F(s-a) = \int_0^\infty e^{-st}[e^{at}f(t)]\ dt = \mathscr{L}\{e^{at}f(t)\} \tag{8.2.9}$$

This indicates that shifting the Laplace transform $F(s)$ along the s axis by a corresponds to multiplying the original function $f(t)$ by e^{at}—that is,

$$F(s-a) = \mathscr{L}\{e^{at}f(t)\} \tag{8.2.10}$$

Note that if the Laplace transform $F(s)$, defined by eq. (8.2.8), exists for $s > \alpha$, from the existence theorem discussed in section 8.1, it readily follows that the shifted Laplace transform (8.2.10) exists for $s > (\alpha + a)$.

8.2.3 Transforms of Simple Functions

The Laplace transforms of some simple functions are reported in Table 8.1 (for a more extensive list, see Erdélyi et al. (1954). The transforms listed in Table 8.1 are obtained directly from the definition (8.1.3), by evaluating the improper integral involved as discussed in section 8.1. For illustrative purposes, we next report the details of the derivation of a few of these transforms.

TABLE 8.1 Laplace Transforms of Elementary Functions

Function, $f(t)$	Laplace Transform, $F(s)$	
1	$\dfrac{1}{s}$	$s > 0$
e^{at}	$\dfrac{1}{s-a}$	$s > a$
$t^n (n = 0, 1, 2, \ldots)$	$\dfrac{n!}{s^{n+1}}$	$s > 0$
$t^p (p > -1)$	$\dfrac{\Gamma(p+1)}{s^{p+1}}$	$s > 0$
$t^n e^{at} (n = 0, 1, 2, \ldots)$	$\dfrac{n!}{(s-a)^{n+1}}$	$s > a$
$\sin at$	$\dfrac{a}{s^2 + a^2}$	$s > 0$
$\cos at$	$\dfrac{s}{s^2 + a^2}$	$s > 0$
$\sinh at$	$\dfrac{a}{s^2 - a^2}$	$s > |a|$
$\cosh at$	$\dfrac{s}{s^2 - a^2}$	$s > |a|$
$e^{at} \sin bt$	$\dfrac{b}{(s-a)^2 + b^2}$	$s > a$
$e^{at} \cos bt$	$\dfrac{(s-a)}{(s-a)^2 + b^2}$	$s > a$
$u(t-a)$	$\dfrac{e^{-as}}{s}$	$s > 0$
$\delta(t-a)$	e^{-as}	

Example 1. Consider $f(t) = t^2$; then from eq. (8.1.3)

$$F(s) = \mathcal{L}\{t^2\} = \int_0^\infty e^{-st} t^2 \, dt \tag{8.2.11}$$

which, after integrating twice by parts, leads to

$$F(s) = -\frac{e^{-st}}{s} t^2 \Big|_0^\infty + 2 \int_0^\infty \frac{e^{-st}}{s} t \, dt = -\frac{e^{-st}}{s} t \Big|_0^\infty + \frac{2}{s^2} \int_0^\infty e^{-st} \, dt$$
$$= -\frac{e^{-st}}{s} \Big|_0^\infty \frac{2}{s^2} = \frac{2}{s^3} \tag{8.2.12}$$

Note that in the derivation above we have assumed that

$$\lim_{t \to \infty} e^{-st} t^n = 0 \qquad \text{for } n = 0, 1, 2 \tag{8.2.13}$$

which is true only for $s > 0$ or $\mathrm{Re}(s) > 0$ if s is complex. This represents the condition for the existence of $\mathcal{L}\{t^2\}$, as reported also in Table 8.1.

Example 2. The Laplace transform of $f(t) = \sin at$ can be obtained using the identity, $\sin at = (1/2i)(e^{iat} - e^{-iat})$, as follows:

$$\mathcal{L}\{\sin at\} = \int_0^\infty e^{-st} \sin at \, dt = \int_0^\infty \frac{1}{2i} [e^{-(s-ia)t} - e^{-(s+ia)t}] \, dt$$
$$= \frac{1}{2i} \left[-\frac{e^{-(s-ia)t}}{(s-ia)} + \frac{e^{-(s+ia)t}}{(s+ia)} \right]_0^\infty = \frac{a}{(s^2 + a^2)} \tag{8.2.14}$$

Note that, similar to the previous case, this transform exists only for $s > 0$ or $\mathrm{Re}(s) > 0$ if s is complex.

Example 3. For the unit step function $u(t - a)$ defined by eq. (8.2.4),

$$\mathcal{L}\{u(t - a)\} = \int_0^\infty e^{-st} u(t - a) \, dt = \int_a^\infty e^{-st} \, dt = \frac{e^{-as}}{s} \tag{8.2.15}$$

which exists for $s > 0$.

Example 4. Some applications of Laplace transform involve discontinuous functions, such as the unit step function or a combination of step functions. As an example, consider a rectangular pulse of intensity K, duration ε and centered at $t = c$, as shown in Figure 8.3. This can be represented as the combination of two unit step functions (8.2.4) as follows:

$$f(t) = K[u(t - a) - u(t - b)] \tag{8.2.16}$$

where $a = c - \varepsilon/2$, $b = c + \varepsilon/2$ and $K = 1/\varepsilon$ so that the pulse is normalized—that is, $\int_0^\infty f(t) \, dt = 1$.

Recalling that the Laplace transform is a linear operator, using eq. (8.2.15) we obtain

$$\mathcal{L}\{f(t)\} = \frac{K}{s} (e^{-as} - e^{-bs}) = \frac{e^{-cs}}{\varepsilon s} (e^{\varepsilon s/2} - e^{-\varepsilon s/2}) \tag{8.2.17}$$

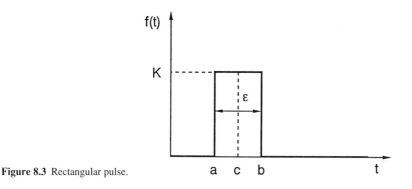

Figure 8.3 Rectangular pulse.

As $\varepsilon \to 0$, the rectangular pulse (8.2.16) approaches the Dirac delta function $\delta(t - c)$—that is, the unit impulse function centered at $t = c$:

$$\lim_{\varepsilon \to 0} f(t) = \delta(t - c) \tag{8.2.18}$$

Thus by letting $\varepsilon \to 0$ in eq. (8.2.17) we obtain the Laplace transform of the Dirac delta function:

$$\mathcal{L}\{\delta(t - c)\} = \frac{e^{-cs}}{s} \lim_{\varepsilon \to 0}\left(\frac{e^{\varepsilon s/2} - e^{-\varepsilon s/2}}{\varepsilon}\right) \tag{8.2.19}$$

The undetermined form of the above limit can be resolved using L'Hôpital's rule, to yield

$$\mathcal{L}\{\delta(t - c)\} = e^{-cs} \tag{8.2.20}$$

This indicates that the Laplace transform of the Dirac delta function centered at the origin of t—that is, $c = 0$—equals unity.

8.2.4 Transforms of Periodic Functions

Let us consider a periodic function $f(t)$, with period $p > 0$, defined for $t \geq 0$:

$$f(t + jp) = f(t) \quad \text{for } t \geq 0 \quad \text{and} \quad j = 1, 2, \ldots \tag{8.2.21}$$

Its Laplace transform is given by

$$\mathcal{L}\{f(t)\} = \frac{1}{1 - e^{-sp}} \int_0^p e^{-st} f(t) \, dt \tag{8.2.22}$$

The proof again follows from the definition (8.1.3), by noting that the interval $t \geq 0$ can be divided into an infinite number of subintervals, $t_{j-1} \leq t \leq t_j$ where $t_j = t_{j-1} + p$ and $j = 1, 2, \ldots$. By taking $t_0 = 0$, we have $t_j = jp$. Thus we can rewrite the expression of the Laplace transform as follows:

$$\mathcal{L}\{f(t)\} = \int_0^\infty e^{-st} f(t) \, dt = \sum_{j=1}^\infty \int_{t_{j-1}}^{t_j} e^{-st} f(t) \, dt$$

$$= \sum_{j=1}^\infty \int_{(j-1)p}^{jp} e^{-st} f(t) \, dt \tag{8.2.23}$$

Introducing in each integral the new independent variable $\tau = t - t_{j-1} = t - (j - 1)p$, using eq. (8.2.21) we obtain

$$
\begin{aligned}
\mathcal{L}\{f(t)\} &= \sum_{j=1}^{\infty} \int_0^p e^{-s[\tau+(j-1)p]} f(\tau + (j-1)p)\, d\tau \\
&= \left[\sum_{j=1}^{\infty} e^{-sp(j-1)}\right] \int_0^p e^{-s\tau} f(\tau)\, d\tau
\end{aligned}
\tag{8.2.24}
$$

Noting that the expression in the square brackets is a geometric series with ratio e^{-sp}, its sum is equal to $1/(1 - e^{-sp})$, so that eq. (8.2.24) reduces to eq. (8.2.22).

8.2.5 Transforms of Derivatives and Integrals

The fundamental property which makes Laplace transform useful for solving differential equations is the relationship

$$
\mathcal{L}\{f'(t)\} = s\mathcal{L}\{f(t)\} - f(0)
\tag{8.2.25}
$$

where $f(t)$ is continuous and $f'(t)$ is piecewise continuous in any finite interval of $t \geq 0$.

This property allows us to evaluate the transform of the derivative of $f(t)$, as a function of the transform of $f(t)$, without requiring the knowledge of $f(t)$, except for its value at $t = 0$, that is, $f(0)$.

In the case where $f'(t)$ is continuous, the proof of eq. (8.2.25) follows directly from the definition of Laplace transform, upon integrating by parts:

$$
\begin{aligned}
\mathcal{L}\{f'(t)\} &= \int_0^{\infty} e^{-st} f'(t)\, dt \\
&= f(t)e^{-st}\Big|_0^{\infty} + s \int_0^{\infty} e^{-st} f(t)\, dt = -f(0) + s\mathcal{L}\{f(t)\}
\end{aligned}
\tag{8.2.26}
$$

where we have assumed that the Laplace transform of $f(t)$ exists and $\lim_{t \to \infty} f(t)\, e^{-st} = 0$.

In the case where $f'(t)$ is only piecewise continuous, and exhibits jump discontinuities at $t = t_j$ with $j = 1, 2, \ldots, m$, the Laplace transform is given by

$$
\mathcal{L}\{f'(t)\} = \int_0^{\infty} e^{-st} f'(t)\, dt = \sum_{j=1}^{m+1} \int_{t_{j-1}}^{t_j} e^{-st} f'(t)\, dt
\tag{8.2.27}
$$

where we have taken $t_0 = 0$ and $t_{m+1} = \infty$.

Integrating by parts yields

$$
\begin{aligned}
\mathscr{L}\{f'(t)\} &= \sum_{j=1}^{m+1} \left[e^{-st} f(t) \Big|_{t_{j-1}}^{t_j} + s \int_{t_{j-1}}^{t_j} e^{-st} f(t) \, dt \right] \\
&= \sum_{j=1}^{m+1} e^{-st} f(t) \Big|_{t_{j-1}}^{t_j} + s \sum_{j=1}^{m+1} \int_{t_{j-1}}^{t_j} e^{-st} f(t) \, dt
\end{aligned}
\tag{8.2.28}
$$

which—recalling that $f(t)$ is continuous, so that alternate terms of the first summation cancel out—reduces to

$$
\mathscr{L}\{f'(t)\} = e^{-st} f(t) \Big|_0^\infty + s \int_0^\infty e^{-st} f(t) \, dt
\tag{8.2.29}
$$

Assuming that the Laplace transform of $f(t)$ exists and that $\lim_{t\to\infty} e^{-st} f(t) = 0$, eq. (8.2.29) leads to eq. (8.2.26).

It may be noted that if we consider a function $f(t)$ which satisfies the sufficient condition (8.1.9), then we are assured that $\mathscr{L}\{f(t)\}$ exists for $s > \alpha$ and that $\lim_{t\to\infty} f(t)e^{-st} = 0$. Accordingly, $\mathscr{L}\{f'(t)\}$ also exists for $s > \alpha$ and is given by eq. (8.2.26).

The above result can be easily extended to derivatives of higher order. Thus for a function $f(t)$ with a continuous second-order derivative, we have

$$
\begin{aligned}
\mathscr{L}\{f''(t)\} &= \int_0^\infty e^{-st} f''(t) \, dt = f'(t) e^{-st} \Big|_0^\infty + s \int_0^\infty e^{-st} f'(t) \, dt \\
&= -f'(0) + s\mathscr{L}\{f'(t)\} \\
&= -f'(0) - sf(0) + s^2 \mathscr{L}\{f(t)\}
\end{aligned}
\tag{8.2.30}
$$

This result can be generalized to the case where $f(t), f'(t), \ldots, f^{(n-1)}(t)$ are continuous for all $t \geq 0$, and $f^{(n)}(t)$ is piecewise continuous in all finite intervals of $t \geq 0$. Then the Laplace transform of $f^{(n)}(t)$ is given by

$$
\begin{aligned}
\mathscr{L}\{f^{(n)}(t)\} = s^n \mathscr{L}\{f(t)\} &- s^{n-1} f(0) - s^{n-2} f'(0) \cdots \\
&- sf^{(n-2)}(0) - f^{(n-1)}(0)
\end{aligned}
\tag{8.2.31}
$$

where we have assumed that

$$
\lim_{t\to\infty} e^{-st} f^{(i)}(t) = 0 \qquad \text{for } i = 0, 1, \ldots, (n-1)
\tag{8.2.32}
$$

and denoted $f^{(0)}(t) = f(t)$. Note that in the case where $f(t)$ and its derivatives $f'(t)$, $f''(t) \cdots f^{(n-1)}(t)$ satisfy the sufficient condition (8.1.9), also eqs. (8.2.32) are satisfied. Then the Laplace transform of $f^{(n)}(t)$ exists for $s > \alpha$ and is given by eq. (8.2.31).

Similarly, it is possible to derive a relation between the transform of the integral of a function $f(t)$ and the transform of $f(t)$, which does not involve the knowledge of $f(t)$ itself. Let us consider a function $f(t)$ which is piecewise continuous in all finite intervals of $t \geq 0$, then

$$
\mathscr{L}\left\{ \int_0^t f(\tau) \, d\tau \right\} = \frac{1}{s} \mathscr{L}\{f(t)\}
\tag{8.2.33}
$$

In order to prove this relation, we introduce the new function

$$g(t) = \int_0^t f(\tau) \, d\tau \tag{8.2.34}$$

Since $f(t)$ is piecewise continuous, $g(t)$ is continuous. Moreover, $g'(t) = f(t)$, except for those t values where $f(t)$ exhibits a jump discontinuity, where $g'(t)$ also becomes discontinuous. Thus $g'(t)$ is piecewise continuous. We can then apply eq. (8.2.25), leading to

$$\mathcal{L}\{g'(t)\} = s\mathcal{L}\{g(t)\} - g(0) \tag{8.2.35}$$

Noting that $g(0) = 0$ from eq. (8.2.34), solving for $\mathcal{L}\{g(t)\}$ gives eq. (8.2.33).

By inspection of eqs. (8.2.25) and (8.2.33) we see that the operations of differentiation and integration of a function $f(t)$ reduce in the transformed space to algebraic operations on $F(s)$—that is, multiplication and division by s, respectively. This clearly indicates the utility of working in the transformed space when dealing with differential equations.

Another useful application of eqs. (8.2.25) and (8.2.33) is in deriving new transforms from known transforms. Let us derive, for example, the Laplace transform of cos at, knowing the transform of sin at from eq. (8.2.14). Thus by noting that

$$\frac{d \sin at}{dt} = a \cos at \tag{8.2.36}$$

we can obtain $\mathcal{L}\{\cos at\}$ using eq. (8.2.25) as follows:

$$\mathcal{L}\{\cos at\} = \frac{1}{a} \mathcal{L}\left\{\frac{d \sin at}{dt}\right\} = \frac{1}{a}[s\mathcal{L}\{\sin at\} - \sin(0)] = \frac{s}{s^2 + a^2} \tag{8.2.37}$$

which is the expression reported in Table 8.1. As a further example, let us derive $\mathcal{L}\{t^n\}$, for $n = 1, 2, \ldots$, knowing from Table 8.1 that $\mathcal{L}\{1\} = 1/s$. Since

$$\int_0^t \tau^n \, d\tau = \frac{1}{n+1} t^{n+1} \tag{8.2.38}$$

eq. (8.2.33) leads to

$$\mathcal{L}\{t^{n+1}\} = (n+1)\mathcal{L}\left\{\int_0^t \tau^n \, d\tau\right\} = \frac{n+1}{s} \mathcal{L}\{t^n\} \tag{8.2.39}$$

This equation can be used recursively to yield

$$n = 0: \quad \mathcal{L}\{t\} = \frac{1}{s} \mathcal{L}\{1\} = \frac{1}{s^2} \tag{8.2.40a}$$

$$n = 1: \quad \mathcal{L}\{t^2\} = \frac{2}{s^3} \tag{8.2.40b}$$

$$n = 2: \quad \mathcal{L}\{t^3\} = \frac{6}{s^4} \tag{8.2.40c}$$

Thus generalizing, we have

$$\mathcal{L}\{t^m\} = \frac{m!}{s^{m+1}}, \qquad m = 0, 1, 2, \ldots \tag{8.2.41}$$

in agreement with the corresponding expression reported in Table 8.1.

8.2.6 Differentiation and Integration of Transforms

Let us suppose that $\mathcal{L}\{f(t)\} = F(s)$ exists; then by differentiating both sides of eq. (8.2.8) with respect to s we obtain

$$F'(s) = \frac{d}{ds} \int_0^\infty e^{-st} f(t) \, dt$$
$$= \int_0^\infty e^{-st}[-tf(t)] \, dt = \mathcal{L}\{-tf(t)\} \tag{8.2.42}$$

Thus the differentiation of the transform $F(s)$ corresponds to the multiplication of the original function $f(t)$, by $(-t)$.

Moreover, from eq. (8.2.42) we see that

$$\frac{d}{ds} \mathcal{L}\left\{\frac{f(t)}{t}\right\} = \mathcal{L}\left\{-t\frac{f(t)}{t}\right\} = -\mathcal{L}\{f(t)\} \tag{8.2.43}$$

By integrating both sides from s to ∞, we obtain

$$\mathcal{L}\left\{\frac{f(t)}{t}\right\}\Bigg|_s^\infty = -\int_s^\infty \mathcal{L}\{f(t)\} \, ds \tag{8.2.44}$$

and then

$$\mathcal{L}\left\{\frac{f(t)}{t}\right\} = \int_s^\infty \mathcal{L}\{f(t)\} \, ds = \int_s^\infty F(s) \, ds \tag{8.2.45}$$

which indicates that integration of the transform, $F(s)$ corresponds to division of the original function, $f(t)$ by t. Equations (8.2.42) and (8.2.45) can be used to derive new transforms from known ones. For example, from the transform of $\sin at$ given by eq. (8.2.14), we can evaluate through eq. (8.2.42) the transform of $(t \sin at)$ as follows:

$$\mathcal{L}\{t \sin at\} = -\frac{d}{ds}\left[\frac{a}{(s^2 + a^2)}\right] = \frac{2as}{(s^2 + a^2)^2} \tag{8.2.46}$$

Similarly, from eq. (8.2.45) we can derive the transform of $(1/t)\sin at$:

$$\mathcal{L}\left\{\frac{1}{t}\sin at\right\} = \int_s^\infty \frac{a}{(s^2 + a^2)} \, ds = \arctan\frac{s}{a}\Bigg|_s^\infty = \frac{\pi}{2} - \arctan\frac{s}{a} \tag{8.2.47}$$

8.2.7 Initial-Value and Final-Value Theorems

The *initial-value theorem* allows one to compute the value of a function $f(t)$ as $t \to 0$ from its Laplace transform $F(s)$, without inverting it. In particular, we have that

$$\lim_{t \to 0} f(t) = \lim_{s \to \infty} [sF(s)] \qquad (8.2.48)$$

The proof involves taking the limit as $s \to \infty$ of both sides of eq. (8.2.25), so that

$$\lim_{s \to \infty} \int_0^\infty \frac{df}{dt} e^{-st} \, dt = \lim_{s \to \infty} [sF(s) - f(0)] \qquad (8.2.49)$$

Since s is independent of time, we can rewrite the lhs of eq. (8.2.49) as follows:

$$\lim_{s \to \infty} \int_0^\infty \frac{df}{dt} e^{-st} \, dt = \int_0^\infty \frac{df}{dt} (\lim_{s \to \infty} e^{-st}) \, dt = 0 \qquad (8.2.50)$$

which, substituted in eq. (8.2.49), leads to eq. (8.2.48). Similarly, the *final-value theorem*

$$\lim_{t \to \infty} f(t) = \lim_{s \to 0} [sF(s)] \qquad (8.2.51)$$

permits the evaluation of a function $f(t)$ as $t \to \infty$, from its Laplace transform $F(s)$, without inverting it. The proof originates again from eq. (8.2.25), but now considering the limit of both sides as $s \to 0$—that is,

$$\lim_{s \to 0} \int_0^\infty \frac{df}{dt} e^{-st} \, dt = \lim_{s \to 0} [sF(s) - f(0)] \qquad (8.2.52)$$

Since s is independent of t, this gives

$$\int_0^\infty \frac{df}{dt} (\lim_{s \to 0} e^{-st}) \, dt = [\lim_{s \to 0} sF(s)] - f(0) \qquad (8.2.53)$$

and hence

$$\int_0^\infty df = [\lim_{s \to 0} sF(s)] - f(0) \qquad (8.2.54)$$

which leads to eq. (8.2.51).

8.2.8 The Convolution Theorem

The convolution theorem provides the inverse of Laplace transforms which can be represented as the product of two terms. Thus if $F(s) = \mathcal{L}\{f(t)\}$ and $G(s) = \mathcal{L}\{g(t)\}$, then the inverse of the transform

$$H(s) = F(s)G(s) \qquad (8.2.55)$$

is given by

$$h(t) = \int_0^t f(t - \tau)g(\tau) \, d\tau \qquad (8.2.56)$$

This is a *convolution* of the functions $f(\tau)$ and $g(\tau)$ over the interval $\tau \in (0, t)$, and is generally represented as $h = f * g$. We may note that, by making the change of variable $\eta = t - \tau$, eq. (8.2.56) becomes

$$h(t) = \int_0^t f(\eta)g(t - \eta) \, d\eta \tag{8.2.57}$$

which shows that the convolution is symmetric; that is, $f * g = g * f$.

To prove the theorem, let us evaluate the transform $H(s)$ of the function $h(t)$ defined by eq. (8.2.56):

$$H(s) = \mathcal{L}\{h(t)\} = \int_0^\infty e^{-st}\left[\int_0^t f(t - \tau)g(\tau) \, d\tau\right] dt \tag{8.2.58}$$

Introducing the unit step function defined by eq. (8.2.4), we see that

$$f(t - \tau)g(\tau)u(t - \tau) = \begin{cases} f(t - \tau)g(\tau) & \text{for } t > \tau \\ 0 & \text{for } t < \tau \end{cases} \tag{8.2.59}$$

so eq. (8.2.58) can be rewritten as

$$H(s) = \int_0^\infty e^{-st}\left[\int_0^\infty f(t - \tau)g(\tau)u(t - \tau) \, d\tau\right] dt \tag{8.2.60}$$

By reversing the order of integration, we obtain

$$H(s) = \int_0^\infty \left[\int_0^\infty f(t - \tau)u(t - \tau)e^{-st} \, dt\right]g(\tau) \, d\tau$$
$$= \int_0^\infty \left[\int_\tau^\infty f(t - \tau)e^{-st} \, dt\right]g(\tau) \, d\tau \tag{8.2.61}$$

Finally, since in the inner integral τ is constant, we can introduce the new integration variable $\eta = t - \tau$, so that

$$H(s) = \int_0^\eta \left[\int_0^\infty f(\eta)e^{-s(\eta+\tau)} \, d\eta\right]g(\tau) \, d\tau$$
$$= \int_0^\eta \left[\int_0^\infty f(\eta)e^{-s\eta} \, d\eta\right]e^{-s\tau}g(\tau) \, d\tau \tag{8.2.62}$$
$$= \int_0^\infty F(s)e^{-s\tau}g(\tau) \, d\tau = F(s)G(s)$$

which proves the theorem.

As an example of application of the convolution theorem, consider two functions: $f(t) = t$ and $g(t) = t^2$, whose Laplace transforms are known from Table 8.1:

$$F(s) = \frac{1}{s^2} \tag{8.2.63a}$$

and

$$G(s) = \frac{2}{s^3} \qquad (8.2.63b)$$

From eq. (8.2.56) we see that the inverse of the transform

$$H(s) = F(s)G(s) = \frac{2}{s^5} \qquad (8.2.64)$$

is given by

$$h(t) = \int_0^t f(\tau)g(t - \tau) \, d\tau$$

$$= \int_0^t \tau(t - \tau)^2 \, d\tau = \frac{t^4}{12} \qquad (8.2.65)$$

This matches with the inverse of $H(s)$ computed using Table 8.1 as

$$h(t) = \mathcal{L}^{-1}\{H(s)\} = 2\mathcal{L}^{-1}\left\{\frac{1}{s^5}\right\} = 2\frac{t^4}{4!} = \frac{t^4}{12} \qquad (8.2.66)$$

8.3   THE INVERSION PROBLEM: THE HEAVISIDE EXPANSION THEOREM

As noted earlier, the problem of inverting Laplace transforms arises when solving differential equations and is often quite difficult. However, there exists a specific class of Laplace transforms, which arises frequently in applications, whose inverse can be readily determined analytically. This can be represented as the ratio between two polynomials $p(s)$ and $q(s)$ with real coefficients; that is,

$$F(s) = \frac{p(s)}{q(s)} \qquad (8.3.1)$$

where the degree of $q(s)$ is larger than that of $p(s)$, and the two polynomials do not share any common factor.

The inversion of the transform $F(s)$ is done by partial fraction expansion. Two separate cases arise.

Case 1. $q(s)$ Has No Repeated Roots
Let us assume that $q(s)$ is of degree n with roots $a_1, a_2, \ldots, a_n$; that is,

$$q(a_i) = 0, \qquad i = 1, 2, \ldots, n \qquad (8.3.2)$$

and $a_i \neq a_j$ for $i \neq j$. We can also assume, with no loss of generality, that the coefficient of the highest power of s in $q(s)$ is unity, so that the polynomial can be factored as

$$q(s) = (s - a_1)(s - a_2) \cdots (s - a_n) \qquad (8.3.3)$$

We can represent the transform $F(s)$, given by eq. (8.3.1), in terms of partial fractions

$$\frac{p(s)}{q(s)} = \frac{A_1}{s - a_1} + \frac{A_2}{s - a_2} + \cdots + \frac{A_n}{s - a_n} \tag{8.3.4}$$

where $A_1, A_2, \ldots, A_n$ are as yet unknown constants, which are determined so as to satisfy the above equation for all values of s.

To evaluate the generic constant A_j, for $j = 1, 2, \ldots, n$, multiply both sides of eq. (8.3.4) by $s - a_j$, leading to

$$\frac{p(s)(s - a_j)}{q(s)} = A_1 \frac{s - a_j}{s - a_1} + \cdots + A_j + \cdots + A_n \frac{s - a_j}{s - a_n} \tag{8.3.5}$$

Let us now take the limit $s \to a_j$ on both sides. Recalling that $q(a_j) = 0$ by definition, the lhs leads to an indeterminate form which can be resolved through L'Hôpital's rule as follows:

$$\lim_{s \to a_j} \frac{p(s)(s - a_j)}{q(s)} = \lim_{s \to a_j} \frac{p'(s)(s - a_j) + p(s)}{q'(s)} = \frac{p(a_j)}{q'(a_j)} \tag{8.3.6}$$

Note that since $q(s)$ has no repeated roots, we are assured that $q'(a_j) \neq 0$. Moreover, as $s \to a_j$, the limit of the rhs of eq. (8.3.5) is given by A_j Thus we have

$$A_j = \frac{p(a_j)}{q'(a_j)}, \qquad j = 1, \ldots, n \tag{8.3.7}$$

The partial fraction expansion of $F(s)$ is then given by

$$F(s) = \frac{p(s)}{q(s)} = \sum_{j=1}^{n} \frac{p(a_j)}{q'(a_j)} \frac{1}{s - a_j} \tag{8.3.8}$$

Recalling the linearity property of Laplace transforms (8.2.1) and the fact that $\mathcal{L}\{e^{a_j t}\} = 1/(s - a_j)$, we see that the inverse of $F(s)$ is given by

$$f(t) = \mathcal{L}^{-1}\{F(s)\} = \sum_{j=1}^{n} \frac{p(a_j)}{q'(a_j)} e^{a_j t} \tag{8.3.9}$$

Case 2. $q(s)$ Has a Root Repeated m Times
Suppose that $q(s)$ has a root $s = b$ repeated m times, together with one or more other repeated or nonrepeated roots. Thus $q(s)$ has a factor $(s - b)^m$, and the transform $F(s)$, given by eq. (8.3.1), can be expanded in partial fractions as follows:

$$\frac{p(s)}{q(s)} = \frac{B_m}{(s - b)^m} + \frac{B_{m-1}}{(s - b)^{m-1}} + \cdots + \frac{B_2}{(s - b)^2} + \frac{B_1}{s - b} + R(s) \tag{8.3.10}$$

where $R(s)$ contains all the repeated and unrepeated factors of $q(s)$ other than $s - b$.

Similarly to the previous case, we multiply both sides of eq. (8.3.10) by $(s - b)^m$, to give

$$\frac{(s - b)^m p(s)}{q(s)} = B_m + (s - b)B_{m-1} + \cdots$$
$$+ (s - b)^{m-2}B_2 + (s - b)^{m-1}B_1 + (s - b)^m R(s) \tag{8.3.11}$$

If we define the lhs of eq. (8.3.11) as $W(s)$, that is,

$$W(s) = \frac{(s - b)^m p(s)}{q(s)} \tag{8.3.12}$$

by taking the limit of both sides of eq. (8.3.11) as $s \to b$, we obtain

$$B_m = W(b) = \lim_{s \to b}\left[\frac{(s - b)^m p(s)}{q(s)}\right] \tag{8.3.13}$$

Next, we differentiate eq. (8.3.11) with respect to s:

$$\frac{dW}{ds} = B_{m-1} + 2(s - b)B_{m-2} + \cdots + (m - 2)(s - b)^{m-3}B_2$$
$$+ (m - 1)(s - b)^{m-2}B_1 + m(s - b)^{m-1}R(s) + (s - b)^m R'(s) \tag{8.3.14}$$

and take the limit of both sides as $s \to b$, leading to

$$B_{m-1} = \left(\frac{dW}{ds}\right)_{s=b} = \lim_{s \to b}\left[\frac{d}{ds}\left\{\frac{(s - b)^m p(s)}{q(s)}\right\}\right] \tag{8.3.15}$$

We can iterate this procedure, successively differentiating eq. (8.3.14) and taking the limit as $s \to b$. For the generic jth coefficient we obtain

$$B_j = \frac{1}{(m - j)!}\left(\frac{d^{m-j}W}{ds^{m-j}}\right)_{s=b}$$
$$= \frac{1}{(m - j)!}\lim_{s \to b}\left[\frac{d^{m-j}}{ds^{m-j}}\left\{\frac{(s - b)^m p(s)}{q(s)}\right\}\right], \qquad j = 1, \ldots, m \tag{8.3.16}$$

Since the coefficients of the partial fraction expansion are determined, we can now turn to the problem of inverting the transform (8.3.10), which can be represented in the equivalent form

$$F(s) = \sum_{j=1}^{m}\frac{B_j}{(s - b)^j} + R(s) \tag{8.3.17}$$

From Table 8.1,

$$\mathcal{L}^{-1}\left[\frac{1}{(s - b)^n}\right] = \frac{1}{(n - 1)!}t^{n-1}e^{bt}, \qquad n = 1, 2, \ldots \tag{8.3.18}$$

and further denote

$$\mathcal{L}^{-1}[R(s)] = r(t) \tag{8.3.19}$$

Then, using the linearity property (8.2.1), the inverse Laplace transform is

$$f(t) = \sum_{j=1}^{m}\frac{B_j}{(j - 1)!}t^{j-1}e^{bt} + r(t) \tag{8.3.20}$$

This equation, along with eq. (8.3.9), is known as the *Heaviside expansion theorem*. It allows us to invert transforms of the type (8.3.1) in all cases where $q(s)$ has any number of repeated or unrepeated roots. In particular, if the term $R(s)$ in eq. (8.3.10) does not contain any repeated root, then we can treat it as in Case 1 above. On the other hand, if $R(s)$ contains one or more repeated roots, then we treat it as in Case 2 one or more times, considering each repeated root separately. Eventually, we are left with a residual which does not contain any repeated root. The final result of this procedure is reported next, for the most general case.

Let us suppose that $q(s)$ has n unrepeated roots $(a_1, a_2, \ldots, a_n)$ and k repeated roots $(b_1$ repeated m_1 times, b_2 repeated m_2 times, $\ldots$, b_k repeated m_k times). The partial fraction expansion of $F(s)$ is then given by

$$\frac{p(s)}{q(s)} = \sum_{j=1}^{n} \frac{A_j}{(s - a_j)} + \sum_{j=1}^{m_1} \frac{B_{j1}}{(s - b_1)^j} + \sum_{j=1}^{m_2} \frac{B_{j2}}{(s - b_2)^j}$$

$$+ \cdots + \sum_{j=1}^{m_k} \frac{B_{jk}}{(s - b_k)^j} \tag{8.3.21}$$

where the coefficients A_j, for $j = 1, 2, \ldots, n$, are given by eq. (8.3.7), while the coefficients B_{ji}, for $j = 1, \ldots, m_i$ and $i = 1, \ldots, k$, are given by eq. (8.3.16). Correspondingly, the inverse transform of $F(s)$ is

$$f(t) = \mathcal{L}^{-1}\{F(s)\} = \sum_{j=1}^{n} A_j e^{a_j t} + \sum_{j=1}^{m_1} \frac{B_{j1}}{(j - 1)!} t^{j-1} e^{b_1 t}$$

$$+ \sum_{j=1}^{m_2} \frac{B_{j2}}{(j - 1)!} t^{j-1} e^{b_2 t} + \cdots + \sum_{j=1}^{m_k} \frac{B_{jk}}{(j - 1)!} t^{j-1} e^{b_k t} \tag{8.3.22}$$

Example 1. Consider the Laplace transform

$$F(s) = \frac{1}{s^2(s - 1)^2(s + 2)} \tag{8.3.23}$$

where $q(s) = s^2(s - 1)^2(s + 2)$ has $n = 1$ unrepeated root: $a_1 = -2$, and $k = 2$ repeated roots: $b_1 = 0$ repeated $m_1 = 2$ times and $b_2 = 1$ repeated $m_2 = 2$ times. From eq. (6.3.21) the partial fraction expansion of $F(s)$ is

$$F(s) = \frac{A_1}{s - a_1} + \sum_{j=1}^{m_1} \frac{B_{j1}}{(s - b_1)^j} + \sum_{j=1}^{m_2} \frac{B_{j2}}{(s - b_2)^j}$$

$$= \frac{A_1}{(s + 2)} + \frac{B_{11}}{s} + \frac{B_{21}}{s^2} + \frac{B_{12}}{s - 1} + \frac{B_{22}}{(s - 1)^2} \tag{8.3.24}$$

Since $p(s) = 1$, from eq. (8.3.7) we have

$$A_1 = \frac{p(a_1)}{q'(a_1)} = \frac{1}{36} \tag{8.3.25}$$

For the coefficients of the repeated factors, let us first consider the first repeated root, $b_1 = 0$. From eq. (8.3.12) we obtain

$$W_1(s) = \frac{(s - b_1)^{m_1} p(s)}{q(s)} = \frac{1}{(s - 1)^2(s + 2)} \tag{8.3.26}$$

Next from eqs. (8.3.13) and (8.3.16) we have

$$B_{21} = W_1(b_1) = \frac{1}{2} \tag{8.3.27a}$$

$$B_{11} = \frac{1}{(m_1 - 1)!} W_1'(b_1) = \frac{3}{4} \tag{8.3.27b}$$

For the second repeated root, $b_2 = 1$, and from eq. (8.3.12):

$$W_2(s) = \frac{(s - b_2)^{m_2}p(s)}{q(s)} = \frac{1}{s^2(s + 2)} \tag{8.3.28}$$

then from eqs. (8.3.13) and (8.3.16):

$$B_{22} = W_2(b_2) = \frac{1}{3} \tag{8.3.29a}$$

$$B_{12} = \frac{1}{(m_2 - 1)!} W_2'(b_2) = \frac{7}{9} \tag{8.3.29b}$$

Finally, the inverse transform is obtained from eq. (8.3.22) as follows:

$$
\begin{aligned}
f(t) &= A_1 e^{a_1 t} + (B_{11} + B_{21}t)e^{b_1 t} + (B_{12} + B_{22}t)e^{b_2 t} \\
&= \frac{1}{36} e^{-2t} + \left(\frac{3}{4} + \frac{1}{2} t\right) + \left(\frac{7}{9} + \frac{1}{3} t\right) e^{t}
\end{aligned}
\tag{8.3.30}
$$

If the roots of the polynomial $q(s)$ are complex, they will occur as complex conjugate pairs. Their contributions to the inverse transform can be combined in the form of trigonometric functions. For example, let us consider the complex conjugate pair of roots

$$b = \alpha + i\beta, \qquad \bar{b} = \alpha - i\beta \tag{8.3.31}$$

repeated m times. We can represent the partial fraction expansion of $F(s)$, given by eq. (8.3.17), as follows:

$$F(s) = \sum_{j=1}^{m} \frac{B_j}{(s - b)^j} + \sum_{j=1}^{m} \frac{\bar{B}_j}{(s - \bar{b})^j} + R(s) \tag{8.3.32}$$

where the coefficients B_j and $\bar{B}_j$ according to eq. (8.3.16) are

$$B_j = \frac{1}{(m - j)!} \lim_{s \to b} \left[\frac{d^{m-j}}{ds^{m-j}} \left\{ \frac{(s - b)^m p(s)}{q(s)} \right\} \right] \tag{8.3.33a}$$

and

$$\bar{B}_j = \frac{1}{(m - j)!} \lim_{s \to \bar{b}} \left[\frac{d^{m-j}}{ds^{m-j}} \left\{ \frac{(s - \bar{b})^m p(s)}{q(s)} \right\} \right] \tag{8.3.33b}$$

The inverse transform is then given by eq. (8.3.20):

$$
\begin{aligned}
f(t) &= \mathcal{L}^{-1}\{F(s)\} = \sum_{j=1}^{m} \frac{B_j}{(j - 1)!} t^{j-1} e^{bt} + \sum_{j=1}^{m} \frac{\bar{B}_j}{(j - 1)!} t^{j-1} e^{\bar{b}t} + r(t) \\
&= e^{\alpha t} \left[\sum_{j=1}^{m} \frac{t^{j-1}}{(j - 1)!} (B_j e^{i\beta t} + \bar{B}_j e^{-i\beta t}) \right] + r(t)
\end{aligned}
\tag{8.3.34}
$$

which, by using Euler relations, becomes

$$f(t) = e^{\alpha t} \sum_{j=1}^{m} \frac{t^{j-1}}{(j - 1)!} [(B_j - \bar{B}_j) \cos \beta t + i(B_j - \bar{B}_j) \sin \beta t] \tag{8.3.35}$$

Since B_j and $\bar{B}_j$ are also complex conjugates, we denote them by

$$B_j = \gamma_j + i\delta_j, \qquad \bar{B}_j = \gamma_j - i\delta_j \tag{8.3.36}$$

Substituting eq. (8.3.36) in eq. (8.3.35) gives

$$f(t) = 2e^{\alpha t} \sum_{j=1}^{m} \frac{t^{j-1}}{(j-1)!} (\gamma_j \cos \beta t - \delta_j \sin \beta t) \tag{8.3.37}$$

Example 2. Consider the inversion of the transform

$$F(s) = \frac{p(s)}{q(s)} = \frac{s+1}{s(s^2 - 2s + 5)(s^2 - 4s + 5)^2} \tag{8.3.38}$$

Note that $q(s)$ has three unrepeated roots: one real, $a_1 = 0$, and one pair of complex conjugates, $a_2 = (1 + 2i)$ and $\bar{a}_2 = (1 - 2i)$, and another pair of complex conjugate roots, $b = (2 + i)$ and $\bar{b} = (2 - i)$, which are repeated twice.

From eqs. (8.3.22) and (8.3.34), the inverse transform is given by

$$f(t) = A_1 e^{a_1 t} + A_2 e^{a_2 t} + \bar{A}_2 e^{\bar{a}_2 t} + (B_1 + B_2 t)e^{bt} + (\bar{B}_1 + \bar{B}_2 t)e^{\bar{b}t} \tag{8.3.39}$$

The coefficients of the terms corresponding to the unrepeated roots $a_1 = 0$, $a_2 = (1 + 2i)$, and $\bar{a}_2 = (1 - 2i)$ are given by eq. (8.3.7):

$$A_1 = \frac{p(a_1)}{q'(a_1)} = \frac{1}{125} \tag{8.3.40a}$$

$$A_2 = \frac{p(a_2)}{q'(a_2)} = \frac{1+i}{8(2-11i)} = -\frac{9}{1000} + \frac{13}{1000} i \tag{8.3.40b}$$

$$\bar{A}_2 = \frac{p(\bar{a}_2)}{q'(\bar{a}_2)} = \frac{1-i}{8(2+11i)} = -\frac{9}{1000} - \frac{13}{1000} i \tag{8.3.40c}$$

For the coefficients of the terms corresponding to the repeated complex roots $b = (2 + i)$ and $\bar{b} = (2 - i)$, we use eqs. (8.3.33) with $m = 2$:

$$B_1 = \lim_{s \to b} \left[\frac{d}{ds} \left\{ \frac{(s-b)^2 p(s)}{q(s)} \right\} \right] = \frac{17 + 6i}{8(-7 + 24i)} = \frac{1 - 18i}{200} \tag{8.3.41a}$$

$$B_2 = \lim_{s \to b} \left[\frac{(s-b)^2 p(s)}{q(s)} \right] = -\frac{3+i}{8(3+4i)} = \frac{-13 + 9i}{200} \tag{8.3.41b}$$

$$\bar{B}_1 = \lim_{s \to \bar{b}} \left[\frac{d}{ds} \left\{ \frac{(s-\bar{b})^2 p(s)}{q(s)} \right\} \right] = \frac{17 - 6i}{-8(7 + 24i)} = \frac{1 + 18i}{200} \tag{8.3.41c}$$

$$\bar{B}_2 = \lim_{s \to \bar{b}} \left[\frac{(s-\bar{b})^2 p(s)}{q(s)} \right] = \frac{3-i}{8(3-4i)} = -\frac{13 + 9i}{200} \tag{8.3.41d}$$

Alternatively, we can represent the inverse transform (8.3.39) in real form using eqs. (8.3.37):

$$f(t) = \frac{1}{125} + 2e^t(\gamma \cos 2t - \delta \sin 2t) + 2e^{2t}[(\gamma_1 \cos t$$
$$- \delta_1 \sin t) + t(\gamma_2 \cos t - \delta_2 \sin t)] \tag{8.3.42}$$

where the coefficients γ and δ are derived from eqs. (8.3.40) and (8.3.41) as follows:

$$\gamma = -\frac{9}{1000}, \qquad \delta = \frac{13}{1000} \qquad\qquad (8.3.43a)$$

$$\gamma_1 = \frac{1}{200}, \qquad \delta_1 = -\frac{18}{1000} \qquad\qquad (8.3.43b)$$

$$\gamma_2 = -\frac{13}{200}, \qquad \delta_2 = \frac{9}{200} \qquad\qquad (8.3.43c)$$

8.4   ORDINARY DIFFERENTIAL EQUATIONS

Laplace transforms are frequently applied to solve ODEs. Other techniques, which are described in chapter 3, can also be used and of course provide the same results. However, Laplace transforms present some advantages which we discuss here. We begin with the simple example of a homogeneous ODE, to illustrate the basic principles of the method.

8.4.1   Homogeneous ODEs

Let us consider the following second-order ODE with constant coefficients:

$$\frac{d^2y}{dt^2} - 3\frac{dy}{dt} + 2y = 0 \qquad\qquad (8.4.1)$$

with ICs

$$y = a \qquad \text{at } t = 0 \qquad\qquad (8.4.2a)$$

$$y' = b \qquad \text{at } t = 0 \qquad\qquad (8.4.2b)$$

To find the solution $y(t)$ of the above equation using Laplace transforms, we follow a three-step procedure.

First, apply the Laplace transform operator to both sides of eq. (8.4.1); that is, multiply by e^{-st} and integrate from $t = 0$ to $t = \infty$. Accordingly, using the linearity condition (8.2.1), eq. (8.4.1) reduces to

$$\mathscr{L}\left\{\frac{d^2y}{dt^2}\right\} - 3\mathscr{L}\left\{\frac{dy}{dt}\right\} + 2\mathscr{L}\{y\} = 0 \qquad\qquad (8.4.3)$$

Using eq. (8.2.31) to evaluate the transforms of the derivatives, together with the ICs (8.4.2), we obtain

$$s^2Y(s) - sa - b - 3[sY(s) - a] + 2Y(s) = 0 \qquad\qquad (8.4.4)$$

where $Y(s)$ represents the Laplace transform of $y(t)$—that is, $Y(s) = \mathscr{L}\{y\}$.

The equation above, sometimes referred to as the *subsidiary equation*, represents the original ODE in the transformed space, and its solution provides the Laplace transform of $y(t)$, that is, $Y(s)$.

It may be noted that the subsidiary equation is *algebraic*, while the original equation is *differential*. This is a typical feature of all methods based on integral transforms, which reduce the dimensionality of the original equation by one. Moreover, the subsidiary equation (8.4.4) already includes the ICs (8.4.2), which otherwise could not be carried separately since this is not a differential equation. Thus, in the transformed space, the subsidiary equation represents not only the original ODE (8.4.1), but rather the entire initial value problem as given by eqs. (8.4.1) and (8.4.2).

Next, solve the subsidiary equation to obtain the transform, $Y(s)$:

$$Y(s) = \frac{sa + b - 3a}{s^2 - 3s + 2} \tag{8.4.5}$$

The last step of the procedure involves inversion of the transform $Y(s)$ to recover the solution in the t domain; that is,

$$y(t) = \mathcal{L}^{-1}\{Y(s)\} \tag{8.4.6}$$

This is typically the most difficult step of the procedure; in some cases, it cannot be performed analytically and requires numerical techniques. In this case, however, since the transform (8.4.5) is of the type defined by eq. (8.3.1), we can readily invert it using the Heaviside expansion theorem. Since the denominator of $Y(s)$,

$$q(s) = s^2 - 3s + 2 = (s - 1)(s + 2) \tag{8.4.7}$$

has no repeated roots, the inverse transform is given by eq. (8.3.9), which in this specific case becomes

$$y(t) = (2a - b)e^t + (b - a)e^{2t} \tag{8.4.8}$$

Indeed the above solution could be obtained with other techniques as well. In particular, if we proceed with the standard method of seeking solutions of the form e^{rt}, then the characteristic equation of eq. (8.4.1) is given by

$$Q(r) = r^2 - 3r + 2 = 0 \tag{8.4.9}$$

which is the same polynomial $q(s)$ in the denominator of $Y(s)$ in eq. (8.4.5). Thus we note that the problem of finding the roots of the characteristic equation is encountered also in the method based on Laplace transforms. The two methods of course lead to the same solution.

It is, however, worth noting two advantages of the Laplace transform method. First, the solution $y(t)$ is obtained by solving an algebraic rather than a differential equation. This is a simpler operation, although we should not underestimate the difficulties involved in the inversion problem. Second, the solution of the initial value problem, which satisfies both the ODE and the ICs, is obtained directly. The procedure does not involve evaluation of arbitrary integration constants, as required by the standard method.

8.4.2 Nonhomogeneous ODEs

To illustrate the application of Laplace transforms to the solution of nonhomogeneous ODEs with constant coefficients, let us consider the same ODE as in section 8.4.1 but with a nonhomogeneous term. Thus we have

$$\frac{d^2y}{dt^2} - 3\frac{dy}{dt} + 2y = f(t) \tag{8.4.10}$$

with ICs

$$y = a \quad\quad \text{at } t = 0 \tag{8.4.11a}$$

$$y' = b \quad\quad \text{at } t = 0 \tag{8.4.11b}$$

By applying the Laplace transform operator to both sides of eq. (8.4.10) and using eq. (8.2.31) and the ICs (8.4.11), we obtain

$$[s^2Y(s) - sa - b] - 3[sY(s) - a] + 2Y(s) = F(s) \tag{8.4.12}$$

Solving for the transform $Y(s) = \mathcal{L}\{y(t)\}$, we get

$$Y(s) = \frac{sa + b - 3a}{s^2 - 3s + 2} + \frac{F(s)}{s^2 - 3s + 2} \tag{8.4.13}$$

where the first term corresponds to the solution of the homogeneous ODE associated with eq. (8.4.10) and coincides with eq. (8.4.5). We have already determined the inverse of this transform in eq. (8.4.8); that is,

$$y_h(t) = \mathcal{L}^{-1}\left\{\frac{sa + b - 3a}{s^2 - 3s + 2}\right\} = (2a - b)e^t + (b - a)e^{2t} \tag{8.4.14}$$

The second term in eq. (8.4.13) represents the contribution of the nonhomogeneous term. In order to invert this term, we regard it as the product of two transforms:

$$\frac{1}{s^2 - 3s + 2} F(s) = G(s)F(s) \tag{8.4.15}$$

and apply the convolution theorem. For this, we first need to find the inverse transform of $G(s)$:

$$g(t) = \mathcal{L}^{-1}\{G(s)\} = \mathcal{L}^{-1}\left\{\frac{1}{s^2 - 3s + 2}\right\} \tag{8.4.16}$$

Note that the denominator of $G(s)$ is the same as that in the corresponding homogeneous problem. Its roots having been determined earlier, use of the Heaviside expansion theorem (8.3.9) leads to

$$g(t) = \frac{p(a_1)}{q'(a_1)} e^{a_1 t} + \frac{p(a_2)}{q'(a_2)} e^{a_2 t} = e^{2t} - e^t \tag{8.4.17}$$

Now, using the convolution theorem (8.2.56), we have

$$\mathcal{L}^{-1}\{G(s)F(s)\} = \int_0^t f(t - \tau)g(\tau)\, d\tau$$
$$= \int_0^t f(t - \tau)[e^{2\tau} - e^\tau]\, d\tau \tag{8.4.18}$$

By combining this with eq. (8.4.14), we obtain the full solution of the nonhomogeneous problem (8.4.10) and (8.4.11):

$$y(t) = y_h(t) + \int_0^t f(t - \tau)g(\tau) \, d\tau$$

$$= (2a - b)e^t + (b - a)e^{2t} + \int_0^t f(t - \tau)[e^{2\tau} - e^\tau] \, d\tau \qquad (8.4.19)$$

The solution above applies to any form of the nonhomogeneous term $f(t)$. We can regard the ODE (8.4.10) as describing the response (or output) $y(t)$ of a system to a forcing function (or input) $f(t)$. In this context, it is of particular interest to study the *impulse response* of the system—that is, the behavior of $y(t)$ when the forcing function is a Dirac delta centered at $t = 0$, namely,

$$f(t) = \delta(t) \qquad (8.4.20)$$

By substituting eq. (8.4.20) in eq. (8.4.18), the convolution integral becomes

$$\mathscr{L}^{-1}\{F(s)G(s)\} = \int_0^t \delta(t - \tau)g(\tau) \, d\tau = g(t) \qquad (8.4.21)$$

Thus $g(t)$ represents the impulse response of the system. Note that by virtue of its role in eq. (8.4.19), $g(t)$ is also the one-sided Green's function (cf. section 3.6).

Although we have treated a specific problem here, the procedure is general. Following the Laplace transform method, as illustrated by eq. (8.4.19), the solution is always a combination of the homogeneous and the particular solution—the latter depending on the nonhomogeneous term $f(t)$ in the ODE—as in the standard method.

An advantage of the Laplace transform method over the standard method is that it can be applied when the nonhomogeneous term is only piecewise continuous. As an example let us consider the two tanks in series shown in Figure 8.4. By assuming that the volumetric flowrate q leaving each tank is proportional to the volume of liquid in the tank, v—that is, $q_i = kv_i$ for $i = 1, 2$—the mass balances in the tanks are given by

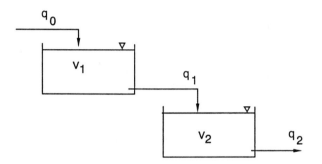

Figure 8.4 Schematic representation of two tanks in series.

$$\frac{dv_1}{dt} = q_0(t) - kv_1 \tag{8.4.22a}$$

$$\frac{dv_2}{dt} = k(v_1 - v_2) \tag{8.4.22b}$$

with ICs

$$v_1 = v_2 = 0 \qquad \text{at } t = 0 \tag{8.4.23}$$

which indicate that the two tanks are initially empty. We are interested in evaluating the volume of liquid in the second tank as a function of time—that is, $v_2(t)$—to prevent it from falling below a minimum or exceeding a maximum value. In particular, consider the case where the feed flowrate is constant, $q_0 = 1$, except for a time interval, say from ($t = \beta$) to ($t = 2\beta$), where it drops to zero. Thus $q_0(t)$ can be represented through the unit step function (8.2.4)—that is,

$$q_0(t) = 1 - u(t - \beta) + u(t - 2\beta) = \begin{cases} 1 & \text{for } 0 < t < \beta \\ 0 & \text{for } \beta < t < 2\beta \\ 1 & \text{for } t > 2\beta \end{cases} \tag{8.4.24}$$

as shown schematically in Figure 8.5.

Since we are interested only in the value of $v_2(t)$, we can eliminate $v_1(t)$ from eq. (8.4.22a) by differentiating both sides of eq. (8.4.22b) with respect to t and then substituting for $v_1'(t)$, as follows:

$$\frac{d^2v_2}{dt^2} + 2k\frac{dv_2}{dt} + k^2v_2 = kq_0 \tag{8.4.25}$$

This is a second-order ODE with constant coefficients and a discontinuous nonhomogeneous term. The proper ICs are given by eqs. (8.4.22b) and (8.4.23):

$$v_2 = 0, \qquad \frac{dv_2}{dt} = 0 \text{ at } t = 0 \tag{8.4.25a}$$

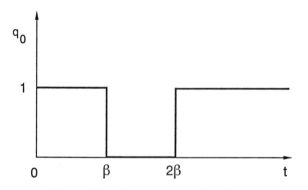

Figure 8.5 Feed flowrate as a function of time.

By applying the Laplace transform operator to both sides of eq. (8.4.25) and following the usual procedure, we obtain

$$\mathscr{L}\{v_2(t)\} = V_2(s) = \frac{kQ_0(s)}{(s + k)^2} \tag{8.4.26}$$

Since the transform of $q_0(t)$ is readily seen from Table 8.1 to be given by

$$Q_0(s) = \frac{1}{s} - \frac{e^{-\beta s}}{s} + \frac{e^{-2\beta s}}{s} \tag{8.4.27}$$

eq. (8.4.26) leads to

$$V_2(s) = \frac{k}{s(s + k)^2} - \frac{ke^{-\beta s}}{s(s + k)^2} + \frac{ke^{-2\beta s}}{s(s + k)^2} \tag{8.4.28}$$

To invert $V_2(s)$, let us first consider the following inverse transform:

$$\mathscr{L}^{-1}\left\{\frac{k}{s(s + k)^2}\right\} = z(t) \tag{8.4.29}$$

Since this transform is of the type (8.3.1), where $p(s) = k$ and $q(s) = s(s + k)^2$, we can use the Heaviside expansion theorem. Noting that $q(s)$ has one unrepeated root ($a_1 = 0$) and one root repeated twice ($b_1 = -k$), eq. (8.3.22) leads to

$$z(t) = A_1 e^{a_1 t} + (B_1 + B_2 t)e^{b_1 t} \tag{8.4.30}$$

where the constant coefficients are obtained from eq. (8.3.7)

$$A_1 = \frac{p(a_1)}{q'(a_1)} = \frac{1}{k} \tag{8.4.31a}$$

and from eqs. (8.3.13) and (8.3.16)

$$B_2 = \lim_{s \to b_1}\left[\frac{(s + k)^2 p(s)}{q(s)}\right] = \lim_{s \to b_1}\left(\frac{k}{s}\right) = -1 \tag{8.4.31b}$$

$$B_1 = \lim_{s \to b_1}\left\{\frac{d}{ds}\left[\frac{(s + k)^2 p(s)}{q(s)}\right]\right\} = \lim_{s \to b_1}\left\{\frac{d}{ds}\left(\frac{k}{s}\right)\right\} = -\frac{1}{k} \tag{8.4.31c}$$

Thus we have

$$z(t) = \frac{1}{k} - \left(\frac{1}{k} + t\right)e^{-kt} \tag{8.4.32}$$

From eqs. (8.2.6) and (8.4.29) we observe that

$$\frac{k}{s(s + k)^2}e^{-\beta s} = e^{-\beta s}\mathscr{L}\{z(t)\} = \mathscr{L}\{u(t - \beta)z(t - \beta)\} \tag{8.4.33a}$$

and similarly

$$\frac{1}{s(s + k)^2}e^{-2\beta s} = e^{-2\beta s}\mathscr{L}\{z(t)\} = \mathscr{L}\{u(t - 2\beta)z(t - 2\beta)\} \tag{8.4.33b}$$

We can now invert the transform (8.4.28), using eqs. (8.4.29), (8.4.32), and (8.4.33), to obtain

$$v_2(t) = \mathcal{L}^{-1}\{V_2(s)\} = z(t) - u(t - \beta)z(t - \beta) + u(t - 2\beta)z(t - 2\beta)$$

$$= \frac{1}{k} - \left(\frac{1}{k} + t\right)e^{-kt} - u(t - \beta)\left[\frac{1}{k} - \left(\frac{1}{k} + t - \beta\right)e^{-k(t-\beta)}\right] \qquad (8.4.34)$$

$$+ u(t - 2\beta)\left[\frac{1}{k} - \left(\frac{1}{k} + t - 2\beta\right)e^{-k(t-2\beta)}\right]$$

To analyze the behavior of the volume of liquid in the second tank as a function of time, $v_2(t)$, let us first assume that $\beta \gg 1/k$. This means that the time interval over which the change in the feed flowrate occurs is much larger than the characteristic time of the tank dynamics. Accordingly, the time interval between the two switches in $q_0(t)$ is sufficiently long for the tank to complete its transient. This behavior is shown in Figure 8.6a. It can be seen that as t approaches β, the transient of the second tank is practically over and the liquid volume reaches its steady-state value corresponding to $q_0 = 1$. From eqs. (8.4.22) this is readily seen to be $v_2 = 1/k$. At time $t = \beta$, the feed flowrate switches to zero and the volume of liquid in the tank begins to decrease. Again, the time interval before the subsequent switch of $q_0(t)$ is long enough for the tank to complete its transient and to approach the new steady-state value. Since now $q_0 = 0$, the new stationary value is $v_2 = 0$. After time $t = 2\beta$, the volume of liquid exhibits a transient similar to that in the first time interval; that is, it increases and approaches the final steady-state value, $v_2 = 1/k$. Thus summarizing, in this case, since the forcing function changes very slowly, the dynamics of the system is given by a sequence of responses to step changes in $q_0(t)$, where after each change the system closely approaches steady-state conditions.

At the other extreme, illustrated in Figure 8.6e, is the case where the switching in the feed flowrate occurs over a time interval much shorter than the characteristic time of the tank dynamics; that is, $\beta \ll 1/k$. Note that for $t < 2\beta$ the system does not change substantially. For $t < 2\beta$ and $\beta \ll 1/k$ we have, in fact, $t \ll 1/k$, and then from eq. (8.4.34) we obtain

$$v_2(t) \simeq 0 \qquad (8.4.35)$$

The system dynamics arises for larger time values (i.e., $t > 2\beta$) where eq. (8.4.34)—recalling that $\beta \ll 1/k$ and hence $e^{\beta k} \simeq 1$—reduces to

$$v_2(t) = \frac{1}{k} - \left(\frac{1}{k} + t\right)e^{-kt} \qquad (8.4.36)$$

Note that eq. (8.4.36) represents the response of the system in the case where $q_0 = 1$ remains constant. This is because the forcing function changes so fast, as compared to the characteristic time of the tank dynamics, that it does not affect the response of the system.

For comparable values of β and $1/k$, the dynamic behavior of the system is intermediate between the limits described above. A few examples are shown in Figures 8.5b to 8.5d. It is seen that near the time values $t = \beta$ and $t = 2\beta$, the behavior of $v_2(t)$

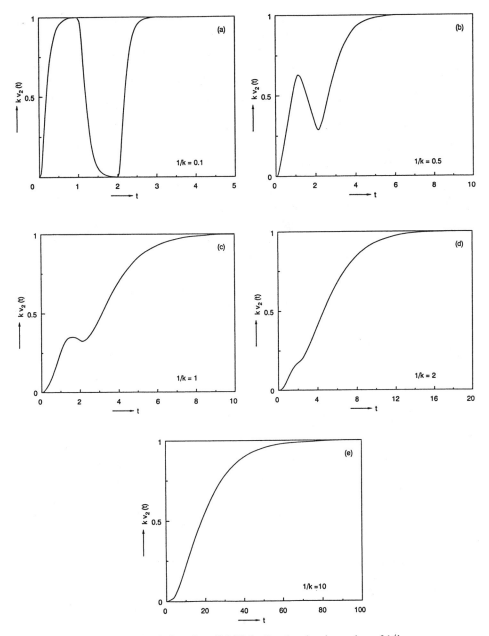

Figure 8.6 Solution of eq. (8.4.25) for $\beta = 1$ and various values of $1/k$.

undergoes some changes as a consequence of the switch in the value of the forcing function q_0. Such changes are, in fact, more evident in $v_1(t)$. From eq. (8.4.22a) we see that the discontinuities in $q_0(t)$ at $t = \beta$ and $t = 2\beta$ cause slope discontinuities in $v_1(t)$, while $v_1(t)$ itself remains continuous. On the other hand, $v_2(t)$ does not exhibit slope discontinuities at all, as seen readily from eq. (8.4.22b). However, from eq. (8.4.25) we observe that the second-order derivative of $v_2(t)$ exhibits discontinuities at $t = \beta$ and $t = 2\beta$.

All of the above dynamic behavior is a consequence of the smoothing effect of the first tank capacity, which introduces some delay in the response of the second tank to changes in the feed flowrate $q_0(t)$. It is worth noting that this is a general behavior for nth-order differential equations with discontinuous nonhomogeneous terms. The derivative of the highest order in the equation exhibits jump discontinuities at the same locations where the nonhomogeneous term becomes discontinuous, while the solution and all its lower-order derivatives remain continuous.

In conclusion, note that Laplace transforms allow us to handle jump discontinuities in the forcing function. In the example above, we derived the solution all at once, while other methods would require to piecewise integrate the ODE by interrupting the integration at all locations where the forcing function exhibits a discontinuity. Thus, in this particular case, we would first integrate eq. (8.4.25) with ICs (8.4.25a) from $t = 0$ to $t = \beta$ with $q_0 = 1$. Then we would change $q_0 = 0$ in eq. (8.4.25) and integrate from $t = \beta$ to $t = 2\beta$. The ICs at $t = \beta$ are selected to make $v_2(t)$ and its first-order derivative continuous. This same procedure is repeated at $t = 2\beta$, where the integration is started again, with new ICs, and $q_0 = 1$ in eq. (8.4.25).

8.5   SYSTEMS OF ORDINARY DIFFERENTIAL EQUATIONS

The method of Laplace transforms can be used conveniently to solve systems of linear ODEs with constant coefficients. The system of linear differential equations in the t domain becomes a system of linear algebraic equations in the s domain, where the unknowns are the Laplace transforms of the original variables.

In this section we discuss some applications of this method, considering first homogeneous and then nonhomogeneous linear systems.

8.5.1   Homogeneous Systems

Let us consider the system of first-order, linear, homogeneous ODEs:

$$\frac{d\mathbf{y}}{dt} = \mathbf{A}\mathbf{y} \qquad (8.5.1)$$

with ICs

$$\mathbf{y} = \mathbf{y}^0 \qquad \text{at } t = 0 \qquad (8.5.2)$$

where $\mathbf{y}$ is the n-column vector containing the unknowns $y_i(t)$, $\mathbf{y}^0$ is the vector of initial values, and $\mathbf{A}$ is the square constant coefficient matrix of order n. A solution of this set of ODEs by the method of matrix analysis has been discussed in section 1.17. We solve the problem here using the method of Laplace transforms.

By applying the Laplace transform operator to both sides of eq. (8.5.1), we obtain

$$s\mathbf{Y}(s) - \mathbf{y}(0) = \mathbf{AY}(s) \tag{8.5.3}$$

which rearranges to

$$(\mathbf{A} - s\mathbf{I})\mathbf{Y}(s) = -\mathbf{y}(0) \tag{8.5.4}$$

where $\mathbf{I}$ is the Identity matrix. This is a linear algebraic system whose solution can be obtained by inverting the matrix $(\mathbf{A} - s\mathbf{I})$:

$$\mathbf{Y}(s) = -(\mathbf{A} - s\mathbf{I})^{-1}\mathbf{y}(0) = -\mathbf{B}\,\mathbf{y}(0) \tag{8.5.5}$$

Following section 1.7, the elements of the inverse of $\mathbf{B}$ are given by

$$b_{ij} = \frac{C_{ji}}{|\mathbf{A} - s\mathbf{I}|} \tag{8.5.6}$$

where C_{ij} is the cofactor of the (i, j) element of the matrix $(\mathbf{A} - s\mathbf{I})$, and $|\mathbf{A} - s\mathbf{I}| = \det(\mathbf{A} - s\mathbf{I})$ represents its determinant. From eq. (8.5.5) the transform of the component $y_i(t)$ is

$$Y_i(s) = -\sum_{j=1}^{n} b_{ij}y_j(0) \tag{8.5.7}$$

Let us note that all the elements on the main diagonal of the matrix $(\mathbf{A} - s\mathbf{I})$ are linear functions of s, while the remaining are independent of s. Accordingly, the cofactors C_{ji} are polynomials of s of degree $(n - 1)$, while $|\mathbf{A} - s\mathbf{I}|$ is a polynomial of degree n. Thus, the coefficients b_{ij} in eq. (8.5.6) are given by the ratio of two polynomials of s, with the numerator being of lower order than the denominator. This is the form of the transforms (8.3.1), which can be inverted using the Heaviside expansion theorem.

By comparing eqs. (8.3.1) and (8.5.7), we note that the denominator of the generic transform $Y_j(s)$ is given by

$$q(s) = |\mathbf{A} - s\mathbf{I}| \tag{8.5.8}$$

which is independent of the index j; that is, it is the same for all the unknowns of the system. From section 8.3, we know that the inverse of the transform (8.3.1), as given by the Heaviside expansion theorem, is the combination of exponential functions (8.3.22) whose exponents are the roots of $q(s)$. From eq. (8.5.8) it follows that these roots are the eigenvalues of the coefficient matrix $\mathbf{A}$. This coincides with the form of the solution of systems of linear ODEs with constant coefficients introduced in section 1.17.

The same procedure can be extended to systems of higher-order ODEs. For example, in the case of differential equations of second order we have

$$\frac{d^2\mathbf{y}}{dt^2} = \mathbf{D}\frac{d\mathbf{y}}{dt} + \mathbf{Ay} \tag{8.5.9}$$

with ICs

$$\mathbf{y} = \mathbf{y}^0, \quad \frac{d\mathbf{y}}{dt} = \mathbf{y}^{0'} \quad \text{at } t = 0 \tag{8.5.10}$$

By applying Laplace transforms we obtain

$$s^2\mathbf{Y}(s) - \mathbf{y}'(0) - s\mathbf{y}(0) = \mathbf{D}[s\mathbf{Y}(s) - \mathbf{y}(0)] + \mathbf{A}\mathbf{Y}(s) \tag{8.5.11}$$

which, using the ICs (8.5.10), reduces to

$$(\mathbf{A} + s\mathbf{D} - s^2\mathbf{I})\mathbf{Y}(s) = (\mathbf{D} - s\mathbf{I})\mathbf{y}^0 - \mathbf{y}^{0'} \tag{8.5.12}$$

This is a linear algebraic system of the same form as (8.5.4) which we can treat following the same procedure discussed above to yield the solution $\mathbf{y}(t)$.

As an example we consider the system of two second-order ODEs:

$$\frac{d^2y_1}{dt^2} - 2\frac{dy_1}{dt} + \frac{dy_2}{dt} - 3y_2 = 0 \tag{8.5.13a}$$

$$\frac{d^2y_2}{dt^2} + \frac{dy_1}{dt} - 2y_1 + y_2 = 0 \tag{8.5.13b}$$

with ICs

$$y_1 = 1, \quad \frac{dy_1}{dt} = 0 \quad \text{at } t = 0 \tag{8.5.14a}$$

$$y_2 = 2, \quad \frac{dy_2}{dt} = 0 \quad \text{at } t = 0 \tag{8.5.14b}$$

This initial value problem can be recast in the matrix form (8.5.9)–(8.5.10) by taking

$$\mathbf{D} = \begin{bmatrix} 2 & -1 \\ -1 & 0 \end{bmatrix}, \quad \mathbf{A} = \begin{bmatrix} 0 & 3 \\ 2 & -1 \end{bmatrix}$$

$$\mathbf{y}^0 = \begin{bmatrix} 1 \\ 2 \end{bmatrix}, \quad \mathbf{y}^{0'} = \begin{bmatrix} 0 \\ 0 \end{bmatrix} \tag{8.5.15}$$

By applying Laplace transforms to eqs. (8.5.13) we obtain

$$(2s - s^2)Y_1 + (3 - s)Y_2 = -s \tag{8.5.16a}$$

$$(2 - s)Y_1 - (s^2 + 1)Y_2 = -(2s + 1) \tag{8.5.16b}$$

which in matrix form is

$$\mathbf{G}\mathbf{Y}(s) = \mathbf{d} \tag{8.5.17}$$

where

$$\mathbf{G} = \begin{bmatrix} (2s - s^2) & (3 - s) \\ (2 - s) & -(s^2 + 1) \end{bmatrix}, \quad \mathbf{d} = \begin{bmatrix} -s \\ -2s - 1 \end{bmatrix} \tag{8.5.18}$$

Note that eq. (8.5.17) is consistent with the general form (8.5.12) since using eqs. (8.5.15) we have

$$(\mathbf{A} + s\mathbf{D} - s^2\mathbf{I}) = \begin{bmatrix} (2s - s^2) & (3 - s) \\ (2 - s) & -(s^2 + 1) \end{bmatrix} = \mathbf{G} \qquad (8.5.19a)$$

and

$$(\mathbf{D} - s\mathbf{I})\mathbf{y}^0 - \mathbf{y}^{0'} = \begin{bmatrix} (2 - s) & -1 \\ -1 & -s \end{bmatrix}\begin{bmatrix} 1 \\ 2 \end{bmatrix} - \begin{bmatrix} 0 \\ 0 \end{bmatrix} = \begin{bmatrix} -s \\ -2s - 1 \end{bmatrix} = \mathbf{d} \qquad (8.5.19b)$$

8.5.2 Dynamics of Linear Systems: The Transfer Function

Linear systems of nonhomogeneous ODEs arise typically in the study of process dynamics. In general, the process is described in terms of one or more output variables, driven by one or more input variables which change in time due to disturbances or following a prescribed transient. The selection of the output and input variables for describing the behavior of a real process is determined by the specific aim of the dynamic analysis. For example, in the case of the two tanks in series discussed in section 8.4.2, we can regard the process as characterized by two output variables, namely, the liquid volumes in each tank: $v_1(t)$ and $v_2(t)$. The process is then described by the nonhomogeneous system of first-order ODEs (8.4.22), which represent the two output variables as a function of one input variable, namely, the feed flowrate $q_0(t)$. On the other hand, if we are interested only in the liquid volume in the second tank, then we can refer to the process description given by the second-order, nonhomogeneous ODE (8.4.25). In this case we have only one output, $v_2(t)$, and one input, $q_0(t)$.

In the above examples we have considered linear processes, where the relation between the selected outputs and inputs is linear. In most cases of interest in applications, the process dynamics is described by nonlinear ODEs. However, linear models remain useful in many instances. As discussed in chapter 2, we are often interested in the behavior of the process in a small neighborhood of the steady state of interest, which represents its normal operation. This is true, for example, when we want to investigate the stability character of the steady state or to control the process close to it. In this case we develop the analysis using a linear model, obtained by linearizing the original model around the steady state. In the linearized model, both the input and the output variables are defined as deviation variables with respect to their steady-state values.

As an example we consider the continuous-flow stirred tank reactor examined in section 2.17. Let us suppose that we are interested in the behavior of the outlet reactant concentration C and temperature T, as a function of concentration C_f and temperature T_f of the feed stream. In the case we can represent the mass and energy balances (2.17.1) and (2.17.2) in the form

$$\frac{dC}{dt} = F_1(C, T, C_f) \qquad (8.5.20a)$$

$$\frac{dT}{dt} = F_2(C, T, T_f) \qquad (8.5.20b)$$

Introducing the deviation variables for the two outputs (i.e., $y_1(t) = C(t) - C_s$ and $y_2(t) = T(t) - T_s$) and for the two inputs (i.e., $f_1(t) = C_f(t) - C_{fs}$ and $f_2(t) = T_f(t) - T_{fs}$) and linearizing eqs. (8.5.20) around the steady state, we obtain

$$\frac{dy_1}{dt} = \left(\frac{\partial F_1}{\partial C}\right)_s y_1 + \left(\frac{\partial F_1}{\partial T}\right)_s y_2 + \left(\frac{\partial F_1}{\partial C_f}\right)_s f_1 \tag{8.5.21a}$$

$$\frac{dy_2}{dt} = \left(\frac{\partial F_2}{\partial C}\right)_s y_1 + \left(\frac{\partial F_2}{\partial T}\right)_s y_2 + \left(\frac{\partial F_2}{\partial T_f}\right)_s f_2 \tag{8.5.21b}$$

Since the partial derivatives are evaluated at the steady state, all coefficients are constant in the above equations. Using matrix notation, these can be written in the general form as follows:

$$\frac{d\mathbf{y}}{dt} = \mathbf{A}\mathbf{y} + \mathbf{B}\mathbf{f} \tag{8.5.22}$$

with ICs

$$\mathbf{y} = 0 \qquad \text{at } t = 0 \tag{8.5.23}$$

where $\mathbf{y}$ is an n-column vector containing the n output variables, $\mathbf{f}$ is an m-column vector containing the m inputs, while the constant coefficient matrices $\mathbf{A}$ and $\mathbf{B}$ have orders $(n \times n)$ and $(n \times m)$, respectively. The above system of linear, nonhomogeneous ODEs describes the dynamic behavior of a generic linear process consisting of n outputs and m inputs.

The linearized CSTR model (8.5.21) includes $n = 2$ outputs and $m = 2$ inputs and can be represented by eq. (8.5.22) by taking

$$\mathbf{A} = \begin{bmatrix} \left(\dfrac{\partial F_1}{\partial C}\right)_s & \left(\dfrac{\partial F_1}{\partial T}\right)_s \\ \left(\dfrac{\partial F_2}{\partial C}\right)_s & \left(\dfrac{\partial F_2}{\partial T}\right)_s \end{bmatrix} \tag{8.5.24a}$$

and

$$\mathbf{B} = \begin{bmatrix} \left(\dfrac{\partial F_1}{\partial C_f}\right)_s & 0 \\ 0 & \left(\dfrac{\partial F_2}{\partial T_f}\right)_s \end{bmatrix} \tag{8.5.24b}$$

Similarly, the linear process (8.4.22) includes $n = 2$ outputs and $m = 1$ input and can be recast in the form of eq. (8.5.22) by taking $\mathbf{y} = \mathbf{v}$, $f_1 = q_0$,

$$\mathbf{A} = \begin{bmatrix} -k & 0 \\ k & -k \end{bmatrix} \quad \text{and} \quad \mathbf{B} = \begin{bmatrix} 1 \\ 0 \end{bmatrix} \tag{8.5.25}$$

By applying Laplace transforms to eq. (8.5.22), we obtain

$$s\mathbf{Y}(s) = \mathbf{A}\mathbf{Y}(s) + \mathbf{B}\mathbf{F}(s) \tag{8.5.26}$$

which can be solved for $\mathbf{Y}(s)$ leading to

$$\mathbf{Y}(s) = -(\mathbf{A} - s\mathbf{I})^{-1}\mathbf{B}\mathbf{F}(s) = \mathbf{G}(s)\mathbf{F}(s) \tag{8.5.27}$$

In the eq. (8.5.27) we have introduced a new $(n \times m)$ matrix $\mathbf{G}(s)$, called the *matrix transfer function*, defined as follows:

$$\mathbf{G}(s) = -(\mathbf{A} - s\mathbf{I})^{-1}\mathbf{B} \tag{8.5.28}$$

This matrix is useful for describing the dynamic behavior of linear systems and is widely used in control theory. To illustrate the significance of the matrix $\mathbf{G}(s)$, let us consider the generic ith output of the model, whose Laplace transform according to eq. (8.5.27) is given by

$$Y_i(s) = \sum_{j=1}^{m} G_{ij}(s)F_j(s) \tag{8.5.29}$$

It is seen that each element $G_{ij}(s)$ of the matrix transfer function represents the effect of the jth input on the ith output of the process. Moreover, since the Laplace transform of the Dirac delta function is $\mathcal{L}\{\delta(t)\} = 1$, $G_{ij}(s)$ represents the impulse response of the ith output of the process to the jth input. It should be noted that the matrix transfer function characterizes the process dynamics independently of the form of the input variable. It is sufficient to multiply the transfer function by the specific input function of interest, $F_j(s)$, to yield the output of the process in the transform space.

As an example, let us consider again the dynamic behavior of two tanks in series. By applying Laplace transforms, eqs. (8.4.22) yield

$$sV_1(s) = Q_0(s) - kV_1(s) \tag{8.5.30a}$$

$$sV_2(s) = kV_1(s) - kV_2(s) \tag{8.5.30b}$$

Solving for $V_1(s)$ and $V_2(s)$, we obtain

$$V_1(s) = \frac{1}{s + k} Q_0(s) \tag{8.5.31a}$$

$$V_2(s) = \frac{k}{s + k} V_1(s) = \frac{k}{(s + k)^2} Q_0(s) \tag{8.5.31b}$$

which, when recast in the form of eq. (8.5.26), where $\mathbf{Y}(s) = \mathbf{V}(s)$ and $F_1(s) = Q_0(s)$, leads to the following expression for the matrix transfer function of the process:

$$\mathbf{G}(s) = \begin{bmatrix} \dfrac{1}{s + k} \\ \dfrac{k}{(s + k)^2} \end{bmatrix} \tag{8.5.32}$$

In section 8.5.1 we noted that each element of the matrix $(\mathbf{A} - s\mathbf{I})^{-1}$ is given by the ratio of two polynomials of s, where the denominator is the same for each element— that is, $q(s) = |\mathbf{A} - s\mathbf{I}|$. Since the elements of the matrix $\mathbf{B}$ in eq. (8.5.26) are constants, from eq. (8.5.28) we observe that the elements of $\mathbf{G}(s)$ are also given by a ratio of two polynomials, with the same denominator $q(s)$. Moreover, by inspection of Table 8.1 we note that most functions $f(t)$, which may be used to describe the time variation of the process input variables, have Laplace transforms given by a ratio of two polynomials. We can then conclude that, in most instances of practical interest, the transforms $Y_i(s)$ given by eq. (8.5.27) are in the form of eq. (8.3.1) and hence can be inverted through the Heaviside expansion theorem. Thus, according to eq. (8.3.22), the generic output, $y_i(t)$ is represented by a combination of exponentials whose arguments are the roots of the polynomial of the denominator of $Y_i(s)$. From eq. (8.5.29), it is seen that $Y_i(s)$ is given by a sum of various terms. The denominator of each of these is the product of the denominator of $G_{ij}(s)$, which is always $|\mathbf{A} - s\mathbf{I}|$, and the denominator of the specific input at hand, $F_j(s)$. Thus the dynamic behavior of $Y_i(t)$ is determined by the roots of the denominator of $G_{ij}(s)$, which are *intrinsic* to the process dynamics and independent of the input considered, and the roots of the denominator of $F_j(s)$, which represent the effect of the *specific* input forcing function.

In general, we are interested in the intrinsic behavior of the process, and hence in the roots of the denominators of the functions $G_{ij}(s)$. As noted above, they are the same for all inputs i and outputs j and are given by

$$|\mathbf{A} - s\mathbf{I}| = 0 \tag{8.5.33}$$

These roots, which are usually referred to as the *poles* of the transfer function, characterize the process dynamics in the neighborhood of the steady state under examination, and in particular its stability character. We have discussed the stability analysis of linear processes in chapter 2 and have found that the stability character is determined by the eigenvalues of the coefficient matrix—that is, the roots of eq. (8.5.33). Using the approach based on Laplace transforms, we recover the same result.

Example. Consider the system of ODEs

$$\frac{dy_1}{dt} = y_1 + 2y_2 + f_1 + 2f_2 \tag{8.5.34a}$$

$$\frac{dy_2}{dt} = y_1 - 2f_1 + f_2 \tag{8.5.34b}$$

with ICs

$$y_1 = 0, \quad y_2 = 0 \quad \text{at } t = 0 \tag{8.5.35}$$

This can be recast in the form of eq. (8.5.22) by taking

$$\mathbf{A} = \begin{bmatrix} 1 & 2 \\ 1 & 0 \end{bmatrix}, \quad \mathbf{B} = \begin{bmatrix} 1 & 2 \\ -2 & 1 \end{bmatrix} \tag{8.5.36}$$

Using eq. (8.5.28) we obtain the matrix transfer function

$$\mathbf{G}(s) = \frac{1}{s^2 - s - 2} \begin{bmatrix} (s - 4) & 2(s + 1) \\ (3 - 2s) & (s + 1) \end{bmatrix} \tag{8.5.37}$$

where note that $|\mathbf{A} - s\mathbf{I}| = s^2 - s - 2 = (s + 1)(s - 2)$. Considering the input functions

$$f_1(t) = e^{-3t} \quad \text{and} \quad f_2(t) = \sin t \tag{8.5.38}$$

with

$$F_1(s) = \frac{1}{s + 3} \quad \text{and} \quad F_2(s) = \frac{1}{s^2 + 1} \tag{8.5.39}$$

the Laplace transforms of the solution are obtained from eq. (8.5.29) as follows:

$$Y_1(s) = \frac{(s - 4)}{(s + 1)(s - 2)(s + 3)} + \frac{2(s + 1)}{(s + 1)(s - 2)(s^2 + 1)} \tag{8.5.40a}$$

$$Y_2(s) = \frac{(3 - 2s)}{(s + 1)(s - 2)(s + 3)} + \frac{(s + 1)}{(s + 1)(s - 2)(s^2 + 1)} \tag{8.5.40b}$$

The above transforms can be inverted using the Heaviside expansion theorem. We observe that all the roots of the involved denominators are unrepeated. In particular, the denominator of the first term has three real roots: $a_1 = -1$, $a_2 = 2$, and $a_3 = -3$, while that of the second term has two real roots: $a_1' = -1$, $a_2' = 2$ and a pair of complex conjugate roots $a_3' = \pm i$. The inverse transforms are given by eqs. (8.3.22) and (8.3.37), which in this specific case reduce to

$$y_1 = \frac{5}{6} e^{-t} + \frac{4}{15} e^{2t} - \frac{7}{10} e^{-3t} - \frac{2}{5} (\cos t + 2 \sin t) \tag{8.5.41a}$$

$$y_2 = -\frac{5}{6} e^{-t} + \frac{2}{15} e^{2t} + \frac{9}{10} e^{-3t} - \frac{1}{5} (\cos t + 2 \sin t) \tag{8.5.41b}$$

8.6 PARTIAL DIFFERENTIAL EQUATIONS

In this section we illustrate the use of Laplace transforms for solving *linear* PDEs through a few examples. When applying this transform to m-dimensional PDE, we obtain a new equation in the transformed space whose dimensionality is decreased by one. Thus in general we obtain an $(m - 1)$-dimensional PDE, where the transform variable s appears as a parameter, which provides the Laplace transform of the solution. Recalling that in the Laplace transform operator the independent parameter, t goes from 0 to ∞, this technique is best suited for problems where at least one of the dimensions is semi-infinite. Since the most typical semi-infinite dimension is time, this technique is particularly useful for studying dynamic systems.

8.6.1 The Diffusion Equation

Let us consider the transient diffusion-reaction problem in a slab, where the generation term $f(t)$ is a function of time and does not depend on composition. Indicating the product concentration by c and the diffusion coefficient by D, the mass balance gives

$$D \frac{\partial^2 c}{\partial x^2} + f(t) = \frac{\partial c}{\partial t}; \quad x \in (0, L) \quad \text{and} \quad t > 0 \tag{8.6.1}$$

where x is the space variable and t is time. The initial and boundary conditions are given by

$$c = 0 \qquad \text{at } 0 \le x \le L \quad \text{and} \quad t = 0 \qquad (8.6.2a)$$

$$c = 0 \qquad \text{at } x = 0 \qquad \text{and} \quad t > 0 \qquad (8.6.2b)$$

$$c = 0 \qquad \text{at } x = L \qquad \text{and} \quad t > 0 \qquad (8.6.2c)$$

As mentioned above, it is convenient to apply the Laplace transform operator along the time dimension, since it is semi-infinite, as follows:

$$\mathcal{L}\{c(x, t)\} = C(x, s) = \int_0^\infty e^{-st} c(x, t) \, dt \qquad (8.6.3)$$

Thus multiplying both sides of eq. (8.6.1) by e^{-st} and integrating from $t = 0$ to $t = \infty$ yields

$$D \int_0^\infty e^{-st} \frac{\partial^2 c}{\partial x^2} \, dt + \int_0^\infty e^{-st} f(t) \, dt = \int_0^\infty e^{-st} \frac{\partial c}{\partial t} \, dt \qquad (8.6.4)$$

By exchanging the integration and differentiation operations in the first term of the lhs and using eq. (8.2.26) for the term on the rhs, we obtain

$$D \frac{\partial^2}{\partial x^2} \int_0^\infty e^{-st} c(x, t) \, dt + F(s) = sC(x, s) - c(x, 0) \qquad (8.6.5)$$

which, by substituting eqs. (8.6.2a) and (8.6.3), leads to

$$D \frac{d^2 C}{dx^2} + F(s) = sC \qquad (8.6.6)$$

This is a linear, second-order ODE whose solution provides the transform $C(x, s)$, which is a function only of the spatial independent variable x while s appears merely as a parameter. The BCs for this equation are obtained directly from eqs. (8.6.2b) and (8.6.2c) by applying the Laplace transform:

$$C(x, s) = 0 \qquad \text{at } x = 0 \qquad (8.6.7a)$$

$$C(x, s) = 0 \qquad \text{at } x = L \qquad (8.6.7b)$$

Since the ODE (8.6.6) has constant coefficients, we can readily obtain the solution of the corresponding homogeneous ODE:

$$C_h(x, s) = d_1 \sinh\left(\sqrt{\frac{s}{D}}\, x\right) + d_2 \cosh\left(\sqrt{\frac{s}{D}}\, x\right) \qquad (8.6.8)$$

and the particular solution

$$C_p(x, s) = \frac{F(s)}{s} \qquad (8.6.9)$$

Then the general solution is given by

$$C(x, s) = d_1 \sinh\left(\sqrt{\frac{s}{D}}\, x\right) + d_2 \cosh\left(\sqrt{\frac{s}{D}}\, x\right) + \frac{F(s)}{s} \qquad (8.6.10)$$

The integration constants d_1 and d_2 are evaluated using the BCs (8.6.7) as follows:

$$d_1 = \frac{F(s)}{s} \frac{\cosh\left(\sqrt{\frac{s}{D}}\, L\right) - 1}{\sinh\left(\sqrt{\frac{s}{D}}\, L\right)} \qquad (8.6.11a)$$

$$d_2 = -\frac{F(s)}{s} \qquad (8.6.11b)$$

which, when substituted in eq. (8.6.10), lead (after some algebraic manipulations) to

$$C(x, s) = \frac{F(s)}{s} \frac{\left\{\cosh\left(\sqrt{\frac{s}{D}}\, L\right)\sinh\left(\sqrt{\frac{s}{D}}\, x\right) - \sinh\left(\sqrt{\frac{s}{D}}\, L\right)\cosh\left(\sqrt{\frac{s}{D}}\, x\right) + \sinh\left(\sqrt{\frac{s}{D}}\, L\right) - \sinh\left(\sqrt{\frac{s}{D}}\, x\right)\right\}}{\sinh\left(\sqrt{\frac{s}{D}}\, L\right)}$$

$$= F(s) \frac{\left\{\sinh\left(\sqrt{\frac{s}{D}}\, L\right) - \sinh\left(\sqrt{\frac{s}{D}}\, x\right) - \sinh\left[\sqrt{\frac{s}{D}}\, (L - x)\right]\right\}}{s \cdot \sinh\left(\sqrt{\frac{s}{D}}\, L\right)}$$

$$(8.6.12)$$

$$= F(s)G(x, s)$$

In order to invert $C(x, s)$ we can use the convolution theorem (8.2.56):

$$c(x, t) = \mathscr{L}^{-1}\{F(s)G(x, s)\} = \int_0^t f(t - \tau)g(x, \tau)\, d\tau \qquad (8.6.13)$$

First, we need to invert the transform

$$G(x, s) = \frac{\left\{\sinh\left(\sqrt{\frac{s}{D}}\, L\right) - \sinh\left(\sqrt{\frac{s}{D}}\, x\right) - \sinh\left[\sqrt{\frac{s}{D}}\, (L - x)\right]\right\}}{s \cdot \sinh\left(\sqrt{\frac{s}{D}}\, L\right)} \qquad (8.6.14)$$

Recalling the series representation of $\sinh a$, namely,

$$\sinh a = a + \frac{a^3}{3!} + \frac{a^5}{5!} + \frac{a^7}{7!} + \cdots \qquad (8.6.15)$$

we can rewrite the transform (8.6.14) in the equivalent form:

$$
G(x, s) = \frac{\left[\sqrt{\dfrac{s}{D}}\,L + \left(\dfrac{s}{D}\right)^{3/2}\dfrac{L^3}{3!} + \cdots\right] - \left[\sqrt{\dfrac{s}{D}}\,x + \left(\dfrac{s}{D}\right)^{3/2}\dfrac{x^3}{3!} + \cdots\right]}{s\left[\sqrt{\dfrac{s}{D}}\,L + \left(\dfrac{s}{D}\right)^{3/2}\dfrac{L^3}{3!} + \cdots\right]}
$$

$$
-\frac{\left[\sqrt{\dfrac{s}{D}}\,(L - x) + \left(\dfrac{s}{D}\right)^{3/2}\dfrac{(L - x)^3}{3!} + \cdots\right]}{s\left[\sqrt{\dfrac{s}{D}}\,L + \left(\dfrac{s}{D}\right)^{3/2}\dfrac{L^3}{3!} + \cdots\right]}
$$

$$
= \frac{\left(\dfrac{s}{D}\right)^{3/2}\left[\dfrac{L^3}{3!} + \left(\dfrac{s}{D}\right)\dfrac{L^5}{5!} + \cdots - \dfrac{x^3}{3!} - \left(\dfrac{s}{D}\right)\dfrac{x^5}{5!} - \cdots - \dfrac{(L - x)^3}{3!} + \left(\dfrac{s}{D}\right)\dfrac{(L - x)^5}{5!} - \cdots\right]}{\left(\dfrac{s}{D}\right)^{3/2}D\left[L + \left(\dfrac{s}{D}\right)\dfrac{L^3}{3!} + \left(\dfrac{s}{D}\right)^2\dfrac{L^5}{5!} + \cdots\right]}
$$

$$
= \frac{\dfrac{1}{3!}[L^3 - x^3 - (L - x)^3] + \left(\dfrac{s}{D}\right)\dfrac{1}{5!}[L^5 - x^5 - (L - x)^5] + \cdots}{D\left[L + \left(\dfrac{s}{D}\right)\dfrac{L^3}{3!} + \left(\dfrac{s}{D}\right)^2\dfrac{L^5}{5!} + \cdots\right]}
\tag{8.6.16}
$$

Thus the transform $G(x, s)$ can be regarded as the ratio between two polynomials of s—that is,

$$
G(x, s) = \frac{p(s)}{q(s)}
\tag{8.6.17}
$$

Note that when taking the same number of terms for all the series involved, the degree of $p(s)$ is lower than that of $q(s)$. Hence we can use the Heaviside expansion theorem to invert $G(x, s)$. For this, we have to consider the roots of the denominator,

$$
q(s) = s\,\sinh\left(\sqrt{\dfrac{s}{D}}\,L\right)
\tag{8.6.18}
$$

which are not common with the numerator $p(s)$. The roots of $q(s)$ are given by the values of s which satisfy either one of the following equations:

$$
s = 0
\tag{8.6.19a}
$$

and

$$
\sinh\left(\sqrt{\dfrac{s}{D}}\,L\right) = 0
\tag{8.6.19b}
$$

We have seen in the derivation of eq. (8.6.16) that the root $s = 0$ is common between $p(s)$ and $q(s)$, because it can be readily confirmed by noting from eq. (8.6.14) that $p(0) = 0$. Thus the only roots that we need to consider are those given by the transcendental eq. (8.6.19b), with the exception of $s = 0$; that is,

$$
\sinh z = 0 \qquad \text{with } z \neq 0
\tag{8.6.20}
$$

Let us assume that there exists a complex solution $z = a + ib$, where a and b are real numbers. Then we have

$$\sinh(a + ib) = \sinh a \cosh(ib) + \cosh a \sinh(ib) = 0 \qquad (8.6.21)$$

Recalling from the Euler relations that

$$\cosh(ib) = \frac{e^{ib} + e^{-ib}}{2} = \cos b \qquad (8.6.22a)$$

and

$$\sinh(ib) = \frac{e^{ib} - e^{-ib}}{2} = i \sin b \qquad (8.6.22b)$$

eq. (8.6.21) becomes

$$\sinh a \cos b + i \cosh a \sin b = 0 \qquad (8.6.23)$$

This implies that a and b should satisfy the two conditions

$$\sinh a \cos b = 0 \qquad (8.6.24a)$$

and

$$\cosh a \sin b = 0 \qquad (8.6.24b)$$

If $a \neq 0$, then $\sinh a \neq 0$ and $\cosh a \neq 0$, and eqs. (8.6.24) imply that

$$\cos b = 0 \qquad (8.6.25a)$$

and

$$\sin b = 0 \qquad (8.6.25b)$$

for which there is no solution. Thus we can conclude that eq. (8.6.20) does not have either real (other than $z = 0$) or complex solutions for the case $a \neq 0$.

Let us now assume that $a = 0$; that is, we consider a purely imaginary solution $z = ib$. Since $\sinh a = 0$ and $\cosh a = 1$, from eqs. (8.6.24) we derive the condition

$$\sin b = 0 \qquad (8.6.26)$$

which is satisfied for $b = \pm n\pi$, with $n = 0, 1, 2, \ldots$. Thus eq. (8.6.20) has an infinite number of purely imaginary solutions, given by

$$z = \sqrt{\frac{s}{D}} L = \pm in\pi, \qquad n = 1, 2, \ldots \qquad (8.6.27)$$

where the value $n = 0$ has been removed since we should not consider the solution $z = 0$. Solving for s we obtain

$$s_n = -\frac{n^2 \pi^2 D}{L^2}, \qquad n = 1, 2, \ldots \qquad (8.6.28)$$

which are an infinite number of negative real values representing the solution of eq. (8.6.19b). Hence these are the roots of the denominator of the transform $G(x, s)$. Since they are unrepeated we can use the Heaviside expansion theorem (8.3.9) which in this case takes the form

$$g(x, t) = \mathcal{L}^{-1}\{G(x, s)\} = \sum_{n=1}^{\infty} \frac{p(s_n)}{q'(s_n)} e^{s_n t} \tag{8.6.29}$$

From eq. (8.6.14), using eq. (8.6.22b), we obtain

$$\begin{aligned}
p(s_n) &= \sinh(in\pi) - \sinh\left(in\pi\, \frac{x}{L}\right) - \sinh\left[in\pi\, \frac{(L - x)}{L}\right] \\
&= i \sin(n\pi) - i \sin\left(n\pi\, \frac{x}{L}\right) - i \sin\left[n\pi\, \frac{(L - x)}{L}\right] \\
&= -i\left[\sin\left(n\pi\, \frac{x}{L}\right) + \sin(n\pi)\cos\left(n\pi\, \frac{x}{L}\right) - \cos(n\pi)\sin\left(n\pi\, \frac{x}{L}\right)\right] \\
&= -i \sin\left(n\pi\, \frac{x}{L}\right)[1 - \cos(n\pi)] \\
&= -i \sin\left(n\pi\, \frac{x}{L}\right)[1 - (-1)^n] \\
&= \begin{cases} 0 & \text{for even } n,\ n = 2, 4, \ldots \\ -2i \sin\left(n\pi\, \frac{x}{L}\right) & \text{for odd } n,\ n = 1, 3, \ldots \end{cases}
\end{aligned} \tag{8.6.30}$$

Similarly, from eq. (8.6.14) using eqs. (8.6.22) and (8.6.28):

$$\begin{aligned}
q'(s_n) &= \sinh\left(\sqrt{\frac{s_n}{D}}\, L\right) + s_n \frac{L}{2\sqrt{s_n D}} \cosh\left(\sqrt{\frac{s_n}{D}}\, L\right) \\
&= i \sin(n\pi) + i \frac{n\pi}{2} \cosh(in\pi) \\
&= i \frac{n\pi}{2} \cos(n\pi) = i \frac{n\pi}{2} (-1)^n
\end{aligned} \tag{8.6.31}$$

Substituting eqs. (8.6.28), (8.6.30), and (8.6.31) in eq. (8.6.29) we have

$$g(x, t) = \frac{4}{\pi} \sum_{n=1,3}^{\infty} \frac{1}{n} \sin\left(n\pi\, \frac{x}{L}\right) e^{-(n^2\pi^2 D/L^2)t} \tag{8.6.32}$$

which represents the *impulse response* for the nonhomogeneity in the PDE. This can be seen from eq. (8.6.13), which, in the case where the nonhomogeneous term in eq. (8.6.1) is given by

$$f(t) = \delta(t) \tag{8.6.33}$$

(the unit impulse function), gives the system response as follows:

$$c(x, t) = \int_0^t \delta(t - \tau)g(x, \tau)\, d\tau = g(x, t) \tag{8.6.34}$$

Similarly, with the knowledge of the impulse response $g(x, t)$, we can obtain the solution of the system for other forms of the nonhomogeneous term $f(t)$ by evaluating the integral in eq. (8.6.13) (see problem 8.10).

8.6.2 The Wave Equation

Let us consider the one-dimensional wave equation discussed in section 5.5.2:

$$\frac{\partial^2 z}{\partial t^2} - c^2 \frac{\partial^2 z}{\partial x^2} = 0; \qquad 0 < x < L \quad \text{and} \quad t > 0 \tag{8.6.35}$$

which describes the displacement $z(x, t)$ normal to the x axis of an elastic string of length L vibrating as a result of an initial perturbation (see Figure 5.10). In particular, we assume that the string is initially at rest with a nonzero displacement having the form of a parabola. Thus the ICs (5.5.7) become

$$z = \phi(x) = x(L - x) \qquad \text{at } 0 \leq x \leq L \quad \text{and} \quad t = 0 \tag{8.6.36a}$$

$$\frac{\partial z}{\partial t} = \psi(x) = 0 \qquad \text{at } 0 \leq x \leq L \quad \text{and} \quad t = 0 \tag{8.6.36b}$$

On the other hand, the two ends of the string are held fixed, so the BCs (5.5.8) reduce to

$$z = z_1(t) = 0 \qquad \text{at } x = 0 \quad \text{and} \quad t > 0 \tag{8.6.37a}$$

$$z = z_2(t) = 0 \qquad \text{at } x = L \quad \text{and } t > 0 \tag{8.6.37b}$$

By applying Laplace transforms and using eqs. (8.2.30) and (8.6.36), eq. (8.6.35) leads to

$$s^2 Z(x, s) - s\phi(x) - \psi(x) - c^2 \frac{d^2 Z}{dx^2} = 0, \qquad 0 < x < L \tag{8.6.38}$$

which takes the form

$$c^2 \frac{d^2 Z}{dx^2} - s^2 Z(x, s) = -sx(L - x), \qquad 0 < x < L \tag{8.6.39}$$

This is a second-order linear ODE in the transformed variable $Z(x, s)$, which we solve next along with the BCs (8.6.37) in the transformed space; that is,

$$Z = 0 \qquad \text{at } x = 0 \tag{8.6.40a}$$

$$Z = 0 \qquad \text{at } x = L \tag{8.6.40b}$$

The solution of the homogeneous ODE associated with eq. (8.6.39) is given by

$$Z_h(x, s) = d_1 \sinh\left(\frac{sx}{c}\right) + d_2 \cosh\left(\frac{sx}{c}\right) \tag{8.6.41}$$

where d_1 and d_2 are integration constants. Using the particular solution

$$Z_p(x, s) = -\frac{2c^2}{s^3} + \frac{x(L - x)}{s} \tag{8.6.42}$$

and the BCs (8.6.40) to evaluate the integration constants, we obtain

$$Z(x, s) = \frac{2c^2}{s^3} \frac{\sinh\left(\dfrac{sx}{c}\right)\left[1 - \cosh\left(\dfrac{sL}{c}\right)\right] + \cosh\left(\dfrac{sx}{c}\right)\sinh\left(\dfrac{sL}{c}\right)}{\sinh\left(\dfrac{sL}{c}\right)}$$
$$- \frac{2c^2}{s^3} + \frac{x(L - x)}{s} \qquad (8.6.43)$$

which can be rearranged to give

$$Z(x, s) = \frac{2c^2}{s^3} \frac{\sinh\left(\dfrac{sx}{c}\right) + \sinh\left[\dfrac{s(L - x)}{c}\right]}{\sinh\left(\dfrac{sL}{c}\right)} - \frac{2c^2}{s^3} + \frac{x(L - x)}{s} \qquad (8.6.44)$$

In order to invert the Laplace transform (8.6.44), let us first consider the series expansion of the ratio:

$$\frac{1}{\sinh\left(\dfrac{sL}{c}\right)} = \frac{2\exp\left(-\dfrac{sL}{c}\right)}{1 - \exp\left(-\dfrac{2sL}{c}\right)} \qquad (8.6.45)$$

Recalling the geometric series

$$\sum_{n=0}^{\infty} q^n = \frac{1}{1 - q} \qquad (8.6.46)$$

with $|q| < 1$, eq. (8.6.45) leads to

$$\frac{1}{\sinh\left(\dfrac{sL}{c}\right)} = 2\exp\left(-\dfrac{sL}{c}\right)\sum_{n=0}^{\infty}\exp\left(-\dfrac{2nsL}{c}\right) \qquad (8.6.47)$$

This relation can be used to rewrite the transform (8.6.44) in the equivalent form:

$$Z(x, s) = \frac{2c^2}{s^3}\sum_{n=0}^{\infty}\left\{\exp\left[-s\frac{(2n + 1)L - x}{c}\right] - \exp\left[-s\frac{(2n + 1)L + x}{c}\right]\right.$$
$$\left. + \exp\left[-s\frac{2nL + x}{c}\right] - \exp\left[-s\frac{2(n + 1)L - x}{c}\right]\right\} - \frac{2c^2}{s^3} + \frac{x(L - x)}{s}$$
$$(8.6.48)$$

Note that all the terms of the summation above have the form

$$F(s) = \frac{1}{s^3}\exp(-sa)$$

Thus, since from Table 8.1 we have

$$\mathcal{L}^{-1}\left\{\frac{1}{s^2}\right\} = t$$

and

$$\mathcal{L}^{-1}\left\{\frac{\exp(-sa)}{s}\right\} = u(t - a)$$

from the convolution theorem (8.2.56) it follows that

$$\mathcal{L}^{-1}\left\{\frac{\exp(-sa)}{s^3}\right\} = \frac{1}{2}(t - a)^2 u(t - a) \tag{8.6.49}$$

We can now invert each of the terms in the rhs of eq. (8.6.48) to yield the desired expression for the displacement:

$$z(x, t) = c^2 \sum_{n=0}^{\infty} \left\{ \left[t - \frac{(2n + 1)L - x}{c} \right]^2 u\left[t - \frac{(2n + 1)L - x}{c} \right] \right.$$

$$- \left[t - \frac{(2n + 1)L + x}{c} \right]^2 u\left[t - \frac{(2n + 1)L + x}{c} \right]$$

$$+ \left[t - \frac{2nL + x}{c} \right]^2 u\left[t - \frac{2nL + x}{c} \right]$$

$$\left. - \left[t - \frac{2(n + 1)L - x}{c} \right]^2 u\left[t - \frac{2(n + 1)L - x}{c} \right] \right\} - c^2 t^2 + x(L - x) \tag{8.6.50}$$

It can be seen by direct substitution that eq. (8.6.50) satisfies the wave equation (8.6.35) together with the ICs (8.6.36) and the BCs (8.6.37).

It is worth noting that for any given t value, the series (8.6.50) is finite, since there is always a finite value of n above which the argument of each of the step functions becomes negative. Accordingly, the corresponding terms of the series vanish. The solution (8.6.50) is illustrated in Figure 8.7, where the displacement $Z(x, t)$ at a fixed location x is represented as a function of time t. It is seen that the solution is periodic in time. A function of this type is often called a *standing wave*.

In general, eq. (8.6.35) can also exhibit solutions having the form of *traveling waves*, which propagate in time. As an example, let us consider eq. (8.6.35) in the semi-infinite domain $x > 0$ with the following ICs:

$$z = 0 \qquad\qquad \text{at } x \geq 0 \quad \text{and} \quad t = 0 \tag{8.6.51a}$$

$$\frac{\partial z}{\partial t} = 0 \qquad\qquad \text{at } x \geq 0 \quad \text{and} \quad t = 0 \tag{8.6.51b}$$

and BCs:

$$z = z_1(t) \qquad\qquad \text{at } x = 0 \quad \text{and} \quad t > 0 \tag{8.6.52a}$$

$$z \rightarrow 0, \quad \frac{\partial z}{\partial x} \rightarrow 0 \qquad \text{as } x \rightarrow \infty \quad \text{and} \quad t > 0 \tag{8.6.52b}$$

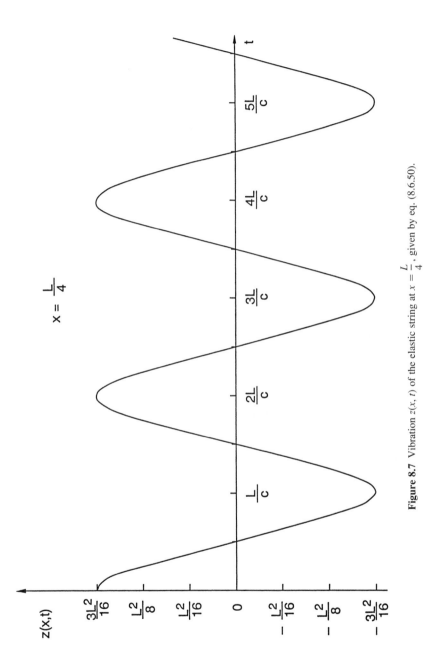

Figure 8.7 Vibration $z(x, t)$ of the elastic string at $x = \dfrac{L}{4}$, given by eq. (8.6.50).

Thus in this case we have a semi-infinite elastic string, initially at rest in the equilibrium position. At time $t = 0$, the end at $x = 0$ is forced to vibrate according to eq. (8.6.52a), and the vibration then propagates through the entire string.

Using Laplace transforms, together with eqs. (8.2.30) and (8.6.51), eq. (8.6.35) reduces to the homogeneous second-order ODE:

$$s^2 Z(x, s) - c^2 \frac{d^2 Z}{dx^2} = 0, \qquad x > 0 \tag{8.6.53}$$

and the BCs (8.6.52) become

$$Z = Z_1(s) \qquad\qquad \text{at } x = 0 \tag{8.6.54a}$$

$$Z \to 0, \quad \frac{dZ}{dx} \to 0 \qquad \text{as } x \to \infty \tag{8.6.54b}$$

where $Z_1(s)$ is the Laplace transform of $z_1(t)$. The general solution of eq. (8.6.53) is given by eq. (8.6.41), which, by evaluating the integration constants through the BCs (8.6.54), leads to

$$Z(x, s) = Z_1(s) \exp\left(-\frac{sx}{c}\right) \tag{8.6.55}$$

This transform can be inverted easily using eq. (8.2.7), which allows us to conclude that $z(x, t)$ is simply the function $z_1(t)$ shifted along the t axis by the quantity (x/c). Since for $t < x/c$ we have $z = 0$, the solution in the (x, t) plane is given by

$$z(x, t) = z_1\left(t - \frac{x}{c}\right) u\left(t - \frac{x}{c}\right) \tag{8.6.56}$$

This indicates that the vibration imposed at the $x = 0$ end of the elastic string propagates unchanged in space as a traveling wave with velocity c.

8.6.3   Moments of the Solution

The solution $y = y(t)$ of a problem can often be regarded as a probability density function, such that $y(t)\, dt$ represents the fraction of the entire population corresponding to a value of the independent variable t, which lies between t and $(t + dt)$. In this case the solution $y(t)$ can be conveniently characterized by its moments. They are integral transforms of the type (8.1.1), where the kernel is given by the power t^j. Thus, we define as *moment of jth order* of $y(t)$:

$$\mu_j = \int_0^\infty t^j y(t)\, dt \tag{8.6.57}$$

In principle, it is possible, although not easy, to retrieve the full distribution $y(t)$ from a suitable number of its moments. In practice, one can introduce some assumptions on the form of the distribution $y(t)$ in order to reduce the number of moments required for this purpose. However, the greatest advantage of the moments of a distribution is their clear physical meaning. The moments are in fact directly related to the most significant characteristics of a distribution such as its average:

$$t_{av} = \frac{\int_0^\infty t y(t)\, dt}{\int_0^\infty y(t)\, dt} = \frac{\mu_1}{\mu_0} \tag{8.6.58a}$$

and its variance:

$$\sigma^2 = \frac{\int_0^\infty (t - t_{av})^2 y(t)\, dt}{\int_0^\infty y(t)\, dt} = \frac{\mu_2}{\mu_0} - \left(\frac{\mu_1}{\mu_0}\right)^2 \tag{8.6.58b}$$

They provide useful global information about the distribution, which in many instances may also be measured experimentally. On the other hand, the experimental evaluation of the entire distribution is generally more difficult. An example of this is given by chromatography, as discussed later in this section.

The above observations provide a motivation for determining the moments of the solution of a PDE, without determining the solution itself fully. For this, Laplace transforms are particularly well suited. Let us consider the Laplace transform $Y(s) = \mathcal{L}\{y(t)\}$ defined by

$$Y(s) = \int_0^\infty e^{-st} y(t)\, dt \tag{8.6.59}$$

Recalling that the exponential function e^{-st} can be expanded in the neighborhood of $s = 0$ as

$$e^{-st} = 1 - st + \frac{1}{2!} s^2 t^2 + \cdots \tag{8.6.60}$$

the transform (8.6.59) can also be expanded, leading to

$$Y(s) = \int_0^\infty y(t)\, dt - s \int_0^\infty t y(t)\, dt + \frac{1}{2!} s^2 \int_0^\infty t^2 y(t)\, dt + \cdots$$
$$= \mu_0 - s\mu_1 + \frac{1}{2!} s^2 \mu_2 + \cdots \tag{8.6.61}$$

This means that the two integral transforms of a function $y(t)$—that is, the moments (8.6.57) and the Laplace transform (8.6.59)—are related. Thus we can determine the moments μ_j from the Laplace transform $Y(s)$ without inverting it. This is of some importance in providing information about $y(t)$, because inversion of Laplace transforms can be cumbersome.

By comparing eq. (8.6.61) with the series expansion of $Y(s)$ in the neighborhood of $s = 0$,

$$Y(s) = Y(0) + Y'(0)s + \frac{1}{2!} Y''(0)s^2 + \cdots \tag{8.6.62}$$

we obtain

$$
\begin{aligned}
\mu_0 &= Y(0) \\
\mu_1 &= -Y'(0) \\
\mu_2 &= Y''(0) \\
&\vdots \\
\mu_n &= (-1)^n \left(\frac{d^n Y}{ds^n} \right)_{s=0}
\end{aligned}
\qquad (8.6.63)
$$

Example. Consider the chromatography model described in chapter 5 and given by eq. (5.4.6), which involves plug flow of the fluid and local equilibrium between the fluid and solid phases. A more realistic representation of fixed-bed adsorbers is obtained by including axial dispersion phenomena and interphase mass transport resistances. Accordingly, we consider the following modified version of the chromatography model:

$$
-\varepsilon E_a \frac{\partial^2 C}{\partial z^2} + v \frac{\partial C}{\partial z} + \varepsilon \frac{\partial C}{\partial \bar{t}} + (1 - \varepsilon) \frac{\partial \Gamma}{\partial \bar{t}} = 0; \qquad 0 < z < L, \quad \bar{t} > 0 \quad (8.6.64)
$$

$$
\frac{\partial \Gamma}{\partial \bar{t}} = k(KC - \Gamma); \qquad 0 < z < L, \quad \bar{t} > 0 \qquad (8.6.65)
$$

where C and Γ represent the concentration of the adsorbable component in the fluid and solid phases, respectively, E_a is the axial dispersion coefficient, k is the interphase mass transport coefficient, and K is the adsorption constant describing the linear equilibrium between the two phases. The remaining quantities retain their meaning as defined in chapter 5. Furthermore, we assume that the fixed bed initially does not contain the adsorbable component; that is

$$
C = 0, \quad \Gamma = 0 \qquad \text{at } 0 \le z \le L \quad \text{and} \quad \bar{t} = 0 \qquad (8.6.66)
$$

As discussed in chapter 5, the proper BCs for eq. (8.6.64) are the Danckwerts conditions given by eqs. (5.2.2). However, in this case, in order to simplify the algebra, but without any loss of generality, we consider the following BCs:

$$
C = C^i(\bar{t}) \qquad \text{at } z = 0 \quad \text{and } \bar{t} \ge 0 \qquad (8.6.67a)
$$

$$
C = \text{finite} \qquad \text{as } z \to \infty \quad \text{and } \bar{t} \ge 0 \qquad (8.6.67b)
$$

Introducing the dimensionless quantities

$$
x = \frac{z}{L}, \quad t = \frac{\bar{t} v}{\varepsilon L}, \quad u = \frac{C}{C_r}, \quad w = \frac{\Gamma}{C_r}
$$

$$
\text{Pe} = \frac{vL}{\varepsilon E_a}, \quad \text{St} = \frac{kL\varepsilon}{v}, \quad \nu = \frac{1 - \varepsilon}{\varepsilon}
$$

$$
\qquad (8.6.68)
$$

where C_r is a reference concentration, eqs. (8.6.64) to (8.6.67) reduce to

$$
-\frac{1}{\text{Pe}} \frac{\partial^2 u}{\partial x^2} + \frac{\partial u}{\partial x} + \frac{\partial u}{\partial t} + \nu \frac{\partial w}{\partial t} = 0, \qquad 0 < x < 1 \quad \text{and} \quad t > 0 \qquad (8.6.69)
$$

$$
\frac{\partial w}{\partial t} = \text{St}(Ku - w), \qquad 0 < x < 1 \quad \text{and} \quad t > 0 \qquad (8.6.70)
$$

with ICs

$$u = 0, \quad w = 0 \qquad \text{at } 0 \le x \le 1 \quad \text{and} \quad t = 0 \qquad (8.6.71)$$

and BCs

$$u = u^i(t) \qquad\qquad \text{at } x = 0 \quad \text{and} \quad t > 0 \qquad (8.6.72a)$$

$$u = \text{finite} \qquad\qquad \text{as } x \to \infty \quad \text{and} \quad t > 0 \qquad (8.6.72b)$$

Introducing the Laplace transforms

$$U(x, s) = \mathcal{L}\{u(x, t)\} = \int_0^\infty e^{-st}u(x, t)\, dt \qquad (8.6.73a)$$

and

$$W(x, s) = \mathcal{L}\{w(x, t)\} = \int_0^\infty e^{-st}w(x, t)\, dt \qquad (8.6.73b)$$

the system of PDEs (8.6.69) to (8.6.72) reduces to a system consisting of one ODE and one algebraic equation—that is,

$$-\frac{1}{\text{Pe}}\frac{d^2U}{dx^2} + \frac{dU}{dx} + sU + sW = 0 \qquad (8.6.74)$$

$$sW = \text{St}(KU - W) \qquad (8.6.75)$$

Solving the algebraic equation for W and substituting in eq. (8.6.74), we obtain the following initial value problem in the transformed space:

$$-\frac{1}{\text{Pe}}\frac{d^2U}{dx^2} + \frac{dU}{dx} + \alpha(s)U = 0 \qquad (8.6.76)$$

where

$$\alpha(s) = s\left[1 + \frac{vKSt}{(s + St)}\right] \qquad (8.6.76a)$$

with BCs

$$U(x, s) = U^i(s) \qquad \text{at } x = 0 \qquad (8.6.77a)$$

$$U(x, s) = \text{finite} \qquad \text{as } x \to \infty \qquad (8.6.77b)$$

The solution of eq. (8.6.76) is

$$U(x, s) = A(s)e^{r_+x} + B(s)e^{r_-x} \qquad (8.6.78)$$

where

$$r_\pm = \frac{1 \pm \sqrt{1 + 4\alpha/\text{Pe}}}{2/\text{Pe}} \qquad (8.6.79)$$

are roots of the characteristic polynomial:

$$-\frac{1}{\text{Pe}}r^2 + r + \alpha = 0 \qquad (8.6.80)$$

From eq. (8.6.79) it is readily seen that $r_+ > 0$ and $r_- < 0$. Thus from the BC (8.6.77b) and eq. (8.6.78) it follows that $A(s) = 0$. Using the BC (8.6.73a) we find that $B(s) = U^i(s)$, and the Laplace transform of the solution $u(x, t)$ is then given by

$$U(x, s) = U^i(s)e^{r_-x} \qquad (8.6.81)$$

In lieu of inverting this transform, which involves a rather complex functional dependence on s, let us use it to find the moments of the solution, that is,

$$\mu_j(x) = \int_0^\infty t^j u(x, t) \, dt \qquad (8.6.82)$$

using eqs. (8.6.63). For this we need to evaluate the derivatives of the transform $U(x, s)$ with respect to s. Noting from eqs. (8.6.76a) and (8.6.79) that r_- is a function of s through the variable $\alpha(s)$—that is, $r_- = r_-(\alpha(s))$—we have

$$\frac{\partial U}{\partial s} = \frac{dU^i}{ds} e^{r_-(s)x} + r'_-(s)xU(x, s) \qquad (8.6.83a)$$

$$\frac{\partial^2 U}{\partial s^2} = \frac{d^2 U^i}{ds^2} e^{r_-(s)x} + r'_-(s)x \frac{dU^i}{ds} e^{r_-(s)x} + r''_-(s)xU(x, s) + r'_-(s)x \frac{\partial U}{\partial s} \qquad (8.6.83b)$$

The values of r_- and its derivatives needed to complete the above relations can be evaluated at $s = 0$, using eqs. (8.6.76a) and (8.6.79), as follows:

$$r_-(0) = 0 \qquad (8.6.84a)$$

$$r'_-(0) = -(1 + \nu K) \qquad (8.6.84b)$$

$$r''_-(0) = \frac{2}{Pe} (1 + \nu K)^2 + \frac{2}{St} \nu K \qquad (8.6.84c)$$

Substituting in eqs. (8.6.63) and (8.6.83), expressions for the first three moments are

$$\mu_0(x) = U(x, 0) = U^i(0) \qquad (8.6.85a)$$

$$\mu_1(x) = -\left(\frac{\partial U}{\partial s}\right)_{s=0} = \frac{dU^i}{ds}(0) + (1 + \nu K)xU^i(0) \qquad (8.6.85b)$$

$$\mu_2(x) = \left(\frac{\partial^2 U}{\partial s^2}\right)_{s=0} = \frac{d^2 U^i}{ds^2}(0) - 2(1 + \nu K)x \frac{dU^i}{ds}(0)$$

$$+ \left\{\left[\frac{2}{Pe}(1 + \nu K)^2 + \frac{2}{St}\nu K\right]x + (1 + \nu K)^2 x^2\right\}U^i(0) \qquad (8.6.85c)$$

In order to get a better physical insight into the solution, let us select a concrete form for the inlet concentration, $u^i(t)$. In particular, we consider a unit impulse of the adsorbable component:

$$u^i(t) = \delta(t) \qquad (8.6.86)$$

Thus we have

$$U^i(s) = \int_0^\infty e^{-st}\delta(t) \, dt = 1 \qquad (8.6.87)$$

and then

$$\frac{dU^i}{ds}(0) = \frac{d^2 U^i}{ds^2}(0) = 0 \qquad (8.6.88)$$

Substituting in eqs. (8.6.85), the first three moments of the concentration of the adsorbable component at various locations x along the fixed bed are

$$\mu_0(x) = 1 \tag{8.6.89a}$$

$$\mu_1(x) = (1 + \nu K)x \tag{8.6.89b}$$

$$\mu_2(x) = \left[\frac{2}{Pe} (1 + \nu K)^2 + \frac{2}{St} \nu K \right] x + (1 + \nu K)^2 x^2 \tag{8.6.89c}$$

These moments relate to the concentration time history at a fixed location, x, which has the qualitative form shown in Figure 8.8. This time history is in fact the *residence-time distribution* of the adsorbable component.

Using eqs. (8.6.58) we can compute two characteristic parameters of the distribution, the average residence time of the adsorbable component:

$$t_{av} = \frac{\mu_1}{\mu_0} = (1 + \nu K)x \tag{8.6.90}$$

and the variance of the concentration distribution:

$$\sigma^2 = \frac{\mu_2}{\mu_0} - \left(\frac{\mu_1}{\mu_0} \right)^2 = \left[\frac{2}{Pe} (1 + \nu K)^2 + \frac{2}{St} \nu K \right] x \tag{8.6.91}$$

As expected, the average residence time of the adsorbable component increases linearly along the fixed-bed axis, and at the bed outlet, i.e., $x = 1$, is equal to:

$$t_{av} = (1 + \nu K) \tag{8.6.92}$$

Note that this value, when transformed into dimensional quantities using eq. (8.6.68), becomes identical to the expression of the average residence time (6.2.18), derived earlier

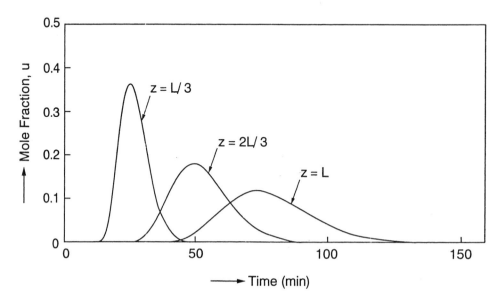

Figure 8.8 Qualitative behavior of the concentration distribution of the adsorbable component along the fixed bed.

in Chapter 6. Since in the previous case axial dispersion and interphase mass transfer resistances were neglected, we can conclude that these phenomena do not influence the average residence time. However, they do affect the width of the distribution around its average value, as seen from the expression of the variance given by eq. (8.6.91). It appears that the concentration pulse spreads while moving along the fixed-bed, and at the outlet its variance is given by:

$$\sigma^2 = \frac{2}{Pe}(1 + vK)^2 + \frac{2}{St}vK \qquad (8.6.93)$$

Both the dissipation processes considered, i.e., the axial dispersion and the interphase mass transport, increase the width of the concentration distribution. When the effects of these processes are negligible, i.e., Pe $\rightarrow \infty$ and St $\rightarrow \infty$, the variance $\sigma \rightarrow 0$, thus indicating that the concentration pulse reaches the fixed-bed outlet without changing its width. This is in agreement with the result obtained in Chapter 6 assuming plug-flow for the fluid phase and local interphase equilibrium along the bed. In this case, from eq. (6.2.24) it is seen that the inlet concentration pulse travels unchanged along the fixed-bed.

The above features are illustrated in Figure 8.9 for a chromatography column filled with zeolite Y. In this case a pulse of a ternary mixture of m-xylene, ethylbenzene and p-xylene was fed to the column using isopropylbenzene as carrier. The experimental data denote the concentrations of each component in the outlet stream of the column as a function of time (Storti et al., 1985). It is seen that the elution curves have the form of broad peaks, which in agreement with eq. (8.6.92), exit the column in the reverse order of the adsorption strengths of the components, i.e., the K values are in the order p-xylene > ethylbenzene > m-xylene. The broadening of the peaks is due to axial dispersion and mass transfer resistances. Note that their overlapping determines the degree of separation of the components—an issue of importance in applications.

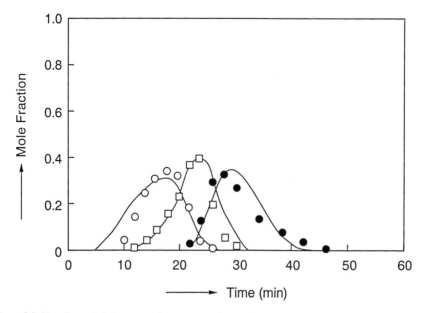

Figure 8.9 Experimental elution curves for a ternary mixture of m-xylene ($\circ$), ethylbenzene ($\square$) and p-xylene ($\bullet$).

References

R. E. Bellman and R. S. Roth, *The Laplace Transform*, World Scientific Publishing, Singapore, 1984.

R. V. Churchill, *Operational Mathematics*, 3rd ed., McGraw-Hill, New York, 1972.

A. Erdélyi, W. Magnus, F. Oberhettinger and F. Tricomi, *Tables of Integral Transforms*, volume 1, McGraw-Hill, New York, 1954.

G. Storti, E. Santacesaria, M. Morbidelli, and S. Carrá, "Separation of Xylenes on Zeolites in the Vapor Phase," *Industrial and Engineering Chemistry, Product Research and Development*, **24**, 89–92 (1985).

Additional Reading

H. S. Carslaw and J. C. Jaeger, *Operational Methods in Applied Mathematics*, Dover, New York, 1963.

P. A. McCollum and B. F. Brown, *Laplace Transform Tables and Theorems*, Holt, Rinehart and Winston, New York, 1965.

F. E. Nixon, *Handbook of Laplace Transformation; Fundamentals, Applications, Tables and Examples*, Prentice-Hall, Englewood Cliffs, NJ, 1965.

E. D. Rainville, *The Laplace Transform*, Macmillan, New York, 1963.

D. V. Widder, *The Laplace Transform*, Princeton University Press, Princeton, NJ, 1941.

Problems

8.1 Prove the following results:

$$\mathcal{L}\{\cos at\} = \frac{s}{s^2 + a^2}, \qquad s > 0$$

$$\mathcal{L}\{\sinh at\} = \frac{a}{s^2 - a^2}, \qquad s > |a|$$

$$\mathcal{L}\{\cosh at\} = \frac{s}{s^2 - a^2}, \qquad s > |a|$$

$$\mathcal{L}\{t^n\} = \frac{n!}{s^{n+1}}, \qquad s > 0 \quad \text{and} \quad n = 0, 1, 2, \ldots$$

8.2 Prove the following results

$$\mathcal{L}\{e^{at} \sin bt\} = \frac{b}{(s - a)^2 + b^2}, \qquad s > a$$

$$\mathcal{L}\{e^{at} \cos bt\} = \frac{s - a}{(s - a)^2 + b^2}, \qquad s > a$$

$$\mathcal{L}\{t^p\} = \frac{\Gamma(p + 1)}{s^{p+1}}, \qquad s > 0 \quad \text{and} \quad p > -1$$

$$\mathcal{L}\{t^n e^{at}\} = \frac{n!}{(s - a)^{n+1}}, \qquad s > a \quad \text{and} \quad n = 0, 1, 2, \ldots$$

8.3 Find the Laplace transform of the periodic function

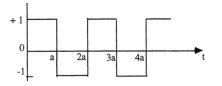

and represent it in terms of simple functions. (*Hint:* $1/(1 + x) = 1 - x + x^2 + \cdots; x^2 <$
∞).

8.4 Solve problem 8.3 using the relation for the Laplace transform of periodic functions
(8.2.22).

8.5 Solve by Laplace transform:

$$y'' + y' - 2y = 3 \cos 3t - 11 \sin 3t$$

with ICs

$$y(0) = 0, \qquad y'(0) = 6$$

8.6 Solve by Laplace transform:

$$y''' + y'' - y' - y = 9e^{2t}$$

with ICs

$$y(0) = 2, \qquad y'(0) = 4, \qquad y''(0) = 3$$

8.7 Solve by Laplace transform the boundary value ODE:

$$y'' + 2y' - y = 2t$$

with BCs

$$y(0) = 1, \qquad y(1) = 0$$

8.8 Solve by Laplace transform the system of first-order ODEs:

$$y_1' + 2y_2' - y_1 = t$$
$$y_1' + y_2 = e^{-t}$$

with ICs

$$y_1(0) = 1, \qquad y_2(0) = 0$$

8.9 Solve by Laplace transform the system of ODEs:

$$y_1'' + y_1' + 2y_2' - y_1 = 1$$
$$y_2''' + y_1'' - 2y_1 + y_2 = e^{-t}$$

with ICs

$$y_1(0) = 2, \ y_1'(0) = 0, \ y_2(0) = 2, \ y_2'(0) = 0, \ y_2''(0) = 1$$

8.10 Solve the diffusion-reaction problem given by eqs. (8.6.1) and (8.6.2), in the case where
the nonhomogeneous term is given by $f(t) = e^{-t}$. Also identify the impulse response.

8.11 Consider the homogeneous parabolic PDE given by eq. (8.6.1) with $f(t) = 0$ and the IC (8.6.2a). Find the impulse responses for the BCs:

$$c = c_1(t) \qquad \text{at } x = 0 \quad \text{and} \quad t > 0$$
$$c = c_2(t) \qquad \text{at } x = L \quad \text{and} \quad t > 0$$

8.12 Consider the transient heat conduction problem in a one-dimensional rod with no internal heat generation, given by eq. (7.1.1) with $f(x) = 0$:

$$\frac{1}{\kappa} \frac{\partial T}{\partial t} = \frac{\partial^2 T}{\partial x^2} - U^2 T; \qquad x \in (0, L), \quad t > 0$$

with the BCs

$$T = T_0; \qquad x = 0, \quad t > 0$$
$$\frac{\partial T}{\partial x} = 0; \qquad x = L, \quad t > 0$$

and the IC

$$T = 0; \qquad x \in [0, L], \quad t = 0$$

Compare the solutions obtained using Laplace and Fourier transform methods.

8.13 Consider the first-order PDE describing the plug-flow, homogeneous reactor with first-order reaction given by eqs. (5.4.2) and (5.4.3). Solve using the Laplace transform, in the case where $u^0(x) = 0$. Compare the solution thus obtained with eq. (5.4.44).

8.14 Determine an expression for the displacement z of a one-dimensional vibrating string, described by eq. (8.6.35), where the ends are held fixed and the ICs are given by

$$z = \phi(x) = \begin{cases} 2x & \text{for } 0 < x < L/4, \quad t = 0 \\ \dfrac{2}{3}(L - x) & \text{for } L/4 < x < L, \quad t = 0 \end{cases}$$

$$\frac{\partial z}{\partial t} = \phi(x) = 0 \qquad \text{for } 0 < x < L, \quad t = 0$$

Compare the solution with that reported in section 5.5.3

8.15 Consider the axial dispersion of a tracer in an empty tube:

$$-\frac{1}{\text{Pe}} \frac{\partial^2 u}{\partial x^2} + \frac{\partial u}{\partial x} + \frac{\partial u}{\partial t} = 0, \qquad 0 < x < 1 \quad \text{and} \quad t > 0$$

with IC

$$u = 0 \qquad \text{at } 0 \leq x \leq 1 \quad \text{and} \quad t = 0$$

and BCs

$$u = \delta(t) \qquad \text{at } x = 0 \quad \text{and} \quad t > 0$$
$$u = \text{finite} \qquad \text{as } x \to \infty \quad \text{and} \quad t > 0$$

Determine the average residence time and variance of the tracer concentration distribution at the tube outlet.

8.16 Find the average residence time and variance of the adsorbable component concentration at the outlet of the fixed-bed adsorber described by the model (8.6.69) to (8.6.72). Consider the case where the inlet concentration is given by the unit step function defined by eq. (8.2.4):

$$u^i(t) = u(t - 0) - u(t - \Delta)$$

8.17 Consider the axial dispersion model of the homogeneous tubular reactor with first-order reaction given by eqs. (5.2.3) and (5.2.4). Determine the first three moments of the solution using Laplace transforms.

8.18 In several cases of practical interest, reacting mixtures involve such a large number of components that it is convenient to treat them as continua rather than as collections of individual species (cf. R. Aris, *AIChE Journal*, **35**, 539, 1989). Thus, by taking x as a label indicating the different species, $c(x) \, dx$ represents the concentration of material whose label lies between x and $x + dx$. In general, the concentration distribution is then given by a continuous function, $c(x) = c_0 f(x)$. Since c_0 is the total concentration, $f(x)$ satisfies the constraint

$$\int_0^\infty f(x) \, dx = 1$$

Consider the case where all chemical species disappear according to a first-order reaction, whose rate constant is a function of the label variable—that is, $k(x) = \bar{k}h(x)$. The function $h(x)$, which is assumed monotonically increasing for $x \in (0, \infty)$, satisfies the constraint

$$\int_0^\infty h(x)f(x) \, dx = 1$$

so that $\bar{k}$ above represents the weighted average value of the reaction rate constant. The disappearance of the various species is governed by the PDE

$$\frac{\partial c(x, t)}{\partial t} = -k(x)c(x, t)$$

with the IC

$$c = c_0 f(x), \qquad t = 0$$

Introducing the dimensionless quantities $u(x, \tau) = c(x, t)/c_0$ and $\tau = \bar{k}t$, the above equation leads to

$$u(x, \tau) = f(x) \exp[-h(x)\tau]$$

and the overall concentration of reacting species is then readily obtained as

$$U(\tau) = \int_0^\infty u(x, \tau) \, dx = \int_0^\infty f(x) \exp[-h(x)\tau] \, dx$$

It is worth noting that, by taking $h(x) = x$ without loss of generality, the overall mass concentration $U(\tau)$ coincides with the Laplace transform of the initial reactants distribution $f(x)$, where τ replaces the usual Laplace variable s:

$$U(\tau) = \mathcal{L}[f(x)]$$

In applications, it is often desired to describe the time evolution of the entire mass of reactants (or of a selected subgroup of them). This corresponds to adopting a so-called *lumped kinetic model* such that

$$\frac{dU}{d\tau} = -F(U)$$

with the IC

$$U = 1 \qquad \text{at } \tau = 0$$

The problem then is to determine the expression for the lumped kinetics $F(U)$ which represents correctly the evolution of the continuous mixture. This is not easy since, even for the simple case being considered, the form of the lumped kinetic model depends upon the form of the initial reactant distribution, $f(x)$. To illustrate this point, we approach the problem from the opposite direction. For the nth-order lumped kinetic model

$$F(U) = U^n$$

derive the corresponding initial reactant distribution $f(x)$ in the two cases: $n = 0$ and $n \geq 1$. Show that even though each component disappears by a first-order kinetics, by properly selecting the form of $f(x)$ the overall mass may disappear according to a lumped kinetics of *any* order $n \geq 1$.

Chapter 9

PERTURBATION METHODS

Perturbation methods are used to provide approximate solutions of nonlinear problems characterized by the presence of a small parameter. These approximations represent an asymptotic series expansion of the true solution truncated typically after a few terms. A condition for the successful application of these methods is that the reduced problem, obtained by the leading order analysis, is amenable to analytic solution.

We first discuss the simple case of regular perturbation expansions and then proceed to analyze the more complex singular perturbation expansions. For the latter, there is no unique procedure and various methods can be applied depending on the specific problem at hand. We describe a few of these, including the method of strained coordinates, matched asymptotic expansions, the WKB method, and the multiple scale method. Several examples are discussed to highlight the potential and limitations of each technique. However, let us first examine the main features of perturbation methods in general and introduce some concepts related to asymptotic expansions.

9.1 THE BASIC IDEA

Equations deriving from mathematical models of physical phenomena usually contain various parameters arising from elementary processes which together constitute the specific phenomenon being studied, such as mass or heat transport, chemical reactions, convective motion, and so on. In most cases such equations are nonlinear and can only be solved numerically. However, it is sometimes possible to obtain analytical approximations of the solution, valid in the entire domain of interest or only in some portion of it, which can be useful in investigating the model at hand. Perturbation methods constitute one important class of such approximation techniques.

Perturbation methods are most conveniently applied when the following two conditions are satisfied by the equation to be solved:

1. One of the parameters of the equation attains only very small or very large values. In fact, since the inverse of a large number is small, without any loss of generality, we may consider only the case of a small parameter, ε.
2. The equation can be solved analytically in the case where $\varepsilon = 0$. The solution thus obtained is usually referred to as the *unperturbed* solution.

Under these circumstances, we can attempt to perturb the solution valid for $\varepsilon = 0$, so as to make it valid also for finite ε values in the neighborhood of zero. This is obtained by developing asymptotic expansions of the solution for small ε values, such as

$$y = y_0 + \varepsilon y_1 + \varepsilon^2 y_2 + \cdots \tag{9.1.1}$$

which can be truncated after one, two, or more terms, leading to the so called zeroth-, first-, or higher-order approximations, respectively. When ε is a naturally small parameter, first- or second-order approximations yield the exact solution in the entire domain of interest with sufficient accuracy for most practical purposes.

The perturbation methods described above are usually referred to as *parameter perturbation* methods. In this chapter we will focus largely on these methods which are *local* with respect to the parameter; that is, the parameter ε takes on small values near zero. However, this local analysis with respect to the parameter ε in fact provides an ideal tool for *global* analysis (i.e., in the entire domain) with respect to the independent variables.

It is worth noting that the two conditions mentioned above are often satisfied in practice, as a consequence of the intrinsic nature of physical phenomena. Usually, each elementary process constituting a given phenomenon does not contribute to the same extent. In particular, one or more among them may provide only marginal contributions. However, these processes with marginal contributions are often responsible for the increased mathematical complexity of the model, which manifests itself in a variety of ways—for example, nonlinearities, higher-order derivatives, or additional independent variables (space or time). This is the ideal situation for applying perturbation methods since the effects of all the elementary processes with marginal contributions are accounted for as perturbation of the basic unperturbed (i.e., $\varepsilon = 0$) solution. The latter is derived from the general model where the mathematical complexities arising from the marginal elementary processes are neglected.

It should also be noted that in some instances perturbation methods can be used to provide analytical solutions of equations which in fact do not contain any small parameter, ε. In these cases, the small parameter is introduced artificially into the equation, in such a way that the zeroth (and possibly higher)-order approximations are obtained in closed form. The desired solution is then obtained from the approximate solution thus developed, by introducing an ε value to recover the original equation (see problem 9.7).

9.2 ASYMPTOTIC SEQUENCES AND ASYMPTOTIC EXPANSIONS

A sequence of functions $u_n(\varepsilon)$ is called *asymptotic* when

$$u_n(\varepsilon) = o[u_{n-1}(\varepsilon)] \qquad \text{as } \varepsilon \to 0 \qquad (9.2.1)$$

where the order symbol o indicates that as $\varepsilon \to 0$, $u_n(\varepsilon)$ approaches zero faster than $u_{n-1}(\varepsilon)$. More rigorously, the statement

$$f(\varepsilon) = o[g(\varepsilon)] \qquad \text{as } \varepsilon \to 0 \qquad (9.2.2)$$

implies that

$$\lim_{\varepsilon \to 0} \left| \frac{f(\varepsilon)}{g(\varepsilon)} \right| = 0 \qquad (9.2.3)$$

The order symbol o should not be confused with the more common symbol O, indicating the order of magnitude—that is,

$$f(\varepsilon) = O[g(\varepsilon)] \qquad \text{as } \varepsilon \to 0 \qquad (9.2.4)$$

which means that

$$\lim_{\varepsilon \to 0} \left| \frac{f(\varepsilon)}{g(\varepsilon)} \right| = \text{finite} \qquad (9.2.5)$$

Examples of asymptotic sequences are

$$\varepsilon^n; \quad n\varepsilon^n; \quad (\sin \varepsilon)^{n/2}; \quad (\sinh \varepsilon)^{2n}; \quad [\ln (1 + \varepsilon)]^n; \quad (\varepsilon \ln \varepsilon)^n \qquad (9.2.6)$$

The asymptotic expansion of any given function, $y(\varepsilon)$ (which we should regard as the exact solution of some specific equation) as $\varepsilon \to 0$, is obtained by using an asymptotic sequence $u_n(\varepsilon)$ and a sequence of coefficients y_n *independent* of ε. In particular, the expression

$$y(\varepsilon) = \sum_{n=0}^{\infty} y_n u_n(\varepsilon) \qquad \text{as } \varepsilon \to 0 \qquad (9.2.7)$$

is called an *asymptotic expansion* if and only if

$$y(\varepsilon) = \sum_{n=0}^{N-1} y_n u_n(\varepsilon) + O[u_N(\varepsilon)] \qquad \text{as } \varepsilon \to 0 \qquad (9.2.8)$$

Similarly, in the case where y is a function of one or more independent variables x, we obtain

$$y(x, \varepsilon) = \sum_{n=0}^{\infty} y_n(x) u_n(\varepsilon) \qquad \text{as } \varepsilon \to 0 \qquad (9.2.9)$$

where the dependence upon the independent variables is confined to the coefficients y_n, thus indicating that the approximation proposed has a *local* validity with respect to the parameter ε, but a *global* one with respect to the independent variable x. Note that even the zeroth-order solution $y_0(x)$ provides an approximation of $y(x)$ valid in the *entire* x domain.

The general procedure for evaluating the coefficients $y_n(x)$ of the asymptotic expansion (9.2.9) of a given function $y(x, \varepsilon)$, once the asymptotic sequence $u_n(\varepsilon)$ has been selected, is based on the following equations:

$$y_0(x) = \lim_{\varepsilon \to 0} \frac{y(x, \varepsilon)}{u_0(\varepsilon)} \tag{9.2.10a}$$

$$y_1(x) = \lim_{\varepsilon \to 0} \frac{y(x, \varepsilon) - y_0(x)u_0(\varepsilon)}{u_1(\varepsilon)} \tag{9.2.10b}$$

$$y_N(x) = \lim_{\varepsilon \to 0} \frac{y(x, \varepsilon) - \sum_{n=0}^{N-1} y_n(x)u_n(\varepsilon)}{u_N(\varepsilon)} \tag{9.2.10c}$$

In the rather frequent case where the asymptotic sequence is given by powers of ε— that is $u_n(\varepsilon) = \varepsilon^n$—this procedure leads to the Taylor series expansion

$$y(x, \varepsilon) = \sum_{n=0}^{\infty} \frac{\varepsilon^n}{n!} \left(\frac{\partial^n y}{\partial \varepsilon^n} \right)_{\varepsilon=0} \tag{9.2.11}$$

This can be readily verified by noting that the nth coefficient of Taylor's expansion, $y_n = (1/n!)(\partial^n y/\partial \varepsilon^n)_{\varepsilon=0}$, can be obtained from eq. (9.2.10c) with $u_n(\varepsilon) = \varepsilon^n$ by applying L'Hôpital's rule for evaluating the limit as $\varepsilon \to 0$.

The series (9.2.8) is a *uniform* asymptotic expansion—that is, an expansion valid uniformly over the entire domain—if it fulfills the condition

$$y(x, \varepsilon) = \sum_{n=0}^{N-1} y_n(x)u_n(\varepsilon) + O[u_N(\varepsilon)] \qquad \text{as } \varepsilon \to 0 \tag{9.2.12}$$

for *all* x values in the domain of interest. It should be stressed that conditions (9.2.8) and (9.2.12) do *not* require the series to be convergent, and in fact asymptotic series can sometimes be divergent.

A simple example of such a situation is given by the asymptotic expansion

$$y(x, \varepsilon) = \sum_{n=0}^{\infty} n!\varepsilon^n x^n \qquad \text{as } \varepsilon \to 0 \tag{9.2.13}$$

valid in the domain $x \in [0, 1]$. By applying the ratio test for series convergence, it is seen that

$$\lim_{n \to \infty} \frac{(n+1)!\varepsilon^{n+1} x^{n+1}}{n!\varepsilon^n x^n} = \lim_{n \to \infty} (n+1)\varepsilon x = \infty \tag{9.2.14}$$

which indicates that the series diverges. On the other hand, from condition (9.2.12), which, using eq. (9.2.5), can be rewritten as

$$\lim_{\varepsilon \to 0} \frac{y(x, \varepsilon) - \sum_{n=0}^{N-1} n!\varepsilon^n x^n}{\varepsilon^N} = \lim_{\varepsilon \to 0} \frac{\sum_{n=N}^{\infty} n!\varepsilon^n x^n}{\varepsilon^N} = N!x^N \tag{9.2.15}$$

it can be concluded that the asymptotic expansion (9.2.13) is uniformly valid in the entire domain of interest $x \in [0, 1]$. From the above we can observe that the only requirement for an asymptotic expansion is that it should approach the original function, as the small parameter approaches zero, with a fixed number of terms [i.e., N terms in eqs. (9.2.8) and (9.2.12)]. In developing an asymptotic expansion, we are not necessarily interested in the behavior of the series when the number of its terms increases—that is, in its convergence characteristics (see problem 9.5).

Nonuniform expansions (i.e., such that condition 9.2.12 is not valid in some portion of the x domain) are quite common in applications. A typical example is given by equations where the independent variable belongs to an infinite domain, $x \in [0, \infty)$, and the perturbation parameter is multiplied by the independent variable x. Thus the asymptotic expansion as $\varepsilon \to 0$ for the function

$$y(x, \varepsilon) = \frac{1}{1 + \varepsilon x} \qquad (9.2.16)$$

can be represented in a Taylor series as follows:

$$y(x, \varepsilon) = 1 - \varepsilon x + \varepsilon^2 x^2 \ldots \qquad (9.2.17)$$

which corresponds to eq. (9.2.12), where $u_n = \varepsilon^n$ and $y_n = (-1)^n x^n$. However, such an asymptotic expansion is nonuniform, since it breaks down for $\varepsilon x > 1$, and thus condition (9.2.12) is certainly violated as $x \to \infty$. In other words, for any fixed ε value, no matter how small, there exists an x value large enough such that any term in the series becomes much larger than the preceding one. However, if the truncated series is applied only in a finite domain (i.e., $x \leq \bar{x}$), then it can certainly be used provided that ε is sufficiently small.

It is worth noting that uniform expansions are possible also in the case of infinite domain for x, as is the case for the function

$$\frac{1}{1 + \varepsilon + x} = \frac{1}{1 + x} - \frac{\varepsilon}{(1 + x)^2} + \frac{\varepsilon^2}{(1 + x)^3} \cdots \qquad (9.2.18)$$

A quite convenient test to verify that a particular expansion is uniform is given by the following condition

$$y_{n+1}(x)u_{n+1}(\varepsilon) = o[y_n(x)u_n(\varepsilon)] \qquad \text{as } \varepsilon \to 0 \qquad (9.2.19)$$

for *all* x values in the domain of interest. This implies that each term of the series is smaller than the preceding one:

$$y_{n+1}(x)u_{n+1}(\varepsilon) \ll y_n(x)u_n(\varepsilon) \qquad \text{as } \varepsilon \to 0 \qquad (9.2.20)$$

which in most cases is easy to verify. It should be noted that eq. (9.2.19) provides a sufficient (but not necessary) condition for (9.2.12) to be satisfied (see problem 9.4).

In the two examples above, the power series ε^n has been used as the asymptotic sequence [i.e., $u_n(\varepsilon) = \varepsilon^n$], thus leading to the asymptotic series (9.2.17) and (9.2.18). This is indeed the most frequent choice, even though it is certainly not the only one. On the other hand, it can be shown that for any given asymptotic sequence there exists a unique asymptotic expansion (see problem 9.2). The proper choice of the asymptotic sequence, and sometimes even of the small parameter ε (whose definition is not always obvious), can be of great importance for the success of the perturbation method. Such choices are usually based on analogy with similar problems or on *a priori* information about some properties of the solution.

In the following sections, we shall address the problem of constructing the asymptotic expansion (9.2.9)—that is, given the asymptotic sequence $u_n(\varepsilon)$—to find the sequence of coefficients, y_n. When the result is a uniform expansion, then the problem is solved, and the only question remaining is where to truncate the series in order to get a satisfactory approximation. On the other hand, if the result is a nonuniform expansion, then other specific techniques must be applied in order to render such an expansion uniformly valid.

9.3 REGULAR PERTURBATION METHOD

To illustrate the general procedure of applying the *regular perturbation* method, let us first consider the following simple *algebraic* equation:

$$y^2 + y + \varepsilon = 0 \qquad (9.3.1)$$

where ε is a small parameter, and we seek an asymptotic expansion of the solution $y(\varepsilon)$ as $\varepsilon \to 0$.

The first step is to select a proper asymptotic sequence, $u_n(\varepsilon)$. Using the power series $u_n = \varepsilon^n$, the following series expansion of $y(\varepsilon)$ as $\varepsilon \to 0$ is obtained:

$$y(\varepsilon) = \sum_{n=0}^{\infty} \varepsilon^n y_n \qquad (9.3.2)$$

where y_n are the coefficients independent of ε, which need to be determined.

The second step is to substitute the proposed approximation in the original equation and to expand for small ε. Thus, substituting eq. (9.3.2) in eq. (9.3.1) it follows that

$$(y_0 + \varepsilon y_1 + \varepsilon^2 y_2 + \cdots)^2 + (y_0 + \varepsilon y_1 + \varepsilon^2 y_2 + \cdots) + \varepsilon = 0 \quad (9.3.3)$$

which, expanding for small ε, reduces to

$$y_0^2 + 2y_0 y_1 \varepsilon + (y_1^2 + 2y_0 y_2)\varepsilon^2 + \cdots + y_0 + \varepsilon y_1 + \varepsilon^2 y_2 + \cdots + \varepsilon = 0 \quad (9.3.4)$$

and grouping coefficients of like powers of ε leads to

$$(y_0^2 + y_0) + (2y_0 y_1 + y_1 + 1)\varepsilon + (y_1^2 + 2y_0 y_2 + y_2)\varepsilon^2 + \cdots = 0 \quad (9.3.5)$$

Since eq. (9.3.5) has to be satisfied for *all* values of ε, and the coefficients y_n do not depend upon ε, we can conclude that *each* of the terms multiplying different powers of ε must be equal to zero. Thus

$$y_0(y_0 + 1) = 0 \qquad (9.3.6)$$
$$2y_0 y_1 + y_1 + 1 = 0 \qquad (9.3.7)$$
$$y_1^2 + 2y_0 y_2 + y_2 = 0 \qquad (9.3.8)$$
$$\vdots$$

The last step of the procedure is to solve the above equations for the unknown coefficients y_n, and then to substitute them into the asymptotic expansion (9.3.2).

In this case, since the first equation (9.3.6) leads to two roots (i.e., $y_0 = 0$ or

-1), there are two distinct sets of values for the sequence of coefficients y_n which satisfy eqs. (9.3.6) to (9.3.8). Each of these can be substituted in eq. (9.3.2), leading to two possible series expansions of $y(\varepsilon)$:

$$y(\varepsilon) = -\varepsilon - \varepsilon^2 + \cdots \qquad (9.3.9a)$$

and

$$y(\varepsilon) = -1 + \varepsilon + \varepsilon^2 + \cdots \qquad (9.3.9b)$$

Since eq. (9.3.1) is a quadratic, the exact solutions are given as

$$y_\pm = \frac{-1 \pm \sqrt{1 - 4\varepsilon}}{2} \qquad (9.3.10)$$

which, by substituting the Taylor series expansion

$$\sqrt{1 - 4\varepsilon} = 1 - 2\varepsilon - 2\varepsilon^2 + \cdots \qquad (9.3.11)$$

become identical to the expansions (9.3.9) generated by the regular perturbation method.

A somewhat more complex problem is given by the following ordinary differential equation

$$\frac{d^2y}{dx^2} = \varepsilon y, \qquad x \in (0, 1) \qquad (9.3.12)$$

with boundary conditions (BCs)

$$\frac{dy}{dx} = 0 \qquad \text{at } x = 0 \qquad (9.3.13a)$$

$$y = 1 \qquad \text{at } x = 1 \qquad (9.3.13b)$$

which represents the concentration profile y in a flat slab of catalyst where an isothermal first-order reaction occurs. The Thiele modulus, defined as $\sqrt{\varepsilon}$, is a small parameter in the case where the rate of internal diffusive mass transport is much faster than the rate of the chemical reaction.

The exact solution of eq. (9.3.12), along with the BCs, is readily obtained following the methods described in chapter 3, leading to

$$y(x) = \frac{\cosh(\sqrt{\varepsilon}x)}{\cosh(\sqrt{\varepsilon})} \qquad (9.3.14)$$

Let us now reproduce this result, in the limit of small ε values, using the regular perturbation method. Realizing that the solution of the problem at $\varepsilon = 0$ corresponding to the situation where no reaction occurs is known as $y(x) = 1$, we are under those circumstances described in section 9.1, where perturbation methods are ideally suited.

Thus, let us first assume the following asymptotic series approximation of the solution

$$y(x, \varepsilon) = \sum_{n=0}^{\infty} \varepsilon^n y_n(x) \qquad (9.3.15)$$

and then substitute it in the original problem (9.3.12) and (9.3.13). By comparing terms of equal orders of magnitude (i.e., containing like powers of the small parameter ε), it follows that

$$O(\varepsilon^0): \qquad \frac{d^2 y_0}{dx^2} = 0, \qquad x \in (0, 1) \qquad\qquad (9.3.16)$$

with BCs

$$\frac{dy_0}{dx} = 0, \qquad \text{at } x = 0$$

$$y_0 = 1, \qquad \text{at } x = 1$$

$$O(\varepsilon^1): \qquad \frac{d^2 y_1}{dx^2} = y_0, \qquad x \in (0, 1) \qquad\qquad (9.3.17)$$

with BCs

$$\frac{dy_1}{dx} = 0, \qquad \text{at } x = 0$$

$$y_1 = 0, \qquad \text{at } x = 1$$

$$O(\varepsilon^2): \qquad \frac{d^2 y_2}{dx^2} = y_1, \qquad x \in (0, 1) \qquad\qquad (9.3.18)$$

with BCs

$$\frac{dy_2}{dx} = 0, \qquad \text{at } x = 0$$

$$y_2 = 0, \qquad \text{at } x = 1$$

and so on.

By solving each of the above problems, it is found that

$$y_0(x) = 1 \qquad\qquad (9.3.19a)$$

$$y_1(x) = -\frac{1}{2}(1 - x^2) \qquad\qquad (9.3.19b)$$

$$y_2(x) = \frac{1}{4}\left(\frac{5}{6} - x^2 + \frac{x^4}{6}\right) \qquad\qquad (9.3.19c)$$

$$\vdots$$

Substituting these in eq. (9.3.15) leads to the desired series expansion

$$y(x, \varepsilon) = 1 - \frac{\varepsilon}{2}(1 - x^2) + \frac{\varepsilon^2}{4}\left(\frac{5}{6} - x^2 + \frac{x^4}{6}\right) + O(\varepsilon^3) \qquad\qquad (9.3.20)$$

which is valid uniformly in the entire domain of interest, $x \in [0, 1]$.

To verify the approximation (9.3.20) produced by the regular perturbation method, we can compare it with the expansion of the exact solution given by eq.

(9.3.14), as $\varepsilon \rightarrow 0$. This can be readily obtained by recalling the following Taylor series expansions as $\varepsilon \rightarrow 0$:

$$\cosh(\varepsilon) = 1 + \frac{\varepsilon^2}{2!} + \frac{\varepsilon^4}{4!} + \frac{\varepsilon^6}{6!} + \cdots \tag{9.3.21a}$$

$$(1 + a\varepsilon)^{-1} = 1 - a\varepsilon + a^2\varepsilon^2 - a^3\varepsilon^3 + \cdots \tag{9.3.21b}$$

from which it follows that

$$[\cosh(\sqrt{\varepsilon})]^{-1} = 1 - \left(\frac{\varepsilon}{2!} + \frac{\varepsilon^2}{4!} + \cdots\right) + \left(\frac{\varepsilon}{2!} + \frac{\varepsilon^2}{4!} + \cdots\right)^2 + \cdots \tag{9.3.22}$$

Substituting these expressions in eq. (9.3.14) leads to

$$\begin{aligned}
\frac{\cosh(\sqrt{\varepsilon}x)}{\cosh(\sqrt{\varepsilon})} &= \left(1 + \frac{\varepsilon x^2}{2!} + \frac{\varepsilon^2 x^4}{4!} + \cdots\right)\left[1 - \left(\frac{\varepsilon}{2!} + \frac{\varepsilon^2}{4!} + \cdots\right)\right. \\
&\quad \left. + \left(\frac{\varepsilon}{2!} + \frac{\varepsilon^2}{4!} + \cdots\right)^2 + \cdots\right] \tag{9.3.23} \\
&= 1 - \frac{\varepsilon}{2}(1 - x^2) + \frac{\varepsilon^2}{4}\left(\frac{5}{6} - x^2 + \frac{x^4}{6}\right) + O(\varepsilon^3)
\end{aligned}$$

which is identical to the solution (9.3.20) obtained by the regular perturbation method.

As a conclusion, we can state that for regular perturbation problems, we can get the solution in terms of an asymptotic series of ε, for $\varepsilon \ll 1$, by substituting

$$y = \sum_{n=0}^{\infty} \varepsilon^n y_n \tag{9.3.24}$$

in the original problem and comparing terms with like powers of ε. The choice of the number of terms to include in the truncated series depends upon the accuracy required and the actual values which the parameter ε can take. Since the basic idea in perturbation methods is to include one or at most a few correction terms (usually, in applications, higher-order terms become much more cumbersome to compute) these methods are best suited when, for physical reasons, the parameter ε actually attains only very small values. In general, perturbation methods should not be regarded as a tool for producing the exact solution through the complete series expansion of ε.

9.4 FAILURE OF REGULAR PERTURBATION METHODS

Regular perturbation methods are not always successful, and in particular they fail when two classes of so-called *singular perturbation* problems are encountered.

9.4.1 Secular-Type Problems

The first class concerns situations where the independent variable belongs to an infinite domain, as is usually the case for time. To illustrate this point, let us consider the time variation of the concentration y of a reactant which is consumed by a slow irreversible second-order reaction in a well-mixed batch reactor, described by the following equation:

$$\frac{dy}{dt} = -\varepsilon y^2 \tag{9.4.1}$$

with IC

$$y = 1 \qquad \text{at } t = 0 \tag{9.4.2}$$

By substituting the usual power series expansion

$$y(t) = \sum_{n=0}^{\infty} \varepsilon^n y_n(t) \tag{9.4.3}$$

it follows that

$$\left(\frac{dy_0}{dt} + \varepsilon \frac{dy_1}{dt} + \varepsilon^2 \frac{dy_2}{dt} + \cdots\right) + \varepsilon[y_0^2 + 2y_0 y_1 \varepsilon \\ + (y_1^2 + 2y_0 y_2)\varepsilon^2 + \cdots] = 0 \tag{9.4.4}$$

with IC

$$y_0 + \varepsilon y_1 + \varepsilon^2 y_2 + \cdots = 1 \qquad \text{at } t = 0 \tag{9.4.5}$$

By grouping terms with like powers of ε, the following set of equations is obtained:

$$O(\varepsilon^0): \quad \frac{dy_0}{dt} = 0, \qquad \text{IC:} \quad y_0(0) = 1 \tag{9.4.6a}$$

$$O(\varepsilon^1): \quad \frac{dy_1}{dt} + y_0^2 = 0, \qquad \text{IC:} \quad y_1(0) = 0 \tag{9.4.6b}$$

$$O(\varepsilon^2): \quad \frac{dy_2}{dt} + 2y_0 y_1 = 0, \qquad \text{IC:} \quad y_2(0) = 0 \tag{9.4.6c}$$

$$\vdots$$

whose solution is

$$y_0 = 1, \qquad y_1 = -t, \qquad y_2 = t^2, \ldots \tag{9.4.7}$$

which, when substituted in eq. (9.4.3), gives the series expansion:

$$y = 1 - \varepsilon t + \varepsilon^2 t^2 + \cdots \tag{9.4.8}$$

It can be readily seen that this expression represents the Taylor expansion of the function

$$y = \frac{1}{1 + \varepsilon t} \tag{9.4.9}$$

as $\varepsilon \to 0$, which indeed represents the exact solution of the original problem (9.4.1) and (9.4.2).

However, since in general the summation of the asymptotic series is not possible, let us regard eq. (9.4.8) as the approximating series which should be truncated after a certain number of terms in order to provide the desired accuracy, and investigate its behavior. In Figure 9.1 the exact solution (9.4.9) is compared with the approximate solutions given by the truncated series expansion $\Sigma_{n=0}^{N} \varepsilon^n y_n(t)$, for various values of N. It can be seen that, for a given ε, for any value of N used, there always exists a sufficiently large value of time t beyond which the approximation breaks down. Thus the series approximation (9.4.8), as previously discussed in section 9.2, is valid non-uniformly over the domain of interest. This is due to the presence of coefficients in the

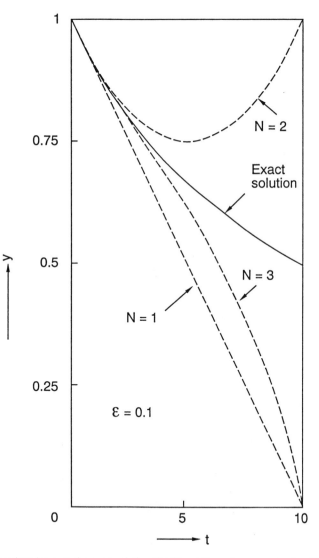

Figure 9.1 Comparison between the exact solution (9.4.9) and the series approximation (9.4.8) truncated after the term of $O(\varepsilon^N)$, with $N = 1, 2, 3$.

series [i.e., t^n in eq. (9.4.8)] which grow unbounded with t, so that each term of the series approaches infinity, no matter how small (but fixed) the parameter ε is. These are called *secular* terms, and consequently these types of singular perturbation problems are referred to as *secular-type* problems. They require special perturbation techniques which, through proper rescaling of the independent variable, remove the secular terms.

It is worth stressing that the secular type of singular behavior is strictly connected with the independent variable t belonging to an *infinite* domain. However, when t belongs to a *finite* interval, then the series expansion (9.4.8) is valid uniformly over such a domain.

9.4.2 Boundary Layer-Type Problems

The second class of problems where regular perturbation methods cannot be applied is where the *order* or *dimensionality* or *type* of the original equation changes when the small parameter vanishes—that is, when $\varepsilon = 0$. This is the case, for example, where the small parameter multiplies the highest-order derivative in an ordinary differential equation, as in

$$\varepsilon \frac{d^2 y}{dx^2} = y, \qquad x \in (0, 1) \tag{9.4.10}$$

with BCs

$$y = 1 \qquad \text{at } x = 1 \tag{9.4.11a}$$

$$\frac{dy}{dx} = 0 \qquad \text{at } x = 0 \tag{9.4.11b}$$

or when it multiplies the highest-order derivative with respect to some independent variable in a partial differential equation, as for example

$$\varepsilon \frac{\partial y}{\partial t} + \alpha \frac{\partial y}{\partial x} + \frac{\partial^2 y}{\partial x^2} = 1, \qquad x \in (0, 1), \quad t > 0 \tag{9.4.12}$$

with BCs

$$y = 0 \qquad \text{at } x = 0, t > 0 \tag{9.4.13a}$$

$$y = 1 \qquad \text{at } x = 1, t > 0 \tag{9.4.13b}$$

and IC

$$y = 0.5 \qquad \text{at } t = 0, 0 < x < 1 \tag{9.4.13c}$$

In both cases the zeroth-order solution, y_0 obtained by setting $\varepsilon = 0$ in the original equation, is incompatible with one of the associated conditions. Specifically, in the first case, y_0 cannot satisfy the boundary condition (9.4.11a), and in the second case it cannot satisfy condition (9.4.13c), since the problem becomes not well-posed, owing to the fact that the specified conditions are too many.

A more detailed analysis of singular problems of this type can be carried out with reference to eq. (9.4.10), which can be regarded as describing the concentration profile of a reactant undergoing irreversible first-order reaction in an isothermal slab catalyst. This is the same problem described by eq. (9.3.12), but in this case the reciprocal of the Thiele modulus, rather than the Thiele modulus itself, is the small parameter $\sqrt{\varepsilon}$. Thus we are now studying the other extreme of the asymptotic behavior of the problem—that is, very large Thiele modulus values.

By applying the regular perturbation procedure to eq. (9.4.10), we find for the zeroth-order approximation

$$y_0 = 0 \qquad\qquad (9.4.14)$$

which satisfies *only one* of the two boundary conditions—that is, eq. (9.4.11b). Since for all the other terms of the series expansion it is readily found that $y_n = 0$, it can be concluded that the approximation of the solution provided by the regular perturbation method is

$$y = 0 \qquad\qquad (9.4.15)$$

The exact analytic solution of eqs. (9.4.10) and (9.4.11) is given by

$$y = \frac{\cosh(x/\sqrt{\varepsilon})}{\cosh(1/\sqrt{\varepsilon})} \qquad\qquad (9.4.16)$$

which is shown in Figure 9.2 for various values of the small parameter ε. It may be seen that for decreasing ε values, the exact solution remains close to zero from $x = 0$ up to x values approaching 1, and finally jumps to $y = 1$ in order to satisfy the BC (9.4.11a). This region, where the solution changes sharply, is generally referred to as a *boundary layer*.

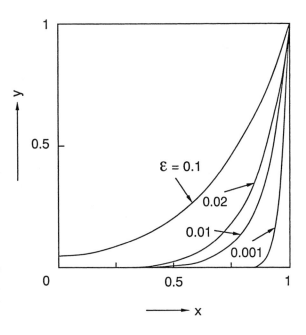

Figure 9.2 Qualitative behavior of the concentration profile $y(x)$ given by eq. (9.4.16), for various values of ε.

The limiting behavior of the exact solution (9.4.16) as $\varepsilon \to 0$ is given by

$$y = \exp[-(1 - x)/\sqrt{\varepsilon}] \qquad \text{as } \varepsilon \to 0 \qquad (9.4.17)$$

which correctly reproduces the two regions shown in Figure 9.2. Thus $y \simeq 0$ for x sufficiently away from 1, and $y = 1$ at $x = 1$, with a consequent sharp change in the boundary layer close to $x = 1$, which is narrower the smaller ε is. An estimate of the width δ of the boundary layer can be obtained from eq. (9.4.17), by imposing the argument of the exponential to be of the order of one—that is,

$$\delta = O[\sqrt{\varepsilon}] \qquad (9.4.18)$$

The approximation (9.4.15) given by the regular perturbation method provides a uniform approximation only in the region away from $x = 1$, but it completely ignores the jump of y in the region close to $x = 1$. This indicates that such an approximation, similar to the case described in section 9.4.1 above, provides an expansion which is not uniformly valid over the entire domain of interest, namely, $x \in [0, 1]$. Moreover, the behavior of the solution $y(x)$ displayed in Figure 9.2 suggests the need for developing *two* distinct expansions, each one valid within a specific interval of x, and then connecting them at the common boundary. This is the basic idea of the method of *matched asymptotic expansions*, usually adopted for solving such problems, often referred to as *boundary layer-type* problems.

Before proceeding on to describe the methods which can be used for solving the types of problems considered above, it is worth discussing some characteristics which differentiate regular and singular perturbation problems.

When an asymptotic expansion obtained through the procedure described in section 9.3 produces an approximation which is uniform in the entire domain of interest of the independent variable, then we are dealing with a regular perturbation problem. For the two types of singular perturbation problems described above, we have shown that such a condition is in fact not satisfied. An important feature in singular perturbation problems is that the zeroth-order solution (i.e., the leading part of the limiting behavior of the exact solution as $\varepsilon \to 0$) does *not* coincide with the solution of the unperturbed problem—that is, where $\varepsilon = 0$. In other words, the exact solution as $\varepsilon \to 0$, and the one computed from the original problem by substituting $\varepsilon = 0$, differ in character. Thus for example, in the case of the secular-type problem given by eqs. (9.4.1) and (9.4.2), the latter solution is simply

$$y = 1 \qquad (9.4.19)$$

which approaches the behavior of the exact solution (9.4.9) as $\varepsilon \to 0$, but not uniformly for all t values. In fact, for very large t values, the difference may become substantial. In particular, as $t \to \infty$, no matter how small (as long as finite) ε is, the exact solution approaches zero. Similarly, for the boundary-layer-type perturbation problem given by eqs. (9.4.10) and (9.4.11), the unperturbed solution does not exist, since $y = 0$ does not satisfy the boundary condition (9.4.11a). On the other hand, the exact solution as $\varepsilon \to 0$ always attains the value unity at $x = 1$. This clearly appears by considering the limiting behavior of the solution as $\varepsilon \to 0$, as given by eq. (9.4.17), from which it follows that

$$\lim_{x \to 1}[\lim_{\varepsilon \to 0} y] = 0 \qquad (9.4.20a)$$

while

$$\lim_{\varepsilon \to 0}[\lim_{x \to 1} y] = 1 \qquad (9.4.20b)$$

9.5   THE METHOD OF STRAINED COORDINATES

In section 9.4.1, singular perturbation problems of the secular type were introduced. In these cases, asymptotic expansions obtained using the regular perturbation method contain secular terms, and thus are not valid uniformly. The method of *strained coordinates* allows for removing such terms, thus leading to uniform asymptotic expansions. Before discussing the application of this method, let us further analyze the occurrence of secular terms in series expansions.

9.5.1   An Introductory Example

As an example, let us consider the free oscillation of a mass with no weight on a nonlinear spring, whose friction with the surrounding medium is negligible. The position of the mass y is given by the equation

$$\frac{d^2y}{dt^2} + y - \varepsilon y^3 = 0, \qquad t > 0 \qquad (9.5.1)$$

with ICs

$$y = a \qquad \text{at } t = 0 \qquad (9.5.2a)$$

$$\frac{dy}{dt} = 0 \qquad \text{at } t = 0 \qquad (9.5.2b)$$

where ε is a small parameter indicating the nonlinearity (i.e., deviation from Hooke's law) of the spring.

Let us first apply the regular perturbation method, thus introducing the asymptotic expansion

$$y(t) = \sum_{n=0}^{\infty} \varepsilon^n y_n(t) \qquad (9.5.3)$$

and substituting it in eqs. (9.5.1) and (9.5.2). Then, expanding the nonlinear term y^3 for small ε and grouping together all terms multiplying like powers of ε, it follows that

$$\left(\frac{d^2y_0}{dt^2} + y_0\right) + \varepsilon\left(\frac{d^2y_1}{dt^2} + y_1 - y_0{}^3\right) + \cdots = 0 \qquad (9.5.4)$$

with ICs

$$y_0 + \varepsilon y_1 + \cdots = a \qquad \text{at } t = 0 \qquad (9.5.5a)$$

$$\frac{dy_0}{dt} + \varepsilon\frac{dy_1}{dt} + \cdots = 0 \qquad \text{at } t = 0 \qquad (9.5.5b)$$

By equating each term to zero, the expansion coefficients $y_0, y_1, \ldots$ can be computed. In particular, the zeroth-order approximation is given by the solution of the equation

$$\frac{d^2 y_0}{dt^2} + y_0 = 0 \tag{9.5.6a}$$

with ICs

$$y_0 = a, \quad \frac{dy_0}{dt} = 0 \quad \text{at } t = 0 \tag{9.5.6b}$$

Thus we have

$$y_0 = a \cos t \tag{9.5.7}$$

indicating that the zeroth-order solution is periodic. Similarly, for the first-order approximation:

$$\frac{d^2 y_1}{dt^2} + y_1 = a^3 \cos^3 t = \frac{a^3}{4} (\cos 3t + 3 \cos t) \tag{9.5.8a}$$

with ICs

$$y_1 = 0, \quad \frac{dy_1}{dt} = 0 \quad \text{at } t = 0 \tag{9.5.8b}$$

The solution of the homogeneous equation corresponding to (9.5.8a) is

$$y_{1h} = C_1 \cos t + C_2 \sin t \tag{9.5.9}$$

To obtain the general solution, we need a particular solution of (9.5.8a), which is given by

$$y_{1p} = -\frac{a^3}{32} \cos 3t + \frac{3a^3}{8} t \sin t \tag{9.5.10}$$

so that the general solution is given by $y_{1h} + y_{1p}$, where the integration constants C_1 and C_2 are computed from the ICs (9.5.8b). This leads to

$$y_1 = \frac{a^3}{32} (\cos t - \cos 3t) + \frac{3a^3}{8} t \sin t$$

which together with eq. (9.5.7) provides the following two-term asymptotic expansion:

$$y = a \cos t + \varepsilon \frac{a^3}{32} [\cos t - \cos 3t + 12t \sin t] + O(\varepsilon^2) \tag{9.5.11}$$

This expression is not valid uniformly in the entire domain $t \in [0, \infty)$, since it contains the secular term ($t \sin t$), which grows unbounded with t.

It is worth noting that the secular term ($t \sin t$) arises in the *particular* solution (9.5.10) of eq. (9.5.8a). This is because the term ($3 \cos t$) in the inhomogeneous part of eq. (9.5.8a) is a solution of the associated homogeneous equation, as is clear from eq. (9.5.9). This is the typical source of secular terms in differential equations with constant coefficients.

Since we seek an approximation valid in the entire domain $t \in [0, \infty)$, we need to render the asymptotic expansion (9.5.11) uniform. In principle, since we know from the existence and uniqueness theorem of section 2.7 that the solution of eq. (9.5.1) is indeed bounded in time, we could compute more terms in the expansion (9.5.11) and then obtain the sum of the series. In fact, even though the truncated series diverges for $t \rightarrow \infty$ because of the secular terms, the series converges in the entire domain. This procedure was successful in the previous case of the asymptotic series (9.4.8), whose sum was given by eq. (9.4.9). However, except for special cases, such a procedure cannot be applied in general because of the difficulty in summing the series.

9.5.2 Development of the Method

A more efficient and practical way of removing secular terms is the method of *strained coordinates*, which is based on expanding, in addition to the dependent variable y, also the independent one, t, according to

$$t = \tau + \varepsilon f_1(\tau) + \varepsilon^2 f_2(\tau) + \cdots \tag{9.5.12}$$

$$y(t) = y_0(\tau) + \varepsilon y_1(\tau) + \varepsilon^2 y_2(\tau) + \cdots \tag{9.5.13}$$

where the coordinate change given by eq. (9.5.12) is such that the original coordinate is recovered for the unperturbed problem (i.e., $\tau \rightarrow t$ as $\varepsilon \rightarrow 0$), while the as yet unknown functions $f_n(\tau)$ are determined in order to *remove* the secular terms. In other words, the idea is to modify the original scale of the independent variable [using eq. (9.5.12) or other expansions suited to the particular problem at hand] to cancel out all secular terms.

Before substituting eqs. (9.5.12) and (9.5.13) in eqs. (9.5.1) and (9.5.2), we need to first compute the derivatives of y with respect to t in terms of the new independent variable τ. Using the chain rule, it follows that

$$\frac{dy}{dt} = \frac{dy}{d\tau}\frac{d\tau}{dt} \tag{9.5.14}$$

From eq. (9.5.13) we have

$$\frac{dy}{d\tau} = \frac{dy_0}{d\tau} + \varepsilon\frac{dy_1}{d\tau} + \cdots \tag{9.5.15}$$

while from eq. (9.5.12) we obtain

$$\frac{dt}{d\tau} = 1 + \varepsilon\frac{df_1}{d\tau} + \cdots \tag{9.5.16}$$

so that

$$\frac{d\tau}{dt} = \frac{1}{1 + \varepsilon\dfrac{df_1}{d\tau} + \cdots}$$

$$= 1 - \varepsilon\frac{df_1}{d\tau} + \cdots \tag{9.5.17}$$

upon expanding by Taylor series as $\varepsilon \rightarrow 0$. Substituting eqs. (9.5.15) and (9.5.17) into eq. (9.5.14) gives

$$\frac{dy}{dt} = \left(\frac{dy_0}{d\tau} + \varepsilon \frac{dy_1}{d\tau} + \cdots \right) \left(1 - \varepsilon \frac{df_1}{d\tau} + \cdots \right) \tag{9.5.18}$$

Similarly, the second-order derivative is computed as follows:

$$\frac{d^2 y}{dt^2} = \left[\left(\frac{d^2 y_0}{d\tau^2} + \varepsilon \frac{d^2 y_1}{d\tau^2} + \cdots \right) \left(1 - \varepsilon \frac{df_1}{d\tau} + \cdots \right) \right.$$
$$\left. + \left(\frac{dy_0}{d\tau} + \varepsilon \frac{dy_1}{d\tau} + \cdots \right) \left(-\varepsilon \frac{d^2 f_1}{d\tau^2} + \cdots \right) \right] \left(1 - \varepsilon \frac{df_1}{d\tau} + \cdots \right)$$

which, by grouping terms with like powers of ε, leads to

$$\frac{d^2 y}{dt^2} = \frac{d^2 y_0}{d\tau^2} + \varepsilon \left(\frac{d^2 y_1}{d\tau^2} - 2 \frac{df_1}{d\tau} \frac{d^2 y_0}{d\tau^2} - \frac{d^2 f_1}{d\tau^2} \frac{dy_0}{d\tau} \right) + \cdots \tag{9.5.19}$$

Now substituting eqs. (9.5.13) and (9.5.19) in eq. (9.5.1) and expanding for small ε, we obtain

$$\frac{d^2 y_0}{d\tau^2} + \varepsilon \left(\frac{d^2 y_1}{d\tau^2} - 2 \frac{df_1}{d\tau} \frac{d^2 y_0}{d\tau^2} - \frac{d^2 f_1}{d\tau^2} \frac{dy_0}{d\tau} \right) + \cdots$$
$$+ (y_0 + \varepsilon y_1 + \cdots) - \varepsilon(y_0^3 + 3\varepsilon y_0^2 y_1 + \cdots) = 0 \tag{9.5.20}$$

To apply the independent variable change (9.5.12) to the ICs (9.5.2), we need to establish a correspondence between the original and the strained time scales. Since it is good to have $t = 0$ equivalent to $\tau = 0$, from eq. (9.5.12) the following constraints for the functions $f_n(\tau)$ arise:

$$f_n(\tau) = 0 \quad \text{at } \tau = 0, \quad \text{for } n = 1, 2, \ldots \tag{9.5.21}$$

Thus, the ICs (9.5.2) reduce as follows:

$$y_0 + \varepsilon y_1 + \cdots = a \quad \text{at } \tau = 0 \tag{9.5.22a}$$

$$\frac{dy_0}{d\tau} + \varepsilon \left(\frac{dy_1}{d\tau} - \frac{df_1}{d\tau} \frac{dy_0}{d\tau} \right) + \cdots = 0 \quad \text{at } \tau = 0 \tag{9.5.22b}$$

where the variable change in the derivative of y from t to τ has been performed using eq. (9.5.18). Equating to zero terms with like powers of ε in (9.5.20) and (9.5.22) leads to the following problems:

$$O(\varepsilon^0): \quad \frac{d^2 y_0}{d\tau^2} + y_0 = 0$$

with ICs

$$y_0 = a, \quad \frac{dy_0}{d\tau} = 0 \quad \text{at } \tau = 0$$

whose solution is

$$y_0 = a \cos \tau \tag{9.5.23}$$

as before, except that the time variable is now τ instead of t.

$$O(\varepsilon^1): \qquad \frac{d^2 y_1}{d\tau^2} + y_1 = 2 \frac{df_1}{d\tau} \frac{d^2 y_0}{d\tau^2} + \frac{d^2 f_1}{d\tau^2} \frac{dy_0}{d\tau} + y_0^3$$

which substituting eq. (9.5.23) and after some algebraic manipulations reduces to

$$\frac{d^2 y_1}{d\tau^2} + y_1 = \frac{a^3}{4} \cos 3\tau - a\left(\frac{d^2 f_1}{d\tau^2} \sin \tau + 2 \frac{df_1}{d\tau} \cos \tau - \frac{3a^2}{4} \cos \tau \right) \tag{9.5.24}$$

with ICs

$$y_1 = 0, \qquad \frac{dy_1}{d\tau} + a \frac{df_1}{d\tau} \sin \tau = 0 \qquad \text{at } \tau = 0 \tag{9.5.25}$$

It is clear that the homogeneous equation associated with eq. (9.5.24) has a general solution in terms of $\sin \tau$ and $\cos \tau$, and further since the inhomogeneous term contains these same functions, it follows that the first-order correction y_1 will introduce secular terms. In fact, this is exactly what happened when we solved eq. (9.5.8a) via the regular perturbation method. However, in the case of eq. (9.5.24) we have one additional degree of freedom—that is, the as yet unknown function $f_1(\tau)$, which we can now select in order to cancel in the inhomogeneous part of the equation all terms which are solutions of the associated homogeneous equation. Thus we set

$$\frac{d^2 f_1}{d\tau^2} \sin \tau + 2 \frac{df_1}{d\tau} \cos \tau = \frac{3a^2}{4} \cos \tau \tag{9.5.26}$$

along with the condition given by eq. (9.5.21)—that is, $f_1(0) = 0$. Rewriting eq. (9.5.26) as

$$\frac{d}{d\tau}\left[\sin^2 \tau \frac{df_1}{d\tau} \right] = \frac{3a^2}{4} \sin \tau \cos \tau$$

integrating once and setting equal to zero the integration constant for convenience, we get

$$\frac{df_1}{d\tau} = \frac{3a^2}{8}$$

which can be integrated with the IC above, that is, $f_1(0) = 0$, leading to

$$f_1 = \frac{3a^2}{8} \tau \tag{9.5.27}$$

With this choice for f_1, eq. (9.5.24) reduces to

$$\frac{d^2 y_1}{d\tau^2} + y_1 = \frac{a^3}{4} \cos 3\tau$$

whose general solution is

$$y_1 = C_1 \sin \tau + C_2 \cos \tau - \frac{a^3}{32} \cos 3\tau$$

where the integration constants are evaluated from the ICs (9.5.25), leading to

$$y_1 = \frac{a^3}{32} (\cos \tau - \cos 3\tau) \tag{9.5.28}$$

Substituting this in eq. (9.5.13), together with eq. (9.5.23), leads to the asymptotic expansion

$$y(t) = a \cos \tau + \varepsilon \frac{a^3}{32} (\cos \tau - \cos 3\tau) + O(\varepsilon^2) \tag{9.5.29}$$

where the strained time scale is given by eqs. (9.5.12) and (9.5.27) as

$$t = \tau + \varepsilon \frac{3a^2}{8} \tau + \cdots$$

and therefore

$$\tau = \frac{t}{1 + \varepsilon(3a^2/8) + \cdots} = \left(1 - \varepsilon \frac{3a^2}{8} + \cdots\right)t = \omega(\varepsilon, a)t \tag{9.5.30}$$

which shows how the frequency of oscillations, ω, depends upon the parameter ε and the amplitude of the initial perturbation, a.

As expected, the asymptotic expansion (9.5.29) produced by the method of strained coordinates is valid uniformly in the entire domain $t \in [0, \infty)$ since secular terms are not present. The method of strained coordinates is certainly not the only one for dealing with secular-type singular perturbation problems. Others, such as the WKB method and the method of multiple scales, which are in general more powerful but more complex, can also be used and are described in forthcoming sections.

9.5.3 Phase Plane Analysis

In order to establish the region where the asymptotic expansion (9.5.29) can be used, it is convenient to develop a phase plane analysis of the original problem (9.5.1)–(9.5.2). In particular, we have seen that both the zeroth-order (9.5.23) and the first-order (9.5.29) approximations are periodic. Thus we now investigate the region of parameter values where the exact solution exhibits a periodic behavior.

Toward this end, we introduce the two new variables:

$$y_1 = y \tag{9.5.31a}$$

$$y_2 = \frac{dy}{dt} = \frac{dy_1}{dt} \tag{9.5.31b}$$

Thus the original problem can be rewritten in the form of a pair of ODEs:

$$\frac{dy_1}{dt} = y_2 \tag{9.5.32a}$$

$$\frac{dy_2}{dt} = -y_1 + \varepsilon y_1^3 \tag{9.5.32b}$$

with ICs

$$y_1 = a, \quad y_2 = 0 \quad \text{at } t = 0 \tag{9.5.32c}$$

Upon dividing eq. (9.5.32b) by eq. (9.5.32a), we get

$$\frac{dy_2}{dy_1} = \frac{-y_1 + \varepsilon y_1^3}{y_2} \tag{9.5.33}$$

which can be integrated using the separation of variables method to yield the relationship

$$y_2^2 = C - y_1^2 + \frac{\varepsilon}{2} y_1^4 = C - g(y_1) \tag{9.5.34}$$

where C is the integration constant, which we will determine through the IC (9.5.32c). Before doing this, let us keep the analysis general, using C as a parameter, and investigate the shape of the solution trajectories in the (y_1, y_2) phase plane. For this, we need to first analyze the shape of the function $g(y_1)$:

$$g(y_1) = y_1^2 \left(1 - \frac{\varepsilon}{2} y_1^2 \right) \tag{9.5.35}$$

With reference to Figure 9.3, we note that the function $g(y_1)$ has three zeroes at $y_1 = 0, \pm\sqrt{2/\varepsilon}$. Also, the first derivative

$$g'(y_1) = 2y_1(1 - \varepsilon y_1^2) \tag{9.5.36}$$

vanishes for three values of y_1 (i.e., $y_1 = 0, \pm 1/\sqrt{\varepsilon}$), where the function $g(y_1)$ exhibits three extrema. In particular, there is a minimum at $y_1 = 0$ [(since $g''(0) = 2 > 0$)] and two maxima at $y_1 = \pm 1/\sqrt{\varepsilon}$ [(since $g''(\pm 1/\sqrt{\varepsilon}) = -4 < 0$)]. Finally, in the limit of very large positive or negative values of y_1, we can approximate $g(y_1) = -(\varepsilon/2)y_1^4$, and so the following asymptotic behavior is obtained:

$$\lim_{y_1 \to \pm\infty} g(y_1) = -\infty \tag{9.5.37}$$

in agreement with the qualitative shape of the function $g(y_1)$ shown in Figure 9.3.

Let us now establish the directions of the solution trajectories in the (y_1, y_2) phase plane. For this, we need to determine the signs of the time derivatives of y_1 and y_2. From eq. (9.5.32a) we see that

$$\frac{dy_1}{dt} > 0 \quad \text{for } y_2 > 0 \tag{9.5.38a}$$

$$\frac{dy_1}{dt} < 0 \quad \text{for } y_2 < 0 \tag{9.5.38b}$$

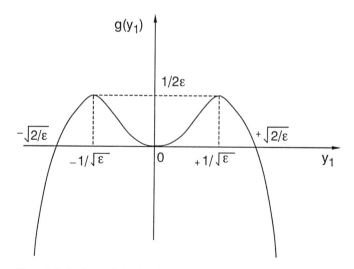

Figure 9.3 Qualitative behavior of the function $g(y_1)$ defined by eq. (9.5.35).

while from eq. (9.5.32b) we can conclude that

$$\frac{dy_2}{dt} > 0 \qquad \text{for } y_1 > 1/\sqrt{\varepsilon} \quad \text{and} \quad -1/\sqrt{\varepsilon} < y_1 < 0 \qquad (9.5.39a)$$

$$\frac{dy_2}{dt} < 0 \qquad \text{for } y_1 < -1/\sqrt{\varepsilon} \quad \text{and} \quad 0 < y_1 < 1/\sqrt{\varepsilon} \qquad (9.5.39b)$$

Combining eqs. (9.5.38) and (9.5.39) we can identify in the (y_1, y_2) phase plane eight regions where the directions of the solution trajectories are uniquely determined. These are shown in Figure 9.4, where it should be noted that dy_1/dt vanishes on the y_1 axis (i.e., $y_2 = 0$), while dy_2/dt vanishes on the three vertical axes defined by the equations $y_1 = 0, \pm 1/\sqrt{\varepsilon}$. Thus there are three stationary points of the system (9.5.32), given by

$$(0, 0); \qquad (-1/\sqrt{\varepsilon}, 0); \qquad (1/\sqrt{\varepsilon}, 0) \qquad (9.5.40)$$

In order to investigate the behavior of the solution trajectories in the vicinity of these stationary points, we consider the original system (9.5.32) linearized around each steady state. We can then evaluate the linearized stability character of these states through the stability theory described in section 2.14 for two-dimensional autonomous systems. Accordingly, we first derive the stability matrix $\mathbf{A}$—that is, the matrix of the coefficients of the system (9.5.32) linearized around the generic steady state (y_1^s, y_2^s):

$$\mathbf{A} = \begin{bmatrix} 0 & 1 \\ -1 + 3\varepsilon y_1^2 & 0 \end{bmatrix}_{(y_1^s, y_2^s)} \qquad (9.5.41)$$

We can now compute tr $\mathbf{A}$ and det $\mathbf{A}$ for each of the steady states (9.5.40) and then establish their stability nature. Thus the steady state $(0, 0)$ is a *center* because tr $\mathbf{A} = 0$ and det $\mathbf{A} = 1 > 0$. On the other hand, the steady states $(\pm 1/\sqrt{\varepsilon}, 0)$ are both *saddle*

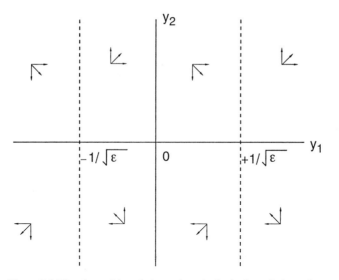

Figure 9.4 Directions of the solution trajectories in the (y_1, y_2) phase plane.

points since tr $\mathbf{A} = 0$ and det $\mathbf{A} = -2 < 0$. It is worth recalling that, as far as the linearized analysis is concerned, the saddle points are unstable while the center is marginally stable. Using Liapunov's theorem we can extend this result to the original nonlinear system (9.5.32) only for the two saddle points, but not for the center.

Using all previous results, we can now proceed to identify the shape of the solution trajectories in the (y_1, y_2) phase plane. To this end, we should return to eq. (9.5.34) and consider various values of C, both positive and negative, corresponding to various possible ICs for the system (9.5.32a,b).

As a preliminary observation we may note that both y_1 and y_2 exhibit even powers in eq. (9.5.34), so that the trajectories are symmetric to both the y_1 and y_2 axes. Moreover, whenever for a range of y_1 values, $C - g(y_1) < 0$, then y_2 becomes imaginary and we should expect discontinuities in the trajectories.

Depending upon the value of the integration constant C, we can identify three qualitatively different shapes of the trajectories.

Case 1. C < 0
In this case, we can readily see from Figure 9.3 that $C - g(y_1) > 0$ only in two separate intervals of y_1 values; that is, $|y_1| > K$, where K is the positive solution of the implicit equation $C = g(K)$ and in all cases is larger than $\sqrt{2/\varepsilon}$. Thus the trajectories exist (i.e., y_2 is real) only for $y_1 < -K$ and $y_1 > K$, and they intersect the y_1 axis at $\pm K$. This, coupled with the indications about the direction of the solution trajectories shown in Figure 9.4, allows us to identify the qualitative phase plane behavior reported in Figure 9.5*a* for $C = -2$.

Case 2. 0 < C < 1/2ε
In this case, there are three intervals of y_1 where $C - g(y_1) > 0$ and the trajectory exists. We indicate with K_1 and K_2 the two positive solutions of the implicit equation $C - g(y_1) = 0$, which, as can be seen in Figure 9.3, belong to the following intervals:

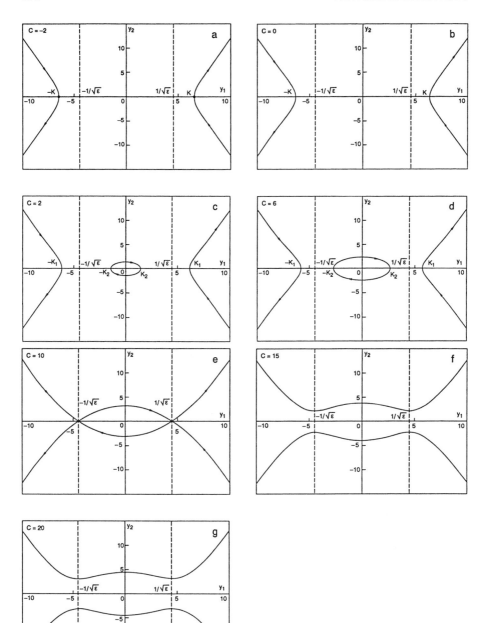

Figure 9.5 Solution trajectories in the (y_1, y_2) phase plane as given by eq. (9.5.34) for various values of C and $\varepsilon = 0.05$.

$\sqrt{1/\varepsilon} < K_1 < \sqrt{2/\varepsilon}$ and $0 < K_2 < 1/\sqrt{\varepsilon}$. Then, the trajectories exist in the three intervals $y_1 < -K_1$, $y_1 > K_1$, and $-K_2 < y_1 < K_2$ and intersect the y_1 axis four times at the points $\pm K_1$ and $\pm K_2$. From the knowledge about the directions of the trajectories shown in Figure 9.4, we can identify the qualitative phase plane behavior shown in Figures 9.5c and 9.5d for two values of C. In both cases, we have two isolated trajectories along with a periodic solution represented by a closed orbit. Note that the case corresponding to $C = 0$, which represents the transition from the previous type of phase plane behavior to the new one, is shown in Figure 9.5b.

Case 3. $C > 1/2\varepsilon$
In this case, $C - g(y_1) > 0$ for all y_1 values, and then the solution trajectories do not exhibit any discontinuity. Moreover, for the same reason, they do not intersect the y_1 axis. Based upon the trajectory directions shown in Figure 9.4, we can identify a phase plane behavior of the type shown in Figures 9.5f and 9.5g for two values of C. Again, the situation shown in Figure 9.5e, corresponding to $C = 1/2\varepsilon$, represents the transition between the last two types of behavior examined.

Let us now go back to analyze the specific initial conditions for our problem— that is, eq. (9.5.32c). By substituting these in eq. (9.5.34), we obtain the following value for the integration constant:

$$C = a^2 - \frac{\varepsilon}{2} a^4 \qquad\qquad (9.5.42)$$

where recall that a represents the initial position, positive or negative, of the mass. From eq. (9.5.42) it can be readily seen that

$$C < 0 \qquad \text{for } a < -\sqrt{\frac{2}{\varepsilon}} \quad \text{or} \quad a > \sqrt{\frac{2}{\varepsilon}} \qquad\qquad (9.5.43a)$$

$$0 < C < \frac{1}{2\varepsilon} \qquad \text{for } -\sqrt{\frac{2}{\varepsilon}} < a < \sqrt{\frac{2}{\varepsilon}} \qquad\qquad (9.5.43b)$$

Thus, when considering the IC (9.5.32c), only the phase plane behavior of type 1 and 2 above can actually be achieved by the system.

Note that the solution can be periodic only in the case where $0 < C < 1/2\varepsilon$, that is, the phase plane behavior is of type 2 above. It is also clear from Figures 9.5c and 9.5d that the closed trajectory is obtained only for a values satisfying

$$a < \frac{1}{\sqrt{\varepsilon}} \qquad\qquad (9.5.44)$$

Recall that the series expansions (9.5.23) and (9.5.29) were also periodic solutions. In deriving these series approximations, we tacitly assumed $a = O(\varepsilon^0)$, as can be seen for example in eq. (9.5.5a). Accordingly, when comparing the asymptotic expansions with the true solution, we should consider only those values of a which are of $O(\varepsilon^0)$, and these indeed satisfy condition (9.5.44) as $\varepsilon \to 0$.

In order to evaluate the accuracy of the series approximation, let us consider the maximum velocity value in the periodic movement of the mass, $y_{2,\text{max}}$, which is equal to the value of y_2 corresponding to $y_1 = 0$. Using eqs. (9.5.34) and (9.5.42), this is given by

$$y_{2,\max} = \pm \sqrt{a^2 - \frac{\varepsilon}{2} a^4} \qquad (9.5.45)$$

On the other hand, from the first-order series expansion (9.5.29) we have

$$y_1 = 0 \quad \text{for } \tau = \pm \frac{\pi}{2} + 2n\pi; \qquad n = 0, 1, 2, \ldots \qquad (9.5.46)$$

The corresponding value of y_2 is obtained by differentiating eq. (9.5.29) with respect to t, which, using eq. (9.5.30), leads to

$$y_2 = \frac{dy}{dt} = \left[-a \sin \tau - \varepsilon \frac{a^3}{32} (\sin \tau - 3 \sin 3\tau) + O(\varepsilon^2) \right]$$

$$\left[1 - \varepsilon \frac{3a^2}{8} + O(\varepsilon^2) \right] \qquad (9.5.47)$$

Using the τ values given by eq. (9.5.46) and neglecting terms of order higher than $O(\varepsilon^1)$ leads to

$$y_{2,\max} = \pm a \mp \varepsilon \frac{a^3}{4} + O(\varepsilon^2) \qquad (9.5.48)$$

It can be readily seen that eq. (9.5.48) is identical to the Taylor series expansion of the exact $y_{2,\max}$ value given by eq. (9.5.45), thus confirming the correctness of the procedure used for deriving the asymptotic expansion (9.5.29).

Finally, in Figures 9.6a and 9.6b, the first-order approximation (9.5.29) is compared with the exact solution (9.5.34) in the (y_1, y_2) phase plane. As expected, it can be seen that the accuracy of the asymptotic expansion improves for decreasing values of ε.

9.6 BOUNDARY LAYER THEORY AND THE METHOD OF MATCHED ASYMPTOTIC EXPANSIONS

The main features of boundary-layer-type singular perturbation problems have been described in section 9.4.2. These are characterized by sharp changes in the dependent variable occurring in a small region of the domain of the independent variable. For example, in the case shown in Figure 9.2, as $\varepsilon \to 0$, the dependent variable y changes relatively smoothly and remains essentially constant in almost the entire domain $x \in [0, 1]$, except for a small region (or boundary layer) close to $x = 1$ where it changes rapidly. Situations of this type are common in various branches of science and engineering, such as fluid mechanics (the classical Prandtl treatment of viscous flow past an immersed object), kinetic theory, combustion theory, and so on.

We have shown previously (section 9.4.2) that in this case a single asymptotic expansion cannot be valid uniformly in the entire domain of x. Thus we develop two asymptotic expansions: the *outer expansion*, which represents the smoothly changing

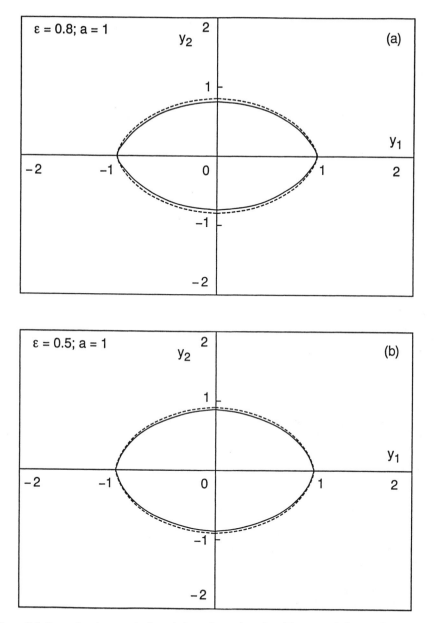

Figure 9.6 Comparison between the (y_1, y_2) phase plane trajectories of the exact solution (continuous curve) and the first-order series approximation (broken curve), as given by eqs. (9.5.34) and (9.5.29), respectively.

portion of the solution, valid in most of the domain of x, and the *inner expansion*, which represents the rapidly changing portion of the solution in the boundary layer. Since the width of such a layer tends to vanish as $\varepsilon \to 0$, we need to magnify it in order to make the variation of y smooth also in this region. This is done by introducing a *stretching transformation* of the independent variable, such as $s = (1 - x)/\varepsilon^{\lambda}$, where λ is a positive parameter, whose value depends upon the problem being studied. For example, in the case of the diffusion-reaction problem described in section 9.4.2, it was shown that the dimension of the boundary layer is of the order of $\sqrt{\varepsilon}$ [eq. (9.4.18)]. This suggests that the stretching transformation

$$s = \frac{1 - x}{\sqrt{\varepsilon}} \tag{9.6.1}$$

should be adopted for the inner expansion. Thus for the problem mentioned above, we know where the boundary layer is located (i.e., $x = 1$ as shown in Figure 9.2) and what the natural stretching transformation is for the inner expansion. This is all that is needed for applying the method of *matched asymptotic expansions*.

9.6.1 An Introductory Example

Let us now apply the method of matched asymptotic expansions to the diffusion-reaction problem of section 9.4.2 by deriving the outer $[y^{o}(x, \varepsilon)]$ and the inner $[y^{i}(x, \varepsilon)]$ expansions.

Outer Expansion.   As mentioned above, due to the singular nature of the problem, the outer expansion cannot satisfy both boundary conditions, so the one corresponding to where the boundary layer is located [i.e., eq. (9.4.11a)] is omitted. In order to derive the outer expansion of the solution, valid in the region outside the boundary layer, denoted as the *outer region*, the following asymptotic series is considered:

$$y^{o}(x, \varepsilon) \sum_{n=0}^{\infty} \varepsilon^{n} y_{n}^{o}(x) \tag{9.6.2}$$

By substituting in the original problem (9.4.10) and (9.4.11b) and by comparing terms with like powers of ε, the following subproblems are obtained:

$$O(\varepsilon^{0}): \quad y_{0}^{o} = 0, \qquad \frac{dy_{0}^{o}}{dx} = 0 \quad \text{at } x = 0 \tag{9.6.3a}$$

$$O(\varepsilon^{1}): \quad \frac{d^{2}y_{0}^{o}}{dx^{2}} - y_{1}^{o} = 0, \quad \frac{dy_{1}^{o}}{dx} = 0 \quad \text{at } x = 0 \tag{9.6.3b}$$

$$O(\varepsilon^{2}): \quad \frac{d^{2}y_{1}^{o}}{dx^{2}} - y_{2}^{o} = 0, \quad \frac{dy_{2}^{o}}{dx} = 0 \quad \text{at } x = 0 \tag{9.6.3c}$$

whose solution can be easily generalized to give $y_{n}^{o} = 0$ for $n \geq 0$. Substituting this in eq. (9.6.2) leads to the outer expansion of the solution

$$y^{o} = 0 \tag{9.6.4}$$

This conclusion is expected from the profiles of y shown in Figure 9.2, where it may be seen that as $\varepsilon \rightarrow 0$, y approaches a constant zero value throughout the entire domain of x away from the boundary layer—that is, the outer region.

Inner Expansion. To investigate the behavior of y in the thin boundary layer, it is necessary to stretch the scale of the independent variable. Thus by applying the stretching transformation (9.6.1) to eqs. (9.4.10) and (9.4.11a), it follows that

$$\frac{d^2y}{ds^2} = y \tag{9.6.5}$$

with BC

$$y = 1 \qquad \text{at } s = 0 \tag{9.6.6}$$

It may be noted that in this case the other boundary condition—that is, the one at $x = 0$—is omitted since the inner expansion is intended to apply only in a narrow region (denoted as the *inner* region) close to the boundary layer and not in the entire domain of x. The stretching transformation makes the second-order derivative of y of the same magnitude as the function itself, as apparent from eq. (9.6.5), while in the original coordinate the second-order derivative was of the order y/ε and therefore much larger than y, as seen from eq. (9.4.10). This shows that the stretching transformation has been successful in making y smooth in the boundary layer.

The inner expansion is obtained in the form of the asymptotic series

$$y^i(s, \varepsilon) = \sum_{n=0}^{\infty} \varepsilon^n y_n^i(s) \tag{9.6.7}$$

where s is now kept fixed while $\varepsilon \rightarrow 0$, by following the same procedure as used above for developing the outer expansion. In particular, the following subproblems are obtained:

$$O(\varepsilon^0): \quad \frac{d^2y_0^i}{ds^2} - y_0^i = 0, \quad y_0^i = 1 \qquad \text{at } s = 0 \tag{9.6.8a}$$

$$O(\varepsilon^1): \quad \frac{d^2y_1^i}{ds^2} - y_1^i = 0, \quad y_1^i = 0 \qquad \text{at } s = 0 \tag{9.6.8b}$$

For the zeroth-order term, solving eq. (9.6.8a) gives

$$y_0^i = (1 - C)\,\exp(s) + C\,\exp(-s) \tag{9.6.9}$$

where C is an integration constant, which needs one additional boundary condition for its evaluation. Here is where the *matching condition* between the outer and inner solutions needs to be introduced.

Specifically, a requirement of the method of matched asymptotic expansions is that there exists a region in the domain of the independent variable where the outer and inner expansions overlap; that is, they are *both* valid. Since the two expansions actually represent the same function, we can force them to be identical in this overlapping region.

For the zeroth-order terms of both expansions, this is obtained through the condition

$$\lim_{x \to 1} y_0^o = \lim_{s \to \infty} y_0^i \qquad (9.6.10)$$

where the overlapping region, wherein the transition from the boundary layer to the bulk of the domain of x occurs, is indicated in the original coordinate by $x \to 1$ while in the stretched coordinate by $s \to \infty$.

Recalling that $y_0^o = 0$ and that y_0^i is defined by eq. (9.6.9), condition (9.6.10) leads to the determination of the integration constant as $C = 1$. Thus we have

$$y_0^i = \exp(-s) \qquad (9.6.11)$$

The existence of an overlapping region allows the method of matched asymptotic expansions to derive a *composite expansion* which represents the solution in the entire domain of x. This is obtained by summing up the outer and inner expansions and subtracting the common value in the overlapping region, as determined by the matching condition (9.6.10). Thus

$$y_0^c(x) = y_0^o(x) + y_0^i(x) - (\text{common part}) \qquad (9.6.12)$$

Since the common part in this case equals zero, the zeroth-order term of the composite solution is given by

$$y_0^c(x) = \exp\left(-\frac{1-x}{\sqrt{\varepsilon}}\right) \qquad (9.6.13)$$

which is identical to the small ε expansion of the exact solution, given by eq. (9.4.17).

9.6.2 Location of the Boundary Layer and the Stretching Transformation

In the example above, the location of the boundary layer and the stretching transformation were both determined by examining the analytical solution of the problem, which obviously will in general not be known *a priori*. Additionally, the matching condition (9.6.10) applies only to the zeroth-order terms of the outer and inner expansions.

To develop these aspects further, let us consider the following example:

$$\varepsilon \frac{d^2 y}{dx^2} + v \frac{dy}{dx} = e^x, \qquad x \in (0, 1) \qquad (9.6.14)$$

with BCs

$$y = \alpha \qquad \text{at } x = 0 \qquad (9.6.15a)$$

$$y = 1 \qquad \text{at } x = 1 \qquad (9.6.15b)$$

where $v = \pm 1$ and α is a constant. The analytical solution of this problem is obtained as the linear combination of the homogeneous and particular solutions:

$$y = C_1 + C_2 e^{-vx/\varepsilon} + \frac{e^x}{v + \varepsilon} \qquad (9.6.16)$$

where C_1 and C_2 are integration constants, determined by the BCs (9.6.15) as follows:

$$C_1 = \frac{[\alpha(\nu + \varepsilon) - 1]e^{-\nu/\varepsilon} + e - (\nu + \varepsilon)}{(\nu + \varepsilon)(e^{-\nu/\varepsilon} - 1)} \tag{9.6.17a}$$

$$C_2 = \frac{(1 - \alpha)(\nu + \varepsilon) + 1 - e}{(\nu + \varepsilon)(e^{-\nu/\varepsilon} - 1)} \tag{9.6.17b}$$

By analyzing the limiting behavior of the exponential term $e^{-\nu/\varepsilon}$ in $y(x)$ as $\varepsilon \to 0$, two distinct expressions are obtained depending upon the value of ν. In particular, for $\nu = +1$ we have

$$y = 1 - \frac{e}{1 + \varepsilon} - \left(1 - \alpha + \frac{1 - e}{1 + \varepsilon}\right)e^{-x/\varepsilon} + \frac{e^x}{1 + \varepsilon} \qquad \text{as } \varepsilon \to 0 \tag{9.6.18}$$

while for $\nu = -1$ we obtain

$$y = \alpha + \frac{1}{1 - \varepsilon} + \left(1 - \alpha - \frac{1 - e}{1 - \varepsilon}\right)e^{-(1-x)/\varepsilon} - \frac{e^x}{1 - \varepsilon} \qquad \text{as } \varepsilon \to 0. \tag{9.6.19}$$

The qualitative behavior of both these functions is illustrated in Figures 9.7 and 9.8, respectively, for various values of the parameters ε and α. It may be seen that the boundary layer, where the solution changes rapidly, is located close to the boundary $x = 0$ in the first case (i.e., $\nu = +1$) and close to the opposite boundary, $x = 1$ in the second case (i.e., $\nu = -1$). Thus the sign of the coefficient of the first derivative in eq. (9.6.14) determines the location of the boundary layer.

The same figures show, by the dashed curves, the solutions obtained by substituting $\varepsilon = 0$ in the original eq. (9.6.14) and neglecting the BC corresponding to the location of the boundary layer (i.e., the zeroth-order terms of the outer expansions). In analytical form, these are given as follows:

$$y = 1 - e + e^x \qquad \text{for } \nu = +1 \tag{9.6.20a}$$

and

$$y = 1 + \alpha - e^x \qquad \text{for } \nu = -1 \tag{9.6.20b}$$

As expected, these zeroth-order solutions approximate rather closely the exact solutions in most of the domain of x, but they break down in the boundary layer region, where they actually ignore the boundary condition. In fact, such a boundary condition does not affect the true solution in the entire domain of x, but only in the narrow region corresponding to the boundary layer. This is another characteristic feature of boundary layer problems.

The example above illustrates that the location of the boundary layer is not always obvious and may require some analysis. For example, in the case of eq. (9.6.14), since both the first and the second derivatives are quite large in the boundary layer, it can be concluded that y' and y'' must have opposite or the same signs, depending upon whether $\nu = +1$ or $\nu = -1$, respectively. By examining Figures 9.7 and 9.8, it can be seen that indeed when the boundary layer is close to $x = 0$, then y' and y'' have opposite signs, and when it is close to $x = 1$, then y' and y'' have the same signs. Therefore, we can

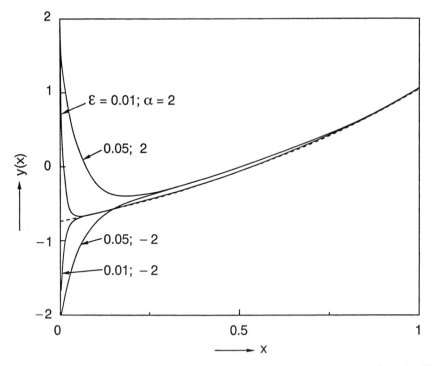

Figure 9.7 Qualitative behavior of the function defined by eq. (9.6.18) for various values of ε and α. The broken curve represents the zeroth-order outer expansion (9.6.20a).

conclude that the boundary layer is located close to $x = 0$ if $\nu = 1$, while it is located close to $x = 1$ if $\nu = -1$.

Unfortunately, simple arguments about the qualitative behavior of the solution, such as those reported above, are not always sufficient to determine *a priori* the location of the boundary layer. Indeed, useful hints can come from a knowledge of the physical phenomenon described by the equation being studied. When no *a priori* answers can be found, then the general procedure described below can be applied. In this procedure, all possible locations of the boundary layer together with various stretching transformations are considered, in order to find those leading to outer and inner expansions having an overlapping region of validity so as to allow for a suitable matching. It is important to note that this procedure provides not only the location of the boundary layer, but also the form of the stretching transformation.

Let us illustrate the procedure with application to eq. (9.6.14).

I. Boundary Layer at $x = 0$. We first *assume* that the boundary layer is located at $x = 0$, and then determine the zeroth-order term of the outer expansion. This is the solution of

$$\nu \frac{dy_0^o}{dx} = e^x \tag{9.6.21a}$$

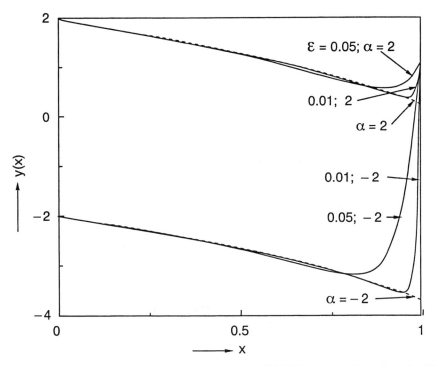

Figure 9.8 Qualitative behavior of the function defined by eq. (9.6.19) for various values of ε and α. The broken curve represents the zeroth-order outer expansion (9.6.20b).

along with the BC (9.6.15b)

$$y_0^o = 1 \qquad \text{at } x = 1 \tag{9.6.21b}$$

which gives

$$y_0^o = 1 + \frac{1}{\nu}(e^x - e) \tag{9.6.22}$$

For the inner expansion, we consider the general stretching transformation

$$s = \frac{x}{\varepsilon^\lambda} \tag{9.6.23}$$

where λ is a positive constant.

Note that the overlapping region for the inner and outer expansions is now near $x = 0$. In this region the two limits $s \to \infty$ and $x \to 0$ hold simultaneously for the inner and outer expansions, respectively.

Substituting in the original eq. (9.6.14) leads to

$$\frac{d^2y}{ds^2} + \nu\varepsilon^{\lambda-1}\frac{dy}{ds} = \varepsilon^{2\lambda-1}\exp(s\varepsilon^\lambda) \tag{9.6.24}$$

Substituting in eqs. (9.6.24) and (9.6.15a) the approximation of y in terms of power series of ε (inner expansion), we get

$$\left[\frac{d^2 y_0^i}{ds^2} + \varepsilon \frac{d^2 y_1^i}{ds^2} + \cdots\right] + \nu\varepsilon^{\lambda-1}\left[\frac{dy_0^i}{ds} + \varepsilon \frac{dy_1^i}{ds} + \cdots\right]$$

$$= \varepsilon^{2\lambda-1}[1 + s\varepsilon^\lambda + \cdots] \tag{9.6.25a}$$

$$y_0^i + \varepsilon y_1^i + \cdots = \alpha \qquad \text{at } s = 0 \tag{9.6.25b}$$

The equation for the zeroth-order term of the inner expansion, y_0^i, will clearly be different depending upon the value of λ; in particular, three cases are possible.

Case 1. $\lambda > 1$
In this case, for the leading order term, eq. (9.6.25) gives

$$\frac{d^2 y_0^i}{ds^2} = 0, \qquad y_0^i = \alpha \qquad \text{at } s = 0 \tag{9.6.26}$$

thus

$$y_0^i = \alpha + Cs \tag{9.6.27}$$

where C is the integration constant, to be determined through matching with the zeroth-order term of the outer expansion (9.6.22). This leads to

$$\lim_{s\to\infty}(\alpha + Cs) = \lim_{x\to 0}\left[1 + \frac{1}{\nu}(e^x - e)\right] \tag{9.6.28}$$

which admits the only solution $C = 0$ and which then implies $\alpha = [1 + (1/\nu)(1 - e)]$. However, since α is an independent parameter and therefore can take any value, the stretching transformation (9.6.23) with $\lambda > 1$ cannot be adopted in general.

Case 2. $\lambda = 1$
In the case of $\lambda = 1$, eq. (9.6.25) leads to

$$\frac{d^2 y_0^i}{ds^2} + \nu \frac{dy_0^i}{ds} = 0, \qquad y_0^i = \alpha \qquad \text{at } s = 0 \tag{9.6.29}$$

whose solution is

$$y_0^i = \alpha - C(1 - e^{-\nu s}) \tag{9.6.30}$$

By applying the matching condition, we obtain

$$\lim_{s\to\infty}[\alpha - C(1 - e^{-\nu s})] = \lim_{x\to 0}\left[1 + \frac{1}{\nu}(e^x - e)\right] \tag{9.6.31}$$

For $\nu = +1$, this admits the solution

$$C = \alpha - 2 + e \tag{9.6.32}$$

while for $\nu = -1$, the only possible solution is $C = 0$, which exists only for the special case where $\alpha = e$.

Thus we can conclude that since the solution (9.6.32) is fully satisfactory, in the case where $\nu = +1$ the boundary layer is indeed located close to $x = 0$. The stretching transformation is then given by eq. (9.6.23) with $\lambda = 1$; that is, $s = x/\varepsilon$. However, the problem remains open for the case $\nu = -1$.

Case 3. $0 < \lambda < 1$
In this case, eq. (9.6.25) gives

$$\frac{dy_0^i}{ds} = 0, \qquad y_0^i = \alpha \qquad \text{at } s = 0 \tag{9.6.33}$$

Thus

$$y_0^i = \alpha \tag{9.6.34}$$

which is clearly not satisfactory since it does not contain any adjustable constant for satisfying the matching condition.

Having exhausted all possible stretching transformations for the boundary layer near $x = 0$, we now proceed by assuming that the boundary layer is located close to $x = 1$.

II. Boundary Layer at $x = 1$.   In this case the outer expansion has to account for the condition at the other boundary (9.6.15a), so that the zeroth-order term now satisfies

$$\nu \frac{dy_0^o}{dx} = e^x, \qquad y_0^o = \alpha \qquad \text{at } x = 0 \tag{9.6.35}$$

whose solution is

$$y_0^o = \alpha - \frac{1}{\nu}(1 - e^x) \tag{9.6.36}$$

For the inner expansion, we now use the stretching transformation

$$s = \frac{1 - x}{\varepsilon^\lambda} \tag{9.6.37}$$

In this case the overlapping region for the inner and outer expansions is near $x = 1$. In this region the corresponding limits for the inner and outer expansions are $s \to \infty$ and $x \to 1$, respectively. Substituting (9.6.37) in the original eq. (9.6.14) for x and using the series (9.6.7) for the inner expansion, we get

$$\left[\frac{d^2 y_0^i}{ds^2} + \varepsilon \frac{d^2 y_1^i}{ds^2} + \cdots \right] - \nu \varepsilon^{\lambda - 1} \left[\frac{dy_0^i}{ds} + \varepsilon \frac{dy_1^i}{ds} + \cdots \right]$$
$$= e \varepsilon^{2\lambda - 1}[1 - \varepsilon^\lambda s + \cdots] \tag{9.6.38a}$$

along with the BC

$$y_0^i + \varepsilon y_1^i + \cdots = 1 \qquad \text{at } s = 0 \tag{9.6.38b}$$

We again have three cases, depending upon the value of λ.

Case 1. λ > 1
For the leading order term, eq. (9.6.38) gives

$$\frac{d^2 y_0^i}{ds^2} = 0, \qquad y_0^i = 1 \qquad \text{at } s = 0$$

whose solution is

$$y_0^i = 1 + Cs \tag{9.6.39}$$

which, following the same arguments for the corresponding case above, can be shown to be unsatisfactory.

Case 2. λ = 1
In this case we have

$$\frac{d^2 y_0^i}{ds^2} - v \frac{dy_0^i}{ds} = 0, \qquad y_0^i = 1 \qquad \text{at } s = 0$$

Thus

$$y_0^i = 1 - C(1 - e^{vs}) \tag{9.6.40}$$

which, upon matching with the zeroth-order term of the outer expansion, leads to

$$\lim_{s \to \infty}[1 - C(1 - e^{vs})] = \lim_{x \to 1}\left[\alpha - \frac{1}{v}(1 - e^x)\right] \tag{9.6.41}$$

In the case $v = -1$, this equation admits the acceptable solution

$$C = e - \alpha \tag{9.6.42}$$

However, in the case $v = +1$, the only possible solution $C = 0$ is valid in the special case where $\alpha = 2 - e$ and is therefore not valid in general.

Case 3. λ < 1
In this case, from eq. (9.6.38) we get

$$\frac{dy_0^i}{ds} = 0, \qquad y_0^i = 1 \qquad \text{at } s = 0 \tag{9.6.43}$$

which gives

$$y_0^i = 1$$

This solution is again not acceptable because the matching is possible only for the special value of $\alpha = 1 + (1/v)(1 - e)$.

Thus summarizing, it is found that the problem given by eqs. (9.6.14) and (9.6.15) admits a boundary layer which is located at $x = 0$ if $v = +1$, and at $x = 1$ if $v = -1$, in agreement with the plots of the exact solution shown in Figures 9.7 and 9.8. Moreover, the stretching transformation required to analyze the boundary layer region with the method of matched asymptotic expansions is indeed $s = x/\varepsilon$ in the first case and $s = (1 - x)/\varepsilon$ in the second case.

The above conclusion about the location of the boundary layer in fact holds even if v is a *function* of x, as long as it does not change sign over the domain of interest.

However, if $v(x) = 0$ for an x value within the domain of interest, often referred to as a *turning point*, then a boundary layer develops on both sides of that particular location, thus giving rise to an *internal boundary layer*. This aspect is discussed in section 9.6.5.

9.6.3   The Matching Principle

We have just seen how the location of the boundary layer is determined in general. After this is done, we need to develop a method to match the inner and outer solutions while retaining any desired number of terms in each expansion. This method is provided by the matching principle, which is illustrated next, in the context of the problem given by eqs. (9.6.14) and (9.6.15) with $v = +1$.

For the outer solution y^o, we introduce the usual series expansion

$$y^o = \sum_{n=0}^{\infty} \varepsilon^n y_n^o(x) \qquad (9.6.44)$$

Substituting this equation in the original problem given by eq. (9.6.14) with $v = +1$ and eq. (9.6.15b) (the boundary layer being at $x = 0$) and comparing terms of like powers of ε leads to

$$O(\varepsilon^0): \qquad \frac{dy_0^o}{dx} = e^x, \qquad y_0^o = 1 \qquad \text{at } x = 1 \qquad (9.6.45)$$

$$O(\varepsilon^1): \qquad \frac{d^2y_0^o}{dx^2} + \frac{dy_1^o}{dx} = 0, \qquad y_1^o = 0 \qquad \text{at } x = 1 \qquad (9.6.46)$$

The solutions of these equations can be substituted in eq. (9.6.44), leading to the *two-term* outer expansion:

$$y^o = (e^x + 1 - e) + \varepsilon(e - e^x) + O(\varepsilon^2) \qquad (9.6.47)$$

Using a similar series expansion for the inner solution, substituting it in the original eqs. (9.6.14) and (9.6.15a) along with the stretching transformation $s = x/\varepsilon$, and comparing terms of like powers of ε leads to the following problems:

$$O(\varepsilon^0): \qquad \frac{d^2y_0^i}{ds^2} + \frac{dy_0^i}{ds} = 0, \qquad y_0^i = \alpha \qquad \text{at } s = 0 \qquad (9.6.48)$$

$$O(\varepsilon^1): \qquad \frac{d^2y_1^i}{ds^2} + \frac{dy_1^i}{ds} = 1, \qquad y_1^i = 0 \qquad \text{at } s = 0 \qquad (9.6.49)$$

whose solutions lead to the following *two-term* inner expansion:

$$y^i = [\alpha - C_1(1 - e^{-s})] + \varepsilon[C_2(1 - e^{-s}) + s] + O(\varepsilon^2) \qquad (9.6.50)$$

where C_1 and C_2 are integration constants to be evaluated through the matching condition. In general, this is provided by the *matching principle*, which can be expressed concisely in the following form:

$$\frac{\text{The } m\text{-term outer expansion}}{\text{of the } k\text{-term inner solution}} = \frac{\text{The } k\text{-term inner expansion}}{\text{of the } m\text{-term outer solution}} \qquad (9.6.51)$$

where the m-term outer expansion of an inner solution u^i is given by

$$\text{the } m\text{-term } \lim_{\substack{\varepsilon \to 0 \\ x \text{ fixed}}} u^i\left(\frac{x}{\varepsilon}\right) \tag{9.6.52a}$$

and similarly, the k-term inner expansion of an outer solution u^o is given by

$$\text{the } k\text{-term } \lim_{\substack{\varepsilon \to 0 \\ s \text{ fixed}}} u^o(s\varepsilon) \tag{9.6.52b}$$

Note that in the expressions above, k and m are integers which are not necessarily equal. Let us now apply the matching principle to the outer and inner expansions determined above, for some specific values of k and m.

Case 1. k = m = 1
From (9.6.47), the 1-term outer solution is $y^o = e^x + 1 - e$, and its 1-term inner expansion, with $x = \varepsilon s$ is

$$1\text{-term } \lim_{\substack{\varepsilon \to 0 \\ s \text{ fixed}}} [e^{\varepsilon s} + 1 - e] = 2 - e \tag{9.6.53}$$

Similarly, from (9.6.50), the 1-term inner solution is $y^i = \alpha - C_1(1 - e^{-s})$, and its 1-term outer expansion, with $s = x/\varepsilon$, is

$$1\text{-term } \lim_{\substack{\varepsilon \to 0 \\ x \text{ fixed}}} [\alpha - C_1(1 - e^{-x/\varepsilon})] = \alpha - C_1 \tag{9.6.54}$$

Next, the matching principle requires equating the right-hand sides of eqs. (9.6.53) and (9.6.54), which gives the value of the first integration constant:

$$C_1 = \alpha - 2 + e \tag{9.6.55}$$

and this coincides with the result (9.6.32) obtained previously for the zeroth-order terms.

Case 2. k = m = 2
The two-term outer solution is given by eq. (9.6.47), and its two-term inner expansion is given by

$$2\text{-term } \lim_{\substack{\varepsilon \to 0 \\ s \text{ fixed}}} [(e^{\varepsilon s} + 1 - e) + \varepsilon(e - e^{\varepsilon s})] = (2 - e) + \varepsilon(s + e - 1) \tag{9.6.56}$$

while the two-term inner solution is given by eq. (9.6.50), and the corresponding two-term outer expansion is

$$2\text{-term } \lim_{\substack{\varepsilon \to 0 \\ x \text{ fixed}}} \left\{ [\alpha - C_1(1 - e^{-x/\varepsilon})] + \varepsilon \left[C_2(1 - e^{-x/\varepsilon}) + \frac{x}{\varepsilon} \right] \right\}$$

$$\tag{9.6.57}$$

$$= (\alpha - C_1 + x) + \varepsilon C_2$$

Equating the two expressions, for any value of ε, it follows that

$$C_1 = \alpha - 2 + e \tag{9.6.58a}$$

$$C_2 = e - 1 \tag{9.6.58b}$$

which are the two constants required to completely define the two-term inner expansion y^i, given in eq. (9.6.50). Note that C_1 given by eq. (9.6.58a) is the same as that given by eq. (9.6.55).

Case 3. k = 2, m = 1
The 2-term inner expansion of the 1-term outer solution is given by

$$\text{2-term } \lim_{\substack{\varepsilon \to 0 \\ s \text{ fixed}}} (e^{\varepsilon s} + 1 - e) = 2 - e + \varepsilon s \tag{9.6.59}$$

On the other hand, the 1-term outer expansion of the 2-term inner solution is readily seen from eq. (9.6.57) to be equal to $(\alpha - C_1 + x)$. Equating this to the right-hand side of eq. (9.6.59) leads only to C_1, and for it the expression is the same as before—that is, eq. (9.6.55) or (9.6.58a).

9.6.4 The Composite Solution

To derive an approximation valid uniformly throughout the entire domain of x, we need to develop the composite solution as defined by eq. (9.6.12). This implies that we should sum up the outer and inner expansions and then subtract the common part in the overlapping region. For example, in the case where $k = m = 2$, this common part is given by the right-hand sides of eqs. (9.6.56) or (9.6.57). Thus the 2-term composite solution is given by

$$y^c(x) = y^o(x) + y^i(x) - [(2 - e + x) + \varepsilon(e - 1)] \tag{9.6.60}$$

which, upon substituting eq. (9.6.47) and eq. (9.6.50) after some algebraic manipulations, leads to

$$y^c(x) = (1 - e + e^x) + (\alpha - 2 + e)e^{-x/\varepsilon} + \varepsilon[(e - e^x) - (e - 1)e^{-x/\varepsilon}] \tag{9.6.61}$$

It is easy to verify that expression (9.6.61) is identical to the two-term expansion of the exact solution, which is readily obtained from eq. (9.6.18) by expanding the term $1/(1 + \varepsilon)$ as follows:

$$\frac{1}{1 + \varepsilon} = 1 - \varepsilon + O(\varepsilon^2) \tag{9.6.62}$$

Thus summarizing, it can be concluded that the method of matched asymptotic expansions is well suited for boundary-layer-type singular perturbation problems. However, it cannot handle situations where secular terms occur (see problem 9.18). It is worth noting that in this section, only those cases where the small parameter ε multiplies the highest-order derivative of a linear ordinary differential equation were considered. As discussed in section 9.4.2, boundary-layer-type singular perturbation problems may arise also from partial differential equations, where the small parameter ε multiplies the highest-order derivative with respect to one particular independent variable. The extension of the method of matched asymptotic expansions to these problems as well as to nonlinear problems is illustrated in section 9.7.

9.6.5 Internal Boundary Layer

In concluding section 9.6.2 we remarked on the possibility that a boundary layer may occur within the domain of integration rather than at one of the boundaries. This occurs for eq. (9.6.14), when the coefficient v is a function of x which vanishes at some point (turning point) within the domain of integration. In this section we discuss this issue by means of a simple example. A more detailed analysis is reported in advanced books dealing with perturbation methods (cf. Bender and Orszag, 1978, section 9.6; Nayfeh, 1973, section 7.3).

Consider the second-order ODE:

$$\varepsilon(x + 2)\frac{d^2y}{dx^2} + x\frac{dy}{dx} - x^2y = 0, \qquad x \in [-1, 1] \tag{9.6.63}$$

with BCs

$$y = -2 \qquad \text{at } x = -1 \tag{9.6.64a}$$

$$y = 1 \qquad \text{at } x = 1 \tag{9.6.64b}$$

which exhibits a turning point at $x = 0$. A boundary layer, referred to as an *internal boundary layer*, develops on both sides of this location.

In this case we develop two outer solutions: y^{o-}, valid to the left of the turning point—that is for $x \in [-1, 0)$—and y^{o+}, which applies to its right—that is, for $x \in (0, +1]$. In the neighborhood of $x = 0$, we develop an inner solution y^i by applying an appropriate stretching transformation of the independent variable x. Note that since in the case of internal boundary layers the solution tends to be complicated, we confine our analysis to the zeroth-order approximation.

Let us first derive the outer solutions—that is, the asymptotic expansions of the solution away from the location $x = 0$:

$$y^o(x) = \sum_{n=0}^{\infty} \varepsilon^n y_n^o(x) \tag{9.6.65}$$

Substituting in eq. (9.6.63) and matching terms of leading order in ε we obtain the following expression for the zeroth-order term for *both* outer expansions:

$$x\frac{dy_0^o}{dx} - x^2y_0^o = 0 \tag{9.6.66}$$

The outer solution valid to the *left* of the internal boundary layer must satisfy the BC (9.6.64a) at $x = -1$; that is,

$$y_0^{o-} = -2 \qquad \text{at } x = -1 \tag{9.6.67}$$

Integrating eq. (9.6.66) with the above IC, we obtain

$$y_0^{o-} = -2 \exp\left[-\frac{(1 - x^2)}{2}\right], \qquad x \in [-1, 0) \tag{9.6.68}$$

On the other hand, the outer solution valid to the *right* of the internal boundary layer must satisfy the BC (9.6.64b) at $x = 1$; that is

$$y_0^{o+} = 1 \qquad \text{at } x = 1 \tag{9.6.69}$$

Integrating eq. (9.6.66) using this IC leads to

$$y_0^{o+} = \exp\left[-\frac{(1 - x^2)}{2}\right], \qquad x \in (0, 1] \tag{9.6.70}$$

We can now turn our attention to deriving the inner solution—that is, the asymptotic expansion of the solution in the neighborhood of $x = 0$—within the internal boundary layer. For this we introduce the stretching transformation

$$s = \frac{x}{\varepsilon^\lambda} \tag{9.6.71}$$

where λ is a positive constant.

Substituting in the original eq. (9.6.63) and using the chain differentiation rule yields

$$\varepsilon^{1-2\lambda}(s\varepsilon^\lambda + 2)\frac{d^2y}{ds^2} + s\frac{dy}{ds} - s^2\varepsilon^{2\lambda}y = 0 \tag{9.6.72}$$

The inner solution is obtained in the form of the asymptotic series:

$$y^i(s) = \sum_{n=0}^{\infty} \varepsilon^n y_n^i(s) \tag{9.6.73}$$

We note that the third term in eq. (9.6.72) is dominated by the second one for any value of $\lambda > 0$. Thus, in order to balance the first two terms, we take $\lambda = 1/2$. Substituting eq. (9.6.73) in eq. (9.6.72), the leading term of the inner solution is given by

$$2\frac{d^2y_0^i}{ds^2} + s\frac{dy_0^i}{ds} = 0, \qquad s \in (-\infty, +\infty) \tag{9.6.74}$$

This equation can be solved by introducing the variable $w(s) = dy_0^i/ds$, leading to

$$2\frac{dw}{ds} + sw = 0$$

which upon integration yields

$$w = \frac{dy_0^i}{ds} = A \exp\left(-\frac{s^2}{4}\right)$$

where A is an integration constant. By a further integration we obtain the zeroth-order approximation of the inner solution:

$$y_0^i(s) = y_0^i(0) + A\int_0^s \exp\left(-\frac{t^2}{4}\right) dt \tag{9.6.75}$$

Note that the integral on the rhs can be expressed in terms of the error function (4.7.1), to give

$$y_0^i(s) = y_0^i(0) + B \operatorname{erf}\left(\frac{s}{2}\right) \tag{9.6.76}$$

where $y_0^i(0)$ and B are constants yet to be determined. For this we have to match the inner solution with the two outer solutions valid on the lhs and the rhs of the internal boundary layer, respectively. Thus for the region $x \in [-1, 0]$ we have

$$\lim_{s \to -\infty} y_0^i(s) = \lim_{x \to 0^-} y_0^{o^-}(x) \tag{9.6.77}$$

which, upon using eqs. (9.6.68) and (9.6.76) and noting that $\operatorname{erf}(-\infty) = -1$, leads to

$$y_0^i(0) - B = -2 \exp(-1/2) \tag{9.6.78}$$

On the other hand, in the interval $x \in [0, 1]$ we have

$$\lim_{s \to +\infty} y_0^i(s) = \lim_{x \to 0^+} y_0^{o^+}(x) \tag{9.6.79}$$

and using eqs. (9.6.70) and (9.6.76) with $\operatorname{erf}(+\infty) = 1$ gives

$$y_0^i(0) + B = \exp(-1/2) \tag{9.6.80}$$

Combining this equation with eq. (9.6.78), we can determine the two constants B and $y_0^i(0)$ and hence complete the expression (9.6.76) for the inner solution:

$$y_0^i(s) = -\frac{1}{2} \exp(-1/2)\left[1 - 3 \operatorname{erf}\left(s/2\right)\right] \tag{9.6.81}$$

In the case of internal boundary layers, where we have more than one outer solution, the composite solution cannot be derived using the procedure outlined in section 9.6.4. However, in this case we note that the leading term of the inner solution $y_0^i(s) \to -2 \exp(-1/2)$ as $s \to -\infty$ and $y_0^i(s) \to \exp(-1/2)$ as $s \to +\infty$. By comparing these limits with the expressions of the leading terms of the outer solutions, $y_0^{o^-}(x)$ and $y_0^{o^+}(x)$, respectively, the following expression for the composite solution can be obtained readily:

$$y_0^c(x) = -\frac{1}{2}\left[1 - 3 \operatorname{erf}\left(\frac{x}{2\sqrt{\varepsilon}}\right)\right] \exp\left[-\frac{(1 - x^2)}{2}\right] \tag{9.6.82}$$

The behavior of the one-term composite solution is shown in Figure 9.9, where it is compared with the numerical solutions of eqs. (9.6.63) and (9.6.64), as well as with the two outer solutions. It can be seen that in the neighborhood of the point $x = 0$, the solution undergoes a rapid change, indicating the presence of the internal boundary layer. Note that at $x = 0$ we obtain $y_0^c = -\frac{1}{2} \exp(-1/2)$, which is different from the value obtained from either one of the outer solutions.

9.7 APPLICATION OF THE METHOD OF MATCHED ASYMPTOTIC EXPANSIONS TO A PROBLEM IN CHEMICAL REACTOR THEORY

In section 5.2, when discussing the issue of properly assessing the boundary conditions of a partial differential equation, we introduced several models with varying degrees of

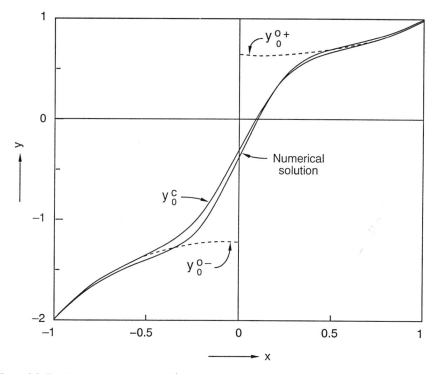

Figure 9.9 Zeroth-order outer (y_0^{o-} and y_0^{o+}), composite (y_0^c), and numerical solutions of eqs. (9.6.63) and (9.6.64); $\varepsilon = 0.02$.

complexity, which are widely used for simulating homogeneous, isothermal, tubular reactors. Let us now focus in particular on one of these, the so-called axial dispersion model, defined by eqs. (5.2.3) and (5.2.4). This model, which we consider here in its transient version, accounts for the concentration changes in the reactor axis direction due to the combined influence of convective transport, axial dispersion, and chemical reaction. The latter, in eq. (5.2.3), was assumed to exhibit first-order kinetics. For a generic reaction kinetics, this model can be rewritten as

$$\frac{\partial u}{\partial t} = \frac{1}{Pe}\frac{\partial^2 u}{\partial x^2} - \frac{\partial u}{\partial x} - Br(u); \qquad 0 < x < 1, t > 0 \tag{9.7.1}$$

with BCs

$$u = \beta(t) + \frac{1}{Pe}\frac{\partial u}{\partial x} \qquad \text{at } x = 0, t > 0 \tag{9.7.2a}$$

$$\frac{\partial u}{\partial x} = 0 \qquad \text{at } x = 1, t > 0 \tag{9.7.2b}$$

and IC

$$u = \alpha(x) \qquad \text{at } t = 0, 0 \le x \le 1 \tag{9.7.2c}$$

where $r(u)$ is a nonlinear function of the reactant concentration u representing the rate of reaction, $\alpha(x)$ is the initial reactant concentration profile along the reactor axis, and $\beta(t)$ is the reactant concentration in the reactor feed, which in general can be a function of time.

A situation which arises in applications is one where mixing in the axial direction is so intense that the reactant concentration becomes spatially uniform in the reactor. In the limit this correspond to the behavior of a well-mixed reactor, which is usually simulated through the ideal CSTR model. Such a situation is indeed described also by the axial dispersion model (9.7.1)–(9.7.2) in the limit where the axial dispersion coefficient becomes very large—that is, where the Peclet number Pe approaches zero. In the present section we investigate the behavior of the axial dispersion model in this limit and derive the asymptotic series expansion of the solution as Pe $\rightarrow$ 0.

However, before this, it is of interest to identify some qualitative features of its behavior. In the limit of Pe $\rightarrow$ 0, eqs. (9.7.1), (9.7.2a), and (9.7.2b) reduce to

$$\frac{\partial^2 u}{\partial x^2} = 0, \qquad 0 < x < 1, t > 0 \tag{9.7.3}$$

and

$$\frac{\partial u}{\partial x} = 0 \qquad \text{at } x = 0, t > 0 \tag{9.7.4a}$$

$$\frac{\partial u}{\partial x} = 0 \qquad \text{at } x = 1, t > 0 \tag{9.7.4b}$$

whose solution is a constant. This indicates, as expected based on the physical arguments given above, that as Pe $\rightarrow$ 0, the reactant concentration u becomes independent of x—that is, spatially uniform within the reactor. However, such a solution does not in general satisfy the IC (9.7.2c).

We can envision a slightly more complex behavior for the solution. At time $t = 0$, the reactant concentration profile in the reactor is that imposed by the IC (9.7.2c); that is, $u = \alpha(x)$. As time increases, due to the very intense mixing the concentration becomes uniform—that is, independent of x. Such a transition from the initial to the uniform profile occurs in a time interval which is inversely proportional to the mixing intensity—that is, directly proportional to the parameter Pe. Accordingly, in the limit of Pe $\rightarrow$ 0, we can expect the solution to undergo a sudden transition, thus exhibiting a *boundary layer* in time. In this time interval, with duration proportional to Pe, the initial profile is made uniform by the intense mixing implied by Pe being very small. This constitutes the *inner solution*, which merges into the *outer solution*, valid for time values much larger than Pe.

The conclusion above suggests that the method of matched asymptotic expansions can be applied to develop the approximate solution, taking the Peclet number as the small parameter (i.e., ε = Pe) and assuming that the boundary layer is located at time $t = 0$.

It is worth noting that this problem provides us the opportunity to apply this method to a problem involving partial derivatives and nonlinear terms—two elements which have not been considered so far.

Let us first derive the *outer expansion*—that is, the asymptotic expansion of the solution in the region away from the time value $t = 0$, where the boundary layer is located:

$$u^o(x, t) = \sum_{n=0}^{\infty} \varepsilon^n u_n^o(x, t) \qquad (9.7.5)$$

Note that the unknown coefficients of the series, u_n^o, are functions of *both* the independent variables x and t, and so we should expect their evaluation to require the solution of subproblems involving partial differential equations—that is, of the same form as the original problem. Following the usual procedure for deriving such subproblems, we substitute the expansion (9.7.5) in the original problem (9.7.1)–(9.7.2) and then compare terms with like powers of ε. As compared to the problems considered earlier, in this case such an operation is further complicated by the presence of the nonlinear term $r(u)$.

This difficulty may be overcome by substituting the nonlinear term by its asymptotic expansion for small ε values, which is most conveniently obtained using the Taylor series expansion. This approximation is clearly appropriate in the framework of this method, where, as in all perturbation methods, we are seeking an approximation of the solution in the form of a series expansion. Thus, by substituting the series approximation (9.7.5) for u, we can regard $r(u(\varepsilon))$ in eq. (9.7.1) as a function of the small parameter ε and can then expand it in a Taylor series as follows:

$$r(\varepsilon) = r(0) + \varepsilon \left(\frac{dr}{d\varepsilon}\right)_{\varepsilon=0} + \frac{1}{2!} \varepsilon^2 \left(\frac{d^2r}{d\varepsilon^2}\right)_{\varepsilon=0} + \cdots \qquad (9.7.6)$$

The coefficients of this series can be evaluated from the specific form of the function $r(u)$ and the series expansion (9.7.5). Thus $r(0) = r(u_0^o)$, where u_0^o is the zeroth-order term of the series expansion (9.7.5). The first-order derivative of r with respect to ε at $\varepsilon = 0$ can be evaluated using the chain rule and eq. (9.7.5) as follows:

$$\left(\frac{dr}{d\varepsilon}\right)_{\varepsilon=0} = \left(\frac{dr}{du}\right)_{u=u_0^o} \left(\frac{du}{d\varepsilon}\right)_{\varepsilon=0} = \left(\frac{dr}{du}\right)_{u=u_0^o} u_1^o \qquad (9.7.7)$$

Similarly, the second-order derivative is given by

$$\begin{aligned}
\left(\frac{d^2r}{d\varepsilon^2}\right)_{\varepsilon=0} &= \left[\frac{d}{d\varepsilon}\left(\frac{dr}{du}\frac{du}{d\varepsilon}\right)\right]_{\varepsilon=0} \\
&= \left(\frac{d^2r}{du^2}\right)_{u=u_0^o} \left(\frac{du}{d\varepsilon}\right)_{\varepsilon=0}^2 + \left(\frac{dr}{du}\right)_{u=u_0^o} \left(\frac{d^2u}{d\varepsilon^2}\right)_{\varepsilon=0} \\
&= \left(\frac{d^2r}{du^2}\right)_{u=u_0^o} u_1^{o2} + 2\left(\frac{dr}{du}\right)_{u=u_0^o} u_2^o
\end{aligned} \qquad (9.7.8)$$

Substituting eqs. (9.7.7) and (9.7.8) in eq. (9.7.6), we obtain the series expansion of the nonlinear term $r(u)$ as $\varepsilon \to 0$:

$$r(u) = r(u_0^o) + \varepsilon \left(\frac{dr}{du}\right)_{u=u_0^o} u_1^o + \frac{1}{2} \varepsilon^2 \left[\left(\frac{d^2r}{du^2}\right)_{u=u_0^o} u_1^{o2} + 2\left(\frac{dr}{du}\right)_{u=u_0^o} u_2^o\right] + \cdots \qquad (9.7.9)$$

We can now proceed to substitute the outer expansion (9.7.5) in the original problem (9.7.1)–(9.7.2), which, using eq. (9.7.9), reduces to

$$
\varepsilon \frac{\partial u_0^o}{\partial t} + \varepsilon^2 \frac{\partial u_1^o}{\partial t} + \varepsilon^3 \frac{\partial u_2^o}{\partial t} + \cdots = \left[\frac{\partial^2 u_0^o}{\partial x^2} + \varepsilon \frac{\partial^2 u_1^o}{\partial x^2} + \varepsilon^2 \frac{\partial^2 u_2^o}{\partial x^2} + \cdots \right]
$$
$$
- \left[\varepsilon \frac{\partial u_0^o}{\partial x} + \varepsilon^2 \frac{\partial u_1^o}{\partial x} + \varepsilon^3 \frac{\partial u_2^o}{\partial x} + \cdots \right]
$$
$$
- \varepsilon B \left\{ r(u_0^o) + \varepsilon \left(\frac{dr}{du} \right)_{u=u_0^o} u_1^o + \frac{1}{2} \varepsilon^2 \left[\left(\frac{d^2 r}{du^2} \right)_{u=u_0^o} u_1^{o2} \right. \right.
$$
$$
\left. \left. + 2 \left(\frac{dr}{du} \right)_{u=u_0^o} u_2^o \right] + \cdots \right\}
$$

(9.7.10)

along with the BCs

$$
\varepsilon u_0^o + \varepsilon^2 u_1^o \cdots = \varepsilon \beta(t) + \frac{\partial u_0^o}{\partial x} + \varepsilon \frac{\partial u_1^o}{\partial x} + \varepsilon^2 \frac{\partial u_2^o}{\partial x} + \cdots \qquad \text{at } x = 0, \, t > 0
$$

(9.7.11a)

$$
\frac{\partial u_0^o}{\partial x} + \varepsilon \frac{\partial u_1^o}{\partial x} + \varepsilon^2 \frac{\partial u_2^o}{\partial x} + \cdots = 0 \qquad \text{at } x = 1, \, t > 0
$$

(9.7.11b)

It is worth stressing that in developing the outer expansion of the solution, we are not considering the IC at $t = 0$ given by eq. (9.7.2c), since the outer solution is valid only away from the boundary layer and hence for $t \gg \varepsilon$.

To evaluate the coefficients u_n^o of the series expansion (9.7.5), we have to now solve the following subproblems which originate from eqs. (9.7.10) and (9.7.11) by comparing terms of like powers of ε.

$$
O(\varepsilon^0): \qquad \frac{\partial^2 u_0^o}{\partial x^2} = 0 \qquad (9.7.12)
$$

with BCs

$$
\frac{\partial u_0^o}{\partial x} = 0 \qquad \text{at } x = 0, \, t > 0 \qquad (9.7.12a)
$$

$$
\frac{\partial u_0^o}{\partial x} = 0 \qquad \text{at } x = 1, \, t > 0 \qquad (9.7.12b)
$$

The solution of this problem is given by

$$
u_0^o = g(t) \qquad (9.7.13)
$$

where $g(t)$ is an as yet undetermined function of time t, independent of the axial coordinate x.

$$
O(\varepsilon^1): \qquad \frac{\partial u_0^o}{\partial t} = \frac{\partial^2 u_1^o}{\partial x^2} - \frac{\partial u_0^o}{\partial x} - Br(u_0^o) \qquad (9.7.14)
$$

with BCs

$$u_0^o = \beta(t) + \frac{\partial u_1^o}{\partial x} \qquad \text{at } x = 0, t > 0 \qquad (9.7.14a)$$

$$\frac{\partial u_1^o}{\partial x} = 0 \qquad \text{at } x = 1, t > 0 \qquad (9.7.14b)$$

Substituting the expression for u_0^o given by eq. (9.7.13), eq. (9.7.14) reduces to

$$\frac{\partial^2 u_1^o}{\partial x^2} = \frac{dg}{dt} + Br(g) \qquad (9.7.15)$$

which, upon integrating both sides from $x = 0$ to $x = 1$ and using eqs. (9.7.14a) and (9.7.14b), leads after some rearrangement to the *consistency condition*:

$$\frac{dg}{dt} = \beta(t) - g - Br(g) \qquad (9.7.16)$$

This is an ordinary differential equation in the only unknown $g(t)$ whose solution allows us to determine, through eq. (9.7.13), the zeroth-order coefficient of the outer expansion. Actually, since there is no initial condition given, eq. (9.7.16) defines the function $g(t)$ only up to an arbitrary constant. The missing condition cannot be given by eq. (9.7.2c), which applies at the opposite end of the boundary layer (i.e., $t = 0$), where the outer expansion is not valid. However, since the *inner solution* is valid within the boundary layer, the missing condition can be obtained by requiring the outer and inner expansions to match in the overlapping region of the boundary layer, where they are both valid simultaneously. As discussed in section 9.6.3, this condition follows directly from the matching principle. Accordingly, the IC for eq. (9.7.16) is provided by the matching condition, which we will derive after developing the inner expansion of the solution.

Before proceeding to determine higher-order terms of the outer expansion, we should underline a particular feature of the procedure that we have just applied to obtain the zeroth-order term of the outer expansion, u_0^o. Unlike the cases illustrated in section 9.6, in order to derive u_0^o we have used not only eq. (9.7.12) which arises by comparing terms of order $O(\varepsilon^0)$ in the original eq. (9.7.10), but also eq. (9.7.14) which arises when comparing terms of the next higher order—that is, $O(\varepsilon^1)$. This is a rather common feature in perturbation problems of this type and, as we shall see next, it occurs also in the evaluation of all the higher-order coefficients of the outer expansion. That is, in order to evaluate the nth coefficient of the series, u_n^o we will have to consider also the subproblem arising when comparing terms of $O(\varepsilon^{n+1})$ in the original eq. (9.7.10), where the subsequent coefficient u_{n+1}^o appears as the unknown.

To better illustrate the point above, let us now proceed to determine the next coefficient of the outer expansion—that is, u_1^o. Using the BC (9.7.14a) and integrating eq. (9.7.15) from $x = 0$ to the generic position x along the reactor axis, we get

$$\frac{\partial u_1^o}{\partial x} = g - \beta(t) + \left[\frac{dg}{dt} + Br(g) \right] x$$

$$= [g - \beta(t)](1 - x) \qquad (9.7.17)$$

where the second equality has been obtained by evaluating the derivative dg/dt through eq. (9.7.16). Integrating eq. (9.7.17) once again, from $x = 1$, we obtain

$$u_1^o = h(t) - \tfrac{1}{2}[g - \beta(t)](1 - x)^2 \tag{9.7.18}$$

where $h(t)$ is an as yet undetermined function of time, which represents the value of u_1^o at $x = 1$. Thus we are in the same situation discussed above when deriving the zeroth-order coefficient u_0^o as given by eq. (9.7.13), which involves the undetermined function $g(t)$. Similarly as before, in order to determine the unknown function $h(t)$, we go on to consider the following subproblem which arises when comparing terms of $O(\varepsilon^2)$ in eqs. (9.7.10)–(9.7.11):

$$O(\varepsilon^2): \qquad \frac{\partial u_1^o}{\partial t} = \frac{\partial^2 u_2^o}{\partial x^2} - \frac{\partial u_1^o}{\partial x} - B\left(\frac{dr}{du}\right)_{u=u_0^o} u_1^o \tag{9.7.19}$$

with BCs

$$u_1^o = \frac{\partial u_2^o}{\partial x} \qquad \text{at } x = 0, \ t > 0 \tag{9.7.19a}$$

$$\frac{\partial u_2^o}{\partial x} = 0 \qquad \text{at } x = 1, \ t > 0 \tag{9.7.19b}$$

Substituting eqs. (9.7.13) and (9.7.18), eq. (9.7.19) reduces to

$$\frac{\partial^2 u_2^o}{\partial x^2} = \frac{dh}{dt} - \frac{1}{2}\left[\frac{dg}{dt} - \frac{d\beta}{dt}\right](1 - x)^2 + [g - \beta(t)](1 - x)$$
$$+ B\left(\frac{dr}{du}\right)_{u=g}\left\{h - \frac{1}{2}[g - \beta(t)](1 - x)^2\right\} \tag{9.7.20}$$

Integrating both sides of eq. (9.7.20) from $x = 0$ to $x = 1$ and using eqs. (9.7.19a) and (9.7.19b), so as to eliminate the new unknown, u_2^o, leads to the consistency condition

$$\frac{dh}{dt} + h = \frac{1}{6}\left(\frac{dg}{dt} - \frac{d\beta}{dt}\right) - B\left(\frac{dr}{du}\right)_{u=g}\left[h - \frac{1}{6}(g - \beta)\right] \tag{9.7.21}$$

which is a first-order ode in the only unknown $h(t)$, whose solution substituted in eq. (9.7.18) yields the first-order coefficient of the outer expansion, u_1^o. Note that, as discussed earlier with reference to eq. (9.7.16), the IC for eq. (9.7.21) will again come from the matching condition between the outer and inner expansions.

Let us leave at this point the derivation of the outer expansion and turn our attention to the *inner expansion*. Its development requires us to first introduce a stretching transformation of the independent variable, t so as to make smooth the variation of the solution within the boundary layer. Since, according to the qualitative behavior of the solution discussed above, we expect the boundary layer to be located at $t = 0$ and its thickness to be proportional to ε, it is reasonable to consider the stretching transformation:

$$\tau = \frac{t}{\varepsilon} \tag{9.7.22}$$

Introducing the stretched time coordinate τ together with the inner series expansion

$$u^i(x, \tau) = \sum_{n=0}^{\infty} \varepsilon^n u_n^i(x, \tau) \qquad (9.7.23)$$

and the series expansion of the nonlinear term $r(u)$ given by eq. (9.7.9), in the original problem (9.7.1)–(9.7.2) leads to

$$\frac{\partial u_0^i}{\partial \tau} + \varepsilon \frac{\partial u_1^i}{\partial \tau} + \cdots = \frac{\partial^2 u_0^i}{\partial x^2} + \varepsilon \frac{\partial^2 u_1^i}{\partial x^2} + \cdots - \varepsilon \frac{\partial u_0^i}{\partial x}$$
$$- \varepsilon^2 \frac{\partial u_1^i}{\partial x} - \cdots - \varepsilon B\left[r(u_0^i) + \varepsilon \left(\frac{dr}{du}\right)_{u=u_0^i} u_1^i + \cdots \right] \qquad (9.7.24)$$

with BCs

$$\varepsilon u_0^i + \varepsilon^2 u_1^i + \cdots = \varepsilon \beta(t) + \frac{\partial u_0^i}{\partial x} + \varepsilon \frac{\partial u_1^i}{\partial x} + \cdots \qquad \text{at } x = 0, \tau > 0 \qquad (9.7.25a)$$

$$\frac{\partial u_0^i}{\partial x} + \varepsilon \frac{\partial u_1^i}{\partial x} + \cdots = 0 \qquad \text{at } x = 1, \tau > 0 \qquad (9.7.25b)$$

and IC

$$u_0^i + \varepsilon u_1^i + \cdots = \alpha(x) \qquad \text{at } \tau = 0, 0 \leq x \leq 1 \qquad (9.7.25c)$$

Grouping together terms of like powers of ε, the following subproblems are obtained:

$$O(\varepsilon^0): \qquad \frac{\partial u_0^i}{\partial \tau} - \frac{\partial^2 u_0^i}{\partial x^2} = 0 \qquad (9.7.26)$$

with BCs

$$\frac{\partial u_0^i}{\partial x} = 0 \qquad \text{at } x = 0, \tau > 0 \qquad (9.7.27a)$$

$$\frac{\partial u_0^i}{\partial x} = 0 \qquad \text{at } x = 1, \tau > 0 \qquad (9.7.27b)$$

and IC

$$u_0^i = \alpha(x) \qquad \text{at } \tau = 0, 0 \leq x \leq 1 \qquad (9.7.27c)$$

$$O(\varepsilon^1): \qquad \frac{\partial u_1^i}{\partial \tau} - \frac{\partial^2 u_1^i}{\partial x^2} = -\frac{\partial u_0^i}{\partial x} - Br(u_0^i) \qquad (9.7.28)$$

with BCs

$$u_0^i = \beta(t) + \frac{\partial u_1^i}{\partial x} \qquad \text{at } x = 0, \tau > 0 \qquad (9.7.29a)$$

$$\frac{\partial u_1^i}{\partial x} = 0 \qquad \text{at } x = 1, \tau > 0 \qquad (9.7.29b)$$

and IC

$$u_1^i = 0 \qquad \text{at } \tau = 0, 0 \leq x \leq 1 \qquad (9.7.29c)$$

Since the solution of the subproblems providing the higher-order coefficients of the series expansion requires intricate manipulations and is rather cumbersome to be reported in detail, we limit our analysis to the zeroth-order term. This is readily evaluated by solving eqs. (9.7.26) and (9.7.27). We may note that the latter is the classical problem of transport of a diffusing species across a slab geometry, in the absence of generation terms and zero flux across either end of the slab. The solution can be obtained readily by the method of finite Fourier transforms (see chapter 7) as

$$u_0^i = a_0 + \sum_{m=1}^{\infty} a_m \cos(m\pi x) \exp(-m^2\pi^2\tau) \tag{9.7.30}$$

where

$$a_0 = \int_0^1 \alpha(x) \, dx \tag{9.7.31a}$$

$$a_m = 2 \int_0^1 \alpha(x) \cos(m\pi x) \, dx \tag{9.7.31b}$$

We can now proceed to derive the composite solution, which needs first to introduce the matching condition between the outer and the inner expansions. Since we are considering here only the zeroth-order terms of both expansions, the matching condition (9.6.51) reduces to

$$\lim_{\substack{\varepsilon \to 0 \\ \tau \text{ fixed}}} u_0^o(x, \varepsilon\tau) = \lim_{\substack{\varepsilon \to 0 \\ t \text{ fixed}}} u_0^i\left(x, \frac{t}{\varepsilon}\right) \tag{9.7.32}$$

Using eqs. (9.7.13), (9.7.22), and (9.7.30) it is readily seen that eq. (9.7.32) implies

$$g(0) = a_0 \tag{9.7.33}$$

which provides the missing initial condition for integrating eq. (9.7.16).

According to eq. (9.6.12), the zeroth-order term of the *composite asymptotic expansion* is given by

$$u_0^c(x, t) = u_0^o(x, t) + u_0^i(x, t) - (\text{common part}) \tag{9.7.34}$$

which, upon using eqs. (9.7.13), (9.7.30), and (9.7.33), reduces to

$$u_0^c(x, t) = g(t) + \sum_{m=1}^{\infty} a_m \cos(m\pi x) \exp(-m^2\pi^2 t/\varepsilon) \tag{9.7.35}$$

Here a_m is given by eq. (9.7.31b), while $g(t)$ is obtained by integrating eq. (9.7.16) with IC (9.7.33) and a_0 is given by eq. (9.7.31a).

To illustrate the solution derived above, let us consider a specific example involving irreversible second-order kinetics, inlet reactant concentration constant with time, and initial reactant concentration profile along the reactor of the exponential type; that is, in eqs. (9.7.1)–(9.7.2)

$$r(u) = u^2, \qquad \beta(t) = 1, \qquad \alpha(x) = e^{-x} \tag{9.7.36}$$

In this case eq. (9.7.16), which provides the zeroth-order outer expansion, reduces to

$$\frac{dg}{dt} = 1 - g - Bg^2 \tag{9.7.37}$$

with IC (9.7.33), which using eqs. (9.7.31a) and (9.7.36) gives

$$g(0) = 1 - e^{-1} \tag{9.7.37a}$$

Using the method of separation of variables, the solution of eq. (9.7.37) is

$$g = \frac{K(1 - \psi)e^{\psi t} - (1 + \psi)}{2B[1 - Ke^{\psi t}]} \tag{9.7.38}$$

where $\psi = (1 + 4B)^{1/2}$ and $K = [2B(1 - e^{-1}) + 1 + \psi]/[2B(1 - e^{-1}) + 1 - \psi]$.

Finally, substituting eqs. (9.7.31b), (9.7.36), and (9.7.38) in eq. (9.7.35), the zeroth-order term of the composite solution is

$$\begin{aligned}
u_0^c(x, t) = {} & \frac{K(1 - \psi)e^{\psi t} - (1 + \psi)}{2B[1 - Ke^{\psi t}]} \\
& + 2 \sum_{m=1}^{\infty} \frac{1 - (-1)^m e^{-1}}{1 + m^2 \pi^2} \cos(m\pi x) \exp\left(-\frac{m^2 \pi^2 t}{\varepsilon}\right)
\end{aligned} \tag{9.7.39}$$

It is instructive at this point to discuss briefly some aspects of the structure of the obtained solution, both in the general form (9.7.35) and in the specific case (9.7.39), and connect them with the observations about the qualitative behavior of the solution reported earlier in this section. In both cases it can be seen that for large time values (i.e. $t \gg \varepsilon$) the contribution of the second term, arising from the inner solution, vanishes and the reactant concentration becomes independent of spatial position x. This behavior is illustrated in Figure 9.10, where the zeroth-order approximation of the reactant concentration profile, as given by eq. (9.7.39), exhibits a transition from the exponential shape at small time values ($u = e^{-x}$ at $t = 0$) to a uniform value at longer times. This uniform value corresponds to the contribution of the outer expansion and is indicated by $g(t)$ in the general form of the solution (9.7.35).

The small time interval during which the transition in the shape of the solution noted above occurs is the boundary layer, and its thickness is proportional to ε—that is, the inverse of the mixing intensity within the reactor. This is also illustrated in Figure 9.10, where the behavior of the solution (9.7.39) is shown for two values of the parameter ε.

From the above we can observe that since the boundary layer occurs in time, the steady-state version of the same problem would not exhibit any boundary layer type of behavior. This can be readily confirmed by noting that the steady-state version of the problem (9.7.1)–(9.7.2), where the time variations in eq. (9.7.1) and the IC (9.7.2c) drop out, leads for Pe $\to 0$ to a *regular* perturbation problem.

A second observation concerns the physical meaning of the zeroth-order term of the outer expansion, $u_0^o = g(t)$, which is given by eq. (9.7.16) with IC (9.7.33). These equations in fact represent the transient mass balance for a well-mixed reactor—that is, the ideal CSTR model, where $g(t)$ represents the outlet reactant concentration. (See section 2.17 for a discussion of the steady state and dynamic behavior of a CSTR.) In

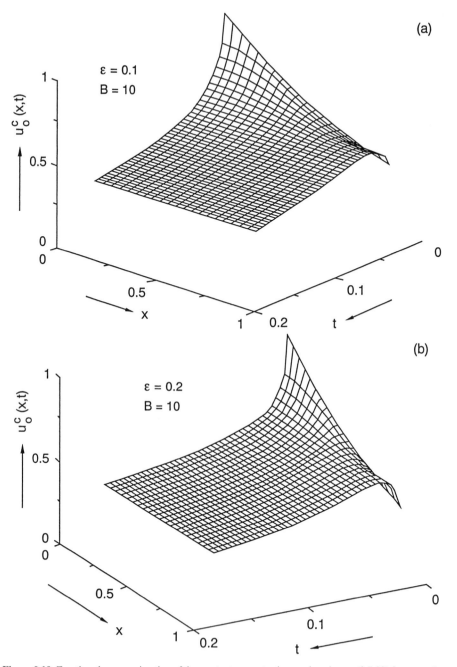

Figure 9.10 Zeroth-order approximation of the reactant concentration as given by eq. (9.7.39) for two values of ε and $B = 10$.

particular, the term in the lhs of eq. (9.7.16) represents the reactant accumulation in the reactor, while the three terms in the rhs represent the reactant inlet and outlet flowrates and the rate of reactant consumption by chemical reaction, respectively. Moreover, eq. (9.7.33) indicates, through the definition of a_0 given by eq. (9.7.31a), that the IC for the CSTR is given by the average value of the initial concentration along the reactor axis $\alpha(x)$.

This corresponds to a well-known result in chemical reactor theory, which states that the ideal CSTR model represents the asymptotic behavior of the axial dispersion model (9.7.1)–(9.7.2) in the limit of intense axial mixing; that is, Pe $\rightarrow$ 0.

A final comment refers to the possibility for the reactor to exhibit multiple steady states, each corresponding to a different solution of the relevant steady-state model. The zeroth-order approximation of the solution at steady-state conditions, u_0^s, can be obtained by taking the limit of the transient solution (9.7.35) as $t \rightarrow \infty$. This leads to

$$u_0^s = g^s \qquad (9.7.40)$$

where g^s indicates the steady-state value of $g(t)$—that is, $g^s = \lim_{t \to \infty} g(t)$, which is obtained by solving the steady-state version of eq. (9.7.16):

$$\beta - g^s - Br(g^s) = 0 \qquad (9.7.41)$$

where β is constant. It is readily seen that, depending upon the specific form of the nonlinear function $r(g)$, this equation may exhibit one or more solutions. Each of these corresponds to the zeroth-order approximation of a possible solution of the full model. In particular, from each of these originates a different series expansion. This can be seen noting that the second coefficient of the outer expansion, u_1^o, is given by eq. (9.7.18), which contains the constant h whose value is given by the following steady-state version of eq. (9.7.21):

$$h^s = \frac{B}{6} \left(\frac{dr}{du} \right)_{u=g^s} (g^s - \beta) \Big/ \left[1 + B \left(\frac{dr}{du} \right)_{u=g^s} \right] \qquad (9.7.42)$$

Thus to each value of g^s corresponds a unique and different value of h^s, and then of u_1^o. This leads to various series expansions, each of which approximates a different steady-state solution of the original model (9.7.1)–(9.7.2).

9.8 THE WKB METHOD

The WKB method (named after Wentzel, Kramers, and Brillouin, who first proposed it) provides a *global* approximation of the solution valid uniformly in the entire domain of the independent variable. The idea is to represent the solution as a linear combination of exponentials of a power series of the small parameter ε. This technique is suited mainly for linear differential equations, where the highest-order derivative is multiplied by a small parameter. In this case the WKB method is likely to provide quite accurate approximations because the exact solution is itself in the form of a linear combination of exponentials of the independent variable. Even though it is limited to linear problems, the WKB method is quite powerful. In fact, for linear problems the method of matched asymptotic expansions can be regarded as a special case of the WKB method.

9.8.1 Description of the Method

To illustrate this method, let us return to the problem of diffusion and reaction in a catalyst slab, in the limit of large Thiele modulus values, as described by eqs. (9.4.10) and (9.4.11). In this case, however, we will assume that the catalyst activity is not constant throughout the slab but rather changes linearly with the independent variable. Thus the original equations change as follows:

$$\varepsilon \frac{d^2 y}{dx^2} = p(x)y, \qquad x \in (0, 1) \tag{9.8.1}$$

where

$$p(x) = a + x, \qquad a > 0 \tag{9.8.1a}$$

The BCs are given by

$$\frac{dy}{dx} = 0 \qquad \text{at } x = 0 \tag{9.8.2a}$$

$$y = 1 \qquad \text{at } x = 1 \tag{9.8.2b}$$

According to the WKB method, the solution is represented by the exponential of a power series of some small parameter δ, to be defined later, as follows:

$$y = \exp\left[\sum_{n=0}^{\infty} \delta^{n-1} S_n(x) \right] \qquad \text{as } \delta \to 0 \tag{9.8.3}$$

This is substituted in the original eq. (9.8.1), leading to

$$\varepsilon \left(\frac{1}{\delta} \frac{d^2 S_0}{dx^2} + \frac{d^2 S_1}{dx^2} + \delta \frac{d^2 S_2}{dx^2} + \cdots \right) + \varepsilon \left(\frac{1}{\delta} \frac{dS_0}{dx} + \frac{dS_1}{dx} \right.$$

$$\left. + \delta \frac{dS_2}{dx} + \cdots \right)^2 = (a + x) \tag{9.8.4}$$

where the exponentials from both sides have been canceled. We can now define the small parameter δ such that the leading order term of the right-hand side is balanced by the leading order term of the left-hand side. Since in eq. (9.8.4), these are given by $O(\varepsilon^0)$ and $O(\varepsilon/\delta^2)$, respectively, this balance is obtained by setting $\delta = \sqrt{\varepsilon}$. Note that this aspect of the WKB method is analogous to the selection of the stretching transformation in the method of matched asymptotic expansions, as given for example by eq. (9.6.1). Thus by substituting $\delta = \sqrt{\varepsilon}$ in eq. (9.8.4), expanding the quadratic term for small ε values, and grouping together terms with like powers of ε, we obtain

$$\left[\left(\frac{dS_0}{dx} \right)^2 - p(x) \right] + \sqrt{\varepsilon} \left[\frac{d^2 S_0}{dx^2} + 2 \frac{dS_0}{dx} \frac{dS_1}{dx} \right]$$

$$+ \varepsilon \left[\frac{d^2 S_1}{dx^2} + \left(\frac{dS_1}{dx} \right)^2 + 2 \frac{dS_0}{dx} \frac{dS_2}{dx} \right] + \cdots = 0 \tag{9.8.5}$$

Since this equation must be satisfied for any value of ε, it follows that each coefficient of the various powers of ε must be identically zero. This leads to the sequence of subproblems for the coefficients S_n described below.

$$O(\varepsilon^0): \qquad \left(\frac{dS_0}{dx}\right)^2 - p(x) = 0$$

thus recalling that $p(x) = a + x$,

$$S_0 = \pm\tfrac{2}{3}p(x)^{3/2} \tag{9.8.6}$$

up to an arbitrary additive constant, since there are no ICs provided for S_0.

$$O(\sqrt{\varepsilon}): \qquad \frac{d^2S_0}{dx^2} + 2\frac{dS_0}{dx}\frac{dS_1}{dx} = 0$$

which similarly gives

$$S_1 = -\tfrac{1}{4}\ln p(x) \tag{9.8.7}$$

$$O(\varepsilon): \qquad \frac{d^2S_1}{dx^2} + \left(\frac{dS_1}{dx}\right)^2 + 2\frac{dS_0}{dx}\frac{dS_2}{dx} = 0$$

Thus

$$S_2 = \pm\frac{5}{48}p(x)^{-3/2} \tag{9.8.8}$$

where it should be noted that since the first subproblem admits two independent solutions for $S_0(x)$, the procedure generates two independent sequences $\{S_n(x)\}$, which we can refer to as $\{S_n^1\}$ and $\{S_n^2\}$, respectively. Each, when substituted in eq. (9.8.3), leads to an independent solution of the original problem, and so the general solution can be written as a linear combination of these, to give

$$y = C_1 \exp\left[\sum_{n=0}^{\infty}\delta^{n-1}S_n^1(x)\right] + C_2 \exp\left[\sum_{n=0}^{\infty}\delta^{n-1}S_n^2(x)\right] \tag{9.8.9}$$

where C_1 and C_2 are integration constants, to be determined through the BCs (9.8.2).

Truncating eq. (9.8.9) at the second term (i.e., $n = 1$) and substituting eqs. (9.8.6) and (9.8.7), we obtain

$$y = p(x)^{-1/4}\left\{C_1 \exp\left[\frac{2}{3\sqrt{\varepsilon}}p(x)^{3/2}\right] + C_2 \exp\left[-\frac{2}{3\sqrt{\varepsilon}}p(x)^{3/2}\right]\right\} \tag{9.8.10}$$

where the integration constants are given by

$$C_1 = \frac{(a + 1)^{1/4}f}{fe^{\phi} + e^{-\phi}} \tag{9.8.11a}$$

$$C_2 = \frac{C_1}{f} \tag{9.8.11b}$$

and

$$f = \frac{4a^{3/2} + \sqrt{\varepsilon}}{4a^{3/2} - \sqrt{\varepsilon}} \exp\left[-\frac{4}{3\sqrt{\varepsilon}} a^{3/2}\right], \qquad \phi = \frac{2}{3\sqrt{\varepsilon}} (a + 1)^{3/2} \quad (9.8.11c)$$

An important aspect concerns the accuracy of the WKB approximation. In general, it is required that the series $\sum_{n=0}^{\infty} S_n(x)\delta^{n-1}$ be an asymptotic series expansion, uniform over the entire domain of interest for the independent variable x. Recalling condition (9.2.12), this means that

$$\ln y = \sum_{n=0}^{N} \delta^{n-1} S_n(x) + O(\delta^N) \qquad \text{as } \delta \to 0 \qquad (9.8.12)$$

which implies that

$$\lim_{\delta \to 0} \left[\frac{\ln y - \sum_{n=0}^{N} \delta^{n-1} S_n(x)}{\delta^N}\right] = \text{finite} \qquad (9.8.13)$$

for any x in the domain of interest—that is, $x \in [0, 1]$. Thus the absolute error involved in approximating $\ln y(x)$ with the series expansion truncated after the Nth term is upper-bounded by $O(\delta^N)$. However, in order to guarantee that also the approximation to $y(x)$ is bounded, the following additional condition needs to be introduced:

$$\delta^N S_{N+1}(x) = o(1) \qquad \text{as } \delta \to 0 \qquad (9.8.14)$$

which indicates that the first term dropped in the truncated series must be much smaller than 1. Thus from eq. (9.8.12) the error can be written as

$$e(\delta, x) = y(x) - \exp\left[\sum_{n=0}^{N} \delta^{n-1} S_n(x)\right]$$
$$= \exp\left[\sum_{n=0}^{N} \delta^{n-1} S_n(x)\right] (\exp\{O[\delta^N S_{N+1}(x)]\} - 1) \qquad (9.8.15)$$

Further, when condition (9.8.14) holds true, the following expansion applies:

$$\exp\{O[\delta^N S_{N+1}(x)]\} = 1 + O[\delta^N S_{N+1}(x)] \qquad (9.8.16)$$

Thus it can be concluded that the relative error is bounded as

$$\frac{e(\delta, x)}{y(x)} = O[\delta^N S_{N+1}(x)] \qquad (9.8.17)$$

In the case of the previous example, since the approximating series (9.8.10) is truncated after two terms (i.e., $N = 1$), the relative error according to (9.8.17) is $O[\delta S_2(x)]$. Using eq. (9.8.8), we obtain

$$\delta S_2(x) = \frac{5\sqrt{\varepsilon}}{48} p(x)^{-3/2} \ll 1 \qquad \text{for } x \in [0, 1] \qquad (9.8.18)$$

thus satisfying condition (9.8.14) for sufficiently small ε values.

It is worth noting that if the above problem is considered in a different domain of x, including the point $x = -a$, then at this point since $p(-a) = 0$, condition (9.8.18) is violated and the WKB approximation breaks down. This occurs for differential equations such as eq. (9.8.1) whenever the coefficient of y [i.e., $p(x)$ in eq. (9.8.1)] vanishes somewhere in the domain of x. As discussed in section 9.6.2, an internal boundary layer develops in the vicinity of these points. In such cases, to develop uniformly valid expansions, various WKB approximations are developed in different subregions of the domain of interest and are then matched together following the method of matched asymptotic expansions.

In the previous example, the WKB method was applied to a linear *homogeneous* ordinary differential equation. This is the type of problem where the WKB method is most useful, so that when handling *nonhomogeneous* problems it is convenient to first apply this method for solving the corresponding homogeneous problem and then to obtain a particular solution through the methods described in chapter 3.

For example, let us consider the homogeneous problem associated with eq. (9.6.14):

$$\varepsilon \frac{d^2y}{dx^2} + v \frac{dy}{dx} = 0, \qquad x \in (0, 1) \tag{9.8.19}$$

Substituting the WKB series expansion (9.8.3) leads to

$$\varepsilon \left(\frac{1}{\delta} \frac{d^2S_0}{dx^2} + \frac{d^2S_1}{dx^2} + \cdots \right) + \varepsilon \left(\frac{1}{\delta} \frac{dS_0}{dx} + \frac{dS_1}{dx} + \cdots \right)^2$$
$$+ v \left(\frac{1}{\delta} \frac{dS_0}{dx} + \frac{dS_1}{dx} + \cdots \right) = 0 \tag{9.8.20}$$

Since the two largest-order terms are $(\varepsilon/\delta^2)(dS_0/dx)^2$ and $(1/\delta)(dS_0/dx)$, for dominant balance we require them to have the same order of magnitude, and so we select $\delta = \varepsilon$. Then, by expanding the squared term, eq. (9.8.20) reduces to

$$\frac{1}{\varepsilon} \left[v \frac{dS_0}{dx} + \left(\frac{dS_0}{dx} \right)^2 \right] + \left[\frac{d^2S_0}{dx^2} + v \frac{dS_1}{dx} + 2 \frac{dS_0}{dx} \frac{dS_1}{dx} \right] + \cdots = 0 \tag{9.8.21}$$

In the usual manner, by equating to zero the coefficients of like powers of ε, we obtain the following subproblems:

$$O(\varepsilon^{-1}): \qquad \frac{dS_0}{dx} \left(v + \frac{dS_0}{dx} \right) = 0$$

which admits the two solutions:

$$S_0 = k_0 \quad \text{and} \quad S_0 = -vx \tag{9.8.22}$$

where k_0 is a constant.

$$O(\varepsilon^0): \qquad \frac{d^2S_0}{dx^2} + v \frac{dS_1}{dx} + 2 \frac{dS_0}{dx} \frac{dS_1}{dx} = 0$$

which yields

$$S_1 = k_1 \tag{9.8.23}$$

a constant, for *both* expressions of S_0. By substituting eqs. (9.8.22) and (9.8.23) in the original series expansion (9.8.3), two solutions of the original problem are obtained, whose linear combination produces the general solution

$$y = C_1 + C_2 \exp\left(-\frac{vx}{\varepsilon}\right) \qquad \text{as } \varepsilon \to 0 \qquad\qquad (9.8.24)$$

where the constants k_0 and k_1 have been incorporated into the constants C_1 and C_2. It is readily seen, by comparison with the exact solution of the original problem given by eq. (9.6.16), that the WKB method has produced the *exact* solution of the associated homogeneous problem. This is *always* the case for linear ordinary differential equations with *constant* coefficients.

Based on the results obtained for the two examples above, it is possible to compare the WKB method with the method of matched asymptotic expansions. For the second example [i.e., eq. (9.8.19)], the two-term WKB expansion coincides with the exact solution, whereas the method of matched asymptotic expansions provides only an approximate solution. Furthermore, it is obtained by directly formulating the global approximation, with no need for developing expansions valid in different regions of the independent variable, and then matching them together.

In the case of the first example, as given by eqs. (9.8.1) and (9.8.2), the method of matched asymptotic expansions cannot produce a two-term approximation, corresponding to the two-term WKB expansion (9.8.10), because of the occurrence of secular terms in the first-order correction to the inner solution (see problem 9.18). The zeroth-order approximation obtained through the boundary layer method, as given by

$$y_0 = \exp[-(1 - x)\sqrt{a + 1}/\sqrt{\varepsilon}] \qquad\qquad (9.8.25)$$

is compared with the two-term WKB approximation and with the exact solution obtained numerically in Figure 9.11. The results confirm the accuracy of the WKB approximation. Of course, it should be reiterated that the WKB method applies *only* to linear differential equations. In the case of nonlinear equations, the exponential terms in the subproblems do not cancel out, and therefore evaluation of the coefficients $S_n(x)$ can generally not be carried out. In these cases the method of matched asymptotic expansions can be applied, as seen in section 9.7.

9.8.2   An Example Involving Rapid Oscillations

Let us now consider a class of problems whose asymptotic behavior can be obtained *only* through the WKB method, whereas the method of matched asymptotic expansions fails. This is the case where the solution changes rapidly, not only in a particular region (i.e., the boundary layer), but rather in the entire domain of interest for the independent variable. In such a situation it is not possible to identify two subregions, where two different expansions of the solution can be developed, using two different scales for the independent variable. Problems of this type can be handled conveniently by the WKB method, as well as by methods based on multiple-scale expansions, to be discussed in the next section.

As an example let us consider the problem of free oscillation of a body on a spring treated in section 9.5.1 where we now refer to a linear spring, and account for

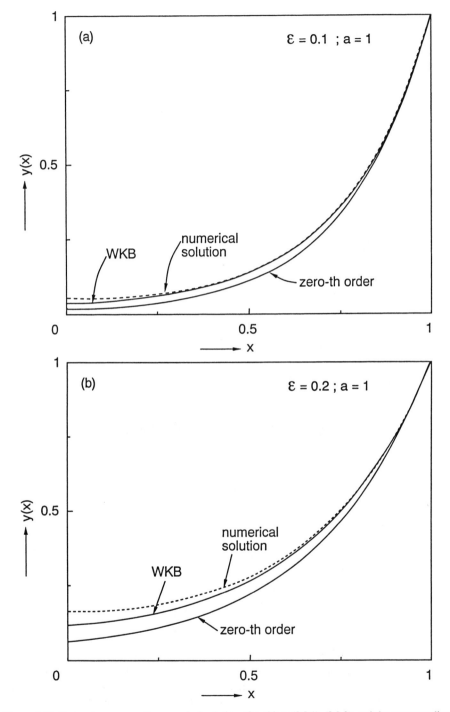

Figure 9.11 Comparison among the numerical solution of problem (9.8.1)–(9.8.2), and the corresponding two-term WKB expansion (9.8.10) and the zeroth-order boundary-layer approximation (9.8.25), for $a = 1$ and two values of ε.

the weight of the body as well as for the friction of the body with the surroundings. The position of the mass y changes with time t according to the equation

$$M \frac{d^2 y}{dt^2} = P - Ky - C \frac{dy}{dt} \qquad (9.8.26)$$

with ICs

$$y = 0, \quad \frac{dy}{dt} = 0 \quad \text{at } t = 0 \qquad (9.8.27)$$

where M is the mass of the body, P is its weight, K is Hooke's elastic constant for the spring, and C is the friction coefficient between the body and the surrounding medium.

Although this is a linear problem with constant coefficients, which readily admits analytical solution, we consider this anyway to illustrate the ability of the WKB method to handle problems of the type noted in the previous paragraph.

The steady-state solution of eq. (9.8.26) is given by

$$y_s = \frac{P}{K} \qquad (9.8.28)$$

which gives the ultimate displacement of the spring due to the weight of the appended body. Let us study the dynamic behavior of this system, in the limit where Hooke's constant is very large—that is, $K = 1/\varepsilon$. Then, substituting in eq. (9.8.26), we have

$$\varepsilon \left[M \frac{d^2 y}{dt^2} + C \frac{dy}{dt} \right] + y = y_s \qquad (9.8.29)$$

This is a linear differential equation with constant coefficients, whose analytical solution is given by

$$y = y_s \left[1 - \frac{\sin(\omega t + \phi)}{\sin \phi} \exp\left(- \frac{Ct}{2M} \right) \right] \qquad (9.8.30)$$

where

$$\omega = \left[\frac{1}{\varepsilon M} \left(1 - \frac{\varepsilon C^2}{4M} \right) \right]^{1/2} \qquad (9.8.31a)$$

and

$$\phi = \tan^{-1} \left[\frac{2M\omega}{C} \right] \qquad (9.8.31b)$$

The qualitative behavior of this solution is shown in Figure 9.12, where it may be seen that while the frequency of the oscillations increases for decreasing ε values, damping of the amplitude is independent of ε. Thus while the duration of the dynamics does not change as ε decreases [since the time constant of the process from eq. (9.8.30) is $2M/C$], the number of oscillations increases, which means that the solution changes relatively

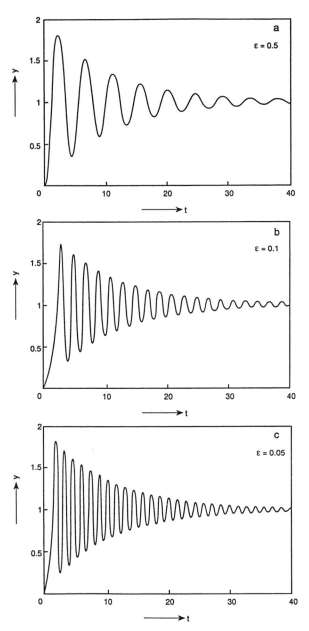

Figure 9.12 Qualitative behavior of the function $y(t)$ as given by eq. (9.8.30) for various values of ε, $C = 0.2$, $M = 1$, and $y_s = 1$.

more rapidly. It is obvious that a boundary-layer-based method cannot be applied in this case.

Let us apply the WKB method by substituting the series expansion (9.8.3) in the homogeneous equation associated with eq. (9.8.29). Grouping terms with like powers of ε, we obtain

$$\frac{\varepsilon M}{\delta^2}\left(\frac{dS_0}{dt}\right)^2 + \frac{\varepsilon}{\delta}\left(M\frac{d^2S_0}{dt^2} + 2M\frac{dS_0}{dt}\frac{dS_1}{dt} + C\frac{dS_0}{dt}\right) + \cdots = -1 \quad (9.8.32)$$

For dominant balance, we select $\delta = \sqrt{\varepsilon}$, so that the following subproblems are obtained:

$$O(\varepsilon^0): \quad M\left(\frac{dS_0}{dt}\right)^2 = -1$$

Thus

$$S_0 = \pm\frac{it}{\sqrt{M}} \quad (9.8.33)$$

$$O(\sqrt{\varepsilon}): \quad M\frac{d^2S_0}{dt^2} + 2M\frac{dS_0}{dt}\frac{dS_1}{dt} + C\frac{dS_0}{dt} = 0$$

which gives

$$S_1 = -\frac{Ct}{2M} \quad (9.8.34)$$

Substituting eqs. (9.8.33) and (9.8.34) in eq. (9.8.3), the following truncated series expansion is obtained:

$$y = A_1 \exp\left[i\frac{t}{\sqrt{M\varepsilon}} - \frac{Ct}{2M}\right] + A_2 \exp\left[-i\frac{t}{\sqrt{M\varepsilon}} - \frac{Ct}{2M}\right] \qquad \text{as } \varepsilon \to 0 \quad (9.8.35)$$

After adding the particular solution $y = y_s$, the constants A_1 and A_2 can be evaluated from the ICs (9.8.27), leading to

$$y = y_s\left[1 - \frac{\sin(\alpha t + \psi)}{\sin \psi}\exp(-Ct/2M)\right] \qquad \text{as } \varepsilon \to 0 \quad (9.8.36)$$

where

$$\alpha = \left[\frac{1}{\varepsilon M}\right]^{1/2} \quad (9.8.37a)$$

and

$$\psi = \tan^{-1}\left(\frac{2M\alpha}{C}\right) \quad (9.8.37b)$$

By comparing with the exact solution (9.8.30), we see that the series expansion (9.8.36) is indeed asymptotically correct, since as $\varepsilon \to 0$, $\omega \to \alpha$, and $\phi \to \psi$.

In all of the problems above, second-order differential equations were considered. The extension of the WKB method to higher-order differential equations is straight-forward (see problem 9.20). It is worth noting here that in the case of an mth-order differential equation, the zeroth-order approximation produces m independent solutions $S_0(x)$. Each of these generates a different sequence of higher-order corrections—that is, $S_1(x)$, $S_2(x)$, and so on—and thus a different asymptotic series (9.8.3). The approxi-mation to the general solution is then obtained by a linear combination of the m ex-ponentials of these asymptotic series.

9.9   THE MULTIPLE SCALE METHOD

The multiple scale method can be applied in various forms. The simplest one, and probably the most widely used, is generally referred to as the *two-variable* expansion method, which will be described in detail in this section. This method can be applied to both secular-type and boundary-layer-type singular perturbation problems, and there-fore it can be regarded as a technique more powerful than both the method of strained coordinates and the method of matched asymptotic expansions. However, the multiple scale method is more complex, and its application usually requires more cumbersome computations. Accordingly, it should be used only in those cases where simpler methods fail.

9.9.1   Development of the Two-Variable Expansion Method

The basic idea behind the multiple scale method can be best illustrated by considering the main features of the solutions obtained through the methods of matched asymptotic expansions and strained coordinates. In the case of boundary-layer-type problems, the solution changes according to two coordinate scales: One of these is given by the original independent variable, which applies in most of the domain of interest; the other is given by the stretched variable, which applies within the boundary layer. In other words, the first coordinate is used to describe the behavior of the solution in the domain away from the boundary layer, while the latter is used to describe it within the boundary layer. For example, in the case of the problem (9.6.14) with $\nu = +1$, we can regard the approximated solution produced by the method of matched asymptotic expansions (9.6.61) as a function of two independent variables: x and x/ε—that is, the original and the stretched variable, respectively.

Also in secular-type problems, the solution changes with the independent variable according to different scales. For example, the approximate solution given by eqs. (9.5.29) and (9.5.30) of problem (9.5.1), obtained through the method of strained co-ordinates, can be regarded as described by two coordinates: t and εt.

Along these lines, in the two-variable expansion version of the multiple scale method, the exact solution is approximated by an asymptotic series whose coefficients are functions of *two independent variables*, each describing a different scale of the original coordinate (e.g., x and x/ε or t and εt in the examples above).

Let us illustrate the two-variable expansion method by considering problem (9.6.14) with $\nu = +1$:

$$\varepsilon \frac{d^2y}{dx^2} + \frac{dy}{dx} = e^x, \qquad x \in (0, 1) \tag{9.9.1}$$

with BCs

$$y = \alpha \qquad \text{at } x = 0 \tag{9.9.2a}$$

$$y = 1 \qquad \text{at } x = 1 \tag{9.9.2b}$$

which has been solved in section 9.6 using the method of matched asymptotic expansions. We introduce the following asymptotic expansion of the solution

$$y = \sum_{n=0}^{N} \varepsilon^n y_n(x, t) + O(\varepsilon^{N+1}) \tag{9.9.3}$$

where the unknown coefficients $y_n(x, t)$ are regarded as functions of the two independent variables x and t. As discussed above, we can take advantage of the solution (9.6.61) obtained earlier and select the second independent variable t so as to reproduce the stretched coordinate

$$t = \frac{x}{\varepsilon} \tag{9.9.4}$$

To substitute the series approximation (9.9.3) in eq. (9.9.1), we need to first compute the total derivatives of y with respect to x. Applying the chain rule and using eqs. (9.9.3) and (9.9.4), we obtain

$$\frac{dy}{dx} = \frac{\partial y}{\partial x} + \frac{\partial y}{\partial t}\frac{dt}{dx} = \sum_{n=0}^{N} \varepsilon^n \frac{\partial y_n}{\partial x} + \frac{1}{\varepsilon} \sum_{n=0}^{N} \varepsilon^n \frac{\partial y_n}{\partial t} + O(\varepsilon^{N+1}) \tag{9.9.5a}$$

and

$$\frac{d^2y}{dx^2} = \sum_{n=0}^{N} \varepsilon^n \frac{\partial^2 y_n}{\partial x^2} + \frac{2}{\varepsilon} \sum_{n=0}^{N} \varepsilon^n \frac{\partial^2 y_n}{\partial x \partial t} + \frac{1}{\varepsilon^2} \sum_{n=0}^{N} \varepsilon^n \frac{\partial^2 y_n}{\partial t^2} + O(\varepsilon^{N+1}) \tag{9.9.5b}$$

Substituting the above derivatives in eq. (9.9.1) and grouping together like powers of ε leads to

$$\frac{1}{\varepsilon}\left(\frac{\partial^2 y_0}{\partial t^2} + \frac{\partial y_0}{\partial t}\right) + \left(\frac{\partial^2 y_1}{\partial t^2} + 2\frac{\partial^2 y_0}{\partial x \partial t} + \frac{\partial y_1}{\partial t} + \frac{\partial y_0}{\partial x}\right) + \varepsilon\left(\frac{\partial^2 y_2}{\partial t^2}\right.$$
$$\left. + 2\frac{\partial^2 y_1}{\partial x \partial t} + \frac{\partial^2 y_0}{\partial x^2} + \frac{\partial y_2}{\partial t} + \frac{\partial y_1}{\partial x}\right) + \cdots = e^x \tag{9.9.6}$$

Thus by proceeding in the usual manner, the coefficients y_n can be determined by solving the subproblems which are obtained by equating to zero the coefficients of like powers of ε in eq. (9.9.6). Each subproblem is completed by the corresponding BCs which are obtained, in similar manner, by substituting the asymptotic expansion (9.9.3) in eq. (9.9.2) and then setting to zero the coefficients of the various powers of ε.

$$O(\varepsilon^{-1}): \qquad \frac{\partial^2 y_0}{\partial t^2} + \frac{\partial y_0}{\partial t} = 0 \tag{9.9.7}$$

along with BCs

$$y_0 = \alpha \quad \text{at } x = 0 \tag{9.9.8a}$$

$$y_0 = 1 \quad \text{at } x = 1 \tag{9.9.8b}$$

The general solution of eq. (9.9.7) is given by

$$y_0 = A_1(x) + A_2(x)e^{-t} \tag{9.9.9}$$

where $A_1(x)$ and $A_2(x)$ are as yet unknown functions which do not depend on t.

$$O(\varepsilon^0): \quad \frac{\partial^2 y_1}{\partial t^2} + \frac{\partial y_1}{\partial t} = -2\frac{\partial^2 y_0}{\partial x \partial t} - \frac{\partial y_0}{\partial x} + e^x \tag{9.9.10}$$

along with BCs

$$y_1 = 0 \quad \text{at } x = 0 \tag{9.9.11a}$$

$$y_1 = 0 \quad \text{at } x = 1 \tag{9.9.11b}$$

Substituting y_0 from eq. (9.9.9), we have

$$\frac{\partial^2 y_1}{\partial t^2} + \frac{\partial y_1}{\partial t} = \left(e^x - \frac{dA_1}{dx}\right) + \frac{dA_2}{dx}e^{-t}$$
$$= a_1(x) + a_2(x)e^{-t} \tag{9.9.12}$$

Since the two inhomogeneous terms in the right-hand side of eq. (9.9.12)—that is, $a_1(x)$ and $a_2(x)e^{-t}$—are both solutions of the associated homogeneous problem, we can expect, as discussed in section 9.5.1, that the first-order correction term y_1 will contain secular terms. We can prevent the occurrence of such terms by selecting $A_1(x)$ and $A_2(x)$ so as to eliminate the two inhomogeneous terms, namely,

$$a_1(x) = e^x - \frac{dA_1}{dx} = 0 \tag{9.9.13a}$$

and

$$a_2(x) = \frac{dA_2}{dx} = 0 \tag{9.9.13b}$$

These lead to

$$A_1 = C_1 + e^x \tag{9.9.14a}$$

and

$$A_2 = C_2 \tag{9.9.14b}$$

where C_1 and C_2 are integration constants to be calculated from the BCs (9.9.8). Thus substituting eq. (9.9.14) in eq. (9.9.9), the zeroth-order approximation of the solution is

$$y_0 = C_1 + e^x + C_2 e^{-x/\varepsilon} \tag{9.9.15}$$

and using the BCs (9.9.8) the values of the two integration constants as $\varepsilon \to 0$ are obtained:

$$C_1 = 1 - e \tag{9.9.16a}$$

and

$$C_2 = \alpha + e - 2 \tag{9.9.16b}$$

Substituting eq. (9.9.16) in eq. (9.9.15) leads to

$$y_0 = 1 - e + e^x + (\alpha + e - 2)e^{-x/\varepsilon} \tag{9.9.17}$$

which indeed represents the zeroth order expansion of the exact solution (9.6.18), as it can be readily seen from eq. (9.6.61).

In order to determine the next term in the series expansion (9.9.3), we return to eq. (9.9.12), whose right-hand side is now equal to zero, since $A_1(x)$ and $A_2(x)$ satisfy eq. (9.9.13). Solving for y_1 gives

$$y_1 = B_1(x) + B_2(x)e^{-t} \tag{9.9.18}$$

where again $B_1(x)$ and $B_2(x)$ are two functions, independent of t, which will be determined so as to prevent the occurrence of secular terms in the next term of the asymptotic series (i.e., y_2 in this case). To this aim, let us consider the subproblem obtained by setting equal to zero the coefficients of ε in eq. (9.9.6):

$$\frac{\partial^2 y_2}{\partial t^2} + \frac{\partial y_2}{\partial t} = -2\frac{\partial^2 y_1}{\partial x \partial t} - \frac{\partial^2 y_0}{\partial x^2} - \frac{\partial y_1}{\partial x} \tag{9.9.19}$$

with BCs

$$y_2 = 0 \quad \text{at } x = 0 \tag{9.9.20a}$$

$$y_2 = 0 \quad \text{at } x = 1 \tag{9.9.20b}$$

Substituting eqs. (9.9.17) and (9.9.18), eq. (9.9.19) reduces to

$$\frac{\partial^2 y_2}{\partial t^2} + \frac{\partial y_2}{\partial t} = -\left(\frac{dB_1}{dx} + e^x\right) + \frac{dB_2}{dx}e^{-t} \tag{9.9.21}$$

Similarly to the case of eq. (9.9.12) above, it can be seen that to prevent the occurrence of secular terms in the expression of y_2, the right-hand side of eq. (9.9.21) must vanish, and accordingly B_1 and B_2 must satisfy the equations

$$B_1 = C_3 - e^x \tag{9.9.22a}$$

$$B_2 = C_4 \tag{9.9.22b}$$

where C_3 and C_4 are integration constants. Substituting eq. (9.9.22) in eq. (9.9.18) leads to

$$y_1 = C_3 - e^x + C_4 e^{-x/\varepsilon} \tag{9.9.23}$$

which can be used, together with the BCs (9.9.11), to evaluate the integration constants as follows:

$$C_3 = e$$

$$C_4 = 1 - e$$

thus leading to the first-order coefficient y_1 of the asymptotic expansion (9.9.3):

$$y_1 = e - e^x + (1 - e)e^{-x/\varepsilon} \tag{9.9.24}$$

Substituting eqs. (9.9.17) and (9.9.24) in the asymptotic series (9.9.3), the following first-order approximation is obtained:

$$y = 1 - e(1 - \varepsilon) + e^x(1 - \varepsilon) + [\alpha - 1 + (e - 1)(1 - \varepsilon)]$$
$$\exp\left(-\frac{x}{\varepsilon}\right) + O(\varepsilon^2) \tag{9.9.25}$$

which is identical to the first-order expansion of the exact solution (9.6.16), as given by eq. (9.6.61) obtained with the method of matched asymptotic expansions.

Thus summarizing, the method of two-variable expansion is based on the idea of representing the solution as an asymptotic series whose coefficients depend upon two independent variables, which represent two different scales of the original independent variable of the problem. In the process of evaluating the coefficients of the series, some arbitrary functions arise which can be determined so as to avoid the occurrence of secular terms, in the same manner as in the method of strained coordinates. The introduction of the second independent variable provides more flexibility to the two-variable expansion method and therefore makes it more powerful than all the others previously described. However, the presence of two independent variables is also the source of the increased complexity of the method. This can be readily seen in the example above, where in order to produce an approximation of the solution of the ODE (9.9.1), the method required us to solve the subproblems (9.9.7), (9.9.12), and (9.9.21) which involved the solution of PDEs.

9.9.2 Selection of the Independent Variables

One point which is often not obvious in the application of the multiple scale method is the selection of the scales of the original coordinate to be used as independent variables. In the previous case, the selection of $t = x/\varepsilon$ was suggested by the boundary layer analysis, and indeed such analysis, as well as insight about the expected behavior of the solution, can be of great help in many instances. However, these approaches are not always successful, and therefore we need to develop a general procedure. This is done in the sequel, in the context of the diffusion-reaction problem in a nonuniform catalyst slab, as given by eqs. (9.8.1) and (9.8.2):

$$\varepsilon \frac{d^2y}{dx^2} = p(x)y, \qquad x \in (0, 1) \tag{9.9.26}$$

where

$$p(x) = (a + x) \tag{9.9.26a}$$

along with BCs

$$\frac{dy}{dx} = 0 \qquad \text{at } x = 0 \tag{9.9.27a}$$

$$y = 1 \qquad \text{at } x = 1 \tag{9.9.27b}$$

which we solved in section 9.8.1 using the WKB method. Before describing the general procedure, it is instructive to analyze the problem with the two-variable expansion method, using as the second independent variable the stretching transformation suggested by the method of matched asymptotic expansion, which from eq. (9.8.25) is given by

$$t = \frac{(1 - x)}{\sqrt{\varepsilon}} \tag{9.9.28}$$

In this case it is convenient to take $\sqrt{\varepsilon}$, rather than ε, as the small parameter of the asymptotic expansion of the exact solution, so that eq. (9.9.3) is replaced by

$$y = \sum_{n=0}^{N} (\sqrt{\varepsilon})^n y_n(x, t) + O[(\sqrt{\varepsilon})^{N+1}] \tag{9.9.29}$$

Substituting eq. (9.9.29) in the original problem (9.9.26)–(9.9.27), computing the second derivative on the lhs of eq. (9.9.26) using the chain rule and eq. (9.9.28)—as we did earlier to derive eq. (9.9.5b)—and finally grouping terms of like powers of ε, we obtain

$$\left[\frac{\partial^2 y_0}{\partial t^2} - p(x)y_0\right] + \sqrt{\varepsilon}\left[\frac{\partial^2 y_1}{\partial t^2} - 2\frac{\partial^2 y_0}{\partial x \partial t} - p(x)y_1\right] + \cdots = 0 \tag{9.9.30}$$

along with the BCs

$$\frac{\partial y_0}{\partial x} + \sqrt{\varepsilon}\frac{\partial y_1}{\partial x} + \cdots = 0 \qquad \text{at } x = 0 \tag{9.9.31a}$$

$$y_0 + \sqrt{\varepsilon}\, y_1 + \cdots = 1 \qquad \text{at } x = 1 \tag{9.9.31b}$$

From eqs. (9.9.30) and (9.9.31), the usual subproblems are derived:

$$O(\varepsilon^0): \qquad \frac{\partial^2 y_0}{\partial t^2} - p(x)y_0 = 0 \tag{9.9.32}$$

along with BCs

$$\frac{\partial y_0}{\partial x} = 0 \qquad \text{at } x = 0 \tag{9.9.33a}$$

$$y_0 = 1 \qquad \text{at } x = 1 \tag{9.9.33b}$$

whose solution is

$$y_0 = A_1(x) \exp[p(x)^{1/2}t] + A_2(x) \exp[-p(x)^{1/2}t] \qquad (9.9.34)$$

$$O(\sqrt{\varepsilon}): \qquad \frac{\partial^2 y_1}{\partial t^2} - p(x)y_1 = 2\frac{\partial^2 y_0}{\partial x \partial t} \qquad (9.9.35)$$

along with BCs

$$\frac{\partial y_1}{\partial x} = 0 \qquad \text{at } x = 0 \qquad (9.9.36a)$$

$$y_1 = 0 \qquad \text{at } x = 1 \qquad (9.9.36b)$$

Substituting eq. (9.9.34) in eq. (9.9.35), recalling the definition $p(x) = a + x$, leads to

$$\frac{\partial^2 y_1}{\partial t^2} - p(x)y_1 = \exp[p(x)^{1/2}t]\left[2p(x)^{1/2}\frac{dA_1}{dx} + \frac{A_1}{p(x)^{1/2}} + A_1 t \right]$$
$$-\exp[-p(x)^{1/2}t]\left[2p(x)^{1/2}\frac{dA_2}{dx} + \frac{A_2}{p(x)^{1/2}} - A_2 t \right] \qquad (9.9.37)$$

From the above equation it appears that in order to prevent the occurrence of secular terms in the expansion of y_1, the functions $A_1(x)$ and $A_2(x)$ should be determined so as to make the quantities in square brackets equal to zero for any value of x and t. However, because such quantities depend both on x and t, it is readily seen that this result cannot be achieved using functions $A_1(x)$ and $A_2(x)$ which depend only on x and not on t.

It should be noted that this is the same difficulty which arises when using the method of matched asymptotic expansions with the stretching transformation (9.9.28), where as mentioned in section 9.8.1 the first-order approximation of the inner solution contains secular terms (see problem 9.18). To overcome this difficulty, we change the definition of the second independent variable t previously adopted [i.e., eq. (9.9.28)] and replace it with

$$t = F(x, \varepsilon) \qquad (9.9.38)$$

At this stage we leave F undetermined. We will use this degree of freedom later, so as to prevent the occurrence of secular terms in the asymptotic expansion.

Using the chain rule and eq. (9.9.38), the following new expressions for the total derivatives of y with respect to x are obtained:

$$\frac{dy}{dx} = \sum_{n=0}^{N}(\sqrt{\varepsilon})^n \frac{\partial y_n}{\partial x} + \frac{dF}{dx}\sum_{n=0}^{N}(\sqrt{\varepsilon})^n \frac{\partial y_n}{\partial t} + O[(\sqrt{\varepsilon})^{N+1}] \qquad (9.9.39a)$$

$$\frac{d^2 y}{dx^2} = \sum_{n=0}^{N}(\sqrt{\varepsilon})^n \frac{\partial^2 y_n}{\partial x^2} + 2\frac{dF}{dx}\sum_{n=0}^{N}(\sqrt{\varepsilon})^n \frac{\partial^2 y_n}{\partial x \partial t}$$
$$+ \frac{d^2 F}{dx^2}\sum_{n=0}^{N}(\sqrt{\varepsilon})^n \frac{\partial y_n}{\partial t} + \left(\frac{dF}{dx}\right)^2 \sum_{n=0}^{N}(\sqrt{\varepsilon})^n \frac{\partial^2 y_n}{\partial t^2} + O[(\sqrt{\varepsilon})^{N+1}] \qquad (9.9.39b)$$

which, when substituted in eq. (9.9.26) and considering only the first two terms of the series, leads to

$$
\begin{aligned}
\varepsilon &\left[\frac{\partial^2 y_0}{\partial x^2} + \sqrt{\varepsilon}\, \frac{\partial^2 y_1}{\partial x^2} + \cdots + 2\, \frac{dF}{dx} \left(\frac{\partial^2 y_0}{\partial x \partial t} + \sqrt{\varepsilon}\, \frac{\partial^2 y_1}{\partial x \partial t} + \cdots \right) \right. \\
&+ \frac{d^2 F}{dx^2} \left(\frac{\partial y_0}{\partial t} + \sqrt{\varepsilon}\, \frac{\partial y_1}{\partial t} + \cdots \right) \\
&\left. + \left(\frac{dF}{dx} \right)^2 \left(\frac{\partial^2 y_0}{\partial t^2} + \sqrt{\varepsilon}\, \frac{\partial^2 y_1}{\partial t^2} + \cdots \right) \right] \\
&= p(x)(y_0 + \sqrt{\varepsilon}\, y_1 + \cdots)
\end{aligned}
\tag{9.9.40}
$$

Let us now return to eq. (9.9.37) and recall why the method failed. This was because it was not possible, using two functions only of x [i.e., $A_1(x)$ and $A_2(x)$], to make identically equal to zero (for any x and t) the expressions in the square brackets which were functions of both x and t. This was instead possible in the case of eq. (9.9.12), where the analogous terms, indicated by $a_1(x)$ and $a_2(x)$, were functions of x alone. Accordingly, we should now select F such that the inhomogeneous part of the new PDE, which will provide the coefficient y_1 in place of eq. (9.9.37), is constituted by the *product* of two functions, one only of x and the other only of t, as was the case for eq. (9.9.12) in the previous problem. From eqs. (9.9.34) and (9.9.35) we see that this requires the exponentials present in the expression of the zeroth-order coefficient y_0 to be functions of t alone. This in turn can happen only if the PDE (9.9.32) which provides y_0 does not contain the variable x. The latter is the condition which we need to satisfy by properly selecting F.

From inspection of eq. (9.9.40), it is clear that to generate a subproblem for y_0 where the variable x is not involved, the function $F(x, \varepsilon)$ should be defined as

$$
\varepsilon \left(\frac{dF}{dx} \right)^2 = p(x)
\tag{9.9.41}
$$

With this, the subproblem for the zeroth-order approximation can be derived from eq. (9.9.40) by balancing terms of $O(1)$:

$$
\frac{\partial^2 y_0}{\partial t^2} = y_0
\tag{9.9.42}
$$

which gives

$$
y_0 = A_1(x)e^t + A_2(x)e^{-t}
\tag{9.9.43}
$$

where the arguments of the exponentials are now independent of x, as desired. The expression of the function F is obtained, recalling that $p(x) = a + x$, by integrating eq. (9.9.41):

$$
t = F(x, \varepsilon) = \frac{2}{3\sqrt{\varepsilon}}\, p(x)^{3/2}
\tag{9.9.44}
$$

up to an arbitrary additive constant, since there are no ICs provided for F. To derive the next coefficient of the series expansion (9.9.29), we consider the following subproblem, which is obtained from eq. (9.9.40) by balancing in the usual manner terms of $O(\sqrt{\varepsilon})$:

$$\frac{\partial^2 y_1}{\partial t^2} - y_1 = -\frac{2}{p(x)^{1/2}} \frac{\partial^2 y_0}{\partial x \partial t} - \frac{1}{2} \frac{1}{p(x)^{3/2}} \frac{\partial y_0}{\partial t}$$

which, by substituting eq. (9.9.43), reduces to

$$\frac{\partial^2 y_1}{\partial t^2} - y_1 = -e^t \left[\frac{2}{p(x)^{1/2}} \frac{dA_1}{dx} + \frac{1}{2} \frac{A_1}{p(x)^{3/2}} \right]$$

$$+ e^{-t} \left[\frac{2}{p(x)^{1/2}} \frac{dA_2}{dx} + \frac{1}{2} \frac{A_2}{p(x)^{3/2}} \right] \quad (9.9.45)$$

It is now possible to prevent the occurrence of secular terms by selecting A_1 and A_2 such that the quantities in square brackets vanish; that is,

$$\frac{dA_1}{dx} = -\frac{1}{4} \frac{A_1}{p(x)} \quad \text{and} \quad \frac{dA_2}{dx} = -\frac{1}{4} \frac{A_2}{p(x)}$$

which upon integration lead to

$$A_1 = C_1 p(x)^{-1/4} \quad \text{and} \quad A_2 = C_2 p(x)^{-1/4} \quad (9.9.46)$$

Substituting eqs. (9.9.44) and (9.9.46) in eq. (9.9.43), the zeroth-order approximation is

$$y_0 = p(x)^{-1/4} \left\{ C_1 \exp\left[\frac{2}{3\sqrt{\varepsilon}} p(x)^{3/2} \right] + C_2 \exp\left[-\frac{2}{3\sqrt{\varepsilon}} p(x)^{3/2} \right] \right\} \quad (9.9.47)$$

which is identical to the corresponding expression (9.8.10) obtained using the WKB method.

In the example above, it is seen that by leaving some flexibility in the definition of the second variable, as done in eq. (9.9.38), we have made the two-variable expansion method more powerful. This has increased the chances of constructing a uniformly valid expansion of the solution. Along these lines, we should mention another version of the two-variable expansion method, where the second independent variable definition (9.9.38) is replaced by the following expansion in terms of the small parameter ε:

$$t = F_0(x) + \varepsilon F_1(x) + \varepsilon^2 F_2(x) + \cdots \quad (9.9.48)$$

In this case, the functions $F_0(x)$, $F_1(x)$, ..., are defined so as to obtain a uniform expansion of the approximate solution. It is worth noting that eq. (9.9.48) is based on the same idea as the transformation (9.5.12) used in the method of strained coordinates. The only difference is that in the present case the transformation applies to the *second* independent variable, rather than to the original one, as in the method of strained coordinates.

Other variants of the multiple scale method have also been reported in the literature. In general, these are based on the idea of increasing the flexibility of the method by introducing additional arbitrary functions in the expansion of the solution. This makes these methods more powerful, because it widens the class of singular perturbation problems that can be solved. However, this also makes their application more cumbersome, particularly in the determination of the arbitrary functions. The reader may refer to the book of Nayfeh (1973, Chapter 6) for a treatment of these methods.

References

M. Abramowitz and I. A. Stegun, *Handbook of Mathematical Functions*, Dover, New York, 1970.

C. M. Bender and S. A. Orszag, *Advanced Mathematical Methods for Scientists and Engineers*, McGraw-Hill, New York, 1978.

A. H. Nayfeh, *Perturbation Methods*, Wiley, New York, 1973.

Additional Reading

R. Bellman, *Perturbation Techniques in Mathematics, Physics, and Engineering*, Dover, New York, 1972.

A. W. Bush, *Perturbation Methods for Engineers and Scientists*, CRC Press, Boca Raton, FL, 1992.

J. D. Cole, *Perturbation Methods in Applied Mathematics*, Blaisdell, Waltham, MA, 1968.

W. Eckhaus, *Asymptotic Analysis of Singular Perturbations*, North-Holland, Amsterdam, 1979.

A. Erdélyi, *Asymptotic Expansions*, Dover, New York, 1956.

N. Fröman and P. O. Fröman, *JWKB Approximation: Contributions to the Theory*, North-Holland, Amsterdam, 1965.

A. K. Kapila, *Asymptotic Treatment of Chemically Reacting Systems*, Pitman Advanced Publishing Program, Boston, 1983.

J. Kevorkian and J. D. Cole, *Perturbation Methods in Applied Mathematics*, Springer-Verlag, New York, 1981.

J. D. Murray, *Asymptotic Analysis*, Springer-Verlag, New York, 1984.

R. E. O'Malley, *Introduction to Singular Perturbations*, Academic Press, New York, 1974.

M. Van Dyke, *Perturbation Methods in Fluid Mechanics*, Academic Press, New York, 1964.

Problems

9.1 Find the asymptotic expansion of the function $\ln(x + \varepsilon)$ as $\varepsilon \to 0$ using the asymptotic sequences ε^n, $\sin^n\varepsilon$, and $[\ln(1 + \varepsilon)]^n$.

9.2 Show that for a given asymptotic sequence $u_n(\varepsilon)$, the coefficients $y_n(x)$ of the asymptotic expansion (9.2.9) are uniquely determined. (*Hint:* The proof follows the same procedure adopted in problem 9.1 to evaluate the coefficients of the asymptotic expansions.)

9.3 Determine the domain of x where the asymptotic expansions found in problem 9.1 are uniform.

9.4 Establish whether the following series expansion is asymptotically uniform over the domain $x \in [0, 1]$

$$y = (x + 0.5) + \varepsilon \ln(x + 0.5) + \varepsilon^2(x + 0.5)^{-2} + \varepsilon^3(x + 0.5)^{-3} + \cdots$$

Show that condition (9.2.20) is not a necessary one.

9.5 The series representation of the mth-order Bessel function of the first kind, $J_m(x)$, is given by [according to eq. (4.8.11)]

$$J_m(x) = \sum_{n=0}^{\infty} \frac{(-1)^n(x/2)^{2n+m}}{n! \, (n + m)!}$$

which can be shown to be convergent for any value of x. However, for large x values the convergence is slow so that the following series expansion (which is of the form discussed in section 4.12) is recommended (Abramowitz and Stegun, 1970, section 9.2):

$$J_m(x) \sim \sqrt{\frac{2}{\pi x}} \left[P(m, x) \cos \alpha - Q(m, x) \sin \alpha \right] \qquad \text{as } x \to \infty$$

where

$$\alpha = x - \frac{\pi}{2} m - \frac{1}{4} \pi$$

$$P(m, x) = \sum_{n=0}^{\infty} (-1)^n \frac{1}{(2x)^{2n}} \frac{\Gamma(1/2 + m + 2n)}{(2n)! \Gamma(1/2 + m - 2n)}$$

$$Q(m, x) = \sum_{n=0}^{\infty} (-1)^n \frac{1}{(2x)^{2n+1}} \frac{\Gamma(1/2 + m + 2n + 1)}{(2n + 1)! \Gamma(1/2 + m - 2n - 1)}$$

Show that the series representing P and Q are not convergent, but nevertheless the above expression is an asymptotic expansion of $J_m(x)$ for large x.

9.6 Find the asymptotic expansion for the error function

$$\text{erf } x = 1 - \frac{2}{\sqrt{\pi}} \int_x^{\infty} e^{-t^2} \, dt$$

in the limit of large x values. Establish whether the obtained series is convergent. (*Hint:* Substitute $\tau = t^2$ in the integral and derive the series expansion by repeatedly applying the method of integration by parts. This approach is usual when dealing with integral functions.)

9.7 Find the approximation to the solution of the problem

$$\frac{d^2 y}{dx^2} = xy$$

with BCs $y(0) = 0$, $y(1) = 1$, such that the residual of the differential equation never exceeds 0.01 in the domain $x \in [0, 1]$. (*Hint:* Multiply the right-hand side by $(1 + \varepsilon)$, use regular perturbation theory, and then recover the solution by setting $\varepsilon = 0$.)

9.8 Solve the diffusion-reaction problem in the case of a nonlinear reaction rate expression $f(y)$ using the regular perturbation method:

$$\frac{d^2 y}{dx^2} = \varepsilon f(y), \qquad x \in (0, 1)$$

with BCs

$$y = 1 \qquad \text{at } x = 1$$
$$\frac{dy}{dx} = 0 \qquad \text{at } x = 0$$

9.9 Find an approximate solution of the system of nonlinear ODEs

$$\frac{d^2 u}{dx^2} = \phi^2 uv$$

$$\frac{d^2 v}{dx^2} = \phi^2 \psi uv$$

with BCs

$$u = 1, \quad \frac{dv}{dx} = 0 \qquad \text{at } x = 0$$

$$\frac{du}{dx} = 0, \quad v = 1 \qquad \text{at } x = 1$$

in the limit where $\phi \ll 1$. Establish whether or not the asymptotic expansion is uniform and find an upper bound for the residuals of the equations. Note that the above equations represent the diffusion-reaction problem for two reacting species (u and v) entering from the opposite sides of a slab and being fully depleted in it.

9.10 Consider the following system of ODEs representing the heat and mass balances in a spherical catalyst particle where a first-order reaction occurs:

$$\frac{d^2u}{ds^2} + \frac{2}{s}\frac{du}{ds} = \phi^2 u \rho(v)$$

$$\frac{d^2v}{ds^2} + \frac{2}{s}\frac{dv}{ds} = -\phi^2 \varepsilon u \rho(v)$$

with BCs

$$\frac{du}{ds} = 0, \qquad \frac{dv}{ds} = 0 \qquad\qquad \text{at } s = 0$$

$$\frac{du}{ds} = Bi_m(1 - u), \quad \frac{dv}{ds} = \alpha\varepsilon(1 - v) \quad \text{at } s = 1$$

Here u and v are the dimensionless reactant concentration and temperature, respectively, while

$$\rho(v) = \exp\left[\gamma\left(1 - \frac{1}{v}\right)\right]$$

and the following variables have been introduced:

Bi_m	Biot number for mass, $k_f R/D$
C_0	bulk reactant concentration
D	effective diffusion coefficient
E	reaction activation energy
h	interphase heat transfer coefficient
$-\Delta H$	heat of reaction
$k(T_0)$	first-order reaction constant at the bulk temperature T_0
k_f	interphase mass transfer coefficient
R	particle radius
R_g	universal gas constant
α	$hRT_0/((-\Delta H)DC_0)$
γ	dimensionless activation energy, $E/R_g T_0$
ε	Prater temperature, $(-\Delta H)DC_0/\lambda T_0$
λ	particle thermal conductivity
ϕ^2	Thiele modulus, $k(T_0)R^2/D$

Since the Prater temperature ε is a relatively small parameter for most gas–solid catalytic reactions (about 0.05), it is convenient to seek an approximate solution to the problem, taking ε as the small parameter. Note that the zeroth-order approximation constitutes the so-called *internal isothermal model* for the catalyst particle. Discuss the possibility that this problem may exhibit multiple solutions. A detailed analysis of this problem is reported by C. J. Pereria, J. B. Wang, and A. Varma, *AIChE Journal*, **25**, 1036 (1979).

9.11 If in problem (9.5.1) describing free oscillations, the nonlinearity was εy^2 rather than εy^3, we would have no secular terms in the regular perturbation expansion. Show that in this case the regular perturbation method and the method of strained coordinates provide the same series expansion of the solution.

9.12 The Rayleigh equation

$$\frac{d^2y}{dt^2} + y + (\varepsilon)\left[-\frac{dy}{dt} + \frac{1}{3}\left(\frac{dy}{dt}\right)^3\right] = 0$$

with ICs

$$y = a, \quad \frac{dy}{dt} = 0 \quad \text{at } t = 0$$

has only one periodic solution corresponding to a specific value of the initial condition, a. Using the method of strained coordinates, determine this value of a as well as the periodic solution in the limit of small positive ε. Compute the coefficients of the asymptotic expansion up to the ε^3 term. (*Hint:* Expand a as a power series of ε, and compute the coefficients to avoid the occurrence of secular terms in the solution, which must be periodic.)

9.13 Use the method of matched asymptotic expansions to find the zeroth-order approximation for the solution of the equation

$$\varepsilon\frac{d^2y}{dx^2} + \beta\frac{dy}{dx} = 1, \quad \beta > 0$$

with BCs

$$y = \alpha \quad \text{at } x = 0$$
$$y = 1 \quad \text{at } x = 1$$

Compare the obtained result with the zeroth-order expansion of the analytical solution.

9.14 Reconsider problem 9.13 for $\alpha = 1 - 1/\beta$.

9.15 Solve problem 9.13 for the case where $\beta < 0$.

9.16 For the problem given by eqs. (9.6.14) and (9.6.15), with $\nu = -1$, develop the two-term composite solution and compare it with the corresponding expansion of the exact solution (9.6.19).

9.17 Using the general procedure outlined in section 9.6.2, for the problem given by eqs. (9.4.10) and (9.4.11), show that the boundary layer is indeed located at $x = 0$ and the corresponding stretching transformation is given by eq. (9.6.1)

9.18 Solve eqs. (9.8.1) and (9.8.2) using the method of matched asymptotic expansions. Show that the zeroth-order approximation coincides with the zeroth-order expansion of the approximation (9.8.10) obtained through the WKB method; that is,

$$y_0 = \exp[-(1 - x)\sqrt{a + 1}/\sqrt{\varepsilon}]$$

Next, derive the two-term expansion using the method of matched asymptotic expansions.

Note that the first-order term of the inner expansion,

$$y_1^i = B_1[e^{t\sqrt{a+1}} - e^{-t\sqrt{a+1}}] + \frac{1}{4}\left[\frac{t}{(a+1)} + \frac{t^2}{\sqrt{a+1}}\right]e^{-t\sqrt{a+1}}$$

contains secular terms. Upon application of the matching condition, this leads to the two-term composite solution:

$$y^c = \left[1 + \frac{(1-x)}{4\sqrt{a+1}} + \frac{(1-x)^2\sqrt{a+1}}{4\sqrt{\varepsilon}}\right]\exp[-(1-x)\sqrt{a+1}/\sqrt{\varepsilon}]$$

which is not an asymptotic expansion since the second-order correction term is proportional to $\varepsilon^{-1/2}$ and hence becomes larger than the first-order term as $\varepsilon \to 0$, thus violating condition (9.2.1).

9.19 The axial dispersion model of an isothermal tubular reactor with an irreversible first-order reaction is given by

$$\frac{1}{\text{Pe}}\frac{d^2y}{dx^2} - \frac{dy}{dx} - \alpha y = 0$$

with BCs

$$\frac{dy}{dx} = \text{Pe}(y-1) \qquad \text{at } x = 0$$

$$\frac{dy}{dx} = 0 \qquad \text{at } x = 1$$

where y is the dimensionless concentration, x is dimensionless axial position, and Pe is the Peclet number. Use the WKB method to find an approximation of the solution up to the $O(\varepsilon^1)$ terms, in the limit of large Pe values (i.e., $\varepsilon = 1/\text{Pe}$). Compare this with the corresponding approximation of the analytical solution.

9.20 Find the approximate solution of the third-order differential equation

$$\varepsilon\frac{d^3y}{dx^3} - x^3y = 0$$

using the WKB method.

9.21 Solve the linear oscillation problem (9.8.26)–(9.8.27) using the two-variable expansion version of the multiple scale method. Compare the obtained solution with that provided by the WKB method.

9.22 Solve the nonlinear free oscillation problem (9.5.1)–(9.5.2) using the two-variable expansion version of the multiple scale method. Show that the appropriate definition of the second independent variable is given by

$$t = F(x) = \frac{1}{2}x^2 + c_1 x$$

Compare this solution with that given by the WKB method.

INDEX